Machine Learning for Future Wireless Communications

Machine Learning for Future Wireless Communications

Edited by

Fa-Long Luo, Ph.D., IEEE Fellow
Silicon Valley, California
USA

This edition first published 2020
© 2020 John Wiley & Sons Ltd

All rights reserved. No part of this publication may be reproduced, stored in a retrieval system, or transmitted, in any form or by any means, electronic, mechanical, photocopying, recording or otherwise, except as permitted by law. Advice on how to obtain permission to reuse material from this title is available at http://www.wiley.com/go/permissions.

The right of Fa-Long Luo to be identified as the author of the editorial material in this work has been asserted in accordance with law.

Registered Offices
John Wiley & Sons, Inc., 111 River Street, Hoboken, NJ 07030, USA
John Wiley & Sons Ltd, The Atrium, Southern Gate, Chichester, West Sussex, PO19 8SQ, UK

Editorial Office
The Atrium, Southern Gate, Chichester, West Sussex, PO19 8SQ, UK

For details of our global editorial offices, customer services, and more information about Wiley products visit us at www.wiley.com.

Wiley also publishes its books in a variety of electronic formats and by print-on-demand. Some content that appears in standard print versions of this book may not be available in other formats.

Limit of Liability/Disclaimer of Warranty
While the publisher and authors have used their best efforts in preparing this work, they make no representations or warranties with respect to the accuracy or completeness of the contents of this work and specifically disclaim all warranties, including without limitation any implied warranties of merchantability or fitness for a particular purpose. No warranty may be created or extended by sales representatives, written sales materials or promotional statements for this work. The fact that an organization, website, or product is referred to in this work as a citation and/or potential source of further information does not mean that the publisher and authors endorse the information or services the organization, website, or product may provide or recommendations it may make. This work is sold with the understanding that the publisher is not engaged in rendering professional services. The advice and strategies contained herein may not be suitable for your situation. You should consult with a specialist where appropriate. Further, readers should be aware that websites listed in this work may have changed or disappeared between when this work was written and when it is read. Neither the publisher nor authors shall be liable for any loss of profit or any other commercial damages, including but not limited to special, incidental, consequential, or other damages.

Library of Congress Cataloging-in-Publication Data

Names: Luo, Fa-Long, editor.
Title: Machine learning for future wireless communications / Dr. Fa-Long Luo.
Description: Hoboken, NJ : Wiley-IEEE, 2020. | Includes bibliographical references.
Identifiers: LCCN 2019029933 (print) | LCCN 2019029934 (ebook) | ISBN 9781119562252 (hardback) | ISBN 9781119562276 (adobe pdf) | ISBN 9781119562313 (epub)
Subjects: LCSH: Wireless communication systems. | Machine learning. | Neural networks (Computer science)
Classification: LCC TK5103.2 .L86 2019 (print) | LCC TK5103.2 (ebook) | DDC 621.3840285/631–dc23
LC record available at https://lccn.loc.gov/2019029933
LC ebook record available at https://lccn.loc.gov/2019029934

Cover Design: Wiley
Cover Image: © Balaji Kharat/Shutterstock

Set in 10/12pt Warnock by SPi Global, Chennai, India

10 9 8 7 6 5 4 3 2 1

Contents

List of Contributors

Daniyal Amir Awan
Network Information Theory Group
Technical University of Berlin
Berlin
Germany

Florian Bahlke
Department of Electrical Engineering and Information Technology
TU Darmstadt
Darmstadt
Germany

Alexios Balatsoukas-Stimming
Department of Electrical Engineering
Eindhoven University of Technology
Eindhoven
The Netherlands

Glauber Brante
Graduate Program in Electrical and Computer Engineering
Federal University of Technology - Paraná
Curitiba
Brazil

Alessandro Brighente
Department of Information Engineering
University of Padova
Padova
Italy

Renato Luis Garrido Cavalcante
Wireless Communications and Networks
Fraunhofer Heinrich Hertz Institute &
Network Information Theory Group
Technical University of Berlin
Berlin
Germany

Kobi Cohen
School of Electrical and Computer Engineering
Ben-Gurion University of the Negev
Beer-Sheva
Israel

Luiz A. DaSilva
CONNECT
Trinity College Dublin
Dublin
Ireland

Nghia Doan
Department of Electrical and Computer Engineering
McGill University
Montreal
Quebec
Canada

Ana Flávia dos Reis
Graduate Program in Electrical and Computer Engineering
Federal University of Technology - Paraná
Curitiba
Brazil

Robert Falkenberg
Communication Networks Institute
TU Dortmund University
Dortmund
Germany

Carlo Fischione
Division of Network and Systems Engineering
Royal Institute of Technology, KTH
Stockholm
Sweden

Gabor Fodor
Division of Decision and Control Systems
Royal Institute of Technology, KTH
Stockholm
Sweden

Francesco Formaggio
Department of Information Engineering
University of Padova
Padova
Italy

Hadi Ghauch
COMELEC Department
Telecom ParisTech
Paris
France

Pere L. Gilabert
Dept. of Signal Theory and Communications
UPC-Barcelona Tech.
Castelldefels
Barcelona
Spain

Warren J. Gross
Department of Electrical and Computer Engineering
McGill University
Montreal
Quebec
Canada

M. Cenk Gursoy
Department of Electrical Engineering and Computer Science
Syracuse University
Syracuse
New York
USA

Seyyed Ali Hashemi
Department of Electrical and Computer Engineering
McGill University
Montreal
Quebec
Canada

Song-Nam Hong
Electrical and Computer Engineering
Ajou University
Suwon
South Korea

Yuxiu Hua
College of Information Science & Electronic Engineering
Zhejiang University
Hangzhou
China

Muhammad Ali Imran
James Watt School of Engineering
University of Glasgow
Glasgow
Scotland
UK

Anu Jagannath
Marconi-Rosenblatt AI/ML Innovation Laboratory
ANDRO Computational Solutions LLC
New York
USA

Jithin Jagannath
Marconi-Rosenblatt AI/ML Innovation Laboratory
ANDRO Computational Solutions LLC
New York
USA

Wei Jiang
Intelligent Networking Group
German Research Center for Artificial Intelligence
Kaiserslautern
Germany

Zhiyuan Jiang
School of Communications and Information Engineering
Shanghai University
Shanghai
China

Seonho Kim
Electrical and Computer Engineering
Ajou University
Suwon
South Korea

Paulo Valente Klaine
James Watt School of Engineering
University of Glasgow
Glasgow
Scotland
UK

Andres Kwasinski
Department of Computer Engineering
Rochester Institute of Technology
New York
USA

Rongpeng Li
College of Information Science & Electronic Engineering
Zhejiang University
Hangzhou
China

Zening Liu
SHIFT
School of Information Science and Technology
ShanghaiTech University
Shanghai
China

David Lopez-Bueno
CTTC/CERCA & Dept. of Signal Theory and Communications
UPC-Barcelona Tech.
Castelldefels
Barcelona
Spain

Elie Ngomseu Mambou
Department of Electrical and Computer Engineering
McGill University
Montreal
Quebec
Canada

Andrei Marinescu
CONNECT
Trinity College Dublin
Dublin
Ireland

Tommaso Melodia
Department of Electrical and Computer Engineering
Northeastern University
Boston
USA

Fatemeh Shah Mohammadi
Kate Gleason College of Engineering
Rochester Institute of Technology
New York
USA

Gabriel Montoro
Dept. of Signal Theory and Communications
UPC-Barcelona Tech.
Castelldefels
Barcelona
Spain

Zhisheng Niu
Department of Electronic Engineering
Tsinghua University
Beijing
China

Marius Pesavento
Department of Electrical Engineering and Information Technology
TU Darmstadt
Darmstadt
Germany

Thi Quynh Anh Pham
Dept. of Signal Theory and Communications
UPC-Barcelona Tech.
Castelldefels
Barcelona
Spain

Nicholas Polosky
Marconi-Rosenblatt AI/ML Innovation Laboratory
ANDRO Computational Solutions LLC
New York
USA

Francesco Restuccia
Department of Electrical and Computer Engineering
Northeastern University
Boston
USA

Gabriele Ruvoletto
Department of Information Engineering
University of Padova
Padova
Italy

Hans Dieter Schotten
Institute for Wireless Communication and Navigation
University of Kaiserslautern
Kaiserslautern
Germany

Ziyu Shao
SHIFT
School of Information Science and Technology
ShanghaiTech University
Shanghai
China

Hossein Shokri-Ghadikolaei
Division of Network and Systems Engineering
Royal Institute of Technology, KTH
Stockholm
Sweden

Mikael Skoglund
Division of Information Science and Engineering
Royal Institute of Technology, KTH
Stockholm
Sweden

Benjamin Sliwa
Communication Networks Institute
TU Dortmund University
Dortmund
Germany

Richard Demo Souza
Department of Electrical and Electronics Engineering
Federal University of Santa Catarina
Florianópolis - SC
Brazil

Slawomir Stanczak
Wireless Communications and Networks
Fraunhofer Heinrich Hertz Institute &
Network Information Theory Group
Technical University of Berlin
Berlin
Germany

Stefano Tomasin
Department of Information Engineering
University of Padova
Padova
Italy

Senem Velipasalar
Department of Electrical Engineering and Computer Science
Syracuse University
Syracuse
New York
USA

Chujie Wang
College of Information Science & Electronic Engineering
Zhejiang University
Hangzhou
China

Kunlun Wang
SHIFT
School of Information Science and Technology
ShanghaiTech University
Shanghai
China

Wenbo Wang
School of Computer Science and Engineering
Nanyang Technological University
Singapore

Christian Wietfeld
Communication Networks Institute
TU Dortmund University
Dortmund
Germany

Zhiqiang Wu
Department of Electrical Engineering
Wright State University
Dayton
USA

Ji-ying Xiang
ZTE Ltd.
Shenzhen
China

Shugong Xu
SICS
Shanghai University
Shanghai
China

Lie-Liang Yang
School of Electronics and Computer Science
University of Southampton
Southampton
UK

Yang Yang
SHIFT
School of Information Science and Technology
ShanghaiTech University
Shanghai
China

Masahiro Yukawa
Department of Electronics and Electrical Engineering
Keio University
Yokohama
Japan

Honggang Zhang
College of Information Science & Electronic Engineering
Zhejiang University
Hangzhou
China

Lin Zhang
School of Electronics and Information Technology
Sun Yat-sen University
Guangzhou
China

Shuang Zhao
Interactive Entertainment Group
Tencent Inc.
Shanghai
China

Zhifeng Zhao
College of Information Science & Electronic Engineering
Zhejiang University
Hangzhou
China

Chen Zhong
Department of Electrical Engineering and Computer Science
Syracuse University
Syracuse
New York
USA

Sheng Zhou
Department of Electronic Engineering
Tsinghua University
Beijing
China

Preface

Due to its powerful nonlinear mapping and distribution processing capability, deep NN-based machine learning technology is being considered as a very promising tool to attack the big challenge in wireless communications and networks imposed by the explosively increasing demands in terms of capacity, coverage, latency, efficiency (power, frequency spectrum, and other resources), flexibility, compatibility, quality of experience, and silicon convergence. Mainly categorized into supervised learning, unsupervised learning, and reinforcement learning, various machine learning (ML) algorithms can be used to provide better channel modeling and estimation in millimeter and terahertz bands; to select a more adaptive modulation (waveform, coding rate, bandwidth, and filtering structure) in massive multiple-input and multiple-output (MIMO) technology; to design more efficient front-end and radio-frequency processing (pre-distortion for power amplifier compensation, beamforming configuration, and crest-factor reduction); to deliver a better compromise in self-interference cancellation for full-duplex transmissions and device-to-device communications; and to offer a more practical solution for intelligent network optimization, mobile edge computing, networking slicing, and radio resource management related to wireless big data, mission-critical communications, massive machine-type communications, and tactile Internet.

In fact, technology development of ML for wireless communications has been growing explosively and is becoming one of the biggest trends in related academic, research, and industry communities. These new applications can be categorized into three groups: (i) ML-based spectrum intelligence and adaptive radio resource management; (ii) ML-based transmission intelligence and adaptive baseband signal processing; and (iii) ML-based network intelligence and adaptive system-level optimization. The successful development and deployment of all these new applications will be challenging and will require huge effort from industry, academia, standardization organizations, and regulatory authorities.

From a practical application and research development perspective, this book aims to be the first single volume to provide a comprehensive and highly coherent treatment on all the technology aspects related to ML for wireless communications and networks by covering system architecture and optimization, physical-layer and cross-layer processing, air interface and protocol design, beamforming and antennal configuration, network coding and slicing, cell acquisition and handover, scheduling and rate adaption, radio access control, smart proactive caching, and adaptive resource allocations.

This book is organized into 21 chapters in 3 parts.

Part 1: Spectrum Intelligence and Adaptive Resource Management

The first part, consisting of eight chapters, presents all technical details on the use of ML in dynamic spectrum access (DSA) and sharing, cognitive radio management and allocation, coverage and capacity optimization, traffic and mobility prediction, energy-efficiency maximization, and intelligent data transfer.

Focusing on advanced ML algorithms for efficient access and adaptive sharing, Chapter 1 presents a comprehensive introduction to spectrum intelligence concepts in different practical application scenarios and real-world settings by covering basic principles, mathematical models, global optimization criterions, learning rules and rates, step-by-step algorithms, and schematic processing flows. More specifically, the key technical aspect presented in this chapter is to address how to employ advanced ML algorithms to dynamically learn the environment by continuously monitoring radio performance and judiciously adjusting the transmission parameters so as to achieve system-wide optimal spectrum usage and resource efficiency for future wireless communications and networks.

Chapter 2 is devoted to the use of various reinforcement learning (RL) algorithms in adaptive radio resource management with emphasis on distributed resource allocation in cognitive radio networks. In comparison with the traditional DSA methods, RL-based techniques can offer the advantages of alignment with the cognition cycle approach and flexibility for the implementation of cross-layer resource allocation without needing prior knowledge of a system model. Taking the mean opinion score as a key quality-of-experience metric, this chapter extensively investigates the impact of individual learning and cooperative learning related to Q-learning and deep Q-networks on resource allocation performance and convergence time.

ML-based spectrum sharing for millimeter-wave communications is the focus of Chapter 3 by proposing a hybrid approach that combines the traditional model-based approach with a data-driven learning approach. The theoretical analyses and experiments presented in this chapter show that the proposed hybrid approach is a very promising solution in dealing with the key technical aspects of spectrum sharing: namely, the choice of beamforming, the level of information exchange for coordination and association, and the sharing architecture. What this chapter presents contributes a new research direction in which the NN-based ML techniques can also be used as an aiding processing component instead of the full replacement of traditional model-based schemes.

Chapter 4 presents an overview of all major state-of-the-art ML-based solutions for network coverage and capacity optimization (CCO) problems in terms of resource reallocation and energy savings. Resource reallocation can be achieved through various load-balancing techniques where either base-station parameters can be adjusted or mobile users are re-allocated among cells. Energy-saving techniques can be accomplished either through antenna power control or by switching base stations on and off depending on demand. More specifically, a deep neural network (DNN) can be used to adaptively configure base-station parameters according to network user's geometry information and resource demand. Furthermore, a data-driven approach on the basis of deep RL is presented in this chapter, which can enable base-station sleeping in an optimized way so as to better address the non-stationarity in real-world traffic.

Radio resource allocation and management play the most important role in wireless network coverage and capacity optimization. Taking ultra-dense heterogeneous wireless networks as illustration examples, the focus of Chapter 5 is on the use of supervised ML technologies in fully decentralized network optimization and radio access control on the basis of the user-centric approach and base-station centric approach. Theoretical analyses and simulation results given in this chapter demonstrate that ML approaches can achieve close-to-optimal network-balancing solutions. Furthermore, it is shown in Chapter 5 that ML-based resource allocation schemes can successfully generalize their knowledge and also become applicable for networks of different size and users with different demand and buffer status.

Chapter 6 serves as a comprehensive overview of the importance and applications of ML algorithms in future wireless networks in terms of energy efficiency (EE) optimization, resource allocation, traffic prediction, self-organizing networks, and cognitive radio networks.

In comparison with the conventional solutions for EE optimization, ML-based approaches can handle complex and dynamic wireless networks in a more efficient, intelligent, and flexible manner. Addressing training-data selection, optimization criterion determination, and learning-rule adjustment, this chapter also provides all possible technical difficulties in designing and implementing adaptive learning solutions for achieving global EE optimization, which can be very useful in bringing ML-based spectrum intelligence technology into practical applications.

As pointed out in Chapter 7, traffic and mobility prediction is the core technology component in designing future mobile communication network architectures and the corresponding signal-processing algorithms. Having overviewed major existing prediction methods in terms of modeling, characterization, complexity, and performance, the emphasis of this chapter moves on to emerging deep learning (DL)-based schemes for traffic and mobility prediction by introducing a random connectivity long short-term memory (RCLSTM) model and combining convolutional NNs (CNNs) with the spatial-domain information estimation of the user's movement trajectory. Moreover, it is shown in Chapter 7 by theoretical analyses and extensive simulations that the RCLSTM model can reduce computational cost by randomly removing some neural connections, and the new CNN-based scheme can deliver improved prediction performance.

Chapter 8 is devoted to the use of ML concepts in anticipatory communications and opportunistic communications, aiming to utilize all the existing resources in a more efficient and intelligent way, which is considered a smart alternative to cost-intense extension of the network infrastructure. Taking mobile crowdsensing as a case study, this chapter presents an opportunistic and context-predictive transmission scheme that relies on ML-based data-rate prediction for channel quality assessment, which is executed online on embedded mobile devices. Through a comprehensive real-world evaluation study, the proposed ML-based transmission scheme is proved to be able to achieve massive increases in the resulting data rate while simultaneously reducing the power consumption of mobile devices.

Part 2: Transmission Intelligence and Adaptive Baseband Processing

Eight chapters in Part 2 focus on various ML algorithms for new radio transmission technologies and new baseband signal-processing, including adaptive modulation waveform and beamforming power selection, intelligent error-correction coding and massive MIMO pre-coding, transmitter classification and signal intelligence, nonlinear channel modeling and parameter estimation, as well as complex-valued signal detection, prediction, and equalization.

Chapter 9 mainly presents three ML-based approaches for adaptive modulation and coding (AMC) schemes so as to make the wireless system able to adapt to variations in channel conditions with fewer model-based approximations, better accuracy, and higher reliability than traditional AMC schemes. Two of them are supervised learning (SL)-based approaches, and the other is implemented by RL in an unsupervised manner. As summarized in Chapter 9, SL-based AMC schemes are suitable for the scenario where training examples are representatives of all the situations that the transmitter might be exposed to. In contrast, RL-based AMC solutions can directly learn from the interacting environment and can gradually achieve satisfactory performance even without any offline training.

As a powerful spatial-domain signal processing tool, massive and nonlinear MIMO technologies are greatly being employed in wireless communications where low-power implementation of nonlinear MIMO signal detection and channel estimation becomes the key to bringing these promising techniques into the practice. To address this problem, Chapter 10 presents SL-based solutions by covering nonlinear MIMO channel modeling, detection problem formulations, lower-bit analog-to-digital conversion (ADC), and parametric and non-parametric learning algorithms. In these reported ML-based solutions, the pilot signals can be used as the training data and further be exploited to directly learn a nonlinear MIMO detector instead of estimating a complex channel transfer function, which can offer a much better compromise among complexity, performance, and robustness in comparison with traditional methods.

In adaptive baseband signal processing, the idea of using ML technology is to mainly replace many building blocks of model-based transceivers with a few learning algorithms, with the intent to drastically reduce the number of assumptions about the system models and the need for complex estimation techniques. However, this reduction in model knowledge brings many technical challenges such as the requirement for large training sets and a long training time. Focusing on symbol detection in multiuser environments, Chapter 11 shows how to combine the theory of reproducing kernel Hilbert spaces (RKHSs) in sum spaces with the adaptive projected sub-gradient method so as to generate a filtering sequence that can be uniquely decomposed into a linear component and a nonlinear component constructed with a Gaussian kernel. By doing so, the corresponding ML solution is able to not only cope with small training sets and sudden changes in the environment but also outperform traditional solutions with perfect channel knowledge.

Chapter 12 provides a comprehensive introduction to the use of various NN architectures and corresponding ML algorithms in channel equalization and signal detection of future wireless communications by covering working principles, channel modeling, training data collection, complexity analyses, and performance comparisons. The NNs presented in this chapter mainly include multilayer perceptron (MLP) networks, radial basis function networks, recurrent NNs (RNNs), functional link artificial NNs, DNNs, CNNs, support vector machines (SVMs), and extreme learning machines as well as their improved versions specifically developed for joint equalization and detection. By using the bit error rate as a key performance index, this chapter demonstrates that NN-based channel equalizers can be very powerful in handling nonlinear and complicated channel conditions, which are the case for future wireless communication systems.

From theory to practice, all technical aspects of ML-based signal intelligence are addressed in Chapter 13, with the emphasis on automatic modulation classification and adaptive wireless interference classification as well as their real-time implementation. More specifically, the automatic modulation classification task involves determining what scheme has been used to modulate the transmitted signal, given the raw signal observed at the receiver. On the other hand, the task of adaptive wireless interference classification essentially refers to identifying what type of wireless emitter exists in the environment. To accomplish these tasks with NN-based approaches in a more efficient and more practical way, this chapter provides a detailed guide for collecting spectrum data, designing wireless signal representations, forming training data, and selecting training algorithms, in particular for the cases of large-scale wireless signal datasets and complexed-valued data representation formats.

Chapter 14 is devoted to the use of various NNs and related learning algorithms in the channel coding (encoder and decoder) of wireless communications and networking including the low-density parity-check (LDPC) codec, polar codec, and Turbo codec. Due to its powerful nonlinear mapping and distributed processing capability, NN-based ML technology could offer a more powerful channel-coding solution than conventional approaches in many aspects including coding performance, computational complexity, power consumption, and processing latency. This chapter presents three different approaches in the use of NNs for channel coding. In the first approach, the NN-based decoder performs all the processing blocks required by a traditional decoder. The second approach uses NNs to perform partial processing of a whole decoding task. The third approach uses NNs to perform the post- or pre-processing of the decoder including noise reduction and log-likelihood ratio estimation.

Due to its capability to achieve the desired capacity for any channel at infinite code length, the polar codec is considered a recent technology breakthrough and is finding wide use in many emerging wireless communications and networking systems. By addressing their working principles, algorithm details, and performance evaluations, three DL-based polar decoding approaches are presented in Chapter 15. More specifically, these three approaches are referred to as the off-the-shelf DL polar decoder, DL-aided polar decoder, and joint-learning approach for both the polar decoder and noise estimator, respectively. This chapter demonstrates that these DL-based polar decoding solutions can reach a maximum-a-posteriori decoding performance for short code lengths and also obtain significant decoding performance gains over their conventional counterparts, while maintaining the same decoding latency.

Chapter 16 focuses on ML and NN approaches for wireless channel state information (CSI) prediction so as to judiciously adapt transmission parameters such as scheduled users, modulation and coding schemes, transmit power, relaying nodes, time slots, sub-carriers, and transmit or receive antennas to instantaneous channel conditions. To make full use of the capability of time-series prediction enabled by RNNs, this chapter first uses RNNs to implement a multi-step predictor for frequency-flat single-antenna channels and then to extend the solution to multi-antenna channels and frequency-selective multi-antenna channels. It can be seen from the theoretical analyses and simulation results reported in Chapter 16 that RNNs exhibit great flexibility, generality, scalability, and applicability for wireless fading-channel prediction applications and can therefore be regarded as a very promising tool to bring transmission intelligence technology into practical implementations.

Part 3: Network Intelligence and Adaptive System Optimization

Organized into five chapters, the third part of this book is devoted to the use of ML and artificial intelligence technologies in system-level and network aspects by covering flexible backhaul and front-haul, cross-layer optimization and coding, full-duplex radio, digital front-end (DFE) and radio-frequency (RF) processing, fog radio access network, and proactive caching.

By focusing on crest-factor reduction, in-phase and quadrature imbalance mitigation, peak-to-average power ratio reduction, and digital pre-distortion linearization, Chapter 17 provides a very comprehensive overview to show how to apply ML technologies to offer an intelligent solution in DFE and software-defined RF processing. Due to its powerful nonlinear mapping and distributed processing capability, NN-based ML technology is proved in this chapter to be able to greatly outperform conventional DFE approaches in many aspects including system performance, programming flexibility, memory access, architecture scalability, computational complexity, power consumption, and processing latency. Moreover, NN-based solutions can also play a very important role in future DFE implementation and deployment including intelligent vector processors and functional IP blocks.

Full-duplex transmission and operation can offer the potential to not only double spectral efficiency (bit/second/Hz) but also improve the reliability and flexibility of dynamic spectrum allocation in future wireless systems. On the other hand, self-interference cancellation (SIC) is the key to bringing full-duplex transmission to reality. Having presented the problem formulations and outlined the shortcomings of conventional SIC approaches, Chapter 18 presents nonlinear NN-based SIC solutions and makes extensive comparisons in terms of cancellation performance, processing latency, computational and implementation complexity, as well as quantitative analyses of logic operations, addition, multiplication, data read/write bandwidth, silicon die area, memory size, and I/O throughput. Open problems for the use of ML in achieving an optimal full-duplex transmission are also discussed at the end of this chapter.

Aiming at ML-based cross-layer and system-level optimization, Chapter 19 investigates online multi-tier operation scheduling in fog-enabled network architectures with heterogeneous node capabilities and dynamic wireless network conditions. On the basis of Lyapunov optimization principles, this chapter proposes a low-complexity online adaptive learning algorithm by covering the trade-off between average network throughput and service delay as well as the resulting benefit from centralized assignment of access node and dynamic online bandwidth allocation. Theoretical analyses and simulation results given in this chapter demonstrate the effectiveness and accuracy of the proposed algorithms. What this chapter addresses also includes the proactive fog access node assignment and resource management problem, given the availability of predictive information in terms of service latency, cost modeling, resource sharing, and predictive scheduling.

As an important technical aspect of network intelligence and security, location verification schemes can be used to verify position-related information independently of that reported by the user or device itself, which is highly desirable because the self-reported information could easily be modified at either the software or hardware level. Taking into account the user-dependent features of physical-layer channels, Chapter 20 discusses ML algorithms for performing in-region location verification tasks with emphasis on multiple-layer nonlinear NNs and SVM approaches. The user-dependent physical-layer features considered in this chapter include multiple-path reflection, path loss, frequency-dependent attenuation, shadowing, and fading. These ML-based verification solutions can achieve optimal performance in a very efficient way from the perspective of computational complexity, resource consumption, and processing latency.

Chapter 21 serves as the last chapter of this book and is devoted to the application of deep RL strategies to edge caching at both small base stations and user equipment. According to different cellular deployment scenarios, this chapter mainly presents a multi-agent actor-critic deep RL framework for edge caching with the goal to increase the cache hit rate and reduce transmission latency. Extensive simulation results, theoretical analyses, and performance comparisons given in this chapter demonstrate that the proposed ML solution can offer a better compromise among latency, complexity, and efficiency than traditional ones for wireless edge-caching and edge-computing applications. As pointed out at the end of this chapter, the proposed concept for multiple agents and multiple tasks can also be used to jointly address the content-caching problem along with other problems including power control and user scheduling so as to obtain an optimal solution at the cross-layer and system levels.

For whom is this book written?

It is hoped that this book serves not only as a complete and invaluable reference for professional engineers, researchers, scientists, manufacturers, network operators, software developers, content providers, service providers, broadcasters, and regulatory bodies aiming at development, standardization, deployment, and applications of ML systems for future wireless communications and networks; but also as a textbook for graduate students in signal and information processing, wireless communications and networks, computer sciences and software engineering, microwave technology, antenna and propagation, circuit theory, and silicon implementation.

Silicon Valley, California
USA

Fa-Long Luo, Ph.D., IEEE Fellow

Part I

Spectrum Intelligence and Adaptive Resource Management

1

Machine Learning for Spectrum Access and Sharing

Kobi Cohen

School of Electrical and Computer Engineering, Ben-Gurion University of the Negev, Beer-Sheva, Israel

1.1 Introduction

Driven by visions of 5G communications and the Internet of Things (IoT), it is expected that tens of billions of wireless devices will be interconnected in the next decade. Due to the increasing demand for wireless communication, along with spectrum scarcity, developing dynamic spectrum access (DSA) algorithms for efficient spectrum sharing among devices and networks that coexist in the same frequency band (enabled by cognitive radio technology) is crucial for obtaining the best communication performance in a complex dynamic environment. The basic idea of DSA algorithms is to learn the environment dynamically by continuously monitoring system performance and judiciously adjusting the transmission parameters for achieving high spectral efficiency. Complete information about the network state typically is not available online for users, which makes the problem challenging: how can we achieve efficient spectrum allocation among users to maximize a global system-wide objective when each user has only partial observation of the system state? Advanced machine learning (ML) algorithms have been developed in the last decade to address this challenging problem.

In general, in this chapter we consider a wireless network consisting of a set $\mathcal{N} = \{1, 2, ..., N\}$ of users and a set $\mathcal{K} = \{1, 2, ..., K\}$ of shared orthogonal channels (e.g. orthogonal frequency-division multiple access [OFDMA]). At the beginning of each time slot, each user selects a channel and transmits its data using a transmission access protocol (e.g. Aloha-type, or carrier-sense multiple access [CSMA]-type narrowband transmission). Transmission on channel k is successful if only a single user transmits over channel k in a given time slot. Otherwise, a collision occurs. The algorithms in this chapter apply to two main models for DSA (Zhao and Sadler, 2007): a hierarchical model that allows secondary (unlicensed) cognitive users to use the spectrum whenever they do not interfere with primary (licensed) users, and the open sharing model among users that acts as the basis for managing a spectral region, where it generally is not assumed that there are primary and secondary users in the networks (e.g. industrial, scientific, and medical radio [ISM] band). Thus, in the hierarchical model, only secondary users implement ML algorithms for learning the environment to improve the spectral usage, whereas primary users are modeled as external processes. This case is illustrated in Figure 1.1. In the open sharing model, all users are cognitive and apply ML algorithms to learn the system dynamics.

The rest of this chapter is organized into three sections. Section 1.2 focuses on online learning algorithms that were developed for DSA, in which a cognitive user aims to learn the occupancy of the spectrum in the presence of external users to improve the spectral usage. The focus is

Machine Learning for Future Wireless Communications, First Edition. Edited by Fa-Long Luo.
© 2020 John Wiley & Sons Ltd. Published 2020 by John Wiley & Sons Ltd.

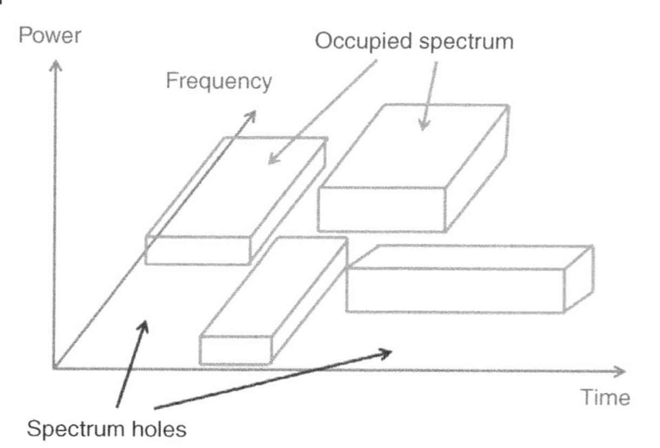

Figure 1.1 An illustration of the hierarchical model that allows secondary (unlicensed) cognitive users to use the spectrum whenever they do not interfere with primary (licensed) users. The secondary users implement ML algorithms for learning the environment to improve spectral usage by exploiting spectrum holes, whereas primary users are modeled as external processes (i.e. the occupied spectrum in the figure).

thus on how quickly the cognitive user can learn the external process, and not on the interaction between cognitive users. Section 1.3 focuses on the more general model where multiple cognitive users share the spectrum, and the goal is to effectively allocate channels to cognitive users in a distributed manner in order to maximize a certain global objective. We first provide an overview of model-dependent solutions, and then focus on the very recent developments of artificial intelligence (AI) algorithms based on deep learning for DSA that can effectively self-adapt to complex real-world settings. Section 1.4 will offer some further discussion and conclusions.

1.2 Online Learning Algorithms for Opportunistic Spectrum Access

1.2.1 The Network Model

We start by providing a representative network model that was investigated in recent years (Tekin and Liu, 2012; Liu et al., 2013; Gafni and Cohen, 2018b, a). Consider K channels indexed by $i = 1, 2, \cdots, K$. The ith channel is modeled as a discrete-time, irreducible, and aperiodic Markov chain with finite state space S^i. This model can be used to model the occupancy of the channels by primary users (i.e. 0 for a busy channel, or 1 for an idle channel or *opportunity*), or by external interference with different levels. At each time, the cognitive user chooses one channel to transmit its data. When transmission is successful, the user receives a certain positive reward (e.g. the achievable data rate) that defines the current state of the channel. Extension to multiple cognitive users will be discussed later. Let $s_i(t)$ denote the state of channel i at time t. We define the maximal reward by $r_{max} \triangleq \max_{1 \le i \le K} \sum_{s \in S^i} s$. Let P^i denote the transition probability matrix and $\vec{\pi}_i = \{\pi_i(s)\}_{s \in S^i}$ be the stationary distribution of channel i. We also define $\pi_{min} \triangleq \min_{1 \le i \le K, s \in S^i} \pi_i(s)$. Let λ_i be the second-largest eigenvalue of P^i, and let $\lambda_{max} \triangleq \max_{1 \le i \le K} \lambda_i$. Also, let $\overline{\lambda}_{min} \triangleq 1 - \lambda_{max}$, and let $\overline{\lambda}_i \triangleq 1 - \lambda_i$ be the eigenvalue gap. Let $M^i_{x,y}$ be the mean hitting time of state y starting at initial state x for channel i, and let $M^i_{max} \triangleq \max_{x,y \in S_i, x \ne y} M^i_{x,y}$. We also define:

$$A_{max} \triangleq \max_i \left(\min_{s \in S^i} \pi_i(s)\right)^{-1} \sum_{s \in S^i} s,$$

$$L \triangleq \frac{30 r^2_{max}}{(3 - 2\sqrt{2})\overline{\lambda}_{min}}. \tag{1.1}$$

The stationary reward mean μ_i is given by $\mu_i = \sum_{s \in S^i} s\pi_i(s)$. We define the channel index permutation of $\{1, ..., K\}$ by σ such that $\mu^* \triangleq \mu_{\sigma(1)} \geq \mu_{\sigma(2)} \geq \cdots \geq \mu_{\sigma(K)}$. Let $t^i(n)$ denote the time index of the nth transmission on channel i, and $T^i(t)$ denote the total number of transmissions on channel i by time t. Given these definitions, we can write the total reward by time t as:

$$R(t) = \sum_{i=1}^{K} \sum_{k=1}^{T^i(t)} s_i(t^i(n)). \tag{1.2}$$

Let $\phi(t) \in \{1, 2, ..., K\}$ be a selection rule that indicates which channel is chosen for transmission at time t, which is a mapping from the observed history of the process (i.e. all past transmission actions and reward observations up to time $(t-1)$ to $\{1, 2, ..., K\}$. A policy ϕ is the time series vector of selection rules: $\phi = (\phi(t), t = 1, 2, ...)$.

1.2.2 Performance Measures of the Online Learning Algorithms

By viewing the cognitive user as a player or agent, and the channels as arms the player can choose so as to maximize the long-term accumulated reward, the network model described can be cast as a restless multi-armed bandit (RMAB) problem, which is a generalization of the classic multi-armed bandit (MAB) problem (Gittins, 1979; Lai and Robbins, 1985; Anantharam et al., 1987). In the classic MAB problem, the states of passive arms remain frozen, which is not suitable to model cognitive radio networks, where the channel states (used for transmission or not) change dynamically. In contrast, in the RMAB setting, the state of each arm (active or passive) can change. Thus, the class of RMAB problems considered here has been studied in recent years in the context of cognitive radio networks in (Tekin and Liu, 2012; Liu et al., 2013; Gafni and Cohen, 2018b, a).

The RMAB problem under the Bayesian formulation with known Markovian dynamics has been shown to be P-SPACE hard in general (Papadimitriou and Tsitsiklis, 1999). Nevertheless, there are a number of studies that obtained optimal solutions for some special cases of RMAB models. In particular, the myopic strategy was shown to be optimal in the case of positively correlated two-state Markovian arms with a unit reward for good state and zero reward for bad state (Zhao et al., 2008; Ahmad et al., 2009; Ahmad and Liu, 2009). In (Liu and Zhao, 2010; Liu et al., 2011), the indexability of a special classes of RMAB has been established. In (Wang and Chen, 2012; Wang et al., 2014), the myopic strategy was shown to be optimal for a family of regular reward functions that satisfy axioms of symmetry, monotonicity and decomposability. In our previous work, optimality conditions of a myopic policy have been derived under arm-activation constraints (Cohen et al., 2014).

Although optimal solutions have been obtained for some special cases of RMAB models as detailed, solving RMAB problems directly is intractable in general (Papadimitriou and Tsitsiklis, 1999). Thus, instead of looking for optimal strategies, it is often desired to develop asymptotically optimal strategies with time. Specifically, a widely used performance measure of an algorithm is the *regret*, defined as the reward loss with respect to a player with side information on the model. An algorithm that achieves a sublinear scaling rate of the regret with time is thus asymptotically optimal in terms of approaching the performance of the player with the side information as time increases. The challenge is to design an algorithm that achieves the best sublinear scaling of the regret with time by learning the side information effectively.

In (Auer et al., 2002b), regret was defined as the reward loss of an algorithm with respect to a player that knows the expected reward of all arms and always plays the arm with the highest expected reward. Since this strategy is known to be optimal in the classic MAB under i.i.d. or rested Markovian rewards (up to an additional constant term (Anantharam et al., 1987)), it is commonly used in RMAB with unknown dynamic settings that allows measuring the algorithm performance in a tractable manner. This approach was used later in (Tekin and Liu, 2012; Liu et al., 2013; Gafni and Cohen, 2018b, a) to design effective learning strategies for DSA. In this chapter, we focus on this definition of regret.

1.2.3 The Objective

As explained in Section 1.2.2, we define the regret $r_\phi(t)$ for policy ϕ as the difference between the expected total reward that can be obtained by using the channel with the highest reward mean, and the expected total reward obtained from using policy ϕ up to time t:

$$r_\phi(t) = t\mu_{\sigma(1)} - \mathbb{E}_\phi[R(t)]. \tag{1.3}$$

The objective is to find a policy that minimizes the growth rate of the regret with time. Achieving this goal requires designing a strategy that effectively addresses the well-known exploration versus exploitation dilemma in online learning problems. On the one hand, the cognitive user should explore all channels to learn their unknown states. On the other hand, it should exploit the inference outcome to use the channel with the highest mean. Since the wireless channels are restless (i.e. both active and passive arms are restless in the context of the RMAB model), the designed strategy should learn the Markovian reward statistics consecutively for a period of time (i.e. epoch). This will avoid the potential reward loss due to the transient effect as compared to steady state when switching channels.

1.2.4 Random and Deterministic Approaches

In this subsection, we overview two algorithms that were developed in recent years to achieve the objective just described. The first uses random epoch lengths, dubbed the regenerative cycle algorithm (RCA) (Tekin and Liu, 2012); and the second uses deterministic epoch lengths, dubbed the deterministic sequencing of exploration and exploitation (DSEE) algorithm (Liu et al., 2013). The idea of the RCA algorithm is to catch predefined channel states each time the algorithm enters a transmission epoch when using a channel. These catching times are called *regenerative cycles* and are defined by random hitting times. The channel selection is based on the upper confidence bound (UCB) index (Auer et al., 2002a). The authors showed that RCA achieves a logarithmic scaling of the regret with time. However, since RCA performs a random regenerative cycle at each epoch, the scaling with the mean hitting time M is of order $O(M \log t)$.

The DSEE algorithm overcomes this issue by avoiding the execution of random regenerative cycles at each epoch. Instead, the idea behind DSEE is to use deterministic sequencing of exploration and exploitation epochs. It has been shown that by judiciously designing these deterministic epochs, a logarithmic regret with time is obtained. However, the design of these deterministic epochs by DSEE requires oversampling bad channels for transmission to achieve the desired logarithmic regret. This oversampling results in a scaling order of $O\left((\frac{1}{\sqrt{\Delta}} + \frac{K-2}{\Delta}) \log t\right)$, where K is the number of channels and $0 < \Delta < (\mu_{\sigma(1)} - \mu_{\sigma(2)})^2$ is a known lower bound on the square difference between the highest reward mean $\mu_{\sigma(1)}$ and the second-highest reward mean $\mu_{\sigma(2)}$.

When the channel state space increases, or when the probability of switching between channel states decreases, then the mean hitting times of catching predefined channel states increases, which decreases performance under the RCA algorithm. On the other hand, increasing the number of channels K when $(\mu_{\sigma(1)} - \mu_{\sigma(2)})$ is small as compared to the differences between $\mu_{\sigma(1)}$ and the reward means of other channels decreases performance under the DSEE algorithm. We next overview a very recent algorithm that we developed to overcome these issues.

1.2.5 The Adaptive Sequencing Rules Approach

In (Gafni and Cohen, 2018b,a), we developed the Adaptive Sequencing Rules (ASR) algorithm for solving the RMAB problem for DSA. The basic idea of ASR is to estimate the desired (unknown) exploration rate of each channel online during the algorithm to achieve efficient learning. This approach avoids oversampling bad channels as in DSEE, and at the same time it significantly reduces the total amount of regenerative cycles as required by RCA. Specifically, we show in (Gafni and Cohen, 2018b,a) that we must explore a bad channel $\sigma(i)$, $i = 2, 3, ..., K$, at least $\overline{D}_i \log t$ times, where

$$\overline{D}_i \triangleq \frac{4L}{(\mu^* - \mu_{\sigma(i)})^2}. \tag{1.4}$$

The intuition is that the smaller the difference between the channel reward means, the more samples we must take to infer which one is better with sufficient accuracy. Since the channel reward means are unknown, however, we replace $\mu_{\sigma(i)}$ with its estimate value, which allows us to estimate $\overline{D}_i$. Then, we can use the estimate of $\overline{D}_i$ (which is updated dynamically during time and controlled by the estimate channel reward means) to design an adaptive sequencing rule for sampling channel i. The designed sequencing rules decide whether to enter an exploration epoch or an exploitation epoch, and are adaptive in the sense that they are updated dynamically and controlled by the current estimated reward means in a closed-loop manner. Interestingly, we found that the size of the exploitation epochs is deterministic and the size of the exploration epochs is random under ASR.

In (Gafni and Cohen, 2018b,a), we showed that ASR achieves a logarithmic scaling of the regret with time as achieved by RCA and DSEE. The scaling with the mean hitting time M under ASR, however, is significantly better than the scaling under RCA. Specifically, ASR achieves a scaling order of $O(M \log\log t)$ as compared to $O(M \log t)$ under RCA. We also showed that scaling with the number of channels and Δ under ASR is significantly better than scaling under DSEE. Specifically, ASR achieves a scaling order of $O\left((\frac{1}{\sqrt{\Delta}} + K - 2)\log t\right)$ as compared to $O\left((\frac{1}{\sqrt{\Delta}} + \frac{K-2}{\Delta})\log t\right)$ under DSEE. Extensive simulation results support the theoretical analysis and demonstrate significant performance gains of ASR over RCA and DSEE. We omit the analysis in this chapter and focus on the description of the algorithm.

1.2.5.1 Structure of Transmission Epochs

As discussed earlier, the designed strategy should use the channels in a consecutive manner for a period of time to learn the restless Markovian reward statistics. The RCA algorithm selects channels based on UCB and uses the channel for a random period of time that depends on hitting times used to catch predefined channel states, while the DSEE algorithm uses channels for a deterministic period of time that grows geometrically with time. The ASR algorithm uses hybrid random/deterministic transmission epoch lengths, while determining the exploration rate for each channel according to judiciously designed adaptive sequencing rules, as discussed later. This approach allows ASR to achieve significant performance improvements both theoretically and numerically.

Figure 1.2 illustrates the structure of the time horizon, which consists of exploration and exploitation epochs. The adaptive sequencing rules decide whether to enter exploration epoch

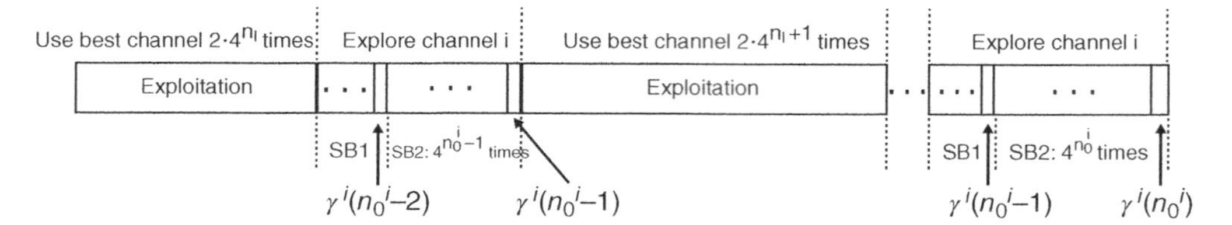

Figure 1.2 An illustration of the exploration and exploitation epochs as implemented by the ASR algorithm. During exploitation epoch, the cognitive user uses the same channel that had the highest sample reward mean at the beginning of the epoch. An exploration epoch is divided into a random-size sub-block SB1 and a deterministic (geometrically growing) size sub-block SB2. SB1 of a channel (say *i*, as in the figure) is random and used to catch the last channel state γ^i observed in the previous exploration epoch. The ASR's selection rule decides which epoch to play at each time. The exploration epoch is illustrated for channel *i* only. In general, an interleaving of exploration epochs for all channels with exploitation epochs (for the channel with the highest sample reward mean) is performed.

and which channel to explore. The design of the adaptive sequencing rules is discussed later. We define $n_O^i(t)$ as the number of exploration epochs in which channel i was used at time t. The exploitation epochs are used to select the channel with the highest sample reward mean, whenever exploration is not being performed. We define $n_I(t)$ as the number of exploitation epochs at time t.

The structure of exploration epoch: As illustrated in Figure 1.2, when entering the $(n_O^i)^{th}$ exploration epoch, the cognitive user starts by using channel i for a random period of time until observing $\gamma^i(n_O^i - 1)$ (i.e. a random hitting time). This random period of time is denoted by sub-block 1 (SB1). After completing SB1, the cognitive user uses the channel for a deterministic period of time with length of $4^{n_O^i}$. This deterministic period of time is denoted by sub-block 2 (SB2). The cognitive user stores the last reward state $\gamma^i(n_O^i)$ observed at the current $(n_O^i)^{th}$ exploration epoch, and so on. The set of time instants during SB2 epochs is denoted by $\mathcal{V}_i$.

The structure of exploitation epoch: The sample reward mean of channel i when entering the $(n_I)^{th}$ exploitation epoch is denoted by $\bar{s}_i$. Then, the cognitive user uses the channel with the highest sample reward mean $\max_i \bar{s}_i$ for a deterministic period of time with length $2 \cdot 4^{n_I - 1}$ (note that the cognitive user does not switch between channels inside epochs). The set of time instants in exploitation epochs is denoted by $\mathcal{W}_i$. The cognitive user computes the sample reward mean $\bar{s}_i$ for each channel based on observations taken at times $\mathcal{V}_i$ and $\mathcal{W}_i$. Observations that have been obtained at the SB1 period are discarded to ensure the consistency of the estimators.

1.2.5.2 Selection Rule under the ASR Algorithm

We next describe the selection rule that the cognitive user applies when deciding whether to explore the channels, or whether to exploit the channel with the highest sample reward mean. Let $\tilde{s}_i(t)$ be the sample reward mean of channel i that the user computes based on observations taken from $\mathcal{V}_i$ only at time t. Let

$$\hat{D}_i(t) \triangleq \frac{4L}{\max\{\Delta, (\max_j \tilde{s}_j(t) - \tilde{s}_i(t))^2 - \epsilon\}}, \tag{1.5}$$

where $0 < \Delta < (\mu_{\sigma(1)} - \mu_{\sigma(2)})^2$ is a known lower bound on $(\mu_{\sigma(1)} - \mu_{\sigma(2)})^2$, and $\epsilon > 0$ is a fixed tuning parameter (we discuss the implementation details later). We also define:

$$I \triangleq \frac{\epsilon^2 \cdot \overline{\lambda}_{min}}{192(r_{max} + 1)^2}. \tag{1.6}$$

As explained earlier, the user must take at least $\overline{D}_i \log t$ samples from each bad channel ($\overline{D}_i$ is given in Eq. (1.4)) to compute the sample means $\bar{s}_i$ with sufficient reliability. Therefore, the user replaces the unknown value $\overline{D}_i$ with its overestimate $\widehat{D}_i(t)$. Furthermore, since $\widehat{D}_i(t)$ is a random variable, we need to guarantee that the desired property holds with a sufficiently high probability. I can be viewed as the minimal rate function of the estimates among all channels and used to guarantee the desired property. Specifically, let $\mathcal{V}_i(t)$ be the set of all time instants during the SB2 period at time t. Then, if there exists a channel (say i) such that the following condition holds

$$|\mathcal{V}_i(t)| \leq \max\left\{\widehat{D}_i(t), \frac{2}{I}\right\} \cdot \log t, \tag{1.7}$$

the cognitive user enters an exploration epoch for channel i. Otherwise, it enters an exploitation epoch (for the channel with the highest sample reward mean). Note that the selection rule for each channel that determines the channel sequencing policy is adaptive in the sense that it is updated dynamically with time and controlled by the random sample reward mean in a closed-loop manner.

1.2.5.3 High-Level Pseudocode and Implementation Discussion

To summarize the discussion, the cognitive user applies the following algorithm:

1) For all K channels, execute an exploration epoch where a single observation is taken from each channel.
2) If condition Eq. (1.7) holds for some channel (say i), then execute an exploration epoch for channel i; and, when completing the exploration epoch, go to Step 2 again. Otherwise, go to Step 3.
3) Execute an exploitation epoch. When completing the exploitation epoch, go to Step 2.

To achieve the theoretical performance, the ASR and DSEE algorithms require the same knowledge of the system parameters. The knowledge of the parameter Δ, however, is not needed under the RCA algorithm. In terms of practical implementations, it is well known that there is often a gap between the sufficient conditions required by theoretical analysis and actual performance. In practice, Δ is not needed, and the parameters can be estimated on the fly. Extensive simulation results demonstrated strong performance when estimating $\overline{D}_i$ directly by setting $\widehat{D}_i(t) \leftarrow \frac{4L}{(\max_j \bar{s}_j(t) - \bar{s}_i(t))^2}$. An example is given in Figure 1.3, where we executed ASR without knowing Δ and by setting ϵ to to zero. $\widehat{D}_i(t)$ was estimated on the fly. It can be seen that ASR significantly outperforms both DSEE and RCA. Additional numerical experiments can be found in (Gafni and Cohen, 2018b,a).

1.3 Learning Algorithms for Channel Allocation

In this section, we consider the more general case where multiple users share the spectrum, and the goal is to effectively allocate channels to users in a distributed manner with the goal of maximizing a certain objective. Since the focus in this section is on the interaction among users (and not between the cognitive user and external processes as discussed in the previous section), we do not assume that there are primary and secondary users in the networks (e.g. as in the open sharing model for DSA (Zhao and Sadler, 2007)). Nevertheless, the model can be extended by adding external processes that are not affected by other users' actions to model the coexistence of primary users in the network.

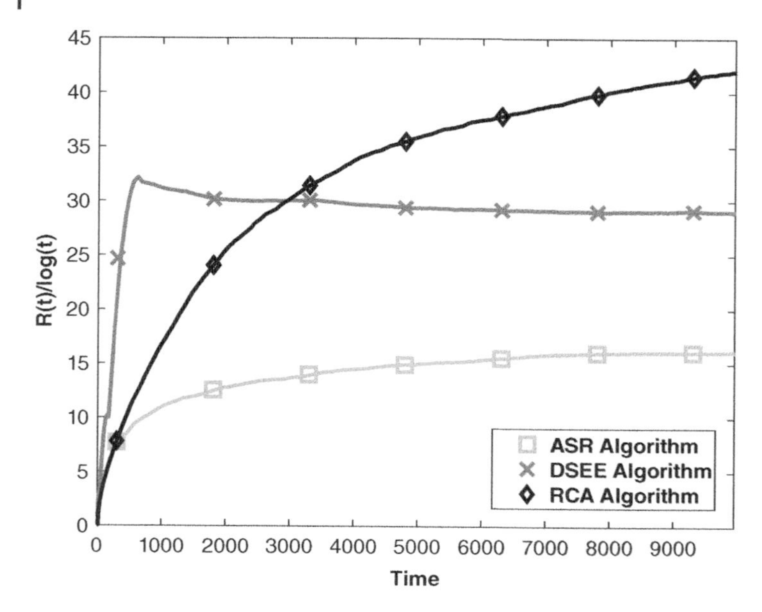

Figure 1.3 The regret (normalized by log t) for the RMAB channel model with eight Markovian channels under ASR, DSEE, and RCA as a function of time. Each channel has two states, good and bad, with rewards 1, 0.1, respectively. The transition probability vectors from good state to bad state and from bad state to good state are [0.09,0.9,0.08,0.7,0.06,0.5,0.04,0.3] and [0.01,0.1,0.02,0.3,0.04,0.5,0.06,0.7], respectively.

1.3.1 The Network Model

Consider a wireless network consisting of a set $\mathcal{N} = \{1, 2, ..., N\}$ of users and a set $\mathcal{K} = \{1, 2, ..., K\}$ of shared orthogonal channels. The users transmit their data on the shared channels using a random access protocol. We mainly focus on Aloha-type narrowband transmission, in which at each time slot, each user is allowed to choose a single channel for transmission with a certain transmission probability. In Section 1.3.2, we discuss other types of communication protocols. We assume that users are backlogged, i.e. all users always have packets to transmit. Transmission on channel k is successful if only a single user transmits over channel k in a given time slot. Otherwise, a collision occurs. Note that in the case where $N \leq K$, it is possible that all users transmit at the same time without collisions. Otherwise, a certain transmission schedule should be applied to reduce the number of collisions. Each user receives an acknowledgement signal (ACK) after transmission to indicate whether its packet was successfully delivered. The ACK signal at time slot t is denoted by the binary observation $o_n(t)$, where $o_n(t) = 1$ indicates successful transmission and $o_n(t) = 0$ indicates that the transmission has failed (i.e. a collision occurred).

The action for user n at time slot t is denoted by $a_n(t) \in \{0, 1, ..., K\}$. We say that $a_n(t) = 0$ when user n does not transmit a packet at time slot t, and $a_n(t) = k$, where $1 \leq k \leq K$, when user n transmits a packet on channel k at time slot t. Note that the actions $a_j(t), j \neq n$, which are taken by other users, are unknown to user n. As a result, the network state (i.e. the entire action profile of all users) at time t is only partially observed by user n through the local signal $o_n(t)$, which makes the problem of obtaining efficient DSA strategies among users challenging. Let $\mathcal{H}_n(t) = (\{a_n(i)\}_{i=1}^t, \{o_n(i)\}_{i=1}^t)$ be the history of user n at time slot t, which is the set of all its past actions and observations. A strategy $\sigma_n(t)$ of user n at time slot t maps from history $\mathcal{H}_n(t-1)$ to a probability mass function over actions $\{0, 1, ..., K\}$. Thus, $\sigma_n(t)$ can be written as $1 \times K$ row vector: $\sigma_n(t) = (p_{n,0}(t), p_{n,1}(t), ..., p_{n,K}(t))$, where $p_{n,k}(t) = \Pr(a_n(t) = k)$ is the probability that user n takes action $a_n(t) = k$ at time t. The time series vector of strategies is denoted by $\boldsymbol{\sigma}_n = (\sigma_n(t), t = 1, 2, ...)$. The strategy profile of all users, and the strategy profile of all users except user n, are denoted by $\boldsymbol{\sigma} = \{\boldsymbol{\sigma}_i\}_{i=1}^n$ and $\boldsymbol{\sigma}_{-n} = \{\boldsymbol{\sigma}_i\}_{i \neq n}$, respectively. The reward for user n at time t, $r_n(t)$ depends on its action $a_n(t-1)$ and the actions taken by all other users at time $t-1$. (i.e. the unknown network state that user n aims to learn). In practice, the reward represents a function of the data rate, latency, etc. We define the accumulated discounted reward by

$$R_n = \sum_{t=1}^{T} \gamma^{t-1} r_n(t), \tag{1.8}$$

where $0 \leq \gamma \leq 1$ is a discounted factor, and T is the time-horizon. Let $\mathbf{E}[R_n(\boldsymbol{\sigma}_n, \boldsymbol{\sigma}_{-n})]$ denote the expected accumulated discounted reward when user n performs strategy $\boldsymbol{\sigma}_n$ and the rest of the users perform strategy profile $\boldsymbol{\sigma}_{-n}$. Then, the objective of each user (say n) is to find strategy $\boldsymbol{\sigma}_n$ that maximizes its expected accumulated discounted reward:

$$\max_{\boldsymbol{\sigma}_n} \quad \mathbf{E}[R_n(\boldsymbol{\sigma}_n, \boldsymbol{\sigma}_{-n})]. \tag{1.9}$$

1.3.2 Distributed Learning, Game-Theoretic, and Matching Approaches

In this subsection, we discuss related works that use game-theoretic models, distributed optimization and learning, and matching techniques, and that were proposed in past and recent years for related models. Very recent developments based on deep learning will be discussed later.

Aloha-Based Protocols and Cross-Layer Optimization. Aloha-based protocols have been widely used for spectrum access primarily because of their easy implementation and random nature. Related work on Aloha-based protocols can be found in (Pountourakis and Sykas, 1992; MacKenzie and Wicker, 2001; Jin and Kesidis, 2002; Shen and Li, 2002; Altman et al., 2004; Bai and Zhang, 2006; Menache and Shimkin, 2008; To and Choi, 2010; Cohen et al., 2012, 2013; Cohen and Leshem, 2013, 2016; Wu et al., 2013) for fully connected networks, where all users interfere with each other, and in (Kar et al., 2004; Wang and Kar, 2006; Baccelli et al., 2006; Kauffmann et al., 2007; Baccelli et al., 2009; Gupta and Stolyar, 2012; Chen and Huang, 2013; Hou and Gupta, 2014; Cohen et al., 2015, 2017) for spatially connected networks, where each user interferes with its neighbors only. Opportunistic Aloha schemes that use cross-layer techniques, in which the design of the medium access control (MAC) layer is integrated with physical layer (PHY) channel information to improve spectral efficiency, have been studied under both the single-channel (Bai and Zhang, 2006; Menache and Shimkin, 2008; Baccelli et al., 2009) and multi-channel (Bai and Zhang, 2006; To and Choi, 2010; Cohen et al., 2012, 2013; Cohen and Leshem, 2013, 2016) cases. Other recent related studies considered opportunistic carrier sensing in a cross-layer design to exploit the channel diversity to optimize a certain objective (e.g. maximizing the data rate, minimizing the invested energy) (Cohen and Leshem, 2009, 2010b,a, 2011; Leshem et al., 2012).

Game-Theoretic Analysis and Distributed Learning. The question of how multiple users can effectively interact and converge to good strategies when applying Aloha-based access has attracted much attention in past and recent years under various network models. A convenient way to address this question is to use game-theoretic analysis in the development of algorithms and convergence analysis because it provides useful tools for modeling and analyzing the multi-user dynamics. Since users in a wireless network are active entities that take actions, modeling the users as players as in a game-theoretic framework allows one to develop game-theoretic tools to establish design principles for the algorithm and to gain insights about its operating points under different objective functions. In (Jin and Kesidis, 2002; Menache and

Shimkin, 2008), the authors considered multiple users that access a single channel using an Aloha-based protocol under a target rate demand for each user (assuming that the demands are feasible), and equilibria analysis has been successfully established. In (Baccelli et al., 2006, 2009), spatial single-channel Aloha networks have been studied under interference channels using stochastic geometry. More recently, the case where multiple users access multiple channels using an Aloha-based protocol as in OFDMA systems, was considered in (Chen and Huang, 2013) and our previous works (Cohen et al., 2012, 2013; Cohen and Leshem, 2013, 2016). In (Chen and Huang, 2013), the authors have developed a distributed algorithm in which a mixed strategy was applied to obtain local information in a spatially distributed network. In (Cohen et al., 2012, 2013; Cohen and Leshem, 2013, 2016), pure strategies were applied, where the local information was obtained by sensing the spectrum in a fully connected network.

A number of studies have considered the problem of achieving proportionally fair rates in spatial random access networks. This problem has been studied under the cases where multiple users access a single channel in (Kar et al., 2004; Wang and Kar, 2006; Gupta and Stolyar, 2012) and multiple channels in (Hou and Gupta, 2014; Cohen et al., 2015, 2017) using Aloha-based protocols. It was shown that by using altruistic plus selfish components in the algorithm design when users update their strategies, proportional fairness can be attained (asymptotically with time). Practically, such algorithms require a small number of message exchanges between users. Furthermore, convergence to the optimal solution requires the judicious application of log-linear learning techniques (see (Young, 1998; Marden and Shamma, 2012) for more details on the theory of log-linear learning). Other related studies that use log-linear learning, and altruistic plus selfish components in the algorithm design under different spectrum-access models and objectives, can be found in (Xu et al., 2012; Herzen et al., 2013; Singh and Chen, 2013; Jang et al., 2014; Herzen et al., 2015).

Cooperative game-theoretic optimization has been studied under frequency-flat interference channels in the single-input and single-output (SISO) (Leshem and Zehavi, 2006; Boche et al., 2007), multiple-input and single-output (MISO) (Jorswieck and Larsson, 2008; Gao et al., 2008) and multiple-input and multiple-output (MIMO) cases (Nokleby et al., 2007). The frequency-selective interference channels case has been studied in (Han et al., 2005; Leshem and Zehavi, 2008). The collision channels case has been studied under a fully connected network and without information sharing between users in (Cohen and Leshem, 2016), where the global optimum was attained under the asymptotic regime (i.e. as the number of users N approaches infinity) and the i.i.d assumption on the channel quality. Other related game-theoretic models have been used in cellular, OFDMA, and 5G systems (Scutari et al., 2006; El-Hajj et al., 2012; Zhao et al., 2014; Zhang et al., 2014; Wang et al., 2016). In (Scutari et al., 2006), the authors modeled and analyzed the problem of power control using potential game theory. Potential games have been used in (Wang et al., 2016; Zhao et al., 2014) as well to analyze effective channel allocation strategies for the downlink operation of multicell networks. In (El-Hajj et al., 2012), the problem of channel allocation in OFDMA systems has been investigated using a two-sided stable matching game formulation. In (Zhang et al., 2014), the authors investigated channel utilization via a distributed matching approach.

Spectrum Access as Matching, and Graph-Coloring Problems. Another set of related works is concerned with modeling spectrum access problem as a matching problem between channels and users. In (Naparstek and Leshem, 2014), the authors developed a fully distributed auction algorithm for optimal channel assignment in terms of maximizing the user sum rate. The expected time complexity was analyzed in (Naparstek and Leshem, 2016). A low-complexity distributed algorithm has been developed in (Leshem et al., 2012) that attains a stable matching solution, where the achievable data rates were assumed known. In (Bistritz and Leshem, 2018a, b; Avner and Mannor, 2016, 2018), channel-assignment algorithms have been developed under partial observation models using distributed learning. Another related approach models the spectrum-access problem as a graph-coloring problem, in which users and channels are represented by vertices and colors, respectively. Thus, coloring vertices such that two adjacent vertices do not share the same color is equivalent to allocating channels such that interference between neighbors is avoided (see (Wang and Liu, 2005; Wang et al., 2009; Checco and Leith, 2013, 2014) and references therein for related works). However, generally, spectrum-access problems are more involved when the number of users is much larger than the number of channels (thus, coloring the graph may be infeasible). Furthermore, users may select more than one channel, and may prefer some channels over others, as well as optimize their rates with respect to the attempt probability.

1.3.3 Deep Reinforcement Learning for DSA

The spectrum-access algorithms that we discussed in the previous section mainly focused on model- and objective-dependent problem settings, often requiring more complex implementations (e.g. carrier sensing, wideband monitoring), and thus the solutions cannot effectively adapt in general for handling more complex real-world models. Therefore, in this section we overview the very recent developments of a deep learning based approach to overcome these issues. Specifically, we now focus on model-free distributed learning algorithms to solve (1.9) that can effectively adapt to changes in network topology, objective functions, time horizons (in which solving dynamic programming becomes very challenging, or often impossible for large T), etc. Obtaining an optimal solution for the spectrum-access problem considered here, however, is a combinatorial optimization problem with a partially observed network state that is mathematically intractable as the network size increases (Zhao et al., 2007). Therefore, in this section we overview methods that use a deep reinforcement learning (DRL) approach, due to its capability to provide good approximate solutions while dealing with a very large state and action spaces. We first describe the basic idea of Q-learning and DRL. We then overview recent developments of DRL to solve the DSA problem.

1.3.3.1 Background on Q-learning and Deep Reinforcement Learning (DRL):

Q-learning is a reinforcement learning method that aims at finding good policies for dynamic programming problems. It has been widely applied in various decision-making problems, primarily because of its ability to evaluate the expected utility from among available actions without requiring prior knowledge about the system model, and its ability to adapt when stochastic transitions occur (Watkins and Dayan, 1992). The algorithm was originally designed for a single agent who interacts with a fully observable Markovian environment (in which convergence to the optimal solution is guaranteed under some regularity conditions in this case). It has been widely applied to more involved settings as well (e.g. multi-agent, non-Markovian environments) and demonstrated strong performance, although convergence to the optimal solution is open in general under these settings. Assume that we can encode the entire history of the process up to time t to a state $s(t)$ that is observed by the system. By applying Q-learning to our setting, the algorithm updates a Q-value at each time t for each action-state pair as follows:

$$\begin{aligned} Q_{t+1}(s_n(t), a_n(t)) = {} & Q_t(s_n(t), a_n(t)) \\ & + \alpha[r_n(t+1) + \gamma \max_{a_n(t+1)} Q_t(s_n(t+1), a_n(t+1)) \\ & - Q_t(s_n(t), a_n(t))], \end{aligned} \tag{1.10}$$

where

$$r_n(t+1) + \gamma \max_{a_n(t+1)} Q_t(s_n(t+1), a_n(t+1)) \tag{1.11}$$

is the learned value obtained by getting reward $r(t+1)$ after taking action $a(t)$ in state $s(t)$, moving to next state $s(t+1)$, and then taking action $a(t+1)$ that maximizes the future Q-value seen at the next state. The term $Q_t(s(t), a(t))$ is the old learned value. Thus, the algorithm aims at minimizing the time difference (TD) error between the learned value and the current estimate value. The learning rate α is set to $0 \leq \alpha \leq 1$, which typically is set close to zero. Typically, Q-learning uses a sliding window to encode the recent history when the problem size is too large.

While Q-learning performs well when dealing with small action and state spaces, it becomes impractical when the problem size increases, mainly for two reasons: (i) a stored lookup table of Q-values for all possible state-action pairs is required, which makes the storage complexity intolerable for large-scale problems; and (ii) as the state space increases, many states are rarely visited, which significantly decreases performance.

In recent years, DRL methods that combine a deep neural network with Q-learning, referred to as deep Q-Network (DQN), have shown great potential for overcoming these issues. Using DQN, the deep neural network maps from the (partially) observed state to an action, instead of storing a lookup table of Q-values. Furthermore, large-scale models can be represented well by the deep neural network so that the algorithm can preserve good performance for very large-scale models. Although convergence to the optimal solution of DRL is an open question, strong performance has been reported in various fields as compared to other approaches. A well-known algorithm was presented in DeepMind's recently published *Nature* paper (Mnih et al., 2015) for teaching computers how to play Atari games directly from the onscreen pixels. A survey of recent studies of DRL in other fields can be found in (Li, 2017).

1.3.4 Existing DRL-Based Methods for DSA

We now overview DRL-based algorithms for DSA that have been developed very recently. We then discuss in detail the deep Q-Learning for spectrum access (DQSA) algorithm that we proposed in (Naparstek and Cohen, 2017, 2018). In (Wang et al., 2017, 2018), the authors developed a spectrum-sensing policy based on DRL for a single user that interacts with an external environment. The multi-user setting that we will discuss in detail in Section 1.3.5, however, is fundamentally different in environment dynamics, network utility, and algorithm design. In (Challita et al., 2017), the authors studied a non-cooperative spectrum-access problem for the case where multiple agents (i.e. base-stations in their model) compete for channels and try to predict the future system state. They use a long short-term memory (LSTM) layer with a REINFORCE algorithm in the algorithm design, and the neural network was trained at each agent. The idea is to let each agent to maximize its own utility, while the predicted state is used to reach a certain fair equilibrium point. From a better comparison point of view, it is worth mentioning here that the DQSA algorithm in Section 1.3.5 as well as the problem setting are fundamentally different. First, in terms of the neural network architecture, DQSA uses LSTM with DQN, which is different from the DQN architecture used in (Challita et al., 2017). Second, in DQSA, the DQN is trained for all users at a single unit (e.g. multi-access edge computing [MEC], cloud, etc.). This setting is more suitable to wireless networking applications implemented by cheap SDRs that only need to update their DQN weights by communicating with the central unit, without implementing extensive training for each device. Examples are cognitive radio networks and the Internet of Things (IoT), which use a large number of cheap SDRs for communications. Third, the DQSA setting allows training the DQN in both cooperative and non-cooperative settings, which leads to fundamentally different operating points depending on the desired objective. Furthermore, in (Challita et al., 2017) the focus was on matching channels to base stations, whereas the focus in DQSA is to share the limited spectrum among a large number of users (i.e. matching might be infeasible). Other related work considered radio control and signal-detection problems, in which a radio signal search environment based on gym reinforcement learning was developed (O'Shea and Clancy, 2016) to approximate the cost of search, as opposed to asymptotically optimal search strategies (Cohen and Zhao, 2015a, b; Huang et al., 2018). Other surveys of recent developments and generalizations of DRL to DSA can be found in (Luong et al., 2018; Mao et al., 2018; Di Felice et al., 2018). Other related works on the general topic of deep learning in mobile and wireless networking can be found in a recent comprehensive survey (Zhang et al., 2018) as well as the other chapters of this book.

1.3.5 Deep Q-Learning for Spectrum Access (DQSA) Algorithm

In this subsection, we discuss in detail the DQSA algorithm that we proposed recently in (Naparstek and Cohen, 2017, 2018), based on deep multi-user reinforcement learning to solve Eq. (1.9). The DQSA algorithm applies for different complex settings and does not require online coordination or message exchanges between users.

We start by introducing the architecture of the DQN used in DQSA in Section 1.3.5.1. The offline and online phases of the algorithm are described in Section 1.3.5.2. The training is done offline at a central unit, whereas the spectrum access is done online in a fully distributed manner (after each user has updated its DQN). More details of the specific setting of the objective function and the design principles of the training phase can be found in (Naparstek and Cohen, 2018).

1.3.5.1 Architecture of the DQN Used in the DQSA Algorithm

We illustrate the architecture of the multi-user DQN used in the DQSA algorithm in Figure 1.4. We next explain each block in the DQN:

1) *Input layer:* The input $\mathbf{x}_n(t)$ to the DQN is a vector of size $2K + 2$. The first $K + 1$ input entries indicate the action (i.e. selected channel) taken at time $t - 1$. Specifically, if the user has not transmitted at time slot $t - 1$, the first entry is set to 1 and the next K entries are set to 0. If the user has transmitted its data on channel k at time $t - 1$ (where $1 \leq k \leq K$), then the $(k + 1)^{th}$ entry is set to 1 and the remaining K entries are set to 0. The following K input entries are the achievable data rate of each channel (when the channel is free), which is proportional to the channel bandwidth. The last input is 1 if an ACK signal has been received. Otherwise, if the transmission has failed or no transmission has been executed, it is set to 0.
2) *LSTM layer:* The complex network model that we handle involves partially observed network states by the users, and non-Markovian dynamics determined by the interactions between the multiple users. Therefore, classical DQNs do not perform well in these cases. To tackle this problem, we add an LSTM layer ((Hausknecht and Stone, 2015)) to the DQN

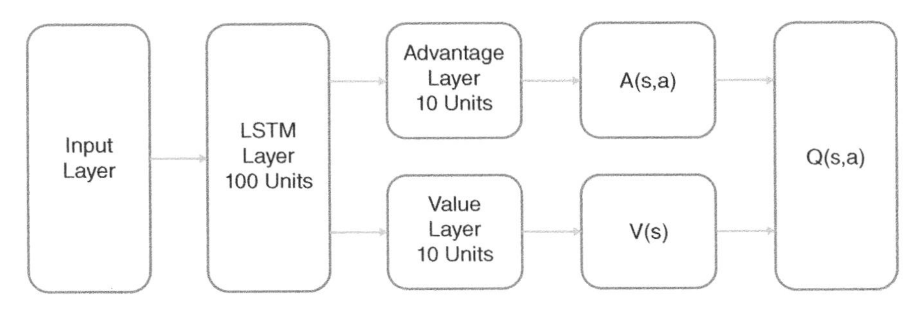

Figure 1.4 An illustration of the architecture of the multi-user DQN used in DQSA algorithm.

that maintains an internal state and aggregates observations over time. The LSTM layer is responsible for learning how to aggregate experiences over time, which makes it possible to infer the true network state based on the history of the process.

3) *Value and advantage layers:* Another improvement that we incorporated is the use of dueling DQNs, as suggested in (Wang et al., 2015), to solve the observability problem in DQNs. The idea of using dueling DQNs is to estimate the average Q-value of the state $V(s_n(t))$ (i.e. the value) independently from the advantage of each action. This operation mitigates the effect of states that are good or bad regardless of the taken action. Specifically, when we input $\mathbf{x}_n(t)$ to the dueling DQNs, we update the Q-value for selecting action $a_n(t)$ at time t by:

$$Q(a_n(t)) \leftarrow V + A(a_n(t)). \tag{1.12}$$

Note that both V and $A(a_n(t))$ depend on the hidden network state $s_n(t)$. The term V is the value of the state, and it estimates the expected Q-value of the state with respect to the chosen action. The term $A(a_n(t)$ is the advantage of each action, and it estimates the Q-value minus its expected value.

4) *Block output layer:* The output of the DQN is a vector of size $K+1$. The first entry of the output vector is the estimated Q-value if the user chooses not to transmit at time t. The $(k+1)^{th}$ entry, where $1 \leq k \leq K$, is the estimated Q-value if the user chooses to transmit on channel k at time t.
5) *Double Q-learning:* Finally, standard Q-learning and DQN that use a max operator (see Eq. (1.10)) are based on the same values for both selecting and evaluating actions. This operation tends to select overestimated values and causes performance reductions. Hence, another component that we add when training the DQN is the use of double Q-learning (Van Hasselt et al., 2016) to decouple the selection of actions from the evaluation of the Q-values. Specifically, we use two neural networks, referred to as DQN_1 and DQN_2. The selection of actions is done by DQN_1, and the estimation of the corresponding Q-value is done by DQN_2.

1.3.5.2 Training the DQN and Online Spectrum Access

We start by describing the pseudocode for training the DQN. Training the DQN is done for all users at a central unit in an offline manner.

Note that the training phase is rarely required to be updated by the central unit (only when the environment characteristics have been significantly changed and no longer reflect the training experiences). The users communicate with the central unit and update their DQN weights from time to time. In real-time, each user (say, n) operates in a fully distributed manner by making autonomous decisions based on the trained DQN. This allows users to learn efficient spectrum-access policies in a distributed manner from local ACK signals only. Specifically, in real-time, each user (say, n) accesses the spectrum as follows:

1) At time slot t, obtain observation $o_n(t)$ and feed input $\mathbf{x}_n(t)$ to the trained DQN_1. Generate output Q-values $Q(a)$ for all actions $a \in \{0, 1, ..., K\}$.
2) Play the following strategy $\sigma_n(t)$: Draw action $a_n(t)$ according to the following distribution:

$$\Pr(a_n(t) = a) = \frac{(1-\alpha)e^{\beta Q(a)}}{\sum\limits_{\tilde{a}\in\{0,1,...,K\}} e^{\beta Q(\tilde{a})}} + \frac{\alpha}{K+1} \quad \forall\ a \in \{0, 1, ..., K\}, \tag{1.13}$$

for small $\alpha > 0$, and β is the temperature (as in simulated annealing techniques). In practice, we take α to zero with time, so that DQSA algorithm tends to choose actions with higher estimated Q-values as time increases.

DQSA Algorithm: Training Phase

1. **for** iteration $i = 1, ..., I$ **do**
2. **for** episode $m = 1, ..., M$ **do**
3. **for** time slot $t = 1, ..., T$ **do**
4. **for** user $n = 1, ..., N$ **do**
5. Observe an input $\mathbf{x}_n(t)$ and feed it into the neural network DQN_1
6. Generate an estimation of the Q-values $Q(a)$ for all available actions $a \in \{0, 1, ..., K\}$ by the neural network
7. Take action $a_n(t) \in \{0, 1, ..., K\}$ (according to Eq. (1.13)) and obtain a reward $r_n(t+1)$
8. Observe an input $\mathbf{x}_n(t+1)$ and feed it into both neural networks DQN_1 and DQN_2
9. Generate estimations of the Q-values $\widetilde{Q}_1(a)$ and $\widetilde{Q}_2(a)$, respectively, for all actions $a \in \{0, 1, ..., K\}$ by the neural networks
10. Form a target vector for the training by replacing the $a_n(t)$ entry by:

$$Q(a_n(t)) \leftarrow r_n(t+1) + \widetilde{Q}_2\left(\arg\max_a\left(\widetilde{Q}_1(a)\right)\right)$$

11. **end for**
12. **end for**
13. **end for**
14. Train DQN_1 with inputs **x**s and outputs Qs.
15. Every ℓ iterations set $Q_2 \leftarrow Q_1$.
16. **end for**

1.3.5.3 Simulation Results

In this subsection, we provide numerical examples to demonstrate the effectiveness of the DQSA algorithm[1]. We refer the reader to (Naparstek and Cohen, 2018) for more details regarding the training mechanisms. We start by investigating the channel throughput of the DQSA algorithm and compare it to slotted-Aloha with an optimal attempt probability. Note that both idle time slots, in which no user accesses the channel, as well as collisions, in which more than two users access the channel at the same time slot, decrease performance. The channel throughput in this simulation is the fraction of time that packets are successfully delivered, where no collisions or idle time slots occur. Note that the slotted-Aloha with optimal attempt probability delivers packets successfully only 37% of the time. By contrast, extensive experimental results that we have performed in (Naparstek and Cohen, 2018) demonstrated that the DQSA algorithm was able to deliver packets successfully almost 80% of the time, about twice the channel throughput obtained by the slotted-Aloha with optimal attempt probability. Furthermore, note that computing the optimal attempt probability when implementing the slotted-Aloha protocol requires knowing the number of users at each time slot. In contrast, the DQSA algorithm was implemented when each user learns only from its ACK signals, without online coordination, message exchanges between users, or carrier sensing.

---

1 We thank Dr. Oshri Naparstek for generating the figures used in this section.

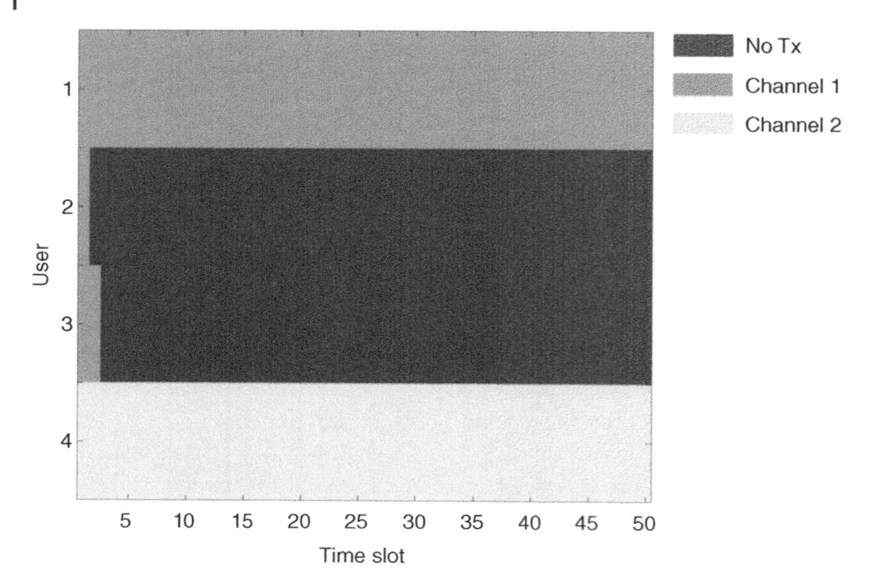

Figure 1.5 A representative channel selection (observed about 80% of the time) when maximizing the user sum rate.

Next, we examined the rate allocation among users obtained under DQSA algorithm. We trained the DQN with different utility functions and demonstrated that users can learn policies that converge to good rate allocations depending on the desired objective. We first considered a network with four users and two channels, and trained the DQN to maximize the user sum rate. In Figure 1.5, we show a representative example of the channel-selection behavior among users. The presented policy after convergence was such that a single user transmits on each channel 100% of the time (users 1, 4 in the figure) while the other users choose not to transmit at all (users 2, 3 in the figure). In terms of the user sum rate, such policies perform very well. Since each user contributes equally to the user sum rate, the users have learned a simple and efficient policy that achieves this goal. Furthermore, the channel throughput was greater than 0.9. We observed such rate allocations in about 80% of the Monte-Carlo experiments.

Although it achieves good performance in terms of user sum rate, the resulting policy in Figure 1.5 performs poorly from a fairness perspective. Therefore, we next examined the rate allocation among users when we trained the DQN such that each user aims at maximizing its own individual rate (i.e. competitive reward). In Figure 1.6, we show a representative example of the channel-selection behavior among users observed in about 80% of the Monte-Carlo experiments. We considered a network with three users and two channels. In this case, we observed that a single user (user 3 in the figure) transmits on a single channel (channel 2 in the figure) 100% of the time, while the two other users (users 1, 2 in the figure) equally share the second channel (channel 1 in the figure) using TDMA-type scheduling. Which one of the users receives a higher data rate depends on the initial conditions and the randomness of the algorithm.

Finally, we examined the performance of the DQSA algorithm when we trained the network to maximize the user sum log-rate, known as *proportionally fair rates*. In Figure 1.7, we show a representative example of the channel-selection behavior among users observed in about 80% of the Monte-Carlo experiments. We considered a network with four users and two channels. Note that the users effectively learn to equally share the channels, which is optimal in terms of proportionally fair rates. It can be seen that the users effectively learn to transmit their data over channels during a batch of time slots and then stop transmitting data. As a result, they (approximately) equally share the number of time slots.

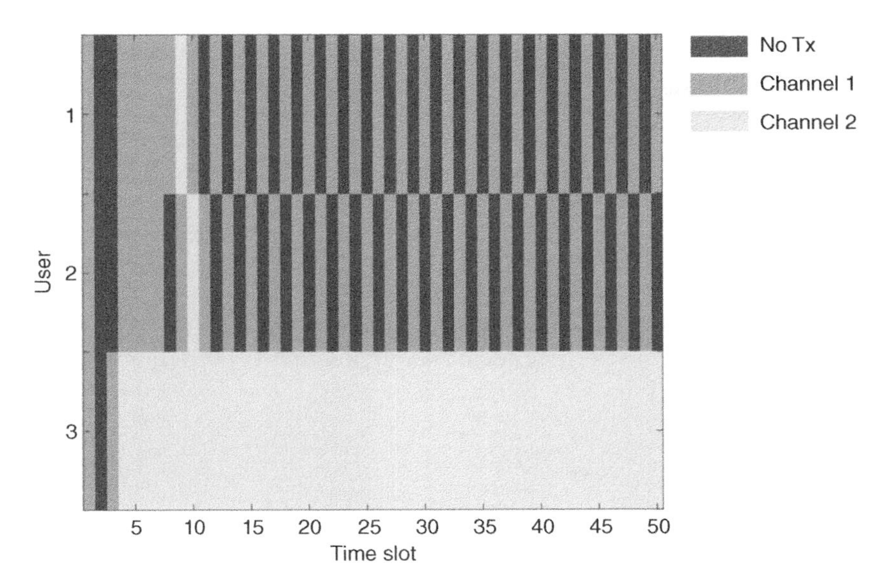

Figure 1.6 A representative channel selection (observed about 80% of the time) under individual utility maximization.

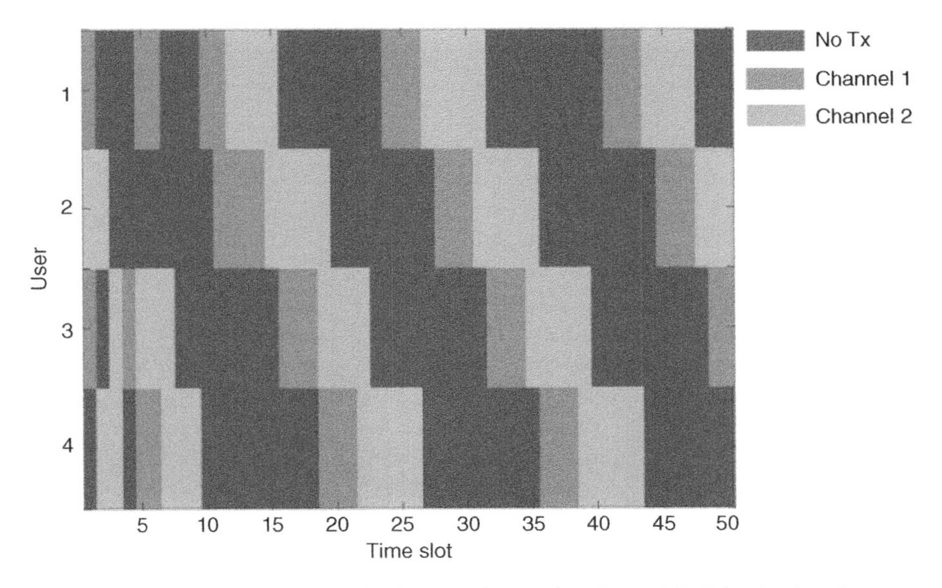

Figure 1.7 A representative channel selection (observed in about 80% of the time) under user sum log-rate maximization.

1.4 Conclusions

In this chapter, we presented an overview of DSA algorithms for efficient spectrum sharing among users that coexist in the same frequency band. The operation of DSA algorithms is based on learning the environment dynamically based on partial observations by continuously monitoring system performance, and judiciously adjusting the transmission parameters to achieve high spectral efficiency.

We started by focusing on online learning algorithms that were developed for learning the occupancy of the spectrum in the presence of external users to improve spectral usage. Then, we focused on algorithms used to effectively allocate channels to users in a distributed manner, with the goal of maximizing a certain global objective. Finally, we focused on very recent developments of artificial intelligence (AI) algorithms based on deep learning for DSA that can effectively self-adapt to complex real-world settings.

Existing AI algorithms for DSA are based mainly on value learning (i.e. Q-learning) methods to obtain the DSA policy. A future research direction in this respect is to develop policy-based learning and hybrid value/policy learning that aim to learn the policy directly, to improve convergence when the state space increases. Another research direction is to develop DRL-based algorithms when users transmit over interference channels with multipacket reception.

Acknowledgments

This work was supported by the Cyber Security Research Center at Ben-Gurion University of the Negev, and by the U.S.-Israel Binational Science Foundation (BSF) under grant 2017723.

Bibliography

Sahand Haji Ali Ahmad and Mingyan Liu. Multi-channel opportunistic access: A case of restless bandits with multiple plays. In *IEEE Annual Allerton Conference on Communication, Control, and Computing*, pages 1361–1368, 2009.

Sahand Haji Ali Ahmad, Mingyan Liu, Tara Javidi, Qing Zhao, and Bhaskar Krishnamachari. Optimality of myopic sensing in multichannel opportunistic access. *IEEE Transactions on Information Theory*, 55(9):4040–4050, 2009.

Eitan Altman, Rachid El Azouzi, and Tania Jiménez. Slotted aloha as a game with partial information. *Computer Networks*, 45(6):701–713, 2004.

Venkatachalam Anantharam, Pravin Varaiya, and Jean Walrand. Asymptotically efficient allocation rules for the multiarmed bandit problem with multiple plays-part II: Markovian rewards. *IEEE Transactions on Automatic Control*, 32(11):977–982, 1987.

Peter Auer, Nicolo Cesa-Bianchi, and Paul Fischer. Finite-time analysis of the multiarmed bandit problem. *Machine learning*, 47(2-3):235–256, 2002a.

Peter Auer, Nicolo Cesa-Bianchi, Yoav Freund, and Robert E Schapire. The nonstochastic multiarmed bandit problem. *SIAM Journal on Computing*, 32 (1):48–77, 2002b.

Orly Avner and Shie Mannor. Multi-user lax communications: a multi-armed bandit approach. In *the 35th Annual IEEE International Conference on Computer Communications (INFOCOM)*, pages 1–9, 2016.

Orly Avner and Shie Mannor. Multi-user communication networks: A coordinated multi-armed bandit approach. *arXiv preprint arXiv:1808.04875*, 2018.

François Baccelli, Bartlomiej Blaszczyszyn, and Paul Muhlethaler. An Aloha protocol for multihop mobile wireless networks. *IEEE Transactions on Information Theory*, 52(2):421–436, 2006.

François Baccelli, Bartlomiej Blaszczyszyn, and Paul Muhlethaler. Stochastic analysis of spatial and opportunistic Aloha. *IEEE Journal on Selected Areas in Communications*, 27(7):1105–1119, 2009.

K. Bai and J. Zhang. Opportunistic multichannel Aloha: distributed multiaccess control scheme for OFDMA wireless networks. *IEEE Transactions on Vehicular Technology*, 55(3):848–855, 2006.

Ilai Bistritz and Amir Leshem. Approximate best-response dynamics in random interference games. *IEEE Transactions on Automatic Control*, 63(6): 1549–1562, 2018a.

Ilai Bistritz and Amir Leshem. Game theoretic dynamic channel allocation for frequency-selective interference channels. *IEEE Transactions on Information Theory*, 2018b.

Holger Boche, Martin Schubert, Nikola Vucic, and Siddharth Naik. Non-symmetric Nash bargaining solution for resource allocation in wireless networks and connection to interference calculus. In *Proc. European Signal Processing Conference*, 2007.

Ursula Challita, Li Dong, and Walid Saad. Proactive resource management in LTE-U systems: A deep learning perspective. *arXiv preprint arXiv:1702.07031*, 2017.

A. Checco and D.J. Leith. Learning-based constraint satisfaction with sensing restrictions. *IEEE Journal of Selected Topics in Signal Processing*, 7(5): 811–820, Oct 2013.

Alessandro Checco and Douglas J Leith. Fast, responsive decentralised graph colouring. *arXiv preprint arXiv:1405.6987*, 2014.

Xu Chen and Jianwei Huang. Distributed spectrum access with spatial reuse. *IEEE Journal on Selected Areas in Communications*, 31(3):593–603, 2013.

Kobi Cohen and Amir Leshem. Time-varying opportunistic protocol for maximizing sensor networks lifetime. In *IEEE International Conference on Acoustics, Speech and Signal Processing (ICASSP)*, pages 2421–2424, 2009.

Kobi Cohen and Amir Leshem. Likelihood-ratio and channel based access for energy-efficient detection in wireless sensor networks. In *IEEE Sensor Array and Multichannel Signal Processing Workshop (SAM)*, pages 17–20, 2010a.

Kobi Cohen and Amir Leshem. A time-varying opportunistic approach to lifetime maximization of wireless sensor networks. *IEEE Transactions on signal processing*, 58(10):5307–5319, 2010b.

Kobi Cohen and Amir Leshem. Energy-efficient detection in wireless sensor networks using likelihood ratio and channel state information. *IEEE Journal on Selected Areas in Communications*, 29(8):1671–1683, 2011.

Kobi Cohen and Amir Leshem. Distributed throughput maximization for multi-channel aloha networks. In *IEEE 5th International Workshop on Computational Advances in Multi-Sensor Adaptive Processing (CAMSAP)*, pages 456–459, 2013.

Kobi Cohen and Amir Leshem. Distributed game-theoretic optimization and management of multichannel aloha networks. *IEEE/ACM Transactions on Networking*, 24(3):1718–1731, 2016.

Kobi Cohen and Qing Zhao. Active hypothesis testing for anomaly detection. *IEEE Transactions on Information Theory*, 61(3):1432–1450, 2015a.

Kobi Cohen and Qing Zhao. Asymptotically optimal anomaly detection via sequential testing. *IEEE Transactions on Signal Processing*, 63(11): 2929–2941, 2015b.

Kobi Cohen, Amir Leshem, and Ephraim Zehavi. A game theoretic optimization of the multi-channel aloha protocol. In *International Conference on Game Theory for Networks (GameNets)*, pages 77–87, 2012.

Kobi Cohen, Amir Leshem, and Ephraim Zehavi. Game theoretic aspects of the multi-channel ALOHA protocol in cognitive radio networks. *IEEE Journal on Selected Areas in Comm.*, 31(11):2276 – 2288, 2013.

Kobi Cohen, Qing Zhao, and Anna Scaglione. Restless multi-armed bandits under time-varying activation constraints for dynamic spectrum access. In *48th Asilomar Conference on Signals, Systems and Computers*, pages 1575–1578, 2014.

Kobi Cohen, Angelia Nedic, and R. Srikant. Distributed learning algorithms for spectrum sharing in spatial random access networks. In *International Symposium on Modeling and Optimization in Mobile, Ad Hoc, and Wireless Networks (WiOpt)*, pages 513–520, May 2015. doi: https://doi.org/10.1109/WIOPT.2015.7151113.

Kobi Cohen, Angelia Nedić, and R Srikant. Distributed learning algorithms for spectrum sharing in spatial random access wireless networks. *IEEE Transactions on Automatic Control*, 62(6):2854–2869, 2017.

Marco Di Felice, Luca Bedogni, and Luciano Bononi. Reinforcement learning-based spectrum management for cognitive radio networks: A literature review and case study. *Handbook of Cognitive Radio*, pages 1–38, 2018.

Ahmad M El-Hajj, Zaher Dawy, and Walid Saad. A stable matching game for joint uplink/downlink resource allocation in ofdma wireless networks. In *Proc. IEEE International Conference on Communications (ICC)*, pages 5354–5359, June 2012.

Tomer Gafni and Kobi Cohen. Learning in restless multi-armed bandits via adaptive arm sequencing rules. *submitted to IEEE Transactions on Automatic Control*, 2018a.

Tomer Gafni and Kobi Cohen. Learning in restless multi-armed bandits using adaptive arm sequencing rules. In *Proc. of the IEEE International Symposium on Information Theory (ISIT)*, pages 1206–1210, Jun. 2018b.

Jie Gao, Sergiy A Vorobyov, and Hai Jiang. Game theoretic solutions for precoding strategies over the interference channel. In *IEEE Global Telecommunications Conference*, pages 1–5, 2008.

John C Gittins. Bandit processes and dynamic allocation indices. *Journal of the Royal Statistical Society*, 41(2):148–177, 1979.

Piyush Gupta and Alexander L Stolyar. Throughput region of random-access networks of general topology. *IEEE Transactions on Information Theory*, 58 (5):3016–3022, 2012.

Zhu Han, Zhu Ji, and KJ Ray Liu. Fair multiuser channel allocation for OFDMA networks using Nash bargaining solutions and coalitions. *IEEE Transactions on Communications*, 53(8):1366–1376, 2005.

Matthew Hausknecht and Peter Stone. Deep recurrent Q-learning for partially observable MDPs. *arXiv preprint arXiv:1507.06527*, 2015.

Julien Herzen, Ruben Merz, and Patrick Thiran. Distributed spectrum assignment for home WLANs. In *IEEE INFOCOM*, pages 1573–1581, 2013.

Julien Herzen, Henrik Lundgren, and Nidhi Hegde. Learning Wi-Fi performance. In *IEEE International Conference on Sensing, Communication, and Networking (SECON)*, pages 118–126, 2015.

I-Hong Hou and Puneet Gupta. Proportionally fair distributed resource allocation in multiband wireless systems. *IEEE/ACM Transactions on Networking*, 22(6):1819–1830, 2014.

B. Huang, K. Cohen, and Q. Zhao. Active anomaly detection in heterogeneous processes. *IEEE Transactions on Information Theory*, pages 1–1, 2018. ISSN 0018-9448. doi: https://doi.org/10.1109/TIT.2018.2866257.

Hyeryung Jang, Se-Young Yun, Jinwoo Shin, and Yung Yi. Distributed learning for utility maximization over CSMA-based wireless multihop networks. In *Proceedings IEEE INFOCOM*, pages 280–288, 2014.

Y. Jin and G. Kesidis. Equilibria of a noncooperative game for heterogeneous users of an ALOHA network. *IEEE Communications Letters*, 6(7):282–284, 2002.

Eduard A Jorswieck and Erik G Larsson. The MISO interference channel from a game-theoretic perspective: A combination of selfishness and altruism achieves pareto optimality. In *IEEE International Conference on Acoustics, Speech and Signal Processing*, pages 5364–5367, 2008.

Koushik Kar, Saswati Sarkar, and Leandros Tassiulas. Achieving proportional fairness using local information in aloha networks. *IEEE Transactions on Automatic Control*, 49(10):1858–1863, 2004.

Bruno Kauffmann, François Baccelli, Augustin Chaintreau, Vivek Mhatre, Konstantina Papagiannaki, and Christophe Diot. Measurement-based self organization of interfering 802.11

wireless access networks. In *In Proc. IEEE International Conference on Computer Communications (INFOCOM) 2007*, pages 1451–1459, 2007.
Tze Leung Lai and Herbert Robbins. Asymptotically efficient adaptive allocation rules. *Advances in applied mathematics*, 6(1):4–22, 1985.
A. Leshem, E. Zehavi, and Y. Yaffe. Multichannel opportunistic carrier sensing for stable channel access control in cognitive radio systems. *IEEE Journal on Selected Areas in Communications*, 30(1):82–95, Jan. 2012.
Amir Leshem and Ephraim Zehavi. Bargaining over the interference channel. In *IEEE International Symposium on Information Theory*, pages 2225–2229, 2006.
Amir Leshem and Ephraim Zehavi. Cooperative game theory and the Gaussian interference channel. *IEEE Journal on Selected Areas in Communications*, 26 (7):1078–1088, 2008.
Yuxi Li. Deep reinforcement learning: An overview. *arXiv preprint arXiv:1701.07274*, 2017.
Haoyang Liu, Keqin Liu, and Qing Zhao. Learning in a changing world: Restless multiarmed bandit with unknown dynamics. *IEEE Transactions on Information Theory*, 59(3):1902–1916, 2013.
Keqin Liu and Qing Zhao. Indexability of restless bandit problems and optimality of whittle index for dynamic multichannel access. *IEEE Transactions on Information Theory*, 56(11):5547–5567, 2010.
Keqin Liu, Richard Weber, and Qing Zhao. Indexability and whittle index for restless bandit problems involving reset processes. In *IEEE Conference on Decision and Control (CDC)*, pages 7690–7696, Dec. 2011.
Nguyen Cong Luong, Dinh Thai Hoang, Shimin Gong, Dusit Niyato, Ping Wang, Ying-Chang Liang, and Dong In Kim. Applications of deep reinforcement learning in communications and networking: A survey. *arXiv preprint arXiv:1810.07862*, 2018.
A.B. MacKenzie and S.B. Wicker. Selfish users in ALOHA: a game-theoretic approach. *IEEE Vehicular Technology Conference*, 3:1354–1357, 2001.
Qian Mao, Fei Hu, and Qi Hao. Deep learning for intelligent wireless networks: A comprehensive survey. *IEEE Communications Surveys & Tutorials*, 20(4): 2595–2621, 2018.
J. R. Marden and J. S. Shamma. Revisiting log-linear learning: Asynchrony, completeness and a payoff-based implementation. *Games and Economic Behavior*, 75:788–808, july 2012.
I. Menache and N. Shimkin. Rate-based equilibria in collision channels with fading. *IEEE Journal on Selected Areas in Communications*, 26(7):1070–1077, 2008.
Volodymyr Mnih, Koray Kavukcuoglu, David Silver, Andrei A Rusu, Joel Veness, Marc G Bellemare, Alex Graves, Martin Riedmiller, Andreas K Fidjeland, Georg Ostrovski, S. Petersen, C. Beattie, A. Sadik, I. Antonoglou, H. King, D. Kumaran, D. Wierstra, S. Legg, and D. Hassabis. Human-level control through deep reinforcement learning. *Nature*, 518(7540):529–533, 2015.
Oshri Naparstek and Kobi Cohen. Deep multi-user reinforcement learning for dynamic spectrum access in multichannel wireless networks. *in Proc. of the IEEE Global Communications Conference (GLOBECOM), available at arXiv*, pages 1–7, Dec. 2017.
Oshri Naparstek and Kobi Cohen. Deep multi-user reinforcement learning for distributed dynamic spectrum access. *IEEE Transactions on Wireless Communications*, 2018. doi: https://doi.org/10.1109/TWC.2018.2879433.
Oshri Naparstek and Amir Leshem. Fully distributed optimal channel assignment for open spectrum access. *IEEE Transactions on Signal Processing*, 62(2):283–294, 2014.
Oshri Naparstek and Amir Leshem. Expected time complexity of the auction algorithm and the push relabel algorithm for maximum bipartite matching on random graphs. *Random Structures & Algorithms*, 48(2):384–395, 2016.
Matthew Nokleby, A Lee Swindlehurst, Yue Rong, and Yingbo Hua. Cooperative power scheduling for wireless MIMO networks. In *IEEE Global Telecommunications Conference*, pages 2982–2986, 2007.

Timothy J O'Shea and T Charles Clancy. Deep reinforcement learning radio control and signal detection with kerlym, a gym rl agent. *arXiv preprint arXiv:1605.09221*, 2016.

Christos H Papadimitriou and John N Tsitsiklis. The complexity of optimal queuing network control. *Mathematics of Operations Research*, 24(2): 293–305, 1999.

IE Pountourakis and ED Sykas. Analysis, stability and optimization of ALOHA-type protocols for multichannel networks. *Computer Communications*, 15(10):619–629, 1992.

Gesualdo Scutari, Sergio Barbarossa, and Daniel P Palomar. Potential games: A framework for vector power control problems with coupled constraints. In *Proc. IEEE International Conference on Acoustics, Speech and Signal Processing (ICASSP)*, volume 4, pages 241–244, May 2006.

D. Shen and V.O.K. Li. Stabilized multi-channel ALOHA for wireless OFDM networks. *The IEEE Global Telecommunications Conference (GLOBECOM)*, 1:701–705, 2002.

Charan Kamal Singh and Chung Shue Chen. Distributed downlink resource allocation in cellular networks through spatial adaptive play. In *25th IEEE International Teletraffic Congress (ITC)*, pages 1–9, 2013.

Cem Tekin and Mingyan Liu. Online learning of rested and restless bandits. *IEEE Transactions on Information Theory*, 58(8):5588–5611, 2012.

T. To and J. Choi. On exploiting idle channels in opportunistic multichannel ALOHA. *IEEE Communications Letters*, 14(1):51–53, 2010.

Hado Van Hasselt, Arthur Guez, and David Silver. Deep reinforcement learning with double Q-learning. In *AAAI*, pages 2094–2100, 2016.

Jiao Wang, Yuqing Huang, and Hong Jiang. Improved algorithm of spectrum allocation based on graph coloring model in cognitive radio. In *WRI International Conference on Communications and Mobile Computing*, volume 3, pages 353–357, 2009.

Kehao Wang and Lin Chen. On optimality of myopic policy for restless multi-armed bandit problem: An axiomatic approach. *IEEE Transactions on Signal Processing*, 60(1):300–309, 2012.

Kehao Wang, Lin Chen, and Quan Liu. On optimality of myopic policy for opportunistic access with nonidentical channels and imperfect sensing. *IEEE Transactions on Vehicular Technology*, 63(5):2478–2483, 2014.

Shangxing Wang, Hanpeng Liu, Pedro Henrique Gomes, and Bhaskar Krishnamachari. Deep reinforcement learning for dynamic multichannel access. In *International Conference on Computing, Networking and Communications (ICNC)*, 2017.

Shangxing Wang, Hanpeng Liu, Pedro Henrique Gomes, and Bhaskar Krishnamachari. Deep reinforcement learning for dynamic multichannel access in wireless networks. *to appear in the IEEE Transactions on Cognitive Communications and Networking*, 2018.

Wei Wang and Xin Liu. List-coloring based channel allocation for open-spectrum wireless networks. In *IEEE Vehicular Technology Conference*, volume 62, page 690, 2005.

Xidong Wang, Wei Zheng, Zhaoming Lu, Xiangming Wen, and Wei Li. Dense femtocell networks power self-optimization: an exact potential game approach. *International Journal of Communication Systems*, 29(1):16–32, January 2016.

Xin Wang and Koushik Kar. Cross-layer rate optimization for proportional fairness in multihop wireless networks with random access. *IEEE Journal on Selected Areas in Communications*, 24(8):1548–1559, 2006.

Ziyu Wang, Tom Schaul, Matteo Hessel, Hado van Hasselt, Marc Lanctot, and Nando de Freitas. Dueling network architectures for deep reinforcement learning. *arXiv preprint arXiv:1511.06581*, 2015.

Christopher JCH Watkins and Peter Dayan. Q-learning. *Machine learning*, 8 (3-4):279–292, 1992.

Huasen Wu, Chenxi Zhu, Richard J La, Xin Liu, and Youguang Zhang. FASA: Accelerated S-ALOHA using access history for event-driven M2M communications. *IEEE/ACM Transactions on Networking*, 21(6):1904–1917, 2013.

Yuhua Xu, Jinlong Wang, Qihui Wu, Alagan Anpalagan, and Yu-Dong Yao. Opportunistic spectrum access in cognitive radio networks: Global optimization using local interaction games. *IEEE Journal of Selected Topics in Signal Processing*, 6(2):180–194, 2012.

H Peyton Young. *Individual strategy and social structure. Princeton University Press*, 1998.

Chaoyun Zhang, Paul Patras, and Hamed Haddadi. Deep learning in mobile and wireless networking: A survey. *arXiv preprint arXiv:1803.04311*, 2018.

Yanru Zhang, Yunan Gu, Miao Pan, and Zhu Han. Distributed matching based spectrum allocation in cognitive radio networks. In *Proc IEEE Global Communications Conference (GLOBECOM)*, pages 864–869, Dec. 2014.

Jun Zhao, Haijun Zhang, Zhaoming Lu, Xiangming Wen, Wei Zheng, Xidong Wang, and Zhiqun Hu. Coordinated interference management based on potential game in multicell OFDMA networks with diverse QoS guarantee. In *Proc. IEEE Vehicular Technology Conference (VTC Spring)*, pages 1–5, May 2014.

Qing Zhao and Brian M. Sadler. A survey of dynamic spectrum access. *IEEE Signal Processing Magazine*, 24(3):79–89, 2007.

Qing Zhao, Lang Tong, Ananthram Swami, and Yunxia Chen. Decentralized cognitive MAC for opportunistic spectrum access in ad hoc networks: A POMDP framework. *IEEE Journal on selected areas in communications*, 25 (3), 2007.

Qing Zhao, Bhaskar Krishnamachari, and Keqin Liu. On myopic sensing for multi-channel opportunistic access: structure, optimality, and performance. *IEEE Transactions on Wireless Communications*, 7(12):5431–5440, 2008.

2

Reinforcement Learning for Resource Allocation in Cognitive Radio Networks

Andres Kwasinski[1], *Wenbo Wang*[2], *and Fatemeh Shah Mohammadi*[3]

[1] *Department of Computer Engineering, Rochester Institute of Technology, New York, USA*
[2] *School of Computer Science and Engineering, Nanyang Technological University, Singapore*
[3] *Kate Gleason College of Engineering, Rochester Institute of Technology, New York, USA*

In this chapter, we will be discussing the use of machine learning (ML) to perform distributed resource allocation in cognitive radio networks. With the idea of framing the presentation within a general scenario, we will consider a setup as illustrated on the left side of Figure 2.1, consisting of two wireless networks sharing a band of the radio spectrum. The primary network (PN) is incumbent to the radio spectrum band. The secondary network (SN) makes use of the radio spectrum band as permitted by the PN following one of the existing dynamic spectrum access (DSA) approaches. For a better presentation, but without loss of generality, this chapter will concentrate on a setup based on the underlay DSA approach, where nodes from the PN and the SN are allowed to transmit simultaneously over the same frequency band as long as interference from the SN to the PN remains below a threshold (Goldsmith et al. (2009)). To access the radio spectrum and operate using underlay DSA, the nodes in the SN, called the secondary users (SUs), instantiate machine learning and signal processing software that collectively form was is called a *cognitive engine*. It is this cognitive engine that effectively enables the SU to become a cognitive radio (CR).

From its early instances, the paradigm of CRs has been based on the cognition cycle as a framework to structure the operation of a wireless device to autonomously gain awareness of its state within a wireless network environment and learn how to adapt its actions as a communication device. When operating based on the cognition cycle, a CR interacts with the wireless network environment by following a cyclical sequence of states, starting with the measurement of variables from the environment (seen as stimuli) in the *Observe* state, from which the CR interprets the state of the environment. The result of the Observe state is fed to the *Decide* state, where the CR uses the current state of the environment and information learned from past experiences to adaptively decide on actions to be used to communicate through a radio waveform. The chosen actions are translated into parameter settings and applied to the transmission of a message in the last state of the cognition cycle, the *Act* state. The complete cognition cycle starts with the CR measuring variables from the environment and ends with the CR interacting with the environment by realizing the chosen actions, effectively having the environment as the medium through which the cycle is closed.

At a deeper level, the use of the CR paradigm, with its cognition cycle, presents opportunities for more effective resource allocation solutions beyond what is apparent from the direct application of ML algorithms. A case to this point is the realization of cross-layer techniques for resource allocation. It is well known that a cross-layer approach to resource allocation leads to

Machine Learning for Future Wireless Communications, First Edition. Edited by Fa-Long Luo.
© 2020 John Wiley & Sons Ltd. Published 2020 by John Wiley & Sons Ltd.

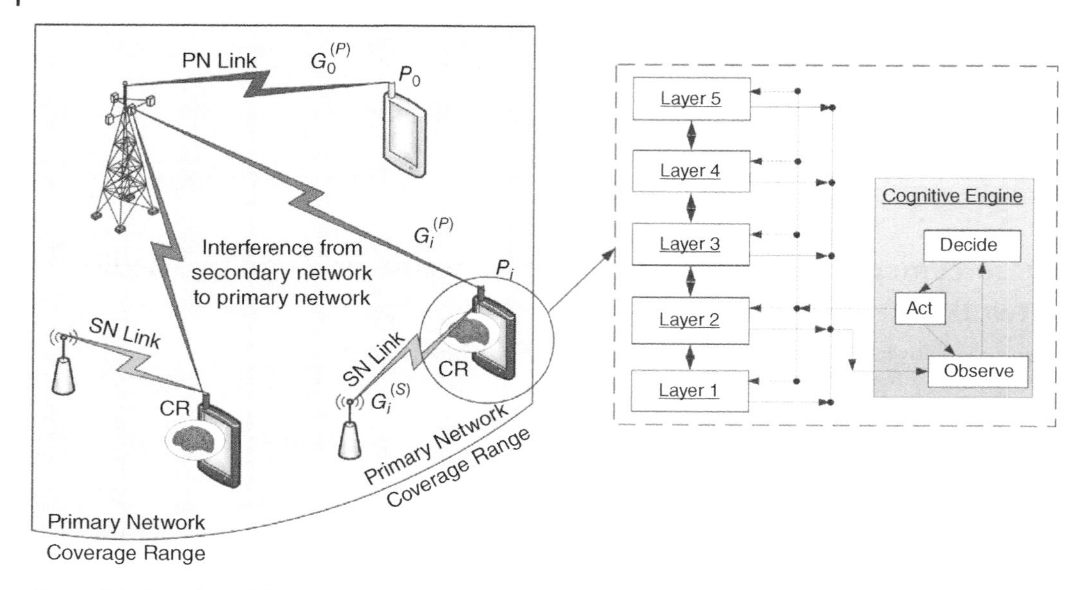

Figure 2.1 Overview of the network setup that is the focus of this chapter.

better-performing solutions. However, commercial wireless device architectures remain layered, not only because a cross-layer design may lead to compromises on a modular network protocol implementation, but, more importantly, because the development process over the complete networking stack is divided into different teams, each specialized in implementing one layer or a sublayer in a specific processor (e.g. main CPU or baseband radio processor). This results in software modules that lack the capability to exchange the data and commands needed to implement cross-layer solutions. Nevertheless, as illustrated on the right part in Figure 2.1, the cognition cycle enables seamless modular cross-layer wireless networking technology by implementing a cognitive engine that senses the environment (in the Observe state) and issues actions on the environment (in the Act state) with variables and parameters that correspond to all layers of the networking protocol stack. This is a framework that gracefully allows cross-layer operations while retaining a modular architecture, and it will be used in this chapter when appropriate.

As indicated, in this chapter we will focus on the use of ML to perform distributed resource allocation in cognitive radio networks. Among the existing ML techniques, we are interested in those that reflect the cognition cycle and that are apt for their application in cross-layer solutions. Moreover, in adopting the CR paradigm, we are interested in considering ML techniques that do not presume an a priori model for the effect of actions on the environment but that derive this knowledge as part of the learning process in the CR. An essential group of ML techniques that agree with these requirements is *reinforcement learning*.

In a high-level view, with reinforcement learning, an agent learns the best action to take by following a process of trying the different possible actions and evaluating their fitness with respect to achieving a goal. The competence of actions to achieve a goal is evaluated by calculating a reward (or cost) from the measured effect that each taken action has on the environment. The name *reinforcement learning* comes from the fact that an agent learns by trying out all actions multiple times, gradually favoring those that result in a larger reward, and eventually deciding on the action with the largest reward. Typically, the learning process gradually evolves from the *exploration phase*, where the agent tries actions mostly at random, with little knowledge of their effect on the environment, to the *exploitation phase*, where the agent prioritizes the trial of those actions with larger rewards. There are many reinforcement learning techniques; one of the most common is Q-learning, which is adopted in this chapter.

While reinforcement learning techniques provide the means for an agent to learn the effects of its actions on the environment, the process of trying the actions multiple times tends to result in a slow learning process. A recent development to address this issue has been to consider the effect of the action on the environment and its reflection on the intended goal as a function, with arguments being the taken action and the state of the environment. With the function being initially unknown, the agent learns it by applying ML techniques. Learning speed is increased by avoiding having to test all actions multiple times. This is the central idea of the deep Q-learning technique that is discussed later in this chapter. Moreover, in the context of having cognitive agents that can communicate with each other, another approach to accelerate learning is to use the experience gained by some agents and transfer it to less-experienced agents so that they can shorten their exploration phase. Later in this chapter, we will address this idea in more detail when discussing cooperative learning and transfer learning.

The rest of this chapter is organized into four sections. Section 2.1 explains the use of Q-learning for cross-layer resource allocations, and Section 2.2 describes resource allocation based on the deep Q-learning technique. Section 2.3 focuses on how different CRs can cooperate during the learning process. Finally, Section 2.4 presents some concluding remarks.

2.1 Use of Q-Learning for Cross-layer Resource Allocation

For ease of exposition, we begin by considering a scenario consisting of a single primary channel, where a group of SUs in the SN try to inject real-time (subject to strict delay constraints) multimedia traffic while avoiding interference to the primary transmission that exceeds a predetermined threshold. Without loss of generality, we consider that the channel is a quasi-static one and that the single PU transmits with fixed power. In addition, the transmissions are subject to additive white Gaussian noise (AWGN) with power σ^2. Let P_0 denote the PU's transmit power and P_i denote SU i's transmit power ($1 \leq i \leq N$). Then the signal-to-interference-plus-noise ratio (SINR) of the PU is expected to satisfy the following condition:

$$\gamma_0^{(P)} = \frac{G_0^{(P)} P_0}{\sum_{i=1}^{N} G_i^{(P)} P_i + \sigma^2} \geq \gamma_0, \tag{2.1}$$

where $G_0^{(P)}$ is the channel gain from the PU to the primary base station (BS), $G_i^{(P)}$ is the channel gain from SU i to the primary BS, and γ_0 is the PU's required SINR threshold as determined from the underlay DSA configuration. Similarly, for SU i, the corresponding SINR is expected to satisfy the condition

$$\gamma_i^{(S)} = \frac{G_i^{(S)} P_i}{G_0^{(S)} P_0 + \sum_{i=1}^{N} G_i^{(S)} P_i + \sigma^2} \geq \gamma_i, \tag{2.2}$$

where $G_0^{(S)}$ is the channel gain from the PU to the secondary BS, $G_i^{(S)}$ is the channel gain from SU i to the secondary BS, and γ_i is SU i's required SINR threshold determined from quality of service (QoS) goals. Assuming that the transmission achieves the performance given by Shannon's channel capacity, we can express the PU throughput as

$$R_0^{(P)} = W \log_2(1 + \gamma_0^{(P)}), \tag{2.3}$$

and that of SU i $(1 \leq i \leq N)$ as

$$R_i^{(S)} = W\log_2(1 + \gamma_i^{(S)}), \tag{2.4}$$

where W is the channel bandwidth.

Conventionally, from the perspective of a single SU i, when transmitting real-time multimedia traffic (e.g. videos), the link performance can be objectively measured at the application layer by the expected end-to-end distortion, denoted as D_i. Being an end-to-end metric, the expected distortion D_i is composed of two components: a source-encoding distortion, D_i^s, and a channel-induced distortion, D_i^c (Kwasinski and Liu (2008)). The source-encoding distortion D_i^s depends on the source-encoding rate $x_i^{(S)}$, which is subject to the limit of the channel throughput $R_i^{(S)}$. For conciseness of discussion, we can adopt a synthetic Gaussian source that provides the worst distortion case for all possible source sample statistics. Then, when adopting the mean squared error as the distortion metric, we can measure D_i^s as

$$D_i^s(x_i^{(S)}) = \eta 2^{-2x_i}, \tag{2.5}$$

where η is the maximum distortion (equal to the variance of the synthetic Gaussian source).

On the other hand, the channel-induced distortion D_i^c is due to errors introduced during transmission over the wireless channel that cannot be corrected by the error-correction mechanism. When assuming that these errors are concealed by replacing the erred source samples with their expected value, the channel-induced distortion is a constant equal to the variance of the source samples. The impact of D_i^c on the end-to-end distortion is determined by the probability that an uncorrectable transmission error occurs, which we denote by Pr_i^e. Furthermore, Pr_i^e depends on both the channel SINR $\gamma_i^{(S)}$ and the channel coding rate $R_i^{(S)}$, which determine the strength of the channel coding. Also, it is worth noting that given a channel throughput $R_i^{(S)}$ and a fixed channel-coding rate $R_i^{(S)}$, the best-effort source-encoding rate can be determined as $x_i^{(S)} = r_i^{(S)} R_i^{(S)}$. Considering all these observations, the end-to-end distortion can be expressed as a function of the parameters $(\gamma_i^{(S)}, r_i^{(S)}, R_i^{(S)})$ in the following form:

$$D_i(\gamma_i^{(S)}, r_i^{(S)}, R_i^{(S)}) = D_i^c \Pr\left(\gamma_i^{(S)}, r_i^{(S)}\right) + \eta 2^{-2x_i(\gamma_i^{(S)}, r_i^{(S)}, R_i^{(S)})}\left(1 - \Pr(\gamma_i^{(S)}, r_i^{(S)})\right). \tag{2.6}$$

Note from Eqs. (2.2) and (2.4) that $\gamma_i^{(S)}$ and $R_i^{(S)}$ are jointly determined by the transmit powers of all the SUs, i.e. $\mathbf{P}^{(S)} = [P_1^{(S)}, \dots, P_N^{(S)}]^\top$. Therefore, Eq. (2.6) can be rewritten as

$$D_i\left(\mathbf{P}^{(S)}, r_i^{(S)}\right) = D_i^c \Pr\left(\mathbf{P}^{(S)}, r_i^{(S)}\right) + \eta 2^{-2x_i(\mathbf{P}^{(S)}, r_i^{(S)})}\left(1 - \Pr\left(\mathbf{P}^{(S)}, r_i^{(S)}\right)\right). \tag{2.7}$$

The goal of each SU will be to perform resource allocation so as to minimize the expected end-to-end distortion during transmission of its real-time multimedia source. Note that this involves a cross-layer approach, since the resources to be allocated by each SU are $(P_i^{(S)}, r_i^{(S)})$, with the goal of minimizing the expected end-to-end distortion, which depends on parameters from multiple layers. Due to the distributed nature of the SN in a practical scenario, we focus on finding for the SUs a decentralized resource-allocation mechanism. In other words, an individual transmitting SU i independently searches for its optimal cross-layer choice of power and channel-coding rate that minimizes end-to-end distortion while keeping the information exchange with the other SUs at a minimum. Here, the requirement of minimum information exchange is due to both the limited signaling capability over a non-dedicated channel and the privacy concerns of revealing local transmitting parameters (e.g. channel-coding parameters) to peer devices. We note that such a limit will cause difficulty coordinating among the SUs. As noted in (Han et al. (2012)), the simple best response-based strategy searching scheme will end up in an arms race, where each SU keeps increasing its transmit power and consequently causes interference to both other SUs and the PU to increase, finally exceeding the SINR limits. This suggests that it is necessary to introduce repeated play among the SUs to help them out of the trap of the prisoner dilemma-like equilibrium. Therefore, considering the strategy coupling between SUs, it is natural to establish our cross-layer strategy-searching scheme based on the framework of reinforcement learning as presented in the following.

Without loss of generality, we consider that both the power level and the channel-coding rate of an SU are selected from a discrete set. Assume that the parameter sets of each SU are the same and are denoted by $\mathcal{P} = \{P_0, \dots, P_L\}$ and $\mathcal{R} = \{r_0, \dots, r_M\}$. Then we can map the problem of transmitting-strategy searching into a general reinforcement learning (RL) process, which consists of three major elements: a set of system state $s \in \mathcal{S}$, a set of system actions $a \in \mathcal{A}$, and a vector of individual cost functions $\mathbf{c} = [c_1, \dots, c_N]^\top$, which measures the instantaneous quality of a local action given the other SUs' choices and the current system state. Specifically, given the current system state s, the transition to a new system state is (probabilistically) determined by the joint actions of all the SUs. Meanwhile, each SU relies on its own observation of the state changes and the evaluated cost level (i.e. feedback) to guide its further selection of transmitting actions.

To avoid explicitly coordinating the power allocation upon the interference conditions given in Eqs. (2.1) and (2.2), we define the system state at a discrete time interval t as $s_t = (I_{i,t}, L_t)$, where $I_{i,t}$ reflects whether the interference condition of an individual SU i is satisfied as per Eq. (2.2), and L_t reflects whether the interference condition of the PU is satisfied according to the underlay DSA approach. Namely,

$$I_{i,t} = \begin{cases} 0, & \text{if } \gamma_{i,t}^{(S)} \geq \gamma_i, \\ 1, & \text{otherwise,} \end{cases} \tag{2.8}$$

and

$$L_t = \begin{cases} 0, & \text{if } \gamma_{0,t}^{(P)} \geq \gamma_0, \\ 1, & \text{otherwise.} \end{cases} \tag{2.9}$$

where $\gamma_{0,t}^{(P)}$ and $\gamma_{i,t}^{(S)}$ are given in Eqs. (2.1) and (2.2), respectively. The action of SU i is defined by $a_{i,t} = (P_{i,t}^{(S)}, r_{i,t}^{(S)})$. By Eq. (2.7), we define the cost function of SU i at time instance t as follows:

$$c_{i,t} = \begin{cases} C, & \text{if } L_t + I_{i,t} > 0, \\ D_{i,t}(P_{i,t}^{(S)}, r_{i,t}^{(S)}), & \text{otherwise,} \end{cases} \tag{2.10}$$

where C is an arbitrarily large constant (larger than the maximum possible valuable for $D_{i,t}(P_{i,t}^{(S)}, r_{i,t}^{(S)})$) indicating a failure for a secondary transmission to meet the target SINR or the violation of the DSA rule for the PU. Also, $D_{i,t}(P_{i,t}^{(S)}, r_{i,t}^{(S)})$ follows from Eq. (2.7).

With the system state, SU actions, and individual cost functions defined by Eqs. (2.8) through (2.10), we are ready to approximate the decentralized strategy-searching process by a group of standard single-agent Q-learning processes (Sutton and Barto (2011), Wang et al. (2016)). Here, by approximation, we assume that the SUs do not observe the joint SU actions and are unaware of their effect on state transitions. Namely, each SU treats the other SUs as if they are part of the stationary environment. Let $\pi_i(a|s_i)$ denote the probability of SU i selecting action $a \in \mathcal{A}$ given the observed state $s_i = (I, L)$. We consider that SU i repeatedly adjusts its strategy π_i in order to finally obtain an optimal strategy that minimizes the state value, i.e. the expected sum of costs over time with a discounting factor β_i $(0 < \beta_i < 1)$:

$$V_{\pi_i}(s_i) = E_\pi \left[c_{t+1}(s_i, a_i) + \beta_i V_{\pi_i}(s_{i,t+1}) | s_{i,t} = s_i \right]. \tag{2.11}$$

Algorithm 1 - Basic strategy-learning based on standard Q-learning.

Require: $\forall i = 1, \dots, N, \forall s_i \in \mathcal{S}, \forall a_i \in \mathcal{A}$, set $t = 0$ and $Q_{i,t}(s_i, a_i) = 0$.
1: **while** $\| \sum_{(s_{i,t}, a_{i,t})} Q_i(s_{i,t}, a_{i,t}) - \sum_{(s_{i,t+1}, a_{i,t+1})} Q_i(s_{i,t+1}, a_{i,t+1}) \| > \chi$ for a given precision $\chi > 0$ and $t \leq T$ **do**
2: **for all** $i = 1, \dots, N$ **do**
3: Select the new action $a_{i,t+1} = \min_{a_i \in \mathcal{A}} Q^*_{i,t}(s'_i = s_{i,t}, a_i)$.
4: **end for**
5: **for all** $i = 1, \dots, N$ **do**
6: Update the states and costs according to Eqs. (2.8)-(2.10).
7: Update the Q-value $Q_{i,t+1}(s_i, a_{i,t})$ according to Eq. (2.13).
8: **end for**
9: $t \rightarrow t + 1$.
10: **end while**

According to Bellman's principle of optimality, the solution for finding the optimal state value can be obtained in a recursive manner. Namely, it can be obtained by taking the current optimal action, provided that the optimal strategies thereafter are available:

$$V^*_{\pi^*_i}(s_i) = \min_{a_i \in \mathcal{A}} \left[c(s_i, a_i) + \beta_i \sum_{s'_i \in \mathcal{S}} p(s'_i | s_i, a_i) V^*_{\pi^*_i}(s'_i) \right], \tag{2.12}$$

where we remove the time instance suffixes t or $t + 1$ to unclutter the notation. In Eq. (2.12), we denote π^*_i as the optimal strategy of SU i and $p(s'_i | s_i, a_i)$ as the state transition probability from s_i to s'_i given the chosen action a_i. Observing Eqs. (2.8) and (2.9), we note that the system state at the next time instance depends only on the joint actions of the SUs. Therefore, by assuming that Eq. (2.12) holds, we are able to apply the standard, single-agent Q-learning scheme for approximately finding the optimal deterministic strategy of each SU. Then, letting $Q_{i,t}(s_i, a_i)$ denote the Q-value (i.e. the estimated state-action value) of SU i in its strategy-updating process, we can update the Q-values with the iterative scheme of state-action value backups as follows:

$$Q_{i,t+1}(s_i, a_{i,t}) \leftarrow (1 - \alpha_{i,t}) Q_{i,t}(s_i, a_{i,t}) + \alpha_{i,t} \left[c_{i,t}(s_i, a_{i,t}) + \beta_i Q^*_{i,t}(s'_i = s_{i,t+1}) \right], \tag{2.13}$$

where $\alpha_{i,t}$ is the learning rate, $0 < \alpha_{i,t} < 1$, and $Q^*_{i,t}$ is SU i's minimum Q-value at stage t:

$$Q^*_{i,t+1}(s_i) = \min_{a_i \in \mathcal{A}} Q^*_{i,t}(s'_i = s_{i,t+1}, a_i), \tag{2.14}$$

which approximates the optimal state-action value after the state is updated to $s'_i = s_{i,t+1}$. The corresponding action a^*_i in Eq. (2.14) also provides the action update at the next time instance $t + 1$. Then, the basic strategy-learning scheme based on the standard, single-agent Q-learning scheme can be summarized in Algorithm 1.

It is worth noting that the standard off-policy Q-learning algorithm has an exponential time complexity. Therefore, Algorithm 1 is only computationally tractable when the state-action space of each SU is sufficiently small. Consider that by ignoring the interaction between SUs and assuming the underlying state-transitions of the SUs are stationary, the basic individual learning scheme in Algorithm 1 only provides an approximated process of optimal strategy searching. Moreover, a coarse discrete state-action space with actions represented in relatively small sets may severely undermine the representativeness of the considered model for optimal strategy searching. As a result, the use of Q-learning faces the challenge that large state or action spaces lead to long convergence times. This is already a challenge in many wireless resource allocation scenarios but could become exacerbated in cross-layer approaches due to states and actions depending on more variables. This issue will be revisited over the following sections of this chapter.

2.2 Deep Q-Learning and Resource Allocation

We now expand the ongoing discussion on resource allocation along multiple dimensions. The first such dimension follows on the note at the end of the previous section, where we said that a challenge encountered with Q-learning is the long convergence time. With a table-based algorithm (referred to this way because it estimates the Q-values for each possible pair (s_i, a_i)), such as Algorithm 1, one reason for the slow convergence is that it estimates each Q-value for each possible pair of a state and an action through an stochastic gradient descent method. An alternative approach is to estimate together the action-value function (also called the *Q-function*) that maps the Q-values as a function of the (state, action) pair of variables. To follow this approach, neural networks can be used as function approximators (of the action-value function) and compensate for the limitations seen with the standard table-based Q-learning algorithm in terms of generalization and function approximation capability. A *deep Q-network* (DQN) is an emerging class of deep reinforcement learning algorithms that is capable of combining the process of RL with a class of neural networks known as *deep neural networks* to approximate the action-value function. The team at Google DeepMind pioneered the use of this technique and demonstrated superhuman performance for a single learning agent playing Go and video games (Mnih et al. (2015), Silver et al. (2016)).

A second dimension of the resource-allocation problem we will expand upon pertains to new foci in network design that are brought about by the development of 5G wireless systems. The evolution of wireless communications toward the 5G era involves a transformation in network design and evaluation that aims at placing the end user at the center of any decision. As a result, resource management techniques for 5G networks need to be based on quality of experience (QoE) performance assessment (Wu et al. (2014)). The shift in performance assessment from objective QoS metrics to subjective end user QoE metrics has been aided by a number of studies in multimedia quality assessment that have contributed techniques to estimate the QoE from objective measurements while retaining high correlation with the subjective perception of quality. Among these techniques, mean opinion score (MOS), a metric rating from 1 (bad) to 5 (excellent), is the most widely used QoE metric (Chen et al. (2015)).

As a third dimension to delve deeper into the resource-allocation problem, we will consider that different SN links may be carrying different types of traffic. For example, one link may be carrying delay-sensitive real-time video, while another link may be seeing regular data traffic. In the context of distributed resource allocation, the challenge in this case is that, because SUs are interacting through actions taken on their shared wireless environment, the rewards or cost functions that is seen by the SUs has to share a common yardstick for all the different types of traffic. Importantly, by providing a single common measuring scale for different types of traffic, the MOS metric provides the means to perform this integrated seamless traffic management and resource allocation across traffic of dissimilar characteristics (Dobrijevic et al. (2014)).

To advance in the presentation along these dimensions, we will continue to assume the same network setup described earlier in this chapter. We will further assume that all primary and secondary links in the network transmit using adaptive modulation and coding (AMC). In AMC, the modulation scheme and channel coding rate are adapted according to the state of the transmission link, usually measured in terms of SINR. Under this assumption, SUs can infer the state of the primary link and maintain their caused interference to the PN below a tolerable threshold

(Mohammadi and Kwasinski (2018)). When using AMC, from (Qiu and Chawla (1999)), the relation between transmit bit rate and the SU threshold SINR is

$$R_i^{(S)} = W \log_2(1 + k\beta_i^{(S)}), \tag{2.15}$$

where $M(\beta_i) = (1 + k\beta_i)$ is in fact the number of bits per modulation symbol (and thus, in practice, takes only a small number of integer values) and $k = \frac{1.5}{-\ln(5BER)}$ is a constant that depends on a target maximum transmit bit error rate (BER) requirement. For the underlay DSA scheme, the SUs will perform resource allocation by selecting their target SINRs γ_i, with the goal of meeting the SINR constraints in Eqs. (2.2) and (2.1). Under the present transmission settings, the selection of β_i is equivalent to the selection of modulation scheme, channel-coding rate and $R_i^{(S)}$.

As mentioned earlier, QoE-based performance assessment is gaining significant attention as we move toward the 5G era. Consequently, we will now use QoE as the network performance metric to assess the quality of the delivered traffic. Among the metrics used to model QoE, we chose MOS because of its popularity and widespread use. Thus, resource allocation will be carried on for all types of traffic using a reward function that is based on average QoE. Since we will concentrate on real-time video and regular data traffic, we summarize next for this type of traffic how MOS metrics are derived from QoS measurements:

- **Data MOS model:**
  From (Dobrijevic et al. (2014)), the MOS Q_D for data traffic is calculated as

  $$Q_D = a \log_{10}\left(bR_i^{(S)}(1 - p_{e2e})\right), \tag{2.16}$$

  where Q_D and p_{e2e} are the data traffic MOS and end-to-end packet loss probability, respectively. The parameters a and b are calculated using the maximum and minimum perceived data quality by the end user. If the transmit rate of a user is $R_i^{(S)}$ and the effective receive rate is also $R_i^{(S)}$, the packet loss rate is zero, and the quality perceived by the end-user when measured in terms of MOS should be maximum (that is, 5). The minimum MOS value of 1 is assigned to the minimum transmission rate. In this chapter $a = 1.3619$ and $b = 0.6780$.
- **Video MOS model:**
  The peak signal-to-noise ratio (PSNR) is commonly accepted as an objective video quality assessment metric to measure, for example, coding performance. However, it is known that PSNR does not accurately reflect subjective human perception of video quality (Sheikh et al. (2006)). A wide variety of techniques have been proposed to estimate user satisfaction for video applications, among which (Khan et al. (2007)) proposed a simple linear mapping between PSNR and MOS, as shown in Figure 2.2. The work in (Piamrat et al. (2009)) presented the heuristic mappings from PSNR to MOS, as shown in Table 2.1, while, according to the recommendation ITU-R BT.500-13 (Assembly (2003)), the relationship between MOS and an objective measure of picture distortion have a sigmoid shape. Consequently, (Hanhart and Ebrahimi (2014)) claimed that if picture distortion is measured with an objective metric, e.g. PSNR (dB), then a logistic function can be used to characterize the relation between MOS and PSNR, as follows:

  $$Q_V = \frac{c}{1 + exp(d\ (PSNR - f))}, \tag{2.17}$$

  where Q_V denotes the MOS for video, and c, d, and f are the parameters of the logistic function. In this chapter, we selected the logistic function to evaluate the quality of video traffic. To compute the parameters of the logistic function in Eq. (2.17), a set of PSNRs and their corresponding MOS values is needed, while at the same time we would like to relate the PSNR to the bit rate $R_i^{(S)}$ so that Eq. (2.17) itself characterizes the video MOS as a function of the QoS variable of bit rate. To first obtain the PSNR-bit rate function representative of video sequences with different resolutions and levels of activity, we averaged over the PSNR-bit rate functions for multiple MPEG-4 coded video sequences at different resolutions (240p, 360p, and 480p), which were combined in the proportions of 39% (240p), 32% (360p), and 28% (480p) so as to follow the mix of resolutions seen in actual traffic (Mobile (2011)). The video sequences used were "Flowervase" and "Race Horses," at 30 frames per second (fps) and resolution 480p, while for resolution 360p, "Tennis" and "Park Scene" at 24 fps were selected. As a result, it was observed that a function of the form $PSNR = k \log R_i^{(S)} + p$ can be used to very closely approximate the average PSNR-bit rate curve, where $k = 10.4$ and $p = -28.7221$ are constants. Next, to obtain the parameters of the logistic function in Eq. (2.17), we computed PSNR values using the PSNR-bit rate curve for bit rates obtained from Eq. (2.15) for all candidate resource allocation actions. To get the MOS values that correspond to the computed PSNR values, we used the average of the linear mapping in Figure 2.2 and the conversion in Table 2.1. After obtaining the MOS-PSNR pairs, the resulting parameters in Eq. (2.17) were $c = 6.6431$, $d = -0.1344$, and $f = 30.4264$.

In the present case, following the user-centric approach, each SU will perform resource allocation aimed at maximizing its QoE, measured as MOS for all traffic types. For the reasons discussed earlier, resource allocation is done through a Q-learning process. However, in order to reduce the convergence time, in this case we will follow a DQN approach instead of table-based Q-learning. Nevertheless, as in any Q-learning process, we first need to design the state space, action space, and the cost or reward function. Because the network setup has not changed from

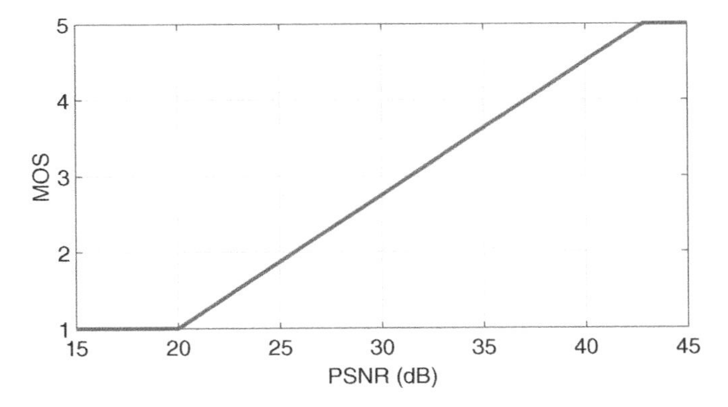

Figure 2.2 MOS versus PSNR.

Table 2.1 PSNR to MOS conversion from (Piamrat et al. (2009)).

PSNR[dB]	MOS
>37	5 (Excellent)
31-37	4 (Good)
25-31	3 (Fair)
20-25	2 (Poor)
<20	1 (Bad)

the previous section, the states are still defined as $s_t = (I_{i,t}, L_t)$ following Eqs. (2.9) and (2.8). Since the goal now is to maximize MOS, the Q-learning process is designed with a reward given by

$$\rho_{i,t} = \begin{cases} M, & \text{if } L_t + I_{i,t} > 0, \\ Q_D^{(i)} \text{ or } Q_V^{(i)}, & \text{otherwise,} \end{cases} \tag{2.18}$$

where M is a constant smaller than the minimum MOS. As can be seen, the MOS for either video (Q_V) or data (Q_D) traffic is the reward obtained when satisfying the interference constraints, and is sent from the SU receiver to its transmitter through the AMC feedback channel. Finally, for the action space, each SU conducts a search into the finite discrete space of candidate target SINRs, $\mathcal{A} = \{\gamma_1^{(i)}, \dots, \gamma_n^{(i)}\}$. At the same time, when choosing one strategy from $\mathcal{A}$, each SU adapts its transmit power as well as its modulation and channel coding rate, which, in turn, determines $R_i^{(S)}$.

As previously indicated, in the DQN approach, a function estimator is used to estimate the optimal action-value function, with a neural network being used as an efficient nonlinear approximator to estimate action-value function $Q_i(s_i, a_i; \theta_i) \approx Q_i^*(s_i, a_i)$ (Mnih et al. (2015)), where the notation of the bivariate function $Q_i(s_i, a_i; \theta_i)$ has now been expanded to include the parameters (the internal weights) θ_i of the neural network (which was designed as a fully connected feed-forward multilayer perceptron [MLP] network). Moreover, the design of the DQN includes a technique known as *experience replay* to improve learning performance. In experience replay, at each time step, the experience of each agent with the environment is stored as the tuple $e_{i,t} = (a_{i,t}, s_{i,t}, \rho_{i,t}, s_{i,t+1})$ into a replay memory $D_{i,t} =< e_{i,1}, \dots, e_{i,t} >$ for the ith agent. Also, each agent utilizes two separate MLP networks as Q-function approximators: one as action-value function approximator $Q_i(s_i, a_i; \theta_i)$ and another as target action-value function approximator $\hat{Q}_i(s_i, a_i; \theta_i^-)$. At each time step of DQN learning, the parameters θ_i of each agent's action-value function approximator are updated through a mini-batch of random samples of experience entries from the replay memory D_i following a gradient descent backpropagation algorithm based on the cost function,

$$L(\theta_i) = E\left[\left(\rho_i(s_i, a_i) + \beta_i \max_{\hat{a}_i \in \mathcal{A}}(\hat{Q}_i(\hat{s}_i, \hat{a}_i; \theta_i^-)) - Q_i(s_i, a_i; \theta_i)\right)^2\right]. \tag{2.19}$$

The parameters θ_i^- of the target action-value function approximator are updated with a lower time scale by doing $\theta_i^- \leftarrow \theta_i$ every C steps of the learning process.

Algorithm 2 summarizes the steps that need to be taken by each SU in order to implement the multi-agent DQN mechanism. It should be noted that the action selection procedure in Algorithm 2 follows the ϵ-greedy policy, which means that an action is randomly chosen from the action set $\mathcal{A}$ with probability ϵ; otherwise, an action with the highest action-value is chosen.

2.3 Cooperative Learning and Resource Allocation

The cognitive SUs seen in this chapter perform resource allocation after learning about the wireless environment and the rewards/penalties of the different actions by running a Q-learning algorithm. Whether the algorithm is the Q-table based or the DQN approach, we can think with a broad perspective that the function of these algorithms is to learn the relation between actions and the wireless environment, a relation that gets represented with more accuracy as the learning algorithm progresses. This relation is represented either as the Q values stored

Algorithm 2 - Multi-agent DQN-based learning framework ([Mnih et al.2015]).

for all SU_i, $i = 1, \dots, N$ **do**
- Initialize replay memory
- Initialization of the neural network for action-value function Q_i with random weights θ_i
- Initialization of the neural network target action-value function $\hat{Q}_i$ with $\theta_i^- = \theta_i$

end for

1: **for** $t < T$ **do**
2: **for** all SU_i, $i = 1, \dots, N$ **do**
3: Select a random action with probability ϵ.
4: Otherwise select the action $a_t^{(i)} = \arg\max_{a_{i,t}} Q_i(s_{i,t}, a_{i,t}; \theta_i)$.
5: Update the state $s_t = (I_{i,t}, L_t)$ following Eqs. (2.9) and (2.8) and the reward $\rho_{i,t}$, using Eq. (2.18).
6: Store $e_{i,t} = (a_{i,t}, s_{i,t}, \rho_{i,t}, s_{i,t+1})$ in experience replay memory of SU_i, D_i.
7: Update parameters θ of action-value function $Q_i(s_i, a_i; \theta_i)$, through mini-batch process of experiences from D_i and backpropagation based on Eq. (2.19).
8: Every C steps update the parameters of the target action-value function $\theta_i^- \leftarrow \theta_i$.
9: **end for**
10: **end for**

in the Q-table or as the neural network parameters that characterize the Q function in the DQN. Because part of the wireless environment involves background noise and the interference created by each SU to the rest of the system, the Q-table or the DQN parameters will reflect both the individual local wireless environment for each SU and the collective interrelation between the system components. It is, therefore, worthwhile to study the differences and similarities between how different SUs represent the relation between actions and the environment.

For the Q-table approach to Q-learning, Figure 2.3 shows a comparison between Q-tables from two SUs as a function of the distance between them. The Q-tables are compared by calculating the average of the relative difference between elements of Q-tables corresponding to the same action. The comparison was done over 10,000 runs of the Q-learning algorithm for a network with 16 SUs, each run corresponding to a different random location of SUs (further network setup details will be presented next). Results are presented in the form of statistical frequency values for the average relative differences. The figure shows the somewhat expected result that Q-tables become more similar as SUs become closer. For distances less than 100 meters, the probability of Q-tables differing on average by less than 10% is close to 50%. Even more interesting, the figure shows that even for SUs separated by large distances, the probability of their Q-tables differing on average by 10% or less is still significant. This is because the Q-tables incorporate information not only about each SU's local environment, but also about the interaction of all network components.

For the DQN approach to Q-learning, Figure 2.4 compares through the mean squared error (MSE) the DQN parameters θ of the action-value function of SUs separated by different distances (we selected one node as a reference SU, and consider different distances by comparing it to the its nearest SU, its second nearest, and its farthest other SU). The comparison was done over 4000 runs of the DQN learning algorithm for a network with 4 SUs, each run corresponding to a different random location of SUs. The statistical frequency of different MSE results in the figure shows that parameters of the Q action-value function are very similar between nearby SUs, and that the similarity decreases as the distance between SU increases but not to a level that makes the DQNs in two SUs remarkably different.

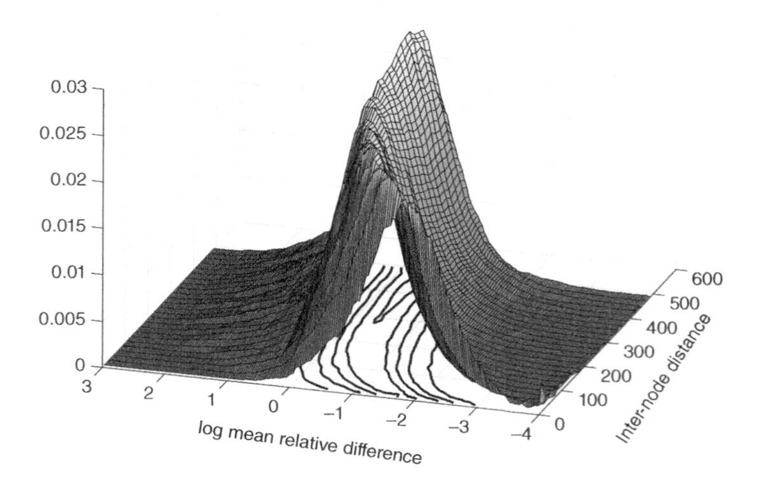

Figure 2.3 Differences and similarities between how different SUs represent the relation between actions and the environment: distribution of the average relative difference between Q-tables of SUs separated by a given distance.

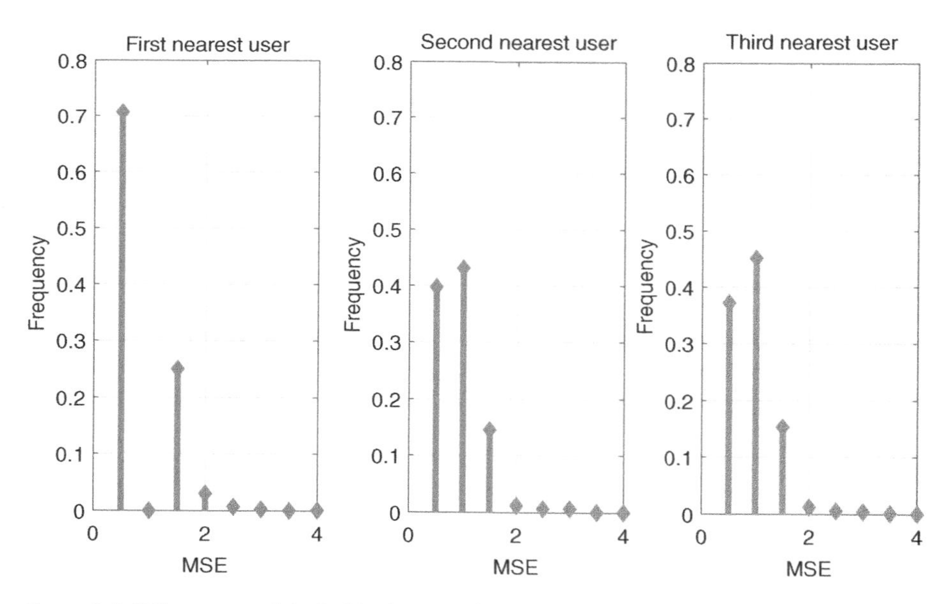

Figure 2.4 Differences and similarities between how different SUs represent the relation between actions and the environment: distribution of the MSE of the action-value function parameters θ between a first SU and other SUs in a secondary network of four links.

Reflecting on Figures 2.3 and 2.4, it is clear that the sharing of the wireless environment leads the SUs to experience significant similarities in how the learned effect of actions on the environment are represented. In a practical scenario, nodes do not join a network all at the same time. Usually, the network will be initially formed by a few nodes, and then other nodes join at a relatively slow time scale. In the case of a new SU joining an existing SN, it makes sense

Algorithm 3 - Cooperative learning in table-based Q-learning.

(Run as a new SU joins the network)
- Add one new SU as SU_{N+1} and initialize $Q_0^{(N+1)}$ with $Q_c = \frac{1}{N}\sum_{i=1}^{N} Q^{(i)}$, where $Q^{(i)}$ is the Q-table from the i SU already active in the SN.

for all SU_i, $i = 1, \dots, N+1$ **do**
  - Individually run Q-learning with the existing $N+1$ Q-tables.

end for

for the newcomer node to avoid learning from scratch (experiencing the complete convergence time for Q-learning) and leverage the experience already learned by the nodes that have been active in the network and have encoded this experience as converged Q-tables. Therefore, Figures 2.3 and 2.4 indicates that learning complexity and the time for Q-learning to converge can be reduced by collaboratively sharing learned experience between nodes. In the rest of this chapter, we discuss how to realize this idea for the two Q-learning approaches discussed earlier.

In the case of table-based Q-learning, the exploration of actions by a newcomer SU can be significantly sped up by resorting to the paradigm of *docitive radio* (Giupponi et al. (2010)), where the Q-table is no longer initialized at random, but instead uses the average of the Q-tables from the other nodes in the network that have already completed their learning process. The implementation of this idea is described in Algorithm 3 (which assumes the availability of a low bit-rate control channel used by SUs when a newcomer join the network to share the Q-tables).

Figures 2.5 and 2.6 illustrate the performance of the table-based Q-learning algorithm for cross-layer resource allocation and the performance impact when implementing cooperative learning. The figures compare the results from simulations of three different systems: a system performing joint cross-layer CR adaptation, called *individual learning*; a system called *docitive* that also performs joint cross-layer CR adaptation but considers a SU joining the network that

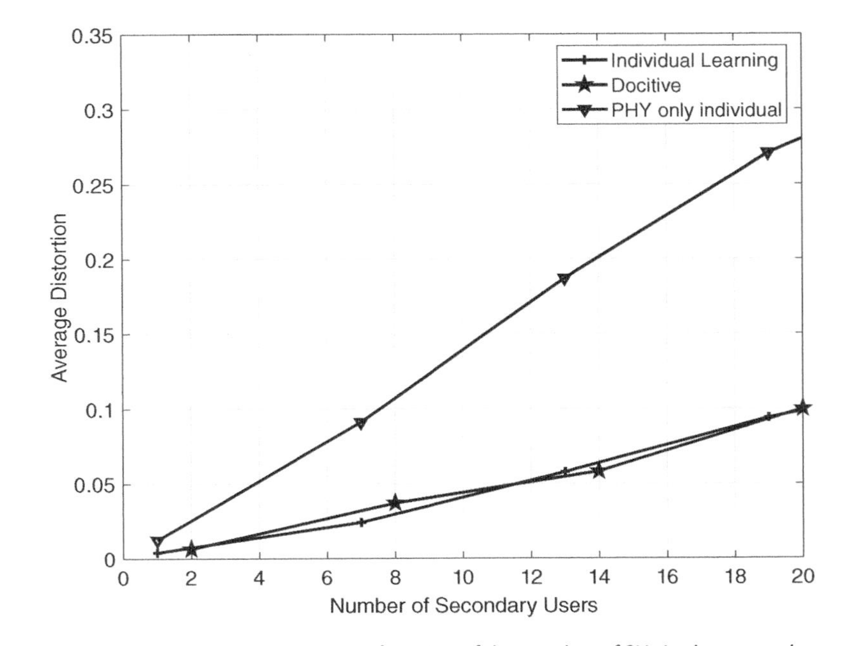

Figure 2.5 Average distortion as a function of the number of SUs in the network.

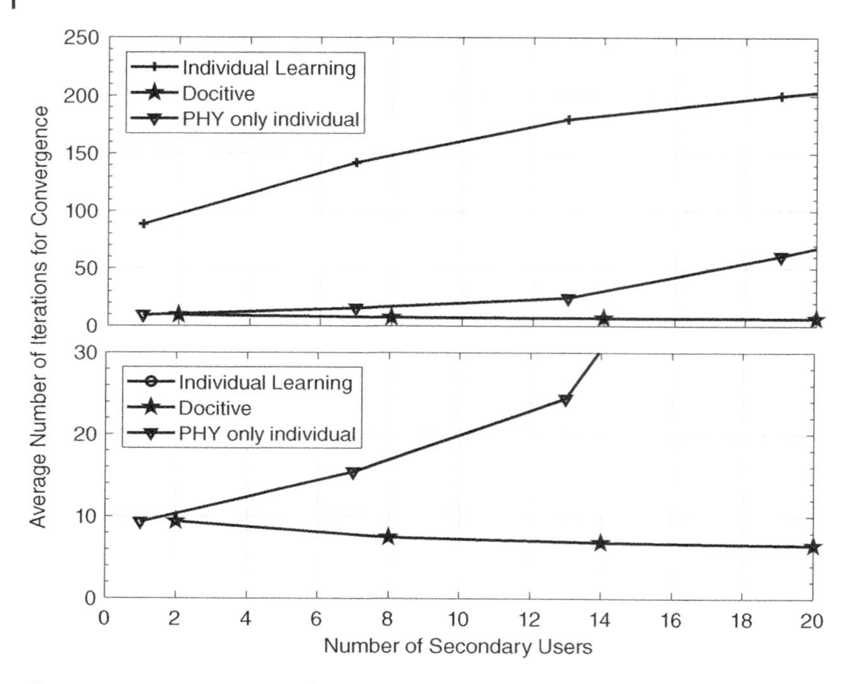

Figure 2.6 Average number of iterations to achieve convergence (lower plot is a detailed version focused on a small number of iterations).

learns through the cross-layer docitive approach implemented in Algorithm 3; and a system identified as *physical layer only*, which is identical to the individual learning system except that it does not implements cross-layer CR, as it only adapts the transmit bit rate. The simulations, based on the Monte Carlo method, were set up for a system with bandwidth 10MHz and system noise power equal to -60 dBm. The PU limited SINR to 10 dB and had a transmit power of 10 dBm. The PU and SUs were placed randomly around their respective base stations within a circle of radius 250 m. Channel gains followed a log-distance path-loss model with path-loss exponent equal to 2.8. For a single SU, its transmit rate could be chosen from the finite rate set $\{9.7655, 19.5313, 39.0625, 78.125, 156.25, 312.5\}$ Kbps. The SUs transmit using binary phase shift keying (BPSK) modulation and used a rate-compatible punctured convolutional (RCPC) channel code with mother code rate 1/4, $K = 9$, puncturing period 8, and available channel coding rates $\{8/9, 4/5, 2/3, 4/7, 1/2, 4/9, 4/10, 4/11, 1/3, 4/13, 2/7, 4/15, 1/4\}$. With respect to the learning algorithm, all the SUs had the same parameter setup with learning rate $\alpha = 0.1$ and discounting factor $\beta = 0.4$.

Figure 2.5 shows the average distortion that the SUs achieve as a function of the number of SUs in the network. It can be seen how with cross-layer resource allocation, the average network distortion is significantly reduced when compared to the physical-layer-only scheme. Moreover, the cross-layer approach exhibits an increase in distortion that is smoother than the physical-layer-only scheme. Figure 2.6 shows the average number of iterations to achieve convergence using table-based Q-learning. The result shows first how the implementation of a cross-layer CR approach, as in the individual learning case, increases the number of iterations necessary to learn compared to the physical-layer-only scheme. However, it can also be seen how the cooperative teaching of SUs already in the network allows the initialization of a new-comer SU with such an accurate sharing of experience that the number of iterations needed to achieve convergence is reduced by as much as 30 times (and more than approximately 10

Algorithm 4 - Transfer learning.

(Run as a new SU joins the network)
- Add one new SU as SU_{N+1}
- Assign the nearest node $n(i)$ to the new SU
- Initialize $Q_{(N+1)}$ with parameters of the action-value function of its nearest neighbor $\theta_{(N+1)} \leftarrow \theta_{(n(i))}$.

for all SU_i, $i = 1, \cdots, N+1$ **do**
- Restart individual DQN learning (Algorithm 2) with the existing $N+1$ action-value function.

end for

times in all cases). The docitive approach even achieves a number of iterations that is always less than the physical-layer-only scheme, even when the latter scheme has an action space 13 times smaller. Even more, because the docitive approach benefits from the knowledge of experienced SUs, the number of iterations actually decreases when the number of SUs in the network increases. Importantly, Figures 2.5 and 2.6 show how cooperation during learning speeds up the learning process without a sacrifice in performance.

In the case of DQN, cooperative learning when a newcomer SU joins the network takes the form of the well-known technique in ML called *transfer learning*. Transfer learning refers to the mechanism where what has been learned in one setting is exploited in order to improve the learning process and generalization in another setting (Goodfellow et al. (2016)). In the present case, transfer learning is realized by having the newcomer SU initialize the parameters of its action-value function neural network approximator with the same parameters of the nearest node. Note that the technique exposes another advantage of using MOS as performance (and reward) metric because the choice of the node from where to transfer the DQN parameters is based on the distance to the newcomer SU but not based on the traffic type being carried by either node. As such, the common yardstick for all types of traffic presented by MOS allows such a seamless integration of dissimilar traffic that experience can be transferred between nodes even when their traffic type is different. This technique is shown in Algorithm 4.

Figure 2.7 illustrates the effect of applying transfer learning to the DQN cognitive engine of a newcomer SU through the results obtained from 500 Monte Carlo simulation runs. The setup for the simulations consisted of a single-link primary network accessing a single channel. The target SINR for the PU was set at 1 dB. The Gaussian noise power and the transmit power of PU were set to be -60 dBm and 10 dBm, respectively. All SUs and PUs were distributed randomly around their respective base stations within a circle of radius 300 m. The distance between the primary base station and the secondary base station was 2 km. Channel gains followed a log-distance path loss model with path loss exponent equal to 2.8. For a single SU, its target SINR could be chosen from the finite discrete set $\{-12, -10, -8, -6, -4, -2, 0, 1, 2, 3, 4, 5\}$ dB. Regarding to the learning algorithm, the same learning rate $\alpha = 0.01$ and discounting factor $\beta = 0.9$ were assumed for all SUs. In the ϵ-greedy exploration, ϵ is initially set to be 0.8, as the number of iteration increases it decreases to a value of 0. As action-value function and target action-value function approximators, we used two separate feed-forward neural networks for each SU. Each neural network was configured with two fully connected hidden layers with three and two neurons. The capacity of replay memory and mini-patch was set to 100 and 10, respectively. The input layer consisted of three nodes representing the state and the selected action to be taken. The output layer had one neuron.

The simulation results in Figure 2.7 compare three resource allocation algorithms in terms of the MOS and the average number of learning steps necessary to achieve convergence. All three

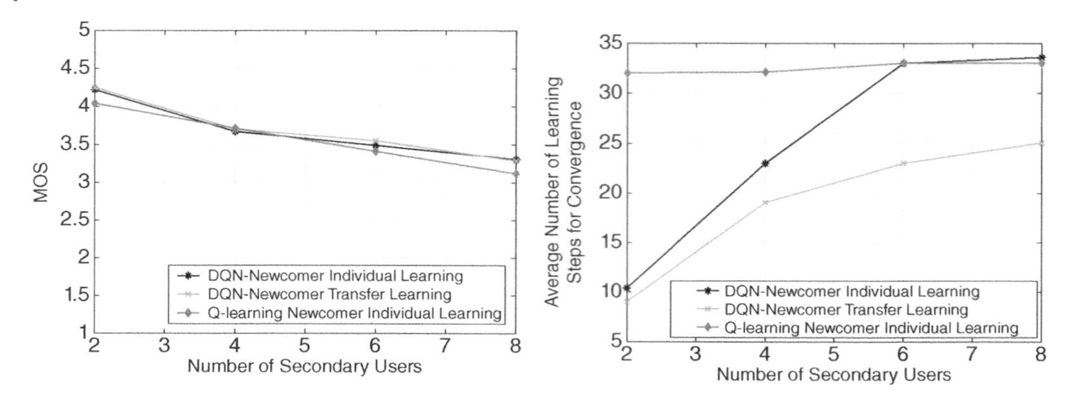

Figure 2.7 (Left): Average MOS in the secondary network. (Right): Average number of learning steps to achieve convergence.

algorithms consider the scenario when one newcomer SU joins the network while the rest of the SUs have already finished learning through individually running the DQN algorithm. The three algorithms considered are: *DQN-Newcomer Individual Learning,* where all SUs perform Algorithm 2 and a newcomer SU initializes its action-value function parameters randomly; *DQN-Newcomer Transfer Learning,* which exploits the transfer learning mechanism for the newcomer SU; and *Q-learning Newcomer Individual Learning,* which implements standard table-based Q-learning for all SUs and initializes the Q-table of new joined SU with zeros. The right side of Figure 2.7 shows how the DQN approach performs faster learning than the table-based Q-learning and how transfer learning accelerates learning in the DQN approach without negative effects on the MOS performance.

2.4 Conclusions

In this chapter, we have discussed the use of ML to perform resource allocation in cognitive radio networks. We framed the presentation within a scenario consisting of a number of CRs that establish wireless links by sharing through underlay DSA a portion of the radio spectrum with an incumbent primary network. The chapter focuses on reinforcement learning because it is a machine learning technique that displays the advantages of alignment with the cognition cycle approach, it is flexible for the implementation of cross-layer resource allocation, and there is no need for prior knowledge of a system model.

We first explained the use of table-based Q-learning, a popular implementation of RL, to allocate resources following a cross-layer approach in a CRN; following this, we noted that if left unaddressed, this approach is at risk of exhibiting long convergence times. We then expanded the presentation in various directions. We explained how a current approach to address the long convergence time in table-based Q-learning is to approximate the correspondence of Q-values (expected rewards) as a function of the actions and environment states variables. The approximation of this Q-function is done through an artificial neural network called a deep Q-network. We also explored resource allocation following the tenets in fifth-generation wireless systems development, which place the end user at the center of design. With this approach, resource allocation is based on the quality of experience as a performance metric. We highlighted that adopting the mean opinion score as a QoE metric enables seamless integration of dissimilar traffic during distributed resource allocation.

In the later part of the chapter, we revisited the question of how to accelerate learning. We noted that in a wireless network, it is rare to see all nodes join the network at the same (initial) moment; instead, nodes enter and leave the network individually at different times. It is then expected that when a node joins a network, it will find other nodes that are already operating and, for CRs, have already gone through the learning process, making them experienced agents. Then, we explained how the learning of CRs that join a network can be accelerated by having those CRs already in the network act as teachers by sharing their experience. This idea takes the form of the described cooperative learning and transfer learning techniques.

Concluding with a broader perspective, we remark that many of the recent developments in CRs leverage significant new advances in ML that are often applied to technical domains that are outside wireless communications. Visible examples are autonomous vehicles, computer vision, and natural language processing. However, the application of these new ML techniques to CRs presents the interest and challenges that are uniquely distinct for the environment in this case, i.e. the wireless network. Particular to a wireless networking environment are how the agents are coupled through the interference they create on each other, and that agents seek to and become able to communicate with each other. Exploring how these unique features of the environment affect ML techniques is where the challenge, potential, and appeal reside for studying the application of ML in this unique environment.

Bibliography

ITU Radiocommunication Assembly. *Methodology for the subjective assessment of the quality of television pictures*. International Telecommunication Union, 2003.

Yanjiao Chen, Kaishun Wu, and Qian Zhang. From qos to qoe: A tutorial on video quality assessment. *IEEE Communications Surveys and Tutorials*, 17 (2):1126–1165, 2015.

Ognjen Dobrijevic, Andreas J Kassler, Lea Skorin-Kapov, and Maja Matijasevic. Q-point: Qoe-driven path optimization model for multimedia services. In *Int. Conf. on Wired/Wireless Internet Comm.*, pages 134–147. Springer, 2014.

L. Giupponi, A. Galindo-Serrano, P. Blasco, and M. Dohler. Docitive networks: an emerging paradigm for dynamic spectrum management. *Wireless Communications, IEEE*, 17(4):47 –54, 2010.

Andrea Goldsmith, Syed Ali Jafar, Ivana Maric, and Sudhir Srinivasa. Breaking spectrum gridlock with cognitive radios: An information theoretic perspective. *Proceedings of the IEEE*, 97(5):894–914, 2009.

Ian Goodfellow, Yoshua Bengio, and Aaron Courville. *Deep Learning*. Adaptive Computation and Machine Learning series. Cambridge, MA: MIT Press, 2016.

Zhu Han, Dusit Niyato, Walid Saad, Tamer Başar, and Are Hjørungnes. *Game Theory in Wireless and Communication Networks: Theory, Models, and Applications*. Cambridge University Press, 2012.

Philippe Hanhart and Touradj Ebrahimi. Calculation of average coding efficiency based on subjective quality scores. *Journal of Visual Communication and Image Representation*, 25(3):555–564, 2014.

Shoaib Khan, Svetoslav Duhovnikov, Eckehard Steinbach, and Wolfgang Kellerer. Mos-based multiuser multiapplication cross-layer optimization for mobile multimedia communication. *Advances in Multimedia*, 2007.

A. Kwasinski and K.J.R. Liu. Source-channel-cooperation tradeoffs for adaptive coded communications [transactions papers]. *IEEE Transactions on Wireless Communications*, 7(9):3347–3358, September 2008. ISSN 1536-1276. https://doi.org/10.1109/TWC.2008.060286.

Volodymyr Mnih, Koray Kavukcuoglu, David Silver, Andrei A Rusu, Joel Veness, Marc G Bellemare, Alex Graves, Martin Riedmiller, Andreas K Fidjeland, Georg Ostrovski, et al. Human-level control through deep reinforcement learning. *Nature*, 518(7540):529–533, 2015.

Byte Mobile. Mobile analytica report. June 2011.

Fatemeh Shah Mohammadi and Andres Kwasinski. Neural network cognitive engine for autonomous and distributed underlay dynamic spectrum access. *arXiv preprint arXiv:1806.11038*, 2018.

Kandaraj Piamrat, Cesar Viho, Jean-Marie Bonnin, and Adlen Ksentini. Quality of experience measurements for video streaming over wireless networks. In *Information Technology: New Generations, 2009. ITNG'09. Sixth International Conference on*, pages 1184–1189. IEEE, 2009.

Xiaoxin Qiu and Kapil Chawla. On the performance of adaptive modulation in cellular systems. *IEEE transactions on Communications*, 47(6):884–895, 1999.

Hamid R Sheikh, Muhammad F Sabir, and Alan C Bovik. A statistical evaluation of recent full reference image quality assessment algorithms. *IEEE Transactions on image processing*, 15(11):3440–3451, 2006.

David Silver, Aja Huang, Chris J Maddison, Arthur Guez, Laurent Sifre, George Van Den Driessche, Julian Schrittwieser, Ioannis Antonoglou, Veda Panneershelvam, Marc Lanctot, et al. Mastering the game of go with deep neural networks and tree search. *Nature*, 529(7587):484–489, 2016.

Richard S. Sutton and Andrew G. Barto. *Reinforcement Learning: An Introduction*. Cambridge, MA: MIT Press, 2011.

W. Wang, A. Kwasinski, D. Niyato, and Z. Han. A survey on applications of model-free strategy learning in cognitive wireless networks. *IEEE Communications Surveys Tutorials*, 18(3): 1717–1757, thirdquarter 2016. ISSN 1553-877X. https://doi.org/10.1109/COMST.2016.2539923.

Yeqing Wu, Fei Hu, Sunil Kumar, Yingying Zhu, Ali Talari, Nazanin Rahnavard, and John D Matyjas. A learning-based qoe-driven spectrum handoff scheme for multimedia transmissions over cognitive radio networks. *IEEE J. on Sel. Areas in Comm.*, 32(11):2134–2148, 2014.

3

Machine Learning for Spectrum Sharing in Millimeter-Wave Cellular Networks

Hadi Ghauch[1], Hossein Shokri-Ghadikolaei[2], Gabor Fodor[3], Carlo Fischione[2], and Mikael Skoglund[4]

[1] COMELEC Department, Telecom ParisTech, Paris, France
[2] Division of Network and Systems Engineering, Royal Institute of Technology, KTH, Stockholm, Sweden
[3] Division of Decision and Control Systems, Royal Institute of Technology, KTH, Stockholm, Sweden
[4] Division of Information Science and Engineering, Royal Institute of Technology, KTH, Stockholm, Sweden

This chapter is devoted to the use of machine learning (ML) tools to address the spectrum-sharing problem in cellular networks. The emphasis is on a hybrid approach that combines the traditional model-based approach with a (ML) data-driven approach. Taking millimeter-wave cellular network as an application case, the theoretical analyses and experiments presented in this chapter show that the proposed hybrid approach is a very promising solution in dealing with the key technical aspects of spectrum sharing: the choice of beamforming, the level of information exchange for coordination and association, and the sharing architecture.

This chapter is organized into the following four sections. Section 3.1 focuses on motivation and background related to spectrum sharing. Section 3.2 presents the system model and problem formulation, the focus of Section 3.3 is on all technical aspects of the proposed hybrid approach. Finally, Section 3.4 discusses further issues and conclusions.

3.1 Background and Motivation

3.1.1 Review of Cellular Network Evolution

The evolution of cellular networks and their enabling technologies is driven by the insatiable demand for mobile broadband and machine-type communication services. This unprecedented growth in data traffic is fueled by the rapidly increasing number of mobile data subscriptions along with a continuous increase in the average data volume per subscription. On average, more than 1 million new mobile broadband subscribers will be added every day up to the end of 2022, resulting in 8.5 billion mobile broadband subscriptions and 6.2 billion unique mobile subscribers worldwide by 2023 (5G Americas White Paper, 2018).

According to Ericsson, overall worldwide data traffic grew by 55% between 2016 and 2017. Data traffic is forecast to grow at a compounded annual rate of 42% from 8.8 exabytes to 71 exabytes between 2016 and 2022, and it is expected that 75% of the worlds mobile data traffic will be video by 2020. In 2017, mobile technologies and services generated 4.5% of the global gross domestic products and will reach 4.6 trillion USD (Ericsson, 2018).

Machine Learning for Future Wireless Communications, First Edition. Edited by Fa-Long Luo.
© 2020 John Wiley & Sons Ltd. Published 2020 by John Wiley & Sons Ltd.

To a large extent, this tremendous growth in mobile data traffic has been enabled by the globally harmonized specifications and technologies developed by the Third Generation Partnership Project (3GPP), which has brought together the entire mobile industry and enabled a worldwide ecosystem of infrastructure, devices, and services. Due to this harmonized global ecosystem, the global uptake of long-Term evolution (LTE) networks by the 3GPP reached a milestone of 3 billion out of the total of 8 billion total cellular connections worldwide at the end of the first quarter of 2018 and will continue to 4 billion by the end of 2019.

As the demand for wireless data traffic in terms of higher subscriber density, higher throughput, and data capacity continues to grow, the research and standardization communities have started to define the enabling technologies for the next generation of wireless and cellular networks: 5G. 5G networks, founded on a new set of physical and higher-layer specifications commonly referred to as *3GPP new radio* (NR), are designed to meet the demands set by mobile broadband services as well as new vertical markets, including massive machine-type communications, ultra-reliable low latency communication, and a broad range of Internet of Things (IoT) applications (Jiang et al., 2018).

In terms of enabling technology components, NR incorporates scalable orthogonal frequency division modulation, where transmission time intervals are optimized for wide bandwidths and fast beam-switching, beam-steering, and beam-tracking to focus energy on highly mobile users. These technologies enable directional communications not only to ensure high signal-to-noise ratios at the desired users, but to also eliminate undesired signals (interference) at the non-intended receivers. Indeed, directional three-dimensional antenna arrays hosting massive multiple-input and multiple-output (MIMO) techniques are key to increasing both the coverage and capacity of 5G networks (Fodor et al., 2016); these systems have been demonstrated to boost coverage and capacity in diverse spectrum bands, including low-bands (< 1 GHz), mid-bands (1 to 6 GHz), and high-bands (above 30 GHz, millimeter-wave).

3.1.2 Millimeter-Wave and Large-Scale Antenna Systems

Although 5G research and standards development are underway to achieve the throughput rates of multiple Gbps, radio carriers of at least 1 GHz will be required, according to the ITU. Such bandwidths are typically only available at frequencies above 6 GHz. Thus, high-frequency spectrum options will be important to 5G, including both centimeter-wave frequencies of 3–30 GHz and millimeter wave (mmWave) frequencies of 30–300 GHz.

Recognizing the need for exploiting mmWave bands, the World Radio Conference 2015 (WRC-15) approved the study of 11 high-band spectrum allocations for WRC-19. It can be expected that many governments will follow similar polices. For example, the 2016 allocation by the FCC in the United States of 28 GHz, 37 GHz, and 39 GHz licensed spectrum for 5G is the first important high-band licensed spectrum for 5G in North America (5G Americas White Paper, 2018). By means of these technologies, including the 3GPP specifications for 5G NR and capitalizing on mmWave bands, 5G is expected to serve as a unified connectivity platform for future innovation and to embrace the evolving cellular ecosystem of diverse services, devices, deployments, and spectrum.

Large-scale antenna systems – sometimes referred to as full-dimension MIMO systems – boost the coverage and capacity of 4G networks deployed in the low-bands and are key enablers of deploying 5G networks in the mid- and high-bands. Indeed, it is expected that in the mmWave bands, a large number of transmit and receive antennas will be deployed at cellular BSs, which will enable directional communications using very narrow beams. Narrow beams are effective for ensuring proper coverage by boosting the link budget in mmWave bands, and they also minimize interference at unintended receivers.

It is therefore clear that mmWave MIMO – enabling highly directional beamforming – is a critical solution for many operators facing the challenges of meeting increasing traffic demands and serving vertical industries. mmWave MIMO channels are characterized by a small number of eigenmodes (represented by a sparsity in the eigenvalues), a high percentage of blockage/penetration loss, and severe pathloss. *Highly directional communications*, consisting of beamforming and combining at base stations (BS) and user equipment (UE), are a prevalent design, to counteract the severe pathloss attenuation with array gain. Additionally, the large number of BS and UE antennas is made possible by using a reduced number of radio frequency (RF) chains, thus reducing the complexity and power consumption of the system. These ideas gave rise to the so-called hybrid analog-digital architecture, where the precoding and combining are performed by separate analog and digital stages (Ardah et al., 2018).

3.1.3 Review of Spectrum Sharing

As discussed, mmWave communications are an essential part of 5G networks, due to their potential to support extremely high data rates and low-latency services. Although mmWave bands offer a much wider spectrum than the commonly used low-and mid-bands, achieving high spectral efficiency and high spectrum utilization by making the best use of licensed and unlicensed spectrum is important due to the large capacity demands and high user densities in many deployment scenarios. Spectrum sharing addresses this goal by allowing multiple service providers and mobile network operators to access the same spectrum band for the same or different sets of services. At the same time, the newly opened bands for 5G services, including the 37 GHz and 57–71 GHz bands, present an opportunity to improve spectrum utilization by sharing valuable spectrum resources.

A precursor to sharing spectrum among multiple mobile network operators is the 3GPP-defined NR-based license-assisted access. It enables the deployment of cellular networks in unlicensed spectrum that co-exist with non-3GPP technologies, thereby facilitating spectrum sharing between multiple radio access technologies. However, extending spectrum sharing facilitated by license-assisted access to sharing even licensed bands opens up new paradigms that are particularly suitable for mmWave deployments exploiting the spatial domain. This new paradigm can be built on the key observation that in mmWave networks, large antenna arrays, directional communications, and the unique propagation environment substantially simplify the problem of managing interference in shared spectrum; see (S. Ghadikolaei et al., 2016), (Boccardi et al., 2016).

Over-the-air demonstrations of 5G NR spectrum-sharing technologies have already been performed. The use of 5G spectrum sharing technologies is expected to improve the performance of mobile broadband and IoT services using unlicensed and shared mmWave spectrum (Tehrani et al., 2016).

The early research reveals that four prominent aspects have a major impact on the performance and feasibility of spectrum-sharing schemes: the choice of beamforming, the amount of information exchange (coordination), the amount of spectrum shared, and the sharing architecture:

- Beamforming and directional communications at both the transmitter and the receiver are the key technology enabler of spectrum sharing in mmWave networks. Digital, analog, and hybrid beamforming are key enablers to realize directional communications. The appropriate beamforming technique depends on the required flexibility in beam formation, processing and constraints on power consumption, cost, and channel state information acquisition.
- The level of coordination among different mobile network operators (inter-operator coordination) has a substantial influence on the feasibility and effectiveness of spectrum sharing.

Without any coordination, spectrum sharing may not even be beneficial, especially in traditional sub-6 GHz networks. Full coordination may bring substantial performance gains in terms of throughput, especially because it enables complementary techniques such as joint precoding and load balancing. From a technical perspective, the challenge is the high overhead of channel estimation and the complexity in the core networks in a multi-operator scenario.

- The amount of spectrum that is shared among the operators is a factor because increasing the bandwidth may improve the achievable capacity by a prelog factor (that is, a factor outside the log function of the achievable data-rate), but usually reduces the signal-to-interference-plus-noise ratio, due to higher noise and interference powers. Spectrum sharing is beneficial if the contribution of the prelog factor outweighs the resulting SINR reduction.
- The supporting architecture is another aspect that affects the performance of spectrum sharing. Infrastructure sharing is an example of an architecture that increases the gain of spectrum sharing. In general, the trade-off between performance gain and protocol overhead determines whether spectrum sharing is beneficial. Reference (Boccardi et al., 2016) discusses various supporting architectures for spectrum sharing in mmWave networks, including interfaces at the radio access networks and at the core network.

3.1.4 Model-Based vs. Data-Driven Approaches

Although the problem of spectrum sharing in millimeter-wave MIMO cellular networks is well-motivated, it is essentially open due to the signaling and computational complexities. Indeed, similar problems such as coordination in cellular networks and multi-cell beamforming have been traditionally addressed using *model-based approaches*. However, these approaches always suffer from elevated signaling and computational complexity, thereby hindering their implementation in practical cellular networks. We will show in Section 3.2.2 that the problem considered here is intractable using these model-based approaches.

Although model-based approaches serve as the foundations of communication systems (Tse and Viswanath, 2005), the simplified models and multiple approximations often required for the mathematical tractability may lead to protocols that do not work well in practice (Ghadikolaei et al., 2018). On the other hand, data-driven approaches address this disadvantage by learning and optimizing from the data – usually acquired by measurements – and making minimal assumptions about the system model. These approaches, however, may need a large number of training samples to perform well, which are not available in most wireless networks due to their time variations (Sevakula et al., 2015).

Recognizing the strengths and limitations of model-based and data-driven approaches, in this chapter we present the use of a hybrid approach, in which the *model-based component* operates on a small time scale, while the *data-driven component* operates on a coarser time scale and refines the models used in the model-based part. We apply this approach to the important (but complicated-to-solve) problem of spectrum sharing in a multi-operator cellular network. As seen in this chapter, the paradigm of hybrid (model-based and data-driven) approaches is applicable to a plethora of problems in the area of ML for wireless communication. We note that a similar conclusion was put forth by Recht (2018), namely that a hybrid approach may still give many of the benefits of a pure model-based approach in the area of optimal control.

Notations: Capital bold letters denote matrices, and lowercase bold letters denote vectors. The superscripts $(\boldsymbol{X})^{\mathrm{T}}$, $(\boldsymbol{X})^{\mathrm{H}}$, and $(\boldsymbol{X})^{\dagger}$ stand for the transpose, transpose conjugate, and Moore-Penrose pseudo-inverse of $\boldsymbol{X}$, respectively. The subscript $[\boldsymbol{X}]_{mn}$ denotes entry of $\boldsymbol{X}$ at row m and column n, and $[\boldsymbol{X}]_n$ represents column n of $\boldsymbol{X}$. $\boldsymbol{I}$ and $\mathbf{1}$ are the identity matrix and all-one matrix of appropriate dimension, respectively. blkdiag(·) the mapping of the arguments to a block diagonal matrix.

3.2 System Model and Problem Formulation

The model presented here is generic and embraces distinct model elements for the network, the association scheme, the antenna and channel models, and models for beamforming and coordination.

3.2.1 Models

3.2.1.1 Network Model

We consider a general multi-user, multi-cell, multi-operator network. The network consists of Z operators, each operating a (sub)set $\mathcal{B}_z$ of the BSs in the network, where it is assumed that each BS belongs to an operator, and no BS is shared among operators. The set of all UEs is denoted by $\mathcal{U}$, and $\mathcal{U}_z$ is the set of all UEs of operator z. Finally, we let W be the total bandwidth available in the network and W_z be that of operator z. We further assume universal frequency reuse.

3.2.1.2 Association Model

We let $a_{bu} = 1$ model that UE $u \in \mathcal{U}$ is served by (or associated with) BS $b \in \mathcal{B}$. Moreover, we denote by $\boldsymbol{A}_z$ the $|\mathcal{B}| \times |\mathcal{U}|$ the *binary association matrix* of operator z, namely, $[\boldsymbol{A}]_{bu} = 1$ if $u \in \mathcal{U}_z$ and $a_{bu} = 1$. We also define the aggregate association matrix $\boldsymbol{A}$ as $\boldsymbol{A} = \sum_{z \in [Z]} \boldsymbol{A}_z$. We assume that no national roaming is allowed, i.e. each BS servers UEs of one operator; see (S. Ghadikolaei et al., 2018), where this assumption is relaxed. For simplicity, let us define $\mathcal{F}_A$ as the set of feasible assignments for $\boldsymbol{A}$. Simply stated, $\mathcal{F}_A$ encodes constraints that each UE is associated with a single BS (since no joint transmission is assumed), that each BS is serving no more than N_b users simultaneously (for the sake of load balancing), and that the UEs of operator z can be only served by BSs of the same operator, thereby preventing national roaming.

An association period is defined as a succession of *coherence intervals (CIs)* during which the association matrix is essentially constant; see Figure 3.1. In that sense, we assume that association is done over a longer time period (many CIs) compared to beamforming, which has to be done during each CI. Such an assumption is natural due to the high overhead incurred by reassociation. After finding the optimal $\boldsymbol{A}_z$ for each operator, each BS and UE recalculates its beamforming vectors every CI. To simplify the presentation, we further assume that each BS can simultaneously serve all of its associated users (in multi-user MIMO fashion).

3.2.1.3 Antenna and Channel Model

Each BS (resp. UE) is equipped with N_{BS} (resp. N_{UE}) antennas, assumed to be uniform linear arrays for simplicity. The $N_{\text{UE}} \times N_{\text{BS}}$ MIMO channel between BS b and UE u is denoted as $\boldsymbol{H}_{bu}$. In a prevalence of mmWave MIMO models, the channel depends on $\boldsymbol{a}_{\text{BS}}$ and $\boldsymbol{a}_{\text{UE}}$, the vector

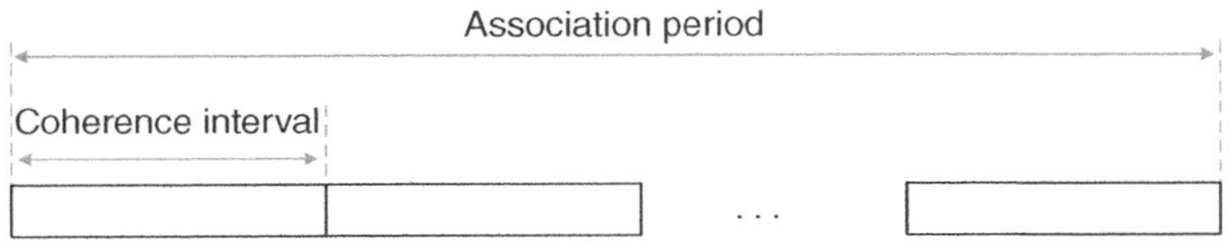

Figure 3.1 A UE-BS association period. Beamforming vectors are fixed for only one CI and should be recomputed afterward. The UE-BS association is fixed over a block of many CI intervals, denoted as an association period.

response functions of the BSs' and UEs' antenna arrays to the angles of arrival and departure (AoAs and AoDs), $\theta^{\mathrm{BS}}_{bun}$, the AoD of the n-th path, $\theta^{\mathrm{UE}}_{bun}$, the AoA of the n-th path, N_{bu}, the number of paths between BS $b \in \mathcal{B}$ and UE $u \in \mathcal{U}$, and g_{bun}, the complex gain of the nth path that includes both path loss and small-scale fading.

3.2.1.4 Beamforming and Coordination Models

Analog Combiners We assume that each UE has a single RF chain, where combiners at the UE consist of phase shifters only. This design is commonly used in mmWave literature due to its simple implementation. Letting $\boldsymbol{w}^{\mathrm{UE}}_u \in \mathbb{C}^{N_{\mathrm{UE}}}$ denote the combining vector of UE u, each combiner is designed to maximize the link budget (Ayach et al., 2012) to its serving BS. Formally, $\boldsymbol{w}^{\mathrm{UE}}_u = \boldsymbol{a}_{\mathrm{UE}}(\theta^{\mathrm{UE}}_{bun^\star})$, where $n^\star = \arg\max_n |g_{bun}|$, i.e. the combiner is matched to the receive array response having the highest path gain. Evidently, this design assumes that channel gains and angles of arrival (AoAs) are available at each UE.

Precoders We assume that each BS uses fully digital precoding, without loss of generality. Furthermore, the UEs associated with each BS are simultaneously served with multi-user MIMO. Moreover, we employ *regularized zero-forcing (RZF)* precoding, as it directly minimizes intra-BS interference (within and among different operators).[1] Let $\boldsymbol{w}^{\mathrm{BS}}_{bu} \in \mathbb{C}^{N_{\mathrm{BS}}}$ be the precoding vector that BS b uses for user u. Using RZF, the precoding vector of UE u is

$$\boldsymbol{w}^{\mathrm{BS}}_{bu} = \lambda_b^{-1}[(\overline{\boldsymbol{H}}_b + \delta \boldsymbol{I})^{\dagger}]_u, \quad 0 < \delta \ll 1, \tag{3.1}$$

where $\overline{\boldsymbol{H}}_b$ is the effective channel for BS b (see Section 3.2.1.5), and λ_b is a normalization to satisfy a maximum transmit power at each BS. The assumption of fully digital precoding is in no way limiting. The precoding method presented here can be applied to the co-called hybrid analog-digital architecture, prevalent in mmWave literature; see (S. Ghadikolaei et al., 2018).

3.2.1.5 Coordination Model

Given the precoding/combining method, we let $\overline{\boldsymbol{h}}_{iu} := (\boldsymbol{w}^{\mathrm{UE}}_u)^{\mathrm{H}} \boldsymbol{H}_{iu}$ be the effective channel between BS $i \in \mathcal{B}$ and UE u, i.e. a cascade of the actual channel, $\boldsymbol{H}_{iu}$, and the combiner, $\boldsymbol{w}^{\mathrm{UE}}_u$ (not to be confused with $\overline{\boldsymbol{H}}_b$). We let $\boldsymbol{C} \in \{0,1\}^{|\mathcal{B}|\times|\mathcal{U}|}$ be the *binary coordination matrix*, where $[\boldsymbol{C}]_{iu} = 1$ denotes BS $i \in \mathcal{B}$ estimating the effective channel $\overline{\boldsymbol{h}}_{iu}$. We define the effective channel $\overline{\boldsymbol{H}}_i \in \mathbb{C}^{\sum_u [\boldsymbol{C}]_{iu} \times N_{\mathrm{BS}}}$ as a matrix whose rows correspond to the effective channels $\overline{\boldsymbol{h}}_{iu}$ for $\{u \mid [\boldsymbol{C}]_{iu} = 1\}$. In this, b (resp. i) denotes the index of the serving (resp. coordinating) BS.

Note that acquiring this effective channel incurs a lower cost when two BSs belong to the same operator than when they belong to different operators. In view of modeling this asymmetry in the usage of the network resources and promoting an efficient use of the backhaul, we include a penalty (cost) for coordination. More specifically, we let $[\boldsymbol{P}]_{iu}$ denote the penalty/cost of BS i estimating the (effective) channel of UE u, corresponding to element $[\boldsymbol{C}]_{iu}$ of the coordination matrix. To keep the exposition simple, we consider hereafter a constant penalty matrix $\boldsymbol{P}$, i.e. having equal entries. In practice, however, the penalty should be a function of distance, operator load, number of antennas, and operators (reflecting various billing policies), etc.; see Section 3.3.3.3 for specific instantiations of the penalty matrix.

Note that $\boldsymbol{P}$ will depend on $\boldsymbol{A}$, since the cost of coordination depends on assignment of UEs to BSs. We express this dependence in general form (concrete choices are discussed in Section 3.3.3.3),

$$[\boldsymbol{P}]_{bu} = g(a_{bu}), \ \forall b \in \mathcal{B}, \ \forall u \in \mathcal{U}, \tag{3.2}$$

1 The use of RZF is in no way restrictive. Moreover, joint transmission is not assumed here, to avoid the resulting latency in signaling through the corresponding core networks.

where $g()$ is affine in its argument. Moreover, the total coordination cost (per CI) for operator z is constrained to be less the a predetermined coordination budget, $P_z^{\max}$, i.e.

$$\sum_{u\in\mathcal{U}_z}\sum_{b\in\mathcal{B}}[\boldsymbol{C}]_{bu}[\boldsymbol{P}]_{bu} := \sum_{u\in\mathcal{U}_z}\sum_{b\in\mathcal{B}}[\boldsymbol{C}]_{bu}\, g(a_{bu}) \le P_z^{\max}\,. \tag{3.3}$$

Ultimately, we aim to find an optimal coordination policy, i.e. one that maximizes the sum-rate of the network while satisfying the coordination budget.

Example 3.1 Figure 3.2 illustrates the intuition behind the approach using a network with two operators, each having two BSs and five UEs – a running example in this chapter. In this example, BS 2 estimates the channel of UE 5 using intra-operator coordination. Moreover, BSs 1 and 2 (of operator black) estimate their channel to UE 7 (of operator red). For this topology, $\mathcal{B}_1 = \{1,2\}, \mathcal{B}_2 = \{3,4\}, N_b = 2$ (for all b), $\mathcal{U}_1 = \{1,\dots,5\}$, and $\mathcal{U}_2 = \{6,\dots,10\}$. The cost of coordination with associated UEs is 1, and those of intra-operator and inter-operator coordination are 10 and 100, respectively. Then we have

$$\boldsymbol{A} = \begin{pmatrix} 1&1&0&0&1&0&0&0&0&0\\ 0&0&1&1&0&0&0&0&0&0\\ 0&0&0&0&0&1&1&0&0&0\\ 0&0&0&0&0&0&0&1&1&1 \end{pmatrix}, \boldsymbol{C} = \begin{pmatrix} 1&1&0&0&1&0&1&0&0&0\\ 0&0&1&1&1&0&1&0&0&0\\ 0&0&0&0&0&1&1&0&0&0\\ 0&0&0&0&0&0&0&1&1&1 \end{pmatrix},$$

$$\boldsymbol{P} = \begin{pmatrix} 1&1&0&0&1&0&100&0&0&0\\ 0&0&1&1&10&0&100&0&0&0\\ 0&0&0&0&0&1&1&0&0&0\\ 0&0&0&0&0&0&0&1&1&1 \end{pmatrix}, \overline{\boldsymbol{H}}_1 = \begin{pmatrix} (\boldsymbol{w}_1^{\mathrm{UE}})^{\mathsf{H}}\boldsymbol{H}_{11}\\ (\boldsymbol{w}_2^{\mathrm{UE}})^{\mathsf{H}}\boldsymbol{H}_{12}\\ (\boldsymbol{w}_3^{\mathrm{UE}})^{\mathsf{H}}\boldsymbol{H}_{13}\\ (\boldsymbol{w}_7^{\mathrm{UE}})^{\mathsf{H}}\boldsymbol{H}_{17} \end{pmatrix}. \square$$

Given an association $\boldsymbol{A}$, we find the combiner. Then, for a given coordination matrix, every BS computes its effective channel and the beamformer.

3.2.2 Problem Formulation

Here, we first show that a pure model-based approach to optimize beamforming/ combining, association, and coordination is computationally intractable, using the models developed in

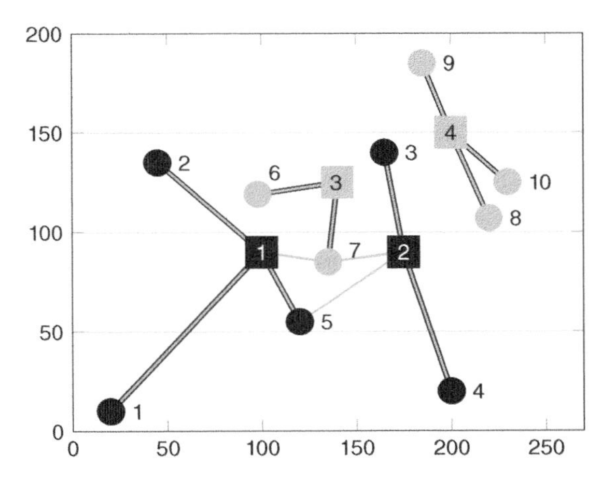

Figure 3.2 An example topology with two operators, gray and black. BSs and UEs are marked by squares and circles, respectively. Black lines show association. Gray lines show coordination.

Section 3.2. Moreover, after reformulating the problem as a multi-armed bandit, we highlight the massive signaling overhead of a pure data-driven method.

3.2.2.1 Rate Models

The SINR of each UE in the network is a function of the desired signal from its serving BS, ρ_{bu}^{Rx}; intra-cell interference, $I^{(1)}$ (signals transmitted to other UEs by the same BS); inter-cell interference, $I^{(2)}$ (interference from the signals transmitted by other BSs of the same network operator); inter-operator interference, $I_{bu}^{(3)}$ (interference from the signals transmitted by all BSs of other operators toward their own UEs); and noise power spectral density. We will skip over the exact expressions and refer the reader to (S. Ghadikolaei et al., 2018).

With that in mind, the average rate achieved by UE u,

$$r_u = \sum_{b \in \mathcal{B}} a_{bu} W_z \mathbb{E}\left[\log\left(1 + \frac{\rho_{bu}^{\mathrm{Rx}}}{I_{bu}^{(1)} + I_{bu}^{(2)} + I_{bu}^{(3)} + W_z \sigma^2}\right)\right], \tag{3.4}$$

where the expectation is over all random channel gains. Moreover, the assignment variable are taken from the feasible set, i.e. $\boldsymbol{A} \in \mathcal{F}_A$.

3.2.3 Model-based Approach

Here we present a (pure) model-based approach to the spectrum sharing. When the association and coordination matrices are given, each BS estimates the effective channel to its UEs, computes the precoding and combining vectors (see Section 3.2.1.4), and computes the average rate for each of the UEs it is serving from Eq. (3.4). A cloud server (logical controller) collects $\{r_u\}$ from all the BSs, computes the coordination cost per CI from Eq. (3.3), and evaluates a network utility $f_z(\boldsymbol{A}, \boldsymbol{C})$ for operator z.[2] Given $\mathcal{B}$ and $\mathcal{U}$, the controller formulates the following optimization problem to find the optimal association and coordination strategies:

$$\mathcal{P}_1 : \underset{\boldsymbol{A},\boldsymbol{C}}{\text{maximize}} \sum_{z=1}^{Z} \alpha_z f_z(\boldsymbol{A}, \boldsymbol{C}) := \sum_{z=1}^{Z} \left(\sum_{u \in \mathcal{U}_z} \log r_u \right), \tag{3.5a}$$

$$\text{subject to } \boldsymbol{A} \in \mathcal{F}_A, \ \boldsymbol{C} \in \{0,1\}^{|\mathcal{B}| \times |\mathcal{U}|}, \tag{3.5b}$$

$$g(a_{bu}) = [\boldsymbol{P}]_{bu}, \forall (b,u) \in (\mathcal{B} \times \mathcal{U}), \tag{3.5c}$$

$$\sum_{u \in \mathcal{U}_z} \sum_{b \in \mathcal{B}} [\boldsymbol{C}]_{bu}\, g(a_{bu}) \leq P_z^{\max}, \forall 1 \leq z \leq Z, \tag{3.5d}$$

where $\{\alpha_z\}_z$ are convex weights, and $\{f_z\}_z \in Z$ is the set of objective functions of the operators. While Eq. (3.5b) denotes the feasible set for $\boldsymbol{A}$ and $\boldsymbol{C}$, constraint Eq. (3.5d) ensures that the coordination cost does not exceed a maximum budget predetermined as $P_z^{\max}$ for each operator.

In the following, we discuss why a direct solution of $\mathcal{P}_1$ would be infeasible with the signaling and time limitations of the current radio access and core networks. Solving $\mathcal{P}_1$ implies that each BS in the network sends (or receives) pilot signals to each UE in the network (including other operators) and exchanges a huge amount of information with a central controller, which then proceeds to solve $\mathcal{P}_1$. Moreover, the complexity and signaling overhead grows with the number of BSs and UEs, and becomes overwhelming for mmWave networks with dense BS deployment. Additionally, there is the large burden of *inter-operator signaling* to compute $I_{bu}^{(3)}$. Furthermore, *channel aging* may render the exchanged information outdated before it serves

2 The use of log is to promote fairness among UEs as well as large sum-rates Andrews et al. (2014).

its purpose. This is particularly problematic for mmWave channels due to their inherently low coherence time. While earlier approaches dealt with these issues by neglecting interference and assuming noise-limited systems, recent works have shown that a few links may observe strong interference (Park and Gopalakrishnan, 2009). These limitations have hindered the application of optimal spectrum sharing to wireless systems. Nonetheless, the solution of $\mathcal{P}_1$ is interesting since it gives a theoretical upper bound for the performance of spectrum sharing (a *benchmark*).

Example 3.2 ***(Illustrative Numerical Results)***
Following the same setup as Example 3.1, we numerically test the impact of the number of antennas, network topology, association, and coordination levels. Here, we restrict our attention to the pure model-based approach, i.e. a *brute-force* solution of $\mathcal{P}_1$. The take-home messages from these results will be the basis of the proposed hybrid approach. We generate 100 random channels, find the optimal beamforming in every realization, and evaluate the interference terms under 2 antenna settings: small ($N_{\mathrm{BS}} = 8, N_{\mathrm{UE}} = 2$) and large ($N_{\mathrm{BS}} = 64, N_{\mathrm{UE}} = 16$).

Figure 3.3 shows different examples' association and coordination matrices: in Figure 3.3(a), we assume no coordination among UE and unintended BS, i.e. $\boldsymbol{C} = \boldsymbol{A}$. In Figure 3.3(b), the same setting is used, while forcing BS 1 to coordinate with UE 6 ($[\boldsymbol{C}]_{16} = 1$). In Figure 3.3(c), full coordination is performed, i.e. $[\boldsymbol{C}]_{bu} = [\boldsymbol{A}]_{bu}$. Although this high level of coordination may improve the rate, the resulting coordination cost will be quite elevated.

The performance of all of these scenarios is shown in Table 3.1. It becomes clear that the sum-rate/minimum UE rate is significantly improved by coordination, notably for UE 6 served by BS 3: indeed, coordinating with BS 1 (by setting $[\boldsymbol{C}]_{16} = 1$) improves the achievable sum-rate, due to a significant reduction in $I^{(3)}$.

As discussed in Section 3.3.3.2, an increase in the number of BS/UE antennas reduces the interference and thereby the need for coordination. This is readily seen in Table 3.1 by comparing the achievable sum-rates for both the small and large antenna settings. Finally, we underline that full coordination is only better than the no-coordination scenario.

3.2.4 Data-driven Approach

Data-driven approaches are attractive because they do not need the models to be accurate reflections of the corresponding physical processes. We refer the reader to (S. Ghadikolaei et al., 2018), where the authors reformulate the spectrum-sharing problem using the MAB

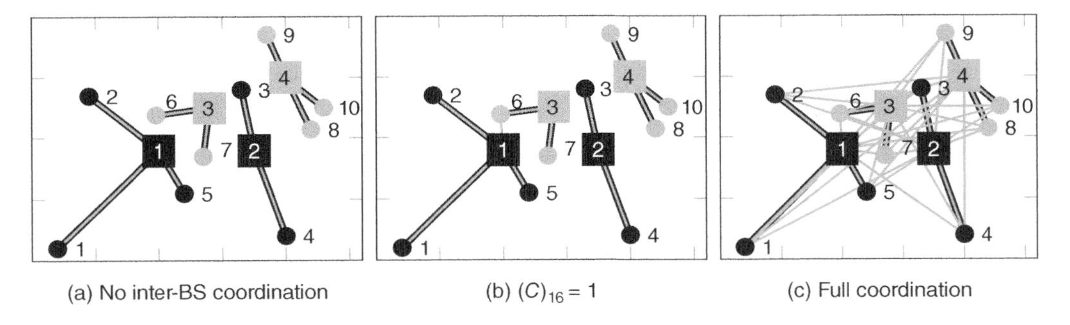

Figure 3.3 Illustration of the association and coordination for a network with two operators, gray and black, distributed in a 200×200 m^2 area. BSs and UEs are marked by squares and circles, respectively. A black (similarly gray) line from BS b to UE u indicates that $[\boldsymbol{A}]_{bu} = 1$ (similarly $[\boldsymbol{C}]_{bu} = 1$). In (a), every BS estimates only the channel of its associated UEs. The setting of (b) is identical to that of (a) except for an extra coordination $[\boldsymbol{C}]_{16} = 1$ to reduce inter-operator interference of UE 1. In (c), every BS estimates the channel of every UE.

Table 3.1 Performance of association and coordination in Figure 3.3. O1 and O2 stand for operator 1 (black) and operator 2 (gray). Rates are in Gbps. Normalized rate of UE 6 shows the rate improvement with respect to scenario (a), baseline, with the same number of antennas. Rate of UE 6 is 0.301 Gbps with ($N_{BS} = 8, N_{UE} = 2$), 1.099 Gbps with ($N_{BS} = 64, N_{UE} = 2$), and 1.884 Gbps with ($N_{BS} = 64, N_{UE} = 16$). Optimal,x corresponds to the solution of $\mathcal{P}_1$ with the coordination budget $P_z^{max} = x$

# Antennas	Scenario	Sum rates of UEs [O1,O2]	Min rate of UEs [O1,O2]	Rate improvement of UE 6 (%)	Coordination cost [O1,O2]
$N_{BS} = 8$, $N_{UE} = 2$	(a)	$[2.120\ 2.463]$	$[0.247\ 0.301]$	0	$[5\ 5]$
	(b)	$[1.968\ 2.896]$	$[0.222\ 0.440]$	148	$[5\ 105]$
	(c)	$[4.591\ 6.297]$	$[0.710\ 1.180]$	346	$[1055\ 1055]$
	Optimal,120	$[2.534\ 3.126]$	$[0.327\ 0.518]$	156	$[115\ 115]$
$N_{BS} = 64$, $N_{UE} = 16$	(a)	$[10.476\ 11.393]$	$[1.864\ 1.884]$	0	$[5\ 5]$
	(b)	$[10.477\ 11.836]$	$[1.912\ 2.223]$	25	$[5\ 105]$
	(c)	$[13.733\ 15.287]$	$[2.642\ 3.018]$	65	$[1055\ 1055]$
	Optimal,120	$[11.512\ 13.028]$	$[2.483\ 2.709]$	47	$[115\ 115]$
	Optimal, 1055	$[14.263\ 15.908]$	$[2.651\ 2.966]$	68	$[1055\ 1055]$

framework. In particular, an *agent* (decision maker) explores an unknown *environment* by taking different *actions* (each action corresponding to a choice of the beamforming vectors, association, and coordination) and observes the corresponding *rewards* (the sum-rate). The aim is to find the optimal set of actions that maximizes the collected reward; see Appendix A.1 for an overview of reinforcement learning.

While this purely data-driven approach offers the distinct advantage of relying on minimal assumptions regarding the model, it is practically infeasible due to the huge dimensions of the action space (which grows exponentially with $\boldsymbol{A}$, $\boldsymbol{C}$, and the number of choices of precoding/combining vectors), as well as the time constraints on the network. More specifically, the number of pilots for CSI acquisition should be limited to a few per CI, which is much lower than what is needed to efficiently run a multi-armed bandit; see (S. Ghadikolaei et al., 2018).

3.3 Hybrid Solution Approach

We have argued so far regarding the practical limitations of a purely model-based approach (i.e. a direct solution to $\mathcal{P}_1$), due to the massive complexity, signaling overhead, and lack of CSI. On the other hand, the curse of dimensionality prevents the pure data-driven approach from finding the optimal coordination and association decisions given the limited time available for the learning task. In the proposed hybrid approach, the learning task continuously refines the rate model of every UE rather than optimizing the decision variables. The model-based part then uses the updated rate models to find the optimal association and coordination strategies.

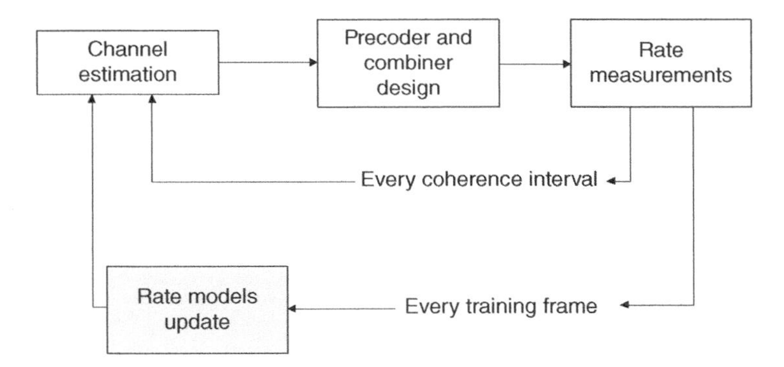

Figure 3.4 Overview of the hybrid solution. Channel estimation and precoding/combining are used to update the rate measurements (per CI). These measurements are in turn used to update the rate models during each training frame.

The so-called hybrid approach consists of two frame types: training and operation frames designed to balance the trade-off between exploration and exploitation. During a *training frame*, the BSs and UEs use a randomized policy to explore the space of feasible solutions for $(\boldsymbol{A}, \boldsymbol{C})$ and improve the rate models (see Section 3.3.2.1). During an *operation frame*, the operators apply a previously found good solution to protect the UE's performance from potential degradation due to a candidate $(\boldsymbol{A}, \boldsymbol{C})$. The solutions are only updated after passing a predefined confidence bound on their rate performance, measured over several training frames; see the illustration in Figure 3.4.

3.3.1 Data-Driven Component

Developing an efficient solution for $\mathcal{P}_1$ is challenging. First, the lack of closed-form solutions implies that iterative approaches are needed: the objective function must be evaluated for several values of associations/coordination matrix, until convergence. This results in sending additional pilots to evaluate the updated combining vectors at the UEs and estimate some new channels. These additional pilot transmissions and channel estimations can be very expensive, as we may need many iterations until convergence, and we may end up estimating almost all the channels; clearly this is impractical in a mmWave cellular network. Moreover, only parts of $I_{bu}^{(2)}$ and $I_{bu}^{(3)}$ for which the respective entry of $\boldsymbol{C}$ is 1 can be computed. Consequently, the cloud is not able compute either $I_{bu}^{(2)}$ nor $I_{bu}^{(3)}$, and therefore the objective function.

The *data-driven component* of the hybrid solution takes as input the network topology (i.e. location/number of BSs and UEs, N_{BS}, N_{UE}, etc.), the association matrix $\boldsymbol{A}$, the coordination matrix $\boldsymbol{C}$, and the effective channels $\overline{\boldsymbol{H}}_b$, and outputs an approximation of the rate of UE u, denoted by $\hat{r}_u$. More specifically, it consists of a dataset and a learning method. The entries of the dataset are $(\boldsymbol{A}, \boldsymbol{C}, \overline{\boldsymbol{H}}_b, \{r_u\}_{u\in\mathcal{U}})$, and the learning method approximates the rate function. The dataset is maintained at at the cloud and updated before/after each training frame; see Section 3.3.3.1.

BS b measures r_u for each of its associated UEs, during each CI, via feedback from the UEs of its rate in this CI. These values are collected at BS b and reported to the cloud before each training frame. The dataset and the mapping, $\hat{r}_u(\boldsymbol{A}, \boldsymbol{C})$ for all $u \in \mathcal{U}$, are updated at the cloud, and the next tuple $(\boldsymbol{A}, \boldsymbol{C})$ to be tested in the following training frame is computed (using the EXPLORE function in Algorithm 1). Then, the cloud updates the dataset and the rate models, and subsequently decides whether the new association and coordination solutions should be applied in subsequent operation frames. The cloud server then gradually updates these models with any new entry in the dataset through the UPDATE procedure. The other functions of this algorithm available to all operators are outlined in Algorithm 1, as follows.

Algorithm 1 Cloud server.

procedure UPDATE ($\boldsymbol{A}_0$,$\boldsymbol{C}_0$,$\{r_u\}$)
// add new entry to dataset and update $\{r_u\}$
procedure INITIALIZE ($\{L_{bu}\}$ if available)
// return $\boldsymbol{A}^{(0)}$,$\boldsymbol{C}^{(0)}$ *(see Section V-D in [Ghadikolaei et al., 2018])*
function OPTIMIZE ($\boldsymbol{A}_0$,$\boldsymbol{C}_0$)
  // initialize $\boldsymbol{A}^{(0)}$, $\boldsymbol{C}^{(0)}$ *(Section 3.3.2)*
  for $k = 1, 2, 3, \ldots$
    Run A-step and find $\boldsymbol{A}^{(k+1)}$ using Eq. (3.7)
    Run C-step and find $\boldsymbol{C}^{(k+1)}$ using Eq. (3.8)
    // break if convergence criteria met
  end for

function EXPLORE ($\boldsymbol{A}^\star$,$\boldsymbol{C}^\star$)
  set ($\boldsymbol{A}^{\text{tf}}$, $\boldsymbol{C}^{\text{tf}}$) at random from $\mathcal{F}_A \times \{0,\ 1\}^{|\mathcal{B}|\times|\mathcal{U}|}$ with probability ϵ
  $(\boldsymbol{A}^{\text{tf}}, \boldsymbol{C}^{\text{tf}}) = (\boldsymbol{A}^\star, \boldsymbol{C}^\star)$ otherwise

---

3.3.2 Model-Based Component

Given the updated rate models, the cloud formulates a variant of $\mathcal{P}_1$ and finds the new association and coordination solutions. Here we state the optimization problem and outline the solution method.

Every operator proceeds to approximate the rate of its UEs (via the DOWNLOAD function in Algorithm 1) and constructs an approximation of $f_z(\boldsymbol{A}, \boldsymbol{C})$, denoted by $\widehat{f}_z(\boldsymbol{A}, \boldsymbol{C})$, where $\widehat{f}_z = \sum_{u\in\mathcal{U}_z} \log \widehat{r}_u$. The modified optimization problem is written as

$$\mathcal{P}_{1M} : \max_{\boldsymbol{A},\boldsymbol{C}} \sum_{z=1}^{Z} \alpha_z \widehat{f}_z(\boldsymbol{A}, \boldsymbol{C}) \tag{3.6a}$$

$$\text{s.t. Eq. } (3.5b), \text{Eq. } (3.5c), \ \text{Eq. } (3.5d)$$

Note that solving $\mathcal{P}_{1M}$ is quite complex, since $\widehat{f}_z$ generally is not jointly convex in $\boldsymbol{A}$ and $\boldsymbol{C}$, and the feasible set is combinatorial. To circumvent such limitations, we leverage the block-coordinate descent (BCD) framework (also known as *alternating optimization*). The BCD method (Razaviyayn et al., 2013) consists of splitting $\mathcal{P}_{1M}$ into two subproblems (steps) that are iteratively updated. In the A-step, $\boldsymbol{A}$ is optimized for a given/known $\boldsymbol{C}$ matrix. Then, in the C-step, $\boldsymbol{C}$ is optimized for a given/known $\boldsymbol{A}$ matrix. Letting $\boldsymbol{A}^{(k)}$ and $\boldsymbol{C}^{(k)}$ be the values for $\boldsymbol{A}$ and $\boldsymbol{C}$ at iteration k, the two steps are formalized as

(A-step)

$$\boldsymbol{A}^{(k+1)} \in \underset{\boldsymbol{A}\in\mathcal{F}_A}{\operatorname{argmax}} \sum_{z=1}^{Z} \alpha_z \widehat{f}_z(\boldsymbol{A}, \boldsymbol{C}^{(k)}),$$

$$\text{s.t.} \sum_{b\in\mathcal{B}} \sum_{u\in\mathcal{U}_z} [\boldsymbol{C}^{(k)}]_{bu}\, g(a_{bu}) \le P_z^{\max}, \quad \forall 1 \le z \le Z,$$

(C-step)

$$C^{(k+1)} \in \underset{C \in \{0,1\}^{|\mathcal{B}| \times |\mathcal{U}|}}{\operatorname{argmax}} \sum_{z=1}^{Z} \alpha_z \hat{f}_z(A^{(k+1)}, C) ,$$
$$\text{s.t.} \sum_{b \in \mathcal{B}} \sum_{u \in \mathcal{U}_z} [C]_{bu} \, g(a_{bu}^{(k+1)}) \leq p_z^{\max} , \quad \forall 1 \leq z \leq Z ,$$

In spite of the combinatorial nature of these subproblems, many techniques may still be used to develop efficient solutions, e.g. a binary program solver and branch-and-bound solvers Bertsekas (1999). Moreover, further complexity reductions may be possible. For instance, when the learning function is bilinear in A and C, the linear program relaxation of these subproblems is optimal (Bertsekas, 1999), and each subproblem reduces to a linear program. More importantly, the use of BCD drastically reduces the size of the search space, from $\mathcal{O}(2^{|\mathcal{B}|^2|\mathcal{U}|^2})$ for $\mathcal{P}_{1M}$ to $\mathcal{O}(2^{|\mathcal{B}||\mathcal{U}|})$ for each of the subproblems. It can be shown that the sequence of updates is monotonically increasing with both the A-step and C-step update, and converges to a limit point provided that $\hat{f}_z$ is biconcave in A and C (see (S. Ghadikolaei et al., 2018) for details).[3]

Algorithm 2 is pseudocode describing our solution. We denote by A^{of} and C^{of} (resp. A^{tf} and C^{tf}) the association and coordination matrices for an operation frame (resp. training frame).

Algorithm 2 Hybrid model-based and data-driven spectrum sharing.

intput CI index n; An indexed sequence of training and operation frames; a feasibility space for the association and coordination $\mathcal{F}$
output Optimal beamforming in every CI, optimal A and C
run $(A^{\text{of}}, C^{\text{of}})$= INITIALIZE () at the cloud
set $A \leftarrow A^{\text{of}}$ and $C \leftarrow C^{\text{of}}$
for $n = 1, 2, 3, \ldots$
//every BS estimates channels and designs combiner to each of its UEs (Section 3.2.1.4)
//every BS estimates effective channels and computes precoding matrices (from Eq. (3.1))
//test this precoder and combiner pair, and record r_u
  if CI n is a *training frame*
    run UPDATE $(A, C, \{r_u\})$ at the cloud
    set $A^{(0)} \leftarrow A^{\text{of}}, \quad C^{(0)} \leftarrow C^{\text{of}}$
    run $(A^\star, C^\star)$= OPTIMIZE $(A^{(0)}, C^{(0)})$ at the cloud
    run $(A^{\text{new}}, C^{\text{new}})$= EXPLORE $(A^\star, C^\star, \mathcal{F})$
    //clear recorded rates at every BS
    if confidence criteria satisfied for $(A^{\text{new}}, C^{\text{new}})$
      set $A^{\text{of}} \leftarrow A^{\text{new}}$ and $C^{\text{of}} \leftarrow C^{\text{new}}$
    end if
    set $A \leftarrow A^{\text{of}}$ and $C \leftarrow C^{\text{of}}$
  end if
end for

---

3 However, showing that this limit point is a stationary point of $\mathcal{P}_{1M}$ is harder to establish due to the coupling between A and C in constraint Eq. (3.5d).

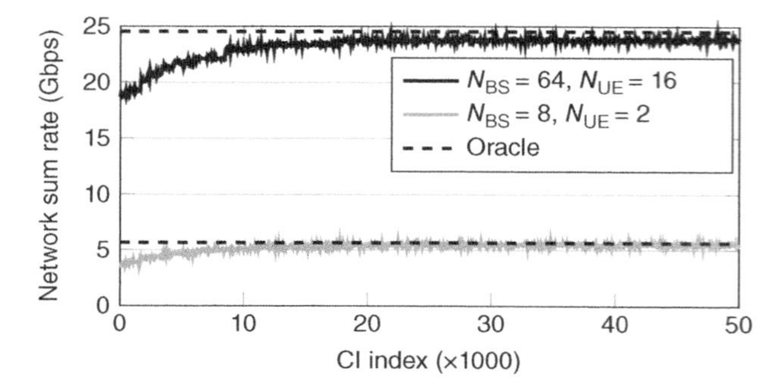

Figure 3.5 Illustration of the rate performance of the hybrid approach with $P_z^{max} = 115$. The dashed black line corresponds to the solution of the pure model-based approach, shown in Table 3.1.

3.3.2.1 Illustrative Numerical Results

We numerically evaluate the performance of Algorithm 2, for the same (stationary) deployment of Example 3.1. We also benchmark against the optimal solution for $\mathcal{P}$ in Table 3.1. We train a fully connected deep neural network with 1 input layer having $2|\mathcal{B}||\mathcal{A}|$ nodes, 5 hidden layers each having 20 nodes, and 1 output layer having $|\mathcal{U}|$ nodes, using backpropagation. More specifically, the input is a vectorization of $\boldsymbol{A}$, $\boldsymbol{C}$, and the output returns the regression results for $\{\hat{r}_u\}_u$. We further assume that the cloud has full topological information, and exploit the inherent properties of mmWave systems such as high path loss and a dense BS deployment to reduce the search space for $\boldsymbol{A}$, $\boldsymbol{C}$. We gradually decay exploration parameter ϵ by setting $\epsilon \leftarrow 0.9 \times \epsilon$ after 1000 CIs, to improve the exploitation.

The instantaneous network sum-rate from Algorithm 2 is shown in Figure 3.5, where the optimal solution of $\mathcal{P}_1$ – requiring the cloud to know all channels in the network perfectly – is also plotted. Despite the fluctuations in the sum-rate (due to the inherent randomness of the channels and CSI estimation errors), the average sum-rate of Algorithm 2 is increasing. Moreover, we observe that the iterations converge to the globally optimal solution of $\mathcal{P}_1$, thereby implying that the algorithm is asymptotically optimal in this case. This convergence behavior is enabled by a good initialization; see Section 3.3.3.2.

3.3.3 Practical Considerations

3.3.3.1 Implementing Training Frames

Training frames are designed to successively refine the current rate models and thereby find a better association and coordination solution. Note that the OPTIMIZE function is called before/after each training frame. Naturally, we expect a high frequency of training frames in the first few association periods, as we assume no a priori knowledge of the network. However, this frequency will gradually decay as more knowledge about the rate models is obtained. If the distribution of the rate is changing over time, more training frames may be required to allow for tracking.

The server obtains the updated rate measurements, updates its models, and runs the BCD procedure prior to each training frame. Then, a randomized exploration is performed on the set of feasible solutions, $\mathcal{F}_A \times \{0,\ 1\}^{|\mathcal{B}|\times|\mathcal{U}|}$, and one association coordination matrix is selected for exploration in the next training frame. This solution is applied in the subsequent operation frames if it passes a reliability check, in terms of a confidence bound on the "perturbation" it causes objective function. Thus, the UE is protected from potential service interruption due to the inherent randomness of the training frame.

3.3.3.2 Initializations

We underline the importance of initializing both the UPDATE procedure and OPTIMIZE functions (Algorithm 1). We thus overview some strategies for picking a "good" starting point to speed up learning $\{r_u\}$, and initial solutions $\boldsymbol{A}^{(0)}$, $\boldsymbol{C}^{(0)}$ to the BCD algorithm.

Rate model: Note that inherent aspects of the system allow for a "good" initialization of learning. When the number of BS and UE antennas is large enough, $I_{bu}^{(2)}$ is small and may be safely neglected. Thus, inter-cell coordination is not needed in large antenna systems – which is the case here – and yields a sparse coordination matrix. Moreover, the high penetration loss and directional communications substantially reduce the intra-cell and inter-operator interference components, compared to sub-6 GHz systems; see (Di Renzo, 2015). Additionally, one may use initializations such as Gaussian approximation for the interference (Verdu, 1998), or set all three interference terms to zero for all links in the network (i.e. interference-free network (S. Ghadikolaei et al., 2016)) when no topological knowledge is assumed. We refer the interested reader to (S. Ghadikolaei et al., 2018) for these discussions.

BCD solver: We use the INITIALIZE function (Algorithm 1) to initialize the BCD method, according to one of the following two strategies:

- *Full/partial topological knowledge available*: $\boldsymbol{A}^{(0)}$ is selected such that each UE is associated with the strongest BS. Then we set $\boldsymbol{C}^{(0)} = \boldsymbol{A}^{(0)}$.
- *No topological knowledge available*: We randomly allocate UEs to BSs within the same operator. We then set $\boldsymbol{C}^{(0)} = \boldsymbol{A}^{(0)}$.

The BCD method is initialized using the current association and coordination solution for the following CIs.

3.3.3.3 Choice of the Penalty Matrix

In general, the cost of estimating the effective channel of a user belonging to another BS or another operator is larger than that of a UE of the same BS. This can be encoded in the penalty matrix, $\boldsymbol{P}$, as follows: $0 \leq [\boldsymbol{P}]_{bu} < [\boldsymbol{P}]_{iu} < [\boldsymbol{P}]_{ju}$, where $i \in \mathcal{B}_z \backslash \{b\}$ and $j \in \mathcal{B}_k$, $k \neq z$. To implement this penalty matrix, let p_b and $\overline{p}_b$ denote the penalty of intra-operator and inter-operator coordination for users served by BS b (assuming the latter is identical for all non-serving operators). Furthermore, we denote by $\boldsymbol{P}_0$ the template penalty that models the cost of channel estimation, where $[\boldsymbol{P}_0]_{bu}$ is the penalty when BS b estimates the channel of UE u. One particular choice for $\boldsymbol{P}_0$ that we adopt here is as follows: $[\boldsymbol{P}_0]_{bu} = p_b$, for $(b, u) \in \mathcal{B}_z \times \mathcal{U}_z$, and $[\boldsymbol{P}_0]_{bu} = \overline{p}_b$, for $(b, u) \in (\mathcal{B} \backslash \mathcal{B}_z) \times \mathcal{U}_z$. Given $\boldsymbol{P}_0$, we then set $[\boldsymbol{P}]_{bu} = [\boldsymbol{P}_0]_{bu} + a_{bu}(p_b - [\boldsymbol{P}_0]_{bu})$, $\forall (b, u) \in (\mathcal{B} \times \mathcal{U})$. We used this instance of the penalty matrix for all the numerical results in this chapter.

3.4 Conclusions and Discussions

In this chapter, we motivated spectrum sharing in millimeter-wave cellular networks as an enabling technology for future cellular networks. After posing the problem, we argued the huge signaling and computational overhead for a (pure) *model-based approach*, i.e. a brute-force solution of the resulting optimization problem. Moreover, we discussed briefly the infeasibilty of a (pure) *data-driven approach* (using reinforcement learning) due to the well-known curse of dimensionality problem in learning. We then derived our so-called hybrid (model-based and data-driven approach) to circumvent these problems, which comprises two components. The *model-based component* finds "good" solutions for coordination, association, and beamforming/combining. Moreover, the *data-driven component* refines these solutions using any of the plethora of available ML methods. We discussed some numerical results showing that, with a proper initialization point, this approach converges to the globally optimal solution of the model-based approach, with a fraction of the signaling and computational overhead.

We underline that although this hybrid approach was presented in the context of spectrum sharing in millimeter-wave networks, the take-home messages are valid for a wide class of "machine leaning for wireless communications" problems: beamforming/combining designs for multi-cell networks, multi-cell coordination, resource allocations, etc. It is well known that model-based solutions for these problems fail due to the huge computational/signaling overhead. On the other hand, acquiring samples – needed to build the training set – is generally done with pilot transmissions, which are scarce resources in most wireless communication systems. This in turn implies that the size of the training set is small, and most ML tools will fail. Thus, we believe that the so-called hybrid approach will offer a new paradigm for solving these problems, often encountered in future wireless communications.

Appendix A

Appendix for Chapter 3

A.1 Overview of Reinforcement Learning

Reinforcement learning (RL) is an area of machine learning concerned with optimizing some learnable utility function in an unknown environment. An RL model is defined by (Sutton and Barto, 2011): a set of environment states $\mathcal{S}$, a set of agent actions $\mathcal{A}$, and a set of immediate reward signals $\mathcal{R} \subset \mathbb{R}$. States describe the environment. A policy is a mapping from the perceived states of the environment to actions to be taken. And a reward is a scalar function that quantifies the benefit of following a policy from an environment state onward.

The agent and environment interact at sequence of time steps $t = 0, 1, 2, \ldots$. At each time t, the agent receives some representation of the environment, encoded in $s_t \in \mathcal{S}$, along with the immediate reward from its last action. The agent then takes a new action $a_t \in \mathcal{A}(s_t) \subseteq \mathcal{A}$, where $\mathcal{A}(s_t)$ is the set of actions available in state s_t. One time step later, the agent receives the reward of its action a_t, which is denoted by $r_t \in \mathcal{R}$, and finds itself in a new state $s_{t+1} \in \mathcal{S}$. It then updates its knowledge of the environment and re-evaluates the optimal policy. The evolution of the agent in the environment can therefore be described by tuple (s_t, a_t, r_t, s_{t+1}) and the one-step dynamics of the environment $p(s', r|s, a) = \mathbb{P}(s_{t+1} = s', r_t = r|s_t = s, a_t = a)$. The sole objective of the RL agent is to find the optimal policy, namely the optimal sequence of actions $\pi^\star(s)$ that maximizes the accumulated rewards it receives over a long-run starting from state s Sutton and Barto (2011) $\pi^*(s) \in \arg\max_\pi \mathbb{E}_\pi \left[\sum_{k=0}^{\infty} \gamma^k r_{t+k+1} \mid s_t = s\right]$, where $0 \leq \gamma \leq 1$ is a discount factor. Solving the optimization involves a trade-off between learning the environment (explore) and taking the optimal policy so far (exploit).

Bibliography

5G Americas White Paper. Wireless technology evolution: Transition from 4g to 5g: 3gpp releases 14 to 16, November 2018.

Jeffrey G Andrews, Sarabjot Singh, Qiaoyang Ye, Xingqin Lin, and Harpreet S Dhillon. An overview of load balancing in HetNets: Old myths and open problems. *IEEE Wireless Commun.*, 21(2):18–25, Apr. 2014.

K. Ardah, G. Fodor, Y. C. B. Silva, W. C. Freitas, and F. R. P. Cavalcanti. A unifying design of hybrid beamforming architectures employing phase shifters or switches. *IEEE TranIEEE Transactions on VTechnology*, 67(11): 11243–11247, 2018.

O. E. Ayach, R. W. Heath, S. Abu-Surra, S. Rajagopal, and Z. Pi. The capacity optimality of beam steering in large millimeter wave MIMO systems. In *Proc. IEEE International Workshop on Signal Processing Advances in Wireless Communications*, pages 100–104, 2012.

D.P. Bertsekas. *Nonlinear Programming*. Athena Scientific, 2 edition, 1999.

F. Boccardi, H. Shokri-Ghadikolaei, G. Fodor, E. Erkip, C. Fischione, M. Kountouris, P. Popovski, and M. Zorzi. Spectrum pooling in mmwave networks: Opportunities, challenges, and enablers. *IEEE Commun. Mag.*, pages 33–39, November 2016.

Marco Di Renzo. Stochastic geometry modeling and analysis of multi-tier millimeter wave cellular networks. *IEEE Trans. Wireless Commun.*, 14(9): 5038–5057, Sept. 2015.

Ericsson. Mobility report 2018. https://www.ericsson.com/en/mobility-report, June 2018.

G Fodor, N. Rajatheva, W. Zirwas, L. Thiele, M. Kurras, K. Guo, A Tölli, J. H Sorensen, and E. De Carvalho. An overview of massive mimo technology components in metis. *IEEE Communications Magazine*, 55(6):155–161, July 2016.

H. S. Ghadikolaei, C. Fischione, and E. Modinao. Interference model similarity index and its applications to millimeter-wave networks. *IEEE Trans. Wireless Commun.*, 17(1):71–85, Jan. 2018.

X. Jiang, H. Shokri-Ghadikolaei, G. Fodor, E Modiano, Z. Pang, M. Zorzi, and C. Fischione. Low-latency networking: Where latency lurks and how to tame it. *Proceedings of the IEEE*, 2018.

M. Park and P. Gopalakrishnan. Analysis on spatial reuse and interference in 60-GHz wireless networks. *IEEE J. Sel. Areas Commun.*, 27(8):1443–1452, Oct. 2009.

Meisam Razaviyayn, Mingyi Hong, and Zhi-Quan Luo. A unified convergence analysis of block successive minimization methods for nonsmooth optimization. *SIAM Journal on Optimization*, 23(2):1126–1153, June 2013.

Benjamin Recht. A Tour of Reinforcement Learning: The View from Continuous Control. *arXiv e-prints*, art. arXiv:1806.09460, June 2018.

Hossein S. Ghadikolaei, Federico Boccardi, Carlo Fischione, Gabor Fodor, and Michele Zorzi. Spectrum sharing in mmWave cellular networks via cell association, coordination, and beamforming. *IEEE J. Sel. Areas Commun.*, 34(11):2902–2917, Nov. 2016.

Hossein S. Ghadikolaei, Hadi Ghauch, Gabor Fodor, Carlo Fischione, and Mikael Skoglund. A hybrid model-based and data-driven approach to spectrum sharing in mmwave cellular network. *In Preparation*, Nov. 2018.

Rahul K. Sevakula, Mohammed Suhail, and Nishchal K. Verma. Fast data sampling for large scale support vector machines. In *IEEE Workshop on Computational Intelligence: Theories, Applications and Future Directions*, Kanpur, India, Dec. 2015.

Richard S Sutton and Andrew G Barto. *Reinforcement learning: An introduction*. Cambridge Univ Press, 2011.

R. H. Tehrani, S. Vahid, D. Triantafyllopoulou, H. Lee, and K. Moessner. Licensed spectrum sharing schemes for mobile operators: A survey and outlook. *IEEE Communications Surveys Tutorials*, 18(4):2591–2623, Fourthquarter 2016. ISSN 1553-877X. doi: 10.1109/COMST.2016.2583499.

David Tse and Pramod Viswanath. *Fundamentals of wireless communication*. Cambridge University Press, 2005.

Sergio Verdu. *Multiuser detection*. Cambridge university press, 1998.

4

Deep Learning–Based Coverage and Capacity Optimization

Andrei Marinescu[1], *Zhiyuan Jiang*[2], *Sheng Zhou*[3], *Luiz A. DaSilva*[1], *and Zhisheng Niu*[3]

[1] *CONNECT, Trinity College Dublin, Dublin, Ireland*
[2] *School of Communications and Information Engineering, Shanghai University, Shanghai, China*
[3] *Department of Electronic Engineering, Tsinghua University, Beijing, China*

4.1 Introduction

Mobile networks have expanded in recent decades with the primary objective of providing coverage for as much of the population as possible. Traditionally, urban locations, due to the high density of users, are the areas that have the better coverage, while sparsely populated rural areas that are not in the vicinity of cities tend to have limited coverage. However, as constant Internet availability becomes more of a necessity for many users, mobile operators are expanding their networks to provide better coverage of users even in remote locations. At the same time, even in locations with typically good coverage, the unexpected appearance of crowds of people decreases the available bandwidth per user due to limited network capacity in the area. The tendency in the latter case is to increase the density of cells in order to boost the total network capacity in particularly busy regions for situations like these, through over-provisioning.

In effect, besides the improvements brought by technical advances in mobile communications, the traditional means of addressing the problems of coverage and capacity are through the addition of extra resources to the network. In the long term, this is an expensive solution that becomes unsustainable economically and environmentally. These extra resources are represented by both physical elements, such as base stations and core network upgrades, and increased energy requirements.

In order to decrease costs associated with commissioning these resources while still providing good quality of service (QoS) to users, several network coverage and capacity optimization (CCO) techniques have been proposed. These typically try to address a trade-off between the two, mainly capacity and coverage. The authors in (Phan et al., 2017) provide a good review of the literature with respect to CCO. Techniques used for this purpose involve turning base stations on and off to save energy (Wu et al., ; Guo et al., 2016), adjusting antenna tilt angles or power depending on demand (Fehske et al., 2013; Gao et al., 2013), load balancing by means of handing over mobile users from more crowded cells to less crowded ones (Li et al., 2011), and antenna placement optimization techniques (Hafiz et al., 2013).

Based on this, we can distinguish two general types of approaches to the CCO problem:

1. Resource reallocation through various load-balancing techniques where either base station parameters can be adjusted or mobile users are reallocated between cells
2. Energy-saving techniques through either antenna power control or switching base stations on and off depending on demand

Machine Learning for Future Wireless Communications, First Edition. Edited by Fa-Long Luo.
© 2020 John Wiley & Sons Ltd. Published 2020 by John Wiley & Sons Ltd.

Both of these measures are dependent on network demand, specifically on spatio-temporal characteristics of mobile users and their traffic patterns. Due to the multitude of factors that determine the state of a mobile network (e.g. network user geometry, mobility patterns, traffic load, etc.) and network parameters to be controlled (e.g. antenna tilt, transmission power levels, handover thresholds, etc.), this represents a complex optimization problem with a large search space. Recent work attempts to leverage the significant advances made in machine learning (ML) in order to address this large search space.

This chapter presents two state-of-the-art ML-based techniques that tackle the CCO problem from each of the main aspects:

- Configuring base-station parameters to address current demand through a deep neural network architecture, where suitable configurations actions are taken on the basis of the inference from current network user geometry information
- Enabling base-station sleeping via a data-driven approach by using deep reinforcement learning (RL), which leverages network traffic models to address the non-stationarity in real-world traffic

We present these two techniques due to their complementary nature. Base station sleeping is a technique that can be used on longer time frames, which can benefit of long-term demand predictions (e.g. hour/day ahead). Base-station parameter adjustment is a more reactive technique, which can operate at the second level depending on current demand patterns, and can be used to help currently operating base stations compensate for sleeping base stations.

The remaining sections of this chapter are divided as follows. Section 4.2 introduces a set of widely used ML techniques and provides an overview of their application to CCO problems in the wireless network domain. Section 4.3 describes the used and achieved result of the deep RL approach in solving the problem of base-station sleeping. Section 4.4 presents the application and evaluation of the multi-agent deep neural network framework on the dynamic frequency reuse problem in mobile networks. And the last section of this chapter, Section 4.5 presents our concluding remarks and potential directions of research in this area in the future.

4.2 Related Machine Learning Techniques for Autonomous Network Management

ML techniques have already found applications in autonomous network management and operation. These applications range from wireless network coverage and capacity optimization to higher-level functions such as network-slicing management and network fault identification and prediction (ETSI, 2018). A significant part of the works in literature that employ ML techniques to optimize radio access network (RAN) functions rely on reinforcement learning (Moysen and Giupponi, 2018). This technique is introduced in the following section, followed by a brief presentation of artificial neural networks, the main driver of the deep learning revolution.

4.2.1 Reinforcement Learning and Neural Networks

Reinforcement learning (RL) is a ML approach in which an autonomous entity (i.e. agent) learns how to perform well in an unknown environment through a process of trial and error. The agent learns how to perform according to the received rewards for each action taken in each given state. The agent can transition from one state to another depending on the action it takes in the current state. This process is pictured in Figure 4.1. The objective of the agent

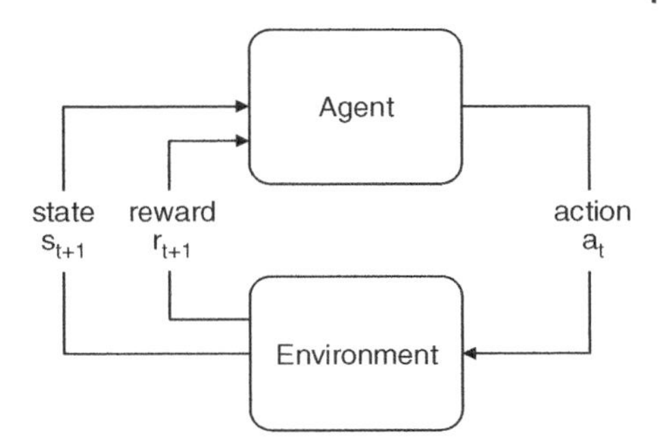

Figure 4.1 Reinforcement learning agent.

is to maximize its long-term reward (i.e. cumulative reward), where the maximum cumulative reward is achieved once the agent finds the optimal policy. However, in order to learn this policy, an agent needs to explore the environment where it is situated. Even though sometimes immediate actions can lead to higher rewards, maximizing the immediate reward can lead to local optima situations; therefore, an agent has to explore each state and action combination to find out the highest-rewarding sequence of actions that can be taken from a particular state.

Formally, for an agent to learn how to achieve optimal performance, it needs to try each state-action combination an infinite number of times (Sutton and Barto). However, in practice, an agent's performance can converge to the optimum if it tries each state-action pair sufficiently often. The number of times that each state-action pair should be visited depends on the stochasticity of the environment, which can insert noise in the rewards observed after each action is taken. The effect of this noise can only be alleviated with a sufficiently large sample size. Note that the formal guarantees of RL hold for stationary environments, where transition probabilities from one state to another as a function of an action are stable. As such, the environment in which an agent operates can be defined as a Markov decision process (MDP).

We can distinguish two main types of RL algorithms: model-free and model-based. Model-free techniques learn the potential long-term reward that can be received from each state and action combination. In addition, model-based techniques also try to keep track of the transition probabilities between states by building a model of the environment, thereby also learning a model of state transitions. Model-based techniques can more accurately address a stationary environment, but will adapt more slowly to an environment that is dynamic and continuously changing/evolving.

RL is a powerful technique, but it is negatively impacted when the state-action space is very large, as the exploration period for an agent takes too long. In particular, with continuous state spaces, the application of RL becomes intractable unless the state space is quantized, which can lead to sub-optimal behavior of the agent since information from the environment is lost through this process. To address this, RL has been combined with function approximation techniques, to approximate the value of a state or state-action pair, even if the specific state that is evaluated has not been previously visited in the exploration period. This has found successful applications in large state-space domains such as backgammon (Tesauro, 1995), the game of Go (Silver et al., 2017), and Atari console games (Mnih et al., 2015). Most of these success stories involve neural networks as the function-approximation technique.

An artificial neural network (ANN) is a nonlinear *supervised* learning technique. This comprises a set of input-output neuron pairs, known as the *input layer* and *output layer*, each consisting of a predefined set of neurons, where each neuron is holding a value. In between these two layers, a neural network has one or more intermediate (or *hidden*) layers of neurons. An example of a very basic neural network is illustrated in Figure 4.2. Neurons from each layer are typically linked with neurons from the next layer through a set of links of different weights.

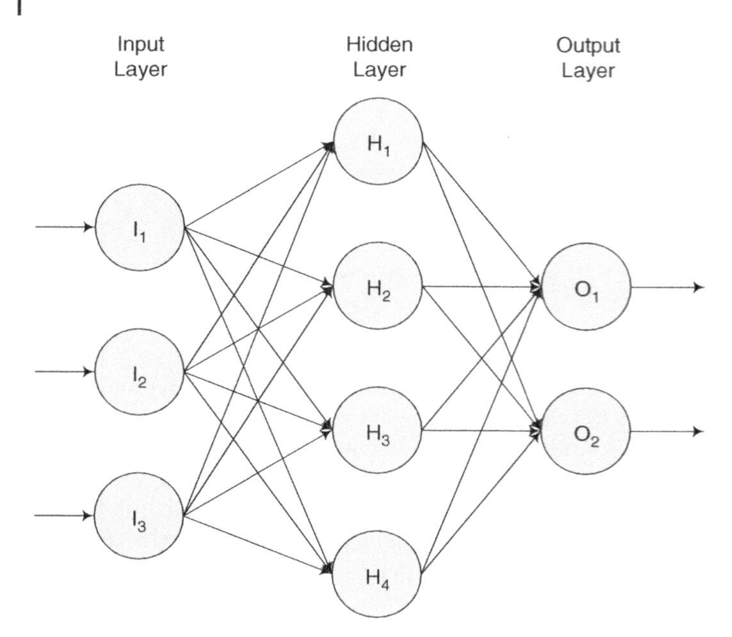

Figure 4.2 Artificial neural network.

Each linked neuron from the next layer combines these weighted values through a nonlinear process known as the *activation function*, to compute its own value. The ANN is trained with a training set of provided input-output pairs, through the process of backpropagation, which calculates the gradient of a loss function while taking into account all the weights in the network. This gradient is used to update the weights, in an attempt to minimize the error between the output of the ANN and the expected value. Once a neural network is trained, it can be used to approximate outputs from provided input values.

Neural networks were introduced in the first half of the last century (McCulloch and Pitts, 1943). However, they have found mainstream success only in the last decade as breakthroughs in training processes combined with advances in hardware have allowed the development of more complex neural network architectures. Specifically, deep neural network (DNN) architectures (architectures with many hidden layers and neurons) have been at the lead of the progress in the fields of computer vision, handwritten text recognition, language translation, and voice recognition.

4.2.2 Application to Mobile Networks

Solutions for the CCO problem employ RL to control network parameters such as antenna tilt and transmission power levels (Razavi et al., 2010; ul Islam and Mitschele-Thiel, 2012; Li et al., 2012; Fan et al., 2014). A related problem, inter-cell interference coordination (ICIC), is also tackled via RL (Galindo-Serrano et al., 2011; Dirani and Altman, 2010; Bennis et al., 2013; Simsek et al., 2011). However, since the domains in which these algorithms operate depend on a large number of parameters, it becomes difficult to quantify all the necessary information from the environment while avoiding an explosion in the state space of the RL agents. Neural networks are a way to address both the large state space, through function approximation, and the nonlinearity aspects of the wireless network parameters. As such, neural networks can be either used in combination with RL – through what is now called deep RL, to address the limitations of tabular RL in such problems– or applied directly as a supervised learning technique. In this chapter, we show how each type of solution (i.e. neural networks, or neural networks in conjunction with RL) can be applied to problems related to coverage and capacity optimization in wireless networks.

4.3 Data-Driven Base-Station Sleeping Operations by Deep Reinforcement Learning

This section presents a data-driven algorithm for dynamic sleeping control of base stations (BSs), called DeepNap. Conventional methods usually adopt queuing theory to derive the optimal sleeping control policies, assuming certain conditions are met. For example, Poisson traffic assumptions are often used to derive analytical solutions. However, real-world traffic usually exhibits self-similarity and non-stationarity, rendering the Poisson assumptions problematic.

On the other hand, deep learning techniques have achieved tremendous success in computer vision, speech recognition, and computer games. It is further discovered that deep learning can help machines make sequential decision making, and hence deep RL is introduced. The biggest advantage of deep RL, compared with conventional methods, is that it is model-free, and therefore robust against modelling error. That is, even if the real-world traffic is non-Poisson, the performance of deep RL methods usually is not affected.

Based on this thinking, we apply deep RL to dynamic BS sleeping control problems. We use a deep Q-network (DQN) to accomplish the learning task of representing high-dimensional raw observations or system belief vectors. Several enhancements of the original DQN method are necessary to make it work in the considered system setting, including an action-wise experience replay and adaptive reward scaling, which will be explained in the following sections.

4.3.1 Deep Reinforcement Learning Architecture

The deep RL architecture usually consists of an agent interacting with the environment. To better fit the BS sleeping control setting we are considering, the architecture is adjusted as shown in Figure 4.3. The system is abstracted by a collection of four components: a traffic emulator (E) that generates traffic based on historical data; a traffic server (S) that has the functionality

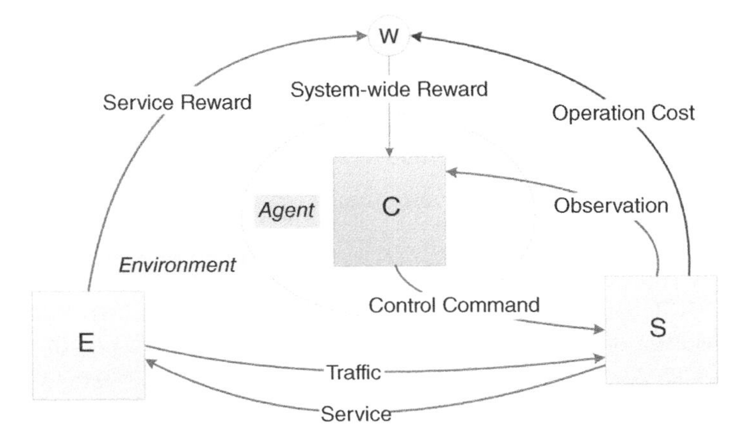

Figure 4.3 RL formulation of BS sleeping control. The controller (C) servers as the agent, while the traffic emulator (E), traffic server (S), and reward combiner (W) together serve as the environment.

of the data plane of the BS; a sleeping controller (C) that has the functionality of the control plane of the BS; and a reward combiner (W) that outputs the reward. In this setting, the sleeping controller can be regarded as the RL agent, and the RL environment is the combination of the traffic emulator, the traffic server, and the reward combiner.

4.3.2 Deep Q-Learning Preliminary

In this section, we introduce some preliminaries about deep Q-learning. Q-learning, without the "deep" concept, is a model-free learning technique to solve sequential decision problems. In essence, it uses a slightly modified value iteration to evaluate each action-state pair. Specifically, the following equation is used:

$$Q^{(i+1)}(s,a) = r + \gamma \max_{a'} Q^{(i)}(s',a'))$$

where s is the state, a is the action at the state s, the reward is r, and the next state after state transition is denoted by s'. The evaluation of each action-state pair is done by continuously iterations based on the obtained data. After the optimal action-value function, denoted by $Q^*(s,a)$, is obtained, it is straightforward that the optimal policy is to greedily select the best action at the current state.

Traditionally the Q-learning method requires that each state-action pair is evaluated. This presents significant challenges in practice, given limited data and convergence time. Hence comes the deep RL approach, which adopts DNNs for action-state space approximation. In particular, the state-action pair is approximated by a set of parameters, i.e. $Q(s,a;\boldsymbol{\theta}) \approx Q^*(s,a)$, and the DNN is trained by minimizing the difference between the predicted state-action pair value and the experienced value, i.e.

$$L(\boldsymbol{\theta}^{(i)}) = \mathbb{E}[(y^{(i)} - Q(s,a;\boldsymbol{\theta}^{(i)}))^2], \tag{4.1}$$

with gradient-based optimization methods, where

$$y^{(i)} = \mathbb{E}[r + \gamma \max_{a'} Q(s',a';\boldsymbol{\theta}^{(i-1)})|s,a] \tag{4.2}$$

Several techniques are widely adopted to stabilize the training process of the DQN: (i) a replay buffer, which is used to store experience in a certain period;(ii) a separate, target Q-network for updating the parameters without changing them on the fly; and (iii) reward clipping (normalization).

4.3.3 Applications to BS Sleeping Control

Now, we will present how to apply the deep RL approach to BS sleeping controls.

From a better comparison point of view, we first introduce the traditional model-based approaches for BS sleeping control. BS sleeping is an effective way to save energy for BS operations. Prior work on this topic usually adopted a model-based approach, e.g. most work has assumed that the incoming traffic at the BS obeys the Poisson distribution. Thereby, a double-threshold hysteretic policy is proved to be optimal in (Kamitsos et al. 2010). Furthermore, it is proved in (Jiang et al. 2018) that under Markov modulated Poisson process (MMPP) modeled traffic and multi-BS scenario, the optimal policy is still threshold-based. When the arrival state is unknown to the sleeping controller, a belief-state value iteration can be applied to solve for the optimal threshold (Leng et al. 2016). However, these approaches are sensitive to modelling error, i.e. when the arrival traffic is not modelled exactly, the sleeping control is not guaranteed to have good performance, and in reality, the traffic pattern is usually insufficiently modeled.

In contrast, RL-based approaches have the advantage of robustness against modelling error. Therefore, with the development of deep RL, it is promising to apply it to BS sleeping control. However, in order to do so, several adjustments have to be made.

If the environment is non-stationary, naive experience replay can become problematic. Shown in Figure 4.4 is an example taken from our experiment with the original DQN algorithm. The top curve shows the average number of requests per time step smoothed over one-minute time windows. The traffic pattern is clearly non-stationary. Driven by the traffic variation, the experience distribution in the replay memory (middle curve) oscillates between waking- and sleeping-dominating regimes. Since the loss in Eq. (4.1) is related with one action (thus one network output) per sample, only a single network output can be updated in these dominating regimes, and the other "free-running" output may drift away from its expected value. For example, observe in the beginning of the experiment that the memory is dominated by waking experiences, and thus the Q value for the free-running sleeping action drifts to around -1. It is only until $16:00$ that the dominance of waking experience is broken by random exploration and the Q value for sleeping action starts to amble toward the expected 0 value. The balance is once more broken by the traffic peak at around $23:00$, with sleeping Q values again pushed to around -1.

4.3.3.1 Action-Wise Experience Replay

With non-stationary traffic patterns, which are often encountered in the real world, the naive experience replay approach may be problematic. As seen in Figure 4.4, a real-world traffic trace is depicted, which is obviously non-stationary. In this case, the naive experience replay may face the problem of premature convergence to the current traffic patterns and not be suitable for future changes. In this regard, we propose an action-wise experience replay method. The experiences from different phases of the traffic are stored in the same experience replay buffer, such that the learning procedure is immune from overfitting to one specific traffic phase.

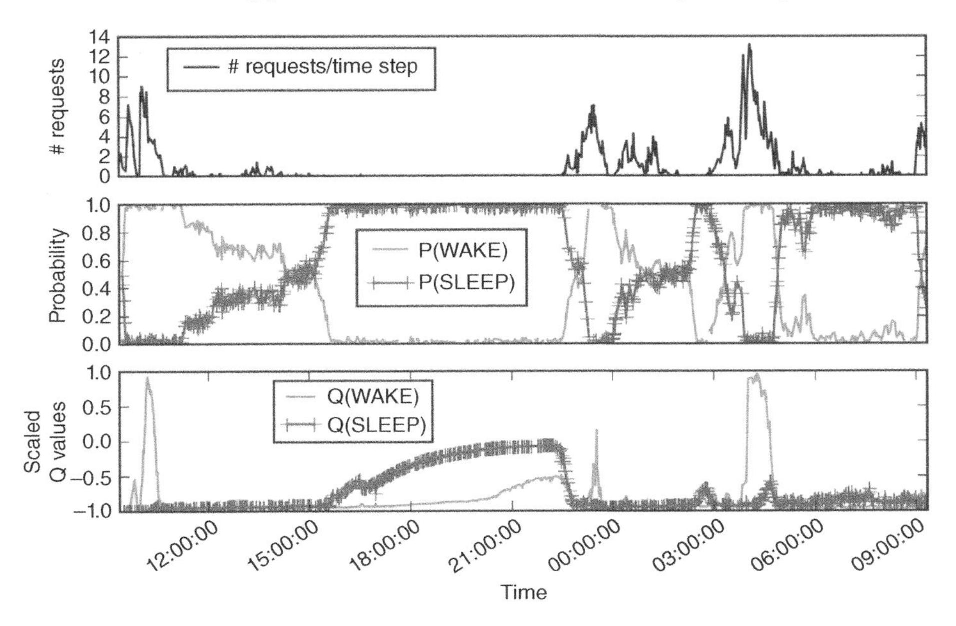

Figure 4.4 Experimental results for the original DQN algorithm. The figures show the number of requests per time step smoothed over a one-minute time window (top), the percentage of experiences with waking and sleeping action in the replay memory (middle), and the average Q values for waking and sleeping actions over one-minute time window (bottom). The data used is taken from September 25–26, 2014.

4.3.3.2 Adaptive Reward Scaling

Another tweak to the original DQN to fit mobile networks is adaptive reward scaling. The traffic profile in mobile networks often has a large dynamic range. The traditional way of dealing with this is to clip the reward to a certain interval, e.g. $[-1, 1]$. Reward values outside this interval are thrown away. This is problematic for mobile traffic profiles given their time-varying dynamic range, especially in BS sleeping scenarios, since the reward value is usually far beyond +1 in high-traffic periods.

Therefore, it is necessary to perform reward rescaling in this scenario. By shrinking the received reward value by a constant factor, the action values can be proportionally scaled down, but this does not take into account the time-varying nature of the traffic profiles. As seen in Figure 4.5, finding an appropriate scaling factor is difficult. To deal with this, the loss function is revised as

$$L'(\theta^{(i)}) = \mathbb{E}[(y'^{(i)} - Q(s, a; \theta^{(i)}))^2 + U(Q(s, a; \theta^{(i)}))], \tag{4.3}$$

where $y'^{(i)} = \frac{r}{R^{(i)}} + \gamma \max_{a'} Q(s', a'; \theta^{(i)-})$ is the rescaled target at iteration i by a scaling factor $R^{(i)} > 0$ and

$$U(Q) = \frac{\kappa}{(Q - 1 - \delta)^2} + \frac{\kappa}{(Q + 1 + \delta)^2}, \tag{4.4}$$

wherein κ and δ are constants. The adaptation process is repeated until the $R^{(i)}$ gradient is smaller than a given tolerance.

4.3.3.3 Environment Models and Dyna Integration

To facilitate the training of the DQN, we adopt the Dyna framework, which combines model-free learning and model-based planning. Specifically, an environmental model is trained to generate artificial synthetic traffic. The online Baum-Welch algorithm is adopted to train a learned interrupted Poisson process (IPP), with parameters to be learned from real data. In this way, the generated data of the learned IPP are mixed with real data in the training phase of the DQN, such that the training process is accelerated.

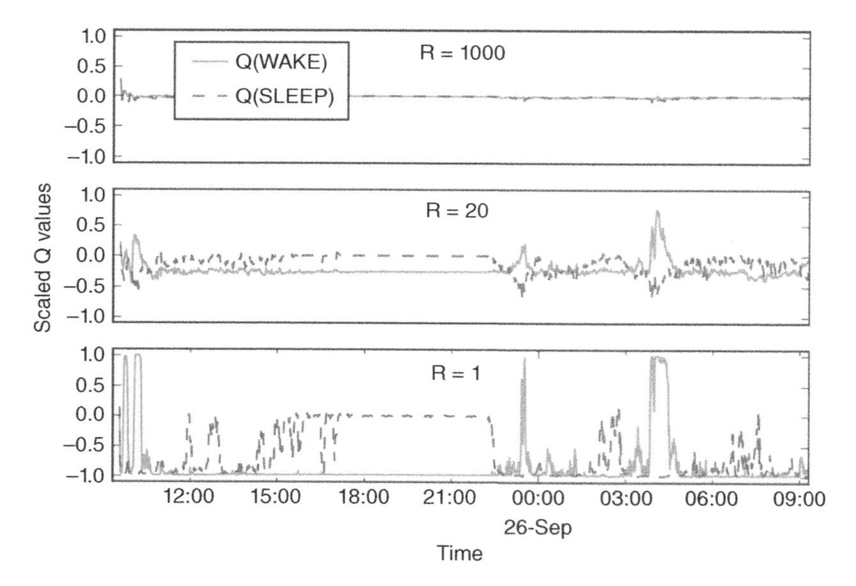

Figure 4.5 Q values of waking and sleeping actions smoothed over a one-minuted time window with fixed reward scaling factors of 1, 20, and 1000.

4.3.3.4 DeepNap Algorithm Description

The procedure of DeepNap is as follows. A feed-forward neural network is used as a DQN and initialized using the Glorot initialization method. An IPP model is periodically fitted with the latest traffic observations from the real world, and the filtered system state is used as input for the DQN. These observations, together with the actions taken, rewards received, and next state observed, are further stored in the action-wise experience replay buffer. Once the buffer reaches capacity, the oldest observations are discarded and replaced with new ones. The DQN is periodically trained with mini-batches taken from the action-wise experience replay buffer. The target Q-network is updated at a longer interval (e.g. after many mini-batch updates) in order to stabilize it.

4.3.4 Experiments

Experimental results are presented in this section to validate the performance advantages. The experiment parameters used are given as follows. The discount factor is set to 0.9, and the exploration probability is 0.02 for the agent. The input to the DQN is of dimensionality 3 and the output is 2. The DQN has 2 hidden layers and 500 units in each layer with the ReLu activation function. The output layer adopts the tanh activation function. The weight and offset of saturation penalties are respectively $\kappa = 10^{-5}$ and $\delta = 10^{-2}$. Network parameters are optimized using mini-batch Nesterov-momentum updates with 0.9 momentum, 10^{-2} step size, and 100 batch size.

4.3.4.1 Algorithm Comparisons

We compare the performance of different algorithm configurations. The investigated algorithms include:

- **Baseline:** Always-on policy agent
- **DQN:** Enhanced DQN using stacked raw observations
- **DQN-m:** Enhanced model-assisted DQN using continuous belief states
- **DQN-d:** Enhanced model-assisted DQN using continuous belief states and Dyna simulation
- **QL-d:** Table-based Q-learning agent using quantized belief states and Dyna simulation

Table 4.1 shows that deep RL-based methods outperform the baseline table-based Q-learning method consistently, due to the generalization capability of DNNs. The reason that DQN-d outperforms DQN most of the time is twofold. (i) The DQN-d method can utilize the human expert knowledge of the underlying models. (ii) DQN-d can use the model-generated synthetic data for training, thus accelerating the training process.

Table 4.1 Method comparison.

Algorithm	Location					
	L1	L2	L3	L4	L5	L6
	Per-step reward					
Baseline	−3.96	−2.49	−4.71	−4.28	−2.91	−4.42
	Gain above baseline					
QL-d	3.280	1.384	3.678	3.014	2.695	3.420
DQN-m	3.443	1.616	3.898	3.228	2.673	3.578
DQN-d	**3.490**	1.879	**3.912**	**3.242**	**2.875**	**3.617**
DQN	3.481	**1.903**	3.880	3.238	2.863	3.600

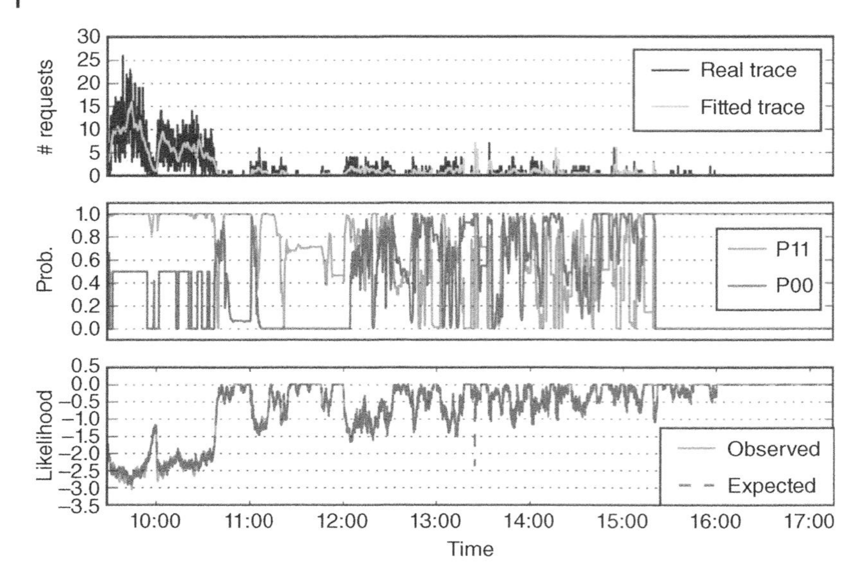

Figure 4.6 Fitting results of the IPP model. Top: traffic and learned emission rates; middle: estimated transition probabilities; bottom: per-step likelihood values.

Figure 4.6 shows the fitness of the learned IPP model, where the learned traffic rate closely matches the the varying traffic volume. On the other hand, the performance advantage of DQN-d over traditional DQN comes at the price of additional computational complexity due to fitting the IPP model. This may constitute a major obstacle in real-world resource-limited scenarios.

4.3.5 Summary

In this section, we presented a data-driven algorithm for dynamic BS sleeping control using deep RL. It is shown that, in face of the non-stationarity in mobile network traffic, ML-based DeepNap enhances the system performance with model robustness compared with conventional methods. Moreover, the integration of IPP traffic model and the use of simulated experience also gives slight improvement over end-to-end DQN at the cost of more computation.

4.4 Dynamic Frequency Reuse through a Multi-Agent Neural Network Approach

This section presents a decentralized learning-based approach for real-time radio resource allocation in mobile networks. Even though network-wide decision-making is optimal when computed centrally, the type of coordination required is unfeasible in practice due to the decentralized nature of mobile networks, which are impacted by communication delays and potential faults along the communication lines. Therefore, decentralized control solutions are more appropriate for this type of problem. As a result, this work employs a multi-agent system (MAS) paradigm as its functional framework. MAS represents a suitable approach for mobile networks, since many elements in the network architecture can be modelled as agents (e.g. eNodeBs, mobile management entities, gateways, etc.).

We further describe the multi-agent solution for resource allocation in mobile networks. This solution employs two different types of agents, mainly at base-level cell agents, and at higher-level coordinator agents that ease the decision-making task and provide higher levels of efficiency at the network level. While a completely decentralized approach is also possible, this hierarchical structure ensures performance that is closer to optimal through the aggregation of information at the coordinator agent level. This is the main role of the coordinator elements, which target the maximization of global network metrics.

The key elements of the architecture are the cell agents, which use function approximation to decide what BS configurations are most suitable for their users. In effect, since the actions depend on local environment information that is affected by nonlinear factors such as noise, pathloss, interference, and user geometry, the cell agents employ DNNs for the inference process required by the agent decision making process. Specifically, the DNNs use cell environment information (e.g. user geometry) together with a list of available actions and constraints imposed by the coordinator agents in order to decide which cell configuration would help achieve the best performance for users at the network level.

4.4.1 Multi-Agent System Architecture

In this subsection, we provide further details on the multi-agent architecture and the design of its underlying agents. This architecture is represented through a hierarchical multi-agent framework. Here, base-level agents have full control over a set of actuators (e.g. BS parameters), while higher-level agents (i.e. coordinators) can impose constraints over another set of the base agent's actuators. This architecture is illustrated in Figure 4.7, from the perspective of an LTE

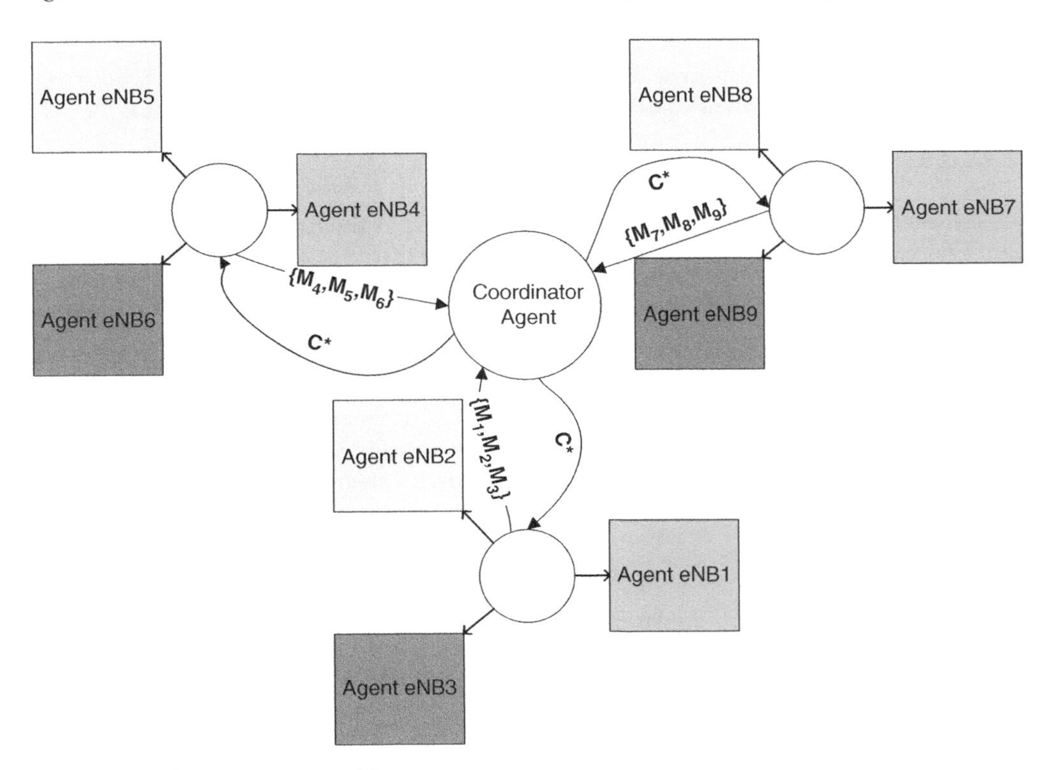

Figure 4.7 Multi-agent system architecture.

network, where base-level agents are eNodeBs (eNBs) cells. While abstracted in the figure, the coordinator agent here can be another eNB cell or an entity higher up in the hierarchy of the network, such as the mobility management entity (MME).

The two types of agents target different performance metrics. At eNB agent level, agents attempt to maximize a local performance metric, which is a function of the users within their coverage; at coordinator agent level, agents attempt to maximize global performance metrics, with regard to all users covered by the eNB agents under their control.

The multi-agent system operates in a three-step process:

1. eNB agents compute local achievable performance under each type of constraint that can be imposed by the coordinator agent, and forward the maximum achievable performance to coordinator agents.
2. The coordinator agent uses this information to compute the most suitable constraint under which global performance is maximized, and imposes this constraint over underlying agents.
3. eNB agents select the local action that maximizes their performance metric under the imposed constraint.

We further describe this decision process formally. We denote the following:

A_j: The set of actions $\mathbf{a}_j$ that cell j can independently select
C: The set of possible values for the constraint imposed by the coordinator to the eNBs

The performance of each cell j is then a function of the composite action $[\mathbf{a}_j, c]$ and the state of the environment $\mathbf{s}_j$:

$$\mu_j = f_j(\mathbf{a}_j, c, \mathbf{s}_j), \tag{4.5}$$

where $c \in C$.

For each possible $c \in C$, an agent selects the local action that maximizes Eq. (4.5) while considering the current environment condition:

$$\mathbf{a}^*_{j,c} = \arg\max_{\mathbf{a}_j \in A_j} f_j(\mathbf{a}_j, c, \mathbf{s}_j). \tag{4.6}$$

Each cell agent provides the coordinator with an estimate of the maximum achievable performance for each of the N constraints, in the form of a vector:

$$M_j = < \mu^*_j(c_1), \mu^*_j(c_2), \ldots \mu^*_j(c_N) >, \tag{4.7}$$

where $\mu^*_j(c)$ represents the maximum value of the cell metric under constraint/policy c, i.e. $\mu^*_j(c) = f_j(\mathbf{a}^*_{j,c}, c, \mathbf{s}_j)$.

When the coordinator agent receives the computed performance vectors from all its underlying cells, it computes the maximum performance achievable for each constraint policy, which represents the **global** metric at the coordinator agent level. Afterward, it selects and imposes the constraint that maximizes the global metric:

$$c^* = \arg\max_{c \in C} < \pi(c_1), \pi(c_2), \ldots \pi(c_N) >, \tag{4.8}$$

where $\pi(c) = \sum_{j=1..m} \mu^*_j(c)$. As such, constraint c^* is selected as the global metric-maximizing constraint, and imposed to the cell agents. Then each agent determines the optimal local action given c^*:

$$\mathbf{a}^*_{j,c^*} = \arg\max_{\mathbf{a}_j \in A_j} f_j(\mathbf{a}_j, c^*, \mathbf{s}_j). \tag{4.9}$$

4.4.1.1 Cell Agent Architecture

The underlying cell agents employ a DNN-based architecture for their own decision-making process. This architecture is presented in abstracted form in Figure 4.8.

The cell agent uses as input local environment information (e.g. number and geometry of served users, traffic information, etc.) and actions that can be taken (both those under full control of the agent and those imposed by the coordinator agent) to estimate the local performance achievable. While the environment information is temporarily fixed and outside the control of the agent, the agent can infer which combinations of constrained actions and local actions can achieve the best performance given the current conditions. Thus, it can select the best action under each type of constraint that is imposed by the coordinator agent, in order to maximize its own performance.

4.4.2 Application to Fractional Frequency Reuse

The previously presented multi-agent framework is applied to the problem of fractional frequency reuse (FFR) in LTE networks. FFR is a technique used to improve coverage for network users by minimizing interference between cells. In particular, cell-edge users are affected by increased levels of interference between neighboring cells in edge areas. FFR techniques are a way of addressing this through the allocation of orthogonal blocks of the spectrum between neighboring cells. Partial frequency reuse (PFR) (also known as *strict frequency reuse*, or *fractional frequency reuse with full isolation*) is one such scheme, where the LTE spectrum is divided into the following four orthogonal blocks:

- A common block to be used by center users, with full reuse between cells and sectors
- Three orthogonal blocks, one per each of the three sectors of a cell

Three types of parameters can typically be controlled in PFR:

- The proportion of bandwidth allocated to the common block versus edge blocks;
- The threshold value (e.g. SINR level) that separates center users from edge users
- The power allocated to the common block versus edge blocks.

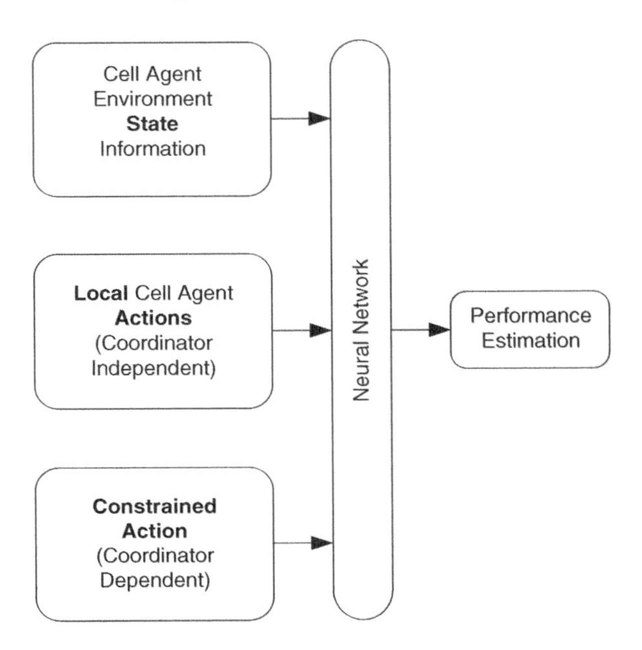

Figure 4.8 General neural network architecture.

This results in neighboring cells having orthogonal parts of the spectrum allocated for edge users. However, this type of interference minimization comes at a cost: since the overall cell's capacity is downsized, sometimes static FFR schemes can actually decrease performance for cell-edge users. As such, dynamic FFR schemes are desirable in such situations,to adapt to current environment conditions on the fly.

In the following section, we will present the application of the previously introduced multi-agent architecture to the problem of dynamic FFR. We show how the problem is addressed with minimal exchange of information, and how the learning-based heuristic solution to be presented achieves near-optimal results in a timely manner. Cell agents are placed in charge of controlling FFR parameters, under bandwidth constraints that can be imposed by coordinator agents in order to maintain spectrum-allocation orthogonality.

The cell agents report estimated gains under each possible bandwidth to the coordinator agent, based on the predictions obtained from neural networks. The coordinator agent obtains local performance estimates from all of its underlying cell agents, and computes and broadcasts back to cell agents the best global orthogonal division of spectrum between edge and center users, i.e. the bandwidth configuration that minimizes interference between cells while maximizing global network performance. After the orthogonal bandwidth division is imposed by the coordinator agent, cell agents locally optimise their own PFR configuration to best suit their own users' geometry with respect to the imposed constraint.

4.4.3 Scenario Implementation

The solution is implemented by modifying ns-3's Strict Frequency Reuse scheme within the LTE system-level simulator module (Baldo et al., 2011). Two parameters are controlled by the implemented cell agent:

- Bandwidth allocation, in number of resource blocks (RBs) allocated to the common block (with the remaining RBs being evenly divided between edge blocks)
- Reference signal received quality (RSRQ) threshold that is used to separate center users from edge users (where UEs with a value greater than the threshold are considered center users)

This dynamic strict frequency reuse scheme is illustrated in Figure 4.9. Note that the default ns-3 implementation represents a static scheme, while the one presented here operates in a dynamic manner, with parameters that can be configured at every second of real-time network operation.

4.4.3.1 Cell Agent Neural Network

The neural network used in this case performs predictions of achievable performance metrics. Since the performance metric is a function of UEs' achievable throughput – and as such is a nonlinear function that depends on factors like user geometry, pathloss, noise, interference, and bandwidth available – a neural network is a suitable approach for this type of estimation.

The features of the cell agent's environment are abstractions of the number of UEs in each RSRQ bin and are served as an input to the neural network. In addition, the type of actions available (both local and coordinator imposed) are also fed as input to the neural network. The overall neural network input layer has the following structure:

- Ten input neurons abstracting information about the environment, with each neuron containing the number of UEs for each RSRQ bin (where the overall RSRQ interval is split into 10 contiguous sub-intervals or bins).
- One input neuron for the bandwidth action (with possible values that the bandwidth used exclusively for center users can take). Only 3 actions are available, to limit the state space: 40, 64, or 88 RBs can be used by the common block (out of a total of 100).

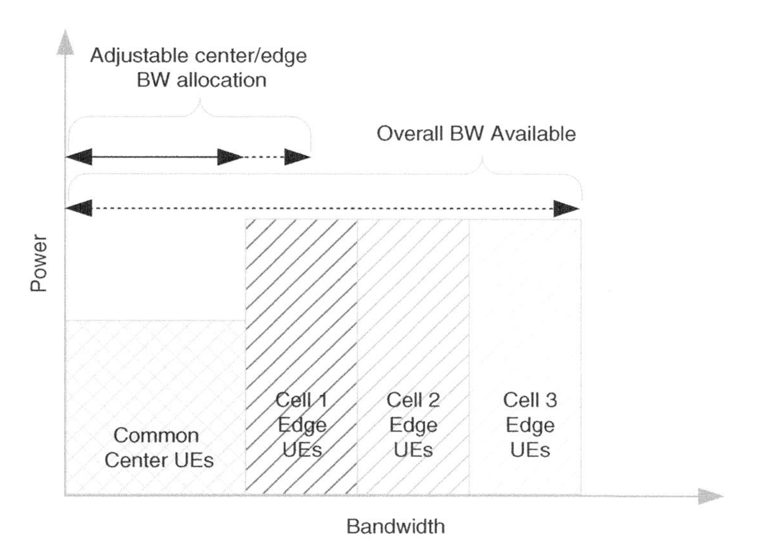

Figure 4.9 Strict frequency reuse and bandwidth action.

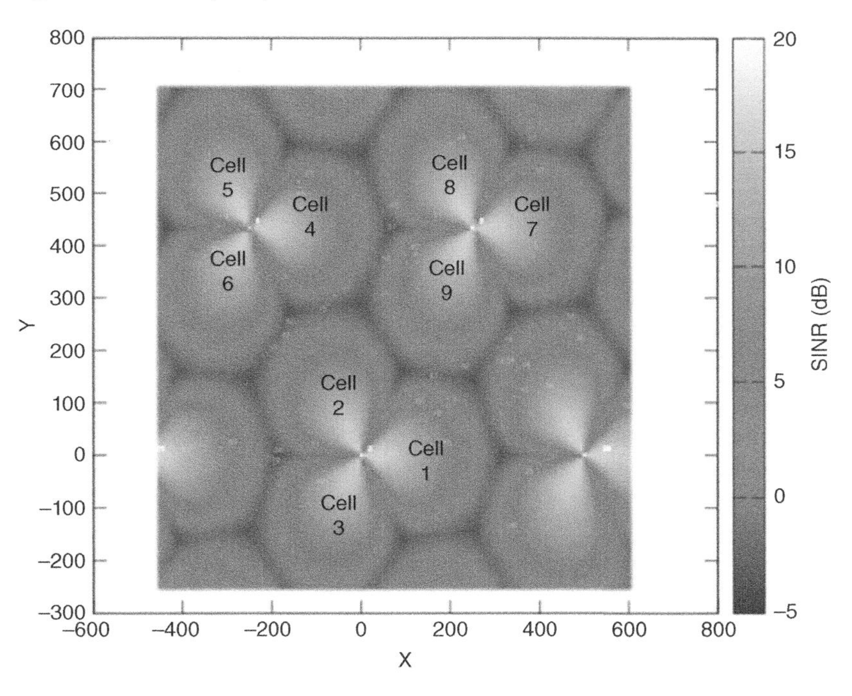

Figure 4.10 Cell setup in ns-3: inner 9 cells.

- One input for the RSRQ threshold (where the threshold can be set between any of the 10 bins from the RSRQ interval). As such, there are 9 actions available (9 delimiters among 10 contiguous intervals).

This neural network is pictured in Figure 4.11. It contains 5 hidden layers, each with 32 fully connected neurons. All layers have *tanh* activation functions, except for the last one, which uses *sigmoid* activations. This is because the output neuron deals exclusively with positive values

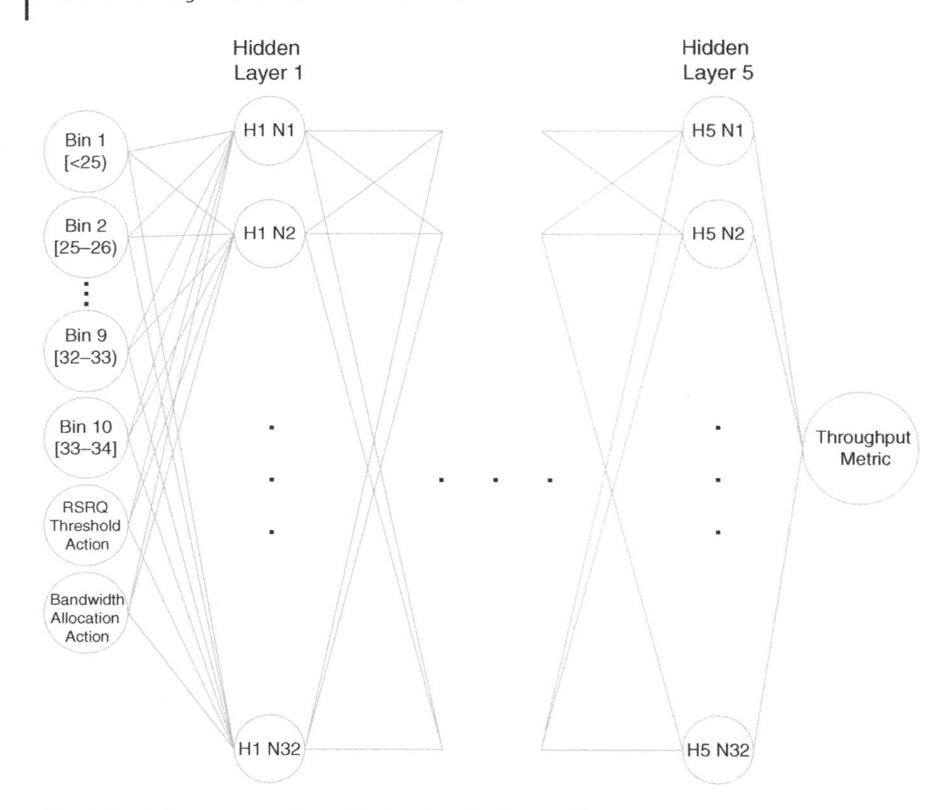

Figure 4.11 Neural network architecture for the SFR problem.

(functions of UE throughput). The architecture is sufficiently deep to be able to estimate more complex functions.

4.4.4 Evaluation

In this section, we present the performance evaluation of both the neural network's prediction abilities and of the level of efficiency achieved by the multi-agent system.

4.4.4.1 Neural Network Performance

The DNN is trained through supervised learning, based on data generated with the ns-3 simulator. The data is acquired using a previously validated setup (Marinescu et al., 2017), in line with the 3GPP requirements of TR 36.814 (TR36.814, 2010). This setup provides input and output pairs based on observed environment information and computed outcomes of the throughput-related function. The pairs are generated using a broad range of UE realizations, which were in turn generated using stochastic geometry tools to simulate realistic user distributions (e.g. clustered behavior, clusters of clusters, sparse UE placements, uneven UE density in cells, etc.), and contain a total of $2e5$ samples for training purposes. This dataset includes all the possible actions that can be taken for each cell realization, in order to have the corresponding output for this in the training set. A separate test set generated using different UE realizations was used to evaluate the accuracy of the DNN. The DNN training was performed using the Keras framework with TensorFlow as a backend (Chollet et al., 2015).

Table 4.2 Neural network performance.

Metric Type	Correlation	MAE	MAPE	RMSE
Minimum TP	0.92	0.020	0.029	0.034
Mean TP	0.96	0.031	0.074	0.040
Harmonic mean TP	0.94	0.031	0.061	0.045
Geometric mean TP	0.95	0.030	0.064	0.041

Table 4.2 presents the results achieved by the DNN when estimating various throughput-related functions/metrics. The values for these metrics were generated according to the throughputs obtained by UEs in a cell. The same network architecture was used in each case, except that during the training phase, the output part of the input-output pair was modified in accordance to the type of function metric to be estimated. It can be observed that the DNN achieves a high accuracy, obtaining up to only 2.9% mean absolute percentile error (MAPE) in the case of minimum throughput. For the convenience of the reader, in Table 4.2 we provide several types of error in addition to MAPE, such as mean absolute error (MAE) and root mean square error (RMSE), together with the correlation between the function outputs and the DNN estimations.

4.4.4.2 Multi-Agent System Performance

The performance of the MAS is evaluated considering the minimum throughput maximization objective, with the agent neural networks being trained as presented in Section 4.4.4.1. We chose this objective for further investigation as it represents a feasible QoS policy for UEs, which attempts to maximize the cell's coverage. As given in the following, three types of MAS are evaluated, to investigate the effect of coordinator agents and of online learning:

1. *Without coordination between agents*: This is a purely decentralized approach, where agents optimize their actions purely based on local environment information, without any constraints imposed by a coordinator entity. As such, they have full control over both bandwidth and RSRQ threshold.
2. *With coordination between agents*: The initially presented hierarchical approach from Figure 4.7, where agents have constraints imposed by a coordinator entity. The agents can only control the RSRQ threshold values, while the bandwidth action is imposed by a coordinator agent.
3. *With online learning abilities*: Similar to the previous scheme, except that after agents take an action, they evaluate the observed outcome against the expected estimates (i.e. values inferred through the DNN) and can choose another action if the actual encountered value is ranked lower than the estimates of other actions. This can be useful when the initially estimated second-ranked action turns out to be more effective than the first-ranked action.

As a baseline, we choose the full-frequency reuse. This is because, as observed in (Marinescu et al., 2018), static-frequency reuse schemes tend to perform worse on average than full-frequency reuse, with respect to edge users. Even though some static-frequency reuse schemes can improve performance for specific cells when particular UE distributions are encountered, expanding these schemes to a large group of neighboring cells when involving realistic UE distributions at network level turns out to be detrimental to the overall performance of the network.

We summarise the performance achieved by the three types of MAS in Table 4.3, considering the cells previously presented in Figure 4.10. We chose the inner three cells for evaluation since

Table 4.3 Inner three cells (2, 4, 9): maximizing minimum throughput policy.

Improv. over FR	A. W/o Coord	B. W/ Coord	C. Online Learning	Optimal
Bottom 10% UEs	10%	15%	16%	18%
Bottom 5% UEs	13%	23%	23%	27%
Bottom 1% UEs	28%	56%	59%	66%
Worst UE	133%	231%	263%	264%

only the nine cells are controlled by agents, and as such we can best notice the impact of actions within the inner three cells. This performance can be also visualized in Figure 4.12.

It can be noticed how each addition to the MAS improves performance. Even a MAS without a coordinator entity achieves better performance when compared to the full-frequency reuse baseline, with performance improvements most noticeable for edge users. Coordination significantly improves the performance of edge users, while learning abilities bring the efficiency

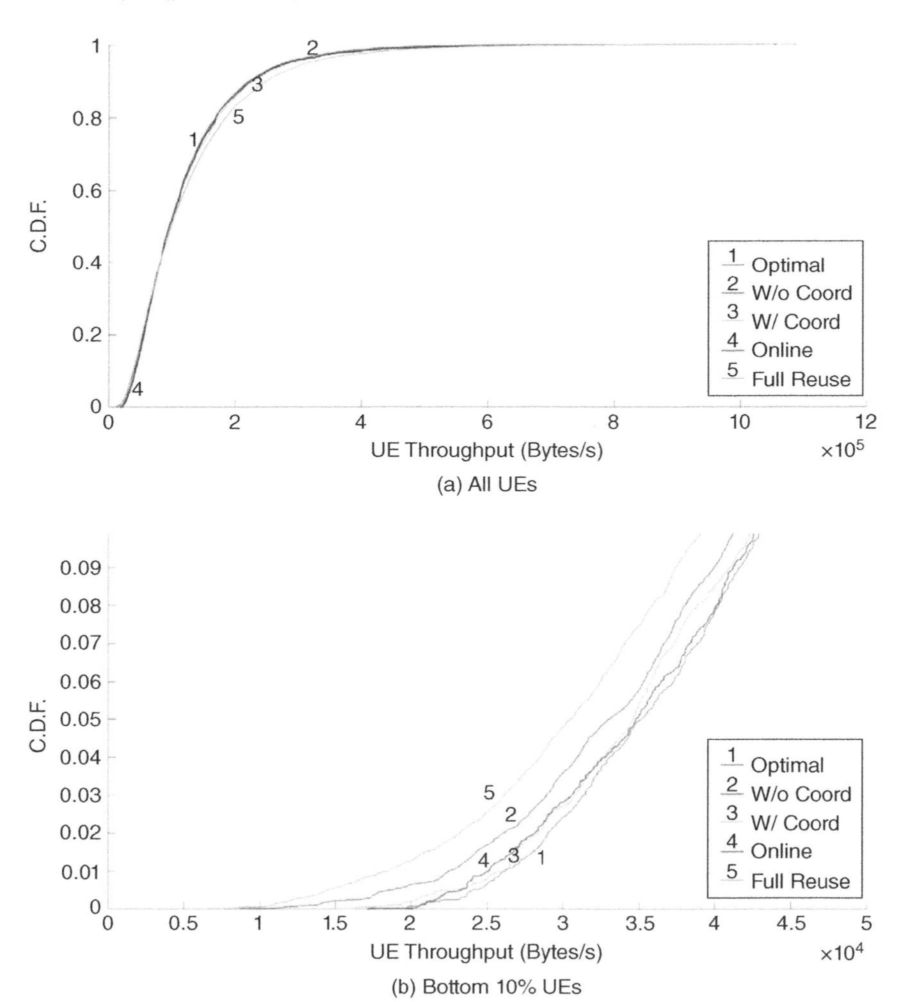

Figure 4.12 Performance over the inner three cells: maximizing minimum throughput policy.

of the MAS to near-optimal levels, after only five or six iterations of the agent action selection process.[1] The performance for the worst user in a cell (in terms of proximity to edge of cell) is improved almost threefold with this type of MAS, providing a 263% improvement, which is very close to the 264% performance that can be achieved optimally. Optimal performance can only be computed after all agents iterate through all the available actions (both coordinator imposed and locally available), a process that can be highly detrimental to the performance of the network, as many of the actions negatively impact the operation of the network. Here, we iterate through these via ns-3 simulations, for the purpose of computing the optimal solution *a posteriori*.

4.4.5 Summary

This section presented the improvements that neural network-based multi-agent systems can bring to LTE network performance. The framework presented is a highly adaptable solution that adjusts to a broad variety of user distributions in a network, addresses imbalances in load between cells, and includes learning capabilities in order to improve performance in the situations that might differ too much from the training base of the neural networks. This chapter also showcases how ML-based agents can achieve improved performance at the global network level by cooperating via a coordinator entity. The neural network proposed in this chapter can assist wireless networks in making optimal choices through inference, without needing to explore sub-optimal actions, which can negatively impact the LTE network at runtime. With this ML-based dynamic frequency reuse scheme, the increase in coverage can be almost threefold for edge users, while the network was shown to retain 95% of its capacity from the full-frequency reuse case.[2]

4.5 Conclusions

In this chapter, we have shown how machine learning–assisted solutions can improve performance in coverage and capacity optimization–related problems in wireless networks. Two different approaches were presented, targeting long-term and short-term adaptations to traffic demands and network environment conditions. The long-term solution involved turning BSs on or off through a deep RL approach in order to save energy while still catering to user needs under non-stationary traffic conditions. The performance of the solution here improves when traffic models are also employed. On the other hand, the short-term solution, complementary in our vision to the long-term one, redistributes currently available resources in the network to maximize global network performance depending on the conditions in the network. Here, a multi-agent system framework based on neural networks' guided cell agents is used to implement dynamic fractional frequency reuse, to improve network coverage while minimizing potential capacity loss. Both of these techniques take advantage of the massive amounts of data/KPIs available in wireless networks to assist the decision-making processes. This way, the presented ML heuristics manage to successfully address complex problems in a timely manner.

Real-time data availability and modelling are essential to these algorithms' decision-making processes. The more historical data is available, the better the algorithms can infer potential future behavior of the wireless network. Predictions of future network behavior can help

1 Even though at cell-agent level there might be only two or three candidates for the best action (in order of ranking under a specific constraint), selections of other actions by agents could force the MAS coordinator entity to select another constraint and as such in turn force the cell agents to iterate through other candidate actions suitable for the new constraint imposed.

2 Full-frequency reuse utilizes the entire spectrum available to the LTE network for each cell, as opposed to fractional frequency reuse solutions, but this comes at higher interference costs.

improve the solution's performance, as agents can preemptively take actions to address upcoming changes, in either the short term or the long term. As such, we expect accurate ML-based traffic and mobility modelling to play a critical role in the future in addressing coverage and capacity optimization issues of wireless networks.

Bibliography

Nicola Baldo, Marco Miozzo, Manuel Requena-Esteso, and Jaume Nin-Guerrero. An open source product-oriented lte network simulator based on ns-3. In *Proceedings of the 14th International conference on Modeling, analysis and simulation of wireless and mobile systems*, pages 293–298. ACM, 2011.

Mehdi Bennis, Samir M. Perlaza, Pol Blasco, Zhu Han, and H. Vincent Poor. Self-organization in small cell networks: A reinforcement learning approach. *Transactions on Wireless Communications*, 12(7):3202–3212, 2013.

François Chollet et al. Keras. https://github.com/fchollet/keras, 2015.

Mariana Dirani and Zwi Altman. A cooperative reinforcement learning approach for inter-cell interference coordination in ofdma cellular networks. In *Modeling and Optimization in Mobile, Ad Hoc and Wireless Networks (WiOpt), 2010 Proceedings of the 8th International Symposium on*, pages 170–176. IEEE, 2010.

GR ETSI. Experiential Networked Intelligence (ENI); ENI use cases. 2018.

Shaoshuai Fan, Hui Tian, and Cigdem Sengul. Self-optimization of coverage and capacity based on a fuzzy neural network with cooperative reinforcement learning. *EURASIP Journal on Wireless Communications and Networking*, 2014(1):57, 2014.

Albrecht J. Fehske, Henrik Klessig, Jens Voigt, and Gerhard P. Fettweis. Concurrent load-aware adjustment of user association and antenna tilts in self-organizing radio networks. *Transactions on Vehicular Technology*, 62(5): 1974–1988, 2013.

Ana Galindo-Serrano, Lorenza Giupponi, and Gunther Auer. Distributed learning in multiuser ofdma femtocell networks. In *VTC Spring*, pages 1–6, 2011.

Minghui Gao, Lianfen Huang, Xiaonan Cui, Hongxiang Cai, and ZhiBin Gao. Intelligent coverage optimization with multi-objective genetic algorithm in cellular system. In *Computer Science & Education (ICCSE), 2013 8th International Conference on*, pages 859–863. IEEE, 2013.

Xueying Guo, Zhisheng Niu, Sheng Zhou, and PR Kumar. Delay-constrained energy-optimal base station sleeping control. *Journal on Selected Areas in Communications*, 34(5):1073–1085, 2016.

Hammad Hafiz, Harjeet Aulakh, and Kaamran Raahemifar. Antenna placement optimization for cellular networks. pages 1–6. IEEE, 2013.

Zhiyuan Jiang, Bhaskar Krishnamachari, Sheng Zhou, and Zhisheng Niu. Optimal sleeping mechanism for multiple servers with MMPP-based bursty traffic arrival. *Wireless Communications Letters*, 7(3):436–439, June 2018.

Ioannis Kamitsos, Lachlan Andrew, Hongseok Kim, and Mung Chiang. Optimal sleep patterns for serving delay-tolerant jobs. In *Proceedings of the 1st International Conference on Energy-Efficient Computing and Networking*, pages 31–40. ACM, 2010.

Bingjie Leng, Bhaskar Krishnamachari, Xueying Guo, and Zhisheng Niu. Optimal operation of a green server with bursty traffic. In *2016 Global Communications Conference (Globecom)*. IEEE, Dec 2016.

Jingyu Li, Jie Zeng, Xin Su, Wei Luo, and Jing Wang. Self-optimization of coverage and capacity in lte networks based on central control and decentralized fuzzy q-learning. *International Journal of Distributed Sensor Networks*, 8(8):878595, 2012.

Zhihang Li, Hao Wang, Zhiwen Pan, Nan Liu, and Xiaohu You. Joint optimization on load balancing and network load in 3gpp lte multi-cell networks. In *Wireless Communications and Signal Processing (WCSP), 2011 International Conference on*, pages 1–5. IEEE, 2011.

Andrei Marinescu, Irene Macaluso, and Luiz A. DaSilva. System level evaluation and validation of the ns-3 lte module in 3gpp reference scenarios. In *Proceedings of the 13th Symposium on QoS and Security for Wireless and Mobile Networks*, Q2SWinet '17, pages 59–64. ACM, 2017.

Andrei Marinescu, Irene Macaluso, and Luiz A. DaSilva. A multi-agent neural network for dynamic frequency reuse in lte networks. In *2018 International Conference on Communications Workshops (ICC Workshops)*, pages 1–6. IEEE, May 2018.

Warren S. McCulloch and Walter Pitts. A logical calculus of the ideas immanent in nervous activity. *The bulletin of mathematical biophysics*, 5(4):115–133, Dec 1943.

Volodymyr Mnih, Koray Kavukcuoglu, David Silver, Andrei A Rusu, Joel Veness, Marc G Bellemare, Alex Graves, Martin Riedmiller, Andreas K Fidjeland, Georg Ostrovski, et al. Human-level control through deep reinforcement learning. *Nature*, 518(7540):529, 2015.

Jessica Moysen and Lorenza Giupponi. From 4g to 5g: Self-organized network management meets machine learning. *Computer Communications*, 2018.

NhuQuan Phan, ThiOanh Bui, Huilin Jiang, Pei Li, Zhiwen Pan, and Nan Liu. Coverage optimization of lte networks based on antenna tilt adjusting considering network load. volume 14, pages 48–58. IEEE, 2017.

Rouzbeh Razavi, Siegfried Klein, and Holger Claussen. A fuzzy reinforcement learning approach for self-optimization of coverage in lte networks. *Bell Labs Technical Journal*, 15(3):153–175, 2010.

David Silver, Julian Schrittwieser, Karen Simonyan, Ioannis Antonoglou, Aja Huang, Arthur Guez, Thomas Hubert, Lucas Baker, Matthew Lai, Adrian Bolton, et al. Mastering the game of go without human knowledge. *Nature*, 550(7676):354, 2017.

Meryem Simsek, Andreas Czylwik, Ana Galindo-Serrano, and Lorenza Giupponi. Improved decentralized q-learning algorithm for interference reduction in lte-femtocells. In *Wireless Advanced (WiAd), 2011*, pages 138–143. IEEE, 2011.

Richard S. Sutton and Andrew G. Barto. *Reinforcement Learning: An Introduction*. MIT Press.

Gerald Tesauro. Td-gammon: A self-teaching backgammon program. In *Applications of Neural Networks*, pages 267–285. Springer, 1995.

TR36.814. 3GPP evolved universal terrestrial radio access (e-utra); further advancements for e-utra physical layer aspects. http://www.3gpp.org/ftp/Specs/html-info/36814.htm, 2010. Rel-9 v9.0.0.

Muhammad Naseer ul Islam and Andreas Mitschele-Thiel. Cooperative fuzzy q-learning for self-organized coverage and capacity optimization. In *Personal Indoor and Mobile Radio Communications (PIMRC), 2012 IEEE 23rd International Symposium on*, pages 1406–1411. IEEE, 2012.

Jian Wu, Sheng Zhou, and Zhisheng Niu. Traffic-aware base station sleeping control and power matching for energy-delay tradeoffs in green cellular networks. volume 12, pages 4196–4209. IEEE, 2013.

5

Machine Learning for Optimal Resource Allocation

Marius Pesavento and Florian Bahlke

Department of Electrical Engineering and Information Technology, TU Darmstadt, Darmstadt, Germany

5.1 Introduction and Motivation

To achieve the postulated performance gains for 5G and subsequent generations of wireless communication network, dramatic enhancements in overall network operation are required (Andrews et al., 2014; Boccardi et al., 2014). Novel services associated with 5G (ITU, 2017; Shafi et al., 2017; Iwamura, 2015) exhibit strict quality-of-service (QoS) and network connectivity requirements even for mobile users at the cell edges and under severe interference. In addition to an expansion of the utilized frequencies to the millimeter-wave range (Ghosh et al., 2014; Rappaport et al., 2013) and the deployment of massive multiple-input and multiple-output (MIMO) antenna arrays (Larsson et al., 2014), a third important resource domain has gathered significant attention: an increase in the spatial density of the network architecture. The overall network capacity increase achieved with the aforementioned approaches critically depends on whether the individual performance gains associated with each technology add up synergetically and over large coverage areas. Finding affirmative answers to this question requires advances in the radio resource management of heterogeneous multi-cell networks. This has been identified as a challenge in the operation of ultra-dense wireless networks (Andrews et al., 2016), especially if the corresponding network architecture requires decentralized network control based on local channel state information (CSI) and limited inter-cell coordination (Ye et al., 2013; You and Yuan, 2017).

Network management and resource optimization in multi-cell networks are generally associated with binary or integer decision-making where, e.g. users are allocated to base stations, and discrete time-frequency resources are assigned to mobile devices (Cheng et al., 2013; Cheng and Pesavento, 2015; Liu and Fan, 2018; Alfa et al., 2016). As today's cellular networks are fundamentally limited by interference, the associated integer programming problems are generally of a combinatorial nature where optimization is carried out with the goal of trading off conflicting interests among players in the network (i.e. mobile users, base stations, subnetworks, etc.), and solutions are at best locally pareto-optimal. Optimal network management, in the strictly mathematical sense, is known to be practically infeasible for large networks, as the corresponding integer programming problems scale poorly with the network dimensions, often exhibit prohibitive computational complexity, and are thus not suitable for online operation. Moreover, due to latency requirements and limited control information exchange among network entities, decision-making in practical multi-cell networks needs to be carried out in a decentralized manner and with partial CSI only. Therefore, strictly optimal resource allocation

Machine Learning for Future Wireless Communications, First Edition. Edited by Fa-Long Luo.
© 2020 John Wiley & Sons Ltd. Published 2020 by John Wiley & Sons Ltd.

as a solution of integer programming formulations is only used as a benchmark for small to medium network scenarios under centralized control.

In practice, large-scale multi-cell networks resource management is generally carried out strictly sub-optimally and based on engineering heuristics. In the past, engineers have developed sophisticated decision rules for optimized network operation based on engineering intuition, real network data analysis, and extensive system-level modeling and simulations. In this context, ML emerges as a promising tool with the potential to bridge the gap between the two worlds: one is from the practical application, the requirements for decentralized decision-making and online network optimization; the other is from a performance perspective, the desire to obtain theoretically optimal network management solutions with potentially large performance gains.

The general idea of the supervised learning approaches for network optimization problems introduced in this chapter in the example of user and resource allocation in heterogeneous networks is intuitive. Instead of manually devising engineering heuristics and decision rules for the combinatorial resource optimization problems, the idea is to train a machine to carry out decentralized decision-making based on statistical classification. In this approach, engineering knowledge about the network optimization problem is only used indirectly through selection of proper classification features. Computationally intractable integer programming problem formulations for optimal resource allocation are employed offline in the generation of training data for a large number of instances of medium-size networks. For these instances, resource allocation is carried out ideally by optimally solving the corresponding optimization problems. While this is computationally demanding, it is important to note that training problems can be solved offline and on central machines with full network-wide CSI available. The optimal decisions obtained for these instances provide the data labels that are used in the training along with the corresponding feature vectors that can be computed decentralized according to local network information for each instance. The machines that are trained with this data are capable of carrying out decentralized decision-making based on local CSI and network information. As in other ML applications, the crucial question that arises in the context of learning-based network management and resource optimization is how well the machines that have been trained with labeled data of given network instances generalize their knowledge to networks of different sizes, topologies, or underlying channel characteristics.

5.1.1 Network Capacity and Densification

A promising and currently very popular technology direction for wireless network expansion is aiming at a joint utilization of massive-MIMO arrays and millimeter wave radio access. The deployment of (indoor and outdoor) small cells is understood as a necessary supplement of this technology to provide widespread network coverage (Xiao et al., 2017; Rappaport et al., 2015) and capacity increase (Jungnickel et al., 2014). It is, however, well established that there exist fundamental limits for the densification of wireless communication networks (Andrews et al., 2016; Nguyen and Kountouris, 2017), beyond which an increase in capacity cannot be achieved. Naturally, there are practical reasons for the limitation of wireless network densification, such as the associated increases in hardware costs and energy consumption (Cavalcante et al., 2014; Hoydis et al., 2011), as well as limited availability of deployment sites and backhaul (Ge et al., 2016). However, the primary reason for this saturation effect in densification is the increase of interference in the network, causing deterioration of the signal-to-noise-ratios (SINR) of network entities. This limits the achievable data rates in the network or, conversely, increases the demand for resources in time, frequency, space, and power to achieve requested QoS beyond an affordable level.

Network optimization and radio access control is required to control and balance the resource consumption in the network, and therefore maximizing network capacity and agility. Simultaneously, strict QoS requirements such as minimum guaranteed data rates and SINR levels of each wireless link need to be provided. An important approach to achieve low resource consumption and a balanced distribution of loads in the network is to employ load balancing between the cells (Majewski and Koonert, 2010; Siomina and Yuan, 2012a). The *load* is defined for each cell as the ratio of its consumed to its available resources. The required spectral resources of a network with multiple cells are lower-bounded by the cell exhibiting the highest demand for such resources, i.e. the most "overloaded" cell. This further emphasizes the need for decreasing resource consumption by balancing the load between cells: for example, through interference management or optimized resource allocation (Lopez-Perez et al., 2011; Hossain et al., 2014; Hu and Qian, 2014). A fundamental challenge, however, is posed by the coordination and information exchange between network entities that is required to perform network optimization, which is addressed in the following.

5.1.2 Decentralized Resource Minimization

Optimized allocation of users to cells in a wireless multi-cell network becomes particularly important in the context of ultra-dense large-scale wireless networks with massive connectivity. It has been the subject of extensive research (You and Yuan, 2017; Athanasiou et al., 2015). Finding the allocation that optimizes the entire network operation naturally requires information about all network entities and all possible wireless links. Network optimization may be performed online during operation, while allocation rules may be devised beforehand based on network information and demand forecasts. A decentralized approach to this problem has been proposed as *range expansion* (Siomina and Yuan, 2012b; Ye et al., 2013). To decrease the resource consumption of the typically overloaded macro cells, mobile devices are offloaded to low-power small cells. Usually, this only happens if users receive the strongest signal from the neighboring small cell; however, if range expansion is utilized, users can be offloaded even if the signal from the small cell is only up to a certain *bias value* weaker than that of the neighboring macro cell. This scheme therefore allows expansion of the size of the coverage areas independently from the transmit power. For predefined bias values, this approach operates fully decentralized, as mobile devices can autonomously select their access point based on received signal strength measurements. Critical for network performance under this allocation scheme is finding the appropriate bias values for all cells in the network, and a common drawback of established methods for computing these values is their significant communication and coordination overheads.

This chapter addresses decentralized approaches for minimizing network resource consumption through both user allocation and range expansion. ML is employed as a tool to provide a decentralized solution for the combinatorial network optimization problem. Multi-class support vector machines (SVMs) and artificial neural networks (ANNs) are used as classifiers to perform offloading based on user allocation and bias values. The application of statistical learning methods in optimizing wireless communications networks has only being considered recently (Jiang et al., 2017; O'Shea and Hoydis, 2017; O'Shea et al., 2017; Xu et al., 2017; Ye et al., 2018; Bahlke and Pesavento, 2018a,b). For the training of these classifiers, local attributes characterizing channel conditions, cell load levels, cell types, and network topologies are locally extracted and used as features. Based on demand forecasts and simulated or historic network data, a resource-optimal configuration that satisfies QoS constraints is determined from a mathematical optimization of network parameters. With this information, the aforementioned classifiers can be trained and subsequently used in a decentralized fashion, while the network is in operation.

5.1.3 Overview

The rest of this chapter is organized as follows: A system model for downlink transmissions in a heterogeneous wireless communication network is introduced in Section 5.2, where Subsection 5.2.2 discusses the model of the cellular network entities and Subsection 5.2.2 outlines basic user-allocation rules. Optimal resource allocation is treated in Section 5.3. The optimal user allocation solution is described in Section 5.3.1, followed by the corresponding feature extraction for the ML-based system in Section 5.3.2. Similarly, the setting of optimal range-expansion parameters and the corresponding feature extraction are discussed in Sections 5.3.3 and 5.3.4, respectively. Subsection 5.3.5 revises multi-class extensions for systems based on SVM and ANNs. Section 5.4 follows with a numerical evaluation of methods based on simulations of a heterogeneous wireless network. Final remarks and a concluding assessment are given in Section 5.5.

5.2 System Model

In this section, a system model for a heterogeneous wireless network with macro cells (MCs) and small cells (SCs) is introduced. The cells in the network serve multiple demand points (DPs) that may represent single mobile users, aggregated data demand from hotspots, or machine-like entities, e.g. in sensor networks. In practical networks, the allocation of DPs to cells is subject to common predefined allocation rules, which are introduced in the following.

5.2.1 Heterogeneous Wireless Networks

The cellular network under consideration is assumed to consist of K cells in total, with the set of all cells denoted as $\mathcal{C} = \{1, \dots, K\}$. The subsets $\mathcal{C}^{\mathrm{MC}} \subset \mathcal{C}$ and $\mathcal{C}^{\mathrm{SC}} \subset \mathcal{C}$ with $\mathcal{C} = \mathcal{C}^{\mathrm{SC}} \cup \mathcal{C}^{\mathrm{MC}}$ and $\mathcal{C}^{\mathrm{SC}} \cap \mathcal{C}^{\mathrm{MC}} = \emptyset$ indicate MCs and SCs, respectively. The network area under consideration contains M DPs, with the set of all DPs $\mathcal{M} = \{1, \dots, M\}$. DP $m \in \mathcal{M}$ exhibits the data rate demand d_m in bits per second. Between cell $k \in \mathcal{C}$ and DP $m \in \mathcal{M}$, an attenuation factor g_{km} is determined by the antenna gains of the base station and the DP antennas, respectively, the path attenuation factor, and the processing gain achieved at the receiver by coherent multiantenna processing schemes such as maximum ratio combining (MRC) and zero forcing (ZF) (Goldsmith, 2004; Tse and Viswanath, 2005). The transmit power of cell k is denoted as p_k, and all radio links to all DPs in the network are subject to additive white Gaussian noise with power σ^2. Based on the attenuation factor g_{km}, the SINR of cell k serving DP m can be computed as

$$\gamma_{km} = \frac{p_k g_{km}}{\sum_{j \in \{\mathcal{C} \setminus \{k\}\}} p_j g_{jm} + \sigma^2}. \tag{5.1}$$

The SINR model (5.1) is commonly used in multiple access networks such as e.g. orthogonal frequency-division multiple access system (OFDMA) (Cimini, 1985; Wong et al., 1999; Majewski and Koonert, 2010), commonly used in LTE and Wi-Fi standards. In the following, we assume that all cells are utilizing the same pool of time-frequency resources and there is full frequency reuse among cells. Based on the considerations in (Mogensen et al., 2007;

Siomina and Yuan, 2012a), the transmission rate for the wireless link between cell k and DP m is upper-bounded by

$$R_{km}(\gamma_{km}) = \eta^{\mathrm{BW}} W \log_2(1 + \gamma_{km}) \tag{5.2}$$

where W is the total system bandwidth in Hz and η^{BW} is the bandwidth efficiency corresponding to the selected modulation and coding scheme. To satisfy the data demands of DP m, cell k needs to utilize at least the fraction d_m/R_{km} of its available resources. The allocation of DPs to cells is in the following indicated by the binary matrix $\boldsymbol{A} \in \{0, 1\}^{K \times M}$ with entries defined as

$$A_{km} = \begin{cases} 1 & \text{if DP } m \text{ is allocated to cell } k \\ 0 & \text{otherwise.} \end{cases} \tag{5.3}$$

The resource consumption Φ_k of a cell k is a measure for the fraction of total resources consumed by the cell to serve the demands of all its allocated DPs. It is defined as

$$\Phi_k = \sum_{m \in \mathcal{M}} A_{km} \frac{d_m}{\eta^{\mathrm{BW}} W \log_2(1 + \gamma_{km})}. \tag{5.4}$$

A cell k is considered overloaded if the resource consumption exceeds one, i.e. $\Phi_k \geq 1$. The violation of this condition for a cell k indicates that the total amount of time-frequency resources available to the cell is insufficient to serve all of its user demands. In the case that the available time-frequency resources exceed the resources requested to fulfill user demands, the cell can either reduce the transmit power on all resources or mute transmission on a selected subset of resources. This reduces interference in the network. Furthermore, depending on the user allocation, there may be cells that are not serving any DP. In this case, the entire base station can be muted, leading to complete elimination of interference caused by the cell on both data and control channels. In this chapter, we consider for simplicity a worst-case design that is commonly used in literature (Majewski and Koonert, 2010; Caballero et al., 2017), where the following worst-case interference assumptions are made:

(A1) Always active: Cells are assumed to be always active, hence the on/of switching of cells is not considered.

(A2) Full load: We assume that DPs have full buffers and always use all their available resources.

Under assumptions (A1) and(A2), the interference that a cell creates to the DPs in the network is independent of the load of the base station. The *always-on* assumption is e.g. a requirement for LTE and LTE-A base stations that are requested to transmit control channels even in the absence of user connections. The more complicated interference model where cells can be switched on and off has been addressed under a similar framework in (Bahlke et al., 2018). The full-load assumption also simplifies the model and is motivated by the desire to have a robust worst-case design that leaves sufficient margins for instantaneous fluctuations of channel conditions and network traffic. The non-full-load case has been studied in (You and Yuan, 2017; Siomina and Yuan, 2012a). It is important to note that the supervised learning–based resource allocation approach can be extended to less conservative network operation designs where the worst-case interference assumptions (A1) and(A2) do not apply.

5.2.2 Load Balancing

A straightforward rule for allocating DPs to cells, referred to as *maximum receive power allocation*, is to minimize the load imposed by each connection individually, which is achieved by allocating each DP to the cell that provides the strongest signal (Siomina and Yuan, 2012a).

This, however, reduces network performance as it leaves the typically low-power SCs underutilized. *Range expansion* can mitigate this problem (3GP, 2012; Siomina and Yuan, 2012b; Ye et al., 2013). In range expansion, the total received power $p_k g_{km}$ from cell k is scaled with a weighting factor θ_k ($\theta_k \geq 1$), the *bias value*, and the resulting biased total receive power is used for the base station allocation decision for DP m. The allocation rule for each DP can be formulated as follows:

$$\textit{max.-receive-power allocation:} \qquad A_{km} = \begin{cases} 1 & \text{if } k = \arg\max_{\ell} \theta_\ell p_\ell g_{\ell m} \\ 0 & \text{otherwise.} \end{cases} \tag{5.5}$$

Considering Eq. (5.4), the sum resource consumption of all cells can be expressed as

$$\sum_{k \in C} \Phi_k = \sum_{k \in C} \sum_{m \in \mathcal{M}} A_{km} \frac{d_m}{\eta^{\mathrm{BW}} W \log_2(1 + \gamma_{km})}. \tag{5.6}$$

To ensure that for each DP m exactly one serving cell k is selected, i.e. $A_{km} = 1$ and $A_{kn} = 0$ for $n \neq m$, the single choice condition $\sum_{k \in C} A_{km} = 1$ must apply. To minimize the sum resource consumption of all cells, each DP m has to be served by the cell k for which it induces the lowest additional load, i.e.:

$$A_{km} = \begin{cases} 1 & \text{if} \quad k = \arg\min_{\ell} \dfrac{d_m}{\eta^{\mathrm{BW}} W \log_2 (1 + \gamma_{\ell m})}; \\ 0 & \text{otherwise.} \end{cases} \tag{5.7}$$

As the demands d_m and the SINR values γ_{km} in Eq. (5.7) are fixed for DP m, it is easy to see that the sum resource consumption of all cells is minimized if each individual DP is served by the base station associated with the largest SINR value γ_{km}. A direct consequence of this observation and the interference assumptions (A1) and (A2) is that the maximum received power allocation rule (5.5) minimizes the resources required by each individual cell if the bias values are chosen as $\theta_k = 1 \; \forall k$. For general bias values, the allocation rule in Eq. (5.5) can equivalently be expressed in form of the inequality

$$\sum_{k \in C} A_{km} \theta_k p_k g_{km} \geq (1 - A_{jm}) \theta_j p_j g_{jm} \quad \forall j, m, \tag{5.8}$$

which is used as a constraint in subsequent network-optimization problems.

5.3 Resource Minimization Approaches

In this section, the DP to base station allocation is addressed on the basis of optimal design of the allocation matrix and the bias values for range expansion. For this purpose, a mixed-integer linear program (MILP) is formulated where allocations and bias values are selected to minimize total network resource consumption while satisfying QoS constraints for each DP. The proposed optimal design is computationally demanding and scales exponentially with the problem network size and is therefore intractable for online optimization of large-scale networks. Furthermore, the problem has to be solved on a central computer with network-wide CSI information available. This information includes, for example, network-wide channel conditions, load levels, and achievable data rates. The usage of centralized CSI is a major problem as it requires global exchange of control information in the network, which is practically intractable. Therefore, we consider in Subsections 5.3.2 and 5.3.4 a fully decentralized supervised learning–based approach to obtain close-to-optimal DP allocation solutions with low complexity and local CSI.

Based on the optimal solution of the integer programming problems introduced in this section, data labels are generated. The labels are used in combination with locally available network information from which selective features are extracted such that classifiers are trained to replicate the behavior of the optimization.

5.3.1 Optimized Allocation

Wireless links between DPs and cells are generally subject to QoS requirements that can be expressed in terms of user guarantees to achieve a minimum SINR threshold γ^{MIN} associated e.g. with a given requested modulation and coding order. Considering Eq. (5.1), these SINR constraints can be reformulated as the inequalities

$$p_k g_{km} \geq \gamma^{\text{MIN}} \left(\sum_{j \in \{C \setminus \{k\}\}} p_j g_{jm} + \sigma^2 \right) \quad \forall (m,k) : A_{km} = 1. \tag{5.9}$$

The required amount of resources Φ is lower-bounded by the maximum amount of resources any cell in the network requires such that $\Phi \geq \arg\max_k \Phi_k$. For the network scenario to be feasible, $\Phi \leq 1$ needs to hold, i.e. the required time-frequency resources cannot exceed those available to the system. Specifically, the following MILP is designed to optimize the allocation of DPs to cells such that resource minimization is achieved:

$$\underset{\Phi, A}{\text{minimize}} \quad \Phi \tag{5.10a}$$

$$\text{subject to} \quad \Phi \geq \sum_{m=1}^{M} A_{km} \frac{d_m}{\eta^{\text{BW}} W \log_2(1+\gamma_{km})} \quad \forall k \tag{5.10b}$$

$$\sum_{k=1}^{K} A_{km} = 1 \ \forall m \tag{5.10c}$$

$$\sum_{k} A_{km} p_k g_{km} \geq \gamma^{\text{MIN}} \left(\sum_{j \in C} (1 - A_{jm}) p_j g_{jm} + \sigma^2 \right) \quad \forall m \tag{5.10d}$$

$$\Phi \in \mathbb{R}_{0+} \tag{5.10e}$$

$$A_{km} \in \{0,1\} \ \forall k, m. \tag{5.10f}$$

In problem (5.10), the parameter Φ in Eq. (5.10b) is the maximum amount of required time-frequency resources. Constraints (5.10c) cause each DP to be allocated to exactly one cell. The minimum SINR condition Eq. (5.10d) is a linear reformulation of problem (5.9). Problem (5.10) is a MILP with $K \times M$ binary optimization parameters. Problems of this type can be solved with high efficiency by general-purpose optimization software (Grant and Boyd, 2008, 2014; Gurobi Optimization, 2018; ApS, 2017) using state-of-the-art approaches such as *branch-and-bound* (Dakin, 1965; Schrijver, 1998; Linderoth and Savelsbergh, 1999).

5.3.2 Feature Selection and Training

The process of allocating the DPs optimally according to problem (5.10) can be approximated by each DP making a local, potentially suboptimal, allocation decision. It has been shown that typically, the three closest cells in terms of signal strength can provide suitable SINR ratios for serving a given DP (Määttänen et al., 2012; Ramos-Cantor et al., 2017; Gulati et al., 2015). The local decision of each DP is therefore whether it should connect to the neighboring cell from which it receives the first-, second-, or third-strongest signal. This can be modeled as a multiclass classification problem, for which each DP locally extracts features from the network, which is discussed in the following.

For each DP m, the three cells, i.e. the respective base stations, that can provide the first-, second-, and third-largest receive power $p_k g_{km}$ at the location of the DPs are in the following referred to as the primary, secondary, and tertiary allocation candidates. Their indices are listed in the vectors $\boldsymbol{\kappa}^{\mathrm{P}}, \boldsymbol{\kappa}^{\mathrm{S}}, \boldsymbol{\kappa}^{\mathrm{T}} \in \{\mathbb{Z}\}^{M\times 1}$, respectively, with their elements determined as

$$\kappa_m^{\mathrm{P}} = \arg\min_{k} p_k g_{km}, \tag{5.11}$$

$$\kappa_m^{\mathrm{S}} = \arg\min_{k\setminus\{\kappa_m^{\mathrm{P}}\}} p_k g_{km}, \tag{5.12}$$

$$\kappa_m^{\mathrm{T}} = \arg\min_{k\setminus\{\kappa_m^{\mathrm{P}},\kappa_m^{\mathrm{S}}\}} p_k g_{km}. \tag{5.13}$$

Recall that according to Subsection 5.2.2, the SINR of DP m is maximized in unbiased networks if $A_{km} = 1$ for $k = \kappa_m^{\mathrm{P}}$. We remark, however, that applying the *max.-receive-power allocation* scheme in Eq. (5.5) to all DPs m does not necessarily prevent overloading of single cells.

For a given network instance, the optimal allocation matrix $\boldsymbol{A}^{\star}$ is obtained from solving problem (5.10). In the classifier training, the label vector $\boldsymbol{y} \in \mathbb{N}^{M\times 1}$ with elements $y_m \in \{c_1, c_2, c_3\}$ corresponding to the classes c_1, c_2, c_3 is determined as follows:

$$y_m = \begin{cases} c_2 & \text{if } A^{\star}_{\kappa_m^{\mathrm{S}} m} = 1, \\ c_3 & \text{if } A^{\star}_{\kappa_m^{\mathrm{T}} m} = 1, \\ c_1 & \text{otherwise.} \end{cases} \tag{5.14}$$

For the proposed learning-based resource-minimization approach, each DP extracts system attributes corresponding to three allocation candidate cells as features. These features must be sufficiently selective for the classification and are chosen according to engineering experience. The first feature in our supervised learning-based approach is a cell-type indicator:

$$F^{\mathrm{TYPE}}(k) = \begin{cases} 1 & \text{if cell } k \text{ is a small cell,} \\ 0 & \text{otherwise.} \end{cases} \tag{5.15}$$

The second attribute describes the additional load that user m would cause to cell k if the user was allocated to it:

$$F^{\mathrm{LOAD}}(k, m) = \frac{d_m}{\eta^{\mathrm{BW}} \log_2(1 + \gamma_{km})}. \tag{5.16}$$

The third attribute is the sum load that would be caused to cell k by all DPs in its coverage area under *max.-receive-power allocation*:

$$F^{\mathrm{COV}}(k) = \sum_{m|\kappa_m^{\mathrm{P}}=k} \frac{d_m}{\eta^{\mathrm{BW}}\log_2(1 + \gamma_{km})}. \tag{5.17}$$

The feature vector used in a classification is constructed using the three aforementioned attributes for each of the three candidate cells associated with DP m:

$$\begin{aligned} \boldsymbol{h}_m = [&F^{\mathrm{TYPE}}(\kappa_m^{\mathrm{P}}), F^{\mathrm{TYPE}}(\kappa_m^{\mathrm{S}}), F^{\mathrm{TYPE}}(\kappa_m^{\mathrm{T}}), F^{\mathrm{LOAD}}(\kappa_m^{\mathrm{P}}, m), \dots \\ &F^{\mathrm{LOAD}}(\kappa_m^{\mathrm{S}}, m), F^{\mathrm{LOAD}}(\kappa_m^{\mathrm{T}}, m), F^{\mathrm{COV}}(\kappa_m^{\mathrm{P}}), F^{\mathrm{COV}}(\kappa_m^{\mathrm{S}}), F^{\mathrm{COV}}(\kappa_m^{\mathrm{T}})]^{\mathrm{T}}. \end{aligned} \tag{5.18}$$

Using the feature vectors Eq. (5.18) and the class labels obtained from Eq. (5.14) and the solution of problem (5.10), a classifier can be trained and subsequently used to optimize the network while it is in operation, which will be discussed in Subsection 5.3.5.

5.3.3 Range Expansion Optimization

Let $\boldsymbol{\theta}$ denote a k-element vector of the bias values θ_k selected for cell k, which determines the allocation of DPs to cells as defined in Eq. (5.8). The set of S available bias values for all cells is in the following denoted as $\mathcal{S} = \{\delta_1, \ldots, \delta_S\}$. For example, if all SCs operate with any of the available bias values in $\mathcal{S}$ and MCs do not apply biasing, then $\theta_k = 1 \quad \forall k \in \mathcal{C}^{\text{MC}}$ and $\theta_k \in \mathcal{S}\ \forall k \in \mathcal{C}^{\text{SC}}$. Denote the optimal allocation matrix $\boldsymbol{A}$ obtained from applying Eq. (5.5) with $\theta_k = 1\ \forall k$ (no bias) as $\boldsymbol{A}^{\text{max.-power}}$.

In the following, problem (5.10) is extended to simultaneously find the optimal bias values for each cell to minimize the maximum load of any cell in the network. The optimal bias values are computed as the solution of an extended problem:

$$\underset{\Phi,\boldsymbol{A},\boldsymbol{\theta}}{\text{minimize}} \quad \Phi \tag{5.19a}$$

$$\text{subject to} \quad (5.10b), (5.10d), (5.10c), (5.8) \tag{5.19b}$$

$$\Phi \in \mathbb{R}_{0+} \tag{5.19c}$$

$$A_{km} \in \{0,1\}\ \forall k, m \tag{5.19d}$$

$$\theta_k \in \mathcal{S}_k\ \forall k. \tag{5.19e}$$

Due to the bilinear product terms $A_{km}\theta_k$ of optimization variables A_{km} and θ_k in constraint (5.8), the problem (5.19) is a mixed integer nonlinear program (MINLP) for which currently no efficient solvers are available. To solve the problem efficiently with general-purpose optimization software, we apply a lifting technique that converts the problem into an equivalent MILP (Adams et al., 2004; Gupte et al., 2013). Let the constant

$$\overline{\theta} = \underset{s,k}{\arg\max} \quad \delta_{s,k} \tag{5.20}$$

denote the largest bias value in set $\mathcal{S}$. An auxiliary parameter Δ_{km} is introduced such that $\Delta_{km} = A_{km}\theta_k\ \forall k, m$ holds. This can be enforced by the following set of linear inequalities:

$$\Delta_{km} \leq A_{km}\overline{\theta} \tag{5.21a}$$

$$\Delta_{km} \leq \theta_{km} \tag{5.21b}$$

$$\Delta_{km} \geq \theta_{km} - (1 - A_{km})\overline{\theta} \tag{5.21c}$$

$$\Delta_{km} \geq 0. \tag{5.21d}$$

With this, problem (5.19) can be equivalently reformulated as the following MILP

$$\underset{\Phi,\boldsymbol{A},\boldsymbol{\theta},\boldsymbol{\Delta}}{\text{minimize}} \quad \Phi \tag{5.22a}$$

$$\text{subject to} \quad \sum_k \Delta_{km} p_k g_{km} \geq (\theta_j - \Delta_{jm}) p_j g_{jm}\ \forall j, m \tag{5.22b}$$

$$(5.10b), (5.10c), (5.10d), (5.21) \quad \forall k, m \tag{5.22c}$$

$$\Phi \in \mathbb{R}_{0+} \tag{5.22d}$$

$$A_{km} \in \{0,1\}\ \forall k, m \tag{5.22e}$$

$$\theta_k \in \mathcal{S}\ \forall k \tag{5.22f}$$

$$\Delta_{km} \in \mathbb{R}_{0+}, \tag{5.22g}$$

which can be solved using conventional general-purpose solvers (Grant and Boyd, 2008, 2014; Gurobi Optimization, 2018; ApS, 2017). The extraction of features for classifier training based on the optimal bias values $\theta_k^\star$ for problem (5.22) is discussed in the following.

5.3.4 Range Expansion Classifier Training

A vector of class labels $\boldsymbol{y}$ is computed from the solutions of problem (5.22) with its element $y_k = \{c_s \mid \theta_k^\star = \delta_s\}$ representing the label corresponding to cell k. These can take values in the discrete set of classes $C = \{c_1, c_2, \dots, c_{|S|}\}$. Hence, there is one class associated with each discrete bias value in $\mathcal{S}$. Similar to Eq. (5.15), the first attribute is denoted as $G^{\mathrm{TYPE}}(k)$, which is determined as $G^{\mathrm{TYPE}}(k) = 1$ if cell k is a SC and $G^{\mathrm{TYPE}}(k) = 0$ otherwise.

For the second set of attributes, we define the index set

$$\mathcal{M}_k^{\{s\}} = \{m | \delta_s p_k g_{km} \geq p_j g_{jm} \forall j \in C\} \tag{5.23}$$

of DPs connected to cell k if it is utilized with a bias value δ_s. These sets can be precomputed for all bias values in $\mathcal{S}$. The corresponding resource consumption in that case forms the second feature, and is calculated as

$$G^{\mathrm{LD}}(k, s) = \sum_{m \in \mathcal{M}_k^{\{s\}}} \frac{d_m}{\eta^{\mathrm{BW}} \log_2(1 + \gamma_{\kappa_k^{\mathrm{P}} m})}. \tag{5.24}$$

In the following, denote as κ_k^{P} and κ_k^{S} the first- and second-strongest neighboring cell to cell k, in terms of receive signal power at the location of base station k. For each assumed bias values in $\delta_s \in S$, the corresponding sum load is computed as

$$G^{\mathrm{PSL}}(k, s) = \sum_{m \in \mathcal{M}_k^{\{s\}}} A_{\kappa_k^{\mathrm{P}} m}^{\mathrm{max.\text{-}power}} \frac{d_m}{\eta^{\mathrm{BW}} \log_2(1 + \gamma_{\kappa_k^{\mathrm{P}} m})} \tag{5.25}$$

and

$$G^{\mathrm{SSL}}(k, s) = \sum_{m \in \mathcal{M}_k^{\{s\}}} A_{\kappa_k^{\mathrm{S}} m}^{\mathrm{max.\text{-}power}} \frac{d_m}{\eta^{\mathrm{BW}} \log_2(1 + \gamma_{\kappa_k^{\mathrm{P}} m})}, \tag{5.26}$$

respectively, that DPs in the coverage area of cell m add to their first and second neighboring cell if the *maximum-receive-power allocation* rule ($A = \boldsymbol{A}^{\mathrm{max.\text{-}power}}$) is applied. The aforementioned attributes are combined into the following $(3S + 1)$-element feature vector:

$$\begin{aligned} \boldsymbol{h}_k = [&G^{\mathrm{TYPE}}(k), G^{\mathrm{LD}}(k, 1), \dots, G^{\mathrm{LD}}(k, s), \\ &G^{\mathrm{PSL}}(k, 1), \dots, G^{\mathrm{PSL}}(k, s), G^{\mathrm{SSL}}(k, 1), \dots, G^{\mathrm{SSL}}(k, s)]^{\mathrm{T}}. \end{aligned} \tag{5.27}$$

This feature vector is used to train a classifier and utilize it to perform resource allocation in the network, similar to the attribute vector defined in Section 5.3.2. Both systems will be evaluated in the following network simulations.

5.3.5 Multi-Class Classification

In Subsections 5.3.2 and 5.3.4, the network optimization problems of DP allocation and bias value selection have been modeled as multi-class classification problems. An outline of how these classification problems can be solved using SVMs and ANNs is provided in the following. Let $\boldsymbol{h}_t$ be the feature vector obtained by feature extraction for allocation (5.18) and for range expansion (5.27) for the training sample t. The class labels corresponding to all training data samples is given in the vector $\boldsymbol{y} = [y_1, \dots, y_T]^\top$. During the training of a classifier based on support vector machines, a hyperplane $\boldsymbol{\omega}^\top \boldsymbol{h} + b = 0$ is determined that best separates the feature vectors (datapoints) into two classes. SVMs are large-margin classifiers, which means they aim to maximize the margin between the hyperplane and the closest data points. Since for large training sets the feature vectors generally are not linearly separable by a hyperplane, two modifications for SVMs have been established. The first modification is soft-threshold training, where the hyperplane does not have to strictly separate the two classes of data points. The resulting mis-classifications are discouraged during the training of the SVM. The second modification is the kernel trick, where a function $\vartheta(\boldsymbol{h}_t)$ maps the attribute vector $\boldsymbol{h}_t$ onto the higher-dimensional lifted feature space of dimension L. In this feature space, for example, polynomial combinations of the attributes are used as training features.

The SVM classifier between classes c_i and c_j is defined by the separating hyperplane $\boldsymbol{\omega}^{\{c_1c_2\}}\vartheta(\boldsymbol{h}_t^{\{c_1c_2\}}) + b^{\{c_1c_2\}}$, where the parameters $\boldsymbol{\omega}^{\{c_1c_2\}}$ and $b^{\{c_1c_2\}}$ are obtained from solving a training optimization problem (Kressel, 1999; Cortes and Vapnik, 1995). If classification is conducted only between these two classes, the predicted class label $\hat{y}^{\{c_1c_2\}}$ for a new data sample with feature vector $\hat{\boldsymbol{h}}^{\{c_1c_2\}}$ is determined as:

$$\hat{y} = \begin{cases} c_i & \text{if} \quad (\boldsymbol{\omega}^{\{c_ic_j\}})^\top \vartheta(\hat{\boldsymbol{h}}^{\{c_1c_2\}}) + b^{\{c_ic_j\}} \geq 0 \\ c_j & \text{if} \quad (\boldsymbol{\omega}^{\{c_ic_j\}})^\top \vartheta(\hat{\boldsymbol{h}}^{\{c_1c_2\}}) + b^{\{c_ic_j\}} < 0. \end{cases} \tag{5.28}$$

For multi-class problems, SVMs are trained to classify between all possible pairings of classes $(c_i, c_j) \in C^{\in}$, $(i \neq j)$. The estimated class $\hat{y} \in C$ is, e.g. chosen according to the majority rule as the class that "wins" the most one-on-one classifications with all other classes. Specifically,

$$\hat{y} = \arg\max_i \left(\sum_{j=1}^{I} H((\boldsymbol{\omega}^{\{ij\}})^\top \vartheta(\hat{\boldsymbol{h}}) + b^{\{ij\}}) \right) \tag{5.29}$$

where $H(\cdot)$ is the Heaviside step function. SVM training problems are typically solved with high computational efficiency in their Lagrange dual formulation using kernel functions (Muller et al., 2001). This functionality is included in common ML software tools (Chang and Lin, 2011; MAT).

In ANNs, the extension from a binary classifier to a multi-class system is commonly performed by applying the *soft-max* function in the output layer, such that the values of the output nodes sum up to one. Therefore, these output values can be utilized as a probability distribution, and the most "likely" class can be chosen.

An example of such an ANN is shown in Figure 5.1. Multi-class functionality with utilization of the described soft-max function is included in state-of-the-art ANN software tools (Chollet

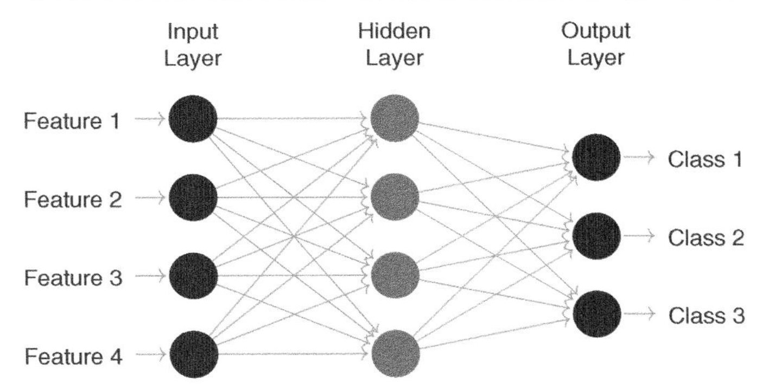

Figure 5.1 Example illustration of a multi-class ANN classifier.

et al., 2015; Pedregosa et al., 2011). From experience, the classification problems for resource allocation presented in this chapter are relatively well linearly separable with ANN classifiers. The ANNs used in this work are therefore designed with one hidden layer with the number of nodes corresponding to the number of input features.

5.4 Numerical Results

A heterogeneous wireless network is simulated with three MCs and six SCs deployed in fixed positions in an area of 1000×1000 meters, as illustrated in Figure 5.2. The MCs utilize a transmit power of $p_k = 46\text{dBmW} \quad \forall k \in C^{\text{MC}}$, and the SCs $p_k = 36\text{dBmW} \quad \forall C^{\text{SC}}$, with an antenna gain of 10dB and 5dB, respectively. The total available bandwidth of the system is set as $W =$ 20MHz, the bandwidth efficiency $\eta^{\text{BW}} = 0.85$, and the noise power spectral density $\sigma^2/W =$ -145dBmW/Hz. The available bias values for range expansion of all cells are either 0 dB (no biasing applied), 3 dB, or 6 dB. The path loss is simulated using the specification in the 3GPP (3GPP, p. 61), with an additional 5 dB log-normal shadow fading. The optimization problems (5.10) and (5.22) are solved using Python and the GUROBI solver v8.1 (Gurobi Optimization, 2018).

The classification problem utilized for resource minimization is solved using support vector machines with linear and quadratic feature mapping (lin. SVM and quad. SVM) and ANNs with a single hidden layer (ANN). The training of the SVMs and ANNs is performed in the scikit-learn toolbox for Python (Pedregosa et al., 2011) using data from 100 network simulations of each 100 DPs that are randomly and uniformly distributed in the network area, and with 9 fixed cell locations as depicted in Figure 5.2. It is observed that generally, only 5% of DPs are allocated to the second- or third-strongest cell in the optimal allocation solution obtained from solving problem (5.10). In a network scenario with $M = 100$ DPs, the number of obtained feature vectors for the allocation- and range-expansion classification according to Subsections 5.3.2 and 5.3.4, respectively, that contribute effectively to the training is therefore roughly the same in both approaches. The algorithms are evaluated using 200 network scenarios with different randomized DP locations. During testing, the result obtained

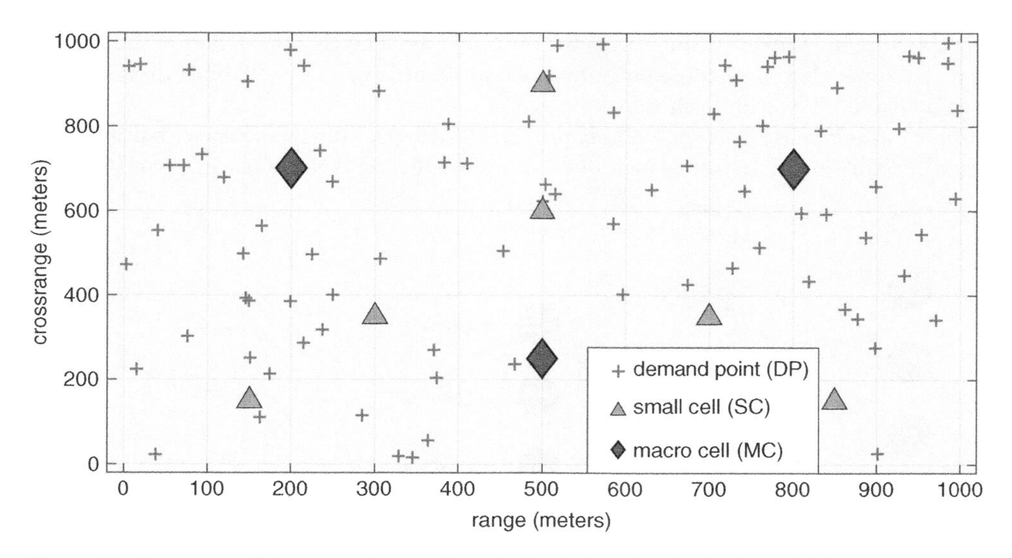

Figure 5.2 Illustration of the network scenario with $M = 100$ randomly distributed DPs.

by the learning-based schemes for allocation is considered a recommendation, where the DP automatically defaults to the *maximum-receive power allocation* if the recommendation of the learning-based system violates the SINR-constraints (5.10d).

The achieved resource-consumption levels over an increasing demand that is uniformly applied for all DPs is depicted in Figure 5.3. The *maximum-receive power allocation* approach allocates each DP to the cell providing the strongest signal, which minimizes the resources necessary to serve each DP. However, this scheme fails in offloading demand between cells, causing the resource consumption of the entire network to be the largest among all approaches. Decreased resource-consumption levels are achieved by the learning-based range expansion schemes, which are lower-bounded by the optimal range-expansion solution, and the learning-based allocation schemes. The optimal allocation solution obtained from Eq. (5.10) marks the lower bound. It can be concluded from Figure 5.3 that optimized allocation can achieve better performance than optimized range expansion. Within both groups, the choice of classifier (SVMs or ANNs) does not make a large difference in performance for this scenario.

Figure 5.4 shows the normalized resource consumption of the network for a varying number of DPs M, but with fixed $d_m = 1\text{Mbit/s}\ \forall\ m$. The resource consumption again increases mostly linearly with M. It is observable that the ANN-based allocation achieves slightly better performance than the other allocation schemes, matching or even outperforming optimized range expansion.

For a single network example scenario, the network consumption of individual cells for all schemes is shown in Figure 5.5. The load of the most resource-consuming MCs, MC1 and MC2, is successfully decreased by all methods. The optimal allocation scheme successfully decreases the maximum load of any cell, which is characterizing the overall resource consumption of the network. It shows, however, unspecific behavior for non-critical cells. The SCs, which for *maximum-receive-power allocation* remain underutilized, take over DPs from other cells for the other schemes. Interestingly, the centermost SC in network SC2 is not utilized at all, implying that no DPs are in close proximity to achieve significant SINR levels.

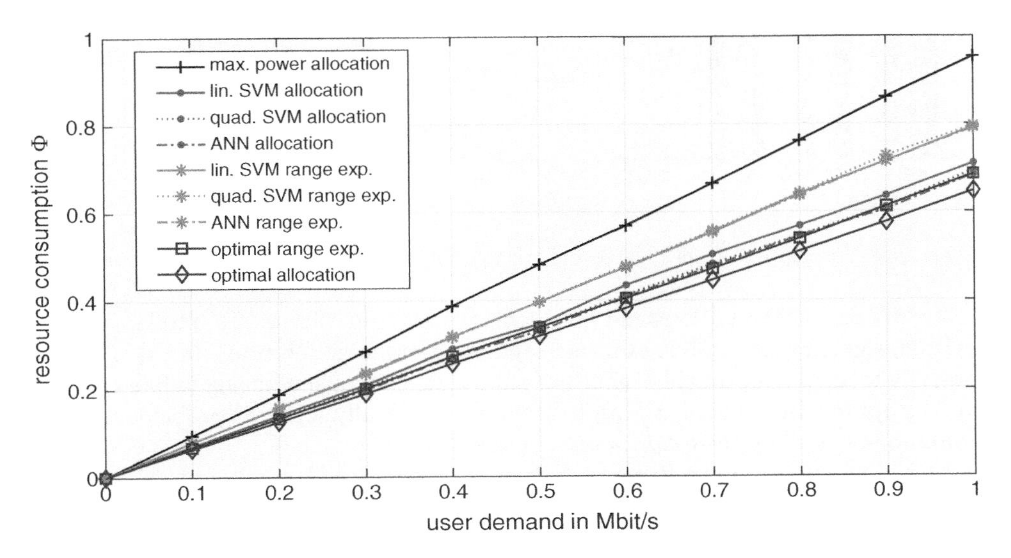

Figure 5.3 Comparison of network resource consumption over data demand for multiple allocation- and range-expansion schemes.

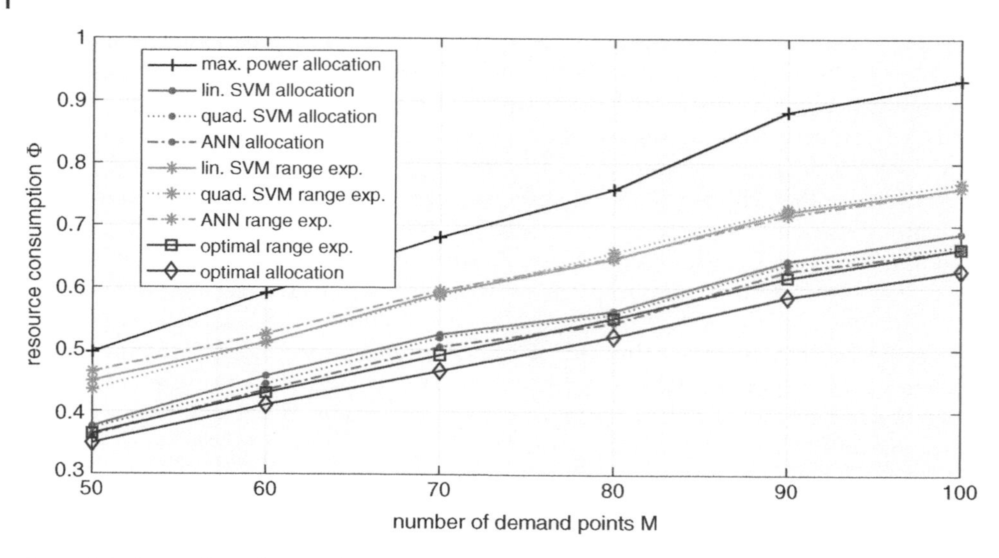

Figure 5.4 Comparison of network resource consumption over a number of DPs for multiple allocation- and range-expansion schemes.

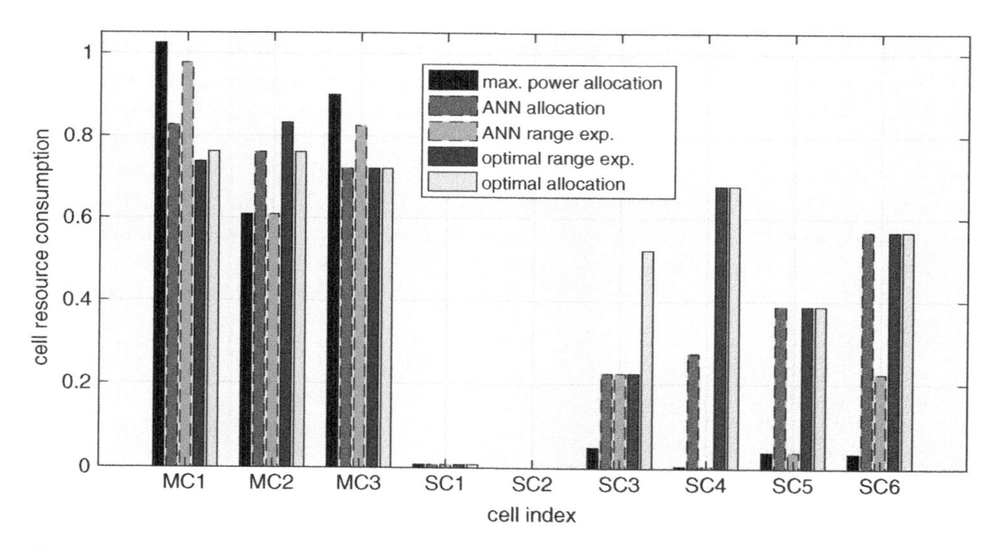

Figure 5.5 Resource consumption of individual cells for a single network scenario example.

In practice, the channel conditions and user demands are time varying. Due to latencies in CSI estimation, signaling, and solving the corresponding optimization problems, it is practically impossible to achieve optimal DP allocation and range expansion during online network operation. The optimization-based approaches therefore generally solve the problem ahead of network operation based on *worst-case demand* forecasts.

To assess the robustness of the schemes with time-varying network conditions, we consider a simulation scenario in which the demand of each DP is drawn from a uniform distribution between 0 and 2Mbit/s. In the following, we refer to this fluctuating demand as the *instantaneous demand*. The optimization-based schemes obtain a solution configuration for allocation and range expansion based on the worst-case demand of $d_m = 2$ Mbit/s $\forall\ m$, while

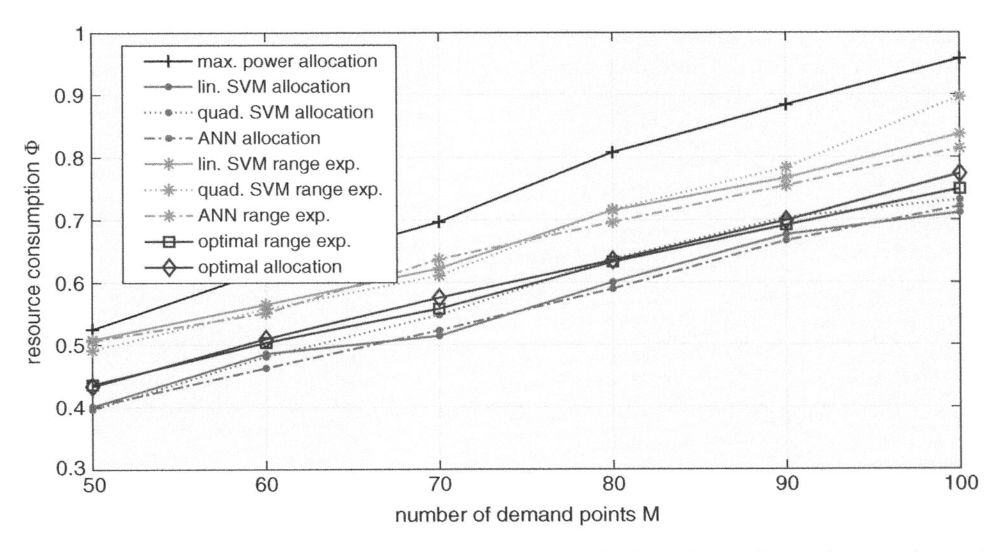

Figure 5.6 Network resource consumption of resource minimization schemes for varying user demand, and worst-case demand information for optimal schemes.

the learning-based approaches operate decentralized on the basis of the instantaneous data demands. The resulting resource consumption, evaluated according to the instantaneous demands for all schemes, is illustrated in Figure 5.6. Interestingly, the learning-based allocation schemes achieve even lower resource consumption than the optimization-based allocation and range-expansion schemes based on problems (5.10) and (5.22), due to the fact that the latter schemes only operate based on worst-case demand forecasts.

5.5 Concluding Remarks

This chapter introduced supervised learning based techniques for fully decentralized network optimization and radio access control in ultra-dense heterogeneous wireless networks. Traditionally, radio access control is performed during network operation based on signal-strength measurements, and the concept of *biasing* is used to reduce resource consumption in the network and offload users from high-power MCs to SC base stations. We formulate optimal user-allocation and bias-value computation schemes based on mixed integer linear programming. Solving these programs requires, however, centralized network-wide combinatorial optimization and thus cannot be carried out online. In turn, we demonstrated that base-station allocation and bias-value solutions obtained from centralized offline optimization are indeed very helpful and suitable for labeling training data used in novel decentralized supervised learning–based approaches. Two alternative classification-based user-allocation schemes were devised: the user-centric approach in which users learn to make allocation decisions only based on local signal strength and network information; and the base station–centric approach in which each base station learns to compute optimized bias values based on local information, and the allocation is carried out according to the biased signal-strength measures. Simulation results illustrated that the proposed ML approaches achieve close to the optimal network-balancing solutions. Furthermore, the results demonstrated that learning-based resource allocation schemes, when trained with optimal network configuration data, successfully generalize their knowledge and become also applicable in networks of different size and users with different demand and buffer status.

Bibliography

MATLAB and Statistics Toolbox Release 2013a, The MathWorks, Inc.

3GPP technical report 36.839, v11.1.0 mobility enhancements in heterogeneous networks, December 2012. URL https://portal.3gpp.org/.

3GPP.

Warren P. Adams, Richard J. Forrester, and Fred W. Glover. Comparisons and enhancement strategies for linearizing mixed 0-1 quadratic programs. *Discrete Optimization*, 1(2):99–120, 2004. doi: 10.1016/j.disopt.2004.03.006.

A.S. Alfa, B.T. Maharaj, S. Lall, and S. Pal. Mixed-integer programming based techniques for resource allocation in underlay cognitive radio networks: A survey. *Journal of Communications and Networks*, 18(5):744–761, October 2016. ISSN 1229-2370. doi: 10.1109/JCN.2016.000104.

J.G. Andrews, S. Buzzi, W. Choi, S.V. Hanly, A. Lozano, A.C.K. Soong, and J.C. Zhang. What will 5G be? *IEEE Journal on Selected Areas in Communications*, 32(6):1065–1082, June 2014. ISSN 0733-8716. doi: 10.1109/JSAC.2014.2328098.

J.G. Andrews, X. Zhang, G.D. Durgin, and A.K. Gupta. Are we approaching the fundamental limits of wireless network densification? *IEEE Communications Magazine*, 54(10):184–190, October 2016. ISSN 0163-6804.

MOSEK ApS. *The MOSEK Optimization Toolbox for MATLAB Manual. Version 8.1.*, 2017.

G. Athanasiou, P.C. Weeraddana, C. Fischione, and L. Tassiulas. Optimizing client association for load balancing and fairness in millimeter-wave wireless networks. *IEEE/ACM Transactions on Networking*, 23(3):836–850, June 2015. ISSN 1063-6692. doi: 10.1109/TNET.2014.2307918.

F. Bahlke and M. Pesavento. Decentralized load balancing in mobile communication networks. In *2018 IEEE International Conference on Acoustics, Speech and Signal Processing (ICASSP)*, pages 3564–3568, April 2018a. doi: 10.1109/ICASSP.2018.8461780.

F. Bahlke and M. Pesavento. Optimized small cell range expansion in mobile communication networks using multi-class support vector machines. In *2018 26th European Signal Processing Conference (EUSIPCO)*, pages 430–434, Sep. 2018b. doi: 10.23919/EUSIPCO.2018.8553221.

F. Bahlke, O.D. Ramos-Cantor, S. Henneberger, and M. Pesavento. Optimized cell planning for network slicing in heterogeneous wireless communication networks. *IEEE Communications Letters*, 22(8):1676–1679, Aug 2018. ISSN 1089-7798. doi: 10.1109/LCOMM.2018.2841866.

F. Boccardi, R. W. Heath, A. Lozano, T.L. Marzetta, and P. Popovski. Five disruptive technology directions for 5g. *IEEE Communications Magazine*, 52 (2):74–80, February 2014. ISSN 0163-6804. doi: 10.1109/MCOM.2014.6736746.

P. Caballero, A. Banchs, G. de Veciana, and X. Costa-Pérez. Multi-tenant radio access network slicing: Statistical multiplexing of spatial loads. *IEEE/ACM Transactions on Networking*, 25(5):3044–3058, Oct 2017. ISSN 1063-6692. doi: 10.1109/TNET.2017.2720668.

R.L.G. Cavalcante, S. Stanczak, M. Schubert, A. Eisenblaetter, and U. Tuerke. Toward energy-efficient 5G wireless communications technologies: Tools for decoupling the scaling of networks from the growth of operating power. *Signal Processing Magazine, IEEE*, 31(6):24–34, Nov 2014. ISSN 1053-5888. doi: 10.1109/MSP.2014.2335093.

Chih-Chung Chang and Chih-Jen Lin. LIBSVM: A library for support vector machines. *ACM Transactions on Intelligent Systems and Technology*, 2: 27:1–27:27, 2011.

Y. Cheng and M. Pesavento. Joint discrete rate adaptation and downlink beamforming using mixed integer conic programming. *IEEE Transactions on Signal Processing*, 63(7):1750–1764, April 2015. doi: 10.1109/TSP.2015.2393837.

Y. Cheng, M. Pesavento, and A. Philipp. Joint network optimization and downlink beamforming for comp transmissions using mixed integer conic programming. *IEEE Transactions on Signal Processing*, 61(16):3972–3987, Aug 2013. ISSN 1053-587X. doi: 10.1109/TSP.2013.2261993.

François Chollet et al. Keras, 2015. URL https://keras.io.

L. Cimini. Analysis and simulation of a digital mobile channel using orthogonal frequency division multiplexing. *IEEE Transactions on Communications*, 33 (7):665–675, jul 1985. doi: 10.1109/tcom.1985.1096357.

Corinna Cortes and Vladimir Vapnik. Support-vector networks. *Machine Learning*, 20(3):273–297, 1995.

R.J. Dakin. A tree-search algorithm for mixed integer programming problems. *The Computer Journal*, 8(3):250–255, 1965. doi: 10.1093/comjnl/8.3.250. URL http://dx.doi.org/10.1093/comjnl/8.3.250.

X. Ge, S. Tu, G. Mao, C.X. Wang, and T. Han. 5G ultra-dense cellular networks. *IEEE Wireless Communications*, 23(1):72–79, February 2016. ISSN 1536-1284. doi: 10.1109/MWC.2016.7422408.

A. Ghosh, T.A. Thomas, M.C. Cudak, R. Ratasuk, P. Moorut, F.W. Vook, T.S. Rappaport, G.R. MacCartney, S. Sun, and S. Nie. Millimeter-wave enhanced local area systems: A high-data-rate approach for future wireless networks. *IEEE Journal on Selected Areas in Communications*, 32(6):1152–1163, June 2014. ISSN 0733-8716. doi: 10.1109/JSAC.2014.2328111.

Andrea Goldsmith. *Wireless Communications*. Cambridge University Press, 2004.

Michael Grant and Stephen Boyd. CVX: MATLAB software for disciplined convex programming, version 2.1, March 2014.

Michael C. Grant and Stephen P. Boyd. Graph implementations for nonsmooth convex programs. In *Recent Advances in Learning and Control*, pages 95–110. Springer, 2008.

S. Gulati, S. Kalyanasundaram, P. Nashine, B. Natarajan, R. Agrawal, and A. Bedekar. Performance analysis of distributed multi-cell coordinated scheduler. In *2015 IEEE 82nd Vehicular Technology Conference (VTC2015-Fall)*, pages 1–5, Sept 2015. doi: 10.1109/VTCFall.2015.7391069.

Akshay Gupte, Shabbir Ahmed, Myun Seok Cheon, and Santanu Dey. Solving mixed integer bilinear problems using milp formulations. *SIAM Journal on Optimization*, 23(2):721–744, 2013. doi: 10.1137/110836183.

LLC Gurobi Optimization. Gurobi Optimizer reference manual, 2018. URL http://www.gurobi.com.

E. Hossain, M. Rasti, H. Tabassum, and A. Abdelnasser. Evolution toward 5G multi-tier cellular wireless networks: An interference management perspective. *IEEE Wireless Communications*, 21(3):118–127, June 2014. ISSN 1536-1284. doi: 10.1109/MWC.2014.6845056.

J. Hoydis, M. Kobayashi, and M. Debbah. Green small-cell networks. *IEEE Vehicular Technology Magazine*, 6(1):37–43, March 2011. ISSN 1556-6072. doi: 10.1109/MVT.2010.939904.

R.Q. Hu and Y. Qian. An energy efficient and spectrum efficient wireless heterogeneous network framework for 5G systems. *IEEE Communications Magazine*, 52(5):94–101, May 2014. ISSN 0163-6804. doi: 10.1109/MCOM.2014.6815898.

ITU. Minimum requirements related to technical performance for IMT-2020 radio interface(s), November 2017. URL https://www.itu.int/pub/R-REP-M.2410-2017.

M. Iwamura. NGMN view on 5G architecture. In *2015 IEEE 81st Vehicular Technology Conference (VTC Spring)*, pages 1–5, May 2015. doi: 10.1109/VTCSpring.2015.7145953.

C. Jiang, H. Zhang, Y. Ren, Z. Han, K.C. Chen, and L. Hanzo. Machine learning paradigms for next-generation wireless networks. *IEEE Wireless Communications*, 24(2):98–105, April 2017. ISSN 1536-1284. doi: 10.1109/MWC.2016.1500356WC.

V. Jungnickel, K. Manolakis, W. Zirwas, B. Panzner, V. Braun, M. Lossow, M. Sternad, R. Apelfrojd, and T. Svensson. The role of small cells, coordinated multipoint, and massive MIMO in 5G. *IEEE Communications Magazine*, 52 (5):44–51, May 2014. ISSN 0163-6804. doi: 10.1109/MCOM.2014.6815892.

Ulrich H.-G. Kressel. Pairwise classification and support vector machines. *Advances in Kernel Methods*, pages 255–268, 1999. URL http://dl.acm.org/citation.cfm?id=299094.299108.

E.G. Larsson, O. Edfors, F. Tufvesson, and T.L. Marzetta. Massive MIMO for next generation wireless systems. *IEEE Communications Magazine*, 52(2): 186–195, February 2014. ISSN 0163-6804. doi: 10.1109/MCOM.2014.6736761.

J.T. Linderoth and M.W.P. Savelsbergh. A computational study of search strategies for mixed integer programming. *INFORMS Journal on Computing*, 11(2):173–187, 1999. doi: 10.1287/ijoc.11.2.173. URL https://doi.org/10.1287/ijoc.11.2.173.

L. Liu and Q. Fan. Resource allocation optimization based on mixed integer linear programming in the multi-cloudlet environment. *IEEE Access*, 6: 24533–24542, 2018. ISSN 2169-3536. doi: 10.1109/ACCESS.2018.2830639.

D. Lopez-Perez, I. Guvenc, G. de la Roche, M. Kountouris, T. Quek, and Jie Zhang. Enhanced intercell interference coordination challenges in heterogeneous networks. *Wireless Communications, IEEE*, 18(3):22–30, June 2011. ISSN 1536-1284. doi: 10.1109/MWC.2011.5876497.

Helka-Liina Määttänen, Kari Hämäläinen, Juha Venäläinen, Karol Schober, Mihai Enescu, and Mikko Valkama. System-level performance of LTE-advanced with joint transmission and dynamic point selection schemes. *EURASIP Journal on Advances in Signal Processing*, 2012(1): 247, Nov 2012. ISSN 1687-6180. doi: 10.1186/1687-6180-2012-247. URL http://dx.doi.org/10.1186/1687-6180-2012-247.

K. Majewski and M. Koonert. Conservative cell load approximation for radio networks with Shannon channels and its application to LTE network planning. In *2010 Sixth Advanced International Conference on Telecommunications (AICT)*, pages 219–225, May 2010. doi: 10.1109/AICT.2010.9.

P. Mogensen, Wei Na, I.Z. Kovacs, F. Frederiksen, A. Pokhariyal, K.I. Pedersen, T. Kolding, K. Hugl, and M. Kuusela. LTE capacity compared to the Shannon bound. In *Vehicular Technology Conference, 2007. VTC2007-Spring. IEEE 65th*, pages 1234–1238, April 2007. doi: 10.1109/VETECS.2007.260.

K.R. Muller, S. Mika, G. Ratsch, K. Tsuda, and B. Scholkopf. An introduction to kernel-based learning algorithms. *IEEE Transactions on Neural Networks*, 12 (2):181–201, Mar 2001. ISSN 1045-9227. doi: 10.1109/72.914517.

V.M. Nguyen and M. Kountouris. Performance limits of network densification. *IEEE Journal on Selected Areas in Communications*, 35(6):1294–1308, June 2017. ISSN 0733-8716. doi: 10.1109/JSAC.2017.2687638.

T. O'Shea and J. Hoydis. An introduction to deep learning for the physical layer. *IEEE Transactions on Cognitive Communications and Networking*, 3 (4):563–575, Dec 2017. ISSN 2332-7731. doi: 10.1109/TCCN.2017.2758370.

T.J. O'Shea, T. Erpek, and T.C. Clancy. Physical layer deep learning of encodings for the MIMO fading channel. In *2017 55th Annual Allerton Conference on Communication, Control, and Computing (Allerton)*, pages 76–80, Oct 2017. doi: 10.1109/ALLERTON.2017.8262721.

F. Pedregosa, G. Varoquaux, A. Gramfort, V. Michel, B. Thirion, O. Grisel, M. Blondel, P. Prettenhofer, R. Weiss, V. Dubourg, J. Vanderplas, A. Passos, D. Cournapeau, M. Brucher, M.

Perrot, and E. Duchesnay. Scikit-learn: Machine learning in Python. *Journal of Machine Learning Research*, 12: 2825–2830, 2011.

Oscar D. Ramos-Cantor, Jakob Belschner, Ganapati Hegde, and Marius Pesavento. Centralized coordinated scheduling in lte-advanced networks. *EURASIP Journal on Wireless Communications and Networking*, 2017(1): 122, 2017.

T.S. Rappaport, S. Sun, R. Mayzus, H. Zhao, Y. Azar, K. Wang, G.N. Wong, J.K. Schulz, M. Samimi, and F. Gutierrez. Millimeter wave mobile communications for 5G cellular: It will work! *IEEE Access*, 1:335–349, 2013. doi: 10.1109/ACCESS.2013.2260813.

T.S. Rappaport, G.R. MacCartney, M.K. Samimi, and S. Sun. Wideband millimeter-wave propagation measurements and channel models for future wireless communication system design. *IEEE Transactions on Communications*, 63(9):3029–3056, Sept 2015. ISSN 0090-6778. doi: 10.1109/TCOMM.2015.2434384.

Alexander Schrijver. *Theory of Linear and Integer Programming*. John Wiley & Sons, 1998. doi: 10.2307/253980.

M. Shafi, A.F. Molisch, P.J. Smith, T. Haustein, P. Zhu, P. De Silva, F. Tufvesson, A. Benjebbour, and G. Wunder. 5G: A tutorial overview of standards, trials, challenges, deployment, and practice. *IEEE Journal on Selected Areas in Communications*, 35(6):1201–1221, June 2017. ISSN 0733-8716. doi: 10.1109/JSAC.2017.2692307.

I. Siomina and D. Yuan. Analysis of cell load coupling for LTE network planning and optimization. *IEEE Transactions on Wireless Communications*, 11(6): 2287–2297, June 2012a. ISSN 1536-1276. doi: 10.1109/TWC.2012.051512.111532.

I. Siomina and D. Yuan. Load balancing in heterogeneous LTE: Range optimization via cell offset and load-coupling characterization. In *2012 IEEE International Conference on Communications (ICC)*, pages 1357–1361, June 2012b. doi: 10.1109/ICC.2012.6364075.

David Tse and Pramoth Viswanath. *Fundamentals of Wireless Communications*. Cambridge University Press, 2005.

Cheong Yui Wong, R.S. Cheng, K.B. Lataief, and R.D. Murch. Multiuser OFDM with adaptive subcarrier, bit, and power allocation. *IEEE Journal on Selected Areas in Communications*, 17(10):1747–1758, 1999. doi: 10.1109/49.793310.

M. Xiao, S. Mumtaz, Y. Huang, L. Dai, Y. Li, M. Matthaiou, G. Karagiannidis, E. Björnson, K. Yang, C.-L. I, and A. Ghosh. Millimeter wave communications for future mobile networks. *IEEE Journal on Selected Areas in Communications*, 35(9):1909–1935, Sept 2017. ISSN 0733-8716. doi: 10.1109/JSAC.2017.2719924.

Z. Xu, Y. Wang, J. Tang, J. Wang, and M.C. Gursoy. A deep reinforcement learning based framework for power-efficient resource allocation in cloud RANs. In *2017 IEEE International Conference on Communications (ICC)*, pages 1–6, May 2017. doi: 10.1109/ICC.2017.7997286.

H. Ye, G.Y. Li, and B. Juang. Power of deep learning for channel estimation and signal detection in OFDM systems. *IEEE Wireless Communications Letters*, 7(1):114–117, Feb 2018. ISSN 2162-2337. doi: 10.1109/LWC.2017.2757490.

Q. Ye, B. Rong, Y. Chen, M. Al-Shalash, C. Caramanis, and J.G. Andrews. User association for load balancing in heterogeneous cellular networks. *IEEE Transactions on Wireless Communications*, 12(6):2706–2716, June 2013. ISSN 1536-1276. doi: 10.1109/TWC.2013.040413.120676.

L. You and D. Yuan. Load optimization with user association in cooperative and load-coupled LTE networks. *IEEE Transactions on Wireless Communications*, 16(5):3218–3231, May 2017. ISSN 1536-1276. doi: 10.1109/TWC.2017.2676762.

6

Machine Learning in Energy Efficiency Optimization

Muhammad Ali Imran[1], *Ana Flávia dos Reis*[2], *Glauber Brante*[2], *Paulo Valente Klaine*[1], *and Richard Demo Souza*[5]

[1] *James Watt School of Engineering, University of Glasgow, Glasgow, Scotland, UK*
[2] *Graduate Program in Electrical and Computer Engineering, Federal University of Technology - Paraná, Curitiba, Brazil*
[5] *Department of Electrical and Electronics Engineering, Federal University of Santa Catarina, Florianópolis - SC, Brazil*

5G wireless networks are expected not only to overcome the limitations of current cellular networks, but also to address new use cases and scenarios, enabling a wide range of new applications. It is envisioned that by 2030, billions of devices will be connected to mobile networks (Cisco, 2017), and, as such, 5G will have to provide connectivity to a huge amount of devices. Moreover, 5G will also have to provide coverage and capacity everywhere, while enhancing user experience and data rates (Huawei, 2013; Valente Klaine et al., 2017). However, a huge concern has been raised, as, for all of these requirements to be possible, it is expected that 5G networks will also consume a thousand times more energy than current systems (Buzzi et al., 2016). In order to overcome that, 5G networks will have to become very energy efficient, pushing the development of greener and more sustainable networks, with respect not only to base stations (BSs), but also to mobile devices (Tullberg et al., 2016). As such, energy efficiency (EE) has been established as one of the primary goals in modern communication scenarios, and several research groups in academia and industry have focused efforts on this topic.

In the literature, several different metrics have been used to define EE. The most common approach is to define it as the ratio between the system throughput and the power consumption, expressed in [bits/Joules] (Feng et al., 2013). Based on this definition, many works in the literature tried to maximise EE depending on a particular application or environment. For instance, in cellular networks, EE has attracted a lot of interest due to the potential of reducing operational costs (Feng et al., 2013). Moreover, the exponential increase of connected devices in wireless communications poses serious sustainable growth concerns, due carbon emissions surge to worrying rates. Therefore, many early studies in green cellular have shown significant EE improvement in order to support user demands and minimize these adverse effects (Gandotra et al., 2017). In these large-scale scenarios, BSs represent the largest share of the total energy consumption: up to 80%, due to power supplies, cooling, and connection to the electrical power grid (Auer et al., 2011).

For example, the work in (Richter et al., 2009) investigates the impact of different BS layouts on the energy consumption of cellular networks, showing that the power saving from deployment of micro-sites is moderate in full-load scenarios and strongly depends on the offset power consumption of both macro- and micro-sites. The extension of this work, presented in (Fehske et al., 2009), employs simulations to evaluate the energy consumption of the deployment of micro-sites in addition to conventional macro-sites, and shows that the deployment

Machine Learning for Future Wireless Communications, First Edition. Edited by Fa-Long Luo.
© 2020 John Wiley & Sons Ltd. Published 2020 by John Wiley & Sons Ltd.

of micro-sites allows a significant decrease in network power consumption. EE has also been applied to other types of networks, such as wireless sensor networks (WSNs). In (Cui et al., 2004, 2005; Pereira et al., 2018), for example, the authors have shown that, when the transmission distance is large (> 100 m), the transmit power dominates total power consumption. On the other hand, for very short-range applications (< 10 m), the power used by the RF circuitry may dominate total power consumption. Therefore, an appropriate power-consumption model for WSNs is required for each scenario, and the approach to optimize the EE may be considerably different.

Traditionally, techniques such as convex optimization (Xu and Qiu, 2013), fractional programming (Zappone et al., 2015), and game theory (AlSkaif et al., 2015) have been used in order to maximize EE in wireless networks. Such approaches attempt to find analytical solutions, optimizing many communication parameters such as transmit power, modulation order, and operational modes. However, many simplifications are usually required to derive these expressions, so that often, only point-to-point scenarios are considered. In addition, several unrealistic assumptions can be a limiting factor of these solutions, as many of them often require information that is not timely or realistically available in a mobile network, such as the number of connected users at a specific time or their exact positions. Furthermore, future cellular networks are expected to become even more complex due to the growth in the number of connected devices and traffic density, as well as their heterogeneous requirements (Valente Klaine et al., 2017). As a result of this complexity, the application of classical optimization techniques (Zappone et al., 2015; AlSkaif et al., 2015; Xu and Qiu, 2013) might not be feasible in future network scenarios.

To overcome these issues, it is clear that conventional solutions will not be optimal, as networks become more and more complex. In addition, more robust and flexible solutions that explore the data generated by the network and make decisions in real time would be preferred. As such, techniques that are able to analyze a huge amount of data and learn from them, such as machine learning (ML) algorithms, can be a viable solution for future networks (Valente Klaine et al., 2017).

In this chapter, an overview of the importance and applications of ML algorithms in future wireless networks in the context of EE is presented. The chapter starts in Section 6.1 by giving a brief definition of self-organizing networks (SONs) and some related solutions involving ML algorithms in the context of EE. Then, an overview of ML techniques applied to more specific topics is presented in Sections 6.2 and 6.3, such as resource allocation, traffic prediction, and cognitive radio networks. Lastly, in Section 6.4, some future trends are presented and conclusions are drawn, highlighting the importance of ML in future networks. Summarizing, the objectives of this chapter can be described as follows:

- Overview ML approaches from the recent literature, with focus on the maximization of the EE of future wireless networks.
- Briefly review the most common ML techniques, as well as highlight application examples and goals of these techniques.
- Identify possible difficulties in current designs, as well as delineate future research directions in terms of ML for energy-efficient wireless networks.
- Present future trends for which ML can be a powerful tool toward energy efficient designs.
- List upcoming challenges raised by the use of ML techniques in order to improve EE.

6.1 Self-Organizing Wireless Networks

As defined in (Valente Klaine et al., 2017; Aliu et al., 2013), a SON is a network that can be adaptive, scalable, stable, agile, and autonomous in order to manage and control certain network objectives. As such, these networks are capable of autonomously making decisions in real time, as well as learn and improve their performance based on previous and historical data. Due to this automation, SONs can provide intelligence inside the network in order to facilitate the work of operators and reduce overall complexity, providing significant cost and energy savings for mobile operators.

SONs can be divided into three main branches (Valente Klaine et al., 2017; Aliu et al., 2013): self-configuration, self-optimization, and self-healing, together denoted as *self-x functions*. Figure 6.1 illustrates some applications of SONs in cellular networks, highlighting the three major self-x branches and the common associated use cases. In this context, ML techniques can certainly improve a SON, allowing the network to adapt by observing its current status, and use such experience to adjust parameters in future actions. In this sense, improving EE could be one of the major guidelines for parameter adjustment.

Another element that can increase network EE, as well as coverage and capacity, is adjustment of the tilt angle of the antennas. Depending on the distribution of the users within a cell, antenna tilt configuration can improve signal reception. Moreover, it can also reduce interference between neighbor BSs, improving the EE of the network (Dandanov et al., 2017). As presented in (Yilmaz et al., 2009), antenna tilting is possible both mechanically and electrically. In mechanical down-tilt (Figure 6.2a), the antenna's main lobe is lowered on one side and the back lobe is raised on the other side, because antenna elements are physically directed toward the ground. On the other hand, electrical down-tilt (Figure 6.2b) lowers both the main and back lobes uniformly by adjusting the phase of antenna elements; this is most commonly used for coverage and capacity optimization.

Once the optimization of the antenna tilt is highly dependent on the cell load, ML techniques become useful in this context. For instance, human mobility and activity patterns could be used to predict network demands, allowing the network to tilt the antennas intelligently and proactively, improving performance. One of the most promising techniques in this predictive context is reinforcement learning (RL), which allows constant adaptive optimization of the network, promoting autonomous actions based on the algorithm observations in order to improve network throughput or EE.

RL for electrical antenna tilting optimization has also been targeted by Dandanov et al. (2017). Unlike other research works that optimize antenna tilting in static network environments only,

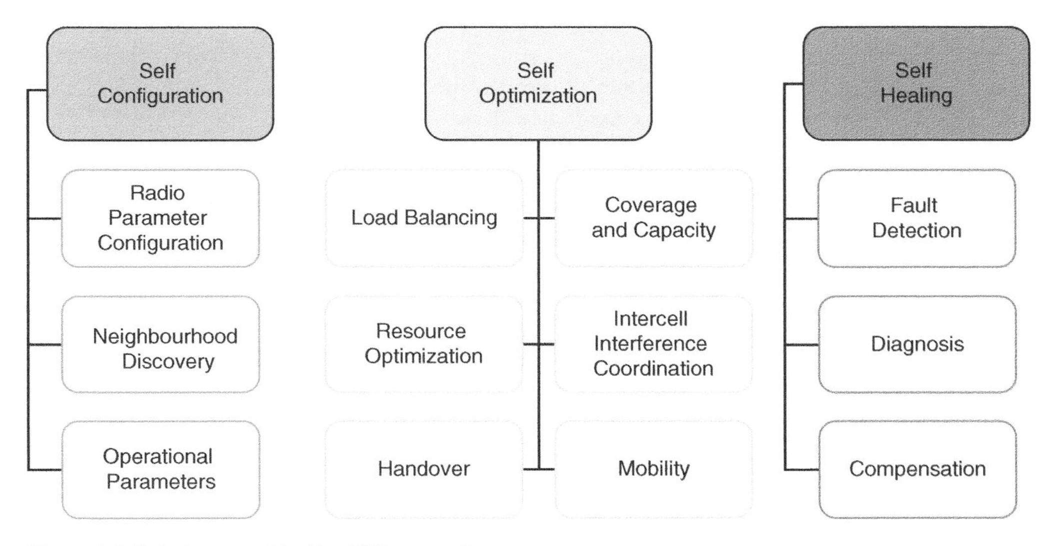

Figure 6.1 Solutions provided by SON categories.

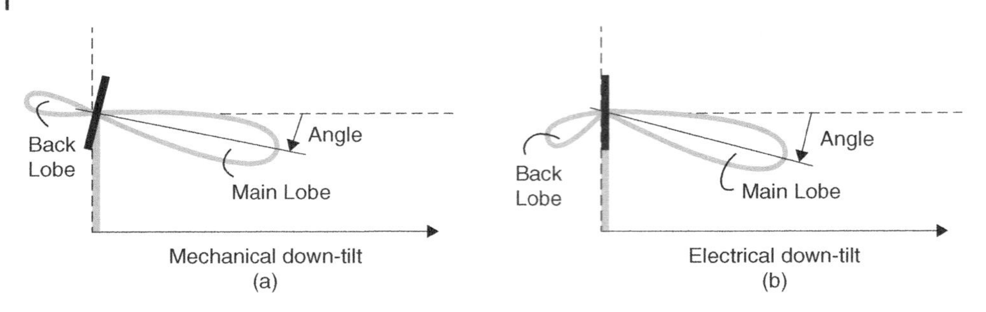

Figure 6.2 Antenna tilting techniques. (Yilmaz et al., 2009).

this work considers the mobile network to be a dynamic environment, adaptive to current user distributions. Simulation results show that the algorithm improves the overall data rate of the network, compared to a scenario without optimization of the antenna tilt.

In addition to antenna tilting, another area that has seen an application of RL algorithms is BS sleeping, in which the authors of (Miozzo et al., 2017) conceive a RL approach based on distributed multi-agent Q-learning to design the on/off switching policies of the network (which BSs to turn on or off). Moreover, the proposed solution is also distributed and has low complexity, allowing it to be performed online. Results show that the EE of the network is improved by up to 15% when compared to a greedy scheme. Finally, (Sinclair et al., 2013) presents another branch of ML useful in SON, this time in the area of unsupervised learning (UL). In this paper, a novel kernel self-organizing map algorithm is proposed in order to minimize unnecessary handovers and, as a consequence, increase network capacity and improve network EE. Simulations demonstrate that the proposed UL algorithm reduces unnecessary handovers by up to 70% in the network.

In (Buzzi et al., 2016), the authors depict economic, operational, and environmental aspects as the main concerns in the last decade of wireless communication systems, while EE emerges as a new prominent figure of merit for cellular networks. In this work, some useful methods to increase EE in 5G networks are classified: resource allocation, network planning and deployment, energy harvesting and transfer, and hardware solutions.

Some examples include a neural network (NN) combined with integer linear programming (ILP) in (Pelekanou et al., 2018) as a means to optimize the backhaul of a 5G communication scenario, aiming at minimizing the overall energy consumption of the network. The results indicate that using a NN can considerably reduce computational complexity when compared to ILP alone while providing similar performance. The efficiency of the proposed technique is also demonstrated in generic joint backhaul and fronthaul service provisioning, in which the overall energy consumption of the 5G infrastructure was minimized.

In Zheng et al. (2016), several ML techniques are used to exploit big data – i.e. data that are usually collected and stored by mobile operators for optimization purposes – are reviewed. As the authors conclude, despite the challenges in terms of data collection, communication overhead, and latency, combining ML tools with big data analysis is a promising solution to constantly optimize 5G network performance. In addition, Kiran et al. (2016) also exploit mobile network big data for radio resource allocation, by implementing a fuzzy controller to efficiently allocate bandwidth among users. As a result, the proposed method is shown to reduce processing latency, which allows lower complexity for distributed solutions. Common to the works in (Zheng et al., 2016; Kiran et al., 2016) is the notion that ML-based approaches are able to follow the rapidly changing conditions of the wireless environment, esspecially in dense scenarios representative of 5G communication systems, and are important tools to continuously improve network performance and operational efficiency.

Very recently, ML is employed in (Zappone et al., 2018a) to enable online power allocation, aiming at maximizing EE in wireless interference networks. As their results show, the model based on deep learning with NNs is able to approach the performance of the analytical method, while requiring lower computational complexity. In (D' Oro et al., 2018), the authors maximize EE given the allocation of transmit powers and subcarrier assignment. Once the optimization is a challenging nonconvex fractional problem, the authors employ a combination of fractional programming, ML using a stochastic learning approach, and game theory. As a result, the proposed ML solution is shown to perform similarly to other algorithms from the literature; but the method has a linear complexity in both the number of users and subcarriers, while other available solutions can only guarantee a polynomial complexity in the number of users and subcarriers. In summary, the proposals from (Zappone et al., 2018a; D' Oro et al., 2018) show that ML can provide the robust and near-optimal solutions required for future cellular networks.

Table 6.1 summarizes some usage examples of ML techniques in SON in the context of EE.

Table 6.1 Machine learning techniques for self-organizing networks.

Literature	Objective
(Dandanov et al., 2017)	Optimize antenna down-tilt using RL, considering the mobile network to be a dynamic environment. The algorithm improves the overall data rate of the network when compared to no antenna tilt optimization.
(Miozzo et al., 2017)	Design the on/off switching policies in the network using RL. The EE of the network is improved by up to 15% when compared to a greedy scheme.
(Sinclair et al., 2013)	Minimize unnecessary handovers in SON through UL. The proposal reduces unnecessary handovers by up to 70% and, as a consequence, increases both network capacity and EE.
(Zheng et al., 2016)	Overview ML techniques for network resource optimization. Despite the challenges in terms of data collection, communication overhead, and latency, combining ML tools with big data analysis is a promising solution to constantly optimize 5G mobile network performance.
(Kiran et al., 2016)	Exploit available big data for radio resource allocation. The proposed fuzzy controller method reduces processing latency, which allows lower complexity for distributed solutions.
(Pelekanou et al., 2018)	Combine NNs with typical ILP solutions in order to minimize overall energy consumption. The proposed approach considerably reduces the computational complexity when compared to ILP alone, while providing similar performance. The efficiency is demonstrated in generic joint backhaul and fronthaul service provisioning, minimizing the overall energy consumption of the 5G infrastructure.
(Zappone et al., 2018a)	Employ deep learning to enable online power allocation, aiming at maximizing EE. The proposed method approaches the performance of the analytical solution, with lower computational complexity.
(D' Oro et al., 2018)	Combine fractional programming, ML, and game theory to maximize EE in wireless interference networks. The proposal performs similarly to other algorithms from the literature, but with linear complexity in both the number of users and subcarriers, while other available solutions have polynomial complexity in the number of users and subcarriers.

6.2 Traffic Prediction and Machine Learning

The quality of the wireless channel for communications varies with time and frequency, so that predicting the amount of data traffic through the network is mandatory in order to maintain good quality of service. Moreover, learning the traffic profile can be a key concept to reduce energy consumption at the network side, since some units could be shut down during low traffic, enabling the network to redistribute resources whenever needed. In other words, the ability to predict data traffic at each BS can be helpful to overall planning of the wireless network, as well as to perform load balancing and providing opportunities to improve network performance. This section deals with the implementation of ML algorithms aimed at predicting network traffic, surveying recently published approaches.

The concept of network densification is seen as one key enabler of future mobile networks in order to address the expected exponential growth in traffic and number of devices (Valente Klaine et al., 2017). However, as a result of this densification process, an increase in the energy consumption of cellular networks is also expected. To tackle this problem, a usual strategy employed in wireless networks is known as *BS sleeping*, which uses temporal traffic variation information to design sleeping periods for BSs. In other words, BSs are switched off during certain time intervals, while traffic from the sleeping cell is accommodated in neighboring cells (Zhang et al., 2017).

Box-Jenkins and auto-regressive integrated moving average (ARIMA) are traditional approaches for data traffic forecasting. Both methods assume that time series are generated from linear processes. Therefore, these methods are not able to model a nonlinear system (Box and Jenkins, 1976), which severely limits their practical applicability, given that real scenarios are often nonlinear. Due to this limitation, substantial research has been performed in the application of more robust ML techniques to traffic load forecasting.

In Nehra et al. (2009), a NN-based energy efficient clustering and routing protocol for WSNs is proposed, whose goal is to maximize network lifetime by minimizing energy consumption. The results show that the proposed scheme has smaller energy consumption and a higher percentage of residual energy when compared to the power-efficient and adaptive clustering hierarchy (PEACH) protocol (Yi et al., 2007).

Another approach, in (Railean et al., 2010), predicts data traffic by associating stationary wavelet transform (SWT) – a powerful tool for processing data sequences at different frequency scales of resolutions – with NNs. Such integration significantly improves the data analysis and data prediction performance, and the obtained results show that the proposal can effectively build prediction models for time series. Similarly, (Zang et al., 2015) proposes a data traffic prediction algorithm for cellular networks based on the combination of a NN with wavelet preprocessing. The prediction of hourly traffic volumes is investigated with the goal of increasing network EE, with results showing that the proposed method outperforms other traditional approaches, such as linear prediction and compressive sensing methods.

Furthermore, predicting mobile traffic in cities is especially challenging, due to the temporal and spatial dynamism introduced by frequent user mobility. In such a context, the work described in (Alvizu et al., 2017) uses NNs to predict tidal traffic variations in a mobile metro-core network. The predicted traffic demand is used to minimize the energy consumption of the network at different hours during the day, to adapt resource occupation to the actual traffic volume. When compared to a common approach in mobile metro-core networks – the virtual wavelength path (VWP) static algorithm – the NN-based proposal results in energy savings up to 31%. A summary of the presented use cases of ML for traffic prediction and their respective characteristics is presented in Table 6.2.

Table 6.2 Applications of ML techniques for traffic prediction.

Literature	Objective
(Nehra et al., 2009)	Design an energy efficient clustering and routing protocol for WSNs. The proposed ANN-based scheme has smaller energy consumption and a higher percentage of residual energy when compared to the PEACH protocol.
(Railean et al., 2010)	Predict data traffic, combining SWT and NNs tools. Data analysis and data prediction performance have been significantly improved, while the proposed prediction models have been used to optimize network resources and reduce energy consumption.
(Zang et al., 2015)	Prodict hourly data traffic for cellular networks, aiming at increasing EE. The proposed combination of NNs with wavelet preprocessing outperforms traditional approaches, such as linear prediction and compressive sensing methods.
(Alvizu et al., 2017)	Predict traffic variations in a mobile metro-core network, in order to reduce energy consumption. The proposed NN solution results in energy savings up to 31% compared to the VWP static algorithm.

6.3 Cognitive Radio and Machine Learning

Spectrum sensing is of great importance in cognitive radio systems, as it is used to predict the availability of a given communication channel and allow secondary (unlicensed) users (SUs) to access frequency bands when primary (licensed) users (PUs) are not communicating. Therefore, spectrum sensing enables the negotiation of network resources, mitigating the under-utilization of the spectrum (Shokri-Ghadikolaei et al., 2012). However, such sensing capability can consume a considerable amount of energy, as the activity of PUs can be highly dynamic in many scenarios. Therefore, EE techniques become highly necessary (Li, 2010). Based on that, this section overviews EE resource allocation techniques based on ML for cognitive radio systems.

ML offers an attractive choice to manage resource allocation in wireless communication systems, since one of the main drawbacks of traditional power allocation algorithms is the need for instantaneous channel state information (CSI), which can be impractical sometimes. Meanwhile, the use of NNs, for example, has the advantage of performing optimization without prior knowledge or assumptions about the environment, using only previously collected data. For instance, a crucial trade-off in spectrum-sensing scenarios is related to the transmission power of the SUs. One the one hand, the higher the transmission power, the better the performance a SU can attain. On the other hand, the interference of both PUs and neighbor SUs also increases (Chen et al., 2013; Ghasemi and Sousa, 2008).

In such a context, the work in (Chen et al., 2013) adopts a RL approach in order to allocate power in a cognitive radio network. The reward function is designed to minimize energy consumption, and as a result, the algorithm is able to improve the network EE. Also in the realm of RL, in (Al-Rawi et al., 2014), the authors attempt to improve routing efficiency and minimize interference in cognitive radio networks. While most spectrum-sensing schemes focus on choosing the best route for SUs, this work focuses on improving the PU network by choosing the route with the minimal interference between PUs and SUs. The proposed method was able to improve EE by minimizing the amount of interference. In addition, a NN solution also aiming at minimizing interference was proposed by (Tumuluru et al., 2010), improving spectrum utilization more than 60%, with a percentage of reduction in sensing energy of up to 51% compared to the case when no spectrum sensing is allowed.

Table 6.3 Applications of ML techniques in cognitive radio systems.

Literature	Objective
(Chen et al., 2013)	Use RL to allocate power for SUs in cognitive radio networks. As a result, overall EE is improved.
(Al-Rawi et al., 2014)	Apply a RL method to improve routing efficiency and minimize interference in cognitive radio networks. The proposed method improves routing efficiency and minimizes interference.
(Tumuluru et al., 2010)	Minimize interference and maximize transmission opportunities using a NN solution. The spectrum utilization improves in more than 60% and the percentage of reduction in sensing energy is up to 51% compared to the case when no spectrum sensing is allowed.
(Xu and Nallanathan, 2016)	Employ a SVM solution to maximize EE, constrained to a maximal amount of interference. This method achieves a trade-off between EE and a satisfaction index of the users.
(Agarwal et al., 2016)	Apply and compare NN and SVM techniques in order to predict PU activity. Results highlight the applicability of ML techniques for enabling dynamic spectrum access.

Finally, energy-efficient resource allocation approaches based on a support vector machine (SVM) technique have been addressed by (Xu and Nallanathan, 2016; Agarwal et al., 2016). In (Xu and Nallanathan, 2016), the authors maximize the EE constrained by a maximum amount of interference and by the total available power, while their results show that this method achieves a trade-off between EE and a satisfaction index of the users. On the other hand, (Agarwal et al., 2016) applies NN and SVM techniques in order to predict PU activity in cognitive radio networks, highlighting the applicability of ML techniques for enabling dynamic spectrum access. The presented use cases of ML techniques in resource allocation for radio systems and their respective features are given in Table 6.3.

6.4 Future Trends and Challenges

This section discusses some trends and challenges in the application of ML techniques for EE improvement of future wireless networks.

6.4.1 Deep Learning

Deep learning algorithms are an important branch of ML tools (Goodfellow et al., 2016). Deep learning differs from other ML techniques in the sense that the employed NNs are very dense, with many layers of neurons between the input and output layers. Moreover, combined with such complex computational structure, a massive amount of data can be processed in order to train the NN. As a consequence, the interest in such techniques for wireless communication networks has recently grown, given that dense 5G scenarios may have a lot of available data, which is a rich training source for deep learning algorithms, allowing the development of powerful supervised methods (Zappone et al., 2018b).

For instance, (Wang et al., 2017) was one of the first to employ a deep NN-based model for cellular traffic prediction at urban scale with massive real-world datasets. Experiment results show that the proposed method outperforms traditional approaches, such as ARIMA, long short-term memory (LSTM), aggregated cellular traffic (GNN-A), and naive forecasting model, in terms of prediction performance, while spatial dependency and the interaction of factors play an important role in the accuracy and robustness of traffic prediction. This demonstrates that in the future, along with the explosive growth of mobile Internet and constantly evolving traffic patterns, new challenges and opportunities will emerge. Zappone et al. (2018b) also proposed a deep learning approach in order to maximize the EE of wireless networks. Results compared the proposal with an analytical optimal solution, indicating near-optimal achievements using the deep learning approach – close to 95% of the optimal value in terms of EE – but with much lower computational complexity.

Furthermore, another interesting research direction involving deep learning has been presented by (O'Shea and Hoydis, 2017; O'Shea et al., 2017). The idea is to use ML tools to redefine the way in which physical layer functions are designed. For instance, in a simple example of a communication system, the transmitter converts the desired information into a data packet, which is corrupted by the channel, while the receiver tries to recover the original information as well as possible. Traditionally, the literature has been concerned with mathematically designing each block belonging to such a communication system. On the other hand, (O'Shea and Hoydis, 2017; O'Shea et al., 2017) use the concept of an *autoencoder*, which is a NN trained to reconstruct the input at the output. For example, the schemes proposed by (O'Shea and Hoydis, 2017; O'Shea et al., 2017) have shown to have a lot of promise in delivering lower-bit error rates and better robustness to wireless channel impairments when compared with traditional schemes, such as space-time block codes (STBCs) and singular value decomposition (SVD) based precoding.

6.4.2 Positioning of Unmanned Aerial Vehicles

Given their adaptability and flexibility, drone small cells (DSCs) are considered a good alternative in order to enable rapid, efficient deployment of an communication network. Typical scenarios include major events, which may congest the deployed network, or disaster situations, where the network infrastructure is compromised. In these situations, the deployment of DSCs through unmanned aerial vehicles (UAVs) is a source of rapid implementation and has a great possibility of reconfiguration compared to terrestrial communications (Zeng et al., 2016). Nevertheless, as also highlighted by (Zeng et al., 2016), although energy efficient communication setups have been extensively studied for terrestrial communications, its systematic investigation for UAV communication systems is still underdeveloped.

In such a context, optimization of UAV placement is of great interest (Alzenad et al., 2017). For instance, (Klaine et al., 2018) recently employed a RL technique in order to optimize the position of UAVs in an emergency scenario. The main goal of the proposed solution is to maximize the number of users covered by the emergency communication network. When compared to different positioning strategies from the literature, the RL solution outperforms the other schemes, decreasing the number of users in outage, which also occurs in a smaller number of episodes. However, EE is still not investigated and is thus a possible future work scenario.

6.4.3 Learn-to-Optimize Approaches

Another research trend in the deep learning community is an approach known as *learn-to-optimize* (Zappone et al., 2018b), which exploits prior information about the problem to solve. In the context of wireless communications, this represents a great research opportunity, since theoretical models are often available, despite their possible simplifications. Therefore, all such available frameworks provide much deeper prior information compared to other fields of science in which ML has been successfully employed.

Such a concept couples with the vision of *machine reasoning* recently presented in (Ericsson, 2018). As stated by the authors, machine reasoning must implement abstract thinking as a computational system, as well as employ logical techniques, such as deduction and induction, to generate conclusions. Therefore, one of the main challenges here is to integrate ML and machine reasoning in an effective way.

6.4.4 Some Challenges

Data collection may involve a big challenge in some cases for ML algorithms. For instance, sleeping-cell-detection scenarios usually lack the necessary data to attest the quality of the network, so that a cell can remain undetected from the point of view of the network while appearing to be fully operational from the user perspective (Fortes et al., 2016; Zoha et al., 2014). Consequently, when data are not available or not reliable, supervised learning techniques cannot be applied, which becomes a challenge for any optimization goal.

In addition, the appropriate training method can be a challenge in wireless network scenarios. As discussed by (Valente Klaine et al., 2017), ML algorithms can be trained either offline or online. Offline training can be applied when a fast response time is not required; hence, it is adequate for algorithms that have a low response time. On the other hand, online training is mostly performed when it is necessary to dynamically adapt to new patterns in the data set: for example, when functions are heavily dependent on time. Deploying algorithms that rely on online training, or that require long training periods with highly time-dependent functions, for example, will result in the inability to generate accurate predictions (Valente Klaine et al., 2017).

In summary, as depicted by (Ericsson, 2018), the acceptance of ML as a viable approach for automation of complex systems such as wireless networks still faces a few important technological challenges. In particular, real-time intelligent decision-making is needed, as well as intelligence for distributed and decentralized systems. Finally, data availability and training are crucial in order to meet the strict demands of future wireless networks.

6.5 Conclusions

In this chapter, ML has been discussed as a means to improve EE of wireless networks. In this sense, we have reviewed the most common ML approaches with focus on the maximization of EE, some application examples, as well as the goals of these techniques.

This work also highlights a few difficulties in current designs in terms of ML for energy efficient wireless networks. Thus, deep learning, machine reasoning, and networks based on UAVs are listed as some of the future trends in the area. In addition, some challenges raised by the use of ML techniques are delineated, particularly related to data collection and network training.

Finally, based on the insights discussed here, we conclude that ML approaches already play a key role for improving EE, optimizing many aspects of network deployments, and allocating resources in an intelligent way where closed-form analytical models are too complex or lack the required flexibility to be employed.

Bibliography

Anirudh Agarwal, Shivangi Dubey, Mohd Asif Khan, Ranjan Gangopadhyay, and Soumitra Debnath. Learning based primary user activity prediction in cognitive radio networks for efficient dynamic spectrum access. In *International Conference on Signal Processing and Communications (SPCOM)*, pages 1–5, 2016.

Hasan A.A. Al-Rawi, Kok-Lim Alvin Yau, Hafizal Mohamad, Nordin Ramli, and Wahidah Hashim. Effects of network characteristics on learning mechanism for routing in cognitive radio ad hoc networks. In *International Symposium on Communication Systems, Networks & Digital Signal Processing (CSNDSP)*, pages 748–753, 2014.

Osianoh Glenn Aliu, Ali Imran, Muhammad Ali Imran, and Barry Evans. A survey of self organisation in future cellular networks. *IEEE Communications Surveys & Tutorials*, 15(1):336–361, 2013.

Tarek AlSkaif, Manel Guerrero Zapata, and Boris Bellalta. Game theory for energy efficiency in wireless sensor networks: Latest trends. *Journal of Network and Computer Applications*, 54:33–61, 2015.

Rodolfo Alvizu, Sebastian Troia, Guido Maier, and Achille Pattavina. Matheuristic with machine-learning-based prediction for software-defined mobile metro-core networks. *Journal of Optical Communications and Networking*, 9(9):D19–D30, 2017.

Mohamed Alzenad, Amr El-Keyi, Faraj Lagum, and Halim Yanikomeroglu. 3-D placement of an unmanned aerial vehicle base station (UAV-BS) for energy-efficient maximal coverage. *IEEE Wireless Communications Letters*, 6 (4):434–437, 2017.

G. Auer, V. Giannini, C. Desset, I. Godor, P. Skillermark, M. Olsson, M.A. Imran, D. Sabella, M.J. Gonzalez, O. Blume, and A. Fehske. How much energy is needed to run a wireless network? *IEEE Wireless Communications*, 18(5):40–49, Oct. 2011. ISSN 1536-1284.

George E.P. Box and Gwilym M. Jenkins. *Time Series Analysis: Forecasting and Control*. San Francisco: Holden-Day, 1976.

S. Buzzi, C.I., T.E. Klein, H.V. Poor, C. Yang, and A. Zappone. A survey of energy-efficient techniques for 5G networks and challenges ahead. *IEEE Journal on Selected Areas in Communications*, 34(4):697–709, April 2016. ISSN 0733-8716.

Xianfu Chen, Zhifeng Zhao, and Honggang Zhang. Stochastic power adaptation with multiagent reinforcement learning for cognitive wireless mesh networks. *IEEE Transactions on Mobile Computing*, 12(11):2155–2166, 2013.

Cisco. Cisco visual networking index: Forecast and methodology, 2016–2021. *White paper*, 2017.

Shuguang Cui, Andrea J. Goldsmith, and Ahmad Bahai. Energy-efficiency of MIMO and cooperative MIMO techniques in sensor networks. *IEEE Journal on Selected Areas in Communications*, 22(6):1089–1098, 2004.

Shuguang Cui, Andrea J. Goldsmith, and Ahmad Bahai. Energy-constrained modulation optimization. *IEEE Transactions on Wireless Communications*, 4 (5):2349–2360, 2005.

Salvatore D' Oro, Alessio Zappone, Sergio Palazzo, and Marco Lops. A learning approach for low-complexity optimization of energy efficiency in multicarrier wireless networks. *IEEE Transactions on Wireless Communications*, 17(5): 3226–3241, 2018.

Nikolay Dandanov, Hussein Al-Shatri, Anja Klein, and Vladimir Poulkov. Dynamic self-optimization of the antenna tilt for best trade-off between coverage and capacity in mobile networks. *Wireless Personal Communications*, 92(1):251–278, 2017.

Ericsson. Artificial intelligence and machine learning in next-generation systems. *White paper*, 2018.

Albrecht J. Fehske, Fred Richter, and Gerhard P. Fettweis. Energy efficiency improvements through micro sites in cellular mobile radio networks. In *IEEE GLOBECOM Workshops*, pages 1–5, 2009.

Daquan Feng, Chenzi Jiang, Gubong Lim, Leonard J. Cimini, Gang Feng, and Geoffrey Ye Li. A survey of energy-efficient wireless communications. *IEEE Communications Surveys & Tutorials*, 15(1):167–178, 2013.

Sergio Fortes, Raquel Barco, and Alejandro Aguilar-Garcia. Location-based distributed sleeping cell detection and root cause analysis for 5G ultra-dense networks. *EURASIP Journal on Wireless Communications and Networking*, 2016(1):149, 2016.

Pimmy Gandotra, Rakesh Kumar Jha, and Sanjeev Jain. Green communication in next generation cellular networks: a survey. *IEEE Access*, 5:11727–11758, 2017.

Amir Ghasemi and Elvino S. Sousa. Spectrum sensing in cognitive radio networks: requirements, challenges and design trade-offs. *IEEE Communications magazine*, 46(4), 2008.

Ian Goodfellow, Yoshua Bengio, Aaron Courville, and Yoshua Bengio. *Deep Learning*, volume 1. MIT Press, 2016.

Huawei. 5G: A technology vision. *White paper*, 2013.

P. Kiran, M.G. Jibukumar, and C.V. Premkumar. Resource allocation optimization in LTE-a/5G networks using big data analytics. In *International Conference on Information Networking (ICOIN)*, pages 254–259, 2016.

Paulo V. Klaine, João P.B. Nadas, Richard D. Souza, and Muhammad A. Imran. Distributed drone base station positioning for emergency cellular networks using reinforcement learning. *Cognitive Computation*, pages 1–15, 2018.

Husheng Li. Multi-agent Q-learning for competitive spectrum access in cognitive radio systems. In *IEEE Workshop on Networking Technologies for Software Defined Radio (SDR) Networks*, pages 1–6, 2010.

Marco Miozzo, Lorenza Giupponi, Michele Rossi, and Paolo Dini. Switch-on/off policies for energy harvesting small cells through distributed Q-learning. In *IEEE Wireless Communications and Networking Conference Workshops (WCNCW)*, pages 1–6, 2017.

Neeraj Kumar Nehra, Manoj Kumar, and R.B. Patel. Neural network based energy efficient clustering and routing in wireless sensor networks. In *International Conference on Networks and Communications (NETCOM'09)*, pages 34–39, 2009.

Timothy O'Shea and Jakob Hoydis. An introduction to deep learning for the physical layer. *IEEE Transactions on Cognitive Communications and Networking*, 3(4):563–575, 2017.

Timothy J. O'Shea, Tugba Erpek, and T. Charles Clancy. Physical layer deep learning of encodings for the MIMO fading channel. In *55th Annual Allerton Conference on Communication, Control, and Computing (Allerton)*, pages 76–80, 2017.

Antonia Pelekanou, Markos Anastasopoulos, Anna Tzanakaki, and Dimitra Simeonidou. Provisioning of 5G services employing machine learning techniques. In *International Conference on Optical Network Design and Modeling (ONDM)*, pages 200–205, 2018.

Zaqueu Cabral Pereira, Thiago Henrique Ton, João Luiz Rebelatto, Richard Demo Souza, and Bartolomeu F. Uchôa-Filho. Generalized network-coded cooperation in OFDMA communications. *IEEE Access*, 6:6550–6559, 2018.

Ion Railean, Cristina Stolojescu, Sorin Moga, and Philippe Lenca. Wimax traffic forecasting based on neural networks in wavelet domain. In *Fourth International Conference on Research Challenges in Information Science (RCIS)*, pages 443–452, 2010.

Fred Richter, Albrecht J. Fehske, and Gerhard P. Fettweis. Energy efficiency aspects of base station deployment strategies for cellular networks. In *IEEE Vehicular Technology Conference Fall (VTC 2009-Fall)*, pages 1–5, 2009.

Hossein Shokri-Ghadikolaei, Younes Abdi, and Masoumeh Nasiri-Kenari. Learning-based spectrum sensing time optimization in cognitive radio systems. In *Sixth International Symposium on Telecommunications (IST)*, pages 249–254, 2012.

Neil Sinclair, David Harle, Ian A. Glover, James Irvine, and Robert C. Atkinson. An advanced som algorithm applied to handover management within LTE. *IEEE Transactions on Vehicular Technology*, 62(5):1883–1894, 2013.

H. Tullberg, P. Popovski, Z. Li, M.A. Uusitalo, A. Hoglund, O. Bulakci, M. Fallgren, and J.F. Monserrat. The METIS 5G system concept: Meeting the 5G requirements. *IEEE Communications Magazine*, 54(12):132–139, Dec. 2016. ISSN 0163-6804.

Vamsi Krishna Tumuluru, Ping Wang, and Dusit Niyato. A neural network based spectrum prediction scheme for cognitive radio. In *IEEE International Conference on Communications (ICC)*, pages 1–5, 2010.

Paulo Valente Klaine, Muhammad Ali Imran, Oluwakayode Onireti, and Richard Demo Souza. A survey of machine learning techniques applied to self organizing cellular networks. *IEEE Communications Surveys and Tutorials*, 2017.

Xu Wang, Zimu Zhou, Zheng Yang, Yunhao Liu, and Chunyi Peng. Spatio-temporal analysis and prediction of cellular traffic in metropolis. In *IEEE 25th International Conference on Network Protocols (ICNP)*, pages 1–10, 2017.

Jie Xu and Ling Qiu. Energy efficiency optimization for MIMO broadcast channels. *IEEE Transactions on Wireless Communications*, 12(2):690–701, 2013.

Lei Xu and Arumugam Nallanathan. Energy-efficient chance-constrained resource allocation for multicast cognitive OFDM network. *IEEE Journal on Selected Areas in Communications*, 34(5):1298–1306, 2016.

Sangho Yi, Junyoung Heo, Yookun Cho, and Jiman Hong. Peach: Power-efficient and adaptive clustering hierarchy protocol for wireless sensor networks. *Computer communications*, 30(14-15):2842–2852, 2007.

Osman N.C. Yilmaz, Seppo Hamalainen, and Jyri Hamalainen. Comparison of remote electrical and mechanical antenna downtilt performance for 3GPP LTE. In *IEEE Vehicular Technology Conference Fall (VTC 2009-Fall)*, pages 1–5, 2009.

Yunjuan Zang, Feixiang Ni, Zhiyong Feng, Shuguang Cui, and Zhi Ding. Wavelet transform processing for cellular traffic prediction in machine learning networks. In *IEEE China Summit and International Conference on Signal and Information Processing (ChinaSIP)*, pages 458–462, 2015.

Alessio Zappone, Eduard Jorswieck, et al. Energy efficiency in wireless networks via fractional programming theory. *Foundations and Trends in Communications and Information Theory*, 11(3-4):185–396, 2015.

Alessio Zappone, Mérouane Debbah, and Zwi Altman. Online energy-efficient power control in wireless networks by deep neural networks. In *IEEE 19th International Workshop on Signal Processing Advances in Wireless Communications (SPAWC)*, pages 1–5, 2018a.

Alessio Zappone, Marco Di Renzo, Mérouane Debbah, Thanh Tu Lam, and Xuewen Qian. Model-aided wireless artificial intelligence: Embedding expert knowledge in deep neural networks towards wireless systems optimization. *arXiv preprint arXiv:1808.01672*, 2018b.

Yong Zeng, Rui Zhang, and Teng Joon Lim. Wireless communications with unmanned aerial vehicles: opportunities and challenges. *IEEE Communications Magazine*, 54(5):36–42, May 2016.

Zhen Zhang, Fangfang Liu, and Zhimin Zeng. A traffic prediction algorithm based on Bayesian spatio-temporal model in cellular network. In *International Symposium on Wireless Communication Systems (ISWCS)*, pages 43–48, 2017.

Kan Zheng, Zhe Yang, Kuan Zhang, Periklis Chatzimisios, Kan Yang, and Wei Xiang. Big data-driven optimization for mobile networks toward 5G. *IEEE network*, 30(1):44–51, 2016.

Ahmed Zoha, Arsalan Saeed, Ali Imran, Muhammad Ali Imran, and Adnan Abu-Dayya. A SON solution for sleeping cell detection using low-dimensional embedding of MDT measurements. In *IEEE 25th Annual International Symposium on Personal, Indoor, and Mobile Radio Communication (PIMRC)*, pages 1626–1630, 2014.

7

Deep Learning Based Traffic and Mobility Prediction

Honggang Zhang, Yuxiu Hua, Chujie Wang, Rongpeng Li, and Zhifeng Zhao

College of Information Science & Electronic Engineering, Zhejiang University, Hangzhou, China

7.1 Introduction

With the proliferation of mobile terminals as well as the astonishing expansion of mobile Internet, the Internet of Things (IoT), and cloud computing, mobile communication networks have become an indispensable social infrastructure, which is bound up with people's lives and various areas of society. Cisco's latest statistics show that mobile data traffic has grown 18-fold over the past 5 years, and it will increase seven-fold between 2016 and 2021 Cisco (2017). In addition, statistics show that mobile traffic has evolved from voice to multimedia, with video traffic accounting for three-quarters of total mobile data traffic worldwide Cisco (2017). In sharp contrast, the theory and methods of performance analysis of mobile communication networks, as well as the corresponding prediction model research, lag behind the rapid growth of mobile services and users. Therefore, it is of crucial importance in terms of efficiency and optimization to acquire the hidden patterns from historical traffic data and predict network traffic.

Generally, most of the decisions that network operators make depend on how the traffic flows in their network. However, although it is very important to accurately estimate traffic parameters, current routers and network devices do not provide the possibility for real-time monitoring; hence, network operators cannot react effectively to traffic changes. To cope with this problem, prediction techniques have been applied to predict network parameters so as to be able to react to network changes in near real time Azzouni and Pujolle (2017). In fact, traffic learning and prediction is acting as a crucial anchor for the design of mobile communication network architecture and embedded algorithms. Moreover, fine traffic prediction on a daily, hourly, or even minutely basis could contribute to the optimization and management of cellular networks like energy savings Li et al. (2014), opportunistic scheduling Li et al. (2014), and network anomaly detection Romirer-Maierhofer et al. (2015). In other words, by contributing to the improvement of network energy efficiency by dynamically configuring network resources according to the practical traffic demand, a precisely predicted future traffic load knowledge can play an important role in designing greener, traffic-aware mobile communication networks Li et al. (2017).

On the other hand, the fifth-generation (5G) mobile communication system is expected to offer a 1000-fold capacity increase compared with the current fourth-generation (4G) deployments, aiming at providing higher data rates and lower end-to-end delay while supporting high-mobility users Andrews et al. (2014). To this end, ultra-dense network deployment has been proposed as a key technology for achieving the capacity goal. The deployment of

Machine Learning for Future Wireless Communications, First Edition. Edited by Fa-Long Luo.
© 2020 John Wiley & Sons Ltd. Published 2020 by John Wiley & Sons Ltd.

a significant number of small cells contributes to boosting the throughput for static users. However, it also leads to several challenging issues for moving users. First, as network density increases, mobile users will inevitably cross more cells, resulting in more frequent handovers. Traditional handovers adopt a passive trigger-based strategy, making the mobile network with no prior preparation. This posterior handover usually incurs negative impacts on both the user side and the network side. On the user hand, since a handover involves intensive signaling interactions between user equipment (UE), serving base station (BS), target BS, and core networks, the UE under a handover usually experiences large delay and obvious throughput reduction Wu and Fan (2016). On the network side, one target BS may reject the mobile UE's access request in its busy period Ulvan et al. (2013). Therefore, frequent handovers caused by user mobility will decrease the quality of service (QoS) and make the network densification in vain.

The other problem is ascribed to load imbalance. The uneven distribution of users and burstiness of services will cause large distinctions in loads of different cells. Moreover, due to user mobility, the continuously varying cell load makes the load-imbalance situation more complicated. For example, unexpectedly high resource utilization in some cells gives rise to a disappointing call-block probability and correspondingly decreases user satisfaction. All of this suggests that simply adding more cells to cellular networks certainly increases some capacity in some areas, but also complicates network management given the current passive event-trigger strategy. Instead, it sounds like a more promising approach is to learn the patterns of human mobility and predict the future location of UEs, so as to proactively reserve some network resources (caching, computing, etc.) and fully reap the gains of network densification.

In this section, we have summarized the background and motivation for why we need to make every effort to predict network traffic and user mobility. On the one hand, the explosive growth of network traffic in the future is bound to greatly increase the difficulty of network resource optimization, whereas learning and predicting traffic, as the cornerstone of designing mobile communication network architecture and embedded algorithms, will contribute to solving this thorny problem. On the other hand, ultra-dense network deployment has been proposed as a key technology for achieving the capacity goal of 5G mobile communication systems, which paradoxically poses a major challenge to guaranteeing the QoS of mobile users; yet learning and predicting user mobility becomes a promising way to alleviate the contradiction.

To address all the technical aspects related to traffic and mobility predictions with the emphasis on the machine learning (ML) based solutions, the rest of this chapter is organized as follows. Sections 7.2 and 7.3 present the problem formation and a brief overview of existing prediction methods in terms of modelling, characterization, complexity, and performance. Section 7.4 presents some deep learning (DL) based schemes for traffic and mobility prediction. We first introduce random connectivity long short-term memory (RCLSTM) – a model that reduces computational cost by randomly removing some neural connections – and its performance in traffic prediction. Then we show three LSTM-based user-mobility prediction schemes, two of which take into account the spatial dependence on the user's movement trajectory, thus combining convolutional neural networks (CNNs) to expect to achieve better prediction performance. Section 7.5 offers further discussions and conclusions.

7.2 Related Work

7.2.1 Traffic Prediction

In the 1970s, researchers used the Poisson model to describe the change of network traffic Sitharaman (2005). Due to the small number of early network applications and limited data transmission, the Poisson model can characterize the network traffic well. However, with the diversification of application types and the continuous expansion of user scale, a new variation of network traffic occurs.

In 1994, Leland et al. (1994) proposed that data traffic is very different from traditional telephone traffic, and mathematically proves that the data traffic arrival decay rate is significantly slower than the Poisson model estimates. Hence, if the network is designed according to the traditional Poisson model, it will cause many serious issues, such as buffer overflow, packet loss, etc. This research reveals that the Poisson model is no longer suitable for the description of network traffic characteristics. Therefore, the Markov model Maheshwari et al. (2013), and autoregressive moving average (ARMA) model Periyanayagi and Sumathy (2014) began to be introduced into the research on network traffic characteristics.

The non-after-effect property of the Markov model indicates that for a system, its future state is only related to the current state and has nothing to do with the past state. Hence, the Markov model can only describe the short-term variation rule of network traffic, and cannot capture the long-term dependencies of network traffic. The ARMA model is linear and has pretty good prediction accuracy, but it can only make short-term predictions with stationary network traffic sequences. For non-stationary sequences, the researchers then proposed the autoregressive integrated moving average (ARIMA) model Raghuveera et al. (2011). Compared with ARMA, the ARIMA model can achieve short-term prediction of non-stationary sequences by smoothing the non-stationary sequences. However, one limitation of the ARIMA model is its natural tendency to concentrate on the mean values of the past series data. Therefore, it remains challenging to capture a rapidly changing process Hong (2012).

These models are all linear; and nonlinear models, such as support vector regression (SVR) WU and WANG (2013) and artificial neural networks (ANNs) Azzouni and Pujolle (2017), are becoming more and more popular for network traffic prediction. The success of SVR lies in four factors: good generation, global optimization solutions, the ability to handle nonlinear problems, and the sparseness of solutions. However, SVR is limited by the lack of structured means to determine some key parameters that are critical to the model's performance, thus incurring a deficiency of knowledge about how to select the key parameters Hong (2012). In recent years, ANN-based models have attracted the attention of many researchers because of their common advantages such as self-adaptation, self-organization, and self-learning, which are not available in the traditional time series prediction model. Nikravesh et al. (2016) investigated the performance of multi-layer perceptron (MLP) (a typical architecture of ANNs) for predicting future behavior of mobile network traffic and their results show MLP has better accuracy than SVR. Wang et al. (2017) used local stacked autoencoders (LSAEs) and global stacked autoencodesr (GSAEs) to extract local traffic characteristics and then utilized LSTM to predict the traffic of cellular networks. However, a large amount of data are required in the training phase of ANNs, and it is almost a consensus that ANNs are difficult to converge and are prone to local minimums. Although these neural network methods can obtain good prediction results, the contradiction between prediction ability and training ability restricts the development of ANN-based models in the field of network traffic prediction Goodfellow et al. (2016).

7.2.2 Mobility Prediction

There has been a substantial body of research toward analyzing and mining the fundamental statistical properties of human mobility. The work in Rhee et al. (2011) studying global positioning system (GPS) traces 44 volunteers in various scenarios and discovers that human walk patterns closely follow Levy walk patterns, with flights and pause-time obeying the heavy-tailed distribution and the moving direction being isotropic. Zhao et al. (2015) verify that log-normal flight distribution can approximate a single movement pattern, while a power-law distribution can approximate human mobility modeled as a mixture of different transportation modes. The authors in Wang et al. (2015) fit the mobility patterns of humans in several cities and find the displacement distributions tend to follow exponential laws. However, based on taxi GPS data and subway transit data, Xia et al. (2018) verify that trip displacement is better fitted to log-normal distribution through an integrated analysis method.

Although some mobility properties have been discovered by data fitting, these results greatly depend on the chosen dataset, leading to inconsistent conclusions in different works. Furthermore, it is even more critical to know the user's specific movement trajectory or the changing trend of the number of mobile users in different cells. However, such information cannot be directly acquired from these statistically significant properties. Fortunately, human mobility has been proved to be predictable to some extent. Song et al. (2010) quantitatively analyze the regularity of the moving trajectory using information theory. They find that user mobility contains 93% potential predictability by calculating the conditional entropy of the position sequence of one user's motion history, which illustrates the feasibility of learning users' mobility patterns through their historical trajectory information. To predict users' future locations, conventional ML techniques like K-nearest neighbors (KNN), linear regression, decision tree, and association rules are obvious candidates to be applied Liao et al. (Maui, USA, Apr., 2012) Tkačík and Kordík (Vancouver, Canada, Jul., 2016). Moreover, it is valuable for us to reference that some DL methods represented by recurrent neural networks (RNNs) have been utilized in time-series problems including predicting traffic Zhang and Patras (LosAngeles, USA, Jun., 2018) and pedestrian trajectories Alahi et al. (LasVegas, USA, Jun.-Jul., 2016). However, the work in Alahi et al. (LasVegas, USA, Jun.-Jul., 2016) can only predict human trajectories through static images in a specific small-range scene such as a hotel or an intersection, which is not the case for the required cellular network scenario.

7.3 Mathematical Background

In this section, we talk about the fundamental mathematical tools that are heavily used in the rest of this chapter. In particular, we give an overview of ANNs and LSTM.

ANNs are constructed as a class of ML models that can eliminate the drawbacks of traditional learning algorithms with rule-based programming LeCun et al. (2015). ANNs can be classified into two main categories: feed-forward neural networks (FFNNs) and RNNs. FFNNs usually consist of an input layer, an output layer, and hidden layers (if necessary). Each layer is composed of a number of neurons and an activation function. A simple diagram of FFNNs is illustrated in Figure 7.1a. In FFNNs, there is no connection between the neurons within the same layer, and all neurons cannot be connected across layers, which means the information flows in one direction from the input layer, through the hidden layers (if any), to the output layer. FFNNs are widely used in various fields like data classification, object recognition, and image processing. However, constrained by their internal structure, FFNNs are unsuitable for handling historical dependencies.

RNNs, as another type of ANNs, are similar to FFNNs in the structure of neural layers, but allow the connections between the neurons within the same hidden layer. An illustration of an RNN is shown on the left side of Figure 7.1b. In addition, the right side of Figure 7.1b is the expanded form of the RNN model, indicating that RNNs calculate the output of the current moment from the input of the current moment $\mathbf{x}_t$ and the hidden state of the previous moment $\mathbf{h}_{t-1}$. Therefore, RNNs allow historical input information to be stored in the network's internal state, and are thereby capable of mapping all of the historical input data to the final

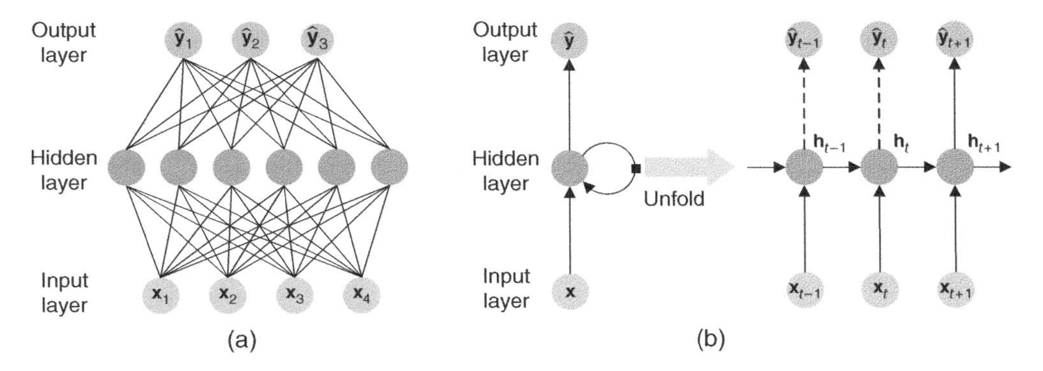

Figure 7.1 An illustration of a FFNN and a RNN.

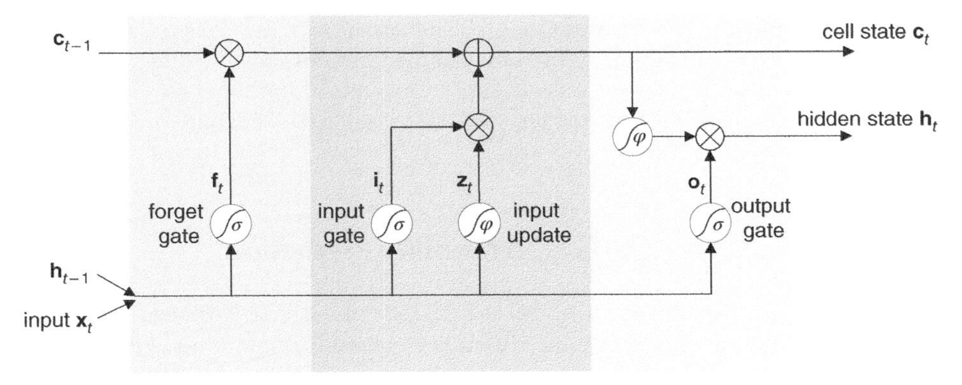

Figure 7.2 An illustration of an LSTM memory block.

output. Theoretically, RNNs are competent to handle such long-range dependencies. However, in practice, RNNs seem unable to accomplish the desired task. This phenomenon has been explored in depth by Hochreiter and Schmidhuber (1997), which explained some pretty fundamental reasons why such learning might be difficult. To tackle this problem, LSTM has been revolutionarily developed by changing the structure of the hidden neurons in traditional RNNs Hochreiter and Schmidhuber (1997).

More specifically, the LSTM neural network is composed of multiple copies of basic memory blocks, and each memory block contains a memory cell and three types of gates (input gate, output gate, and forget gate), as illustrated in Figure 7.2. The memory cell is the key component of LSTM and is responsible for the information transfer at different time-steps. Meanwhile, the three gates, each of which contains a sigmoid layer to optionally pass information, are responsible for protecting and controlling the cell state. As its name implies, the input gate controls which part of the input will be utilized to update the cell state. Similarly, the forget gate controls which part of the old cell state will be thrown away, while the output gate determines which part of the new cell state will be output.

For the memory block at time-step t, we use f_t, i_t, and o_t to represent the forget, input, and output gates, respectively. Assuming that x_t and h_t represent the input and output at the current time-step, h_{t-1} is the output at the previous time-step, σ represents the sigmoid activation function, and $\otimes$ denotes the Hadamard product, the key equations of the LSTM scheme are given here:

$$
\begin{aligned}
f_t &= \sigma(W_{xf}x_t + W_{hf}h_{t-1} + b_f) \\
i_t &= \sigma(W_{xi}x_t + W_{hi}h_{t-1} + b_i) \\
o_t &= \sigma(W_{xo}x_t + W_{ho}h_{t-1} + b_o) \\
\tilde{c}_t &= tanh(W_{xc}x_t + W_{hc}h_{t-1} + b_c) \\
c_t &= f_t \otimes c_{t-1} + i_t \otimes \tilde{c}_t \\
h_t &= o_t \otimes tanh(c_t)
\end{aligned}
\tag{7.1}
$$

where W and b are the corresponding weight matrices and biases of the three gates, and memory cells with subscripts f, i, and o stand for the forget, input, and output gates, respectively, while the subscript c is used for the memory cell. After calculating the values of the three gates, the process of updating information through the gate structure can be divided into three steps. First, multiplying the value of forget gate f_t by the old cell state c_{t-1} decides which part of the previous cell state c_{t-1} should be thrown away. Then, the information in the cell state is updated by multiplying the value of input gate i_t by the new candidate memory cell value $\tilde{c}_t$. Finally, multiplying the output gate o_t by the updated cell state c_t through a *tanh* function leads to the output value h_t. The output h_t and cell state value c_t will be passed to the next memory block at the $t + 1$ time-step.

7.4 ANN-Based Models for Traffic and Mobility Prediction

In the following, we will present the use of RNNs and LSTMs in both network traffic prediction and user-mobility prediction. For a better illustration, related simulation results will be provided as well.

7.4.1 ANN for Traffic Prediction

7.4.1.1 Long Short-Term Memory Network Solution

As we addressed in the previous section, benefiting from the component of the memory cells and three types of gates, the LSTM block has become a powerful tool to make time-series predictions. A LSTM network is a kind of RNN whose hidden units (i.e. the part that cycles over time) are replaced with LSTM blocks. Hua et al. (2018) used a three-layer stacked LSTM network, which is depicted in Figure 7.3, to make traffic prediction. The input data of the LSTM network are from y_1 to y_T, where T denotes the length of the input sequences, and the output of the LSTM network is a prediction of the actual value at time $T + 1$, denoted as $\hat{y}_{T+1}$. Before training the LSTM network, the raw data need to be preprocessed, given that they are so uneven in numerical size. To do so, the authors in this reference first take the logarithm of the raw data and then carry out a normalization process according to $\frac{x - min(x)}{max(x) - min(x)}$, where $\boldsymbol{x}$ is the vector after taking the logarithm of the raw data, and $min(\boldsymbol{x})$ and $max(\boldsymbol{x})$ denote the minimum and maximum value of $\boldsymbol{x}$, respectively. Through this process, the raw data are limited to a range between 0 and 1. Then, the notion of a sliding window is introduced, which indicates a fixed number of previous time slots to learn and then predict the current value. Finally, the processed data is split into two sets (i.e. a training set and a test set). The training set is used to train the LSTM network, and the test set is used to evaluate its prediction accuracy.

After data preprocessing, the training set is used to train the three-layer LSTM network. The training objective is to reduce the value of the loss function, which can be the root mean

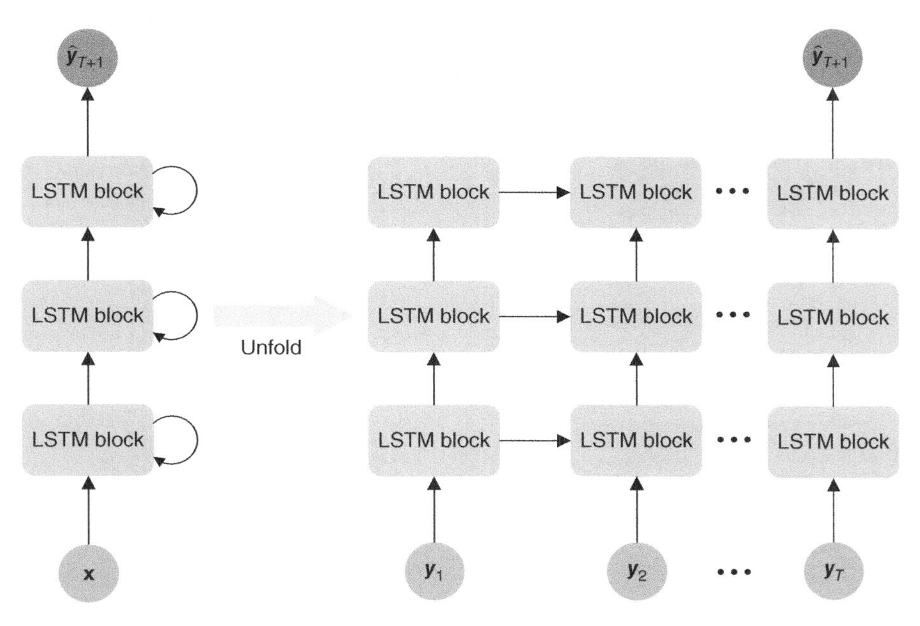

Figure 7.3 An illustration of a three-layer stacked LSTM network.

square error (RMSE) between the actual value and the predicted value. Next, the test set is used to estimate the performance of the LSTM network. The performance of LSTM-based traffic prediction will further be addressed and compared in the next subsection.

7.4.1.2 Random Connectivity Long Short-Term Memory Network Solution

Since the invention of LSTM, a number of scholars have proposed several improvements with respect to its original architecture. Greff et al. (2017) evaluated the aforementioned conventional LSTM and eight different variants thereof (e.g. gated recurrent unit [GRU] Chung et al. (2014)) on three benchmark problems: TIMIT, IAM Online, and JSB Chorales. Each variant differs from the conventional LSTM by a single simple change. They found that the conventional LSTM architecture performs well on the three datasets, and none of the eight investigated modifications significantly improve the performance. This suggests that much more effort is needed to further improve the performance of LSTM solutions.

On the other hand, LSTM's computing time is proportional to the number of parameters if no customized hardware or software acceleration is used. Therefore, given this disappointing characteristic, Hua et al. (2018) present an approach to decrease the number of involved parameters, and thus put forward a new model that reduces the computational cost. As a matter of fact, conventional LSTM (including its variants) follows the classical pattern that the neurons in each memory block are fully connected and this connectivity cannot be changed manually. On the other hand, it has been found that for certain functional connectivity in neural microcircuits, random topology formation of synapses plays a key role and can provide a sufficient foundation for specific functional connectivity to emerge in local neural microcircuits Hill et al. (2012). This discovery is different from conventional cases where neural connectivity is considered to be more heuristic so that neurons need to be connected in a more fully organized manner. It raises a fundamental question as to whether a strategy of forming more random neural connectivity, like in the human brain, might yield potential benefits for LSTM's performance and efficiency. With this conjecture, the RCLSTM model has been proposed by the authors of this chapter and will be presented as follows.

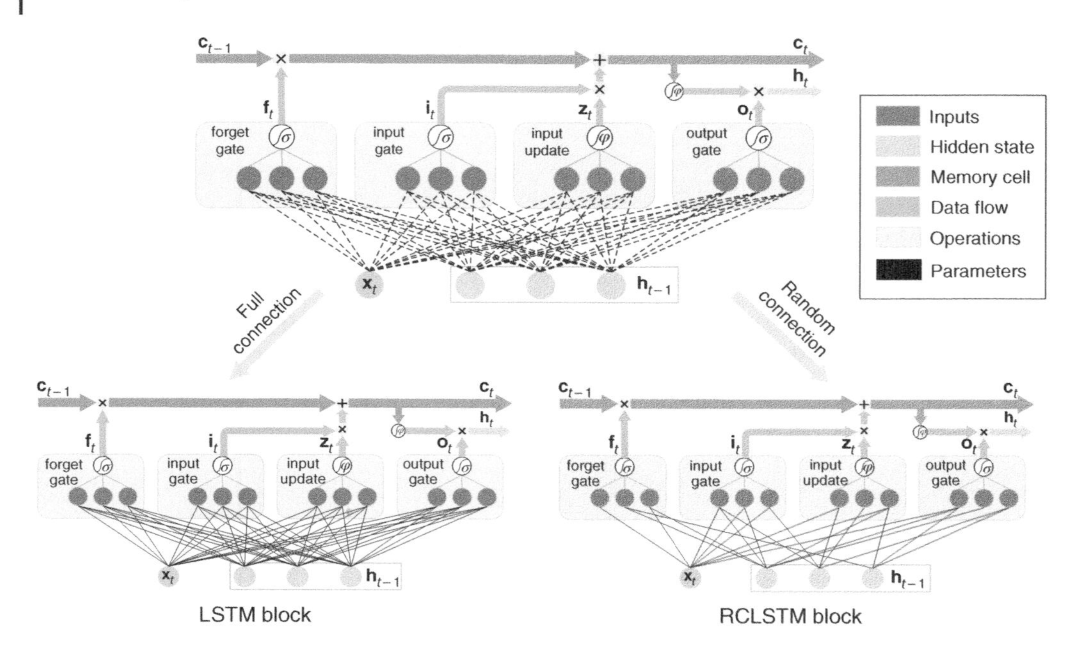

Figure 7.4 The evolution of LSTM to RCLSTM.

In the proposed RCLSTM model, neurons are randomly connected rather than being fully connected as in LSTM. Actually, the trainable parameters in LSTM only exist between the input part – the combination of the input of the current moment (i.e. $\mathbf{x}_t$) and the output of the previous moment (i.e. $\mathbf{h}_{t-1}$), and the functional part – the combination of the gate layers and the input update layer. Therefore, the LSTM architecture can be further depicted in Figure 7.4, which indicates that whether the LSTM neurons are connected or not can be determined by certain randomness. As depicted in the upper part of Figure 7.4, dashed lines are used to denote the neural connections that can be added or omitted. If the neurons are fully connected, then it becomes a standard LSTM model. On the other hand, if the neurons are connected according to the probability values generated at random, then an RCLSTM model is created. The lower-right part of Figure 7.4 shows an example RCLSTM structure in which the neural connections are randomly sparse, unlike the LSTM model.

Based on the proposed RCLSTM block, Hua et al. (2018) construct a three-layer RCLSTM network similar to the LSTM network in Figure 7.3; but the recurrent memory blocks are replaced by RCLSTM blocks, which are used in simulations to verify the RCLSTM performance for traffic prediction. Simulation results are shown in Figures 7.5 and 7.6. Figure 7.5a depicts the RMSE and computing time under different percentages of neural connectivity in the RCLSTM model (note that 100% connectivity means the fully connected LSTM model), which reveals that the performance of the RCLSTM model is slightly less adequate than that of the LSTM model; but the RCLSTM with very sparse neural connections (i.e. 1%) can reduce the computing time by around 30% compared with the baseline LSTM. Figure 7.5b intuitively illustrates the actual and predicted traffic values, from which it can be observed that the predicted values can match the variation trend and features of the actual values very well. Figures 7.5c and d show how the predictive capability of the RCLSTM model is influenced by the number of training samples and the length of input traffic sequences. It can be observed from Figure 7.5c that RCLSTM models are more sensitive to the number of training samples than LSTM, because when the number of training samples increases, the RMSEs of the RCLSTM models vary more

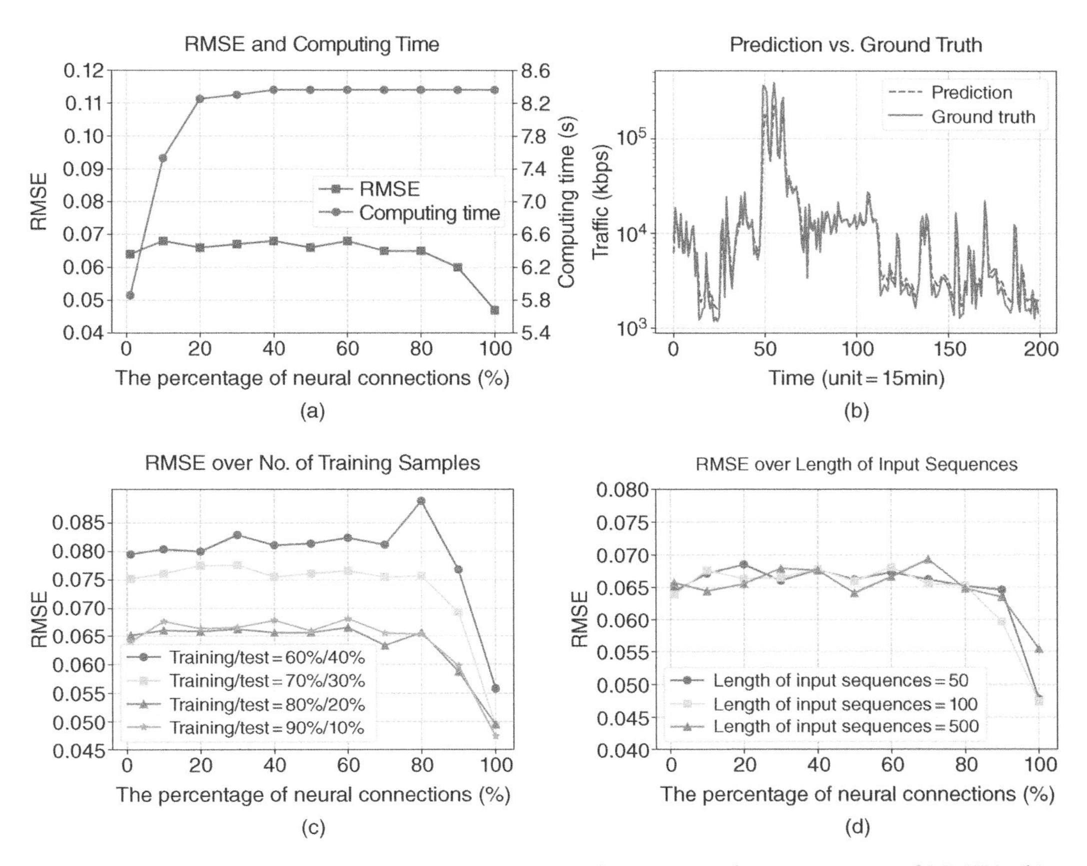

Figure 7.5 Results of traffic prediction using RCLSTM network: (a) RMSE and computing time of RCLSTMs; (b) predicted traffic data and actual traffic data; (c) RMSE of RCLSTMs over different numbers of training samples; (d) RMSE of RCLSTMs over different lengths of input sequences.

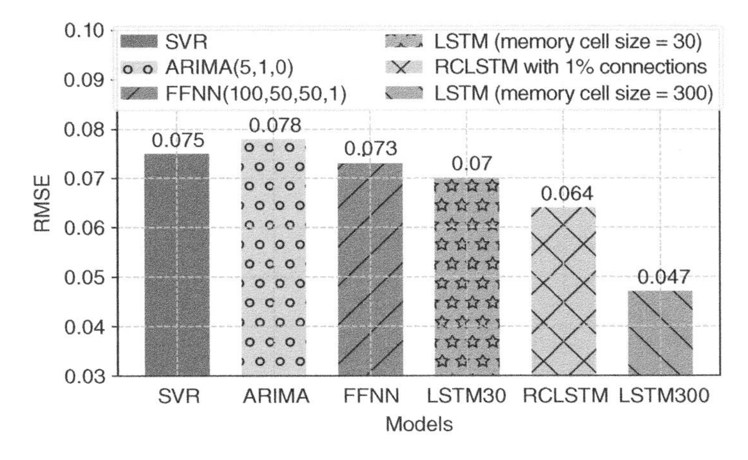

Figure 7.6 Performance comparison of RCLSTM, SVR, ARIMA, FFNN, and LSTM.

significantly than that of the LSTM model. Figure 7.5d gives the related results and shows that RCLSTM models are less susceptible to the length of the input sequences than the LSTM model. Finally, Figure 7.6 is the comparison of the RCLSTM with four well-known prediction techniques: SVR, ARIMA, FFNN, and LSTM. These results reveal that LSTM with a memory cell size of 300 performs much better than the others, followed by the RCLSTM with a memory cell size of 300 and 1% neural connections. In conclusion, the RCLSTM model is a highly competitive traffic-prediction solution in terms of performance and computing time.

7.4.2 ANN for Mobility Prediction

7.4.2.1 Basic LSTM Network for Mobility Prediction

As shown in Figure 7.7, Wang et al. (2018) proposed a basic, generalized mobility model that is based on the combination of multiple LSTM units and, further, contains input and output layers. In their work, they first extract the location points from the user's historical trajectory at a certain sampling interval. Each location point, represented by either a two-dimensional coordinate or a one-dimensional cell ID according to the requirement of spatial granularity, indicates the user's position at one specific time-step. Then a sequence of positions $p = \{p_0, p_1, \cdots, p_{T-1}\}$ can be obtained by augmenting location points at different time-steps, where T is the observation length.

The use of the processed mobility data to train a three-layer LSTM network can be divided into three main steps. (i) The sequence of positions is processed by a fully connected input layer so that each value in the sequence is mapped to a multidimensional feature tensor. (ii) The processed sequence is sent to the main part of the mobility model. (iii) A fully connected output layer maps the output of the last LSTM layer at each time-step i to a location point $\tilde{p}_i$ with the same dimension as p_i, to get the output sequence $\tilde{p} = \{\tilde{p}_0, \tilde{p}_1, \cdots, \tilde{p}_{T-1}\}$, where $\tilde{p}_i$ represents the

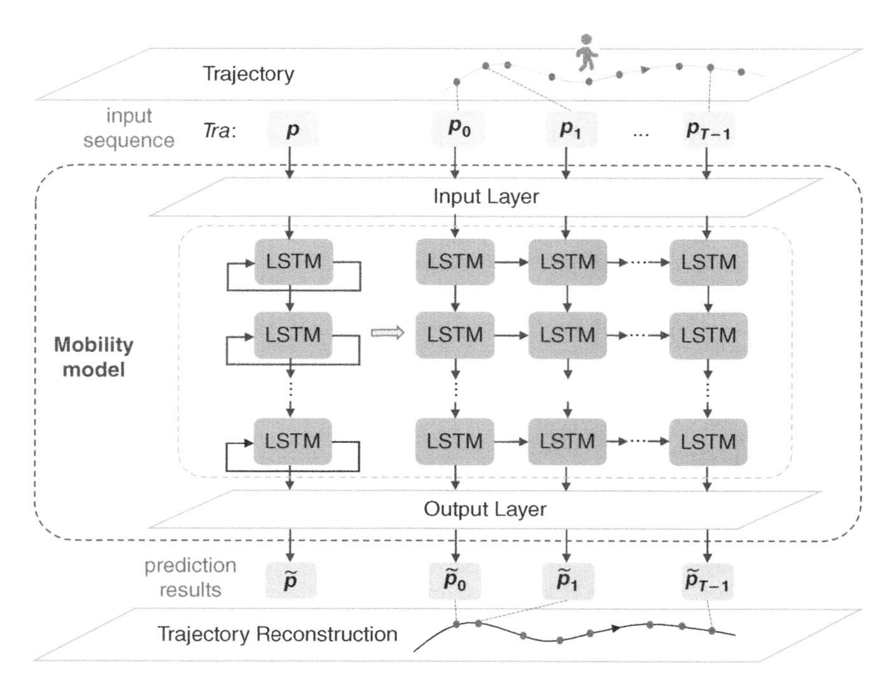

Figure 7.7 LSTM scheme and basic framework for mobility prediction.

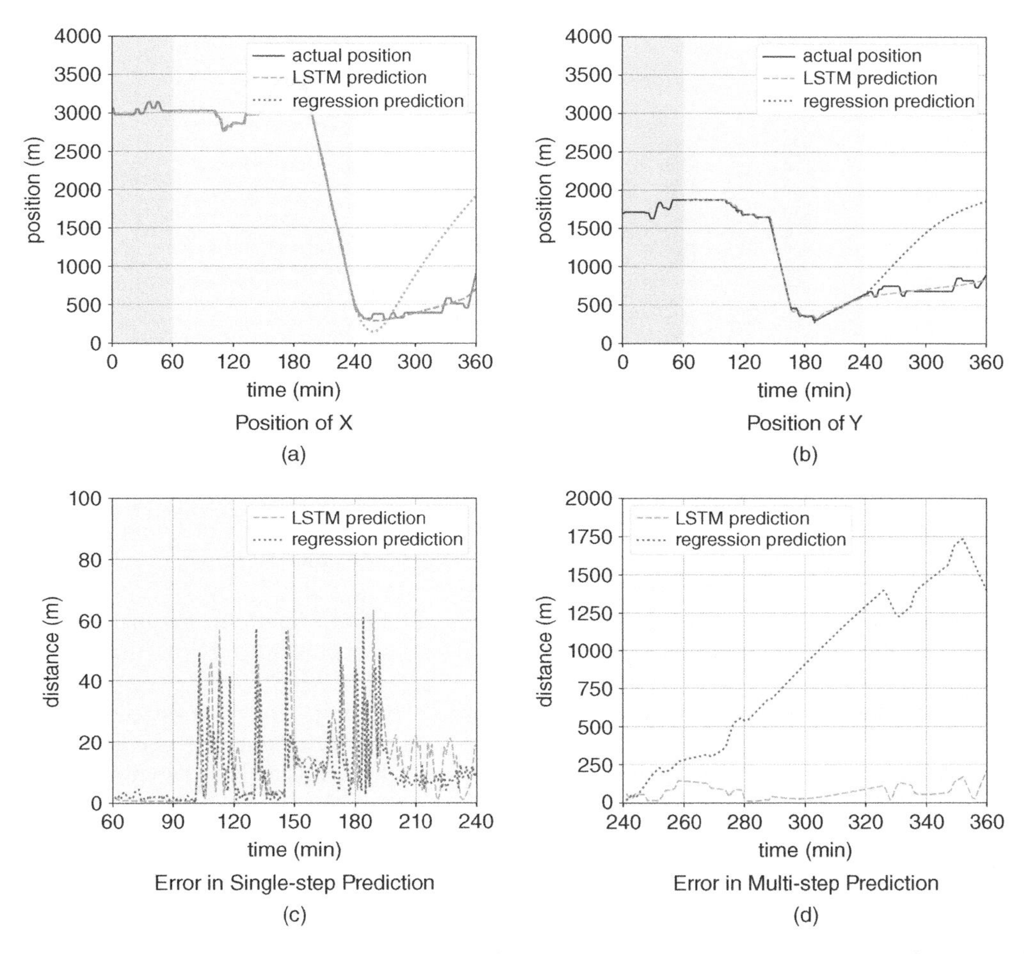

Figure 7.8 Comparison of mobility prediction performance on a coordinate-level dataset over five hours (after a one-hour observation) with two representative feasible algorithms in the ML field: LSTM and linear regression. The first three hours are single-step predictions, and the last two hours are multistep predictions. (a) and (b) show the predicted positions of the x-coordinate and y-coordinate, respectively; (c) and (d) represent the distance error in the single-step prediction and the multistep prediction, respectively.

prediction result of the position at the $i + 1$ time-step. The training objective is to reduce the value of the loss function, which can be either mean square error (MSE) when the input p_i is a two-dimensional coordinate or the cross entropy when p_i is a cell ID over all time-steps. Wang et al. (2018) conduct the relevant experiments with different mobility-prediction methods on two spatial granularities (i.e. coordinate-level and cell-level).

As for coordinate-level prediction, Figure 7.8 presents a performance comparison of the LSTM-based framework and the conventional linear regression. As shown in Figure 7.8a and b, after a one-hour observation (in the left region), the two methods first make single-step predictions given the fully observable ground truth (the user's real position) at each time-step (in the middle region). However, when position measurements become unavailable, the conventional linear regression algorithm fails to follow the actual evolution of the user's trajectory, but the LSTM scheme yields predictions with superior accuracy. Figure 7.8c and d further show the distance error between the prediction results and the ground truth in single-step

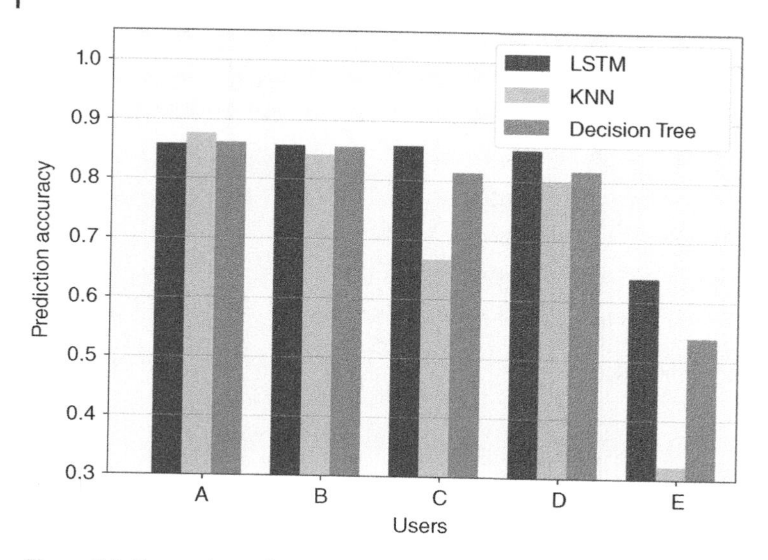

Figure 7.9 Comparison of mobility prediction performance on a cell-level dataset.

and multi-step prediction respectively; it can be observed that both methods perform well in most of single-step prediction cases, yet the LSTM scheme performs much better than linear regression in the multiple-step prediction case.

The results of cell-level prediction are shown in Figure 7.9 by comparing the prediction accuracy for five users with different algorithms, where accuracy is defined as the percentage of values with correct prediction results. The results verify that the LSTM-based framework in general yields superior performance compared with the other methods.

7.4.2.2 Spatial-Information-Assisted LSTM-Based Framework of Individual Mobility Prediction

Figure 7.10 shows an improved version of the basic LSTM-based framework for individual mobility prediction. This new proposed version is mainly composed of three modules: a spatial module for extracting geographical features, an attribute module for providing auxiliary information, and a temporal LSTM-based module for processing time-series features. Since the coordinate-level trajectory can capture the change of position more accurately and is more commonly used in practice compared with the cell-level trajectory, this improved framework only considers the coordinate-level case.

The spatial module is used to get the spatial information between locations over a period of time. In a historical trajectory $Tra = \{p_0, p_1, \dots, p_{T-1}\}$, where each p_i contains the corresponding longitude and latitude values, the location points in a short period of time, such as $\{p_0, p_1, p_2\}$ and $\{p_1, p_2, p_3\}$, often have a strong spatial correlation that can be captured by a one-dimensional convolutional layer. A typical convolutional layer consists of several convolutional filters. In the field of image processing, a filter learns the spatial dependencies in a multi-channel image by applying the convolution operation on each of the two-dimensional local patches. For position sequences, in order to acquire the spatial dependency in a short period of time, the filters need to slide in the time dimension, which is known as 1-D convolution.

The attribute module in Figure 7.10 extracts some auxiliary information that is helpful for mobility prediction, including a userID that represents the user's identity, a workID that represents the user's work type, a weekID that represents the day of week, and a timeID that

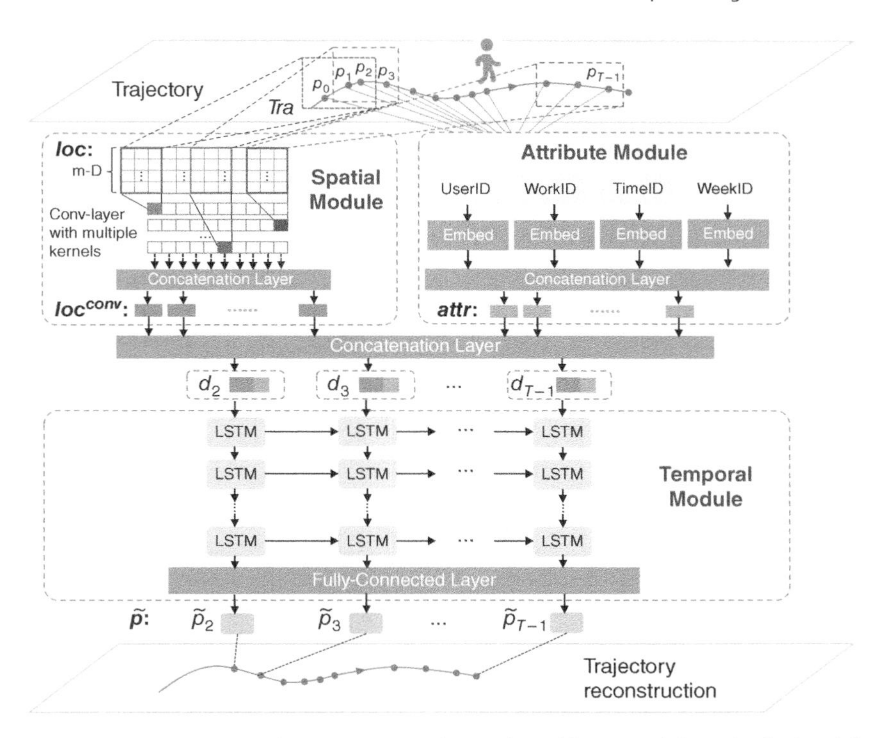

Figure 7.10 The spatial-information assisted LSTM-based framework for individual mobility prediction.

represents the specific time at the location point. Furthermore, the attribute module can alleviate the problem of data missing and low multi-step prediction accuracy. Since these attributes are categorical values, we use the embedding method to transform each categorical attribute into a low-dimensional vector. Then different categorical vectors can be concatenated into one vector at each time-step, yielding $attr = \{attr_2, \dots, attr_{T-1}\}$, where $attr_i$ integrates all attributes at the i time-step.

Concerning the temporal module, the geographically featured vector loc_i^{conv} captured by the spatial module and the corresponding attribute vector $attr_i$ obtained by the attribute module at each time-step are concatenated together as the input sequence $\{d_2, d_3, \dots, d_{T-1}\}$ of the temporal module. The temporal module consists of multiple stacked LSTM layers and a fully connected layer, and has internal operations similar to the corresponding parts in the basic LSTM-based framework. Through this temporal module, we can obtain the output sequence $\tilde{p} = \{\tilde{p}_2, \tilde{p}_3, \dots, \tilde{p}_{T-1}\}$, where $\tilde{p}_i$ represents the prediction result of the position at the $i+1$ time-step.

7.4.2.3 Spatial-Information-Assisted LSTM-Based Framework of Group Mobility Prediction

For group mobility prediction, the objective is to make accurate predictions of the distribution of mobile users in different cells simultaneously in a large area. We express the distribution of group mobile users over time T as a spatio-temporal sequence of data points $D = \{D_0, D_1, \dots, D_{T-1}\}$, where D_i is a snapshot of the number of mobile users at timestamp i in a geographical region represented as an $M \times N$ grid.

Note that the spatio-temporal sequence forecasting problem is different from the conventional time-series forecasting problem, since the prediction target of the former is a sequence that contains both spatial and temporal structures. In order to address this problem, a

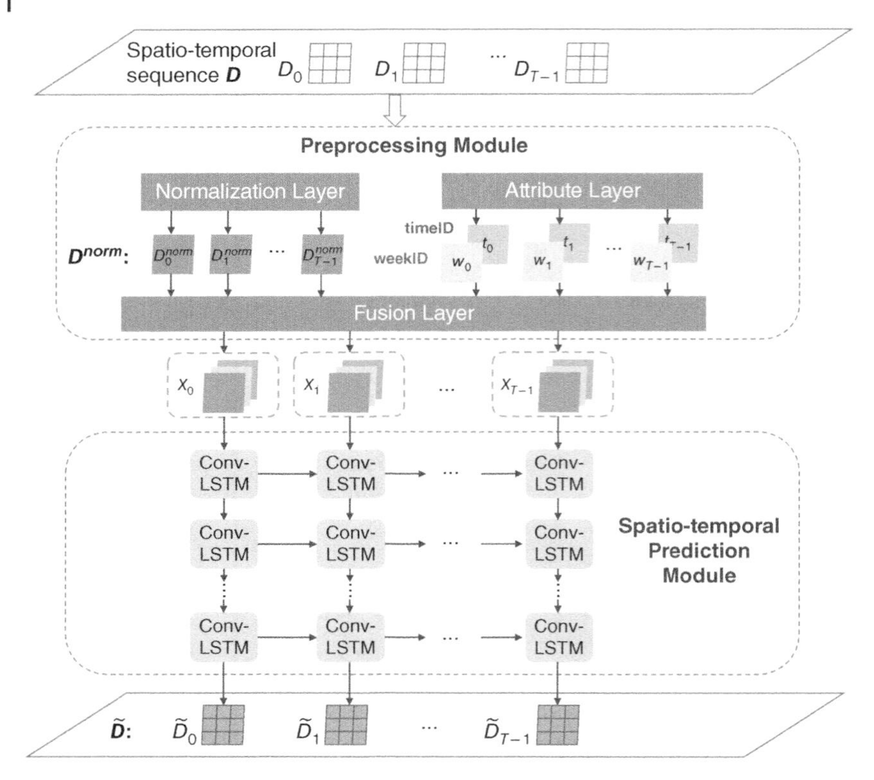

Figure 7.11 The spatial-information-assisted LSTM-based framework of group mobility prediction.

spatio-temporal mobility-prediction framework is given in Figure 7.11, which mainly contains a preprocessing module and a spatio-temporal prediction module.

In the preprocessing module, the raw observation sequence $D = \{D_0, D_1, \dots, D_{T-1}\}$ can be coped with from two perspectives. On the one hand, a norm-layer is used on D, thus making D become a normalized sequence $D^{norm} = \{D_0^{norm}, D_1^{norm}, \dots, D_{T-1}^{norm}\}$. On the other hand, some auxiliary information can be captured, such as weekID, which represents the day of week, and timeID, which represents the specific moment of the current time-step from the raw data. Then weekID and timeID are transformed into scalar values as week offset w_i and time offset t_i at the i time-step, respectively. Furthermore, a fusion layer is used to blend the normalized value D_i^{norm} with the corresponding attribute value at each time-step and obtain the final input sequence $\{X_1, X_2, \dots, X_{T-1}\}$ of the spatio-temporal prediction module, where each X_i is a 3-D tensor in $\mathbb{R}^{3 \times M \times N}$ with the original D_i^{norm}, the fusion of D_i^{norm} and w_i, and the fusion of D_i^{norm} and t_i.

Unlike the previous individual mobility-prediction mechanism, the convolutional LSTM (ConvLSTM) unit is adopted to take the place of the standard LSTM unit in the spatio-temporal prediction module. In ConvLSTM, the inputs X_t, cell memories C_t, hidden states H_t, cell candidates $\tilde{C}_t$, and gates i_x, f_t, o_t are all 3-D tensors, where the first dimension indicates the number of feature maps and the last two form the spatial dimension ($M \times N$). Given a sequence of 3-D inputs denoted as $X = \{X_1, X_2, \dots, X_{T-1}\}$, the operations of a single ConvLSTM are formulated as follows:

$$\begin{aligned} f_t &= \sigma(W_{xf} * X_t + W_{hf} * H_{t-1} + b_f) \\ i_t &= \sigma(W_{xi} * X_t + W_{hi} * H_{t-1} + b_i) \\ o_t &= \sigma(W_{xo} * X_t + W_{ho} * H_{t-1} + b_o) \\ \tilde{C}_t &= tanh(W_{xc} * X_t + W_{hc} * H_{t-1} + b_c) \\ C_t &= f_t \odot C_{t-1} + i_t \odot \tilde{C}_t \\ H_t &= o_t \odot tanh(C_t) \end{aligned} \quad (7.2)$$

where $\odot$ denotes the Hadamard product and $*$ is the 2-D convolution operator. Since each hidden element of this neural network is represented as a 2-D map, the cross-spatial mobility correlations can be effectively captured through the convolution operations. Finally, we obtain the output sequence $\tilde{D} = \{\tilde{D}_0, \tilde{D}_1, \dots, \tilde{D}_{T-1}\}$, where $\tilde{D}_i$ represents the predicted distribution of a group of mobile users in different cells at the $i + 1$ time-step.

By taking advantage of spatial information, the group mobility prediction, essentially the spatio-temporal forecasting problem, can not only use important spatial correlations to improve prediction performance but also predict the distribution of mobile users in different cells simultaneously.

7.5 Conclusion

In this chapter, we summarized the importance of traffic and user-mobility prediction in communication networks and review some commonly used prediction methods with the emphasis on the current mainstream DL-based prediction methods. In particular, we focused on a LSTM-based model. For traffic prediction, an variation named RCLSTM model can be intuitively interpreted as an artificial neural network formed by stochastic connections among neurons. We provided proof of the effectiveness of the RCLSTM under various conditions as well as its comparison with several traditional ML algorithms and plain ANNs. When it comes to mobility prediction, we first investigated whether the pure LSTM network can boost prediction accuracy and then presented a valuable research result that gives an intuitive sense that the LSTM network outperforms traditional algorithms. Furthermore, based on the original LSTM scheme that directly uses the LSTM network to predict the next-moment location from historical trajectory data, this chapter also discussed the two slightly more complex models proposed by Wang et al. (2018), which neatly integrate LSTM, CNN, and other modules (e.g. embedded module for individual mobility prediction and ConvLSTM module for group mobility prediction). Although the performance of the two high-level models has not fully been verified by simulations, the idea behind them is illuminating since the user's movement is not only similar to the state transition in terms of time but also affected by the geographical location in terms of space.

It is envisaged that as the applicability of DL techniques for traffic and mobility prediction increases, traffic and mobility prediction will be improved with respect to computational overhead and accuracy. However, there are still many hard-to-tackle problems. Single models like LSTM may be difficult to use with long time-series data, while the combination of LSTM and other neural networks like CNN may lead to a more complex structure and be hard to train. In addition, long-term prediction and a better understanding of the relationship of both the endogenous and exogenous variables in traffic and mobility prediction Zhang and Dai (2018) are expected to be researched in the future.

Bibliography

Alexandre Alahi, Kratarth Goel, Vignesh Ramanathan, Alexandre Robicquet, Li Fei-Fei, and Silvio Savarese. Social LSTM: Human trajectory prediction in crowded spaces. In *Proc. CVPR 2016*, Las Vegas, USA, Jun.-Jul. 2016.

Jeffrey G. Andrews, Stefano Buzzi, Wan Choi, Stephen V. Hanly, Angel Lozano, Anthony C.K. Soong, and Jianzhong Charlie Zhang. What will 5G be? *IEEE J. Select. Areas Commun.*, 32(6):1065–1082, 2014.

Abdelhadi Azzouni and Guy Pujolle. A long short-term memory recurrent neural network framework for network traffic matrix prediction. *arXiv preprint arXiv:1705.05690*, 2017.

Junyoung Chung, Caglar Gulcehre, KyungHyun Cho, and Yoshua Bengio. Empirical evaluation of gated recurrent neural networks on sequence modeling. *arXiv preprint arXiv:1412.3555*, 2014.

Cisco. Cisco visual networking index: Global mobile data traffic forecast update, 2016-2021 white paper. https://www.cisco.com/c/en/us/solutions/collateral/service-provider/visual-networking-index-vni/mobile-white-paper-c11-520862.html, March 2017.

Ian Goodfellow, Yoshua Bengio, Aaron Courville, and Yoshua Bengio. *Deep Learning*, volume 1. MIT Press Cambridge, 2016.

Klaus Greff, Rupesh K. Srivastava, Jan Koutník, Bas R. Steunebrink, and Jürgen Schmidhuber. LSTM: A search space odyssey. *IEEE transactions on neural networks and learning systems*, 28(10):2222–2232, 2017.

Sean L. Hill, Yun Wang, Imad Riachi, Felix Schürmann, and Henry Markram. Statistical connectivity provides a sufficient foundation for specific functional connectivity in neocortical neural microcircuits. *Proceedings of the National Academy of Sciences*, 109(42):E2885–E2894, 2012.

Sepp Hochreiter and Jürgen Schmidhuber. Long short-term memory. *Neural computation*, 9(8):1735–1780, 1997.

Wei-Chiang Hong. Application of seasonal SVR with chaotic immune algorithm in traffic flow forecasting. *Neural Computing and Applications*, 21(3):583–593, 2012.

Yuxiu Hua, Zhifeng Zhao, Rongpeng Li, Xianfu Chen, Zhiming Liu, and Honggang Zhang. Deep learning with long short-term memory for time series prediction. *arXiv preprint arXiv:1810.10161*, 2018.

Yann LeCun, Yoshua Bengio, and Geoffrey Hinton. Deep learning. *Nature*, 521 (7553):436, 2015.

Will E. Leland, Murad S. Taqqu, Walter Willinger, and Daniel V. Wilson. On the self-similar nature of Ethernet traffic (extended version). *IEEE/ACM Transactions on Networking (ToN)*, 2(1):1–15, 1994.

Rongpeng Li, Zhifeng Zhao, Xuan Zhou, and Honggang Zhang. Energy savings scheme in radio access networks via compressive sensing-based traffic load prediction. *Transactions on Emerging Telecommunications Technologies*, 25 (4):468–478, 2014.

Rongpeng Li, Zhifeng Zhao, Jianchao Zheng, Chengli Mei, Yueming Cai, and Honggang Zhang. The learning and prediction of application-level traffic data in cellular networks. *IEEE Transactions on Wireless Communications*, 16(6): 3899–3912, 2017.

Wen-Hwa Liao, Kuo-Chiang Chang, and Sital Prasad Kedia. An object tracking scheme for wireless sensor networks using data mining mechanism. In *Proc. NOMS 2012*, Maui, USA, Apr. 2012.

Sumit Maheshwari, Sudipta Mahapatra, Cheruvu Siva Kumar, and Kantubukta Vasu. A joint parametric prediction model for wireless internet traffic using Hidden Markov Model. *Wireless Networks*, 19(6):1171–1185, 2013.

Ali Yadavar Nikravesh, Samuel A. Ajila, Chung-Horng Lung, and Wayne Ding. Mobile network traffic prediction using MLP, MLPWD, and SVM. In *Big Data (BigData Congress), 2016 IEEE International Congress on*, pages 402–409. IEEE, 2016.

S. Periyanayagi and V. Sumathy. S-ARMA model for network traffic prediction in wireless sensor networks. *Journal of Theoretical & Applied Information Technology*, 60(3), 2014.

T. Raghuveera, P.V.S. Kumar, and K.S. Easwarakumar. Adaptive linear prediction augmented autoregressive integrated moving average based prediction for VBR video traffic. *Journal of Computer Science*, 7(6):871, 2011.

Injong Rhee, Minsu Shin, Seongik Hong, Kyunghan Lee, Seong Joon Kim, and Song Chong. On the levy-walk nature of human mobility. *IEEE/ACM Trans. on Networking (TON)*, 19(3):630–643, 2011.

Peter Romirer-Maierhofer, Mirko Schiavone, and Alessandro D'alconzo. Device-specific traffic characterization for root cause analysis in cellular networks. In *International Workshop on Traffic Monitoring and Analysis*, pages 64–78. Springer, 2015.

Sai Ganesh Sitharaman. Modeling queues using Poisson approximation in IEEE 802.11 ad-hoc networks. In *2005 14th IEEE Workshop on Local & Metropolitan Area Networks*, pages 6–pp. IEEE, 2005.

Chaoming Song, Zehui Qu, Nicholas Blumm, and Albert-László Barabási. Limits of predictability in human mobility. *Science*, 327(5968):1018–1021, 2010.

Jan Tkačík and Pavel Kordík. Neural turing machine for sequential learning of human mobility patterns. In *Proc. IJCNN 2016*, Vancouver, Canada, Jul. 2016.

Ardian Ulvan, Robert Bestak, and Melvi Ulvan. Handover procedure and decision strategy in LTE-based femtocell network. *Telecommunication Systems*, 52(4):2733–2748, 2013.

Chujie Wang, Zhifeng Zhao, Qi Sun, and Honggang Zhang. Deep learning-based intelligent dual connectivity for mobility management in dense network. *arXiv preprint arXiv:1806.04584*, 2018.

Jing Wang, Jian Tang, Zhiyuan Xu, Yanzhi Wang, Guoliang Xue, Xing Zhang, and Dejun Yang. Spatiotemporal modeling and prediction in cellular networks: A big data enabled deep learning approach. In *INFOCOM 2017-IEEE Conference on Computer Communications, IEEE*, pages 1–9. IEEE, 2017.

Wenjun Wang, Lin Pan, Ning Yuan, Sen Zhang, and Dong Liu. A comparative analysis of intra-city human mobility by taxi. *Physica A: Statistical Mechanics and its Applications*, 420:134–147, 2015.

Jin-qun WU and Wang-xian WANG. Network traffic prediction based on PSO-SVM model. *Journal of Hunan Institute of Humanities, Science and Technology*, 2:023, 2013.

Jingxian Wu and Pingzhi Fan. A survey on high mobility wireless communications: Challenges, opportunities and solutions. *IEEE Access*, 4: 450–476, 2016.

Feng Xia, Jinzhong Wang, Xiangjie Kong, Zhibo Wang, Jianxin Li, and Chengfei Liu. Exploring human mobility patterns in urban scenarios: A trajectory data perspective. *IEEE Communications Magazine*, 56(3):142–149, 2018.

Chaoyun Zhang and Paul Patras. Long-term mobile traffic forecasting using deep spatio-temporal neural networks. In *Proc. MOBIHOC 2018*, Los Angeles, USA, Jun. 2018.

H. Zhang and L. Dai. Mobility prediction: A survey on state-of-the-art schemes and future applications. *IEEE Access*, pages 1–1, 2018. ISSN 2169-3536. doi: 10.1109/ACCESS.2018.2885821.

Kai Zhao, Mirco Musolesi, Pan Hui, Weixiong Rao, and Sasu Tarkoma. Explaining the power-law distribution of human mobility through transportation modality decomposition. *Scientific Reports*, 5:9136, 2015.

8

Machine Learning for Resource-Efficient Data Transfer in Mobile Crowdsensing

Benjamin Sliwa, Robert Falkenberg, and Christian Wietfeld

Communication Networks Institute, TU Dortmund University, Dortmund, Germany

The exploitation of mobile entities (persons, vehicles) as moving sensor nodes is an enabler for data-driven services such as intelligent transportation systems within future smart cities. In contrast to traditional approaches that rely on static sensors, mobile crowdsensing is able to provide large-scale sensing coverage while simultaneously guaranteeing data freshness and cost-efficiency.

With the massive increases in machine-type communication (MTC), resource optimization of data transmission becomes a major research topic as spectrum resources are limited and shared with other users within the cellular network. As an alternative to cost-intensive extension of the network infrastructure, anticipatory communication aims to utilize the existing resources in a more resource-efficient way.

In this chapter, we provide an overview of applications and requirements for vehicular crowdsensing as well as anticipatory data transmission. As a case study, we present an opportunistic, context-predictive transmission scheme that relies on machine learning–based data-rate prediction for channel quality assessment, which is executed online on embedded devices. The proposed transmission scheme is analyzed in a comprehensive real-world evaluation study, where it is able to achieve massive increases in the resulting data rate while simultaneously reducing the power consumption of mobile devices.

This chapter is organized into five sections. Section 8.1 provides an overview of requirements and current work in mobile crowdsensing and anticipatory data transfer. In Section 8.2, the proposed machine learning–based transmission approach is introduced and set into relation to its groundwork. Section 8.3 presents the methodology setup for the real-world performance evaluation. Section 8.4 presents and discusses the results of the empirical performance evaluation about the proposed solutions. Section 8.5 offers further discussion and conclusions.

8.1 Mobile Crowdsensing

In this section, we summarize the intended applications and their respective requirements for data transmissions. Furthermore, an overview of existing methods for anticipatory communication techniques is provided. The section is closed with an overview of the contribution provided by the proposed transmission schemes, which are further analyzed in the remainder of this chapter.

Machine Learning for Future Wireless Communications, First Edition. Edited by Fa-Long Luo.
© 2020 John Wiley & Sons Ltd. Published 2020 by John Wiley & Sons Ltd.

8.1.1 Applications and Requirements

The next generation of smart city-based intelligent transportation systems will be *data-driven* (Zanella et al. (2014)) and closely linked to data analytics (Chen et al. (2017), Djahel et al. (2015)).

As the existing sensor infrastructure will not be able to fulfill the needs for coverage and data freshness, and since extending the existing sensor systems is often highly cost-intensive, e.g. as additional roadwork is required, novel and scalable sensing methods are highly desired.

A highly promising approach relies on using mobile entities (e.g. vehicles and pedestrians) themselves as *mobility sensor nodes* that actively sense their environment and gather information for traffic-monitoring systems (Wang et al. (2016)). While this approach turns vehicles into cyber-physical systems, it also marks a paradigm shift from locally limited sensing to mobile and distributed data acquisition. Figure 8.1 provides an overview of an example application scenario where vehicles sense their environment in order to gather information for crowdsensing-based traffic analysis.

Apart from traffic-related sensing, mobile crowdsensing is also applicable for services such as weather analysis (Calafate et al. (2017)), detection of potholes (Wang et al. (2017)), and air quality measurement (Chen et al. (2018)). While data-exploiting applications benefit greatly from improved sensor coverage and enhanced data freshness, the presence of these services leads to a massive increase in MTC and competition for the available spectrum resources within the cellular communication network. As a straightforward but not at all cost-efficient approach, the capacity of the communication network can be increased by providing more spectrum resources and more network infrastructure. In contrast, a more sophisticated idea with the aim of using the existing infrastructure in a more resource-efficient way is the exploitation of *anticipatory communication principles* (Bui et al. (2017)). Since the intended applications can be considered as *delay-tolerant* within defined limits, data transmissions can be performed opportunistically within the bounds of the application requirements.

In the following, we will focus on the deadline requirements in the lower minute range, which is a realistic assumption for real-world systems for traffic monitoring and optimization (Shi and Abdel-Aty (2015), Chen et al. (2018)) as well as environmental sensing (Vandenberghe et al. (2012)). Furthermore, we consider the resulting data rate as a metric for the resource-efficiency

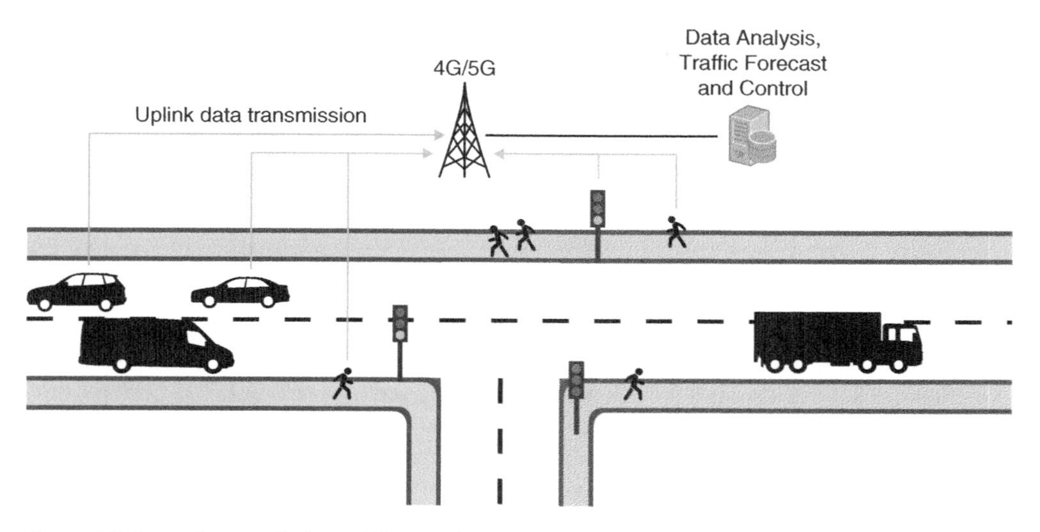

Figure 8.1 Example scenario for mobile crowdsensing.

of a data transmission, as it is closely related to the transmission duration as well as the spectrum occupation time and has a severe impact on the power consumption of the mobile device (Ide et al. (2015)).

8.1.2 Anticipatory Data Transmission

In order to allow opportunistic and anticipatory communication, knowledge about the channel quality is a crucial factor. While long-term evolution (LTE) provides several indicators for downlink channel quality, similar indicators are missing for the intended uplink use case. Nevertheless, assuming bidirectional channel properties, uplink connectivity can be approximated with decent accuracy based on the passive downlink indicators, as the analysis in Ide et al. (2016) shows. Based on literature review, two main anticipatory optimization principles can be identified: infrastructure-based and user equipment (UE) based optimization. Infrastructure-based approaches usually change the resource-scheduling mechanisms based on additional information (Feng et al. (2017)).

While central approaches are widely investigated in scientific evaluations, they have a number of disadvantages. As the authors of Zheng et al. (2015) analyze, the need to communicate locally sensed context information back to the infrastructure leads to a significantly increased signaling overhead. Moreover, research works in these fields are usually limited to simulative analysis or small-scale experiments, as the changed algorithms would need to be deployed by the network operator. In contrast to that, as UE-based approaches only affect single devices, evaluations can be performed in the real world using the public cellular network. However, since the available cell resources are unknown for the UE, it has to rely on predictions, which are usually handled by machine learning (ML) models (Jiang et al. (2017)). The authors of Jomrich et al. (2018) provide a ML-based data rate estimation for uplink and downlink in real-world vehicular scenarios. For the predictions, random forests are applied on a large measurement data set.

Software-defined radio (SDR) based techniques enable access to low-level indicators by sniffing control channel information of the LTE signal. In Bui and Widmer (2018), the authors apply such an approach in order to reduce the consumption of cell resources by 50 %. Similarly, the client-based control-channel analysis for connectivity estimation (C^3ACE) (Falkenberg et al. (2016)) and enhanced C^3ACE (E-C^3ACE) (Falkenberg et al. (2017a)) schemes utilize radio network temporary identifier (RNTI) histograms to determine the number of active users and their individual resource consumption as features for the data-rate prediction. This information is used to achieve highly precise data-rate prediction results using artificial neural networks.

For the purpose of predicting future network behavior, a popular approach is to use crowdsensing-based *connectivity maps* to manage network quality information (Pögel and Wolf (2015)). Given the fact that human mobility is predictable to a high degree (Song et al. (2010)) due to recurring patterns, the trajectory of a mobile entity can be forecasted precisely enough to derive network-quality estimations from connectivity maps. The estimates can then be exploited to perform data transmissions in a context-aware manner. Furthermore, anticipatory communication is used for interface-selection in heterogeneous vehicle-to-everything (V2X) scenarios (Cavalcanti et al. (2018)). In Sepulcre and Gozalvez (2018), Sepulcre et al. (2015), the authors use context-awareness for infrastructure-assisted alternating between LTE and IEEE 802.11p data transmission, by evaluating cost models for each available technology per road segment.

In this work, we define the abstract *context* as a composition of information from multiple domains:

- The **channel context** is described by the available indicators of the respective communication technology. For LTE, the passive downlink indicators reference signal received power

(RSRP), reference signal received quality (RSRQ), signal-to-noise-plus-interference ratio (SINR), and channel quality indicator (CQI) are used to describe the received power and signal quality. Note that some of these indicators are not well-defined and are specific for different modem manufacturers.
- The **mobility context** is based on global navigation satellite system (GNSS) information and includes position, direction, and velocity of the vehicle. This information is utilized to determine the significance of the channel context parameters with respect to the coherence time. In order to enable *context prediction, mobility prediction* is a basic requirement, which is utilized by machine learning predictive channel-aware transmission (ML-pCAT) in Section 8.2.4.
- The **application context** provides high-level information about the application. This includes information about the payload size of the packets as well as deadline requirements of the intended crowdsensing application.

In the following sections, we summarize the work on machine learning CAT (ML-CAT) (Sliwa et al. (2018b)) and ML-pCAT (Sliwa et al. (2018a)), which extend the established transmission schemes channel-aware transmission (CAT) (Ide et al. (2015)) and predictive CAT (pCAT) (Wietfeld et al. (2014)) with ML-based data-rate prediction. The proposed mechanisms implement client-side anticipatory networking and bring together different research directions of anticipatory optimization such as ML-based channel quality assessment, mobility prediction, and connectivity maps.

It should be noted that the proposed data-transmission scheme does not intend to fulfill the real-time requirements of safety-related applications (ETSI (2009)). Nevertheless, anticipatory communication is expected to play an important role in this domain as well. The interested reader is forwarded to recent work on mobility-predictive mesh routing (Sliwa et al. (2016)). Here, mobility prediction is applied to establish robust routing paths.

8.2 ML-Based Context-Aware Data Transmission

In this section, the different evolution stages of the anticipatory data transmission scheme CAT, illustrated in Figure 8.2, are presented and explained.

As groundwork, the pure probabilistic the CAT model is presented, and its channel-predictive extension pCAT is derived. Afterward, ML-CAT introduces a novel approach for channel-quality assessment using ML. Finally, ML-pCAT brings the different individual components together in a unified transmission scheme. All of the discussed variants are *probabilistic* transmission schemes. Acquired data is stored in a local buffer until a transmission decision is made for the whole buffer. Therefore, compared to straightforward direct transmissions, an additional delay is introduced due to the buffering.

8.2.1 Groundwork: Channel-aware Transmission

The general approach of the basic CAT scheme (Ide et al. (2015)) is based on the observation that data transmissions during low channel-quality periods suffer from low transmission and energy efficiencies. Not only are retransmissions often required to cope with high packet-loss probabilities, but high transmission power also needs to be applied in order to achieve a sufficient signal level at the receiver. Since straightforward periodic data transmissions are performed regardless of the current channel situation, they are uniformly distributed among the overall value range of the signal-to-interference-plus-noise ratio (SINR).

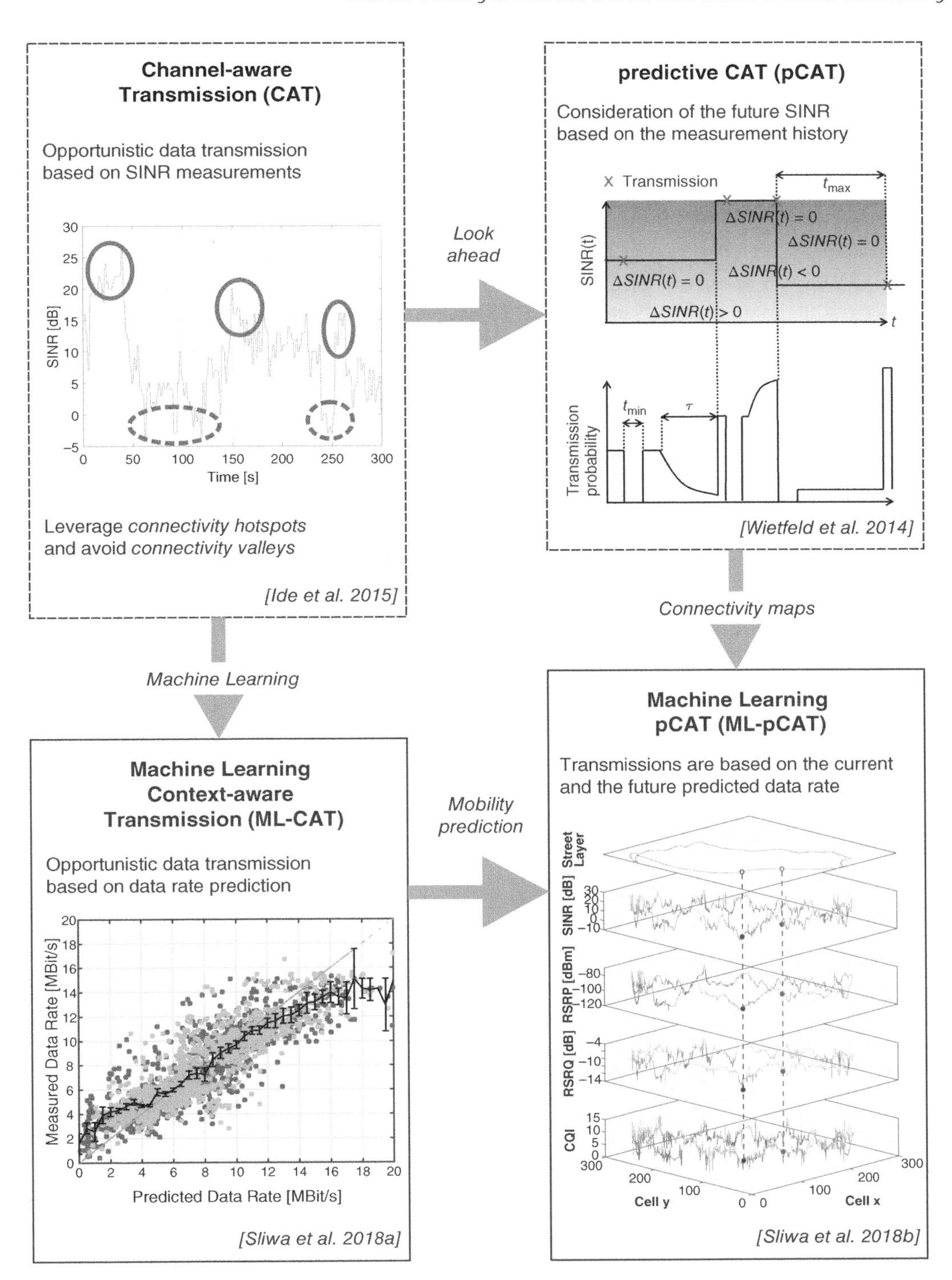

Figure 8.2 Evolution steps and methodological continuity of the CAT scheme.

Figure 8.3 provides an illustration of this behavior that shows data transmissions with respect to the SINR based on real-world experiments. The need to avoid data transmissions during low channel-quality periods (also referred to as *connectivity valleys*) is addressed by the CAT scheme by a probabilistic process with respect to channel-quality measurements. Based on measurements of the SINR and the velocity v, the resulting transmission probability $p(\text{SINR}, v)$ is calculated as

$$p(\text{SINR}, v) = \left(\frac{\text{SINR}}{\text{SINR}_{\max}}\right)^{\alpha} \cdot \left(1 - \frac{v}{v_{\max}}\right)^{\beta} \tag{8.1}$$

The exponents α and β are used to control the impact of the corresponding indicators and their dependencies on high indicator values. As shown in Ide et al. (2016), RSRP and RSRQ can be exploited for deriving estimations for the SINR if the modem does not explicitly report this indicator. The resulting transmission probability is shown in Figure 8.4.

8.2.2 Groundwork: Predictive CAT

While the immediate channel properties of single measurements are highly dynamic in the mobile scenario, their behaviors often follow a general trend as the distance to the respective base station is either decreasing or increasing. Therefore, if the current channel quality

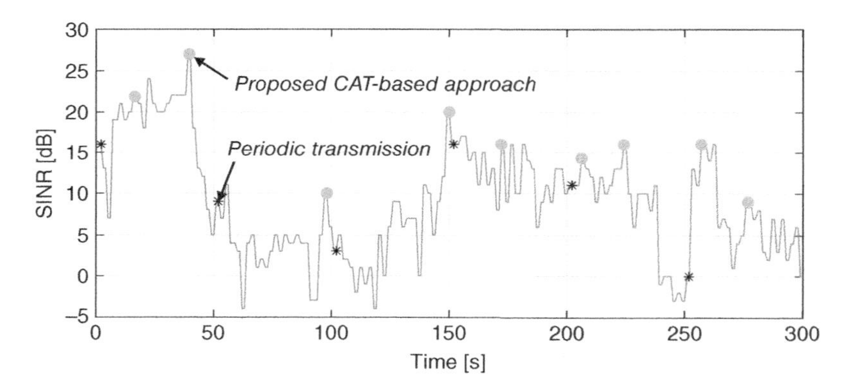

Figure 8.3 Example comparison of straightforward periodic data transfer and SINR-based CAT.

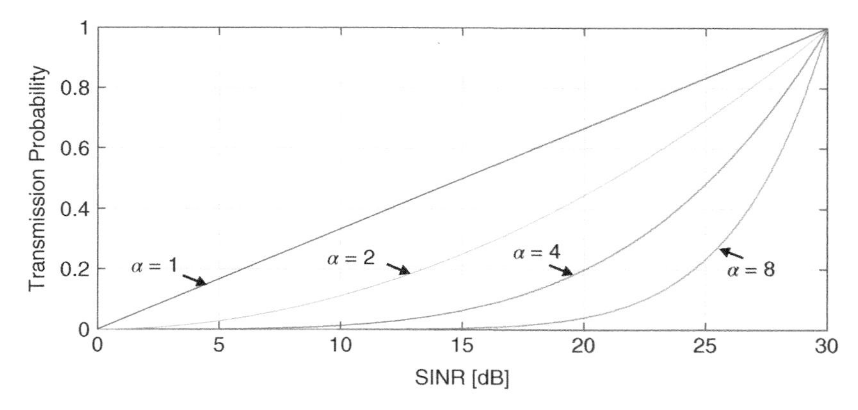

Figure 8.4 Analytical transmission probability for SINR-based CAT with different values for α ($\beta = 0$, $\text{SINR}_{\max} = 30$ dB).

decreases, it will likely decrease further in the near future before it will increase again, e.g. due to a cellular handover. In order to pay attention to this phenomenon, CAT was extended to pCAT by introducing a *temporal lookahead* τ in order to proactively detect low channel-quality periods. The extended transmissions scheme sends data early if it expects the channel quality to decrease and late if the channel quality is anticipated to improve. The prediction is based on the previous measurements that have been performed on the same track.

pCAT uses two timeout parameters in order to provide bounds for transmission efficiency and data freshness. $t_{\mathbf{min}}$ ensures a minimum payload size, and $t_{\mathbf{max}}$ is used to specify a maximum value for the buffering delay. The resulting transmission probability is computed as

$$p(t) = \begin{cases} 0 & : \Delta t \leq t_{\min} \\ \left(\dfrac{\text{SINR}}{\text{SINR}_{\max}}\right)^{\alpha \cdot z_1} & : t_{\min} < \Delta t < t_{\max}, \Delta\text{SINR}(t) > 0 \\ \left(\dfrac{\text{SINR}}{\text{SINR}_{\max}}\right)^{\alpha / z_2} & : t_{\min} < \Delta t < t_{\max}, \Delta\text{SINR}(t) \leq 0 \\ 1 & : \Delta t > t_{\max} \end{cases} \tag{8.2}$$

with

$$z_1 = \max\left(\Delta\text{SINR}(t) \cdot \left(1 - \frac{\text{SINR}(t)}{\text{SINR}_{\max}}\right) \cdot \gamma, 1\right), \tag{8.3}$$

$$z_2 = \max\left(\left|\Delta\text{SINR}(t) \cdot \left(\frac{\text{SINR}(t)}{\text{SINR}_{\max}}\right) \cdot \gamma\right|, 1\right) \tag{8.4}$$

and

$$\Delta\text{SINR}(t) = \overline{\text{SINR}}(t, t+\tau) - \text{SINR}(t) \tag{8.5}$$

The resulting transmission probability for pCAT is shown in Figure 8.5. It can be observed that pCAT formally meets the defined tasks: transmissions are delayed if the channel quality is expected to improve and performed early if a decrease is anticipated. Further results of empirical evaluations are described in Pillmann et al. (2017).

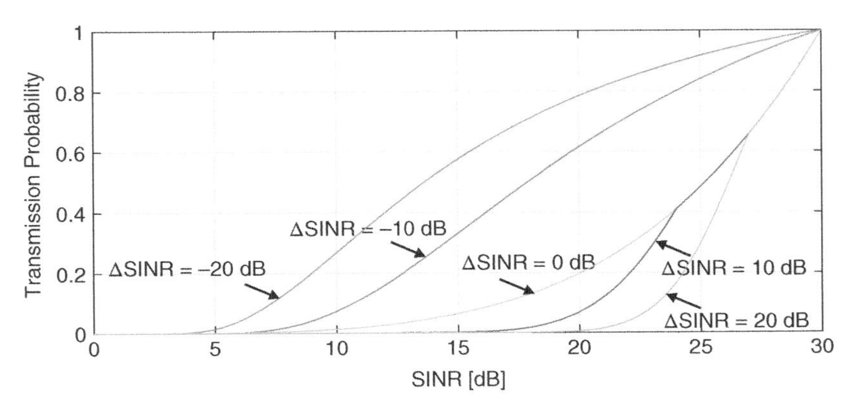

Figure 8.5 Analytical transmission probability for SINR-based pCAT with different values for $\Delta\text{SINR}(t)$ ($\alpha = 4$).

8.2.3 ML-based CAT

Due to the highly dynamic channel characteristics, the channel properties are likely changing during ongoing transmissions. Therefore, the meaningfulness of the channel assessment is limited, especially for larger packet sizes. Moreover, even for high channel coherence times, the resulting data rate is impacted by transport-layer interdependencies such as the slow start of the Transmission Control Protocol (TCP). Although the complexity of these factors exceeds the limits of analytical system descriptions, ML is able to uncover the hidden interdependencies between the different types of context information. Apart from the SINR, LTE offers further network quality indicators such as RSRP, RSRQ, and CQI that can be used to achieve a better channel quality assessment. Therefore, CAT can be extended to ML-CAT, which uses ML-based data-rate prediction as a transmission metric.

In the first step, the probabilistic transmission scheme of CAT is generalized, as for CAT and pCAT the metric calculation is implicitly normed to the value range of the SINR. Therefore, an abstract metric Φ is introduced, for which the value range is specified with defined values for Φ_{min} and Φ_{max}. Each measured value $\Phi(t)$ is converted in a *normed current metric* $\Theta(t)$ with

$$\Theta(t) = \frac{\Phi(t) - \Phi_{\text{min}}}{\Phi_{\text{max}} - \Phi_{\text{min}}} \tag{8.6}$$

The resulting transmission probability $p_{\Phi(t)}$ is then computed as

$$p_{\Phi(t)} = \begin{cases} 0 & : \Delta t \leq t_{\text{min}} \\ \Theta(t)^{\alpha} & : t_{\text{min}} < \Delta t < t_{\text{max}} \\ 1 & : \Delta t > t_{\text{max}} \end{cases} \tag{8.7}$$

The performance of different ML models for uplink data-rate prediction in vehicular scenarios is further analyzed in Sliwa et al. (2018b). The analysis shows that for all considered models, the M5 regression tree is able provide the highest prediction accuracy while still being able to be implemented in a very resource-efficient manner for online execution on embedded devices.

Based on these observations, the processing pipeline for the context information is derived in Figure 8.6. The prediction features are the channel context parameters RSRP, RSRQ, SINR, and CQI. Further information is added by the current vehicle speed and the payload size of the data packet. During the training phase, the resulting data rate of active transmission provides the label for the prediction.

Figure 8.7 shows an example of the temporal behavior of the proposed data-rate prediction mechanism in a real-world evaluation. It can be seen that the payload size has a severe impact on the overall performance, as the anticipated data rate significantly decreases after each transmission, which clears the current transmission buffer. For low payload sizes, the slow start of TCP has a dominant impact; for very high payload sizes, the channel dynamics become the major influence on overall behavior.

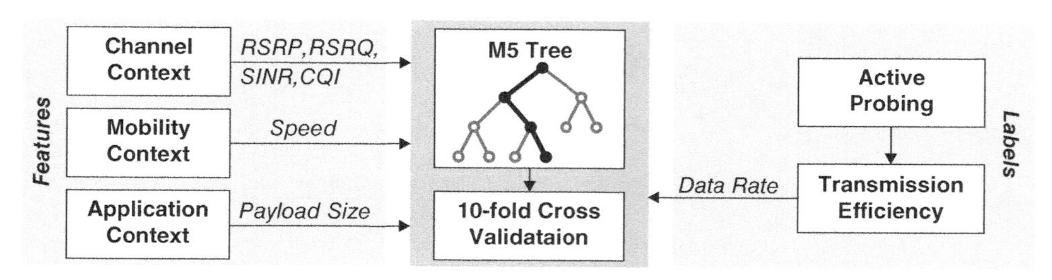

Figure 8.6 Learning pipeline for M5-based data-rate prediction using different types of context parameters.

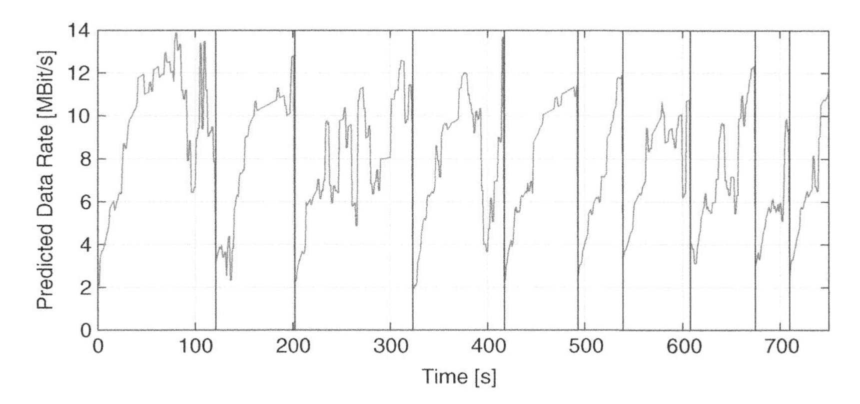

Figure 8.7 Example temporal behavior of the uplink data-rate prediction of a real-world drive test.

The resulting prediction accuracy for suburban and highway scenarios is shown in Figure 8.8. It can be observed that although both tracks have different velocity statistics, the prediction itself shows a similar behavior. It should be noted that, from an application perspective, prediction failures in the upper-left triangle region – referred to as *underestimations* – are not problematic for the considered use case as the end-to-end data rate is even higher than anticipated.

Although the focus of this work is on using ML-based data-rate prediction as a metric, the abstract definition of Θ allows the extension of ML-CAT for flexible use of other metrics as well, e.g. directly using the passive downlink indicators for scheduling the transmission decision. Furthermore, it enables the transfer of the proposed method to different communication technologies, e.g. for IEEE 802.11p-based vehicle-to-infrastructure communication (Sliwa et al. (2018d)).

Figure 8.9 shows the trained and pruned M5 regression tree used for the data-rate prediction. Since it only consists of threshold decisions and a small number of linear regression models, it can be implemented in a very resource-efficient manner using a sequence of `if/else`-statements and executed in real time.

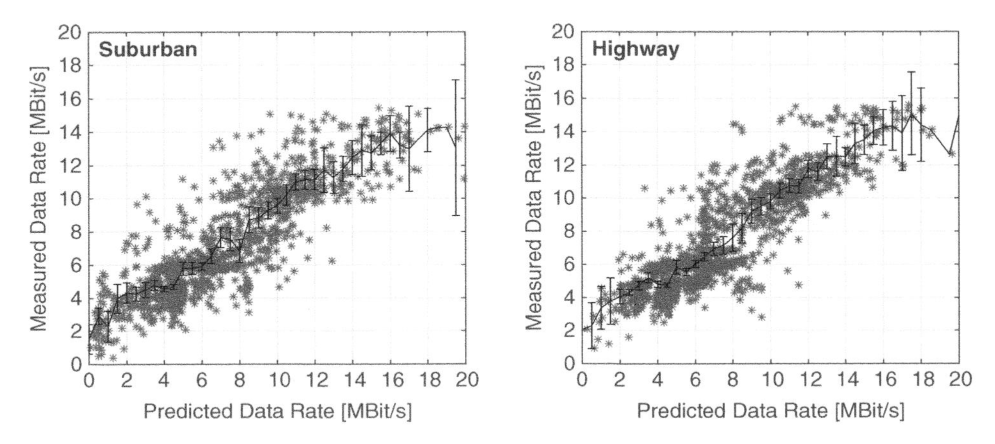

Figure 8.8 Performance of M5 regression tree data-rate predictions on real-world measurement data.

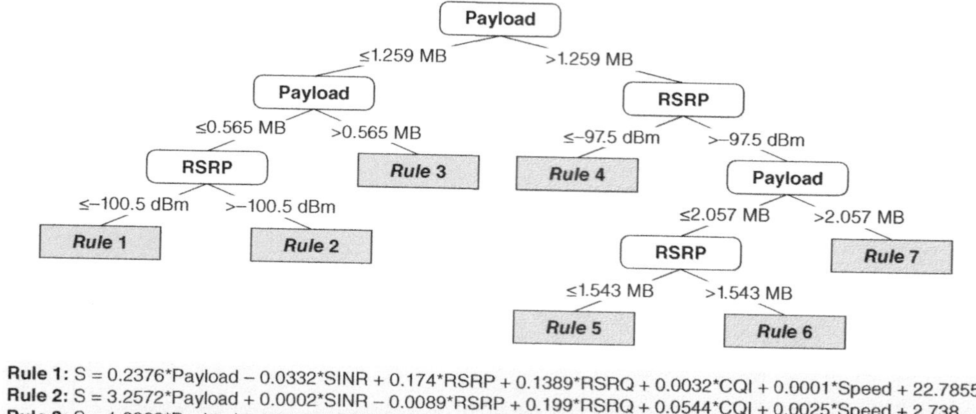

Rule 1: S = 0.2376*Payload – 0.0332*SINR + 0.174*RSRP + 0.1389*RSRQ + 0.0032*CQI + 0.0001*Speed + 22.7855
Rule 2: S = 3.2572*Payload + 0.0002*SINR – 0.0089*RSRP + 0.199*RSRQ + 0.0544*CQI + 0.0025*Speed + 2.738
Rule 3: S = 1.8368*Payload + 0.0002*SINR + 0.0126*RSRP + 0.225*RSRQ + 0.0008*CQI + 0.0048*Speed + 6.8621
Rule 4: S = 0.3136*Payload + 0.0026*SINR + 0.4682*RSRP + 0.0408*RSRQ – 0.0852*CQI + 54.619
Rule 5: S = 0.3383*Payload + 0.052*SINR + 0.0012*RSRP + 0.3109*RSRQ + 0.0009*CQI + 8.9411
Rule 6: S = 1.75*Payload + 0.0297*SINR + 0.0012*RSRP + 0.0436*RSRQ + 0.0489*CQI + 6.535
Rule 7: S = 0.5932*Payload + 0.0583*SINR + 0.001*RSRP + 0.4585*RSRQ – 0.0012*CQI + 12.0811

Figure 8.9 Trained and pruned M5 regression tree model for data-rate prediction, consisting of seven linear regression models.

8.2.4 ML-based pCAT

Finally, ML-pCAT brings together the ideas of pCAT and ML-CAT for context-predictive and ML-based sensor data transmission.

A blueprint for the composition and implementation of the different components is illustrated by the architecture model in Figure 8.10 In order to allow the estimation of the future channel context for all considered network quality indicators, all vehicles manage a *multi-layer connectivity map* that stores aggregated information of previous indicator measurements in a cellular structure. Based on the measured mobility context parameters, vehicles derive their future location by *mobility prediction*. The estimated future channel context is then derived as a lookup operation for the predicted cell within the connectivity map. In Sliwa et al. (2018a), different methods for predicting future vehicle locations and their effects on the accuracy of the obtained context information from connectivity maps are discussed in detail. In the following evaluations, only the trajectory-aware mobility prediction approach is applied, as it is able to provide precise and turn-robust predictions.

While the connectivity map can be generated in a decentralized manner by using only measurements of the respective UE, the approach highly benefits from large-scale, crowdsensing-based map generation. A reasonable approach for data synchronization would be to use a non-cellular data connection whenever it is available.

Figure 8.11 shows an example excerpt of the derived connectivity map within the overall evaluation scenario. Although the different indicators characterize larger road segments similarly, intensity and variance are significantly different. The transmission probability relies on the metric abstraction from Section 8.2.3 and is calculated as

$$p_{\Phi(t)} = \begin{cases} 0 & : \Delta t \leq t_{\min} \\ \Theta(t)^{\alpha \cdot z} & : t_{\min} < \Delta t < t_{\max} \\ 1 & : \Delta t > t_{\max} \end{cases} \tag{8.8}$$

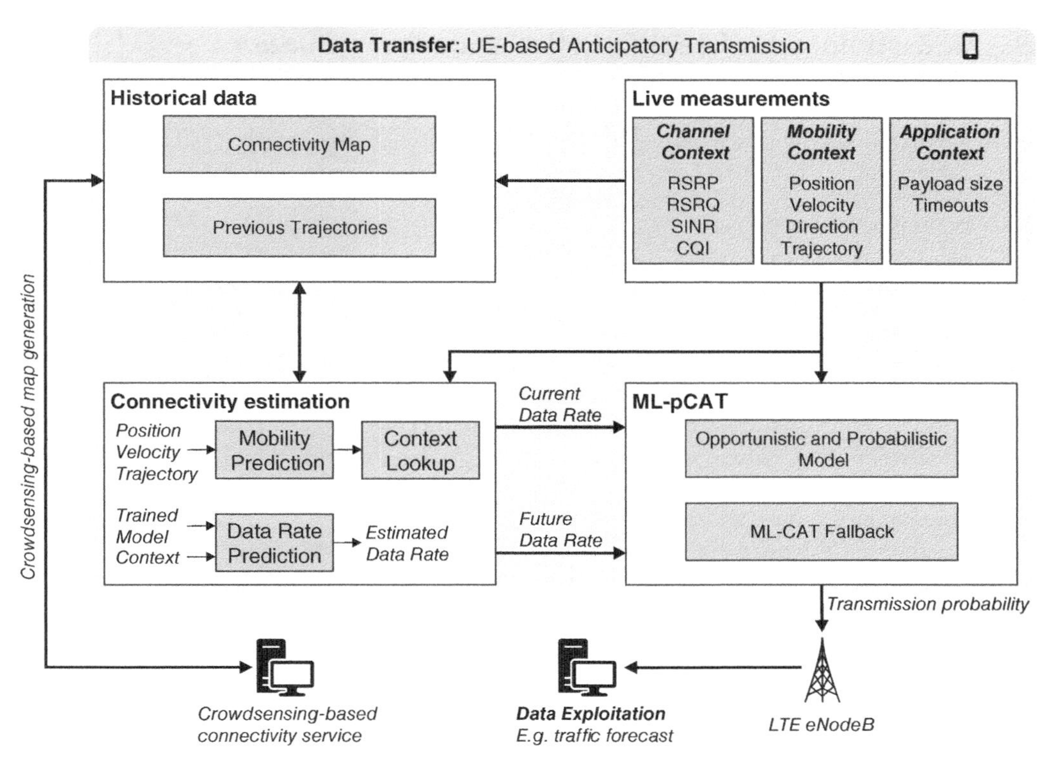

Figure 8.10 Overall architecture model for ML-pCAT-based data transmission.

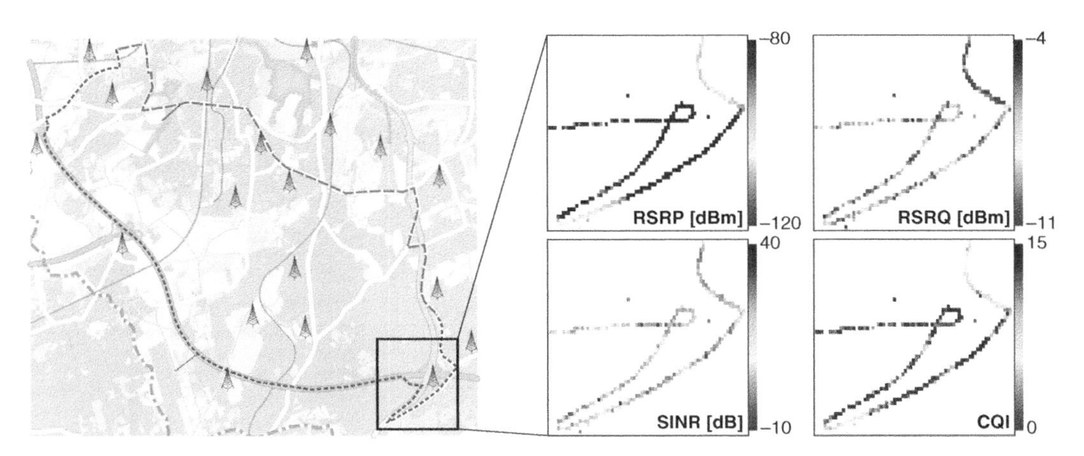

Figure 8.11 Overview of the evaluation scenario and excerpt of the connectivity map with cell width of 25 m. (Map: OpenStreetMap contributors, CC BY-SA.)

with

$$z = \begin{cases} \max(|\Delta\Phi(t) \cdot (1 - \Theta(t)) \cdot \beta|, 1) & : \Delta\Phi > 0 \\ (\max(|\Delta\Phi(t) \cdot \Theta(t) \cdot \beta|, 1))^{-1} & : \Delta\Phi \leq 0 \end{cases}. \tag{8.9}$$

8.3 Methodology for Real-World Performance Evaluation

In this section, the methodological setup for the real-world performance evaluation is presented. Furthermore, the process for the power consumption analysis is described.

8.3.1 Evaluation Scenario

The measurements are performed with off-the-self Android-based UE, which executes the CAT application and performs the online ML-based data-rate prediction. Further information about the framework has been published in Sliwa et al. (2018c). A virtual sensor application generates a fixed amount of sensor data per second, which is stored in a local buffer. The probabilistic transmission schemes assess the channel each t_p and compute a transmission probability for the whole buffer. All raw measurements (Sliwa (2018)) and the developed measurement application are provided in an open source way (available at https://github.com/BenSliwa/MTCApp). In total, the data set consists of more than 7500 transmissions, which were performed within a driving range of more than 2000 km.

The transmission schemes are evaluated in a real-world scenario using the public cellular network infrastructure. Figure 8.11 shows a map of the evaluation scenario. Drive tests are performed on different track types (suburban and highway) with different velocity characteristics.

Table 8.1 shows the parameters of the considered evaluation scenario. Φ_{SINR} represents the transmission metrics of CAT and pCAT. Φ_{ML} is applied for ML-CAT and ML-pCAT. The required information for the creation of the connectivity map is based on the data obtained from the experiments in Sliwa et al. (2018b).

8.3.2 Power Consumption Analysis

Apart from transmission efficiency, which is represented by the achieved data rate in the following section, the changed transmission behavior is likely having a severe impact on energy efficiency. While the communication-related battery drain is nearly negligible for cars, it has a crucial impact on the flight-time potential of small-scale aerial vehicles. As these types of vehicles are becoming more and more integrated into road traffic scenarios (Menouar et al. (2017)), the application of CAT-based data transfer for UAV-based data provisioning is a promising option for increasing the possible flight time.

Table 8.1 Parameters of the evaluation scenario.

Parameter	Value
Sensor data arrival rate	50 kByte / s
Channel assessment interval t_p	1 s
Minimum delay between transmissions t_{min}	10 s
Maximum buffering time t_{max}	120 s
Prediction lookahead τ	30 s
Connectivity map cell width	25 m
$\Phi_{\text{SINR}}\{\text{min, max}, \alpha, \beta\}$	{0 dB, 30 dB, 8, 0.5}
$\Phi_{\text{ML}}\{\text{min, max}, \alpha, \beta\}$	{0 MBit/s, 18 MBit/s, 8, 1}

Measuring the impact of data transmissions on power consumption of a mobile device is a nontrivial task, as it involves the isolation of the running process from the rest of the system. For evaluating the energy efficiency of the considered transmission schemes, we therefore rely on a sequential process that is executed in a post-processing step:

- **ML-based estimation of transmission power**: The model of Falkenberg et al. (2018) is utilized to predict the applied value of P_{TX} for each performed transmission from the passive downlink indicators, analogously to the process described in Section 8.2.3. As P_{TX} has a strong correlation with the distance to the base station, the RSRP provides a meaningful indicator for the learning process.
- **Determination of device-specific characteristics**: The relationship between P_{TX} and the actual power consumption is highly specific for the UE and depends on its different power amplifiers. Therefore, this dependency is captured by controlled variation of P_{TX} in a laboratory environment (Falkenberg et al. (2017b)).
- **Model-based analysis**: Finally, the state-based context-aware power consumption model (CoPoMo) (Dusza et al. (2013)) is applied to compute the average power consumption of the UE with respect to the different possible transmission states.

8.4 Results of the Real-World Performance Evaluation

In this section, the results of the empirical performance evaluation are presented and discussed. First, the behavior of the measured network quality indicators is discussed. Afterward, the results of the considered transmissions schemes are analyzed, and finally, an overall summary is provided.

8.4.1 Statistical Properties of the Network Quality Indicators

Figure 8.12 shows the statistical distributions of the measured channel context parameters during the real-world measurements. It can be observed that the reported values have different dynamics and allow the derivation of different conclusions about the channel quality. The RSRP has a clear Gaussian distribution around the center value of 90 dBm. For the RSRQ, higher values are more frequently reported than lower values. The empirical results of the channel quality indicator (CQI) show a peak for the amount of CQI reports for the value 2. Since this indicator is not standardized, its definition and implementation are specified by the modem manufacturer. Therefore, it can be assumed that the CQI is less significant than the other indicators, which is confirmed by the small CQI coefficients shown in Figure 8.9.

8.4.2 Comparison of the Transmission Schemes

Figure 8.13 shows the resulting data rate for the different considered transmission schemes. Three characteristic areas can be identified: periodic data transfer, SINR-based approaches, and ML approaches. The lowest data rate is achieved by the straightforward approach, which sends the data periodically, regardless of the channel conditions. With the opportunistic approach, SINR-based schemes CAT and pCAT are able to increase the data rate significantly. The probabilistic and channel-aware method is applied for exploiting channel-quality knowledge.

Finally, the highest values are achieved by the ML-based transmission schemes ML-CAT and ML-pCAT. The predicted data rate provides a better metric for channel quality assessment

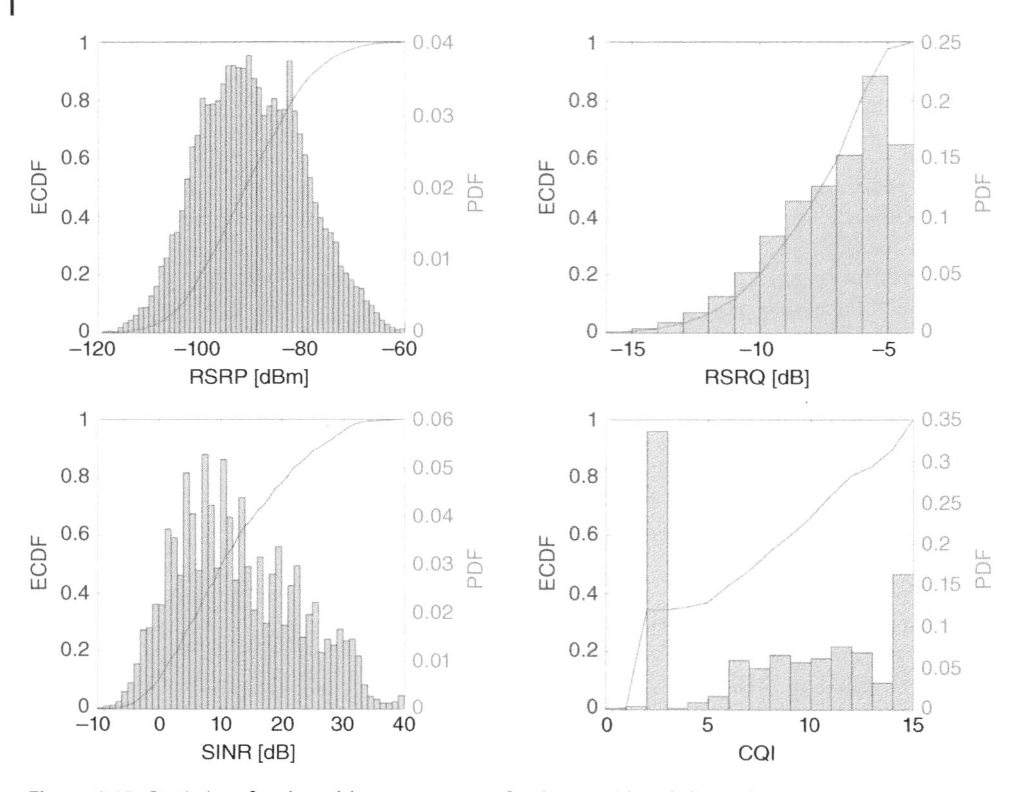

Figure 8.12 Statistics of real-world measurements for the considered channel context parameters.

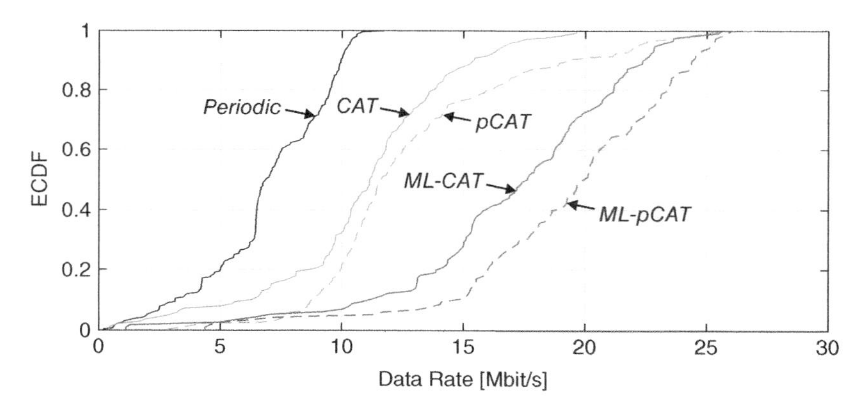

Figure 8.13 Comparison of the resulting data rate for the considered transmission schemes.

than pure SINR measurements. In particular, the integration of the payload size information allows the implicit consideration of hidden interdependencies of the network quality indicators in their significance for different channel coherence times. For both variants, additional benefits are achieved by the context-lookahead of the pCAT approach.

The empirical cumulative distribution function (ECDF) for the age of the sensor information contained in the transmitted packets is shown in Figure 8.14. As expected, the curve of the periodic approach is very steep due to the fixed interval. Rare deviations from the mean are related

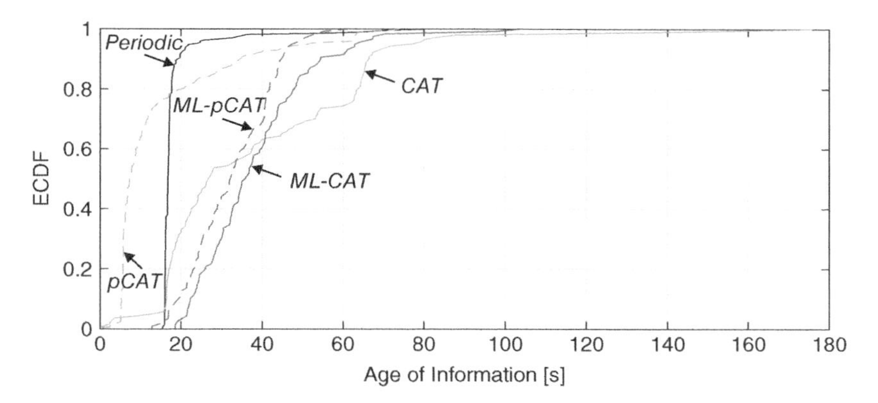

Figure 8.14 Comparison of the resulting age of information for the considered transmission schemes.

to scheduling effects and the need for data retransmissions and cellular handovers. For pCAT, the average age of information is even lower as transmissions are performed more frequently. As the vehicle moves forward on its trajectory, the channel lookahead τ behaves like a moving window. This behavior is intensified by the dynamics of the SINR metric, resulting in a high probability of experiencing SINR-peaks during the transmission evaluation.

Although generally, ML-pCAT is equal to pCAT with a different channel assessment metric, their behaviors show significant differences for the data age. The M5 tree-based prediction is far less dynamic than the SINR assessment (c.f. Figure 8.7). Therefore, it is less influenced by short-term influences (e.g. multipath propagation) that only have a minor significance in the considered scenario, as the vehicle will likely have moved to a different location when the transmission has been initialized. Since the payload size has a severe impact on the achievable data rate, the ML-based approach waits longer until a minimum transmission buffer size is achieved. CAT and ML-CAT, which only rely on the current channel quality measurements, show a similar behavior.

The impact of the optimized transmission behavior on the power consumption of the mobile device is shown in Figure 8.15. Here, a similar behavior as for the data-rate evaluations can be observed. The energy efficiency benefits from the improved transmission behavior in multiple ways. As transmissions are performed during better channel-quality periods, less transmission power needs to be applied in order to deliver the data, which is the most dominant factor for overall power consumption (Dusza et al. (2013)). In addition, as the transmissions are performed with higher data rates, the transmission duration itself is reduced, and the modem is able to stay in the `IDLE` state for longer time periods. As a result, ML-CAT reduces the average power consumption of the UE by 49.51 % and ML-pCAT by 58.29 %.

8.4.3 Summary

Figure 8.16 summarizes the key results of the considered transmission schemes in the domain's data rate, energy efficiency, and data freshness. Overall, it can be seen that there is a trade-off among gains in data rate, power consumption, and additional buffering delay. The flexible definition of the proposed scheme allows its behavior to be configured with respect to application requirements.

The great benefit of using ML for assessing channel quality in comparison to traditional approaches (e.g. SINR measurements) is the implicit consideration of hidden effects that have a

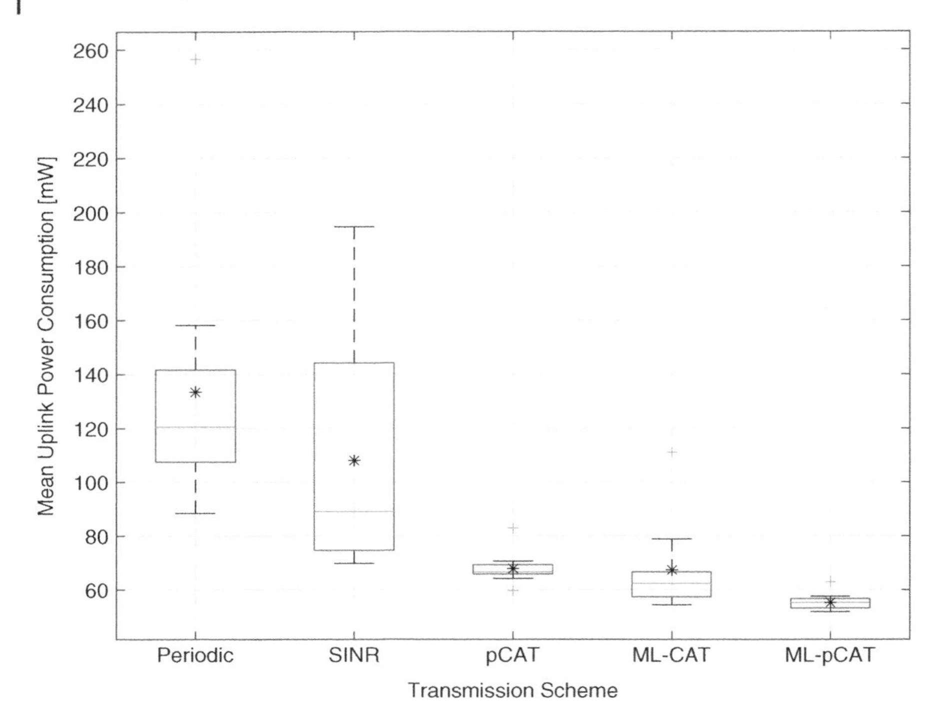

Figure 8.15 Comparison of the resulting power consumption efficiency for the considered transmission schemes.

strong influence on the overall behavior but are too complicated to be covered by a closed analytical description. As the proposed approach integrates the payload size and the velocity of the vehicle into the data-rate prediction, it implicitly considers payload versus overhead ratio, slow start of TCP, and the dependency on the channel coherence time.

8.5 Conclusion

In this chapter, we presented a ML-based approach for anticipatory data transmission for using mobile sensor nodes. Since the spectrum resources are highly limited and the medium is shared between different cell users, the overall goal is to improve transmission efficiency by freeing occupied resources as early as possible.

The presented ML-pCAT approach relies on context-predictive data transmission and the exploitation of favorable channel conditions in order to boost the resulting transmission efficiency. Through anticipatory and opportunistic data transfer, the existing network is utilized in a more efficient way for both individual UE and the network as a whole. As a side effect, the average power consumption of the UE is significantly reduced, which increases the operation lifetime of the mobile device.

The presented approach also shows how communication engineering and ML can benefit from each other to achieve better results. While the foundation of the proposed transmission scheme relies on analytical modeling and exploitation of expert knowledge, the integration of ML allows the consideration of hidden interdependencies between the variables, which are too complex to be described in an analytical way.

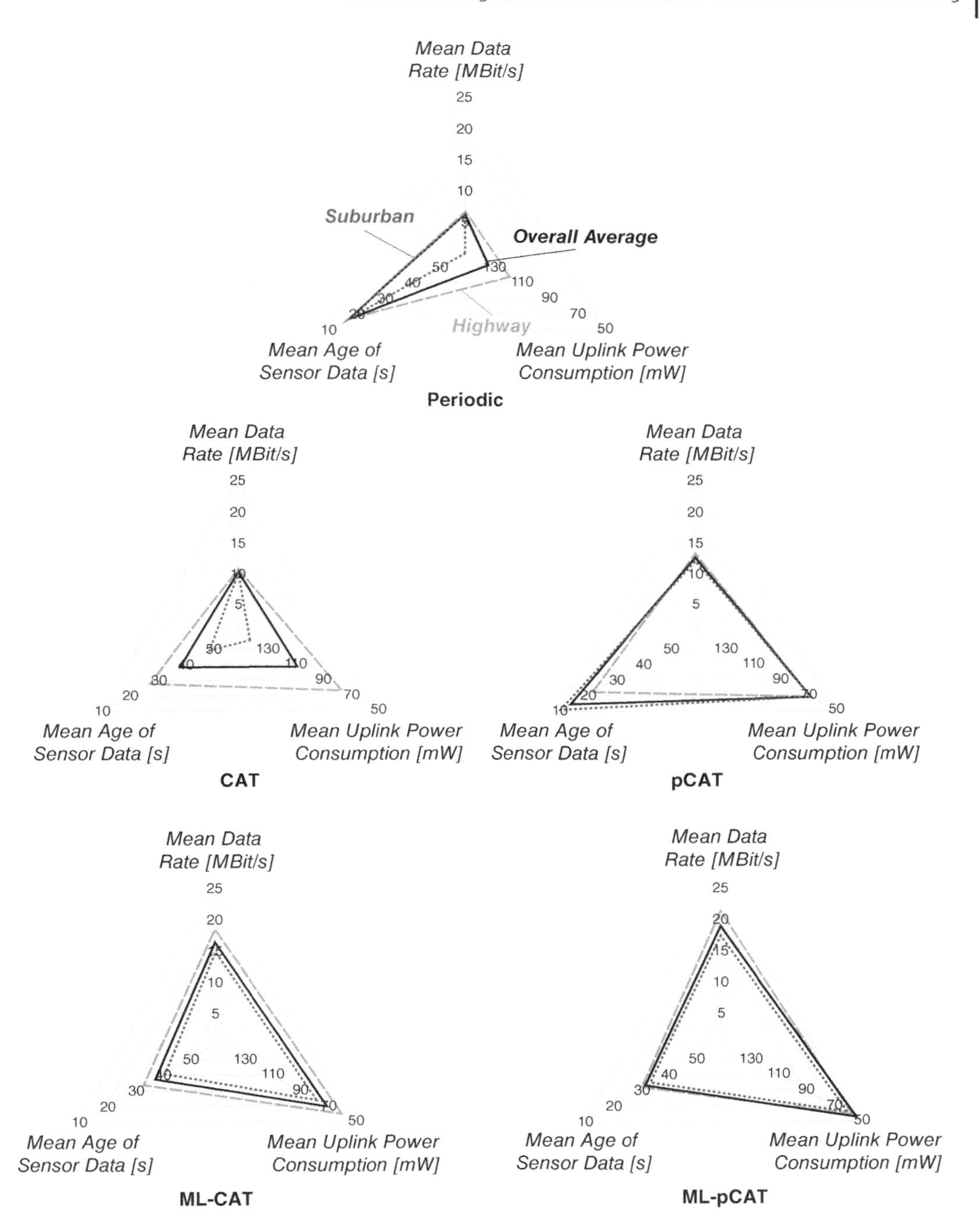

Figure 8.16 Summarizing comparison of the considered transmission schemes in the domain's data rate, energy efficiency, and data freshness. The scales for sensor data age and power consumption are inverted such that a large footprint means better performance.

In future work, we will investigate the applicability of ML-pCAT for multi-interface communication with heterogeneous network technologies. In addition, the data-rate prediction will be improved by considering information about the cell load obtained from passive probing of the LTE control channels.

Acknowledgments

Part of the work on this paper has been supported by Deutsche Forschungsgemeinschaft (DFG) within the Collaborative Research Center SFB 876 "Providing Information by Resource-Constrained Analysis," projects A4 and B4.

Bibliography

N. Bui and J. Widmer. Data-driven evaluation of anticipatory networking in LTE networks. *IEEE Transactions on Mobile Computing*, PP (99): 1–1, 2018. ISSN 1536-1233. doi: 10.1109/TMC.2018.2809750.

Nicola Bui, Matteo Cesana, S Amir Hosseini, Qi Liao, Ilaria Malanchini, and Joerg Widmer. A survey of anticipatory mobile networking: Context-based classification, prediction methodologies, and optimization techniques. *IEEE Communications Surveys & Tutorials*, 2017.

C.T. Calafate, K. Cicenia, O. Alvear, J.C. Cano, and P. Manzoni. Estimating rainfall intensity by using vehicles as sensors. In *2017 Wireless Days*, pages 21–26, March 2017. doi: 10.1109/WD.2017.7918109.

Elmano Ramalho Cavalcanti, Jose Anderson Rodrigues de Souza, Marco Aurelio Spohn, Reinaldo Cezar de Morais Gomes, and Anderson Fabiano Batista Ferreira da Costa. VANETs' research over the past decade: Overview, credibility, and trends. *SIGCOMM Comput. Commun. Rev.*, 48 (2): 31–39, May 2018. ISSN 0146-4833. doi: 10.1145/3213232.3213237.

C. Chen, T.H. Luan, X. Guan, N. Lu, and Y. Liu. Connected vehicular transportation: Data analytics and traffic-dependent networking. *IEEE Vehicular Technology Magazine*, 12 (3): 42–54, Sept 2017. ISSN 1556-6072. doi: 10.1109/MVT.2016.2645318.

L.J. Chen, Y.H. Ho, H.H. Hsieh, S.T. Huang, H.C. Lee, and S. Mahajan. ADF: An anomaly detection framework for large-scale PM2.5 sensing systems. *IEEE Internet of Things Journal*, 5 (2): 559–570, April 2018. doi: 10.1109/JIOT.2017.2766085.

S. Djahel, R. Doolan, G.M. Muntean, and J. Murphy. A communications-oriented perspective on traffic management systems for smart cities: Challenges and innovative approaches. *IEEE Communications Surveys Tutorials*, 17 (1): 125–151, Firstquarter 2015. ISSN 1553-877X. doi: 10.1109/COMST.2014.2339817.

Björn Dusza, Christoph Ide, Liang Cheng, and Christian Wietfeld. CoPoMo: A context-aware power consumption model for LTE user equipment. *Transactions on Emerging Telecommunications Technologies (ETT), Wiley*, 24 (6): 615–632, October 2013.

ETSI. Intelligent transport systems (ITS); vehicular communications; basic set of applications; definitions. Technical report, ETSI TR 102 638 V1.1.1 (2009-06), 2009.

Robert Falkenberg, Christoph Ide, and Christian Wietfeld. Client-based control channel analysis for connectivity estimation in LTE networks. In *IEEE Vehicular Technology Conference (VTC-Fall)*, Montréal, Canada, Sep 2016. IEEE. doi: 10.1109/VTCFall.2016.7880932.

Robert Falkenberg, Karsten Heimann, and Christian Wietfeld. Discover your competition in LTE: Client-based passive data rate prediction by machine learning. In *GLOBECOM 2017 - 2017 IEEE*

Global Communications Conference, pages 1–7, Singapore, Dec 2017a. doi: 10.1109/GLOCOM.2017.8254567.

Robert Falkenberg, Benjamin Sliwa, and Christian Wietfeld. Rushing full speed with LTE-advanced is economical - a power consumption analysis. In *IEEE Vehicular Technology Conference (VTC-Spring)*, Sydney, Australia, jun 2017b.

Robert Falkenberg, Benjamin Sliwa, Nico Piatkowski, and Christian Wietfeld. Machine learning based uplink transmission power prediction for LTE and upcoming 5G networks using passive downlink indicators. In *2018 IEEE 88th Vehicular Technology Conference (VTC-Fall)*, Chicago, USA, Aug 2018.

Z. Feng, Z. Feng, and T.A. Gulliver. Biologically inspired two-stage resource management for machine-type communications in cellular networks. *IEEE Transactions on Wireless Communications*, 16 (9): 5897–5910, Sept 2017.

C. Ide, B. Dusza, and C. Wietfeld. Client-based control of the interdependence between LTE MTC and human data traffic in vehicular environments. *IEEE Transactions on Vehicular Technology*, 64 (5): 1856–1871, May 2015. ISSN 0018-9545. doi: 10.1109/TVT.2014.2337516.

C. Ide, R. Falkenberg, D. Kaulbars, and C. Wietfeld. Empirical analysis of the impact of LTE downlink channel indicators on the uplink connectivity. In *2016 IEEE 83rd Vehicular Technology Conference (VTC Spring)*, pages 1–5, May 2016. doi: 10.1109/VTCSpring.2016.7504211.

C. Jiang, H. Zhang, Y. Ren, Z. Han, K.C. Chen, and L. Hanzo. Machine learning paradigms for next-generation wireless networks. *IEEE Wireless Communications*, 24 (2): 98–105, April 2017. ISSN 1536-1284. doi: 10.1109/MWC.2016.1500356WC.

Florian Jomrich, Alexander Herzberger, Tobias Meuser, Björn Richerzhagen, Ralf Steinmetz, and Cornelius Wille. Cellular bandwidth prediction for highly automated driving - Evaluation of machine learning approaches based on real-world data. In *Proceedings of the 4th International Conference on Vehicle Technology and Intelligent Transport Systems 2018*, number 4, pages 121–131. SCITEPRESS, Mar 2018. ISBN 978-989-758-293-6.

H. Menouar, I. Guvenc, K. Akkaya, A.S. Uluagac, A. Kadri, and A. Tuncer. UAV-enabled intelligent transportation systems for the smart city: Applications and challenges. *IEEE Communications Magazine*, 55 (3): 22–28, March 2017. ISSN 0163-6804. doi: 10.1109/MCOM.2017.1600238CM.

Johannes Pillmann, Benjamin Sliwa, Christian Kastin, and Christian Wietfeld. Empirical evaluation of predictive channel-aware transmission for resource efficient car-to-cloud communication. In *IEEE Vehicular Networking Conference (VNC)*, Torino, Italy, Nov 2017.

T. Pögel and L. Wolf. Optimization of vehicular applications and communication properties with connectivity maps. In *2015 IEEE 40th Local Computer Networks Conference Workshops (LCN Workshops)*, pages 870–877, Oct 2015. doi: 10.1109/LCNW.2015.7365940.

M. Sepulcre, J. Gozalvez, O. Altintas, and H. Kremo. Context-aware heterogeneous V2I communications. In *2015 7th International Workshop on Reliable Networks Design and Modeling (RNDM)*, pages 295–300, Oct 2015. doi: 10.1109/RNDM.2015.7325243.

Miguel Sepulcre and Javier Gozalvez. Context-aware heterogeneous V2X communications for connected vehicles. *Computer Networks*, 136: 13–21, 2018. ISSN 1389-1286. doi: https://doi.org/10.1016/j.comnet.2018.02.024.

Qi Shi and Mohamed Abdel-Aty. Big data applications in real-time traffic operation and safety monitoring and improvement on urban expressways. *Transportation Research Part C: Emerging Technologies*, 58: 380–394, 2015. ISSN 0968-090X. Big Data in Transportation and Traffic Engineering.

Benjamin Sliwa. Raw experimental cellular network quality data. Mar 2018. doi: 10.5281/zenodo.1205778. URL http://doi.org/10.5281/zenodo.1205778.

Benjamin Sliwa, Daniel Behnke, Christoph Ide, and Christian Wietfeld. B.A.T.Mobile: Leveraging mobility control knowledge for efficient routing in mobile robotic networks. In *IEEE*

GLOBECOM 2016 Workshop on Wireless Networking, Control and Positioning of Unmanned Autonomous Vehicles (Wi-UAV), Washington D.C., USA, Dec 2016.

Benjamin Sliwa, Thomas Liebig, Robert Falkenberg, Johannes Pillmann, and Christian Wietfeld. Machine learning based context-predictive car-to-cloud communication using multi-layer connectivity maps for upcoming 5G networks. In *2018 IEEE 88th Vehicular Technology Conference (VTC-Fall)*, Chicago, USA, Aug 2018a.

Benjamin Sliwa, Thomas Liebig, Robert Falkenberg, Johannes Pillmann, and Christian Wietfeld. Efficient machine-type communication using multi-metric context-awareness for cars used as mobile sensors in upcoming 5G networks. In *2018 IEEE 87th Vehicular Technology Conference (VTC-Spring)*, Porto, Portugal, Jun 2018b. Best Student Paper Award.

Benjamin Sliwa, Thomas Liebig, Robert Falkenberg, Johannes Pillmann, and Christian Wietfeld. Resource-efficient transmission of vehicular sensor data using context-aware communication. In *19th IEEE International Conference on Mobile Data Management (MDM)*, Aalborg, Denmark, Jun 2018c.

Benjamin Sliwa, Johannes Pillmann, Maximilian Klaß, and Christian Wietfeld. Exploiting map topology knowledge for context-predictive multi-interface car-to-cloud communication. In *2018 IEEE 88th Vehicular Technology Conference (VTC-Fall)*, Chicago, USA, Aug 2018d.

Chaoming Song, Zehui Qu, Nicholas Blumm, and Albert-László Barabási. Limits of predictability in human mobility. *Science*, 327 (5968): 1018–1021, 2010. ISSN 0036-8075. doi: 10.1126/science.1177170.

W. Vandenberghe, E. Vanhauwaert, S. Verbrugge, I. Moerman, and P. Demeester. Feasibility of expanding traffic monitoring systems with floating car data technology. *IET Intelligent Transport Systems*, 6 (4): 347–354, Dec 2012. ISSN 1751-956X. doi: 10.1049/iet-its.2011.0221.

S. Wang, S. Kodagoda, L. Shi, and H. Wang. Road-terrain classification for land vehicles: Employing an acceleration-based approach. *IEEE Vehicular Technology Magazine*, 12 (3): 34–41, Sept 2017. ISSN 1556-6072. doi: 10.1109/MVT.2017.2656949.

X. Wang, X. Zheng, Q. Zhang, T. Wang, and D. Shen. Crowdsourcing in ITS: The state of the work and the networking. *IEEE Transactions on Intelligent Transportation Systems*, 17 (6): 1596–1605, June 2016. ISSN 1524-9050. doi: 10.1109/TITS.2015.2513086.

Christian Wietfeld, Christoph Ide, and Björn Dusza. Resource-efficient wireless communication for mobile crowd sensing. In *51st ACM/EDAC/IEEE Design Automation Conference (DAC)*, San Francisco, USA, June 2014. IEEE.

A. Zanella, N. Bui, A. Castellani, L. Vangelista, and M. Zorzi. Internet of things for smart cities. *IEEE Internet of Things Journal*, 1 (1): 22–32, Feb 2014. ISSN 2327-4662. doi: 10.1109/JIOT.2014.2306328.

K. Zheng, Q. Zheng, P. Chatzimisios, W. Xiang, and Y. Zhou. Heterogeneous vehicular networking: A survey on architecture, challenges, and solutions. *IEEE Communications Surveys Tutorials*, 17 (4): 2377–2396, Fourthquarter 2015. ISSN 1553-877X. doi: 10.1109/COMST.2015.2440103.

Part II

Transmission Intelligence and Adaptive Baseband Processing

9

Machine Learning–Based Adaptive Modulation and Coding Design

Lin Zhang[1] *and Zhiqiang Wu*[2]

[1] *School of Electronics and Information Technology, Sun Yat-sen University, Guangzhou, China*
[2] *Department of Electrical Engineering, Wright State University, Dayton, Ohio, USA*

9.1 Introduction and Motivation

Adaptive modulation and coding (AMC) technologies have been widely used in wired and wireless communication systems in order to adapt to the variations of channel conditions. AMC schemes can help communication systems to achieve higher spectrum efficiency and a better trade-off between data rate and reliability via adjusting the modulation order and/or coding rate according to real-time channel conditions. Since wireless channels are usually more dramatically time-variant than wired channels, most standards for wireless communication, such as the Wi-Fi standards IEEE 802.11n and IEEE 802.11ac, propose to employ AMC for improving spectrum efficiency in order to meet user demands. In order to further improve spectrum efficiency and reliability, AMC schemes have been combined with multiple-input and multiple-output (MIMO) and orthogonal frequency division multiplexing (OFDM). Owing to the enhanced reliability, these AMC-aided systems allow reduced overhead and latency, while meeting the upper-layer demand for real-time transmissions.

However, due to the complicated operational scenarios and time-varying nature of wireless channels, in wireless communication systems such as IEEE 802.11n-based wireless networks, the physical layer configuration/link adaptation is difficult to implement in practice. Consequently, optimal physical layer configurations have rarely been achieved in practice (Iera et al., 2005; Daniels et al., 2010), making the AMC schemes not well adapted to the channel's states, and thereby yielding performance degradation.

In addition, when AMC schemes are combined with MIMO and OFDM, modeling the joint effect of OFDM modulation, convolutional coding, and MIMO processing is highly challenging due to the following issues. First, in a practical system, there are various imperfections, such as nonlinear distortion of power amplifiers, quantization error of the analog-digital converters (ADCs), and non-Gaussian additive noise. In this case, accurate modeling of a complete telecommunication system is difficult. Second, in the design of communication systems, it is the convention to split the signal processing into multiple independent blocks, with each executing a specific and isolated function. Consequently, jointly optimizing these components leads to computationally complex systems (OShea and Hoydis, 2017). For the reasons as above-mentioned, existing AMC implementations are either inaccurate due to the model-based approximations or cumbersome due to the large-size lookup tables (Peng et al., 2007; Jensen et al., 2010). Therefore, in order to save the computation expense as well as to avoid making impractical approximations, such as the Gaussian approximation of non-Gaussian

Machine Learning for Future Wireless Communications, First Edition. Edited by Fa-Long Luo.
© 2020 John Wiley & Sons Ltd. Published 2020 by John Wiley & Sons Ltd.

distributions, a concise approach is desired to enable AMC schemes to be operated with mathematical models that need as few assumptions as possible.

On the other side, machine learning (ML), or "the programming of computers to optimize a performance criterion using example data" (Alpaydin, 2009), has the potential to jointly optimize the rigid modular blocks using a unified nonlinear framework. Hence, it may serve as a good candidate for the optimization of AMC-aided wireless systems having specific hardware configurations, when communicating over wireless channels.

Based on these considerations, in this chapter, we apply ML techniques to AMC-aided wireless systems to allow them to adapt to the variations of channels. We introduce and analyze two types of ML-assisted AMC schemes in the context of the MIMO and OFDM scenarios, where the conflicting demands for high date rate and high reliability are met by adjusting the modulation order and coding rate. Specifically, a supervised learning approach and a reinforcement learning approach are considered.

Here we first provide an overview of the ML-assisted AMC.

9.1.1 Overview of ML-Assisted AMC

Supervised learning (SL) approaches have the following advantages to facilitate their applications in AMC. First, during the operation of the communication network, training data can be collected online for updating the dataset. Second, there are abundant training datasets, including cyclic redundancy checks (CRCs), channel state information (CSI), the associated modulation and coding scheme (MCS), etc. These data can be used to update the training set for learning AMC in a specific wireless network, or be reused for training in other networks via transfer learning (Pan et al., 2010). Lastly, non-ideal operations that are often approximated or neglected in the model-based approach, such as nonlinear amplifications and finite resolutions in the analog circuit components (Daniels et al., 2008), are welcomed by the SL techniques when performing link adaptation, thanks to the black box approach.

One drawback of the SL approaches is that the sample data obtained can hardly represent accurately all the situations that signals may experience, including time-varying wireless channel conditions, nonlinear behavior of the amplifier, non-Gaussian distributed noises and interference, etc. (Leite et al., 2012). Hence, SL approaches may be infeasible for online learning, where the feature set could be inconstant and collecting large number of training examples could be impractical.

By contrast, reinforcement learning (RL) and Markov decision process (MDP) methods are able to directly learn from the environment, which thus provides solutions for the online learning required by communication systems. Unlike SL approaches, which learn from examples given by external supervisors, RL methods have the capability to learn online. Recently, (Jiang et al., 2017) has introduced RL to the field of wireless communications (Jiang et al., 2017). In (Leite et al., 2012), a Q-learning-based framework has been proposed for selecting the best MCS at a given signal-to-noise ratio (SNR). In (Melián-Gutiérrez et al., 2015), the authors have proposed an algorithm based on a hybrid hidden Markov decision and upper confidence bound (UCB), which is used to optimize the performance of secondary users in cognitive radios. However, a main drawback of using this technique in AMC is that the interacting time with the environment might be too long for it to select the preferable MCS. This is explicitly undesirable for operation in time-variant wireless channels. In order to mitigate this problem, some offline training may be required.

As we know, ML-assisted AMC has so far mainly been implemented in the context of SL or RL. Although unsupervised learning, like principal component analysis (PCA), has been used in (Daniels et al., 2008) to reduce the feature dimensions, its final decision is still accomplished with the aid of the k-nearest neighbor(k-NN) algorithm, which is a SL method. The reason is that unsupervised learning is mainly suitable for data preprocessing, but not suitable for decision-making, such as in AMC.

Considering SL, the most widely used learning algorithms include support vector machines (SVMs), decision trees, k-NN, neural networks, and so on. All these algorithms require labeled data. They can all be operated as classifiers after appropriate training. Therefore, without loss of generality, in the following sections, we will only consider k-NN and SVM as our two examples, in order to show the applications of SL in AMC. For RL, we will address Q-learning, since it has been widely investigated in literature (Venkatraman et al., 2010; Oksanen et al., 2010; Leite et al., 2012).

In the following sections, we will first present the principles of both SL-assisted and RL-assisted AMC. Then the performance of the AMC schemes is investigated. In our investigation, the specifications and parameters defined in the IEEE 802.11n standards are used. Note that IEEE 802.11n is the underlying protocol of the Wi-Fi Alliance, which is the first Wi-Fi protocol to support MIMO-OFDM transmissions.

The rest of this chapter is organized as follows. Section 9.1.2 provides a brief overview of the AMC schemes specified in the IEEE 802.11n, and gives details of the modulation types and coding rates used in the standards. Then, Section 9.2 is devoted to SL-assisted AMC, where both the k-NN and SVM approaches are considered. In Section 9.3, the RL- and MDP-based AMC schemes are addressed. Corresponding simulation results and analyses are provided at the end of Section 9.2 and Section 9.3, respectively. The last section of this chapter provides some further discussion and observations.

9.1.2 MCS Schemes Specified by IEEE 802.11n

In this subsection, we briefly introduce the MCS schemes specified by the IEEE 802.11n standards. Part of the MCS specifications are shown in Table 9.1, where MCS_i indicates the i_{th}

Table 9.1 List of modulation & coding schemes defined in IEEE 802.11n.

MCS_i	Nss	Modulation type	Coding rate	Date rate (Mbps)
0	1	BPSK	1/2	6.5
1	1	QPSK	1/2	13
2	1	QPSK	3/4	19.5
3	1	16QAM	1/2	26
4	1	16QAM	3/4	39
5	1	64QAM	2/3	52
6	1	64QAM	3/4	58.5
7	1	64QAM	5/6	65
8	2	BPSK	1/2	13
9	2	QPSK	1/2	26
10	2	QPSK	3/4	39
11	2	16QAM	1/2	52
12	2	16QAM	3/4	78
13	2	64QAM	2/3	104
14	2	64QAM	3/4	117
15	2	64QAM	5/6	130

MCS, and Nss represents the number of spatial streams that may be transmitted simultaneously via the corresponding air interfaces of IEEE 801.11n transceivers. As shown in Table 9.1, the modulation types include binary phase shift keying (BPSK) and quadrature phase shift keying (QPSK), which transmit, respectively, one bit and two bits per symbol by modulating information using two phase states and four phase states. The 16 quadrature amplitude modulation (QAM) and 64QAM transmit respectively 4 bits and 6 bits per symbol, with the aid of 16 constellation points and 64 constellation points uniformly distributed on the corresponding constellation maps. In Table 9.1, the ratio of the number of information bits and the number of coded bits input to modulate is defined as the *coding rate*. The modulation level and coding rate determine the data rate of the system, expressed as the number of bits per second (bps).

Note that, in the IEEE 802.11n standards, MCSs are indexed from 0–31, corresponding to different MCSs and hence different data rates. In Table 9.1, we only list the parameters of the MCSs with indexes from 0–15. Explicitly, the data rate achieved by these MCSs ranges from 6.5–130 Mbps. In general, when a higher order of modulation type and/or higher coding rate are employed, and/or two spatial streams instead of one are employed, the system can achieve a higher date rate.

Therefore, in practice, when a wireless channel is time-variant, the system may adapt to the corresponding communications situation by employing an appropriate MCS via activating a corresponding MCS index, in order to attain the best possible trade-off between data rate and reliability. However, identifying a near-optimum MCS in responding to time-varying wireless channels is not an easy task. In the following sections, we will introduce a range of AMC schemes based on the SL and RL methods.

9.2 SL-Assisted AMC

In the principle of ML, a SL algorithm aided by some training data generates an inferred function, which is thereby used for mapping new examples. In SL, labeled training data are constituted by a set of training examples, each of which contains an input object and its desired output value. The set of training examples is used to produce an inferred function.

Naturally, it is expected that the class of labels can be correctly determined for the unknown instances by the learning algorithm. Hence, it is required that the learning algorithm is able to adapt to unknown situations by using the training data intelligently and reasonably. In supervised ML, the most widely used learning algorithms include SVM, decision trees, the k-NN algorithm, neural networks, and so on. Here, we will consider k-NN and SVM as two examples to illustrate their applications in AMC schemes.

9.2.1 k-NN-Assisted AMC

With the objective to maximize throughput under the constraint of a given packet error rate (PER), the link adaptation relying on AMC is required to measure a specific feature set in order to determine the near-optimal AMC parameters – the modulation order and coding rate – as defined in Table 9.1. In literature, PER has often been selected as the performance metric to determine an MCS index (Kant et al., 2007), which may be formulated as:

$$i^* = \underset{i}{\text{argmax}}\{R_i : PER_i \leq PER_{target}\}, \tag{9.1}$$

where i^* represents the near-optimal MCS index identified by solving this optimization problem, and the index i corresponding to MCS_i is defined by the IEEE 802.11n standards as shown in Table 9.1. R_i is the throughput when MCS_i is used, PER_i is the corresponding PER, and PER_{target} is the constraint on PER. Therefore, if PER_i is the largest PER that is less than or equal to PER_{target} for a given channel realization, then MCS_i is selected for data transmission.

When k-NN-assisted AMC is considered, the previous optimization problem is transformed into a classification problem, which is then solved using the k-NN algorithm (Daniels et al., 2008; YİĞİT and Kavak, 2013). Note that k-NN is a non-parametric SL algorithm that can be used for both classification and regression, which is detailed as follows.

9.2.1.1 Algorithm for *k*-NN-Assisted AMC

From the previous discussion, we know that AMC in wireless communication is a typical classification problem. In order to fulfill the classification for a desired MCS, the feature set for training should be extracted first. We assume that there are W distinct realizations in the training set, expressed as $\mathcal{W} = \{0, 1, \cdots, W-1\}$. Then, after the training of the k-NN system, each realization $\omega \in \mathcal{W}$ is assigned to a class, i.e. a MCS, $i \in \mathcal{I}$, where $\mathcal{I} = \{0, 1, \cdots, I-1\}$ by solving the problem of Eq. (9.1).

To be more explicit, let z_ω represent the feature set obtained from training, using the realization of $\mathcal{W}$ that can be expressed as

$$\{z_\omega\} \Rightarrow \{i(\omega)\}, \tag{9.2}$$

where $i(\omega)$ is the class MCS_i obtained from the realization of ω. Thus, after training, a mapping table between the feature set and its corresponding MCSs can be obtained. With the aid of this table, once a new z_ω, corresponding to a communication situation, is observed, the k-NN algorithm can determine a corresponding MCS scheme to achieve AMC transmission. It is worth pointing out that in practical AMC classification applications, the k-NN algorithm has been improved to enhance training efficiency in (Daniels et al., 2008).

Algorithm 1 summarizes the k-NN algorithm for AMC by using a feature query $\mathbf{q}$. In this algorithm, k nearest neighbors ω_a $(a \in \{1, 2, \cdots, k\})$ are listed first. Then, the classes that occur most often among $\{\omega_a\}$ are selected. Finally, the class having the smallest i, which leads to the lowest throughput R_i is chosen, in order to guarantee the required PER.

Algorithm 1 k-NN algorithm for AMC (Daniels et al., 2008)

Require: $n_{i(\omega)} \leftarrow 0 \; \forall i(\omega) \in \mathcal{I}$
1: **for** $a = 1 \rightarrow k$ **do**
2: $\quad \omega_a \leftarrow \arg\min_{\omega}\{d(\mathbf{z}_\omega, \mathbf{q}) : \omega \notin \{\omega_1, \cdots, \omega_{a-1}\}\}$
3: $\quad n_{i(\omega)} \leftarrow n_{i(\omega)} + 1$
4: **end for**
5: $modeset = \arg\max_i\{n_i\}$
6: **return** $\min\{\arg\min_i\{R_i : i \in modeset\}\}$

As an example, Figure 9.1 illustrates the k-NN algorithm for AMC, where $k = 5$ is assumed. According to Algorithm 1, the first five nearest neighbors are identified, as shown by the five points together with their labels inside the circle in Figure 9.1. Then, the classes that occur most often among the feature query $\mathbf{q}$ are identified. In this example, the query is represented by the black-filled square. As shown by the classes in the circle, there are two neighbors with MCS_1 $(i = 1)$, two neighbors with MCS_8 $(i = 8)$, and one neighbor with MCS_4 $(i = 4)$. Hence, the selected classes are $i = 1$ and $i = 8$. Finally, the class indexed by the smallest i is chosen,

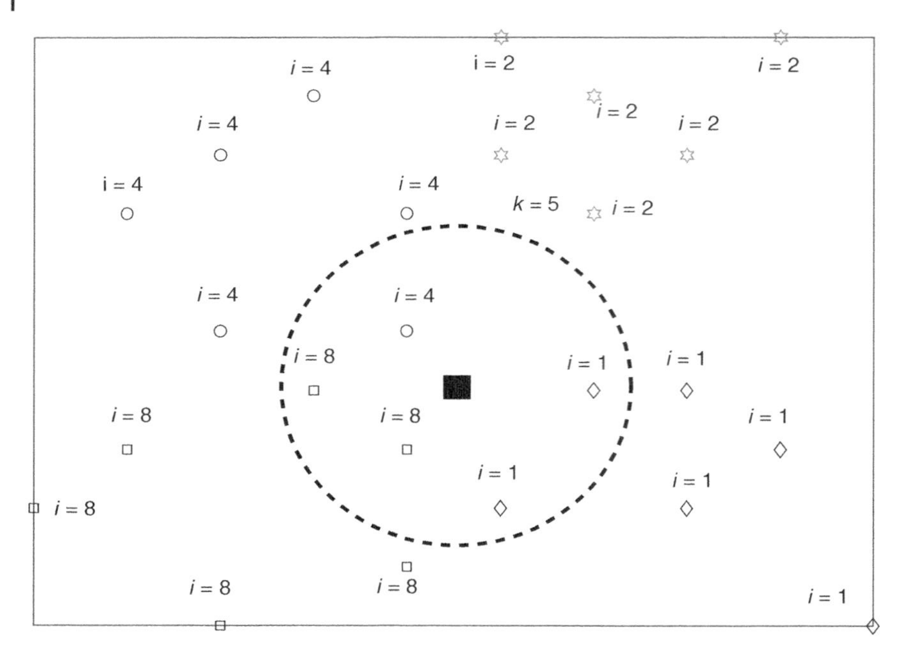

Figure 9.1 Illustration of MCS class selection in a 5-NN algorithm.

which fields the smallest throughput R_i to guarantee the required PER. In this specific example, the throughput $R_1 = R_8$, as shown in Table 9.1, but this is not necessarily true.

Additionally, as shown in Table 9.1, there are classes $i' < i$, but the throughput $R_i < R_{i'}$. In this case, although classes $\{i, i'\}$ are selected from the first step, class i is chosen as the last. In other words, given a selected class set $\{i_1, i_2, \cdots, i_n\}$, the final class selected is the class generating the lowest throughput. This is because the lowest throughput yields the highest reliability. Therefore, the previous selection rule could guarantee the required reliability.

9.2.2 Performance Analysis of *k*-NN-Assisted AMC System

In the MIMO-OFDM-based IEEE 802.11n wireless systems (Daniels et al., 2008; YİĞİT and Kavak, 2013) with N_t transmit antennas and N_r receive antennas, after the discrete Fourier transform (DFT), the signal received from the nth ($n \in \{1, 2, \cdots, N\}$) subcarrier can be represented as:

$$\mathbf{y}_n = \sqrt{E_s}\mathbf{H}_n\mathbf{x}_n + \mathbf{v}_n, \tag{9.3}$$

where N is the DFT size and also the number of subcarriers; $\mathbf{x}_n$ is the transmit symbol vector; $\mathbf{v}_n \sim CN(0, N_0\mathbf{I})$ is the additive complex Gaussian noise vector, with each element having a zero mean and a variance of N_0; $\mathbf{H}_n$ represents the $N_r \times N_t$ channel matrix of the nth subcarrier; and E_s represents the expected total received signal energy. Therefore, we have $E[|\mathbf{x}_n|^2] = 1\ \forall n$.

In the performance analysis, we assume that the wireless channel $\mathbf{H}_n$ undergoes quasi-static block fading, meaning the channel states remain constant across all OFDM symbols within one packet, but are independent for different packets. In addition to this assumption, the following assumptions (Daniels et al., 2008) are also employed in our analysis:

- *Fixed packet length*: All packets are set to 128 bytes. Hence, for a given bit error rate, all packets have the same expected PER.

- *Perfect synchronization*: All packets are synchronized perfectly in both the time domain and frequency domain. Therefore, any time and frequency offsets can be ignored.
- *Estimation*: Unlike (Daniels et al., 2008), we assume that the noise power and channels estimation are estimated at receivers. Corresponding to IEEE 802.11n, legacy long training field (L-LTF) symbols are used to estimate the noise power, while the high-throughput long training field (HT-LTF) is used to estimate the channel state.
- *Linear equalization*: A linear zero-forcing (ZF) equalizer is employed at the receiver to combat inter-symbol interference (ISI).

As described in Algorithm 1, the feature set is required to be extracted first, and is then used to calculate the distances between the input query and the data in the training set so that we can determine the k-nearest neighbors. More explicitly, for the considered MIMO-OFDM AMC system, specifically at the receiver, after post-processing, a link quality metric (LQM) to describe the performance of the link can be evaluated based on $\mathbf{H}_n$, E_s, and N_0. This LQM is regarded as the feature set extracted from $\mathbf{H}_n$, E_s, and N_0.

Specifically, when the ZF post-processing is employed, for each of the spatial streams of $a \in \{1, \cdots, N_s\}$, where N_s is the number of spatial streams and $N_s \leq \min\{N_t, N_r\}$ (Daniels et al., 2010), the SNR of subcarrier $n \in \{1, \cdots, N\}$ is given by (YİĞİT and Kavak, 2013):

$$\gamma[a, n] = \frac{E_s}{N_0 \sum_{a'=1}^{N_s} \left|[\mathbf{G}_{ZF}[n]]_{a,a'}\right|^2}, \tag{9.4}$$

where $\mathbf{G}_{ZF} = (\mathbf{H}_n)^{\dagger} = (\mathbf{H}_n^H \mathbf{H}_n)^{-1} \mathbf{H}_n^H$, with $(\cdot)^{\dagger}$ and $(\cdot)^{-1}$ denote the pseudo-inverse Hermitian transpose and inverse operation, respectively. Note that the SNR dimension for each subcarrier is equal to its spatial stream number.

According to Eq. (9.4), for each subcarrier, we can obtain a feature query from the SNR of each spatial stream. However, as described in (Daniels et al., 2008), a clear-implementation MIMO-OFDM AMC methodology is highly desired. In order to simplify the AMC system, in the following analysis, each packet will only produce one feature query instead of producing a feature query for each subcarrier. Moreover, each subcarrier of a packet will also apply the same MCS according to the feature query. In the following, we will discuss situations for frequency-flat fading and frequency-selective fading, since the way to extract a feature query from a packet is different for these two fading channels.

For transmissions over frequency-flat fading channels, all subcarriers share the same channel state. Therefore, post-processing SNR averaged over all subcarriers is available as the feature query. Then we can obtain a one-dimensional feature set for MCS 0-7 with one spatial stream and a two-dimensional feature set for MCS 8-15 with two spatial steams, according to Eq. (9.4). Since there are two feature sets with different dimensions, we perform Algorithm 1 in each feature set. Once we obtain the suggested MCS for each feature set, the MCS with the highest rate is selected.

However, for transmissions over frequency-selective fading channels, average post-processing SNR cannot effectively reflect the variations of channel conditions, which are uniquely determined by the channel matrix, signal energy, and noise variance. In this case, the feature space of the training set might include $\mathbf{H}_n$, E_s, and N_0, which leads to a higher feature space dimension. On the other hand, due to the curse of dimensionality in SL, dimensions on order of $N_r \times N_t \times N$ require exceedingly large training sets even for typical values of N_r, N_t, and N.

In order to reduce the dimension of the feature space, a subcarrier ordering method is proposed in (Daniels et al., 2008). In this method, under the assumption that subcarrier position does not affect total packet performance, the packet performance can be determined by analysis of the per-subcarrier post-processing SNR distribution. According to this method, define $SNR_a^{ZF}[n]$ as the post-processing SNR for the spatial stream a for the subcarrier $n \in \{1, 2, \cdots, N_{ds}\}$, where N_{ds} is the number of the data-bearing subcarrier and $N_{ds} = 52$ for 20MHz channel bandwidth. All of the SNRs can compose a SNR set, which is expressed as:

$$\left\{ \{SNR_a^{ZF}[n]\}_{a=1}^{N_s} \right\}_{n=1}^{N}. \tag{9.5}$$

Then define $SNR_{(\eta)}^{ZF}$ to be the ηth smallest element in the set, where $\eta \in \{1, 2, \cdots, N_{ds}N\}$. Inspired by empirical observations in IEEE 802.11n channels, this method proposes that $SNR_{(\eta)}^{ZF}$ for a few values of η often determines packet performance and can work as the feature query. In order to obtain appropriate values of η, extensive computer searches are utilized, and a four-dimension feature query with $\eta = 1, 2, 6, 26$ is discovered to be necessary for MCS 0-7 and MCS 8-15-based IEEE 802.11n systems over frequency selective fading channels, which are given by:

$$\mathbf{q} = [SNR_{(1)}^{ZF}, SNR_{(2)}^{ZF}, SNR_{(6)}^{ZF}, SNR_{(26)}^{ZF}]^T. \tag{9.6}$$

Finally, Algorithm 1 is performed in the feature set consisting of these four-dimension feature queries and the suggested MCS is selected.

9.2.3 SVM-Assisted AMC

The k-NN-assisted AMC can effectively help the system to adapt to channel variations and enhance system performance. However, similar to other conventional SL algorithms, this method cannot explicitly construct class boundaries. Additionally, the k-NN method has high computational complexity and requires a large offline database for training, which prevents its efficient use in AMC.

To elaborate a bit further, the conventional methods use an offline database with the PER subjected to different fixed-channel realizations. However, the offline database is usually too large to implement online AMC. Thus we have to collect the training data in an online way instead of constructing an offline database. Since a large amount of PER data for each channel should be collected to decrease the variance of the PER and improve the accuracy of the estimation, how to reduce the required PER training data has attracted a lot of research interest. A good representative in this direction is an online AMC scheme proposed in (Daniels and Heath, 2010), which could achieve a fast AMC by using a support vector machine as given in the following subsections.

9.2.3.1 SVM Algorithm

The SVM-aided algorithm uses a single measurement of the frame like the success/failure and the measurements of current channel states to train SVM classifiers (Daniels and Heath, 2010). Compared with other fast online AMC algorithms based on the nearest neighbor classification (Daniels and Heath, 2009), the SVM-based fast online AMC algorithm can achieve the same performance with fewer excessive memory and processing requirements. In addition, the researchers have presented that the SVM-based fast online AMC algorithm has much lower complexity than the nearest neighborhood classification–based online AMC algorithm. Moreover, the SVM-based fast online AMC algorithm has further been improved to enhance performance, like using kernel functions to construct the nonlinear boundary, selecting appropriate training set and feature set to balance performance and the complexity, and so on.

In this subsection, we discuss how to apply the SVM to learn channel condition variations, with the aim to provide a reliable solution for fast AMC (Xu and Lu, 2006; Yun and Caramanis,

2009). For building binary online AMC classifiers, SVM can significantly discriminate complex nonlinear class region boundaries with lower complexity than many other SL algorithms, including the neural network-based ones. Moreover, the implementation of SVM also alleviates the issues of optimization, overfitting, and interpretation (Daniels and Heath, 2010), which benefit from the dual representations of SVM.

The SVM algorithm maximizes the margin between different classes in the training set, and also places the label $y \in \{+1, -1\}$ on the feature set realizations according to the distance from different margin boundaries of each class. That is to say, the SVM constructs the maximum margin classifier to determine the class region of each feature set. When we apply SVM for AMC, the class regions correspond to the success or failure of transmissions. Namely, they represent whether frame decoding at the receiver is successful or not. Usually, success/failure classes are denoted by $+1$ and -1, respectively. For the feature set realization $\mathbf{x} \in \mathbb{R}^p$, where p is the dimension of the feature vector and $\mathbb{R}$ is the set of real numbers, the margin of SVM is defined by the inference function $h(\mathbf{x}) = \mathbf{w}^T\phi(\mathbf{x}) + b$, where $\mathbf{w}$ can be defined as a linear discriminant vector for training, $\phi(\mathbf{x})$ is the SVM feature transformation performing on the feature set to generate the SVM feature set, and b is denoted as the margin bias.

Then the result of the inference function acts as the input to the discriminant function $g(h(\mathbf{x}))$, which is the compound function of x defined as $g(z) = 1$ when $z \geq 0$, and $g(z) = -1$ when $z < 0$. Each SVM provides the binary classification, so μ SVMs are needed to distinguish μ modulation and coding scheme classes. The classifiers are defined as one-versus-none classifiers (Daniels and Heath, 2010). The mth classifier is used to select the appropriate one among the class m. When the PER constraint $\text{PER}(m, \mathbf{x}) \leq F$ is not met, where F is the threshold of PER, no class is selected. When a frame encoded with MCS i is chosen by the classifier as a successful transmission, the classifier is reinforced to choose MCS i. Otherwise, when a frame transmission is considered a failure transmission, the classifier is reinforced to choose no MCS classes. Thus, the appropriate MCS for the frame transmission can be selected from μ MCS classes after applying the μ one-versus-none classifiers.

As a matter of fact, one-versus-none classification cannot address the PER constraint. In order to fit in with the change of PER of training sets, the training data from training sets are equally weighted. The discriminant functions with PER regression are replaced with regression functions and are respectively set with different MCS classes so as to calculate the posterior probability of each class. Then the regression functions are set as $r_m : \mathbb{R} \to [0, 1]$, which differs depending on m.

Therefore, the SVM-based algorithm used for online AMC to classify different feature-set realizations $\mathbf{x}$ with a two-stage classifier framework – one-versus-none classifiers and regression functions – can be defined as follows (Daniels and Heath, 2010):

1. Compute $h_m(\mathbf{x})$ according to the training data.
2. Compute $r_m(h_m(\mathbf{x}))$ according to $h_m(\mathbf{x})$ and the training data.
3. Map $r_m(h_m(\mathbf{x}))\ \forall m \in \{0, 1, \ldots, \mu - 1\}$.
4. Find the optimized $m^* = \arg\min_m\{(1 - r_m(h_m(\mathbf{x})))/T_m : r_m(h_m(\mathbf{x})) \leq F\}$.
5. If the number of appropriate m^* is equal to 1, $m^* \leftarrow \arg\min_{m \in m^*}$.
6. If the best m^* is not found, randomly choose one.

where T_m is the time consumed to transmit the data bits with MCS m. Steps 5 and 6 guarantee that only one optimal m^* is chosen for the transmission. Steps 3–6 are used to find the most appropriate MCS for an unknown channel realization. It can be concluded from Step 4 that $r_m(h_m(\mathbf{x}))$ also provides an estimate of $\text{PER}(m, \mathbf{x})$. It is worth pointing out that there is no need to complete steps 1–2 for each new frame observation since regression and inference functions need not be evaluated frequently.

Specifically, the nth element of the mth classifier's dataset is $\mathbf{x}_{m,n}$, $\forall n \in \{0, 1, ..., N-1\}$. The subscript m can be omitted since the SVM optimization is done with the aid of each of the μ binary classifiers separately.

In the SVM algorithm, the geometric margin for each $\mathbf{x}_n$ is defined as

$$\gamma_n = y_n \left(\left(\frac{\mathbf{w}^T}{||\mathbf{w}||} \right) \phi(\mathbf{x}_n) + \frac{b}{||\mathbf{w}||} \right). \tag{9.7}$$

The geometric margin of $\mathbf{w}, b$ corresponding to the training set is the smallest margin of geometric margins on $\mathbf{x}_n$, which can be defined as

$$\gamma = \min_{n=0,1,...,N-1} \gamma_n. \tag{9.8}$$

The objective of the SVM algorithm is to maximize the geometric margin, which can be denoted as

$$\begin{aligned} \min_{\mathbf{w}\in\mathbb{R}^P, b\in\mathbb{R}} \quad & \frac{\hat{\gamma}}{||\mathbf{w}||} \\ \text{subject to} \quad & y_n(\mathbf{w}^T\phi(\mathbf{x}_n) + b) \geq \hat{\gamma} \end{aligned} \tag{9.9}$$

where $\gamma = \hat{\gamma}/||\mathbf{w}||$ and $\hat{\gamma}$ is the functional margin. Considering that w and b can be added with an arbitrary scaling constraint: here we use the scaling constraint, which means the functional margin $\hat{\gamma}$ of w, b can be set as 1, i.e. $\hat{\gamma} = 1$.

Therefore, the previous optimization problem can be transformed as

$$\begin{aligned} \min_{\mathbf{w}\in\mathbb{R}^P, b\in\mathbb{R}} \quad & \frac{1}{2}||\mathbf{w}||^2 \\ \text{subject to} \quad & y_n(\mathbf{w}^T\phi(\mathbf{x}_n) + b) \geq 1. \end{aligned} \tag{9.10}$$

Let C denote the penalty for the feature-set realization requiring $\xi_n > 0$ to solve the non-separable case. Then the optimization problem can be rewritten as

$$\begin{aligned} \min_{\mathbf{w}\in\mathbb{R}^P, b\in\mathbb{R}} \quad & \frac{1}{2}||\mathbf{w}||^2 + C\sum_{n=0}^{N-1}\xi_n \\ \text{subject to} \quad & y_n(\mathbf{w}^T\phi(\mathbf{x}_n) + b) \geq 1 - \xi_n \\ & \xi_n \geq 0. \end{aligned} \tag{9.11}$$

The one-versus-none classifier will become non-separable when the feature space realizations cannot lead to the deterministic class outcomes. Hence C acts as the parameter to optimize the performance of the classifier and minimize the probability of choosing the wrong MCS class.

To remove the non-separable classifier and reduce the dimension of the feature set, the kernel function can be applied to transform the feature set. The kernel function for the SVM classifier can be defined as $\kappa(\mathbf{x}_n, \mathbf{x}_{n'}) := \phi(\mathbf{x}_n)^T\phi(\mathbf{x}_{n'})$. In addition, defining $\mathbf{Q} \in \mathbb{R}^{N\times N}$ where $[\mathbf{Q}]_{n,n'} := y_n y_{n'} \kappa(\mathbf{x}_n, \mathbf{x}_{n'})$, with the help of Lagrangian dual and the Karush-Kuhn-Tucker (KKT) conditions, the maximum margin can be constructed by

$$\begin{aligned} \max_{\alpha\in\mathbb{R}^N} \quad & \sum_{n=0}^{N-1}\alpha_n - \frac{1}{2}\sum_{n,n'=0}^{N-1}\alpha_n\alpha_{n'}y_n y_{n'}\mathbf{x}_n^T\mathbf{x}_{n'} \\ \text{subject to} \quad & \sum_{n=0}^{N-1} y_n\alpha_n = 0 \\ & 0 \leq \alpha_n \leq C \end{aligned} \tag{9.12}$$

Then, using the kernel function, the term $\mathbf{x}_n^T\mathbf{x}_{n'}$ can be updated as $\phi(\mathbf{x}_n)^T\phi(\mathbf{x}_{n'})$, and the previous equation becomes

$$\begin{aligned} &\max_{\alpha\in\mathbb{R}^N} \quad \mathbf{e}^T\alpha - \frac{1}{2}\alpha^T\mathbf{Q}\alpha \\ &\text{subject to} \quad \sum_{n=0}^{N-1} y_n\alpha_n = 0 \\ &\qquad\qquad\quad 0 \le \alpha_n \le C \end{aligned} \tag{9.13}$$

where $\mathbf{e}$ is a vector with all elements being "1", α_n is the nth element in the vector $\boldsymbol{\alpha}$, and the inference function becomes $h(\mathbf{x}) = \sum_{n=0}^{N-1} y_n\alpha_n\kappa(\mathbf{x}_n, \mathbf{x}_{n'}) + b$. The kernel function generalizes the boundary shape between classes. The most frequently used kernels, such as the linear kernel $\kappa_{lin}(\mathbf{x}_n, \mathbf{x}_{n'}) := \mathbf{x}_n^T\mathbf{x}_n$, the polynomial kernel $\kappa_{poly}(\mathbf{x}_n, \mathbf{x}_{n'}) := (\mathbf{x}_n^T\mathbf{x}_n)^d$ where d is the polynomial dimension, and the Gaussian kernel $\kappa_{gau}(\mathbf{x}_n, \mathbf{x}_{n'}) := e^{-\gamma||\mathbf{x}_n - \mathbf{x}_{n'}||^2}$ where $\gamma > 0$ defines the boundaries according to the exponentially scaled Euclidean distance (Daniels and Heath, 2010), can provide different kinds of class boundaries.

Notably, in the dual formulation, the dual-optimization variable α_n reveals the concept of the support vector, whose value has three cases. When $\alpha_n = 0$, $\mathbf{x}_n$ is not the support vector and does not contribute to the inference function. When $0 < \alpha_n < C$, $\mathbf{x}_n$ is the support vector lying on the margin. When $\alpha_n = C$, $\mathbf{x}_n$ is also a support vector, but lies inside the margin.

In the SVM-based fast online algorithm, the regression function $r_m(h_m(\mathbf{x}))$, wherein $h_m(\mathbf{x})$ is the inference function, is applied on μ MCS classes to address the PER constraint for each one-versus-none classifier. The prior probabilities of the normally distributed inference function outputs can be expressed in the sigmoid form as (Daniels and Heath, 2010)

$$\Pr[y = 1|h(\mathbf{x})] = (1 + e^{A_1h(\mathbf{x})^2 + A_2h(\mathbf{x}) + A_3})^{-1} \tag{9.14}$$

where A_1, A_2, and A_3 are constants.

Considering that the quadratic function is non-monotonic, the inference output can be simplified, and the corresponding prior probabilities can be described as

$$\Pr[y = 1|h(\mathbf{x})] = (1 + e^{B_1h(\mathbf{x}) + B_2})^{-1} \tag{9.15}$$

where B_1 and B_2 are constants (Platt et al., 1999). The output of the SVM classifier provides reasonable and accurate results of choosing the MCS class with the fitting function given in Eq. (9.15) (Niculescu-Mizil and Caruana, 2005; Keerthi et al., 2001). Thus for each one-versus-none classifier, the regression function to address the PER constraint can be denoted as

$$r_m(h_m(\mathbf{x})) = (1 + e^{B_{1,m}^* h_m(\mathbf{x}) + B_{2,m}^*})^{-1} \tag{9.16}$$

where optimized parameters $B_{1,m}^*$ and $B_{2,m}^*$ can be calculated by the fitting algorithm according to the result of Eq. (9.15). Moreover, the subscript m can be omitted, since the fitting could be implemented with each MCS.

In order to achieve B_1^* and B_2^*, the cost function can be constructed and optimized, and then B_1^* and B_2^* are defined as

$$\begin{aligned} \{B_1^*, B_2^*\} = \arg\min_{B_1,B_2} \Bigg\{ &-\sum_{n=0}^{N-1} \lambda_n \log((1 + e^{B_1h(\mathbf{x}) + B_2})^{-1}) \\ &+ (1 - \lambda_n)\log(1 - (1 + e^{B_1h(\mathbf{x}) + B_2})^{-1}) \Bigg\} \end{aligned} \tag{9.17}$$

Table 9.2 Comparison of complexity. (Daniels and Heath, 2010).

Algorithm	Processing	Memory(bits)
SVM(linear)	$\mu(p+2)$ multiplications	$\mu b(8p+4)$
	μ divisions	
	$\mu(p+4)$ additions	
	μ exponential maps	
	μ-length sort	
SVM(Gaussian)	$\mu(p+303)$ multiplications	$\mu b(300p+5)$
	μ divisions	
	$\mu(p+304)$ additions	
	2μ exponential maps	
	μ-length sort	
kNN	$\mu(pN+1)$ multiplications	$\mu b(300p+1)$
	μ divisions	
	$\mu(2pN+k+1)$ additions	
	μ N-length sorts	
	μ-length sort	

where λ_n is defined as

$$\lambda_n = \begin{cases} (N_p+1)/(N_p+2) & \text{if } y_n = +1 \\ 1/(N_m+2) & \text{if } y_n = -1 \end{cases} \tag{9.18}$$

where $N_p = \sum_{n=0}^{N-1}(1+y_n)/2$ and $N_m = N - N_p$. The optimization for choosing the best B_1^* and B_2^* can be efficiently implemented by the Newton algorithm that can converge quickly to the optimal value (Lin et al., 2007).

Furthermore, Table 9.2 compares the complexity of SL-based AMC schemes. It can be observed that the complexity of the SVM-based fast online AMC algorithm with a linear kernel has the lowest processing complexity and memory cost. Since training data far from the margin should be maintained for the following optimizations, the SVM-based AMC with a Gaussian kernel and k-NN will consume much more processing resources.

9.2.3.2 Simulation and Results

System Parameters Similar to the discussions in Section 9.2.1, the simulations are provided to investigate the performance of SVM-based AMC systems following the IEEE 802.11n standard (802, 2010). The simulation parameter settings are given as follows (Daniels and Heath, 2010): the channel bandwidth is 20 MHz, the number of receive antennas and transmit antennas is 2, the number of available MCS classes is 16 from MCS0–MCS15, the PER constraint is 0.1, and the frame length is 128 bytes. In the simulations, we assume that the synchronization and channel estimation at the receiver are performed perfectly, and that the zero-forcing algorithm is used for equalizations. In addition, the simulated online AMC algorithm is based on the SVM that uses linear or Gaussian kernels and nearest neighbor classifiers with $k = 10$ (Daniels and Heath, 2009). The post-processing SNR feature sets are well ordered (Daniels et al., 2010), 32000 channel realizations with SNRs ranging from 0–30dB are applied, and 4-tap Gaussian channels are employed with severe frequency selectivity and uniform power-delay profiles. The parameters of μ SVM classifiers are determined by the empirical results of the cross-validation set: hereby we set $C = 10$ for both the linear and Gaussian kernels, and we set $\gamma = 0.005$ for the Gaussian kernel. Moreover, it is noticeable that according to (Daniels and Heath, 2010), when the number of training examples is set to $N = 60$, the classifier can provide robust performances with no need to increase the training sample number.

Results Figure 9.2 describes the relationship between the throughput and the average SNR over all preset channels, while Figure 9.3 provides the result of the PER at different SNRs. From these two figures, we can see that SL-based AMC can help the wireless communication system reach a good compromise between high data rate and high reliability by using experience information originated from labeled training data. When N increases, the throughput increases, but the PER also increases. When N is small and SNR is large, the SVM-based algorithm achieves higher throughput than the k-NN-based algorithm. Hence, using the linear kernel can provide satisfactory performances for the system, and we do not need to pay extra to use the Gaussian kernel.

It is worth pointing out that the performance of the k-NN algorithm-based AMC scheme is deeply dependent on the selection of the sampled data and k, since in the k-NN algorithm, the k-nearest neighbor is determined by calculating the distance between the unlabeled object and all labeled objects. Thus, compared with the SVM-based AMC scheme, the robustness of the k-NN algorithm-based AMC scheme is worse.

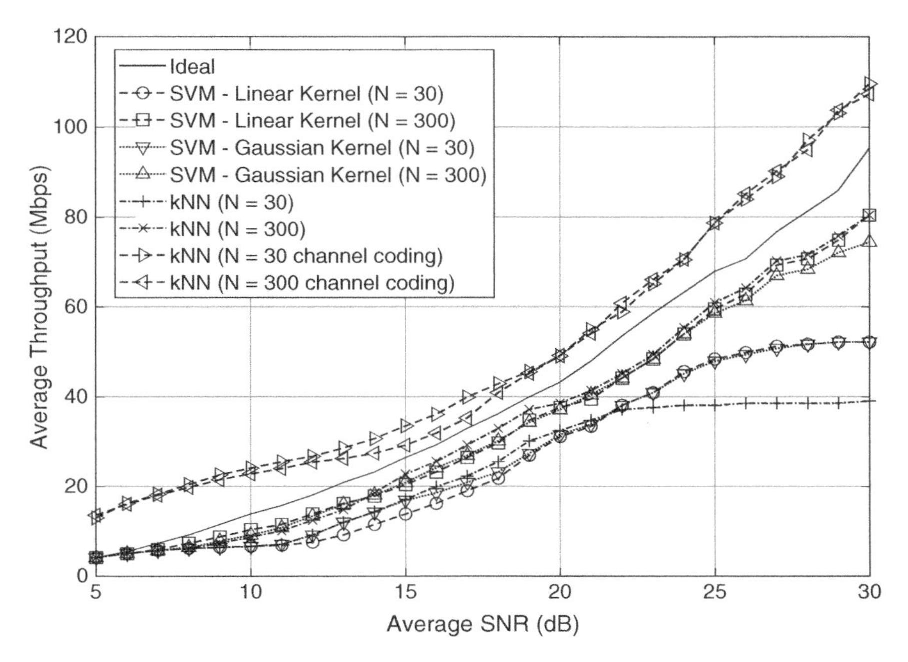

Figure 9.2 Throughput vs. average SNR over different channels. (Daniels and Heath, 2010)

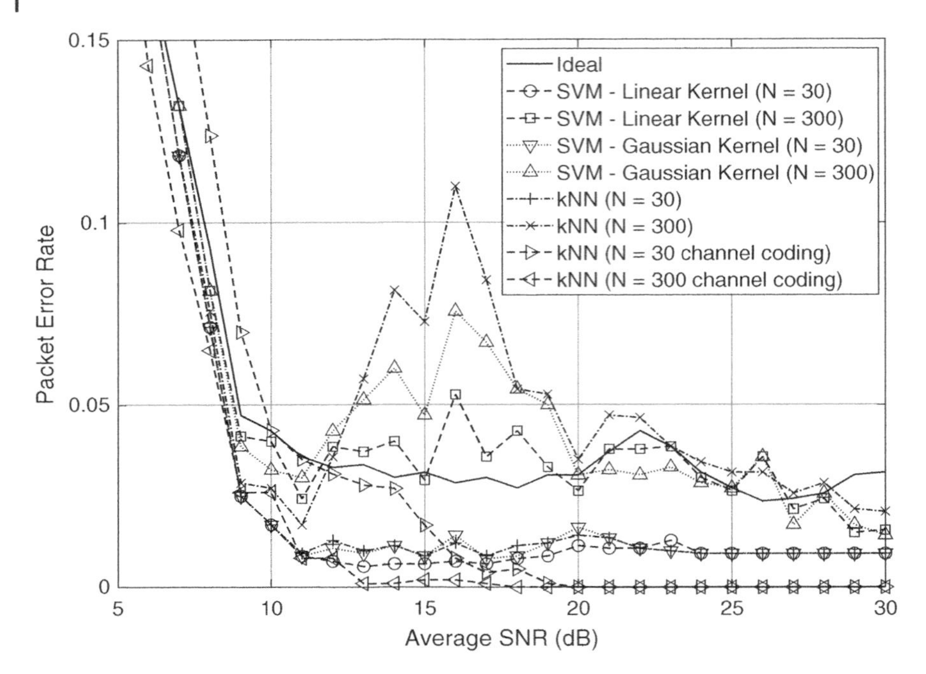

Figure 9.3 PER vs. SNR over different channels. (Daniels and Heath, 2010)

9.3 RL-Assisted AMC

In this section, we introduce the RL algorithm and address how to apply it in AMC. As mentioned earlier, unlike the SL method, the RL approach can directly interact with the environment, which means an agent can choose its actions according to the interactions and rewards obtained from the environment (Sutton et al., 1998). Specifically, the RL approach does not require an external supervisor, since the learning examples are obtained from interactions with the environment. When applying the RL approach to AMC, an agent can learn and formulate the best modulation and coding scheme by using past experiences obtained in real time and channel states, with minimal assumptions about the communication scenario. Accordingly, with the objective of maximizing spectrum efficiency, a Markov decision process is constructed to decide which modulation and coding scheme should be used (Leite et al., 2012). Notably, the RL-aided AMC does not require offline training that needs to consider all possible situations the physical transmissions may experience, and thus this type of AMC can provide more adaptive services than SL-aided AMC.

9.3.1 Markov Decision Process

As mentioned at the beginning of this chapter, the RL approach can be represented by the theory of MDP (Puterman, 2014). There are four key elements in a MDP problem: the states $\boldsymbol{S}$, the actions $\boldsymbol{A}$, the transition function P, and the reward R. Specifically, $\boldsymbol{S} = \{s_1, s_2, ..., s_n\}$ denotes the set of n possible states, which describes the dynamic variations of the environment; $\boldsymbol{A} = \{a_1, a_2, ..., a_m\}$ denotes the set of m possible agent actions; P: $\boldsymbol{S} \times \boldsymbol{A} \times \boldsymbol{S} \rightarrow [0, 1]$ is a transition probability function in which $P(s, a, s')$ is the probability of making a transition from state $s \in \boldsymbol{S}$ to state $s' \in \boldsymbol{S}$ when action $a \in \boldsymbol{A}$ is taken; and R is a reward function, where $R(s, a)$ is the immediate reward of the environment when taking action a at state s.

At the k_{th} stage of the learning process, the agent performs an action $a_k \in \boldsymbol{A}$ at the observed state $s_k \in \boldsymbol{S}$. At the next stage, before the state changes to $s_{k+1} \in \boldsymbol{S}$ with probability $P(s_k,a_k,s_{k+1})$, the agent receive the reward $R_k(s_k,a_k)$ generated by the environment. Later on, a similar process is carried out, and the agent receives a series of subsequent rewards R_{k+1}, R_{k+2}, R_{k+3},... from the environment.

The aim of the agent is to find a policy π that defines the behavior of the mapping from the states to the actions – π: $\boldsymbol{S} \rightarrow \boldsymbol{A}$ – to obtain the discounted accumulative reward as much as possible. Denoting $\pi(s)$ as the action that the agent takes at state s, then the accumulative reward is a function related to the state values, which can be calculated by

$$\begin{aligned} V^{\pi}(s) &= \sum_{k=0}^{\infty} \gamma^k R_k(s_k, \pi(s_k)|s_0 = s) \\ &= R(s, \pi(s)) + \gamma \sum_{s' \in S} P(s'|s, \pi(s)) V^{\pi}(s') \end{aligned} \tag{9.19}$$

where $0 \leq \gamma \leq 1$ is a discount factor used for weighting the long-term accumulative rewards. The discount factor determines how much the future reward will contribute to the accumulative reward function.

As time goes by, the future reward becomes less reliable and predictable, and hence the future reward is less important than the current reward such that $\gamma \leq 1$. More concretely, if the discount factor is close to 0, the agent thinks highly of the current reward and almost neglects the future rewards, while higher values of γ make the agent attach more importance to the long-term future reward.

As given by Eq. (9.19), once the state of the environment at time k is given, the reward only depends on the action taken by the agent. Therefore, for each s, the agent needs to find an optimal policy $\pi^*(s) \in \boldsymbol{A}$ to maximize the accumulative reward as defined in Eq. (9.19). To be more specific, the objective of the agent is to find a policy $V^*(s)$ such that $V^*(s) = max_{\pi} V^{\pi}(s)$.

9.3.2 Solution for the Markov Decision

In the MDP model, the policy iteration and value iteration dynamic programming (Sutton et al., 1998) methods are two well-known methods to obtain the optimal policy. However, these two methods both require prior knowledge of the environmental dynamics, which is hard to estimate beforehand in a practical view. In other words, we can hardly obtain prior knowledge of $R(s, a)$ and $P(s, a, s')$ in the AMC scenario. In this case, exploration of the state-action-reward relationship is required to query the environment. To accomplish this task, Q-learning stands out as an outstanding reinforcement technique due to its simplicity and low computational cost (Watkins and Dayan, 1992).

In the Q-learning method, the state-action value function (Q-function) is used to characterize policies. Specifically, the Q-function represents fitness to perform action a when in state s. It starts from a given state and then accumulates its value according to the rewards that the agent received when taking a series of actions following the policy thereafter. The Q-function is defined as

$$Q^{\pi}(s,a) = E\left(\sum_{k=0}^{\infty} \gamma^k R_k | s_0 = s, a_0 = a\right) \tag{9.20}$$

The optimal Q-function, $Q^*(s,a)$, is the one that maximizes the discounted cumulative rewards, namely $Q^*(s,a) = \max_{\pi} Q^{\pi}(s,a)$. According to Bellman's optimality principle, a greedy policy can be used to solve this optimization problem. In the greedy policy, an action with the largest Q-value is selected at each state. This means that for each state $s \in \mathbf{S}$, once $Q^*(s,a)$ is known, the optimal policy is to take action a with the highest $Q^*(s,a)$ value, i.e.

$$\pi^*(s) = \max_{a \in A} Q^*(s,a) \tag{9.21}$$

Moreover, in the Q-learning algorithm, $Q^*(s,a)$ is updated recursively using the current state s, the next state s', action a taken at s, and reward R obtained from the environment due to taking a at the current state. The updating formula is given as

$$Q(s,a) \leftarrow Q(s,a) + \alpha[R + \gamma \max_{a} Q(s',a) - Q(s,a)] \tag{9.22}$$

where α is the learning rate.

As can be inferred from the previous discussions, the RL algorithm has an exploration versus exploitation dilemma. *Exploration* means collecting informative data about the environment by taking new actions for the given state space. *Exploitation* means preserving the process well enough according to the available knowledge obtained previously. To balance the need for exploration and exploitation, two commonly adopted strategies are the ε-greedy search and the softmax action-selection method (Sutton et al., 1998).

ε-greedy search In the ε-greedy search, the agent uses the probability ε to decide whether to exploit the Q-table or explore the new choices. Since ε is usually small, the agent tends to select the action that satisfies $\max_{a} Q(s,a)$. On rare occasions, with probability ε, the agent takes a random action for the purpose of experiencing as many actions and effects as possible.

softmax action-selection Due to the small probability of random selection, the ε-greedy search can only attain a suboptimal policy. To solve this problem, random selection is abandoned in the softmax action-selection method. Instead, the action a taken at state s is chosen with probability $Pr(a)$ given by

$$Pr(a) = \frac{e^{Q(s,a)/\tau}}{\sum_{i=1}^{m} e^{Q(s,a_i)/\tau}}, \tag{9.23}$$

where τ determines the selection trend. As can be inferred from Eq. (9.23), if τ is close to 0, the agent prefers choosing actions with the highest Q-value. If τ becomes larger, the policy trends toward random selection, as all actions tend to have the same probability of being chosen.

9.3.3 Actions, States, and Rewards

Using the Markov process, ML-assisted AMC systems can take actions based on the collected environment states and rewards to adapt to dynamically changing channel conditions. More details about the actions, states, and rewards are given as follows.

Actions The target of AMC is to maximize throughput for a given state of the environment by adopting the modulation and coding schemes. In practical protocols such as IEEE 802.11n, there are only a finite number of modulation and coding schemes. Each scheme is considered an action, and the agent selects the optimal scheme based on past experiences just before packet transmission.

States The link establishment has enabled the agent to collect environment information including the past and the current channel states, extracted network features, and so on. Although the environment state can be determined by various features, as an example of one possible solution, the received SNR averaged over all subcarriers is used to determine the state of the environment (Leite et al., 2012). Due to the constraint that the state value should belong to a finite set, we consider SNR values in the range from -2–20 dB with step of 1 dB to avoid dealing with an infinite number of states. As a result, a total of $n = 23$ states are used to determine the environment. At each single state s, the agent chooses one action from a finite action set, and different actions lead to different expected rewards. Since the channel link may be time-varying, the agent must continually track and update the value of $Q(s, a)$, embodying the real-time compatibility of the RL approach. To maximize the achievable throughput, the agent must search and find the optimal MCS for a given state of the channel environment.

Rewards To ensure the effectiveness of applying the RL approach to a practical problem, the reward function R must be defined appropriately in a specific context. Given the link-adaptation scenario, the throughput achieved by taking action a at state s is used to define the reward function R, namely

$$R(s, a) = log_2(M_a)\rho_a[1 - PER(s, a)] \tag{9.24}$$

where M_a denotes the modulation order of action a, ρ_a is the coding rate when taking action a, and $PER(s, a)$ is the packet error rate when taking action a at channel state s. Using the CRC field in each packet, the receiver can detect packets with errors, and the PER can be estimated by the error packet measurements. This information is then fed back to the transmitter to update the learning algorithm in order to increase spectrum efficiency.

9.3.4 Performance Analysis and Simulations

Due to the exploration, the RL framework is not as powerful as SL. Next, we perform simulations to investigate the performance of RL-aided AMC.

In the simulations, we only consider the combination of $m = 8$ modulation and coding with $N_{ss} = 1$, as given in Table 9.1. Unless stated otherwise, we have set $\tau = 0.4$ on the softmax action-selection method given by Eq. (9.23). Similarly, the learning rate of the Q-learning algorithm, given by Eq. (9.22), is set to $\alpha = 0.4$ by default. These settings will be further discussed as follows.

Figures 9.4 and 9.5 investigate the effects of tuning parameters, including the softmax parameter and the learning rate, on the convergence behavior of the algorithm. To indicate the differences among the RL policies and different modulation and coding for a given SNR, the mean square error (MSE) is calculated over the frame numbers.

From Figure 9.4, we can see that the lower the value of the softmax parameter, the faster the convergence. The reason is that for $\tau = 0$, actions with a higher Q-value are preferred. However, low τ can lead to bad system performance because the learning procedure finishes its exploration too early. For example, $\tau = 0.2$ leads to a MSE of about $-1dB$, which is larger than that of the scenario $\tau = 0.4$ or $\tau = 0.8$.

Moreover, it can be observed from Figure 9.5 that when the learning rate α has a larger value, the learning engine can get more immediate rewards than the accumulated rewards and vice versa. Hence, the learning rate can be used for the balance of the exploitation and exploration, and the optimal learning rate depends on specific systems. For example, the simulated system performs best when $\alpha = 0.4$.

Next we compare system performance using the RL engine with that using the lookup table approach, which is also known as RawBER mapping (Jensen et al., 2010) used for link adaptations. In RawBER mapping, the probability of uncoded bit errors at each subcarrier is used to calculate the LQM. Specifically, the relationship between RawBER and PER can be determined by a regression generated by simulations performed over the additive white Gaussian noise (AWGN) channel, which can be prepared beforehand (Lamarca and Rey, 2005).

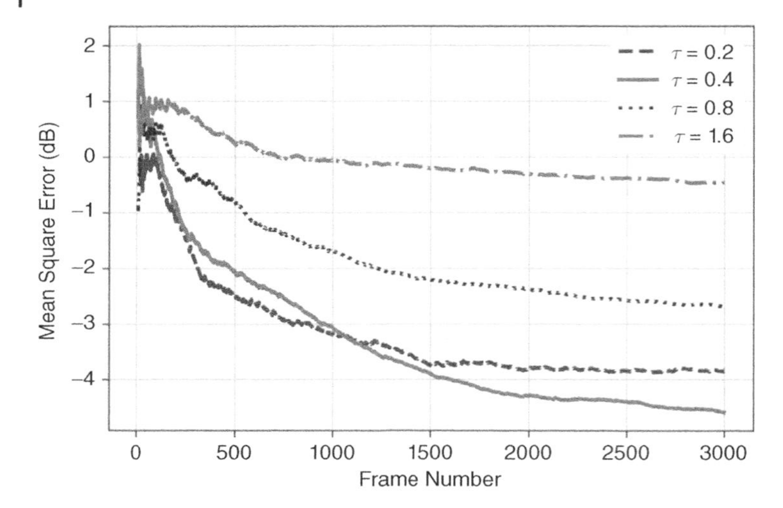

Figure 9.4 Influence of the softmax parameter τ on the convergence of the RL algorithm for $\alpha = 0.4$.

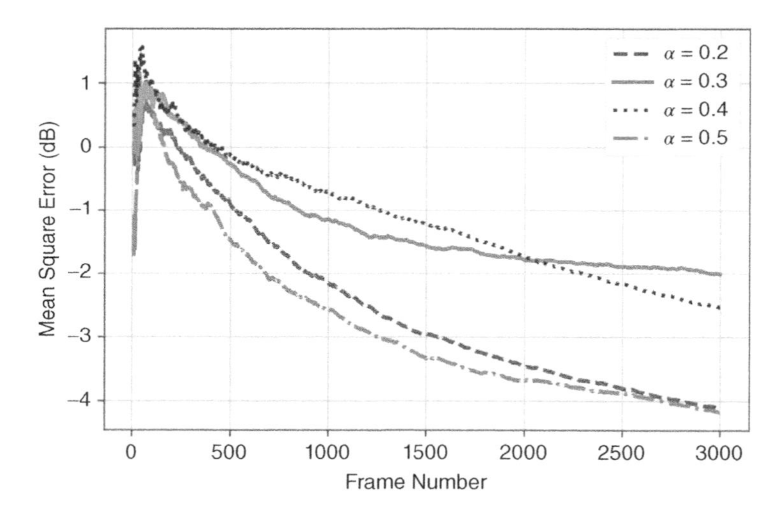

Figure 9.5 Influence of the learning rate α on the convergence of the RL algorithm for $\tau = 0.4$.

The drawback of the lookup table approach is that it requires a large amount of memory and simulation time for each scenario. For example, in some scenarios, the Gaussian assumption – i.e. the interference along with the Gaussian noise can be regarded as a single Gaussian distribution – may be unreasonable (Aljuaid and Yanikomeroglu, 2010). Hence, one kind of generated data cannot guarantee to cover all possible situations. Consequently, lookup tables can potentially lead to suboptimal solutions.

Let us consider the scenario where the channel SNR can be ideally obtained. Figure 9.6 shows that the spectrum efficiency is a function of the average SNR in this scenario. It can be observed that the RL approach achieves lower throughput as compared to the lookup table approach, since the latter approach is optimized in the specific channel, while the RL gradually adapts to the environment.

By contrast, as shown in Figure 9.7, when the channel SNR cannot be ideally obtained due to the estimation error, which is a more common scenario in practical communication systems,

Figure 9.6 Average spectrum efficiency of the lookup table and the RL technique over an AWGN channel with ideal SNR estimation.

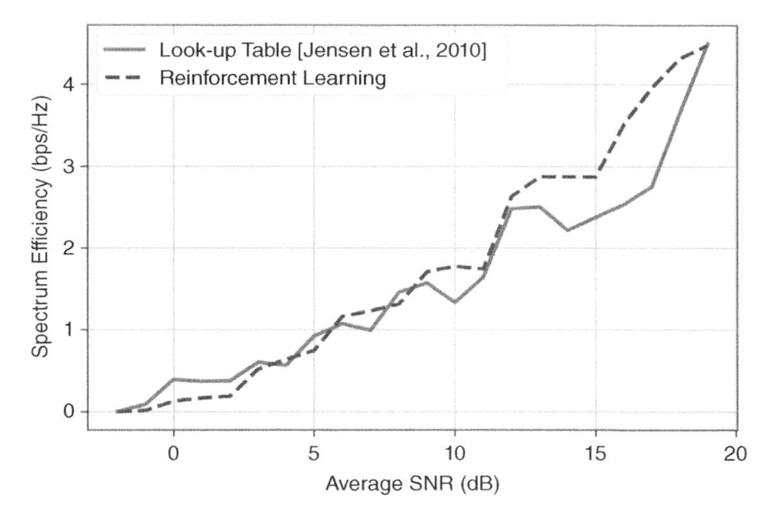

Figure 9.7 Average spectrum efficiency of the lookup table and the RL technique over an AWGN channel with SNR estimation error.

the proposed RL scheme can outperform the lookup table approach. This is because RL is able to learn from the environment and adapts to the environment, including the SNR estimation error, while the lookup table approach must be optimized in a specific environment. Moreover, RL can operate in online mode, and hence it does not require an expert or extensive simulations to adapt to different scenarios. The best MCS selection does not require an exhaustive trial-and-error procedure but a small amount of programming effort to build the system.

It can be seen from the simulation results and analyses that the RL framework can provide a promising solution for the AMC problem. In this method, the maximization of spectral efficiency is regarded as a Markov decision process, and the mean SNR is used to determine the channel state of the radio link. A relation between the SNR values and modulation and coding schemes can be determined by adopting Q-learning for solving the problem of the Markov decision process. From the simulations, we can further see that RL-aided AMC can be used for online radio link establishment, since offline training is not required by RL. Moreover, this method is adaptive to changes of the radio environment. Thus, with the aid of RL, AMC can dynamically adjust the parameters and provide services for users with higher efficiencies and reliability.

9.4 Further Discussion and Conclusions

In this chapter, we have presented three ML-assisted AMC schemes for wireless communication systems to adapt to variations in channel conditions with fewer model-based approximations and potentially higher accuracy than traditional adaptive modulation and coding schemes. Two of them are SL-based approaches, and the other is implemented by RL. Notably, although there are many SL algorithms such as SVM, decision trees, k-NN, neural networks, and so on, they have similar mechanisms, such as the requirement of labeled data and the analogy to classifiers after training has been done. Therefore, without a loss of generality, we have only taken k-NN and SVM as examples to show the application of SL in AMC. For the RL approach in AMC, we have taken Q-learning as an example, since it represents the primary mechanisms of RL and has been widely used due to its conciseness.

To verify the practicability and effectiveness of applying these learning algorithms to AMC, simulations with parameter settings referring to the IEEE 802.11n standard have been done to compare the presented ML-assisted AMC schemes with traditional approaches. It is worth pointing out that SL-based AMC schemes are suitable for the scenario where training examples are representative of all the situations the transmitter might be exposed to. By contrast, the RL algorithm can directly learn from the interacting environment and can gradually achieve satisfactory performance with no offline training as required by SL. Therefore, these two types of learning mechanisms can be selected according to whether training is performed offline or online based on different user demands in different application scenarios.

Thanks to training and learning mechanisms, ML-aided AMC approaches have achieved intelligent and outstanding reliability in wireless systems. Future research directions may include the combination of RL and SL, how to enhance the robustness of ML-assisted AMC systems, and how to achieve better trade-offs between complexity and system performance, with the objective of providing more adaptive, intelligent, and better services for end users.

Bibliography

IEEE draft standard for information technology - telecommunications and information exchange between systems - local and metropolitan area networks - specific requirements - part 11: Wireless LAN medium access control (MAC) and physical layer (PHY) specifications. *IEEE Draft P802.11-REVmb/D6.0, September 2010 (Revision of IEEE Std 802.11-2007, as amended by IEEE Std 802.11k-2008, IEEE Std 802.11r-2008, IEEE Std 802.11y-2008, IEEE Std 802.11w-2009 and IEEE Std 802.11n-2009)*, pages 1–2097, Sept 2010.

M. Aljuaid and H. Yanikomeroglu. Investigating the validity of the gaussian approximation for the distribution of the aggregate interference power in large wireless networks. In *2010 25th Biennial Symposium on Communications*, pages 122–125, May 2010.

Ethem Alpaydin. *Introduction to Machine Learning*. MIT Press, 2009.

R. Daniels and R.W. Heath. Online adaptive modulation and coding with support vector machines. In *2010 European Wireless Conference (EW)*, pages 718–724, April 2010.

R.C. Daniels and R.W. Heath. An online learning framework for link adaptation in wireless networks. In *2009 Information Theory and Applications Workshop*, pages 138–140, Feb 2009.

R.C. Daniels, C. Caramanis, and R.W. Heath Jr. A supervised learning approach to adaptation in practical MIMO-OFDM wireless systems. In *IEEE GLOBECOM 2008 - 2008 IEEE Global Telecommunications Conference*, pages 1–5, Nov 2008.

R.C. Daniels, C.M. Caramanis, and R.W. Heath. Adaptation in convolutionally coded MIMO-OFDM wireless systems through supervised learning and SNR ordering. *IEEE Transactions on Vehicular Technology*, 59 (1): 114–126, Jan 2010.

A. Iera, A. Molinaro, G. Ruggeri, and D. Tripodi. Improving QoS and throughput in single and multihop WLANs through dynamic traffic prioritization. *IEEE Network*, 19 (4): 35–44, July 2005.

T.L. Jensen, S. Kant, J. Wehinger, and B.H. Fleury. Fast link adaptation for MIMO OFDM. *IEEE Transactions on Vehicular Technology*, 59 (8): 3766–3778, Oct 2010.

C. Jiang, H. Zhang, Y. Ren, Z. Han, K. Chen, and L. Hanzo. Machine learning paradigms for next-generation wireless networks. *IEEE Wireless Communications*, 24 (2): 98–105, April 2017.

Shashi Kant, T. Lindstrom Jensen, and B.W. Channel. Fast link adaptation for IEEE 802.11 n. *Agenda*, 2: 1, 2007.

S. Sathiya Keerthi, Shirish Krishnaj Shevade, Chiranjib Bhattacharyya, and Karuturi Radha Krishna Murthy. Improvements to Platt's SMO algorithm for SVM classifier design. *Neural computation*, 13 (3): 637–649, 2001.

M. Lamarca and F. Rey. Indicators for PER prediction in wireless systems: A comparative study. In *2005 IEEE 61st Vehicular Technology Conference*, volume 2, pages 792–796 Vol. 2, May 2005.

J.P. Leite, P.H.P. de Carvalho, and R.D. Vieira. A flexible framework based on reinforcement learning for adaptive modulation and coding in OFDM wireless systems. In *2012 IEEE Wireless Communications and Networking Conference (WCNC)*, pages 809–814, April 2012.

Hsuan-Tien Lin, Chih-Jen Lin, and Ruby C. Weng. A note on Platt's probabilistic outputs for support vector machines. *Machine Learning*, 68 (3): 267–276, 2007.

Laura Melián-Gutiérrez, Navikkumar Modi, Christophe Moy, Faouzi Bader, Iván Pérez-Álvarez, and Santiago Zazo. Hybrid UCB-HMM: A machine learning strategy for cognitive radio in HF band. *IEEE Transactions on Cognitive Communications and Networking*, 1 (3): 347–358, 2015.

Alexandru Niculescu-Mizil and Rich Caruana. Predicting good probabilities with supervised learning. In *Proceedings of the 22nd International Conference on Machine Learning*, pages 625–632. ACM, 2005.

J. Oksanen, J. Lunden, and V. Koivunen. Reinforcement learning-based multiband sensing policy for cognitive radios. In *2010 2nd International Workshop on Cognitive Information Processing*, pages 316–321, June 2010.

T. OShea and J. Hoydis. An introduction to deep learning for the physical layer. *IEEE Transactions on Cognitive Communications and Networking*, 3 (4): 563–575, Dec 2017.

Sinno Jialin Pan, Qiang Yang, et al. A survey on transfer learning. *IEEE Transactions on Knowledge and Data Engineering*, 22 (10): 1345–1359, 2010.

F. Peng, J. Zhang, and W.E. Ryan. Adaptive modulation and coding for IEEE 802.11n. In *2007 IEEE Wireless Communications and Networking Conference*, pages 656–661, March 2007.

John Platt et al. Probabilistic outputs for support vector machines and comparisons to regularized likelihood methods. *Advances in Large Margin Classifiers*, 10 (3): 61–74, 1999.

Martin L. Puterman. *Markov decision processes: discrete stochastic dynamic programming*. John Wiley & Sons, 2014.

Richard S. Sutton, Andrew G. Barto, et al. *Reinforcement Learning: An Introduction*. MIT Press, 1998.

P. Venkatraman, B. Hamdaoui, and M. Guizani. Opportunistic bandwidth sharing through reinforcement learning. *IEEE Transactions on Vehicular Technology*, 59 (6): 3148–3153, July 2010.

Christopher J.C.H. Watkins and Peter Dayan. Q-learning. *Machine Learning*, 8 (3-4): 279–292, 1992.

G. Xu and Y. Lu. Channel and modulation selection based on support vector machines for cognitive radio. In *2006 International Conference on Wireless Communications, Networking and Mobile Computing*, pages 1–4, Sept 2006.

HALİL YİĞİT and Adnan Kavak. A learning approach in link adaptation for MIMO-OFDM systems. *Turkish Journal of Electrical Engineering & Computer Sciences*, 21 (5): 1465–1478, 2013.

S. Yun and C. Caramanis. Multiclass support vector machines for adaptation in MIMO-OFDM wireless systems. In *2009 47th Annual Allerton Conference on Communication, Control, and Computing (Allerton)*, pages 1145–1152, Sept 2009.

10

Machine Learning–Based Nonlinear MIMO Detector

Song-Nam Hong and Seonho Kim

Electrical and Computer Engineering, Ajou University, Suwon, South Korea

10.1 Introduction

Wireless systems make it possible to provide communication links with Gbps data rates by using a massive antenna array (Swindlehurst et al., 2014) and/or by using wide (possibly multi-gigahertz) bandwidth (Pi and Khan, 2011). The common weakness of both approaches is significant power consumption at the receiver, caused by high-precision (e.g. 8∼14-bit precision) analog-to-digital converters (ADCs), because total power consumed by ADCs scales linearly with the number of precision levels, bandwidth, and the number of the ADCs (Murmann; Walden, 1999; Mezghani et al., 2010). For example, power consumption of ADCs is shown to be proportional to both the number of precision levels and the bandwidth, under Nyquist rate sampling (Mezghani et al., 2010). Therefore, the use of high-precision ADCs at the receiver becomes impractical when a massive antenna array and/or wide bandwidth are used.

Low-resolution (e.g. 1∼3-bit precision) ADCs have been regarded as a cost-effective solution to reduce power consumption of future wireless systems that include massive multiple-input and multiple-output (MIMO) systems and wideband communication systems (Nossek and Ivrlač, 2006; Singh et al., 2009; Dabeer and Madhow, 2010; Mo and Heath, 2015). In spite of the benefits, the use of low-resolution ADCs gives rise to numerous challenges. One challenge is that it is difficult to obtain accurate channel-state information at the receiver (CSIR) with conventional pilot-based channel estimation techniques. In addition, conventional data-detection methods, developed for linear MIMO systems, provide poor detection performance due to the nonlinearity at the ADCs.

Extensive research has been performed to resolve the channel-estimation and data-detection problems in uplink massive MIMO systems with one-bit ADCs Jacobsson et al. (2015,2017), Risi et al. (2014), Choi et al. (2016), Li et al. (2017), Mollén et al. (2016,2017), Wang et al. (2015), Liang and Zhang (2016), Wen et al. (2016). For the channel-estimation problem in such systems, numerous methods have been developed to improve the accuracy of CSIR. The developed methods include a least-squares (LS)-based method (Risi et al., 2014), a maximum-likelihood–based method (Choi et al., 2016), and a method using Bussgang decomposition (Li et al., 2017). For the data-detection problem, optimal maximum-likelihood detection (MLD) has been introduced in (Choi et al., 2016), and some other low-complexity methods have also been developed in (Mollén et al., 2016,2017; Jeon et al., 2018a).

These works can be extended into a multihop MIMO system, called a *distributed uplink MIMO system*, where an end-to-end channel transfer function between K sources and the data center is highly nonlinear. Thus, it is extremely challenging to estimate such a channel transfer

Machine Learning for Future Wireless Communications, First Edition. Edited by Fa-Long Luo.
© 2020 John Wiley & Sons Ltd. Published 2020 by John Wiley & Sons Ltd.

function with a limited number of quantized pilot signals. This motivates the consideration of a data-driven supervised learning (SL) based detector where pilot signals (or training data) are exploited to directly learn a MIMO detector, rather than estimating a complex channel transfer function. In this chapter, we will explain various SL-based detectors for nonlinear multihop MIMO communication systems.

The rest of this chapter is organized as follows. Section 10.2 is devoted to the problem formulation and system model of a multihop MIMO system with one-bit ADCs. Section 10.3 explains MIMO detectors with regard to SL Learning for considered system. In this section, these detectors are described as being in two categories: non-parametric learning and parametric learning. Section 10.4 introduces a low-complexity SL detector that applies clustering algorithms on SL detectors with parametric learning. Section 10.5 shows the performance of introduced SL detectors via simulation. Section 10.6 will offer some further discussion and conclusions.

10.2 A Multihop MIMO Channel Model

As shown in Figure 10.1, let us consider a multihop distributed uplink MIMO system in which K sources transmit independent messages to one data center with the aid of intermediate relays. Also, the data center is equipped with $N_r \geq K$ receive antennas with one-bit ADCs in the uplink system.

Let $w_k \in \{0, .., m-1\}$ denote the source k's message for $k \in \{1, ...K\}$, each of which contains $\log m$ information bits. We also denote the m-ary constellation set by $S = \{s_0, ..., s_{m-1}\}$ with power constraint $\frac{1}{m}\sum_{i=0}^{m-1}|s_i|^2 = P_t$. Let sign$(\cdot)$: $R \rightarrow \{-1, 1\}$ represent the one-bit ADC quantizer function with sign$(u) = 1$ if $u \geq 0$ and sign$(u) = -1$, otherwise. Then the transmitted symbol of source k, $\tilde{x}_k$, is obtained by a modulation function $f : W \rightarrow S$ as $\tilde{x}_k = f(w_k) \in S$.

Then, from real and imaginary parts, the data center observes

$$r = \text{sign}(\Phi(\tilde{\boldsymbol{x}}) + \tilde{\boldsymbol{z}}) \in \{-1, 1\}^N, \tag{10.1}$$

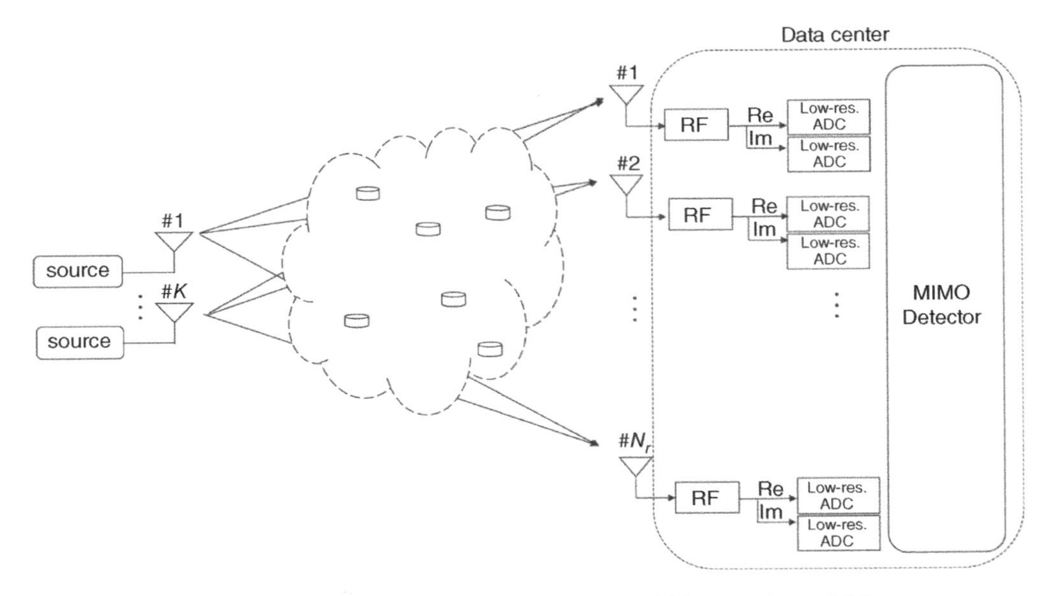

Figure 10.1 Description of a distributed multihop MIMO system with low-resolution ADCs.

where $N = 2N_r$, $\Phi(\cdot)$ represents a complex nonlinear function (called an *end-to-end channel transfer function*) and $\tilde{\mathbf{z}} = [\tilde{z}_1, \dots, \tilde{z}_N] \in \mathbb{R}^N$ denotes the noise vector whose elements are independently identically distributed as circularly symmetric complex Gaussian random variables with zero mean and unit variance: in other words, $\tilde{z}_i \sim \mathcal{N}(0, \sigma_z^2/2)$. Note that $\Phi(\cdot)$ captures all the intermediate relays' operations and all the local wireless channels in the network. For example, each local channel in the multihop can be assumed as Rayleigh fading. It is remarkable that the channel model in Eq. (10.1) is quite general, including an uplink MIMO system with one-bit ADCs.

The communication framework consists of training and data-transmission phases (see Figure 10.2). Note that while undergoing these two phases, the channel is assumed to be fixed within the coherence time:

- *Training phase*: During this phase, K sources transmits "known" sequences (i.e. pilot signals) so that the data center can learn a nonlinear function $\Phi(\cdot)$. From a machine learning (ML) perspective, the data center collects the data and the corresponding labels. Let $\mathcal{M} = \{0, \dots, m-1\}^K$ denote the set of all possible messages of the K sources. For each class $c \in \mathcal{M}$, the K sources transmit T pilot signals $\tilde{\mathbf{x}}_i^c$ for $i = 1, \dots, T$. In other words, T is the number of repetitions of a signal according to each message(class) c. From Eq. (10.1), the data center can collect the labelled data set $\mathcal{D}$ as

$$\mathcal{D} = \{\tilde{\mathbf{r}}_i^c : c \in \mathcal{M}, i = 1, \dots, T\}, \tag{10.2}$$

  where $\tilde{\mathbf{r}}_i^c \in \{-1, 1\}^N$. After sending the pilot signals in Eq. (10.2), the receiver creates empirical conditional probability mass functions (PMFs) by using the received signals observed during the pilot transmission. An empirical conditional PMF for each message (class) c is given by

$$\hat{p}(r|c) = \frac{1}{T}\sum_{t=1}^{T} \mathbf{1}_{\{r=\tilde{r}_t^c\}} \text{ for } c \in 1, \dots, \mathcal{M}, \tag{10.3}$$

  where $\mathbf{1}_{\{\mathcal{A}\}}$ is an indicator function that equals 1 if an event A is true and zero otherwise.
- *Data-transmission phase*: Given the $\mathcal{D}$ and a new observation $\mathbf{r}$, the data center detects the class of $\mathbf{r}$ (i.e. users' messages $\hat{\mathbf{w}} = (\hat{w}_1, \dots, \hat{w}_K)$) as

$$\Psi(\mathbf{r}) = c \in \mathcal{M}, \tag{10.4}$$

  which will be introduced in the next sections.

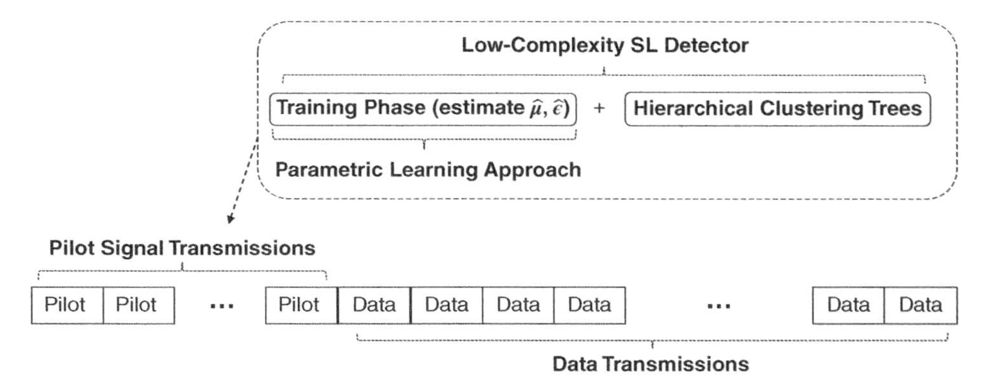

Figure 10.2 Illustration of the training and data-transmission phases within the coherence time.

10.3 Supervised-Learning-based MIMO Detector

From a ML perspective, the detection problem in Eq. (10.4) (a.k.a. the *supervised learning problem*) can be classified into two approaches (Robert, 2014): non-parametric learning and parametric learning. Non-parametric learning – such as k-nearest neighbor (kNN), decision trees, and support vector machines (SVMs) – does not require a priori knowledge of a data set $\mathcal{D}$ (e.g. a distribution of data). On the other hand, in parametric learning – such as logistic regression, naive Bayes, and neural networks – data is assumed to be generated from a given probabilistic model with some parameters (e.g. Gaussian model); then, the parameters are optimized using the given data set $\mathcal{D}$. Therefore, it is very important to choose a proper probabilistic model using a priori knowledge (or domain knowledge) about the data set $\mathcal{D}$.

With this classification, we will briefly explain the existing (parametric or non-parametric) SL detectors. Note that they can be immediately applied to a multihop MIMO system, since SL detectors do not rely on a specific system model.

10.3.1 Non-Parametric Learning

According to (Jeon et al., 2018a), non-parametric learning characterizes empirical conditional probability mass function (PMF) based on a training data set. Non-parametric learning is categorized into empirical-maximum-likelihood detection and minimum-mean-distance detection, as follows:

1) *Empirical-maximum-likelihood detection (eMLD)*: The key idea is the selection of an index for the input symbol vector that maximizes the empirical conditional PMF in Eq. (10.3) as follows:

$$\hat{c} = \underset{c\in\{1,\ldots,\mathcal{M}\}}{\arg\max}\ \hat{p}(r|c). \tag{10.5}$$

When training repetitions T increases to infinity, this detection is equivalent to the optimal maximum-likelihood detection method, since the empirical distribution converges to the corresponding true distribution by the law of large numbers. When T is insufficient, however, the empirical distribution cannot represent the true distribution, resulting in detection errors. More specifically, there is a non-zero probability $r \notin \mathcal{D}$, leading to $\hat{p}(r|c) = 0$ for all messages $c \in \mathcal{M}$. To solve this problem, the receiver finds $\tilde{r} \in \mathcal{D}$ that are the closest vectors to r. Let $\mathcal{N}(r)$ be the set of these closest vectors. Then, the eMLD method $\Psi_{eMLD} : r \rightarrow c$ is given by

$$\Psi_{eMLD}(r) = \underset{r\in\mathcal{N}}{\arg\max} \sum_{\tilde{r}\in\mathcal{N}} \hat{p}(\tilde{r}|c). \tag{10.6}$$

As in Figure 10.3, this is similar to the kNN classifier in the sense that they simply compare the number of neighbor labels. One notable difference is that eMLD uses the neighbor set of elements that are equidistant from the received vector.

2) *Minimum-mean-distance detection (MMD)*: The drawback of eMLD is partially using the empirical conditional PMFs, because the set is limited to the nearest neighbors ($\mathcal{N}$) only. Therefore, to fully exploit the empirical PMFs, we introduce an alternative detection method: MMD. As illustrated in Figure 10.4, the MMD method, $\Psi_{MMD} : r \rightarrow c$, selects the index of the symbol vector that yields the conditional minimum mean distance, i.e.

$$\Psi_{MMD}(r) = \underset{c\in\mathcal{M}}{\arg\min}\ \mathbb{E}_{\tilde{r}_i}[||r - \tilde{r}_i||_2|c] = \underset{c\in\mathcal{M}}{\arg\min} \sum_{\tilde{r}_i} ||r - \tilde{r}_i||_2 \hat{p}(\tilde{r}_i|c). \tag{10.7}$$

MMD finds the message that minimizes the weighted sum of the empirical PMFs, where the weights are the distance between the received vector and the trained vector, $||r - \tilde{r}_i||_2$.

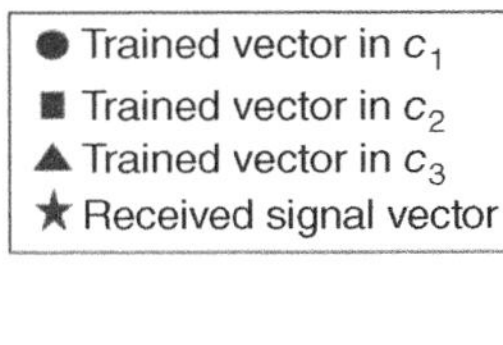

Figure 10.3 Empirical-maximum-likelihood detection (eMLD).

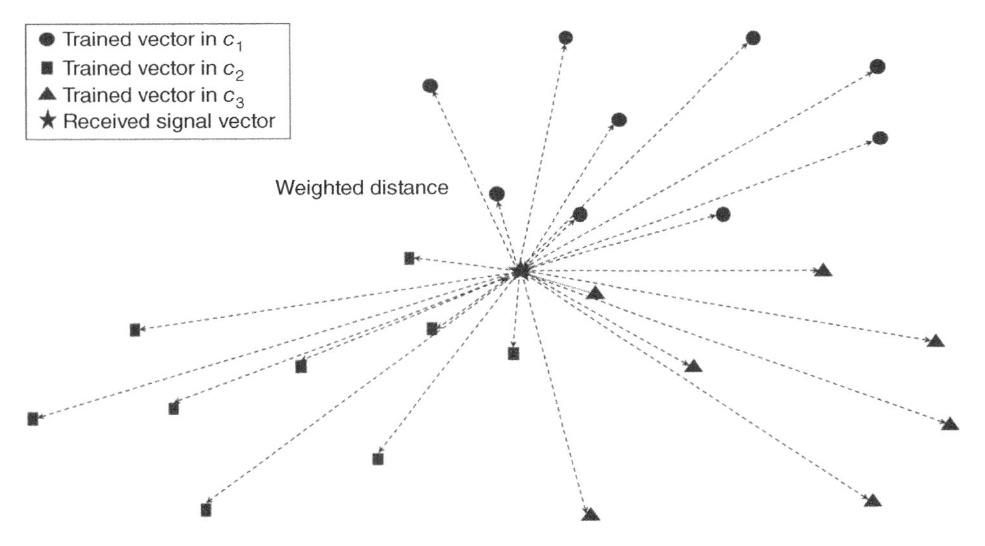

Figure 10.4 Minimum-mean-distance detection (MMD).

Although the optimality of MMD is not guaranteed, it may perform better than eMLD when L is insufficient, because MMD additionally uses reliability information captured by the distance between the received signal and the trained signal.

10.3.2 Parametric Learning

The weakness of non-parametric approaches (eMLD, MMD) is that they cause high computational complexity. In both methods, the receiver should compute all distances in the training set ($|\mathcal{D}| = \mathcal{M} \times T$). In particular, when the size of $\mathcal{D}$ is large (e.g. $T \gg 1$), a non-parametric approach is intractable in practical systems. Therefore, by learning parameters during the training phase, we can alleviate the complexity significantly.

1) *SL based on a Gaussian model (SLGM)*: In this approach, it is very important to identify a proper probabilistic model for a given data set $\mathcal{D}$. A Gaussian model is the most widely used (Jeon et al., 2018a; Robert, 2014) where the data $\boldsymbol{r} \in \mathcal{D}$ is assumed to be generated from the following probability distribution: $P(\boldsymbol{r}|c, \boldsymbol{\theta}_c) = \mathcal{N}(\boldsymbol{\mu}_c, \Sigma_c)$ where $c \in \mathcal{M}$ denotes the class (or message) of the K sources and $\boldsymbol{\theta}_c$ represents the parameter vector for the class c. Using the given data $\{\tilde{\boldsymbol{r}}_t^c : t = 1, \ldots, T\}$, we can optimize $\boldsymbol{\theta}_c = (\hat{\boldsymbol{\mu}}_c, \hat{\Sigma}_c)$ via maximum likelihood estimation as

$$\hat{\boldsymbol{\mu}}_c = \frac{1}{T}\sum_{t=1}^{T} \tilde{\boldsymbol{r}}_t^c \tag{10.8}$$

$$\hat{\Sigma}_c = \frac{1}{T}\sum_{t=1}^{T} (\tilde{\boldsymbol{r}}_t^c - \hat{\boldsymbol{\mu}}_c)(\tilde{\boldsymbol{r}}_t^c - \hat{\boldsymbol{\mu}}_c)^T, \tag{10.9}$$

where $\hat{\boldsymbol{\mu}}_c$ and $\hat{\Sigma}_c$ represent the mean and covariance of the training data corresponding to class c, respectively. As in Figure 10.5, $\hat{\boldsymbol{\mu}}_c$ could be considered representative vectors according to each class c. When the training data is not sufficient, the covariance matrix tends to be rank-deficient and ill-conditioned. This problem can be resolved with a shrinkage estimator Schäfer and Strimmer (2005). Given $\hat{\boldsymbol{\theta}}_c = (\hat{\boldsymbol{\mu}}_c, \hat{\Sigma}_c)$, the optimal maximum-likelihood detector is derived as

$$\Psi_{PGD}(\boldsymbol{r}) = \underset{c \in \mathcal{M}}{\operatorname{argmin}} \ (\boldsymbol{r} - \hat{\boldsymbol{\mu}}_c)^T \hat{\Sigma}_c^{-1} (\boldsymbol{r} - \hat{\boldsymbol{\mu}}_c). \tag{10.10}$$

In particular, the distance measure in this detector is referred to as *Mahalanobis distance*, and the inverse matrix of $\hat{\Sigma}_c$ in Eq. (10.9) is called a *precision matrix*. When $\Sigma_c = \mathbf{I}$ for all c, as a special case, the resulting detector is equivalent to the minimum-centered-distance (MCD) detector proposed in Jeon et al. (2018b). In Jeon et al. (2018b), it was shown that, among the previously discussed SL detectors, MCD and eMLD detectors yield the best performance. Since the complexity of eMLD is higher than MCD, the latter was highly recommended. However, one can argue that the Gaussian model in Eq. (10.10) may not be appropriate as the probability distribution of binary data $\boldsymbol{r} \in \{1, -1\}^N$. This motivates the development of a SL detector using a novel probabilistic model that is suitable for binary data.

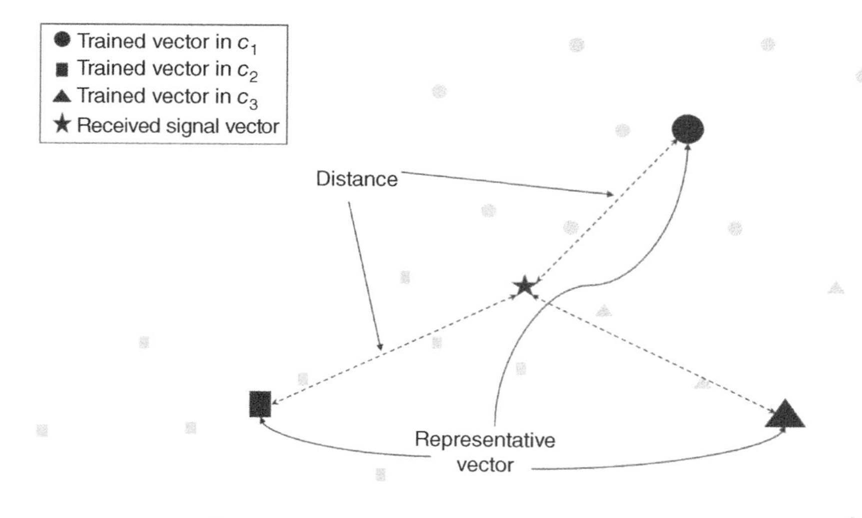

Figure 10.5 Description of a parametric learning.

2) *SL based on a Bernoulli-like model (SLBM)*: A SL detector based on a Bernoulli-like model is introduced. Here, data are assumed to be generated from the following probability distribution:

$$P(\boldsymbol{r}|c,\boldsymbol{\theta}_c) = \prod_{i=1}^{N} \epsilon_{c,i}^{\mathbf{1}_{\{r_i \neq \mu_{c,i}\}}} (1-\epsilon_{c,i})^{\mathbf{1}_{\{r_i = \mu_{c,i}\}}}, \tag{10.11}$$

where $\boldsymbol{\theta}_c = (\boldsymbol{\mu}_c, \boldsymbol{\epsilon}_c)$ for $c \in \mathcal{M}$, $\epsilon_{c,i} < 0.5$ for all i, and $\mathbf{1}_{\{\mathcal{A}\}}$ denotes an indicator function with $\mathbf{1}_{\{\mathcal{A}\}} = 1$ if $\mathcal{A}$ is true and $\mathbf{1}_{\{\mathcal{A}\}} = 0$ otherwise. Given the training data for the class c (e.g. $\mathcal{D} = \{\tilde{\boldsymbol{r}}_t^c : t = 1, \ldots, T\}$), the parameter vector $\boldsymbol{\theta}_c$ is optimized using ML estimation as

$$(\hat{\boldsymbol{\mu}}_c, \hat{\boldsymbol{\epsilon}}_c) = \underset{\boldsymbol{\mu}_c, \boldsymbol{\epsilon}_c}{\operatorname{argmax}} \prod_{t=1}^{T} P(\tilde{\boldsymbol{r}}_t^c | \boldsymbol{\mu}_c, \boldsymbol{\epsilon}_c). \tag{10.12}$$

By inserting Eq. (10.11) into Eq. (10.12), we can get

$$(\hat{\boldsymbol{\mu}}_c, \hat{\boldsymbol{\epsilon}}_c) = \underset{\boldsymbol{\mu}_c, \boldsymbol{\epsilon}_c}{\operatorname{argmax}} \prod_{i=1}^{N} \prod_{t=1}^{T} \epsilon_{c,i}^{\mathbf{1}_{\{\tilde{r}_{t,i}^c \neq \mu_{c,i}\}}} (1-\epsilon_{c,i})^{\mathbf{1}_{\{\tilde{r}_{t,i}^c = \mu_{c,i}\}}}. \tag{10.13}$$

For any $\epsilon_{c,i} < 0.5$, this objective function is maximized by taking

$$\hat{\mu}_{c,i} = \operatorname{sign}\left(\sum_{t=1}^{T} \tilde{r}_{t,i}^c \right), \tag{10.14}$$

for $i = 1, \ldots, N$. Letting $N_\text{d} = \sum_{t=1}^{T} \mathbf{1}_{\{\tilde{r}_{k,i}^c \neq \hat{\mu}_{c,i}\}}$ and $N_\text{s} = \sum_{t=1}^{T} \mathbf{1}_{\{\tilde{r}_{k,i}^c = \hat{\mu}_{c,i}\}}$, we can find an optimal $\epsilon_{c,i}$ independently from the other $\epsilon_{c,j}$'s with $i \neq j$ by taking the solution of $\operatorname{argmax}_{\epsilon_{c,i}} \epsilon_{c,i}^{N_d} (1-\epsilon_{c,i})^{N_s}$.

Taking $\frac{\partial(\epsilon_{c,i}^{N_d}(1-\epsilon_{c,i})^{N_s})}{\partial \epsilon_{c,i}} = 0$, the optimal $\epsilon_{c,i}$ is obtained as

$$\hat{\epsilon}_{c,j} = \frac{\sum_{t=1}^{T} \mathbf{1}_{\{\hat{\mu}_{c,j} \neq \tilde{r}_{j,i}^c\}}}{T}. \tag{10.15}$$

Using the optimal parameter vector $\hat{\boldsymbol{\theta}}_c = (\hat{\boldsymbol{\mu}}_c, \hat{\boldsymbol{\epsilon}}_c)$ in Eq. (10.14) and Eq. (10.15), the optimal maximum-likelihood estimator is derived as

$$\Psi_{\mathcal{D}}(\boldsymbol{r}) = \underset{c \in \mathcal{M}}{\operatorname{argmin}}\ (\boldsymbol{r} - \hat{\boldsymbol{\mu}}_c)^T \begin{bmatrix} -\log \hat{\epsilon}_{c,1} & \cdots & 0 \\ \vdots & \ddots & \vdots \\ 0 & \cdots & -\log \hat{\epsilon}_{c,N} \end{bmatrix} (\boldsymbol{r} - \hat{\boldsymbol{\mu}}_c). \tag{10.16}$$

When one-bit ADCs are employed at receivers, it is shown that SLBM outperforms SLGM since the former is developed for and more proper to the treatment of binary data.

The principle of parametric learning is very close to that of a nearest-centroid classifier (NCC), which is a simple solution of the classification problem in SL. NCC assigns the class label of an unlabelled observed vector by using the centroid vectors that represent their classes. Similarly, in parametric learning, the SL detector works by selecting a message that has the minimum distance from the conditional mean vector of the pilot signals. This resemblance is a good example to show an interesting connection between a data-detection problem in wireless communications and a classification problem in SL.

10.4 Low-Complexity SL (LCSL) Detector

Although a parametric learning detector requires lower complexity than a non-parametric learning detector, the computational complexity is prohibitive as the size of search-space (e.g. $|\mathcal{M}| = m^K$) grows exponentially with K. Thus, as in conventional MIMO systems, sphere decoding, which efficiently finds a reduced search space, can be considered. However, the conventional sphere decoding in Hassibi and Vikalo (2005) cannot be used directly due to the nonlinearity of the considered channel models. So-called one-bit sphere decodings (OSDs), which are suitable for one-bit quantizations, have been proposed in Jeon et al. (2018b), Kim et al. (2017). In this section, OSD based on a hierarchical clustering forest (OSD-HCF) is introduced, which can yield a higher-quality reduced search space. In this method, a fast binary nearest neighbor search algorithm (flann) Muja and Lowe (2012) is used, which finds a reduced search space efficiently by exploiting hierarchical clustering structures. Combining with the SL detector, it is called a low-complexity SL (LCSL) detector. The overall procedures of the LCSL detector are provided as follows:

1. First decompose the (binary) Hamming space hierarchically so as to build a tree structure. It starts with choosing J elements from $\hat{\mathcal{U}}$ randomly, which act as J cluster centroids.
2. The previous step forms the J clusters around these centroids, and the decomposition process is repeated recursively (see Algorithm 1).
3. This process is performed W times to construct $\{\mathcal{T}_1, \dots, \mathcal{T}_W\}$ trees.

The resulting trees in will be used in the data-transmission phase to efficiently reduce the search space according to a new observation $\boldsymbol{r}$. It is noticeable that the algorithm constructs multiple hierarchical trees having possibly different decomposition structures, and thus is able to improve the quality of a resulting reduced search space.

As you see Figure 10.6, we explain this process (Algorithm 1) by giving a simple example ($J = 2, |\mathcal{M}| = 16$) in the case of a single tree. The decomposition procedure is applied into binary space (in this case, 16 $\hat{\boldsymbol{\mu}}_c'$s) hierarchically until the cluster size is less than J, which is called the size of the leaf node. (A cluster size at the last hierarchical level can be called the size of the leaf node because the output structure of this process can be viewed as a tree: Figure 10.7.) Also, the leaf node size equals branch factor J. This is due to the reduction of the number of parameters and the calculation of approximate complexity. However, we skip this as it is not the main point of this section. If we operate this process multiple times, several distinctive trees (forest) are constructed.

Data-transmission phase: Given a current observation $\boldsymbol{r}$, the search algorithm begins with traversing multiple trees in parallel (see Algorithm 2). Note that W multiple trees share a single priority queue ($\mathcal{Q}$), and the nodes in the priority queue are arranged in order of shortest hamming distance with regard to the observation $\boldsymbol{r}$. Then it can efficiently produce the reduced search-space $\mathcal{S}(\boldsymbol{r}) \subseteq \mathcal{M}$, which only contains the nearest $\hat{\boldsymbol{\mu}}_c$'s to the $\boldsymbol{r}$. Leveraging this, the LCSL detector is performed as

$$\Psi_D(\boldsymbol{r}) = \underset{c \in \mathcal{S}(\boldsymbol{r})}{\operatorname{argmin}} \ (\boldsymbol{r} - \hat{\boldsymbol{\mu}}_c)^T \begin{bmatrix} -\log \hat{\epsilon}_{\mathrm{c},1} & \dots & 0 \\ \vdots & \ddots & \vdots \\ 0 & \dots & -\log \hat{\epsilon}_{\mathrm{c,N}} \end{bmatrix} (\boldsymbol{r} - \hat{\boldsymbol{\mu}}_c). \tag{10.17}$$

In order to help you understand Algorithm 2, we provide a simple example (Figure 10.8) that finds a reduced search space based on the structure built in the prior example (Figure 10.6). During the first search, the observation continues selecting the closest centroid at each level

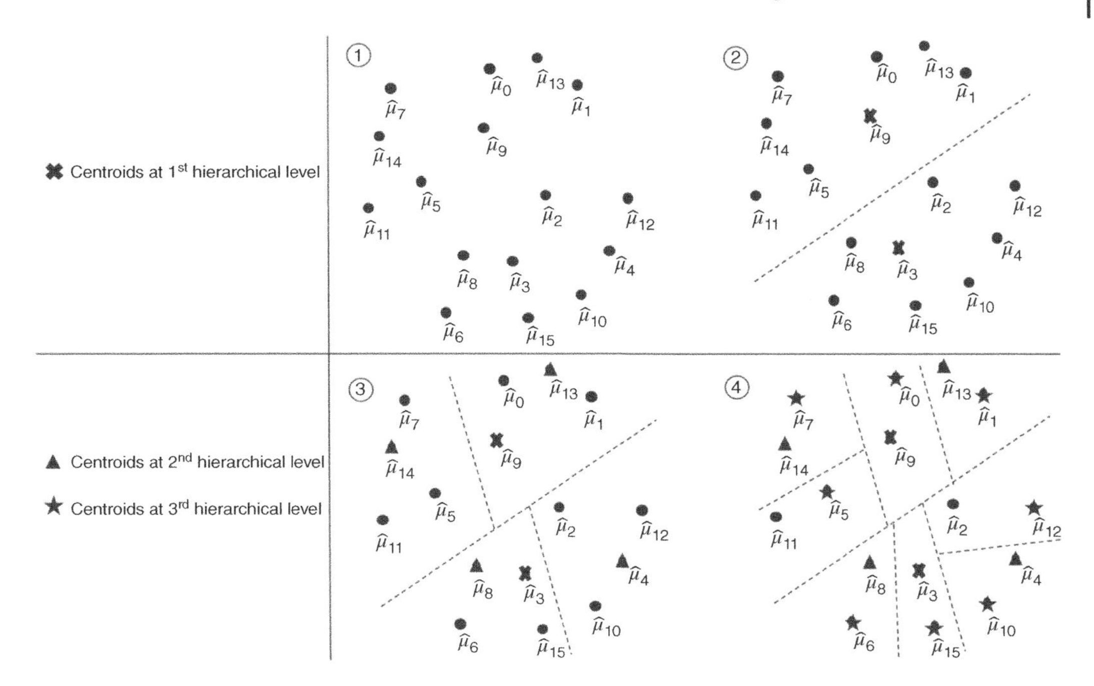

Figure 10.6 Simple example of a decomposition process.

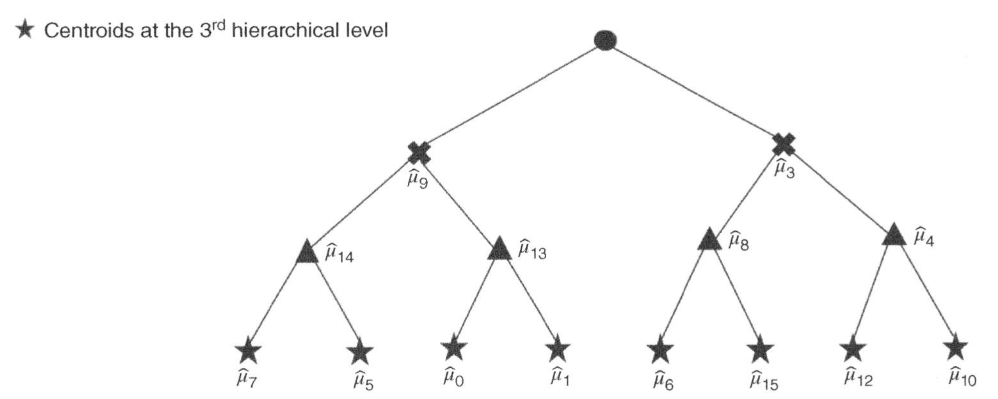

Figure 10.7 Illustration of the view of a tree structure.

until it reaches a leaf node. The elements included in the leaf node are returned into the reduced search space ($S(\boldsymbol{r})$). Moreover, the unselected centroids are put into the priority queue that sorts centroids in such an order that makes an efficient nearest-neighbor search possible. For the second search, the same search method begins from the first node in the priority queue. This algorithm ends when it satisfies the predetermined size of the reduced search space ($|S(\boldsymbol{r})|$).

Beyond a single tree search, a tree search is expanded to a forest search algorithm (Algorithm 2) where several different tree structures are constructed. In Figure 10.9, the search starts by going down from the root node to a leaf node simultaneously at each tree along the

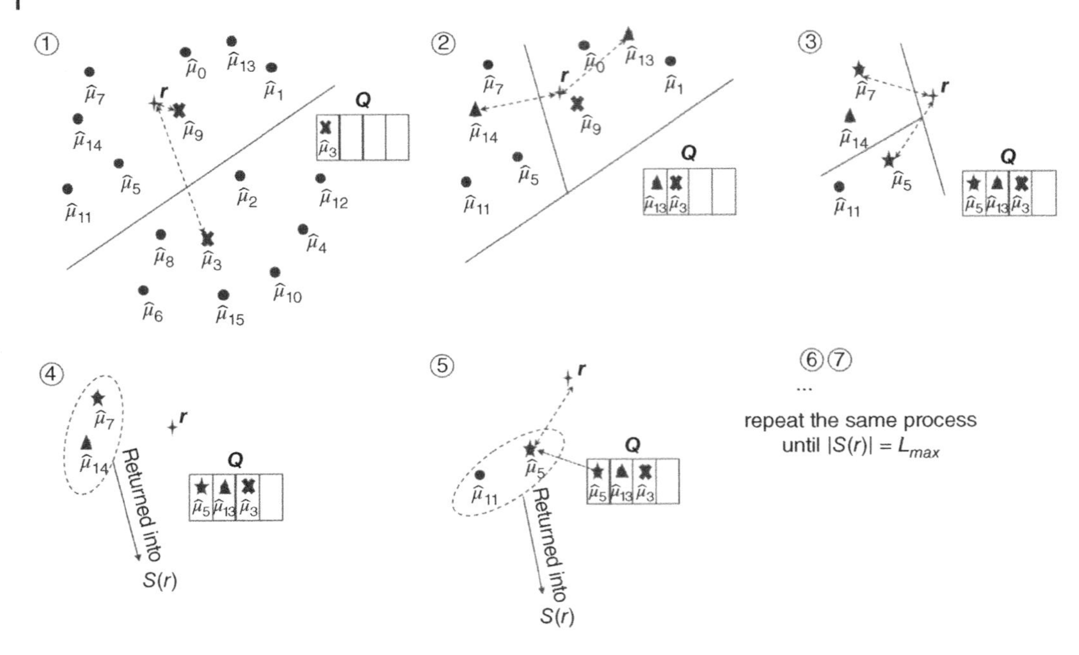

Figure 10.8 Brief example of a tree search algorithm.

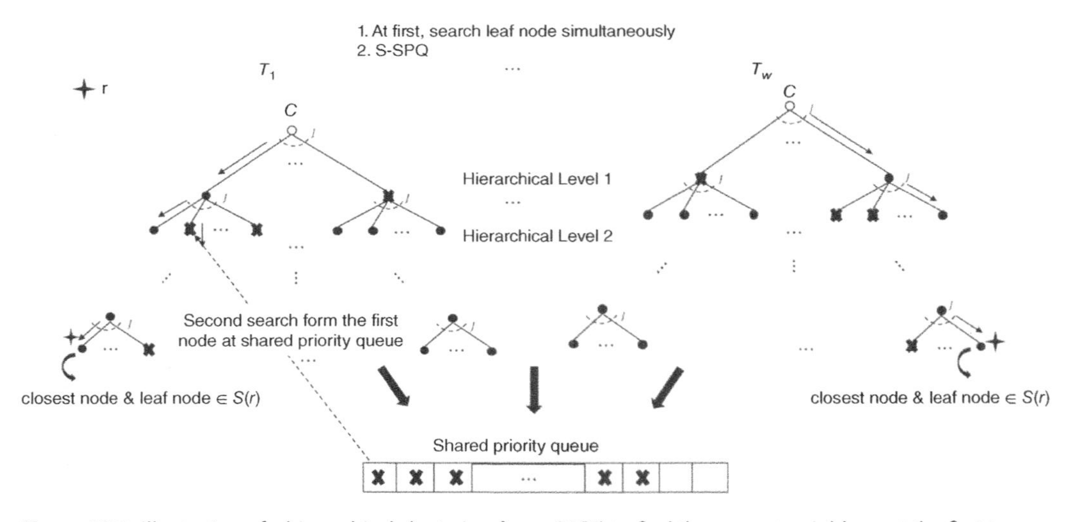

Figure 10.9 Illustration of a hierarchical clustering forest (HCF) to find the nearest neighbors at the first two steps, where the resulting reduced search space for a given current observation $\mathbf{r}$ is denoted by $S(\mathbf{r})$. S-SPQ: Search from the first node at the shared priority queue.

closest centroid to $\mathbf{r}$ at each level. Unselected nodes (**X** icon) are stored in a single shared priority queue in order of the shortest distance from $\mathbf{r}$. The search stops when it reaches a leaf node and it obtains approximately J $\hat{\boldsymbol{\mu}}_c^{\prime}s$ contained in the leaf node of each tree. In the next step, the search starts from the first node of the shared priority queue in the same way. This iterates until it satisfies the parameter $|S(\mathbf{r})|$.

Algorithm 1 Constructing a hierarchical clustering tree $h(C_i, J)$

Input: $\hat{\mathcal{U}} = \{\hat{\boldsymbol{\mu}}_1, \ldots, \hat{\boldsymbol{\mu}}_{|\mathcal{M}|}\}$
Output: hierarchical clustering tree $\mathcal{T}_i$
Parameters: J (branching factor and maximum leaf size)

1: **if** $|\hat{\mathcal{U}}| \leq J$ **then**
2:     *create leaf node with the elements in* $\hat{\mathcal{U}}$
3: **else**
4:     $\mathcal{P} \leftarrow$ *select J elements randomly from* $\hat{\mathcal{U}}$ *(centroids)*
5:     $C \leftarrow$ cluster the elements in $\hat{\mathcal{U}}$ w. r. t. the centroid $i \in \mathcal{P}$
6:     **for** each cluster $C_i \in C = \{C_1, \ldots, C_J\}$ **do**
7:         *create non-leaf node with centroid* $\mathcal{P}_i$
8:         *recursively apply the algorithm* $h(C_i, J)$ *(with the updated* C_i*)*
9:     **end for**
10: **end if**

Algorithm 2 Searching parallel hierarchical clustering trees

Input: hierarchical clustering trees $\{\mathcal{T}_i : i = 1, \ldots, W\}$ and a new observation $\mathbf{r}$
Output: $\mathcal{S}(\mathbf{r})$(reduced search space associated with $\mathbf{r}$)
Parameters: $L_{\max}$(the desired size of a reduced search space, e.g. $|\mathcal{S}(\mathbf{r})| = L_{\max}$)

1: L $\leftarrow$ 0 {L = *number of points* $\hat{\mu}_c$ *searched*}
2: $\mathcal{Q} \leftarrow$ *empty priority queue*
3: $\mathcal{R} \leftarrow$ *empty priority queue*
4: **for** each tree T_i **do**
5:     *call* TraverseTree($\mathcal{T}_i$,$\mathcal{Q}$,$\mathcal{R}$)
6: **end for**
7: **while** $|\mathcal{Q}| \neq 0$ *and* $L < L_{max}$ **do**
8:     $j \leftarrow$ *top index of* $\mathcal{Q}$
9:     call TraverseTree(j, $\mathcal{Q}$, $\mathcal{R}$)
10: **end while**
11: **return** *K top points from* $\mathcal{R}$

1: **procedure** TRAVERSETREE(j, $\mathcal{QQ}$, $\mathcal{R}$)
2:     **if** *node j is a leaf node* **then**
3:         $\mathcal{S} \triangleq$ {all the elements in leaf node j}
4:         $\mathcal{R} = \mathcal{R} \cup \mathcal{S}$
5:         $L \leftarrow L + |\mathcal{S}|$
6:     **else**
7:         $C \leftarrow$ *child nodes of* N
8:         $i \leftarrow$ *closest node of* C *to observation* $\mathbf{r}$
9:         $C_p \leftarrow C \setminus \{i\}$
10:         *add all nodes in* C_p *to* $\mathcal{Q}$
11:         call TraverseTree(i, $\mathcal{Q}$, $\mathcal{R}$)
12:     **end if**
13: **end procedure**

10.5 Numerical Results

In this section, we evaluate the average bit-error rate (BER) performances of the introduced SLGM and SLBM detectors. Furthermore, it is shown that the LCSL detector can achieve the original performance with much lower complexity for a large-scale distributed MIMO system.

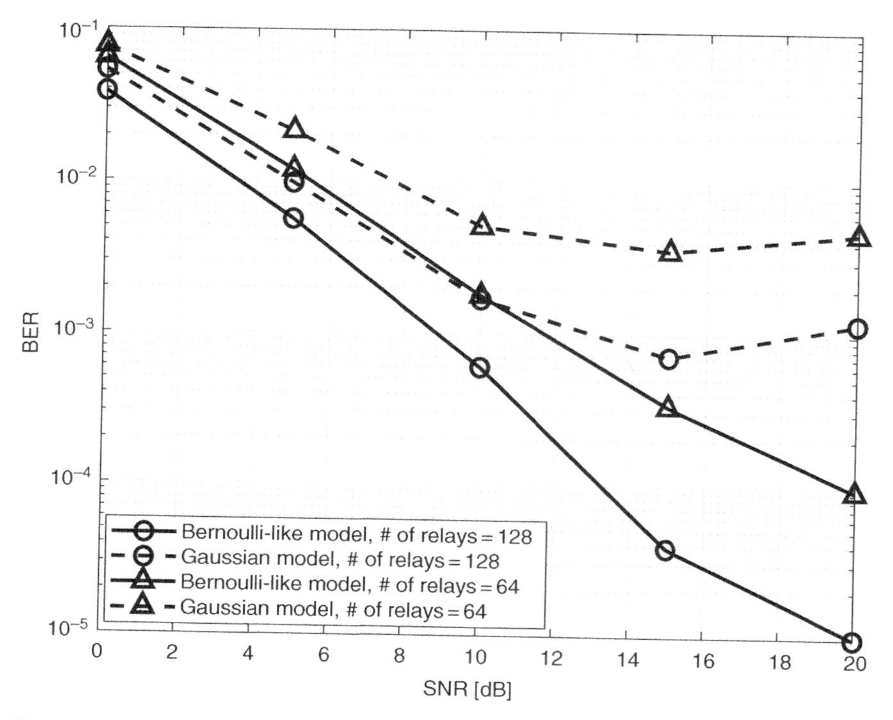

Figure 10.10 $K = 8$, $N_r = 64$, and $T = 15$. Performance comparisons of the SL detectors based on Bernoulli-like and Gaussian models, according to the number of intermediate relays.

For the simulations, a Rayleigh fading channel is considered for local wireless channels where each element of a channel matrix is drawn from an independent and identically distributed (i.i.d.) circularly symmetric complex Gaussian random variable with zero mean and unit variance. In addition, the relays' operations are assumed to be amplify-and-forward (AF), and QPSK modulation is assumed. However, it is remarkable that performance trends are kept for other relays' operations. Note that each relay is equipped with a single antenna. When training overhead is small (e.g. T is small), an empirical error probability (e.g. $\epsilon_{c,i}$) can be underestimated as zero, although it is actually not. This can cause a severe error-floor problem. To prevent this, we assign a minimum value of $\hat{\epsilon}_{c,i}$ as 10^{-3}.

Figure 10.10 shows the BER performances of the parametric learning detectors with training overhead $T = 15$. Also, the following two scenarios are considered: (i) 64 intermediate relays; (ii) 128 intermediate relays. Figure 10.10 shows that the SLBM detector outperforms the SLGM detector, which implies that the Bernoulli-like model is more suitable to binary data than a Gaussian model.

Figure 10.11 shows the BER performance of the low-complexity detector according to L_{max} in Algorithm 2. Also, we set $J = 32$ in Algorithm 1. As seen in Figure 10.11, the complexity is extremely high when the number of users becomes large. From this simulation, we observe that the low-complexity detector can achieve optimal performance perfectly with only 6% of the original complexity. Also, it is expected that the use of a low-complexity technique is more beneficial for a large-scale distributed reception system (e.g. large K).

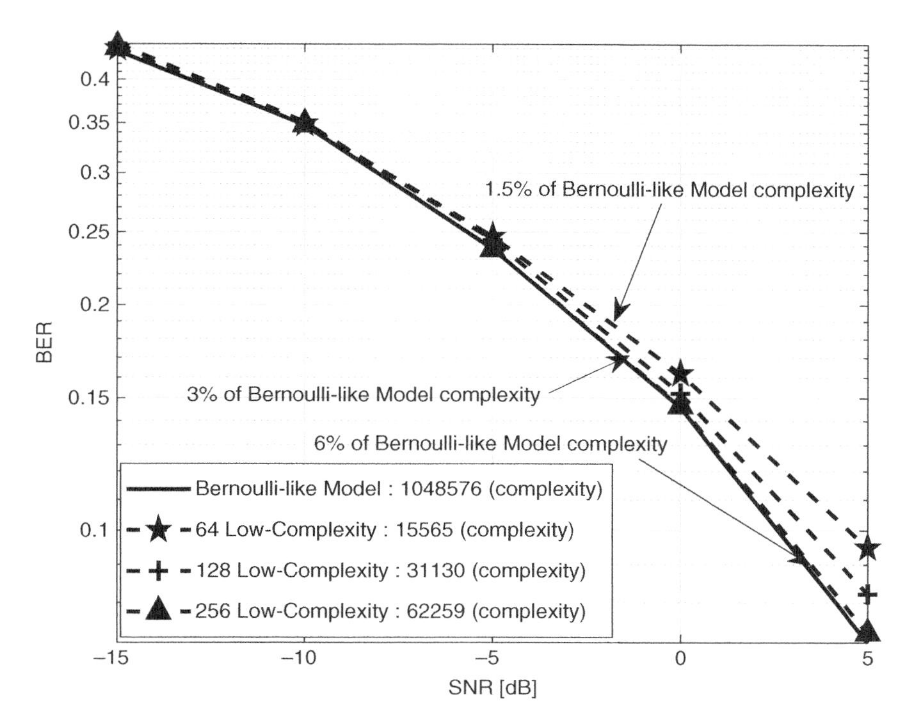

Figure 10.11 $K = 14$, $N_r = 64$, and $T = 15$. Performance of the LCSL detector according to the size of the reduced search space (L_{max}).

10.6 Conclusions

In this chapter, promising SL detectors were described for distributed multihop MIMO systems, especially when receivers are equipped with one-bit ADCs. The SL detectors can be categorized as non-parametric or parametric. The former have the benefit of a large training set (e.g. a large number of pilot signals) since the resulting empirical conditional probability distribution approaches a true one. However, this approach suffers from extremely high complexity. Parametric SL detectors can reduce this complexity by estimating representative vectors (called codewords) corresponding to each class (message) during the training phase. Furthermore, by combining them with one-bit sphere decodings, they can perform very well with reasonable complexity. Thus, the SL-based detectors produced in this chapter can be considered cutting-edge for a nonlinear communication system.

Bibliography

Junil Choi, Jianhua Mo, and Robert W. Heath. Near maximum-likelihood detector and channel estimator for uplink multiuser massive MIMO systems with one-bit ADCs. *IEEE Transactions on Communications*, 64 (5): 2005–2018, 2016.

Onkar Dabeer and Upamanyu Madhow. Channel estimation with low-precision analog-to-digital conversion. In *Communications (ICC), 2010 IEEE International Conference on*, pages 1–6. IEEE, 2010.

Babak Hassibi and Haris Vikalo. On the sphere-decoding algorithm i. expected complexity. *IEEE transactions on signal processing*, 53 (8): 2806–2818, 2005.

Sven Jacobsson, Giuseppe Durisi, Mikael Coldrey, Ulf Gustavsson, and Christoph Studer. One-bit massive MIMO: Channel estimation and high-order modulations. In *Communication Workshop (ICCW), 2015 IEEE International Conference on*, pages 1304–1309. IEEE, 2015.

Sven Jacobsson, Giuseppe Durisi, Mikael Coldrey, Ulf Gustavsson, and Christoph Studer. Throughput analysis of massive MIMO uplink with low-resolution ADCs. *IEEE Transactions on Wireless Communications*, 16 (6): 4038–4051, 2017.

Yo-Seb Jeon, Song-Nam Hong, and Namyoon Lee. Supervised-learning-aided communication framework for MIMO systems with low-resolution ADCs. *IEEE Transactions on Vehicular Technology*, 2018a.

Yo-Seb Jeon, Namyoon Lee, Song-Nam Hong, and Robert W. Heath. One-bit sphere decoding for uplink massive MIMO systems with one-bit ADCs. *IEEE Transactions on Wireless Communications*, 2018b.

Seonho Kim, Namyoon Lee, and S-N Hong. A low-complexity soft-output wmd decoding for uplink MIMO systems with one-bit ADCs. *arXiv preprint arXiv:1707.02868*, 2017.

Yongzhi Li, Cheng Tao, Gonzalo Seco-Granados, Amine Mezghani, A. Lee Swindlehurst, and Liu Liu. Channel estimation and performance analysis of one-bit massive MIMO systems. *IEEE Trans. Signal Process*, 65 (15): 4075–4089, 2017.

Ning Liang and Wenyi Zhang. Mixed-ADC massive MIMO. *IEEE Journal on Selected Areas in Communications*, 34 (4): 983–997, 2016.

Amine Mezghani, Nesrine Damak, and Josef A. Nossek. Circuit aware design of power-efficient short range communication systems. In *Wireless Communication Systems (ISWCS), 2010 7th International Symposium on*, pages 869–873. IEEE, 2010.

Jianhua Mo and Robert W. Heath. Capacity analysis of one-bit quantized MIMO systems with transmitter channel state information. *IEEE Transactions on Signal Processing*, 63 (20): 5498–5512, 2015.

Christopher Mollén, Junil Choi, Erik G. Larsson, and Robert W. Heath. One-bit ADCs in wideband massive MIMO systems with ofdm transmission. In *Acoustics, Speech and Signal Processing (ICASSP), 2016 IEEE International Conference on*, pages 3386–3390. IEEE, 2016.

Christopher Mollén, Junil Choi, Erik G. Larsson, and Robert W. Heath Jr. Uplink performance of wideband massive MIMO with one-bit ADCs. *IEEE Trans. Wireless Communications*, 16 (1): 87–100, 2017.

Marius Muja and David G. Lowe. Fast matching of binary features. In *Computer and Robot Vision (CRV), 2012 Ninth Conference on*, pages 404–410. IEEE, 2012.

B. Murmann. ADC performance survey 1997-2016, accessed on mar. 2017.

Josef A. Nossek and Michel T. Ivrlač. Capacity and coding for quantized MIMO systems. In *Proceedings of the 2006 International Conference on Wireless Communications and Mobile Computing*, pages 1387–1392. ACM, 2006.

Zhouyue Pi and Farooq Khan. An introduction to millimeter-wave mobile broadband systems. *IEEE communications magazine*, 49 (6), 2011.

Chiara Risi, Daniel Persson, and Erik G. Larsson. Massive MIMO with 1-bit ADC. *arXiv preprint arXiv:1404.7736*, 2014.

Christian Robert. *Machine learning, a probabilistic perspective.* Taylor & Francis, 2014.

Juliane Schäfer and Korbinian Strimmer. A shrinkage approach to large-scale covariance matrix estimation and implications for functional genomics. *Statistical Applications in Genetics and Molecular Biology*, 4 (1), 2005.

Jaspreet Singh, Onkar Dabeer, and Upamanyu Madhow. On the limits of communication with low-precision analog-to-digital conversion at the receiver. *IEEE Trans. Communications*, 57 (12): 3629–3639, 2009.

A Lee Swindlehurst, Ender Ayanoglu, Payam Heydari, and Filippo Capolino. Millimeter-wave massive MIMO: The next wireless revolution? *IEEE Communications Magazine*, 52 (9): 56–62, 2014.

Robert H. Walden. Analog-to-digital converter survey and analysis. *IEEE Journal on Selected Areas in Communications*, 17 (4): 539–550, 1999.

Shengchu Wang, Yunzhou Li, and Jing Wang. Multiuser detection in massive spatial modulation MIMO with low-resolution ADCs. *IEEE Transactions on Wireless Communications*, 14 (4): 2156–2168, 2015.

Chao-Kai Wen, Chang-Jen Wang, Shi Jin, Kai-Kit Wong, and Pangan Ting. Bayes-optimal joint channel-and-data estimation for massive MIMO with low-precision ADCs. *IEEE Transactions on Signal Processing*, 64 (10): 2541–2556, 2016.

11

Adaptive Learning for Symbol Detection: A Reproducing Kernel Hilbert Space Approach

Daniyal Amir Awan[1], *Renato Luis Garrido Cavalcante*[2], *Masahario Yukawa*[3], *and Slawomir Stanczak*[2]

[1] *Network Information Theory Group, Technical University of Berlin, Berlin, Germany*
[2] *Wireless Communications and Networks, Fraunhofer Heinrich Hertz Institute & Network Information Theory Group, Technical University of Berlin, Berlin, Germany*
[3] *Department of Electronics and Electrical Engineering, Keio University, Yokohama, Japan*

11.1 Introduction

Solutions to problems in wireless communications are typically based on models that require, for example, channel state information, knowledge of interference patterns, and information about the phase of desired signals, to name a few. As a result, before sending any information over the wireless channel, transceivers traditionally estimate many model parameters. However, this approach has two major drawbacks that can severely impair the performance of communication systems. First, perfect estimation of parameters is impossible in general because of the presence of noise and the limitations of the algorithms being used. Second, the models themselves are only idealizations, and it is often unclear how well they can capture the true behavior of real systems.

To mitigate these handicaps of purely model-based approaches, researchers have been proposing alternatives based on data-driven approaches including machine learning. The idea is to replace some building blocks of model-based transceivers with learning algorithms, with the intent to drastically reduce the number of assumptions about the models and the need for complex estimation techniques. However, this reduction in model knowledge brings many technical challenges, especially in the physical layer of the communication stack. In particular, some state-of-the-art learning tools require large training sets and a long training time. However, in the physical layer, the environment (channel and user distribution, etc.) can be considered roughly constant only for a very short time, which can be all the time available to collect a training set based on pilots, train a machine, and perform the communication task. If this temporal aspect is not taken into account, then by the time enough samples are available to train existing learning algorithms, the propagation environment may have changed so drastically as to render the training set useless for the current propagation conditions, even if we assume that training can be performed instantaneously.

Against this background, this chapter introduces a novel machine learning algorithm for symbol detection in multiuser environments. We consider a challenging multiuser uplink scenario in which the number of antennas available at the base station may be smaller than the number of active users. More specifically, the proposed method is an adaptive (nonlinear) receive filter that learns to detect symbols from data directly, without performing any intermediate estimation tasks (e.g. channel estimation). Furthermore, the method is robust against abrupt changes

Machine Learning for Future Wireless Communications, First Edition. Edited by Fa-Long Luo.
© 2020 John Wiley & Sons Ltd. Published 2020 by John Wiley & Sons Ltd.

of the wireless environment. We build upon recent adaptive learning methods in reproducing kernel Hilbert spaces (RKHSs) that have already been successfully applied to the *robust nonlinear beamforming* problem in Theodoridis et al. (2011), Slavakis et al. (2009). In contrast to these studies, rather than assuming knowledge of the angle of arrivals or channels of the desired user, we provide our algorithm with additional robustness against the vagaries of the wireless environment by considering partially linear filters, as proposed in Yukawa (2015a) in the context of a different application.

The rest of this chapter is organized as follows. We present the relevant mathematical background in Section 11.2. After presenting our system model in Section 11.3, we present our symbol-detection method in detail in Section 11.4. Simulations are presented in Section 11.5, and Section 11.6 summarizes the chapter.

11.2 Preliminaries

In the remainder of this chapter, $\mathbb{R}$ is the set of reals, and $\mathbb{N}$ is the set of natural numbers with the convention that $0 \notin \mathbb{N}$. We define $\overline{N_1, N_2} := \{N_1, N_1 + 1, N_1 + 2, \dots, N_2\}$ for $(N_1, N_2) \in \mathbb{N} \times \mathbb{N}$ with $N_1 \leq N_2$. By $\| \cdot \|_{\mathbb{R}^d}$ we denote the conventional Euclidean norm in $\mathbb{R}^d$. The real and imaginary parts of complex-valued scalars or vectors are given by $\Re(\cdot)$ and $\Im(\cdot)$, respectively. Let $(\mathcal{H}, \langle \cdot, \cdot \rangle_{\mathcal{H}})$ be a real Hilbert space endowed with the inner product $\langle \cdot, \cdot \rangle_{\mathcal{H}}$, which induces the norm $(\forall \boldsymbol{x} \in \mathcal{H})\ \|\boldsymbol{x}\|_{\mathcal{H}} = \langle \boldsymbol{x}, \boldsymbol{x} \rangle_{\mathcal{H}}^{1/2}$. If $C \subset \mathcal{H}$ is a nonempty closed convex set, the projection $P_C : \mathcal{H} \to C$ is defined to be the operator that maps an arbitrary point $\boldsymbol{x} \in \mathcal{H}$ to the uniquely existing vector $\boldsymbol{y} \in C$ satisfying $(\forall \boldsymbol{u} \in C)\ \|\boldsymbol{x} - \boldsymbol{y}\|_{\mathcal{H}} \leq \|\boldsymbol{x} - \boldsymbol{u}\|_{\mathcal{H}}$. By span$(S)$ we denote the set of all finite linear combinations of the elements of the set S. If $\kappa : \mathcal{U} \times \mathcal{U} \to \mathbb{R}$ is a function of two variables, we denote the function of a single variable obtained by keeping the second argument fixed to $\boldsymbol{y} \in \mathcal{U}$ by $\kappa(\cdot, \boldsymbol{y}) : \mathcal{U} \to \mathbb{R} : \boldsymbol{x} \mapsto \kappa(\boldsymbol{x}, \boldsymbol{y})$.

11.2.1 Reproducing Kernel Hilbert Spaces

The proposed algorithms for symbol detection are based on the theory of reproducing kernel Hilbert spaces (RKHSs), which have been extensively used in diverse fields such as statistics, probability, signal processing, and machine learning, among others Berlinet and Thomas-Agnan (2004), Theodoridis et al. (2011), Slavakis et al. (2009), Yukawa (2015a). These spaces can be formally defined as follows:[1]

Definition 11.1 ***(Reproducing kernel Hilbert spaces and reproducing kernels (RKHS))*** Let $\mathcal{U}$ be an arbitrary nonempty set. A Hilbert space $(\mathcal{H}, \langle \cdot, \cdot \rangle_{\mathcal{H}})$ of real-valued functions $f : \mathcal{U} \to \mathbb{R}$ is called a *reproducing kernel Hilbert space* if and only if there exists a function $\kappa : \mathcal{U} \times \mathcal{U} \to \mathbb{R}$, called *reproducing kernel*, such that:

(i) $(\forall x \in \mathcal{U})\ \kappa(\cdot, \boldsymbol{x}) \in \mathcal{H}$; and
(ii) $(\forall x \in \mathcal{U})(\forall f \in \mathcal{H})\ f(\boldsymbol{x}) = \langle f, \kappa(\cdot, \boldsymbol{x}) \rangle_{\mathcal{H}}$ (reproducing property).

For readers who are not familiar with these particular Hilbert spaces, Definition 11.1 may not give any hints about the functions $\kappa : \mathcal{U} \times \mathcal{U} \to \mathbb{R}$ that qualify as reproducing kernels or an intuition on the functions $f : \mathcal{U} \to \mathbb{R}$ in an RKHS. Therefore, it may be enlightening

---

1 In this study, we only deal with real Hilbert spaces of real-valued functions, but the definition of RKHSs shown here can be naturally extended to complex Hilbert spaces Berlinet and Thomas-Agnan (2004).

to construct inner product spaces (or pre-Hilbert spaces) for which RKHSs emerge as their completion (this construction can be seen as part of the Moore-Aronszajn theorem Aronszajn (1950)(Berlinet and Thomas-Agnan, 2004, Ch. 1)). For this purpose, we now introduce the concept of *positive definite kernels*, which we later show to be equivalent to the concept of reproducing kernels:

Definition 11.2 ***(Positive definite kernels):*** Let $\mathcal{U}$ be an arbitrary nonempty set. We say that $\kappa : \mathcal{U} \times \mathcal{U} \to \mathbb{R}$ is a (real) positive definite kernel (or simply kernel for brevity) if the following properties hold:

(Symmetry) $(\forall \boldsymbol{x} \in \mathcal{U})(\forall \boldsymbol{y} \in \mathcal{U})\ \kappa(\boldsymbol{x}, \boldsymbol{y}) = \kappa(\boldsymbol{y}, \boldsymbol{x})$

(Positivity) $(\forall M \in \mathbb{N})(\forall (\alpha_1, \dots, \alpha_M) \in \mathbb{R}^M)(\forall (\boldsymbol{x}_1, \dots, \boldsymbol{x}_M) \in \mathcal{U}^M)$

$$\sum_{k=1}^{M} \sum_{j=1}^{M} \alpha_k \alpha_j \kappa(\boldsymbol{x}_k, \boldsymbol{x}_j) \geq 0.$$

We emphasize that this definition requires no structure on the set $\mathcal{U}$. However, in the application considered in this chapter, we focus on the case $\mathcal{U} \subset \mathbb{R}^d$. We also note that the *positivity* property in the definition is equivalent to the following property: for arbitrary $M \in \mathbb{N}$ and $(\boldsymbol{x}_1, \dots, \boldsymbol{x}_M) \in \mathcal{U}^M$, the matrix $\mathbf{K} \in \mathbb{R}^{M \times M}$ with the element in the ith row and jth column given by $[\mathbf{K}]_{i,j} := \kappa(\boldsymbol{x}_i, \boldsymbol{x}_j)$ has to be positive semi-definite. Note that kernels can be further categorized into positive definite kernels and positive semi-definite kernels depending on whether $\mathbf{K}$ is always positive definite or only positive semi-definite for distinct vectors $(\boldsymbol{x}_1, \dots, \boldsymbol{x}_M) \in \mathcal{U}^M$. However, in this study we do not make this distinction.

To construct a pre-Hilbert space of functions associated with a given positive definite kernel k, we define the following vector space of functions $f : \mathcal{U} \to \mathbb{R}$:

$$\mathcal{H}_0 := \text{span}(\{\kappa(\cdot, \boldsymbol{x}) \mid \boldsymbol{x} \in \mathcal{U}\}).$$

By definition of $\mathcal{H}_0$, any two functions $f \in \mathcal{H}_0$ and $g \in \mathcal{H}_0$ can be written as $f : \mathcal{U} \to \mathbb{R} : \boldsymbol{x} \mapsto \sum_{k=1}^{M} \alpha_k\ \kappa(\boldsymbol{x}, \boldsymbol{x}_k)$ and $g : \mathcal{U} \to \mathbb{R} : \boldsymbol{x} \mapsto \sum_{j=1}^{N} \beta_j\ \kappa(\boldsymbol{x}, \boldsymbol{x}'_j)$ for some $(M, N) \in \mathbb{N} \times \mathbb{N}$, $(\boldsymbol{x}_1, \dots, \boldsymbol{x}_M) \in \mathcal{U}^M$, $(\boldsymbol{x}'_1, \dots, \boldsymbol{x}'_N) \in \mathcal{U}^N$, $(\alpha_1, \dots, \alpha_M) \in \mathbb{R}^M$, and $(\beta_1, \dots, \beta_N) \in \mathbb{R}^N$. Readers can verify that the function $\langle \cdot, \cdot \rangle_{\mathcal{H}_0} : \mathcal{H}_0 \times \mathcal{H}_0 \to \mathbb{R}$ defined by

$$\langle f, g \rangle_{\mathcal{H}_0} := \sum_{k=1}^{M} \sum_{j=1}^{N} \alpha_k \beta_j \kappa(\boldsymbol{x}_k, \boldsymbol{x}'_j) \tag{11.1}$$

is an inner product, so $(\mathcal{H}_0, \langle \cdot, \cdot \rangle_{\mathcal{H}_0})$ is a pre-Hilbert space. We can also prove that a unique completion of $(\mathcal{H}_0, \langle \cdot, \cdot \rangle_{\mathcal{H}_0})$, which is the Hilbert space denoted by $(\mathcal{H}, \langle \cdot, \cdot \rangle_{\mathcal{H}})$, satisfies both: (i) $(\forall x \in \mathcal{U})\ \kappa(\cdot, \boldsymbol{x}) \in \mathcal{H}$ and (ii) $(\forall x \in \mathcal{U})(\forall f \in \mathcal{H})\, f(\boldsymbol{x}) = \langle f, \kappa(\cdot, \boldsymbol{x}) \rangle_{\mathcal{H}}$. As a result, $(\mathcal{H}, \langle \cdot, \cdot \rangle_{\mathcal{H}})$ is a RKHS. Furthermore, positive definite functions are reproducing kernels, and the converse can also be shown to be valid. The Moore-Aronszajn theorem also states that, given a positive definite kernel κ, there is only one Hilbert space for which κ is the reproducing kernel.

11.2.2 Sum Spaces of Reproducing Kernel Hilbert Spaces

We now briefly summarize the ideas originally proposed in Yukawa (2015a). Suppose that we have the task of learning an unknown nonlinear function $f : \mathcal{U} \to \mathbb{R}$ (where $\mathcal{U} \subset \mathbb{R}^d$) that can be decomposed into the sum of Q distinct components, such as high- and low-frequency components, linear and nonlinear components, and periodic and aperiodic components. Assume that each of these components belongs to one of $Q \in \mathbb{N}$ RKHSs $(\mathcal{H}_1, \langle\cdot,\cdot\rangle_{\mathcal{H}_1}), \ldots, (\mathcal{H}_Q, \langle\cdot,\cdot\rangle_{\mathcal{H}_Q})$ of functions mapping $\mathcal{U}$ to the real line. In this case, the unknown function f is a member of the *sum space* defined by

$$\mathcal{H}^+ := \left\{ \sum_{q\in\mathcal{Q}} f_q \mid (\forall q \in \mathcal{Q})\, f_q \in \mathcal{H}_q \right\},\ \mathcal{Q} = \overline{1, Q}.$$

For fixed (strictly) positive weights $[w_1, \ldots, w_Q] =: \boldsymbol{w}$, we can equip the space $\mathcal{H}^+$ with the (weighted) norm

$$(\forall f \in \mathcal{H}^+) \quad \| f \|^2_{\mathcal{H}^+,\boldsymbol{w}} := \min \left\{ \sum_{q\in\mathcal{Q}} w_q^{-1} \| f_q \|^2_{\mathcal{H}_q} \mid f = \sum_{q\in\mathcal{Q}} f_q,\ (\forall q \in \mathcal{Q})\, f_q \in \mathcal{H}_q \right\}, \tag{11.2}$$

and it can be shown that the resulting normed vector space $(\mathcal{H}^+, \|\cdot\|_{\mathcal{H}^+,\boldsymbol{w}})$ is also the RKHS $(\mathcal{H}^+, \langle\cdot,\cdot\rangle_{\mathcal{H}^+,\boldsymbol{w}})$ associated with the reproducing kernel $\kappa := \sum_{q\in\mathcal{Q}} w_q\, \kappa_q$ (Berlinet and Thomas-Agnan, 2004, p. 24)Aronszajn (1950).

If no additional structure is imposed on $(\mathcal{H}^+, \langle\cdot,\cdot\rangle_{\mathcal{H}^+,\boldsymbol{w}})$, then, for a given function $f \in \mathcal{H}^+$, the decomposition $f = \sum_{q\in\mathcal{Q}} f_q$, where $(\forall q \in \mathcal{Q})\, f_q \in \mathcal{H}_q$ is not necessarily unique, which in turn makes the computation of the norm in Eq. (11.2) a particularly challenging task in general. One notable exception for the non-uniqueness of the decomposition is shown here:

Remark 11.1 Assuming that the sum RKHS $(\mathcal{H}^+, \langle\cdot,\cdot\rangle_{\mathcal{H}^+,\boldsymbol{w}})$ is constructed with RKHSs $(\mathcal{H}_1, \langle\cdot,\cdot\rangle_{\mathcal{H}_1}), \ldots, (\mathcal{H}_Q, \langle\cdot,\cdot\rangle_{\mathcal{H}_Q})$ satisfying $\mathcal{H}_j \cap \mathcal{H}_q = \{0\}$ whenever $j \neq q$, then we have both

$$(\forall f \in \mathcal{H}^+)\ \| f \|^2_{\mathcal{H}^+,\boldsymbol{w}} = \sum_{q\in\mathcal{Q}} w_q^{-1}\ \| f_q \|^2_{\mathcal{H}_q} \tag{11.3}$$

and

$$(\forall f \in \mathcal{H}^+)(\forall g \in \mathcal{H}^+)\langle f, g\rangle_{\mathcal{H}^+,\boldsymbol{w}} = \sum_{q\in\mathcal{Q}} w_q^{-1}\ \langle f_q, g_q\rangle_{\mathcal{H}_q}. \tag{11.4}$$

From a practical perspective, with a sum space satisfying the assumption in Remark 11.1, algorithms can perform many operations by simply considering the Hilbert spaces $(\mathcal{H}_1, \langle\cdot,\cdot\rangle_{\mathcal{H}_1}), \ldots, (\mathcal{H}_Q, \langle\cdot,\cdot\rangle_{\mathcal{H}_Q})$ separately and by summing the results, as done in Eqs. (11.3) and (11.4), for example. By doing so, hard-to-solve optimization problems, such as those required for the evaluation of norms in Eq. (11.2), are avoided. The algorithms described later in this chapter use a sum space of this type.

11.3 System Model

We now turn our attention to the adaptive learning method for symbol detection in multiuser environments. For brevity, in the following we focus on the uplink of a wireless network, but we emphasize that the proposed approaches can also be used in the downlink.

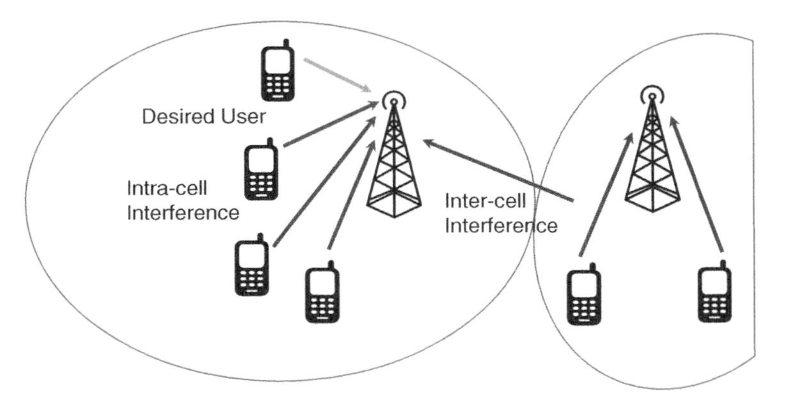

Figure 11.1 Multiuser uplink: The received baseband signal $\mathbf{r}(t)$ consists of the desired signal and noise plus interference from users in the same cell and also from users from other cells.

11.3.1 Symbol Detection in Multiuser Environments

Consider the uplink in Figure 11.1, where a base station with $M \in \mathbb{N}$ antennas receives a noisy superposition of signals from $K \in \mathbb{N}$ users, each of which is equipped with a single antenna. The receive baseband signal at the base station at time $t \in \mathbb{N}$ is given by

$$\mathbf{r}(t) = \sum_{k=1}^{K} \sqrt{p_k(t)} b_k(t) \mathbf{h}_k(t) + \mathbf{n}(t) \in \mathbb{C}^M, \tag{11.5}$$

where $p_k(t) \in]0\ \infty[$, $b_k(t) \in \mathbb{C}$, and $\mathbf{h}_k(t) \in \mathbb{C}^M$ are the power, the modulation symbol, and the channel, respectively, for user $k \in \overline{1, K}$, and where $\mathbf{n}(t) \in \mathbb{C}^M$ denotes noise. As is common in the literature, we assume that the channels between the users and the base station undergo *Rayleigh block fading*. Under this assumption, the channels remain constant for a block of complex-valued channel symbols known as the *coherence block*. More precisely, let $(b \in \mathbb{N})$ $t_b \in \mathbb{N}$ denote the start of the coherence block b, where $|t_b - t_{b+1}| := T_{\text{block}}$ is the coherence block size. Then, $(\forall k \in \overline{1, K})$ $(\forall t \in \overline{t_b, t_{b+1} - 1})(\exists \mathbf{h}_k^b \in \mathbb{C}^M)$ $\mathbf{h}_k(t) = \mathbf{h}_k^b$; i.e., $\mathbf{h}_k^b$ is the fixed channel of user k, lasting from $t = t_b$ to $t = t_{b+1} - 1$, for the coherence block b.

Now, we introduce the general idea of the proposed learning algorithm from a high-level perspective. Without any loss of generality, we outline the learning process for detecting the data symbols of user 1 in Eq. (11.5). In mathematical terms, the algorithm should ideally learn a function $g : \mathbb{C}^M \to \mathbb{C}$ such that $(\forall t \in \mathbb{N})$ $g(\mathbf{r}(t)) = b_1(t)$.[2] To this end, in each coherence block b, user 1 transmits a sequence of pilot symbols $(b_1(t))_{t \in \overline{t_b, t_b + T_{\text{train}} - 1}}$, with $T_{\text{train}} < T_{\text{block}}$. The pilots are also known to the base station, and it uses them, along with the corresponding received signals $(\boldsymbol{r}(t))_{t \in \overline{t_b, t_b + T_{\text{train}} - 1}}$, to learn g. Unlike traditional batch learning methods, the proposed algorithm improves its estimate of g sequentially with each incoming training sample; i.e. it operates in an *online* fashion and does not wait for the acquisition of the entire batch $\mathcal{S} = \{(\mathbf{r}(t), b_1(t))\}_{t \in \overline{t_b, t_b + T_{\text{train}} - 1}}$ to start the learning process. The detection can therefore, in principle, be performed without any delay starting at $t = t_b + T_{\text{train}}$, which is important in high-data-rate systems. Denoting the latest estimate (at $t = t_b + T_{\text{train}} - 1$) of g by $\tilde{f} : \mathbb{C}^M \to \mathbb{C}$, we use $(\tilde{f}(\boldsymbol{r}(t)))_{t \in \overline{t_b + T_{\text{train}}, t_{b+1} - 1}}$ as the estimate of the information symbols $(b_1(t))_{t \in \overline{t_b + T_{\text{train}}, t_{b+1} - 1}}$.

2 A well-known example of a detection function is that of a linear function $h : \mathbb{C}^M \to \mathbb{C}$ given by $(\mathbf{w} \in \mathbb{C}^M)$ $h(\mathbf{r}(t)) := \mathbf{w}^H \mathbf{r}(t)$.

11.3.2 Detection of Complex-Valued Symbols in Real Hilbert Spaces

Before proceeding with the description of the learning algorithm for symbol detection, we need to make a brief technical detour. The reason is that the theory presented in Section 11.2 involves vector spaces of functions mapping members of an abstract set $\mathcal{U}$ (see Definition 11.2) to real numbers. However, in the problem described in Section 11.3.1, we have to learn a function mapping complex-valued vectors in $\mathbb{C}^M$ [the input signals $(r(t))_{t\in\mathbb{N}}$] to complex numbers [the symbols $(b_1(t))_{t\in\mathbb{N}}$]. We could naturally define $\mathcal{U} = \mathbb{C}^M$ because there are no restrictions imposed on $\mathcal{U}$, but the co-domain of the functions being learned requires special attention because we are working with real RKHSs. To deal with this issue, we use the approach described in Slavakis et al. (2009), Yukawa et al. (2013), which exploits the trivial bijection between $\mathbb{C}^N$ and $\mathbb{R}^{2N}$ for arbitrary $N \in \mathbb{N}$ to enable the estimation of complex-valued symbols with real RKHSs. In more detail, we define $\mathcal{U} := \mathbb{R}^{2M}$, and, for each $t \in \mathbb{N}$, we split $\mathbf{r}(t)$ in Eq. (11.5) into two $2M$-dimensional vectors given by $\mathbf{r}_1(t) := [\Re\,(\mathbf{r}(t))^\mathsf{T}\ \Im\,(\mathbf{r}(t))^\mathsf{T}]^\mathsf{T} \in \mathbb{R}^{2M}$ and $\mathbf{r}_2(t) := [\Im\,(\mathbf{r}(t))^\mathsf{T} - \Re\,(\mathbf{r}(t))^\mathsf{T}]^\mathsf{T} \in \mathbb{R}^{2M}$. Similarly, the complex-valued symbols $(b_1(t))_{t\in\mathbb{N}}$ are mapped to vectors $[b_{1,1}(t)\ b_{1,2}(t)]^\mathsf{T} := [\Re(b_1(t))\ \Im(b_1(t))]^\mathsf{T} \in \mathbb{R}^2$. Our task is now to learn a function $f : \mathbb{R}^{2M} \to \mathbb{R}$ that operates on $\mathbf{r}_1(t)$ and $\mathbf{r}_2(t)$ separately, as depicted in Figure 11.2. The relation between f and the ideal complex-valued function g is given by $(\forall t \in \mathbb{N})\, g(\mathbf{r}(t)) = f(\mathbf{r}_1(t)) + if(\mathbf{r}_2(t))$, where i is the solution to the equation $i^2 = -1$.

To simplify notation in the discussion that follows, we define: $(\forall t \in \mathbb{N})\ (\forall l \in \overline{1,2})\ n := 2t + l - 2$, $\mathbf{y}_n = \mathbf{y}_{2t+l-2} := \mathbf{r}_l(t)$, and $s_n = s_{2t+l-2} := b_{1,l}(t)$. The advantage of using this simplified notation is that we have natural mappings from the natural numbers to the real and imaginary parts of the complex-valued symbols $(b_1(t))_{t\in\mathbb{N}}$ and the complex-valued received signals $(r(t))_{t\in\mathbb{N}}$. Note that, in particular, we have the following equivalence between the ideal functions g and f:

$$(\forall t \in \mathbb{N})\, f(\mathbf{r}_1(t)) + if(\mathbf{r}_2(t)) = b_1(t) \Leftrightarrow (\forall n \in \mathbb{N})\, f(\mathbf{y}_n) = s_n. \tag{11.6}$$

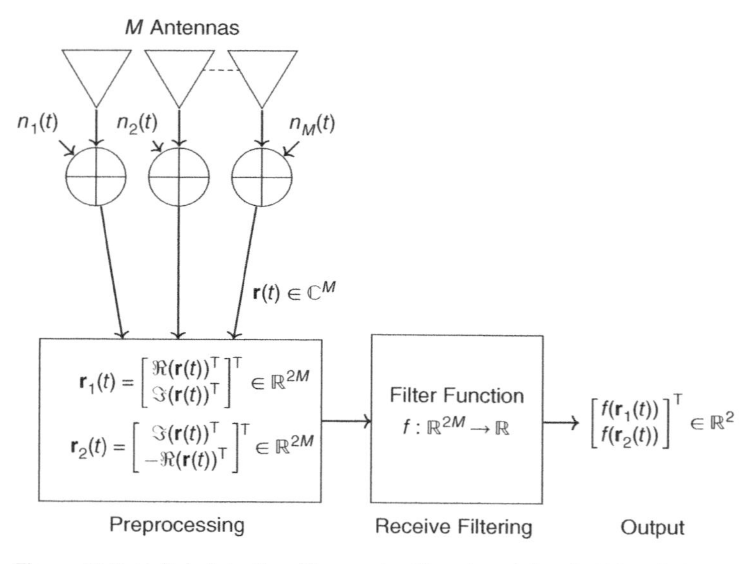

Figure 11.2 Uplink detection: The received baseband signal $\mathbf{r}(t)$ is split into two real parts $\mathbf{r}_1(t)$ and $\mathbf{r}_2(t)$ to enable real processing. Note that the illustration shows the processing for a single user, but the same processing is applied to every user of interest in parallel.

In the remainder of this chapter, to align our terminology with that of previous studies Slavakis et al. (2009), Theodoridis et al. (2011), we will use the words *filter* and *function* interchangeably when referring to the function f and its estimates.

11.4 The Proposed Learning Algorithm

Two important properties of an ideal receive filter are *high spatial resolution* and *robustness* against changes in the environment. In Slavakis et al. (2009), the authors design a high-resolution nonlinear receive filter (which they refer to as a *nonlinear beamformer*) in an infinite-dimensional RKHS associated with the Gaussian kernel. It has been shown that this receive filter outperforms the *linearly constrained minimum variance* receive filter Bourgeois and Minker (2009). However, when compared to linear filters, one of the main drawbacks of nonlinear filters is that they are in general less robust against changes in the environment. For instance, in the case of linear filtering, if a user leaves the system, the signal-to-interference-plus-noise ratios (SINRs) of the remaining users improve. However, this property cannot be ensured in general with nonlinear filters; performance may in fact deteriorate.

To achieve a good trade-off between the robustness of conventional linear filters and the high resolution of nonlinear filters, we incorporate linear and nonlinear components in our filter design by considering the theory in Section 11.2.2. In particular, we propose to work in the sum space of a linear kernel and a Gaussian kernel, defined as $(\forall \mathbf{u} \in \mathbb{R}^{2M})(\forall \mathbf{v} \in \mathbb{R}^{2M})\ \kappa_{\mathrm{L}}(\mathbf{u}, \mathbf{v}) := \mathbf{u}^T\mathbf{v}$ and $\kappa_{\mathrm{G}}(\mathbf{u}, \mathbf{v}) := \exp\left(-\frac{\|\mathbf{u}-\mathbf{v}\|^2_{\mathbb{R}^{2M}}}{2\sigma^2}\right)$, respectively. It is known that κ_{L} and κ_{G} are reproducing kernels associated with RKHSs, which we denote by $(\mathcal{H}_{\mathrm{L}}, \langle\cdot,\cdot\rangle_{\mathcal{H}_{\mathrm{L}}})$[3] and $(\mathcal{H}_{\mathrm{G}}, \langle\cdot,\cdot\rangle_{\mathcal{H}_{\mathrm{G}}})$, respectively. By defining $\mathcal{H} := \mathcal{H}_{\mathrm{L}} + \mathcal{H}_{\mathrm{G}}$, we construct the sum RKHS associated with the kernel $\kappa := w_{\mathrm{L}}\ \kappa_{\mathrm{L}} + w_{\mathrm{G}}\ \kappa_{\mathrm{G}}$, where $w_{\mathrm{L}}, w_{\mathrm{G}} > 0$ are fixed weights. With this particular sum space $(\mathcal{H}, \langle\cdot,\cdot\rangle_{\mathcal{H},\boldsymbol{w}})$, with $\boldsymbol{w} := [w_{\mathrm{L}},\ w_{\mathrm{G}}]$, we have the desirable property $\mathcal{H}_{\mathrm{L}} \cap \mathcal{H}_{\mathrm{G}} = \{0\}$ Berlinet and Thomas-Agnan (2004), Yukawa (2015a), which in particular implies that norms and inner products in $(\mathcal{H}, \langle\cdot,\cdot\rangle_{\mathcal{H},\boldsymbol{w}})$ can be easily computed as shown in Eqs. (11.3) and (11.4). In the rest, we denote the RKHSs $(\mathcal{H}, \langle\cdot,\cdot\rangle_{\mathcal{H},\boldsymbol{w}})$, $(\mathcal{H}_{\mathrm{L}}, \langle\cdot,\cdot\rangle_{\mathcal{H}_{\mathrm{L}}})$, and $(\mathcal{H}_{\mathrm{G}}, \langle\cdot,\cdot\rangle_{\mathcal{H}_{\mathrm{G}}})$ simply by $\mathcal{H}$, $\mathcal{H}_{\mathrm{L}}$, and $\mathcal{H}_{\mathrm{G}}$, respectively.

11.4.1 The Canonical Iteration

We now proceed to pose the learning problem as a special case of the *convex feasibility problem* involving (possibly) countably infinitely many sets. This problem is then solved with a particular version of the *adaptive projected subgradient method* Yamada and Ogura (2005). To simplify the exposition, we assume that all transmitted symbols are pilots, and the coherence block size T_{block} is infinite. By doing so, we highlight that the computational complexity during training is kept at low levels even if the number of pilots is very large, and later we show that the resulting algorithm has good performance with few pilots.

By recalling that an ideal filter $f \in \mathcal{H}$ should satisfy the equalities in Eq. (11.6), we can expect that reasonable estimates of an ideal filter should belong to closed convex sets $(C_n)_{n\in\mathbb{N}}$ given by

$$(\forall n \in \mathbb{N})\ C_n := \{h \in \mathcal{H} : |h(\mathbf{y}_n) - s_n| = |\langle h, \kappa(\mathbf{y}_n, \cdot)\rangle_{\mathcal{H},\boldsymbol{w}} - s_n| \leq \epsilon\}, \tag{11.7}$$

3 Note that $(\mathcal{H}_{\mathrm{L}}, \langle\cdot,\cdot\rangle_{\mathcal{H}_{\mathrm{L}}})$ is nothing but the $2M$-dimensional Euclidean space $\mathbb{R}^{2M}$ in which an inner product is the dot product, i.e. $(\forall \mathbf{u} \in \mathbb{R}^{2M})(\forall \mathbf{v} \in \mathbb{R}^{2M})\ \langle \mathbf{u}, \mathbf{v}\rangle_{\mathcal{H}_{\mathrm{L}}} = \mathbf{u} \cdot \mathbf{v}$.

where $\epsilon \in]0\ \infty[$ is a relaxation parameter used to, for example, combat erroneous assumptions and the detrimental effects of noise in measurements. With these sets, similarly to the problems in Theodoridis et al. (2011), we pose the learning problem as follows:

$$\text{For some } n_o \in \mathbb{N}, \text{find } f^* \in \mathcal{H} \text{such that } f^* \in \bigcap_{n \geq n_o} C_n, \tag{11.8}$$

where we assume that $\bigcap_{n\in\mathbb{N}} C_n \neq \emptyset$. Briefly, we seek filters that belong to all but finitely many sets in Eq. (11.7). Removing a finite number of sets from the intersection in Eq. (11.8) enables us to derive computationally simple learning algorithms based on the adaptive projected subgradient method Yamada and Ogura (2005) (note, however, that the solution to Eq. (11.8) belongs to infinitely many sets in $(C_n)_{n\in\mathbb{N}}$). In particular, with some mild technical assumptions, the problem in Eq. (11.8) can be solved with a particular case of the adaptive projected subgradient method Yamada and Ogura (2005), which we now describe.

Denote by $\mathcal{J}_n \subset \mathbb{N}$ the indices of the sets in $(C_n)_{n\in\mathbb{N}}$ that we intend to process at iteration $n \in \mathbb{N}$ of the training procedure. In particular, note that $\mathcal{J}_n$ can be changed at every iteration n, which is useful to process the sets in $(C_n)_{n\in\mathbb{N}}$ as soon as they become available, or to adjust the computation complexity of the algorithm according to the hardware capabilities of the receiver, or both. Starting from $f_1 = 0$, the canonical training procedure produces a sequence of filters $(f_n)_{n\in\mathbb{N}}$ in $\mathcal{H}$ with the iterations given by

$$(\forall n \in \mathbb{N}) f_{n+1} = \sum_{j\in\mathcal{J}_n} q_j^n P_{C_j}(f_n), \tag{11.9}$$

where $P_{C_j}(f_n) = f_n + \beta_j^n \kappa(\mathbf{y}_n, \cdot) = f_n + \beta_j^n (w_{\mathrm{L}}\ \kappa_{\mathrm{L}}(\mathbf{y}_n, \cdot) + w_{\mathrm{G}}\ \kappa_{\mathrm{G}}(\mathbf{y}_n, \cdot))$ is the projection of f_n onto the set C_j, with β_j^n given by

$$\beta_j^n := \begin{cases} \dfrac{s_j - \langle f_n, \kappa(\mathbf{y}_j, \cdot)\rangle_{\mathcal{H},\boldsymbol{w}} - \epsilon}{\kappa(\mathbf{y}_j, \mathbf{y}_j)}, & \text{if } \langle f_n, \kappa(\mathbf{y}_j, \cdot)\rangle_{\mathcal{H},\boldsymbol{w}} - s_j < -\epsilon \\ 0, & \text{if } |\langle f_n, \kappa(\mathbf{y}_j, \cdot)\rangle_{\mathcal{H},\boldsymbol{w}} - s_j| \leq \epsilon \\ \dfrac{s_j - \langle f_n, \kappa(\mathbf{y}_j, \cdot)\rangle_{\mathcal{H},\boldsymbol{w}} + \epsilon}{\kappa(\mathbf{y}_j, \mathbf{y}_j)}, & \text{if } \langle f_n, \kappa(\mathbf{y}_j, \cdot)\rangle_{\mathcal{H},\boldsymbol{w}} - s_j > \epsilon \end{cases}$$

and where $(q_j^n)_{j\in\mathcal{J}_n}$ are non-negative weights satisfying $\sum_{j\in\mathcal{J}_n} q_j^n = 1$. Note that the iterates produced by this algorithm are steered toward the sets with proportionally large weights, which is a feature that can be useful in scenarios where some sets are known to be more reliable than others.

In particular, the iteration in Eq. (11.9) has the following desirable property. If $f_n \notin \cap_{k\in\mathcal{J}_n} C_k$, then (Yamada and Ogura, 2005, Theorem 2)

$$(\forall f^\star \in \cap_{n\in\mathbb{N}} C_n) \quad \| f_{n+1} - f^\star \|_{\mathcal{H},\boldsymbol{w}} < \| f_n - f^\star \|_{\mathcal{H},\boldsymbol{w}};$$

i.e. the recursion in Eq. (11.9) at time n is guaranteed to move f_n to a point closer to the solution to problem Eq. (11.8). For other properties of the algorithm, including its convergence, we refer readers to Yamada and Ogura (2005)

11.4.2 Practical Issues

Next, we look at issues related to the implementation of Eq. (11.9). The first issue is the selection of an appropriate set $\mathcal{J}_n$. A reasonable and simple way of selecting $\mathcal{J}_n$ is to include the $W \in \mathbb{N}$ most recent training samples. More precisely, at time $n \in \mathbb{N}$, we define $\mathcal{J}_n$ as the set given by

$\mathcal{J}_n := \overline{n - W + 1, n}$ if $n \geq W$, or $\mathcal{J}_n := \overline{1, n}$ otherwise. The window size W is a design parameter chosen according to the available computational power. Larger sizes typically improve the performance at the cost of increased computational complexity.

The second issue pertains to the memory requirement and the complexity of the learning framework. To understand the main challenges, let us look at how the canonical algorithm in Eq. (11.9) proceeds. At time $n \in \mathbb{N}$, we can show that the filter estimate generated by Eq. (11.9) is given by $f_n = \sum_{i=1}^{n-1} \gamma_i^{(n)} \kappa(\mathbf{y}_i, \cdot)$, where $(\gamma_i^{(n)})_{i \in \overline{1,n-1}}$ are real coefficients that depend on the sets $(C_n)_{n \in \mathbb{N}}$ Theodoridis et al. (2011). Since f_n is expressed as a linear combination of the elements in the set $\mathcal{D}_{n-1} := \{\kappa(\mathbf{y}_1, \cdot), \kappa(\mathbf{y}_2, \cdot), \dots, \kappa(\mathbf{y}_{n-1}, \cdot)\}$,

$$f_n \in \mathcal{H}_{n-1} = \text{span}(\mathcal{D}_{n-1}) \subset \mathcal{H}.$$

Note that $\mathcal{H}_{n-1}$ is also a Hilbert space if equipped with the same inner product of the sum RKHS $\mathcal{H}$. In the following, we will refer to $\mathcal{D}_{n-1}$ as the *learning dictionary*. At each iteration $n \in \mathbb{N}$ of the algorithm, a new element $\kappa(\mathbf{y}_n, \cdot) = w_{\text{L}}\, \kappa_{\text{L}}(\mathbf{y}_n, \cdot) + w_{\text{G}}\, \kappa_{\text{G}}(\mathbf{y}_n, \cdot)$ is admitted, and the dictionary is extended to $\mathcal{D}_n = \mathcal{D}_{n-1} \cup \{\kappa(\mathbf{y}_n, \cdot)\}$. It follows that $\mathcal{H}_{n-1} \subset \mathcal{H}_n \subset \mathcal{H}$, and the extended space $\mathcal{H}_n$ is spanned by $\mathcal{D}_n$. Therefore, to evaluate $(\mathbf{y} \in \mathbb{R}^{2M})\, f_{n+1}(\mathbf{y}) = \langle f_{n+1}, \kappa(\mathbf{y}, \cdot)\rangle_{\mathcal{H}}$ (by using the reproducing property discussed in Section 11.2), we need to store $\mathcal{D}_n^{\text{mem}} := \{\mathbf{y}_1, \mathbf{y}_2, \dots, \mathbf{y}_n\}$ along with the coefficients $\gamma_1^{(n+1)}, \gamma_2^{(n+1)}, \dots, \gamma_n^{(n+1)}$ in the memory of the receiver (note: $\mathcal{D}_n \subset \mathcal{H}$ can be trivially recovered from $\mathcal{D}_n^{\text{mem}} \subset \mathbb{R}^{2M}$). Moreover, the coefficients $\gamma_i^{(n)}$, the number of which increases with n, are required at each iteration n by the projections $(j \in \mathcal{J}_n)$ $P_{C_j}(f_{n-1})$ in Eq. (11.9). This fact shows that the memory requirements and the computational complexity may become prohibitive when n becomes sufficiently large. To keep the complexity and the memory requirements of the learning algorithm at manageable levels, we use online dictionary learning techniques, as explained next.

11.4.3 Online Dictionary Learning

It follows from Section 11.2.2 that the filter estimate f_n at time $n \in \mathbb{N}$ can be uniquely decomposed as the sum of its linear and Gaussian components as follows:

$$f_n := \sum_{i=1}^{n-1} \gamma_i^{(n)} \kappa(\mathbf{y}_i, \cdot) := f_{\text{L},n} + f_{\text{G},n} = w_{\text{L}} \sum_{i=1}^{n-1} \gamma_i^{(n)} \kappa_{\text{L}}(\mathbf{y}_i, \cdot) + w_{\text{G}} \sum_{i=1}^{n-1} \gamma_i^{(n)} \kappa_{\text{G}}(\mathbf{y}_i, \cdot),$$

where $f_{\text{L},n} \in \text{span}(\mathcal{D}_{\text{L},n-1})$ and $f_{\text{G},n} \in \text{span}(\mathcal{D}_{\text{G},n-1})$, and where $(\forall k \in \mathbb{N})$ $\mathcal{D}_{\text{L},k} = \{\kappa_{\text{L}}(\cdot, \mathbf{y}_1), \dots, \kappa_{\text{L}}(\cdot, \mathbf{y}_k)\}$ and $\mathcal{D}_{\text{G},k} = \{\kappa_{\text{G}}(\cdot, \mathbf{y}_1), \dots, \kappa_{\text{G}}(\cdot, \mathbf{y}_k)\}$. To curb the growth of the dictionaries $\mathcal{D}_{\text{L},n}$ and $\mathcal{D}_{\text{G},n}$ as a function of n in such a way as to have a minor impact on the performance of the filter f_n, we use *dictionary sparsification*, as explained next.

Dictionary sparsification has its origins in the seminal work in Engel et al. (2004), but here we use an approach similar to that proposed in Yukawa (2015b), which handles the linear and Gaussian components of the sequence $(f_n)_{n \in \mathbb{N}}$ of filters separately. In the proposed approach, we use admission control to verify whether the most recent inputs $\kappa_{\text{L}}(\mathbf{y}_n, \cdot)$ and $\kappa_{\text{G}}(\mathbf{y}_n, \cdot)$ should be added to the dictionaries $\mathcal{D}_{\text{L},n-1}$ and $\mathcal{D}_{\text{G},n-1}$, respectively. Briefly, the idea is to check if $\kappa_{\text{L}}(\mathbf{y}_n, \cdot)$ and $\kappa_{\text{G}}(\mathbf{y}_n, \cdot)$ can be approximated (in some sense) by a linear combination of elements previously admitted in dictionaries $\mathcal{D}_{\text{L},n-1}$ and $\mathcal{D}_{\text{G},n-1}$. Newly arriving elements are added to the dictionary only if such an approximation is not possible. The particular techniques for dictionary sparsification of the linear and Gaussian components of the proposed filter are described in the next two subsections.

11.4.3.1 Dictionary for the Linear Component

Admission control for the linear part can be easily done as follows. Since $\mathcal{H}_\mathrm{L}$ is nothing but the Euclidean space $\mathbb{R}^{2M}$, it is spanned by the Euclidean basis $\mathcal{D}_\mathrm{L} := \{\kappa_\mathrm{L}(\mathbf{e}_1, \cdot), \kappa_\mathrm{L}(\mathbf{e}_2, \cdot), \dots, \kappa_\mathrm{L}(\mathbf{e}_{2M}, \cdot)\}$, where $\mathbf{e}_m \in \mathbb{R}^{2M}$ is a vector having a one at the mth index and zeros elsewhere. So, every $\kappa_\mathrm{L}(\mathbf{y}_n, \cdot)$ can be written in terms of a linear combination given by $\sum_{m=1}^{2M} [\mathbf{y}_n]_m \kappa_\mathrm{L}(\mathbf{e}_m, \cdot)$, with $[\mathbf{y}_n]_m$ the mth entry of $\mathbf{y}_n$. As a result, the linear component $f_{\mathrm{L},n} = w_\mathrm{L} \sum_{i=1}^{n-1} \gamma_i^{(n)} \kappa_\mathrm{L}(\mathbf{y}_i, \cdot) = w_\mathrm{L} \sum_{m=1}^{2M} \gamma_m^{(\mathrm{L},n)} \kappa_\mathrm{L}(\mathbf{e}_m, \cdot)$ at time $n \in \mathbb{N}$ consists of only $2M$ basis functions with their coefficients $\gamma_m^{(\mathrm{L},n)}$ updated by each projection $(j \in \mathcal{J}_n)\ P_{C_j}(f_n)$ in Eq. (11.9), and we also have $(\forall n \in \mathbb{N})\ \mathcal{D}_{\mathrm{L},n} = \mathcal{D}_\mathrm{L}$ and $\mathcal{H}_{\mathrm{L},n} = \mathcal{H}_\mathrm{L}$. With the proposed sparsification technique for the linear component, note that the memory and computational requirements of $f_{\mathrm{L},n}$ do not increase with n.

11.4.3.2 Dictionary for the Gaussian Component

The proposed sparsification technique for the dictionary $\mathcal{D}_{\mathrm{G},n}$ is based on the studies in Engel et al. (2004), Slavakis and Theodoridis (2008), and it can be summarized as follows. Suppose that we start with the dictionary $\mathcal{D}_{\mathrm{G},1} = \{\kappa(\mathbf{y}_1, \cdot)\}$. At time $n \geq 2$, we have the dictionary $\mathcal{D}_{\mathrm{G},n-1}$, and the objective is to determine whether the newly arriving element $\kappa_\mathrm{G}(\mathbf{y}_n, \cdot)$ should be added to $\mathcal{D}_{\mathrm{G},n-1}$ to construct $\mathcal{D}_{\mathrm{G},n}$. Informally, if $\kappa_\mathrm{G}(\mathbf{y}_n, \cdot)$ can be "well approximated" by any vector in the subspace span$(\mathcal{D}_{\mathrm{G},n-1})$, then functions in span$(\mathcal{D}_{\mathrm{G},n-1} \cup \{\kappa_\mathrm{G}(\mathbf{y}_n, \cdot)\})$ can also be "well approximated" by functions in span$(\mathcal{D}_{\mathrm{G},n-1})$, so $\kappa_\mathrm{G}(\mathbf{y}_n, \cdot)$ does not need to be added to $\mathcal{D}_{\mathrm{G},n-1}$. As commonly done in approximation theory in Hilbert spaces, we can define as the best approximation of $\kappa_\mathrm{G}(\mathbf{y}_n, \cdot)$ in the subspace $\mathcal{H}_{\mathrm{G},n-1} := \mathrm{span}(\mathcal{D}_{\mathrm{G},n-1})$ the projection $P_{\mathcal{H}_{\mathrm{G},n-1}}(\kappa_\mathrm{G}(\mathbf{y}_n, \cdot))$. With this definition, the squared norm $d_n := \|\kappa_\mathrm{G}(\mathbf{y}_n, \cdot) - P_{\mathcal{H}_{\mathrm{G},n-1}}(\kappa_\mathrm{G}(\mathbf{y}_n, \cdot))\|^2_{\mathcal{H}_\mathrm{G}}$ of the residual $\kappa_\mathrm{G}(\mathbf{y}_n, \cdot) - P_{\mathcal{H}_{\mathrm{G},n-1}}(\kappa_\mathrm{G}(\mathbf{y}_n, \cdot))$ serves as a measure to indicate how well the best vector $P_{\mathcal{H}_{\mathrm{G},n-1}}(\kappa_\mathrm{G}(\mathbf{y}_n, \cdot))$ in the subspace span$(\mathcal{D}_{\mathrm{G},n-1})$ is able to approximate $\kappa_\mathrm{G}(\mathbf{y}_n, \cdot)$. Therefore, we can update the dictionary as follows:

$$\mathcal{D}_{\mathrm{G},n} = \begin{cases} \mathcal{D}_{\mathrm{G},n-1}, & \text{if } d_n \leq \alpha \\ \mathcal{D}_{\mathrm{G},n-1} \cup \{\kappa_\mathrm{G}(\mathbf{y}_n, \cdot)\} & \text{otherwise,} \end{cases}$$

where $\alpha > 0$ is a design parameter. For completeness, we show the steps required for the computation of d_n in the Appendix.

11.4.4 The Online Learning Algorithm

Applying the previous sparsification techniques to the canonical algorithm in Eq. (11.9), we obtain the following iterations for all $n \in \mathbb{N}$:

$$\begin{aligned} f_{n+1} &= P_{\mathcal{H}_n}\left(\sum_{j \in \mathcal{J}_n} q_j^n P_{C_j}(f_n)\right), \\ &= P_{\mathcal{H}_n}\left(f_n + \sum_{j \in \mathcal{J}_n} q_j^n \beta_j^n \kappa(\mathbf{y}_j, \cdot)\right), \\ &= f_n + \sum_{j \in \mathcal{J}_n} q_j^n \beta_j^n P_{\mathcal{H}_n}(\kappa(\mathbf{y}_j, \cdot)) \text{(linearity of projections)}, \end{aligned} \tag{11.10}$$

where $f_1 = 0$, $(\forall j \in \mathcal{J}_n)\ P_{\mathcal{H}_n}(\kappa(\mathbf{y}_j, \cdot)) = w_\mathrm{L}\ P_{\mathcal{H}_{\mathrm{L},n}}(\kappa_\mathrm{L}(\mathbf{y}_j, \cdot)) + w_\mathrm{G}\ P_{\mathcal{H}_{\mathrm{G},n}}(\kappa_\mathrm{G}(\mathbf{y}_j, \cdot))$. The projections $P_{\mathcal{H}_{\mathrm{L},n}}(\kappa_\mathrm{L}(\mathbf{y}_j, \cdot))$ are given by $P_{\mathcal{H}_{\mathrm{L},n}}(\kappa_\mathrm{L}(\mathbf{y}_j, \cdot)) = \sum_{m=1}^{2M} [\mathbf{y}_j]_m \kappa_\mathrm{L}(\mathbf{e}_m, \cdot)$, and details of the projections $P_{\mathcal{H}_{\mathrm{G},n}}(\kappa_\mathrm{G}(\mathbf{y}_j, \cdot))$ are given in the Appendix.

In the following, we summarize the proposed learning method.

Algorithm 1 Online adaptive filtering algorithm.

Initialization: Fix $\epsilon > 0$, training block length $T_{\text{train}} \in \mathbb{N}$, $W \in \mathbb{N}$, $\alpha > 0$, $\mathcal{D}_0 := \emptyset$, and $f_1 = 0$.
At $n \geq 1$ **Repeat**:

1. **Sample update**: The training samples $\{(\mathbf{y}_j, s_j) : j \in \mathcal{J}_n\}$ are available. Set $(\forall j \in \mathcal{J}_n)$ $q_j^n = 1/|\mathcal{J}_n|$, where $|\mathcal{J}_n|$ is the cardinality of $\mathcal{J}_n$.
2. **Dictionary update**: Follow the procedure in Section 11.4.3 to update $\mathcal{D}_{n-1}$.
3. **Adaptive learning**: Follow the procedure in Section 11.4.4 to calculate f_{n+1}.

11.5 Simulation

For the simulation, we consider $K \in \{3, 4, 5\}$ users and a single base station with $M = 3$ antennas. The modulation scheme is quadrature phase-shift keying (QPSK). The simulation parameters are shown in Table 11.1. The average user performance is shown in terms of the averaged gray coded bit error rate (BER). We simulate two Rayleigh channel blocks, each consisting of $T_{\text{block}} = 500$ complex channel symbols, and we average results over 100 channel realizations. To show the online learning progress of Eq. (11.10), we evaluate the detection performance for every 100 training samples.

In the following, we denote our proposed method by *PLAF*. Figure 11.3 compares the performance of *PLAF* against the *minimum mean squared error* (MMSE) receive filter with symbol-level *successive interference cancellation* (SIC), commonly referred to as the symbol-level *MMSE-SIC* receiver. We assume perfect channel knowledge for the *MMSE-SIC*, which is an unrealistic assumption in practice, but it highlights the advantages of the proposed method (which does not require channel knowledge explicitly). To demonstrate that *PLAF* has a high resolution, we are interested in the number of users that we can detect, with a certain average user BER, for a fixed number of antennas at the base station. Figure 11.3 shows that *PLAF* is able to support a larger number of users than the *MMSE-SIC*-based receiver.

Table 11.1 Simulation parameters.

Parameter	Symbol	Value
Number of BS antennas	M	3
Number of users	K	$\{3, 4, 5\}$
User SNR	SNR	$\{0, 5, 10, 15, 20\}$ dB
User spatial location	θ	$\{30°, 60°, 90°, 120°, 150°\}$
Modulation	$b(t)$	QPSK$[\pm 1 \pm i1]$
Prob. of active users	ρ	$\{1, 0.75, 0.60\}$
Coherence block size	T_{block}	500
Dictionary novelty	α	0.1
Window size	W	50
Precision	ϵ	0.01
Gaussian/Linear weight	w_G, w_L	0.8, 0.2

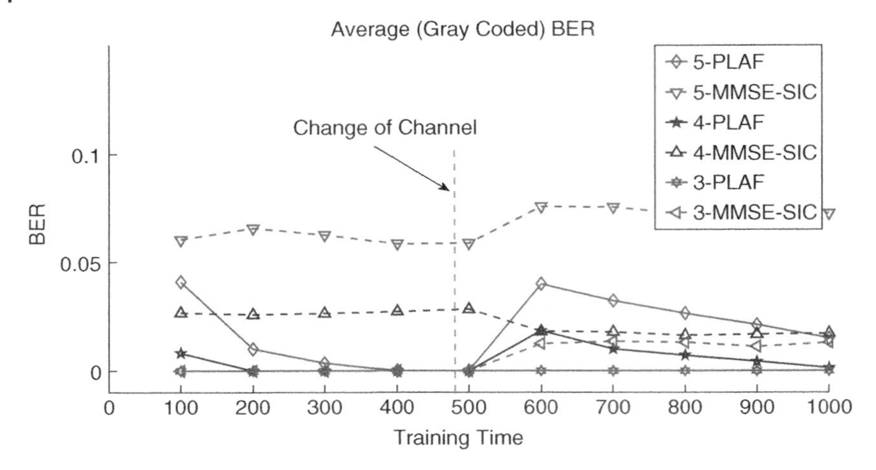

Figure 11.3 Comparison between PLAF and MMSE-SIC for $\rho = 1$.

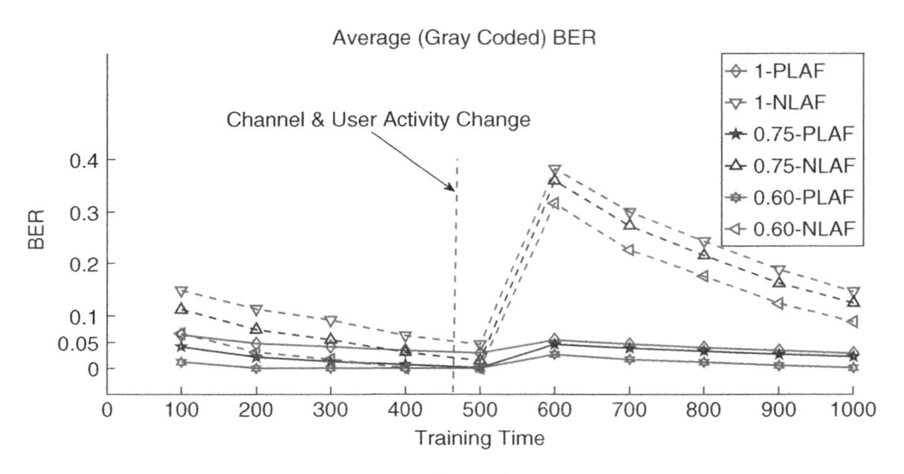

Figure 11.4 Comparison between PLAF and NLAF for $K = 5$.

In addition to *MMSE-SIC*, we also compare the proposed method with the purely nonlinear adaptive receive filter, which we denote by *NLAF*. The difference between *PLAF* and *NLAF* is that *NLAF* employs a Gaussian kernel only. With this comparison, we demonstrate that *PLAF* is more robust than *NLAF* against changes in the environment. Specifically, at the start of each block, we change user channels and the user activity. We denote by $\rho \in [0\ 1]$ the probability with which all users are active during a block. After 500 channel symbols, the channel and user activity are changed, which is a common scenario in communication systems. It is clear from Figure 11.4 that *PLAF* is more robust against changes in the environment than *NLAF*.

11.6 Conclusion

In this chapter, we showed that the theory of RKHS in sum spaces can be applied in the design of robust receive filters for symbol detection. In more detail, we used a particular version of the adaptive projected subgradient method to generate a sequence of filters that can be uniquely decomposed into a linear component and a nonlinear component constructed with a Gaussian kernel. By doing so, we equipped the resulting algorithm with the robustness of linear filters and the high performance of nonlinear filters. Simulations have shown that the learning algorithm is able to cope with small training sets and abrupt changes in the environment. In the simulated scenario, the proposed method outperformed, in particular, the traditional MMSE-SIC with perfect channel knowledge.

Appendix A

Derivation of the Sparsification Metric and the Projections onto the Subspace Spanned by the Nonlinear Dictionary

Let $\mathcal{D}_{G,n-1}$ denote the Gaussian dictionary at time index $n-1$, and let $S_{n-1} := |\mathcal{D}_{G,n-1}|$ denote its cardinality. Denote by $\boldsymbol{\Psi}_l^{n-1} \in \mathcal{D}_{G,n-1}$ the lth element of $\mathcal{D}_{G,n-1}$. We denote by $\mathbf{K}_{n-1} \in \mathbb{R}^{S_{n-1}\times S_{n-1}}$ the standard *Gram* matrix at time $n-1$, with the element in the ith row and jth column given by $(\forall i \in \overline{1, S_{n-1}})$ $(\forall j \in \overline{1, S_{n-1}})$ $\mathbf{K}_{n-1} := \langle \boldsymbol{\Psi}_i^{n-1}, \boldsymbol{\Psi}_j^{n-1} \rangle_{\mathcal{H}_G}$. Note that $\mathbf{K}_{n-1}$ is positive definite because the elements $(\boldsymbol{\Psi}_l^{n-1})_{l\in\overline{1,S_{n-1}}}$ of the Gaussian dictionary $\mathcal{D}_{G,n-1}$ are linearly independent by assumption (see Section 11.4.3.2). As a result, the inverse $\mathbf{K}_{n-1}^{-1}$ exists.

The projection of $\kappa_G(\mathbf{y}_n, \cdot)$ onto the linear closed subspace $\mathcal{H}_{G,n-1} \subset \mathcal{H}_G$ spanned by $\mathcal{D}_{G,n-1}$ is given by Slavakis and Theodoridis (2008)

$$\mathbf{P}_{\mathcal{H}_{G,n-1}}(\kappa(\mathbf{y}_n, \cdot)) = \sum_{l=1}^{S_{n-1}} \zeta_{\mathbf{y}_n,l}^{n} \boldsymbol{\Psi}_l^{n-1}, \tag{A.1}$$

where $\zeta_{\mathbf{y}_n}^n \in \mathbb{R}^{S_{n-1}}$ is given by $\zeta_{\mathbf{y}_n}^n = \mathbf{K}_{n-1}^{-1}\xi_{\mathbf{y}_n}^n$; the vector $\xi_{\mathbf{y}_n}^n$ is given as

$$\xi_{\mathbf{y}_n}^n = \begin{bmatrix} \langle \kappa_G(\mathbf{y}_n, \cdot), \boldsymbol{\Psi}_1^{n-1} \rangle_{\mathcal{H}_G} \\ \vdots \\ \langle \kappa_G(\mathbf{y}_n, \cdot), \boldsymbol{\Psi}_{S_{n-1}}^{n-1} \rangle_{\mathcal{H}_G} \end{bmatrix}.$$

Suppose now that $\mathbf{K}_{n-1}^{-1}$ is given; then the distance of $\kappa_G(\mathbf{y}_n, \cdot)$ from $\mathcal{D}_{G,n-1}$ is the solution to Slavakis and Theodoridis (2008)

$$d_n^2 := \kappa_G(\mathbf{y}_n, \mathbf{y}_n) - (\xi_{\mathbf{y}_n}^n)^\intercal \zeta_{\mathbf{y}_n}^n.$$

Given d_n, $\xi_{\mathbf{y}_n}^n$, and $\zeta_{\mathbf{y}_n}^n$, the inverse $\mathbf{K}_{n-1}^{-1}$ is updated for the next iteration n to $\mathbf{K}_n^{-1}$, which further enables us to calculate $\xi_{\mathbf{y}_{n+1}}^{n+1}$, $\zeta_{\mathbf{y}_{n+1}}^{n+1}$, and d_{n+1}, in that order. In more detail, we initialize the inverse by $\mathbf{K}_1^{-1} := 1/\kappa_G(\mathbf{y}_1, \mathbf{y}_1)$. For $n \geq 2$ if $\kappa_G(\mathbf{y}_n, \cdot)$ is admitted to the dictionary, i.e. if $\mathcal{D}_{G,n} = \mathcal{D}_{G,n-1} \cup \{\kappa_G(\mathbf{y}_n, \cdot)\}$, then

$$\mathbf{K}_n^{-1} := \begin{bmatrix} \mathbf{K}_{n-1}^{-1} + \frac{\zeta_{\mathbf{y}_n}^n (\zeta_{\mathbf{y}_n}^n)^\intercal}{d_n^2} & -\frac{\zeta_{\mathbf{y}_n}^n}{d_n^2} \\ -\frac{(\zeta_{\mathbf{y}_n}^n)^\intercal}{d_n^2} & \frac{1}{d_n^2} \end{bmatrix},$$

otherwise $\mathbf{K}_n^{-1} := \mathbf{K}_{n-1}^{-1}$.

Now we look at how to calculate $\mathbf{P}_{\mathcal{H}_{G,n}}(\kappa(\mathbf{y}_j, \cdot))$ for each $j \in \mathcal{J}_n$. We start by considering the latest training sample $j = n$. If $\kappa_G(\mathbf{y}_n, \cdot) \in \mathcal{D}_{G,n}$ then obviously $\mathbf{P}_{\mathcal{H}_{G,n}}(\kappa_G(\mathbf{y}_n, \cdot)) = \kappa_G(\mathbf{y}_n, \cdot)$, otherwise $\mathbf{P}_{\mathcal{H}_{G,n}}(\kappa_G(\mathbf{y}_n, \cdot)) = \mathbf{P}_{\mathcal{H}_{G,n-1}}(\kappa_G(\mathbf{y}_n, \cdot))$. The projection $\mathbf{P}_{\mathcal{H}_{G,n-1}}(\kappa_G(\mathbf{y}_j, \cdot))$ in (A.1) is already available to us because $\zeta^n_{\mathbf{y}_n}$ and $\mathcal{D}_{G,n-1}$ are both known to us from the dictionary update step (see Section 11.4.3.2 and Algorithm 1). It follows that $(\forall n \in \mathbb{N})$ $(\forall j \in \mathcal{J}_n)$, either $\mathbf{P}_{\mathcal{H}_{G,n}}(\kappa_G(\mathbf{y}_j,\cdot)) = \kappa_G(\mathbf{y}_j,\cdot)$ or $\mathbf{P}_{\mathcal{H}_{G,n}}(\kappa_G(\mathbf{y}_j,\cdot)) = \mathbf{P}_{\mathcal{H}_{G,n-1}}(\kappa_G(\mathbf{y}_j,\cdot))$.

Bibliography

Nachman Aronszajn. Theory of reproducing kernels. *Transactions of the American Mathematical Society*, 68 (3): 337–404, 1950.

Alain Berlinet and Christine Thomas-Agnan. *Reproducing Kernel Hilbert Spaces in Probability and Statistics*. Springer Science & Business Media, 2004.

Julien Bourgeois and Wolfgang Minker. *Linearly Constrained Minimum Variance Beamforming*. Springer US, Boston, MA, 2009.

Y. Engel, S. Mannor, and R. Meir. The kernel recursive least-squares algorithm. *IEEE Transactions on Signal Processing*, 52 (8): 2275–2285, Aug 2004. ISSN 1053-587X. doi: 10.1109/TSP.2004.830985.

K. Slavakis, S. Theodoridis, and I. Yamada. Adaptive constrained learning in reproducing kernel Hilbert spaces: The robust beamforming case. *IEEE Transactions on Signal Processing*, 57 (12): 4744–4764, Dec 2009. ISSN 1053-587X. doi: 10.1109/TSP.2009.2027771.

Konstantinos Slavakis and Sergios Theodoridis. Sliding window generalized kernel affine projection algorithm using projection mappings. *EURASIP Journal on Advances in Signal Processing*, 2008 (1): 735351, Apr 2008. ISSN 1687-6180. doi: 10.1155/2008/735351. URL https://doi.org/10.1155/2008/735351.

Sergios Theodoridis, Konstantinos Slavakis, and Isao Yamada. Adaptive learning in a world of projections. *IEEE Signal Processing Magazine*, 28 (1): 97–123, 1 2011. ISSN 1053-5888. doi: 10.1109/MSP.2010.938752.

Isao Yamada and Nobuhiko Ogura. Adaptive projected subgradient method for asymptotic minimization of sequence of nonnegative convex functions. *Numerical Functional Analysis and Optimization*, 25 (7-8): 593–617, 2005.

M. Yukawa. Adaptive learning in cartesian product of reproducing kernel Hilbert spaces. *IEEE Transactions on Signal Processing*, 63 (22): 6037–6048, Nov 2015a. ISSN 1053-587X. doi: 10.1109/TSP.2015.2463261.

M. Yukawa. Online learning based on iterative projections in sum space of linear and Gaussian reproducing kernel Hilbert spaces. In *2015 IEEE International Conference on Acoustics, Speech and Signal Processing (ICASSP)*, pages 3362–3366, April 2015b. doi: 10.1109/ICASSP.2015.7178594.

M. Yukawa, Y. Sung, and G. Lee. Dual-domain adaptive beamformer under linearly and quadratically constrained minimum variance. *IEEE Transactions on Signal Processing*, 61 (11): 2874–2886, June 2013. ISSN 1053-587X. doi: 10.1109/TSP.2013.2254481.

12

Machine Learning for Joint Channel Equalization and Signal Detection

Lin Zhang[1] *and Lie-Liang Yang*[2]

[1] *School of Electronics and Information Technology, Sun Yat-sen University, Guangzhou, China*
[2] *School of Electronics and Computer Science, University of Southampton, Southampton, UK*

12.1 Introduction

Over the last two decades, wireless communication has faced the demand to provide services anywhere. One of the vital technical challenges to achieve this is to support high reliability and high-data-rate communication in communication environments where signals experience severe inter-symbol interference (ISI) caused by multipath signal propagation embedded in broadband signal transmissions (Ranhotra et al., 2017). Another technical challenge is the mitigation of nonlinear signal distortions caused by imperfect design of wireless transceivers (Olmos et al., 2010).

In wireless communications, channel equalization is the main technique to combat the effect of ISI (Rappaport, 2002). Various equalizers have been proposed and implemented, which can be used to mitigate ISI and even to exploit the dispersive nature of wireless signals, in order to improve reliability. Among the family of channel equalization, adaptive equalizers have been proposed and applied to track the time-varying characteristics of wireless mobile channels.

Adaptive equalization typically includes a training phase and a tracking phase. During the training phase, a given training sequence is sent by a transmitter to a receiver in order to implement a proper setting for the equalizer in the receiver. After training, user data can then be sent. During this stage, the adaptive equalizer may further estimate the channel statistics and use the information to update the adaptive filter's coefficients, in order for the adaptive filter to adapt to time-varying communication environments.

In general, equalization techniques can be classified into two categories – linear and nonlinear equalization – depending on whether the output of an adaptive equalizer is fed back for performance enhancement (Chiu and Chao, 1996). Specifically, in the family of linear equalization, the most commonly used are zero-forcing (ZF) and minimum mean-square error (MMSE) equalizers (Haykin, 1986). In applications where channel distortion is too severe to be handled by linear equalizers, nonlinear equalizers are usually introduced, which can achieve higher reliability than linear equalization, but usually with high complexity. In the family of nonlinear equalization, there are typically three types of nonlinear methods developed: the decision feedback equalizer (DFE), the maximum likelihood symbol detection (MLSD) assisted equalizer, and the maximum likelihood sequence estimation (MLSE) assisted equalizer (Uesugi et al., 1989). Among these three, the MLSE equalizer is optimal in the sense that it minimizes the probability of sequence errors. However, the classic MLSE equalizer requires that the channel statistics are known to the receiver. Furthermore, the MLSE equalizer's complexity grows

Machine Learning for Future Wireless Communications, First Edition. Edited by Fa-Long Luo.
© 2020 John Wiley & Sons Ltd. Published 2020 by John Wiley & Sons Ltd.

exponentially with the number of taps of the channel impulse response (CIR) (Patwary and Rapajic, 2005).

Conventional equalizers deconvolve received signals to combat distortions induced by channels and, ultimately, to recover the received message. Therefore, conventional equalizers can be viewed as inverse filters, and the joint response function of a channel and the corresponding equalizer should be close to an ideal delay function (Lee and Messerschmitt, 1994). Accordingly, the complexity of equalization is dependent on the channels and may be very high in practical communication environments, where multiple reflections, multiple refractions and scattering, etc. exist.

In order to reduce the computational cost of traditional equalizers, machine learning (ML) based equalizers have been proposed for ISI suppression. In channel equalization, typical ML methods include multilayer perceptron (MLP) (Peng et al., 1991), radial basis functions (RBFs) (Guha and Patra, 2009), recurrent RBFs (Cid-Sueiro and Figueiras-Vidal, 1993), and Pao networks (Arcens et al., 1992). In these ML-based equalizers, in principle, each class is defined by the possible output of the symbol alphabet and a proper nonlinear boundary found in a higher dimensional space. Hence, ML-based channel equalization can be viewed as a problem of pattern recognition. Consequently, the complexity of ML-based equalization is mainly determined by the symbol number and the size of the constructed space, but is not determined by the CIRs. In addition to those previously mentioned, other benefits of ML-based channel equalizers include their intelligent and adaptive signal-processing capabilities, which enable them to operate in dynamic and time-varying wireless communication environments.

In this chapter, our focus is on ML-based channel equalization and data detection, which utilize deep learning neural networks (NNs) to learn and formulate the feature sets of time-varying wireless channels in order for them to be efficiently operated in highly dynamic wireless channels.

The rest of this chapter is organized as follows. After a brief overview of ML-based equalizers in Section 12.2, in Section 12.3 we describe three classic equalization algorithms: ZF equalization, MMSE equalization, and MLSE equalization. Section 12.4 provides the details of the NN models to be used for channel equalization, and presents the structure as well as the training process of NN-based equalization. Then, in Section 12.5, we investigate the performance of orthogonal frequency-division multiplexing (OFDM) systems involving NN-based channel equalization. Finally, Section 12.6 concludes this chapter.

12.2 Overview of Neural Network-Based Channel Equalization

By exploiting the nonlinear characteristics of neural networks (NNs), NN-based channel equalizers are capable of extracting the key features of communication channels and combating inter-symbol interference (ISI) intelligently with the aid of high-efficiency NN algorithms, including MLP (Gibson et al., 1989a, b), functional link artificial NN (FLANN) (Patra and Pal, 1995; Patra et al., 1999), radial basis function NN (RBFNN) (Chen et al., 1992, 1993b), self-constructing recurrent fuzzy NN (SCRFNN) (Lin et al., 2005; Chang et al., 2010), recurrent neural networks (RNNs) (Kechriotis et al., 1994; Zhang et al., 2004), deep learning (DL) (Ye and Li, 2017; Ye et al., 2018), extreme learning machines (ELMs) (Tang et al., 2016; Yang et al., 2018), etc. In the following, we provide a detailed overview of research on NN-based channel equalization.

12.2.1 Multilayer Perceptron-Based Equalizers

Channel equalization can be regarded as a classification problem, where the equalizer is constructed as a decision-making device with the motivation to classify the transmitted signals as accurately as possible. In this way, (Gibson et al., 1989a, b) proposed an adaptive equalizer using a NN architecture based on the MLP, to combat ISI over linear channels with white Gaussian noise. Following this work, (Chen et al., 1990; Gibson et al., 1991) applied the MLP-based equalizer for nonlinear channels with colored Gaussian noise, which demonstrated that the MLP-based equalizer is able to achieve bit error rate (BER) performance close to that of the optimal equalizer. However, MLP-based equalizers in these references considered mainly real-valued and bipolar signals.

In the MLP-based equalizers considered previously, the sigmoid function used by the output layer nodes confines network output to the range $[-1, 1]$. Therefore, by relaxing or replacing it with other functions, MLP-based equalizers can be designed to support multiple amplitude signals, such as pulse amplitude modulation (PAM) signals. Furthermore, in order to support complex quadratic-amplitude modulation (QAM) signals, (Peng et al., 1991, 1992) proposed to separately process the real and imaginary parts of the threshold relied weighted sum in the neural unit. They demonstrated that the MLP-based equalizer outperforms the least mean square (LMS) based linear equalizer, when nonlinear channel distortions exist. Similarly, (Chang and Wang, 1995) proposed a MLP-based equalizer to equalize complex-valued phase shift keying (PSK) signals received from nonlinear satellite channels.

In these works, all the MLP equalizers are supported by supervised learning. Similar to the ML algorithms considered in the previous chapters, a large amount of channel input/output data are usually needed to train the equalizer – using, for example, the backward propagation (BP) algorithm or its variants – before the equalizers are used for data detection. In contrast, in blind equalization where training data is unavailable, channel equalizers can also be designed on the basis of the MLP by exploiting the higher-order statistics of the source sequences, as done, for example, in (You and Hong, 1998; Gao et al., 2009). Moreover, (Gao et al., 2009) showed that their proposed blind equalizer is capable of achieving a faster convergence rate and a smaller mean square error (MSE) in comparison with the original MLP-based equalizer requiring training data, when communicating in an underwater acoustic digital communication scenario.

12.2.2 Functional Link Artificial Neutral Network-Based Equalizers

Because a FLANN is typically a single- or flat-layer perceptron, it has low computational complexity and can be readily implemented using hardware (Patra et al., 2008). In a FLANN, functional expansion can not only be applied to enhance the original input pattern but also be exploited to increase the pattern dimension. Specifically, the functional expansion of the input pattern can be implemented with the aid of basis functions such as trigonometric, Gaussian, Chebyshev, and Legendre polynomials (Burse et al., 2010).

In comparison with the MLP, the FLANN has no hidden layers, but nonlinear mappings. Owing to its simplicity, FLANNs have attracted wide research interest and have also applied for channel equalization (Patra and Pal, 1995; Patra et al., 1999). In these FLANN-based equalizers, functional expansions are achieved with the aid of orthogonal trigonometric functions. In (Weng and Yen, 2004), a reduced decision feedback FLANN (RDF-FLANN) based channel equalizer was introduced. Instead of taking the channel output as the input signals to the network, the RDF-FLANN-based equalizer also feeds back its own output signals to the input layer of the network. Moreover, the Chebyshev NN (ChNN) designed according to the FLANN was proposed for channel equalizations where four types of QAM signals were considered (Patra and Kot, 2002; Patra et al., 2005). Studies show that by expanding the input space, ChNNs can provide more efficient computation for the static function approximation than trigonometric polynomials.

FLANN-based equalizers may provide better performance than MLP-based equalizers. However, FLANN-based equalizers have the weakness of a higher-order complexity increase, especially in cases when the input space dimensionality is expanded at lower BER (Burse et al., 2010).

12.2.3 Radial Basis Function-Based Equalizers

RBFs were initially regarded as good alternatives for sigmoidal transfer function networks (Burse et al., 2010). Then, the RBF NN was proposed for data interpolation in the multidimensional space (Mulgrew, 1996). Furthermore, the cyclostationary characteristics of received signals are not required to be known in RBF NNs; hence, RBF NNs can also mitigate distortion caused by co-channel and adjacent channel interference.

In RBF-aided equalizers, received signals are classified according to their center vectors. As the network structure of RBF NNs is designed on the basis of Bayesian principles, the output in RBF-based equalizers is different from that in MLP and FLANN-based equalizers. According to (Chen et al., 1992,1993b,a), in comparison with adaptive MLSE, RBF-based equalizers achieve superior performance over time-varying Rayleigh fading channels and also have lower computational complexity.

Moreover, in (Chen et al., 1994) and (Cha and Kassam, 1995), the real-valued RBF was generalized to the complex-valued RBF (CRBF). Correspondingly, a stochastic-gradient training algorithm was developed to train the CRBF network (Cha and Kassam, 1995). To implement blind equalization with the aid of RBF, many techniques were proposed and studied (Tan et al., 2001; Uncini and Piazza, 2003; Xie and Leung, 2005).

In order to solve practical problems more effectively and to further improve the performance of RBF NN-based equalization, neuron fuzzy systems were developed according to the fuzzy logic methodology (Soria-Olivas et al., 2003; Rong et al., 2006). For example, the sequential fuzzy extreme learning algorithm was proposed in (Rong et al., 2009), which has been regarded as a good alternative to the batch-processing algorithm. This is because the sequential fuzzy extreme learning algorithm does not require a complete dataset for training, and the data can be utilized either in the form of small blocks or one by one during the training process. Therefore, in comparison with batch learning algorithm, retraining new datasets is not necessary in the sequential fuzzy learning algorithm.

12.2.4 Recurrent Neural Networks-Based Equalizers

In principle, a RNN can ideally implement the inverse of a finite memory system, with the result that it can substantially model a nonlinear infinite memory filter. Therefore, RNNs have the capability to effectively mitigate all the interference introduced by channels.

RNNs have been applied to channel equalization in various scenarios. The authors of (Kechriotis et al., 1994) proposed an adaptive RNN-based equalizer that can be operated either in trained mode, aided by a sequence of training data, or in blind equalization mode without training data. This adaptive RNN-based equalizer is suitable for both linear and nonlinear communication channels. Studies and simulation results show that RNN-based equalizers may outperform traditional linear filter-based equalizers in some specific application scenarios.

Subsequently, (Zhang et al., 2004) proposed a blind equalization algorithm on the basis of the bilinear RNN (BLRNN) and demonstrated that it is able to achieve better convergence performance and lower BER than the traditional constant modulus algorithm (CMA) (Zhang et al., 2004). The complex version of BLRNN (Park, 2008) supported by genetic algorithms was applied to channel equalization in wireless asynchronous transfer mode (ATM), and it was shown that better MSE or symbol error rate (SER) performance than conventional equalizers can be achieved. Furthermore, RNNs were applied to implement blind equalization in underwater acoustic communications, showing that a significant performance improvement can be achieved in comparison with traditional feedforward neural networks (FNNs) (Xiao et al., 2008).

In addition to RNNs, in (Li et al., 2017), convolutional neural networks (CNNs) were employed for channel equalization in order to combat nonlinear channel distortions and temporal changes of radio signals. In this joint CNN-RNN network, received radio signals are passed first through a CNN-based subnetwork and then through a RNN-based subnetwork. With the aid of CNN, the previous internal state can be exploited by the current state at each step. Therefore, a RNN with long short-term memory (LSTM) is capable of learning the temporal dependency. Consequently, the RNN has the potential to solve the problem of temporal variation of radio signals.

12.2.5 Self-Constructing Recurrent Fuzzy Neural Network-Based Equalizers

Fuzzy NN employs both the favorable interpretation of fuzzy logic and a high capability of learning (Chang and Ho, 2009). Owing to this, an adaptive network-based fuzzy inference system (ANFIS) was presented in (Jang et al., 1997). Furthermore, as the fuzzy NN is able to construct nonlinear decision boundaries, fuzzy NN-based channel equalizers can achieve high efficiency for ISI mitigation.

As fuzzy NN-based equalizers are capable of directly forming decision boundaries according to received signals, no information about channel characteristics, including channel coefficients and channel order, has to be known. SCRFNN is the combination of a RNN and a self-constructing fuzzy NN. In online learning processes, parameter learning is based on the supervised gradient descent approach with the aid of a delta adaptation law, and structure learning is based on the partition of the input space (Lin et al., 2005). Hence, the parameters and structure of SCRFNNs can be self-adjusted simultaneously. Note that SCRFNNs can restrict the generation of new fuzzy rules by setting constraints and therefore are more compact compared with ANFIS.

In terms of channel equalization, a SCRFNN-based nonlinear equalizer was proposed in (Lin et al., 2005). In contrast, (Chang et al., 2010) focused on time-varying and time-invariant wireless channels, and proposed a fast self-constructing fuzzy NN-based decision feedback equalizer (FSCFNN-DFE). In comparison with the SCRFNN, the FSCFNN-DFE can achieve improvement in terms of BER performance, hardware cost, and computational complexity, while enjoying the merit that channel characteristics are not required to be known to the receiver.

12.2.6 Deep-Learning-Based Equalizers

DL neural networks employing a number of hidden layers have great representation capability (Ye and Li, 2017). With the aid of DL, (Ye and Li, 2017) applied the MLP to joint channel equalization and decoding, and demonstrated robust performance under various channel conditions, including a time-varying frequency selective channel that generates severe ISI. Following this work, in (Ye et al., 2018), the DL neural network (DNN) structure was utilized to recover data symbols conveyed in OFDM principles. Furthermore, studies show that although DNNs are trained using data generated by the statistics of known channel models, the considered DNN-aided equalizer is robust, even when the maximum delay spread and number of paths assumed in training are different from those used at the test stage.

12.2.7 Extreme Learning Machine–Based Equalizers

Extreme learning machines (ELMs) have the capability to provide high learning accuracy by minimizing training error at the training stage as well as the norm of output weights. Since, in general, smaller training error results in a smaller norm of weights, this type of feedforward neural network allows the achievement of a desirable generalization performance with fixed hidden-layer parameters. For this reason, it has attracted increasing research interest in computer version applications (Tang et al., 2016).

Likewise, researchers in wireless communications have applied ELM to design channel equalizers for ISI mitigation and performance enhancement. Thanks to its merits of learning accuracy and efficiency, an ELM-based channel equalizer was demonstrated (Yang et al., 2018) to have the capability to converge fast in short data-packet transmissions where quadrature amplitude modulation (QAM) is used.

In general, when properly trained with the aid of training data, NN-based equalizers can effectively suppress ISI intelligently and significantly improve the reliability of communication. However, we should note that NN-based equalizers are required to extract the key features of wireless channels via training, which may induce concern about convergence, computation complexity, and spectrum efficiency of communication. In order to address these concerns, in recent years, the support vector machine (SVM) and Gaussian processes for regression (GPR) techniques have been introduced in channel equalization in wireless communications, and will be briefly reviewed in the next subsection.

12.2.8 SVM- and GPR-Based Equalizers

According to (Cover, 1965), Cover's theorem states that it's highly possible to create linearly separable clouds of data by projecting the pattern space $\mathcal{P}$ to the higher-dimensional feature space $\mathcal{F}$ via nonlinear mapping. Based on this theorem, the SVM nonlinearly maps the inner products of data in the pattern space, instead of the data themselves, via a kernel, so as to use an optimal hyperplane to separate clouds of data in the feature space (Sebald and Bucklew, 2000).

In the context of applications of SVM for channel equalization, in (Perez-Cruz et al., 2001), the authors presented a SVM-based equalizer for equalizing burst time division multiple access (TDMA) channels. In order to obtain SVM-based equalization, an iterative re-weighted least squares (IRWLS) procedure is used to reduce the computational complexity of the conventional SVM, in which the training procedure is implemented according to quadratic programming (QP). Then, a blind equalization algorithm based on the least squares support vector regressor (LSSVR) was proposed by (Yao et al., 2011), which exploits only characteristics of the transmitted signal and does not have to use training sequences to achieve spectrum efficiency of communications.

However, SVM-based blind equalization does not work well if the actual channel does not fit the channel model (Perez-Cruz et al., 2001). In order to improve the performance of SVM-based blind equalization algorithms, (Wang et al., 2016) proposed to combine the conventional cost function of the support vector regressor (SVR) with the probability density function (PDF) of the error function. Thus the PDF of the equalizer's output signals is formulated so that it can match the PDF of the constellations known to the receiver. Simulation results demonstrated that this equalization method outperforms legacy algorithms, such as the PDF algorithm and the SVR algorithm, when considering communications over ISI channels. However, in SVM-based equalizers, the hyperparameters have to be either prespecified or estimated by cross-validations. Moreover, SVM-based equalizers need a long training sequence and hence have high complexity. Thus, in practice, SVM-based equalizers can usually only tune one or two hyperparameters by cross-validation.

In order to address these issues, a nonlinear channel equalizer designed according to GPR was proposed in (Perez-Cruz et al., 2008). In comparison with SVM-based methods, GPR algorithms have the advantage of optimizing kernel hyperparameters based on maximum likelihood. This allows significant performance improvement in the case of short training sequences, making GPR-based equalizers outperform SVM-aided equalizers. In contrast, in the case of using long training datasets, hyperparameters only slightly affect the solution, with the result that SVM and GPR-based equalizers achieve similar performance. In order to further improve the performance of GPR-based equalizers, (Olmos et al., 2010) combined nonlinear channel equalization with low-density parity checking (LDPC) codes (Yang et al., 2015). This design motivates the use of short training sequences to obtain an accurate posterior probability estimation for the LDPC code, rather than directly working on the error rate of the LDPC decoding, which was considered in the previous research.

Having outlined all the NN- and ML-based channel equalization solutions, next we will address these techniques in more detail by considering the algorithm design, complexity analysis, and performance comparison. First, let us present the principles and problem formulations of channel equalization and related signal detection.

12.3 Principles of Equalization and Detection

Figure 12.1 presents a diagram for the basic communication system with an equalizer employed at the receiver side. Let $d(t)$ denote the baseband data and $f(t)$ represent the combined impulse response of the transmitter, radio channel, radio frequency (RF)/IF processing modules, and matched filter. At the receiver, the received signal can be expressed as

$$y(t) = d(t) \otimes f(t) + n(t) \tag{12.1}$$

where $n(t)$ is the additive noise input to the equalizer, and $\otimes$ denotes the convolution operation.

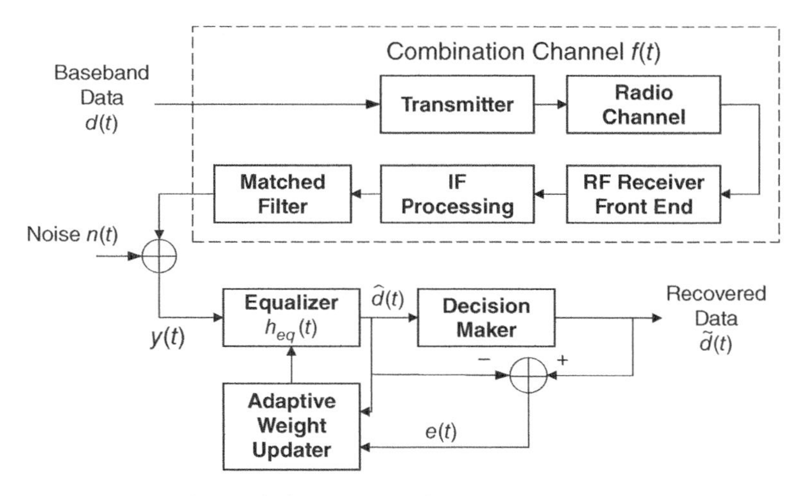

Figure 12.1 Equalizer-aided transmission diagram.

Let $h_{eq}(t)$ denote the impulse response function of the equalizer. Then the output signals of the equalizer denoted by $\hat{d}(t)$ are given by

$$\begin{aligned}\hat{d}(t) &= d(t) \otimes f(t) \otimes h_{eq}(t) + n(t) \otimes h_{eq}(t) \\ &= d(t) \otimes g(t) + n(t) \otimes h_{eq}(t)\end{aligned} \tag{12.2}$$

where $g(t) = f(t) \otimes h_{eq}(t)$.

In order to reliably recover the transmitted data, a desired output of the equalizer is $d(t)$. In this case, we have $\hat{d}(t) = d(t)$, which is obtained when $h_{eq}(t)$ is designed to satisfy

$$g(t) = f(t) \otimes h_{eq}(t) = \delta(t) \tag{12.3}$$

When Eq. (12.3) is satisfied, the ISI induced by the channel can be fully eliminated. Considering the frequency domain, from Eq. (12.3) we have

$$F(f)H_{eq}(f) = 1 \tag{12.4}$$

where $F(f)$ and $H_{eq}(f)$ are, respectively, the Fourier transformations of $f(t)$ and $h_{eq}(t)$. Equation (12.4) means the ideal equalizer is an inverse filter of the combined channel $f(t)$. From Eq. (12.4), we know that when communicating over frequency-selective fading channels, frequency components with low response gains are enhanced, while frequency components with high response gains are reduced after the equalization, so that the spectrum after equalization becomes flat. Consequently, the channel-fading effect can be efficiently mitigated.

Channel equalization is a classic, well-known technique in telecommunication engineering, and considerable research has been done. Typical equalization algorithms include the ZF algorithm, MMSE algorithm, optimum MLSE algorithm, etc. Before introducing the ML-based equalization algorithms, let us first briefly review the principles of these three equalizers.

Zero-Forcing Equalization In ZF equalization, according to Eq. (12.4), the frequency response of the equalizer is designed to be

$$H_{eq,ZF}(z) = \frac{1}{F(z)} \tag{12.5}$$

The ZF-equalizer in Eq. (12.5) can ideally eliminate the ISI introduced by a channel. However, the noise embedded in the received signal varies with the frequency response of the equalizer. To demonstrate this effect, we can analyze the noise power spectral density (PSD) function after the ZF equalization, which can be expressed as

$$N(z) = N_0|H_{eq,ZF}(z)|^2 = \frac{N_0}{|F(z)|^2} \tag{12.6}$$

where N_0 is the noise power before ZF equalization. As shown in Eq. (12.6), if the combined channel response $F(z)$ has a frequency component with a very small amplitude, the corresponding noise power will be amplified significantly with an amplification factor of $\frac{1}{|F(z)|^2}$. In order to achieve a better trade-off between ISI elimination and noise-power amplification, a MMSE algorithm has been introduced (Haykin, 1986), which has the following principles.

Minimum Mean Square Error Equalization The aim of the MMSE-equalizer is to minimize the MSE between the desired equalizer's output and the actual equalizer's output. Referring to Figure 12.1, the detection error is

$$e_k = d_k - \hat{d}_k \tag{12.7}$$

Assume that the equalizer is a casual linear transversal equalizer with a length of $N = L + 1$. Then the frequency response of the equalizer is

$$H_{eq}(z) = \sum_{i=0}^{L} w_i z_i \tag{12.8}$$

Then, when the samples of $y(t)$, which are expressed as $\{y_n\}$, are input to the equalizer, the output of the equalizer given by the convolution of the discrete impulse response of $h_{eq}(t)$ and $\{y_n\}$ can be expressed as

$$\hat{d}_k = \sum_{i=0}^{L} w_i^* y_{k-i} = \mathbf{w}^H \mathbf{y}_k \tag{12.9}$$

where $\mathbf{w}^H = [w_0^*, \ldots, w_L^*]$ and $\mathbf{y}_k = [y_k, \ldots, y_{k-L}]^T$ represent, respectively, the weighted vector of the equalizer and the channel output vector input to the equalizer for detecting d_k.

Thus, given the desired output d_k and based on Eq. (12.7), the MSE is given by

$$J(\mathbf{w}) = E[e_k e_k^*] = E[|d_k|^2] + \mathbf{w}^H E[\mathbf{y}_k \mathbf{y}_k^H]\mathbf{w} - 2E[d_k \mathbf{y}_k^H]\mathbf{w} \tag{12.10}$$

where $E[\bullet]$ represents the numerical expectation operation and $\mathbf{y}_k^H = [y_k^*, \ldots, y_{k-L}^*]$. In order to minimize $J(w)$, we take the differentiation of $J(w)$ with respect to w^*, yielding

$$\frac{\partial J}{\partial \mathbf{w}^*} = -\mathbf{p} + \mathbf{R}\mathbf{w} = 0 \tag{12.11}$$

where $\mathbf{p}$ is the cross-correlation vector between $\mathbf{y}_k$ and d_k expressed as

$$\mathbf{p} = E[y_k d_k^*, \ldots, y_{k-L} d_k^*]^T \tag{12.12}$$

where the expectation operation is performed on each element of the matrix. In Eq. (12.11), R is the auto correlation matrix of $\mathbf{y}_k$ denoted by

$$\begin{aligned} \mathbf{R} = E[\mathbf{y}_k \mathbf{y}_k^H] &= E\left\{ \begin{bmatrix} y_k \\ y_{k-1} \\ \vdots \\ y_{k-L} \end{bmatrix} \begin{bmatrix} y_k^* & y_{k-1}^* & \cdots & y_{k-L}^* \end{bmatrix} \right\} \\ &= E\begin{bmatrix} |y_k|^2 & y_k y_{k-1}^* & \cdots & y_k y_{k-L}^* \\ y_{k-1} y_k^* & |y_{k-1}|^2 & \cdots & y_{k-1} y_{k-L}^* \\ \cdots & \cdots & \cdots & \cdots \\ y_{k-L} y_k^* & y_{k-L} y_{k-1}^* & \cdots & |y_{k-L}|^2 \end{bmatrix} \end{aligned} \tag{12.13}$$

From Eq. (12.11), the optimal weighted vector can be obtained to be

$$\mathbf{w}_{\mathbf{opt}} = \mathbf{R}^{-1}\mathbf{p} \tag{12.14}$$

Note that, in the case where the length of the equalizer is infinite, i.e. $\mathbf{w}^T = [w_0, \ldots, w_\infty]$, it can be shown that the frequency response of the MMSE-equalizer can be expressed as (Stuber, 1999)

$$H_{eq,MMSE}(z) = \frac{1}{F(z) + N_0} \tag{12.15}$$

Comparing Eq. (12.5) with Eq. (12.15), it can be seen that when the noise power is zero, meaning the signal-to-noise ratio (SNR) is infinite, the MMSE-equalizer is equivalent to the ZF-equalizer. By contrast, when noise exists, the MMSE-equalizer is capable of achieving a better balance between ISI mitigation and noise-power reduction, yielding more reliable detection than the ZF-equalizer.

Maximum Likelihood Sequence Estimation–Assisted Equalizers To attain near-optimum error performance, researchers have proposed various nonlinear equalizers for application in mobile communication systems. Among them, there are a range of nonlinear equalizers that are designed according to the classical maximum likelihood detection algorithm. One of them is the MLSE-equalizer. In the principle of MLSE equalization, the posteriori detection probabilities of all the possible data sequences are first calculated. Then the data sequence with the maximum posteriori probability is selected as the desired output. While the MLSE-equalizer is capable of minimizing the probability of the sequence error, it has a main defect of high computational complexity, which is extreme when the delay spread of the channel is large.

Let us consider the detection of L most recent input data symbols. Then, there are M^L states, when the size of the symbol alphabet of the modulation is M. As shown in Figure 12.2, with the aid of the Gram-Schmidt orthogonal operation, the received signals corresponding to the L symbols can be expressed as

$$y(t) = \sum_{n=1}^{N} w_n \phi_n(t) \tag{12.16}$$

where $\{\phi_n(t)\}$ is a set of N orthogonal basis, $t \in [0, LT_s]$, and T_s is the sampling time. Furthermore, in Eq. (12.16), the coefficients w_n are computed by the formulas of

$$w_n = \sum_{k=0}^{L-1} d_k f_{nk} + v_n \tag{12.17}$$

where $f_{nk} = \int_0^{LT_s} f(t - kT_s)\phi_n^*(t)dt$ and $v_n = \int_0^{LT_s} n(t)\phi_n^*(t)dt$.

It can be shown that the coefficient vector $\mathbf{w} = [w_1, w_2, \ldots, w_N]^T$ follows the multidimensional Gaussian distribution with the PDF given by

$$p(\mathbf{w}|\mathbf{d}, f(t)) = \prod_{n=1}^{N} \frac{1}{\pi N_0} \exp\left(-\frac{1}{N_0}\left|w_n - \sum_{k=0}^{L-1} d_k f_{nk}\right|^2\right) \tag{12.18}$$

As shown in Figure 12.2, given the signal $y(t)$ corresponding to the L data symbols and having the corresponding coefficient vector $\mathbf{w}$, the MLSE-equalizer estimates a data sequence

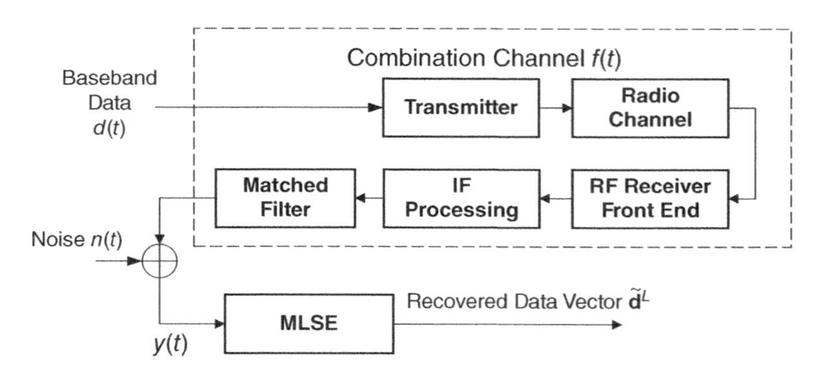

Figure 12.2 The structure of a MLSE equalizer.

$\tilde{\mathbf{d}} = [d_0, d_1, \ldots, d_{L-1}]$ by maximizing a logarithm likelihood function, which can be expressed as

$$\begin{aligned}\tilde{\mathbf{d}} &= \arg\max_{\mathbf{d}} \{\log p(\mathbf{w}|\mathbf{d}, f(t))\} \\ &= \arg\max_{\mathbf{d}} \left\{ -\sum_{n=1}^{N} \left| w_n - \sum_{k=0}^{L-1} d_k f_{nk} \right|^2 \right\} \\ &= \arg\max_{\mathbf{d}} \left\{ 2\mathrm{Re}\left\{ \sum_{k=0}^{L-1} d_k^* \sum_{n=1}^{N} w_n f_{nk}^* \right\} - \sum_{k=0}^{L-1}\sum_{m=0}^{L-1} d_k d_m^* \sum_{n=1}^{N} f_{nk} f_{nm}^* \right\}.\end{aligned} \tag{12.19}$$

It can be shown that in Eq. (12.19),

$$\sum_{n=1}^{N} w_n f_{nk}^* = \int_{-\infty}^{\infty} w(\tau) f^*(\tau - kT_s) d\tau = y[k] \tag{12.20}$$

$$\sum_{n=1}^{N} f_{nk} f_{nm}^* = \int_{-\infty}^{\infty} f(\tau - kT_s) f^*(\tau - mT_s) d\tau = u[k-m] \tag{12.21}$$

where $y[k]$ is the sampled signal of $y(t)$, $u[k]$ is the channel parameter satisfying $u[k-m] = u(kT_s - mT_s)$, and $u(t) = f(t) * f^*(-t)$ (Goldsmith, 2005). Substituting Eq. (12.20) and Eq. (12.21) into Eq. (12.19), we obtain

$$\tilde{\mathbf{d}} = \arg\max_{\mathbf{d}} \left\{ 2\mathrm{Re}\left\{ \sum_{k=0}^{L-1} d_k^* y[k] \right\} - \sum_{k=0}^{L-1}\sum_{m=0}^{L-1} d_k d_m^* u[k-m] \right\}. \tag{12.22}$$

Here we have reviewed the principles of three types of classic equalization schemes. It can be seen that all these equalizers assume that the channel impulse response is a priori known to the equalizer. However, in practice, especially in wideband wireless communication systems, the exact channel response can hardly be retrieved due to the time-varying nature of wireless channels. This results in a significant challenge to the design of traditional equalizers. In order to adapt to channel variation intelligently, NN- and ML-based equalization algorithms have been developed, which will be the focus of the next section.

12.4 NN-Based Equalization and Detection

In this section, we first briefly introduce a quintessential model for the feedback neural network, i.e. the MLP, in order to provide some insight into the construction and training of NNs. Then, based on the MLP and DL, a DNN solution is proposed for the implementation of channel equalizers. Finally, equalizers based on different types of NNs are briefly described.

12.4.1 Multilayer Perceptron Model

Generally, NNs need to be trained with a large set of data in order to extract the key feature parameters. The training process is usually implemented by the gradient descent algorithm, which uses forward and backward propagation in each iteration. This subsection briefly introduces the principles of the MLP as well as its training method.

12.4.1.1 Generalized Multilayer Perceptron Structure

A general neural node, as shown in Figure 12.3, is the basic unit of a MLP. Given the inputs $x_1, x_2, ..., x_n$, the output of the neural node is denoted by $y = f\left(\sum_{i=1}^{n} w_i x_i + b\right)$, where w_i is the weighting of x_i, b is the bias, and $f(\cdot)$ is the nonlinear function, which is called the *activation function*. There are various choices for the activation function (Goodfellow et al., 2016), including the sigmoid function, hyperbolic tangent function, and ReLu function, among which the ReLu function is defined as:

$$f_{\text{ReLu}}(z) = \max\{0, z\}. \tag{12.23}$$

A MLP has the typical structure as shown in Figure 12.4, where multiple neural nodes are arranged to form multiple layers, including an input layer, possibly multiple hidden layers, and an output layer. The neural nodes of two adjacent layers are fully connected, with the inputs to a neural node of the current layer consisting of all the outputs of the neural nodes in the last layer. The input layer is used to feed data into the network, while the output layer is used to output the computation results. The other layers between the input and output layers are called hidden layers because the training data of NNs do not generate the desired output for each of these layers (Goodfellow et al., 2016). The MLP in Figure 12.4 has four hidden layers.

Denote the number of layers of a MLP as L. The lth layer consists of $n^{[l]}$ neural nodes, where $l = 0, 1, ..., L-1$ and layer 0 is referred as the input layer. At the lth layer, the inputs are denoted as a vector $\boldsymbol{x}^{[l]} = [x_1^{[l]}, x_2^{[l]}, ..., x_{n^{[l]}}^{[l]}]^{\mathrm{T}}$; the weighting factors are denoted as a matrix $\boldsymbol{W}^{[l]} = [\boldsymbol{w}_1^{[l]}, \boldsymbol{w}_2^{[l]}, ..., \boldsymbol{w}_{n^{[l]}}^{[l]}]$, where $\boldsymbol{w}_i^{[l]}$ is the column vector corresponding to the ith neural node of layer l and is defined as $\boldsymbol{w}_i^{[l]} = [w_{i,1}^{[l]}, w_{i,2}^{[l]}, ..., w_{i,n^{[l-1]}}^{[l]}]^{\mathrm{T}}$; and the biases are denoted as $\boldsymbol{b}^{[l]} = [b_1^{[l]}, b_2^{[l]}, ..., b_{n^{[l]}}^{[l]}]^{\mathrm{T}}$. Therefore, the input to the ith neural node of layer l is $\sum_{j=1}^{n^{[l-1]}} w_{ij}^{[l]} x_i^{[l]} + b_i^{[l]}$.

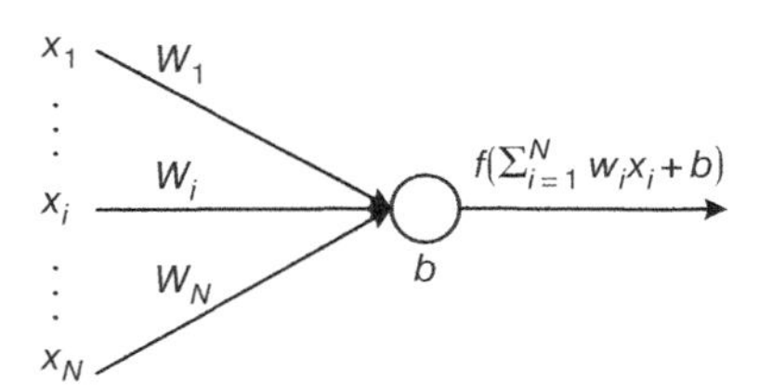

Figure 12.3 Neural node.

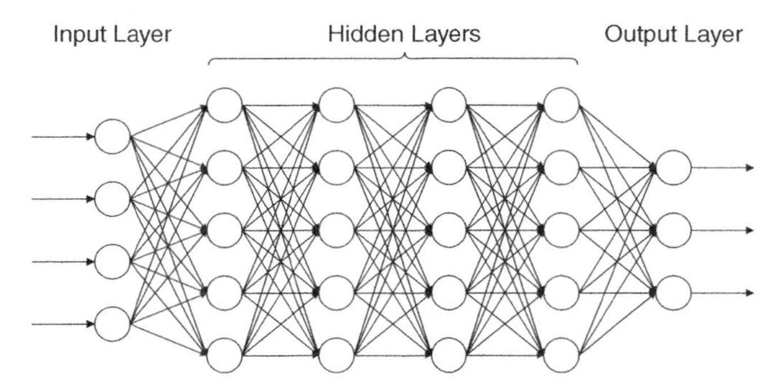

Figure 12.4 An example of the MLP model.

In principle, a MLP implements a nonlinear mapping expressed as $f(\boldsymbol{x}^{[0]}) : \mathbb{R}^{n^{[0]}} \to \mathbb{R}^{n^{[L]}}$, where the mapping function is constructed in an iterative way expressed as:

$$f(\boldsymbol{x}^{[0]}) = f^{[L]}(f^{[L-1]}(...f^{[1]}(\boldsymbol{x}^{[0]}))), \tag{12.24}$$

where $f^{[l]}(\cdot)$ is the activation function at the lth layer. The goal of MLP is to approximate some function f^* in order to generate the desired output denoted by $f^*(\boldsymbol{x}^{[0]})$. To achieve a close approximation, the MLP should be trained by tuning the weighting factors and the bias of each layer in a supervised way. For this purpose, the loss function can be defined to measure the difference between the actual output and the desired output, which should be minimized by the training process.

Considering a training set with M examples, the loss function in the L_2 norm is defined as

$$L_2(\boldsymbol{Y}, \widehat{\boldsymbol{Y}}) = \frac{1}{M} \sum_{m=1}^{M} ||\boldsymbol{y}^{(m)} - \hat{\boldsymbol{y}}^{(m)}||^2 \tag{12.25}$$

where $\boldsymbol{y}^{(m)}$ and $\hat{\boldsymbol{y}}^{(m)}$ are, respectively, the actual output and the desired output of the MLP corresponding to the mth example; $\boldsymbol{Y}$ and $\widehat{\boldsymbol{Y}}$ are defined as $\boldsymbol{Y} = [\boldsymbol{y}^{(1)}, \boldsymbol{y}^{(2)}, ..., \boldsymbol{y}^{(M)}]$ and $\widehat{\boldsymbol{Y}} = [\hat{\boldsymbol{y}}^{(1)}, \hat{\boldsymbol{y}}^{(2)}, ..., \hat{\boldsymbol{y}}^{(M)}]$, respectively.

In MLP, the training process is usually completed with the aid of the gradient descent algorithm or its variants, some of which are introduced as follows.

12.4.1.2 Gradient Descent Algorithm

The basic principle behind the gradient descent algorithm is to use the derivative of the objective function to update the decision variables, so that the value of the objective function (or the cost) can be reduced.

Consider a function $r(x)$ in the real domain, and denote its derivative as $r'(x)$. Then, at the point x, $r(x+\epsilon)$ for small ϵ can be approximated as $r(x) + \epsilon r'(x)$. Furthermore, we can see that, provided $|r'(x)| > 0$, $r(x - \epsilon \text{sign}(r'(x)))$ is smaller than $r(x)$. In other words, the derivative gives the information to reduce the function value by introducing a small change in the variable x. In this way, we may repeatedly update the variable using $x' = x - \epsilon\text{sign}(r'(x))$ to derive the minimal value of the function as well as the point achieving this minimum value.

Return to the MLP training, the objective function is the loss function, and the variables include the weighing factors and the bias in each layer, i.e. $\boldsymbol{W}^{[l]}$ and $\boldsymbol{b}^{[l]}$, $l = 1, 2, ..., L$. According to the principles of the gradient descent algorithm, these variables are updated as

$$\boldsymbol{W}^{[l]\prime} = \boldsymbol{W}^{[l]} - \alpha \frac{\partial L_2(\boldsymbol{Y}, \widehat{\boldsymbol{Y}})}{\partial \boldsymbol{W}^{[l]}}, \quad l = 1, 2, ..., L \tag{12.26}$$

$$\boldsymbol{b}^{[l]\prime} = \boldsymbol{b}^{[l]} - \alpha \frac{\partial L_2(\boldsymbol{Y}, \widehat{\boldsymbol{Y}})}{\partial \boldsymbol{b}^{[l]}}, \quad l = 1, 2, ..., L \tag{12.27}$$

where α is called the *learning rate*. The value of α can be set to a small constant; or, alternatively, it can be adaptively changed with the aid of the line-search strategy (Goodfellow et al., 2016), which evaluates the objective function for several α values and chooses the one resulting in the smallest function value.

In the gradient descent algorithm, the cost of calculating the partial derivatives is $O(M)$, with the result that the loss function is related to all the training examples. This may be expensive when the training set is large, which is necessary for good generalization. To reduce the complexity, the stochastic gradient descent algorithm (Goodfellow et al., 2016) has been developed as a variant of the gradient descent algorithm. In the stochastic gradient descent algorithm,

M' uniformly chosen examples are used to estimate the whole data set, where M' is called the *minibatch size* and is usually much smaller than M. In this way, we may fit a training set with millions of examples using updates computed from a smaller number of examples, while the cost of calculating the partial derivative is reduced to $O(M')$ instead of $O(M)$.

Another drawback of the gradient descent algorithm is that it may be trapped at a local minimum. However, (Goodfellow et al., 2016) argued that many ML models can work well when they are trained with the gradient descent algorithm. Furthermore, the authors pointed out that even though it's not guaranteed to find the global minimum within a reasonable time, the algorithm can usually quickly find the value of the loss function, which is low enough for practical purposes. Hence, this property makes the gradient descent algorithm practically useful.

12.4.1.3 Forward and Backward Propagation

In each iteration of the gradient descent algorithm, forward propagation is utilized to evaluate the loss function value, while backward propagation is utilized to compute the partial derivatives. Considering a stochastic gradient descent algorithm with minibatch size M', the loss function is written as:

$$L_2(Y, \hat{Y}) = \frac{1}{M'} \sum_{m=1}^{M'} (y^{(m)} - \hat{y}^{(m)})^2. \tag{12.28}$$

Starting from the input layer, the forward propagation processes carry out the linear and activation operations layer by layer until the output layer. Specifically, at the lth layer, when denoting, respectively, the results of the linear and activation operations as $\boldsymbol{Z}^{[l]}$ and $\boldsymbol{A}^{[l]}$, whose dimensions are both $n^{[l]} \times M'$, the operations can be formulated as:

$$\boldsymbol{Z}^{[l]} = \boldsymbol{W}^{[l]}\boldsymbol{A}^{[l-1]} + \boldsymbol{b}^{[l]}, \tag{12.29}$$

$$\boldsymbol{A}^{[l]} = f^{[l]}(\boldsymbol{Z}^{[l]}). \tag{12.30}$$

Hence, when given $\boldsymbol{A}^{[l]}$ and $\boldsymbol{b}^{[l]}$, the computational chain of forward propagation is as follows:

$$\begin{aligned} \boldsymbol{A}^{[0]} = \boldsymbol{X}^{[0]} &\rightarrow \boldsymbol{Z}^{[1]}, \boldsymbol{A}^{[1]} \\ &\rightarrow \boldsymbol{Z}^{[2]}, \boldsymbol{A}^{[2]} \\ &\rightarrow \dots \\ &\rightarrow \boldsymbol{Z}^{[L]}, \boldsymbol{A}^{[L]} \end{aligned}$$

where $\boldsymbol{X}^{[0]}$ is the initialization operation.

Finally, the output of the MLP is given by the loss functions after the operations at the Lth layer.

In contrast, backward propagation starts from the output layer and processes in a layer-by-layer manner until the input layer. The chain rule is used for computing the partial derivatives in terms of each of the L layers. Specifically, for the lth layer, the corresponding partial derivatives are calculated as follows:

$$d\boldsymbol{Z}^{[l]} = \frac{\partial L_2}{\partial \boldsymbol{Z}^{[l]}} = \frac{\partial L_2}{\partial \boldsymbol{A}^{[l]}} \frac{\partial \boldsymbol{A}^{[l]}}{\partial \boldsymbol{Z}^{[l]}} = d\boldsymbol{A}^{[l]} \odot f^{[l]'}(\boldsymbol{Z}^{[l]}), \tag{12.31}$$

$$d\boldsymbol{W}^{[l]} = \frac{\partial L_2}{\partial \boldsymbol{W}^{[l]}} = \frac{\partial L_2}{\partial \boldsymbol{Z}^{[l]}} \frac{\partial \boldsymbol{Z}^{[l]}}{\partial \boldsymbol{W}^{[l]}} = \frac{1}{M'} d\boldsymbol{Z}^{[l]} \boldsymbol{A}^{[l-1]\mathrm{T}}, \tag{12.32}$$

$$d\boldsymbol{b}^{[l]} = \frac{\partial L_2}{\partial \boldsymbol{b}^{[l]}} = \frac{\partial L_2}{\partial \boldsymbol{Z}^{[l]}} \frac{\partial \boldsymbol{Z}^{[l]}}{\partial \boldsymbol{b}^{[l]}} = \frac{1}{M'} \sum_{m=1}^{M'} d\boldsymbol{z}^{[l](m)}, \tag{12.33}$$

$$d\boldsymbol{A}^{[l-1]} = \frac{\partial L_2}{\partial \boldsymbol{A}^{[l-1]}} = \frac{\partial L_2}{\partial \boldsymbol{Z}^{[l]}} \frac{\partial \boldsymbol{Z}^{[l]}}{\partial \boldsymbol{A}^{[l-1]}} = \boldsymbol{W}^{[l]\mathrm{T}} d\boldsymbol{Z}^{[l]}, \tag{12.34}$$

where $\odot$ denotes the Hadamard product operation. The computational chain of backward propagation is

$$\begin{aligned} d\boldsymbol{A}^{[L]} = \boldsymbol{Y} - \boldsymbol{A}^{[L]} &\rightarrow d\boldsymbol{Z}^{[L]}, d\boldsymbol{W}^{[L]}, d\boldsymbol{b}^{[L]}, d\boldsymbol{A}^{[L-1]} \\ &\rightarrow d\boldsymbol{Z}^{[L-1]}, d\boldsymbol{W}^{[L-1]}, d\boldsymbol{b}^{[L-1]}, d\boldsymbol{A}^{[L-2]} \\ &\rightarrow \dots \\ &\rightarrow d\boldsymbol{Z}^{[1]}, d\boldsymbol{W}^{[1]}, d\boldsymbol{b}^{[1]}, d\boldsymbol{A}^{[0]}. \end{aligned}$$

Finally, the partial derivatives required by the gradient descent algorithm are given by $d\boldsymbol{Z}^{[1]}$, $d\boldsymbol{W}^{[1]}$, $d\boldsymbol{b}^{[1]}$, $d\boldsymbol{A}^{[0]}$.

12.4.2 Deep-Learning Neural Network-Based Equalizers

As the MLP becomes deeper, i.e. as it has more hidden layers, it becomes more powerful for the representation of complicated functions, e.g. time-varying channel impulse responses. Therefore, with the aid of DL, this subsection extends the MLP to construct a DL NN, based on which a channel equalizer is further presented.

12.4.2.1 System Model and Network Structure

By introducing more fully connected hidden layers, a MLP is extended to a DNN, which has the structure as shown in Figure 12.5. Like other neural networks, DNNs may be trained using different tasks so as to perform different functions. Specifically, when it is trained with the task of recovering signals undergoing channel distortions, a DNN is capable of learning channel features and hence approximating the equalization function, so as to bring the received signals close to the desired signals. In this way, a DNN-based channel equalizer can be constructed.

The architecture of a generalized communication system with a DNN-based channel equalizer is illustrated in Figure 12.6. At the transmitter side, the information message is first encoded and the codeword is digital modulated according to certain baseband modulation scheme. The modulated symbols are then sent to the wireless channel, where different distortions may be experienced. At the receiver side, the DNN-based channel equalizer is utilized to combat channel distortion before data demodulation and decoding. In Figure 12.6, the DNN is usually trained offline before online deployment, i.e. actual real-time application. Details of the network

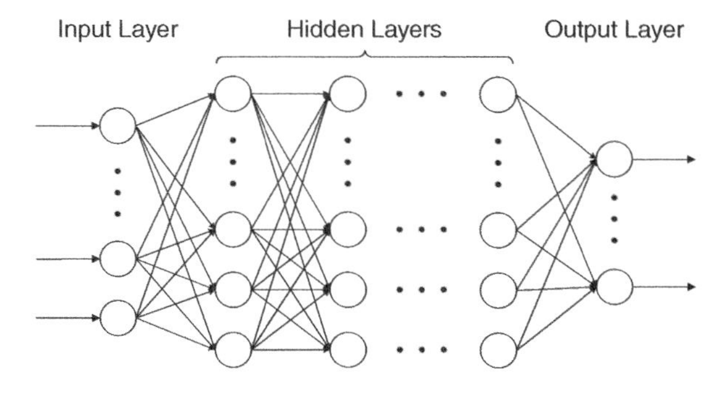

Figure 12.5 The structure of DNN.

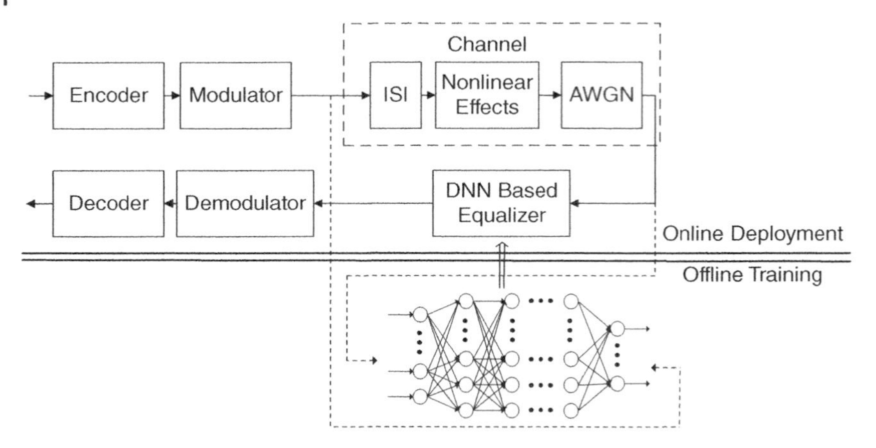

Figure 12.6 The system model for a DNN-based equalizer.

training are provided in the following subsection. Note that, at the online development stage, DNN-based equalizers can recover the desired signals without explicit channel estimations.

As shown in Figure 12.6, channel distortions may consist of ISI, nonlinear distortion, and additive white Gaussian noise (AWGN). Here, nonlinear distortion, denoted as a g-function, is mainly introduced by amplifiers and mixers, at both the transmitter and receiver, which can be viewed as part of the channel effects. Assume that the impulse response of a dispersive channel is formulated as

$$h(z) = \sum_{l=0}^{n_L-1} h[l] \cdot z^{-l}, \tag{12.35}$$

where n_L is the length of the impulse response. Then, the output of this channel can be expressed as

$$r[i] = g\left(\sum_{j=0}^{n_L-1} s[i-j] \cdot h[j]\right) + n[i], \tag{12.36}$$

where $n[i]$ is Gaussian noise, $s[j]$ is the channel input, and $r[i]$ is the channel output after nonlinear distortion contributed by the $g(\cdot)$ operation.

12.4.2.2 Network Training

In order to perform channel equalization, the DNN is first required to be trained in a supervised way using a training set with a great amount of channel input/output data. Conventionally, online training is adopted to train small-size MLP-based channel equalizers, which utilize, for example, transmitted pilot data to adjust network parameters. However, this online training method is not suitable for DNNs, because many parameters in a DNN need to be determined. Therefore, a large training set is required, and a long training period needs to take place in the DNN-based scheme, which can actually degrade overall equalization performance.

In order to circumvent the disadvantages of online training, DNNs are usually trained offline (Ye and Li, 2017; Ye et al., 2018) using various information sequences obtained under diverse channel conditions. After being properly trained, DNNs can efficiently recover the received data online without relying on time-consuming online training.

One merit of offline training of DNNs is that, unlike many other ML tasks where a large size training set is difficult to obtain, the training data for DNN-based equalizers can be directly generated with the aid of simulation approaches, once the channel model and other parameters are known or correctly estimated. In detail, for each simulation, a random message $\boldsymbol{m}$ is generated, based on which the encoder chooses a corresponding codeword $\boldsymbol{c}$ from the codebook. Then, data modulation is executed to yield the symbol sequence $\boldsymbol{s}$; it is transmitted over the wireless channel, which can be modeled as a black box. At the receiver, the channel output $\boldsymbol{r}$ is collected together with the corresponding symbol sequence transmitted, which are assigned a supervising label to form a training example. From this one can see that training examples are generated based on modeling and simulation. Thus, an arbitrarily large training set can be obtained to train a DNN-based equalizer.

As shown in Figure 12.6, at the offline training stage, DNNs are required to be trained to minimize the difference between transmitted symbols $\boldsymbol{s}$ and estimated ones $\hat{\boldsymbol{s}}$. More explicitly, the input to a DNN is the received signal $\boldsymbol{r}$, while $\boldsymbol{s}$ and $\hat{\boldsymbol{s}}$, respectively, denote the targeted output and the actual output of the DNN, and $\hat{\boldsymbol{s}}$ is the estimate of $\boldsymbol{s}$. Therefore, the loss function of the DNN is

$$L_2(\boldsymbol{s}, \hat{\boldsymbol{s}}) = \frac{1}{M'} \sum_{m=1}^{M'} \|\boldsymbol{s}^{(m)} - \hat{\boldsymbol{s}}^{(m)}\|^2, \tag{12.37}$$

where the superscript m denotes the index of the training example, and M' is the minibatch size. Naturally, the vectors $\boldsymbol{s}$ and $\hat{\boldsymbol{s}}$ have size M'. To elaborate a bit further, with the aid of the known input $\boldsymbol{r}$ to the DNN, the actual output can be expressed as $\hat{\boldsymbol{s}} = f(\boldsymbol{r})$, where $f(\cdot)$ represents the nonlinear mapping of the DNN determined by the weights that are unknown but obtained via the known "$\boldsymbol{r}$" and "$\boldsymbol{s}$".

Through the minimization of the loss function $L_2(\boldsymbol{s}, \hat{\boldsymbol{s}})$ of Eq. (12.37), by using the gradient descent algorithm with forward/backward propagation, as shown in Section 12.4.1.2 and Section 12.4.1.3, offline training can be implemented with the aid of known transmitted and received signals. Since the received signals potentially experience various channel variations, after the training stage, the resultant DNN can be used to effectively detect the received signals experiencing various channel variations.

The main concern of offline training of DNNs is that the channels used for generating training data might be mismatched with the actual channels. However, the simulation results in (Ye et al., 2018) show that variations of channel statistics usually do not significantly degrade the achievable performance of signal detection. A possible reason for this may be the generalization capability of DNNs, which can extend a DNN trained according to a specific model to the application scenarios where constant channel changes may occur.

12.4.3 Convolutional Neural Network-Based Equalizers

A convolutional neural network (CNN) is a kind of neural network for processing data with known grid-like topologies. Rather than using fully connected layers, as in MLP, CNN is constituted by convolutional layers, as shown in Figures 12.7 and 12.8: a convolution stage, a detection stage, and a pooling stage are utilized to implement the convolution operations, activation operations, and down-samplings.

Figure 12.9 shows a CNN-based equalizer, which is similar to Figure 12.6 for a DNN-based equalizer from a data-flow point of view. Similar to the discussion in Section 12.4.2 for DNN-based equalizers, the training set can be generated on the basis of the channel model and parameters that are derived from practice. Furthermore, with the similar system model and the same optimization objective, a CNN-based equalizer can be trained in the same way as a DNN-based equalizer. Specifically, the $L_2(\cdot)$ loss function and the stochastic

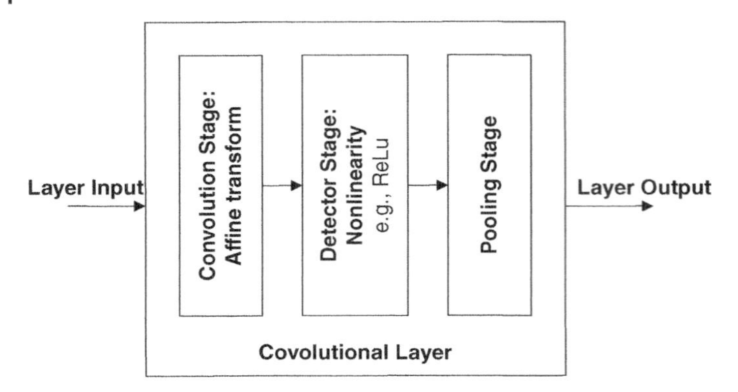

Figure 12.7 A typical convolutional layer.

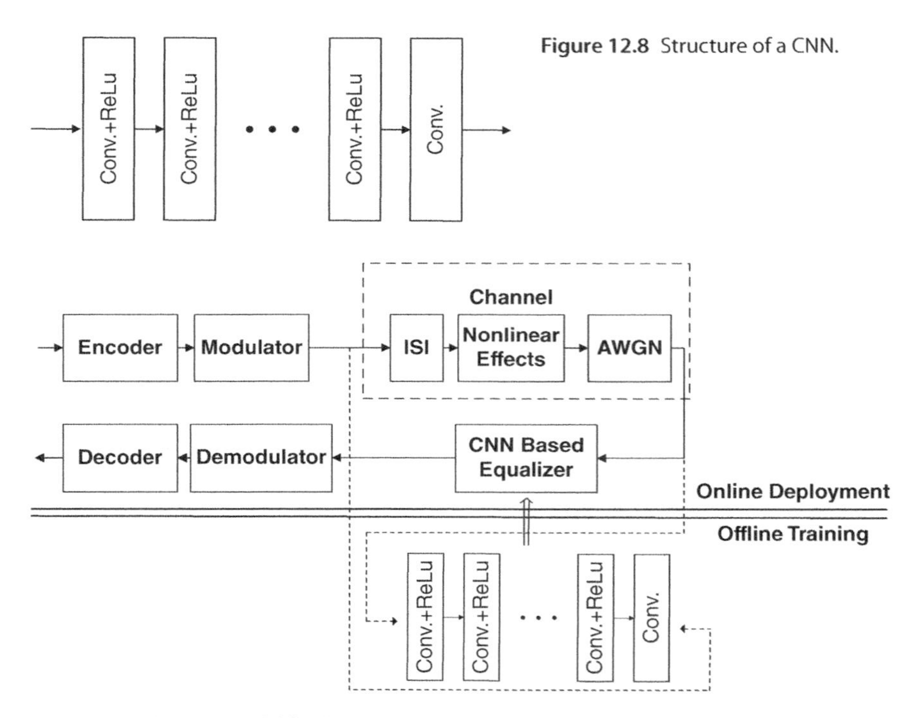

Figure 12.8 Structure of a CNN.

Figure 12.9 System model for CNN-based equalizer.

gradient descent algorithm with forward/backward propagations can be adopted to obtain the appropriate network parameters.

It is worth pointing out that the convolutional layer introduced by CNN-based equalizers has the following merits, which are beneficial to channel equalization:

1. Convolutional neural nodes only process data from a restricted subarea of the previous layer. This property agrees with the channel characteristic that ISI only exists between bits transmitted consecutively, while the nonlinear effect usually influences different bits independently (Xu et al., 2018).
2. CNNs have shift-invariant properties, which can be used to learn for matched filters and reduce temporal variations during signal detection (O'Shea et al., 2016; Li et al., 2017).

This is because the information recovery process in wireless communication systems is usually invariant to time shifting, scaling, rotation, linear mixing, and convolution through random filters.

12.4.4 Recurrent Neural Network-Based Equalizers

Recurrent neural networks (RNN)are another widely used neural network framework, and are powerful for one-dimensional data processing (Goodfellow et al., 2016). In RNNs, the current output of a hidden node depends not only on its current input but also on its past output. In other words, RNNs have memory, which enables them to model an infinite impulse response (IIR) filter. This property is important for the application of RNNs in channel equalization.

The RNN with three hidden recurrent units shown in Figure 12.10, where the neural nodes are represented by circles, was proposed in (Kechriotis et al., 1994) for channel equalization. In Figure 12.10, the left part is a folded computational graph for the RNN, while the right part is an unfolded version corresponding to the time series that guide the computation (Goodfellow et al., 2016).

RNNs can also be trained using the method for DNN- and CNN-based equalizers, described in Sections 12.4.2 and 12.4.3, respectively. Furthermore, the $L_2(\cdot)$ loss function and the stochastic gradient descent algorithm with forward/backward propagation can be adopted to obtain the appropriate network parameters.

In summary, with the aid of these neural networks, equalizers are able to extract the key features of wireless channels and reduce ISI adaptively and intelligently. Based on MLP, we can design DNN-based equalizers having the capability to mitigate ISI, thanks to embedded multiple-layered neural networks. Moreover, as shown in Figure 12.11, the channel-favorite properties of RNN and CNN facilitate equalizer design; owing to this, RNNs and CNNs have been widely used to combat ISI and achieve reliable communications. In the next section, we will present a range of simulation results to demonstrate the achievable performance of NN-based equalizers in OFDM systems.

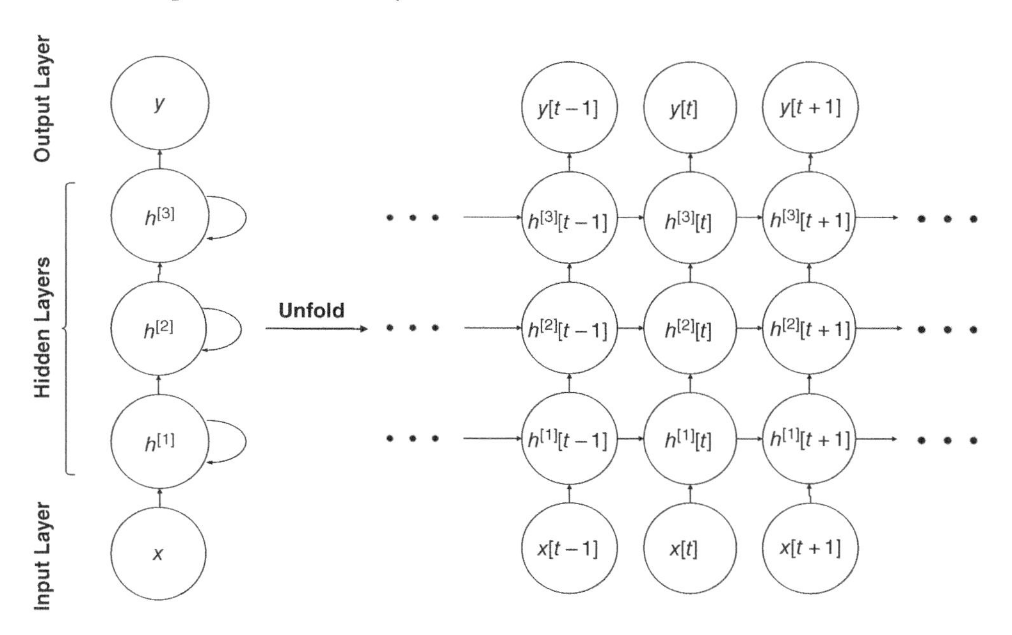

Figure 12.10 Structure of an RNN: folded and unfolded computational graph.

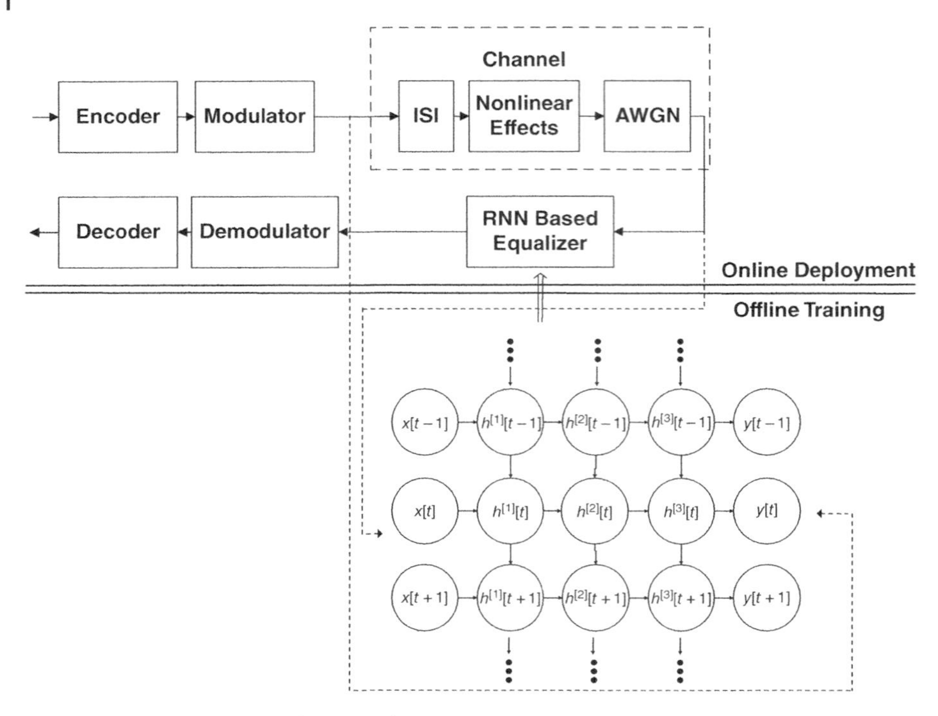

Figure 12.11 System model for a RNN-based equalizer.

12.5 Performance of OFDM Systems With Neural Network-Based Equalization

In this section, we investigate the performance of OFDM systems, which adopt DL techniques, including CNN and DNN, for channel equalization. Before providing the simulation performance, let us first describe the system model and network structure used for this performance investigation.

12.5.1 System Model and Network Structure

Figure 12.12 illustrates the structure of an OFDM system with a NN-based equalizer. The baseband OFDM system is identical to traditional ones (E. Dahiman and Skold, 2014). At the transmitter side, the transmitted QAM modulated symbols associated with the pilots for channel estimation are first converted to a parallel data representation. After that, the inverse fast Fourier transform (IFFT) module converts the data from the frequency domain to the time domain. Then, a cyclic prefix (CP) is inserted in order to avoid inter-block interference. Finally, after a parallel-to serial conversion, the signals are transmitted over the channel, which is usually dispersive.

Let the discrete sample-spaced multiple-path channel be expressed as $\{h(n)\}_{n=0}^{N-1}$, where the complex random variable $h(n)$ is the channel gain at instant n. Then the discrete received signal can be expressed as

$$y(n) = h(n) \otimes x(n) + w(n), \ n = 0, 1, \ldots \tag{12.38}$$

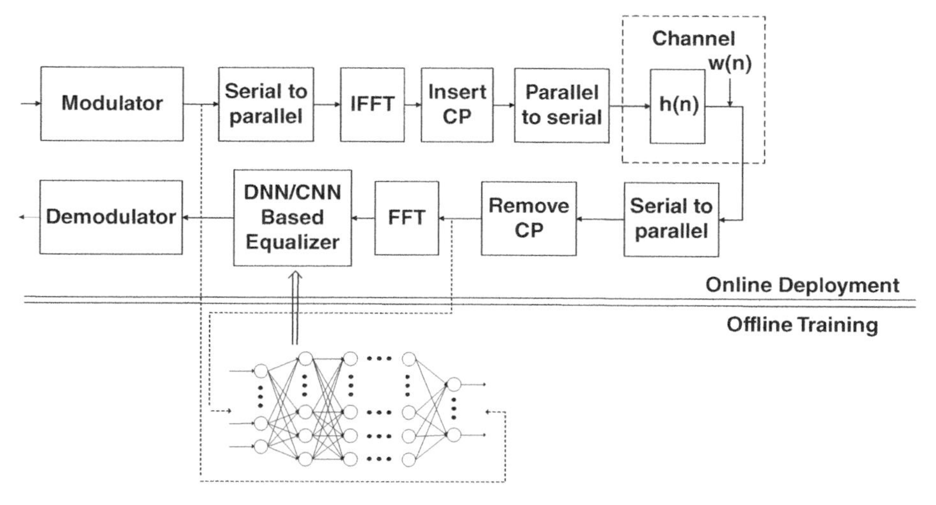

Figure 12.12 System model for OFDM with a DNN- or CNN-based equalizer.

where $\otimes$ denotes the circular convolution, $x(n)$ is the transmitted time-domain signal, and $w(n)$ is the AWGN.

As shown in Figure 12.12, at the receiver side, after the serial-to-parallel conversion and removing the cyclic prefix, the fast Fourier transform (FFT) converts the received signal from the time domain to the frequency domain. Hence, in the OFDM system considered, the DNN- or CNN-based equalizer operates in the frequency domain. Finally, received information is recovered by demodulation based on the output of the equalizer, as seen in Figure 12.12.

12.5.2 DNN and CNN Network Structure

For OFDM systems, Figure 12.13 depicts the specific structure of the DNN, which consists of six layers, four of them being hidden layers marked as ELU. In our study, the numbers of neurons in each layer are set to 256, 600, 400, 300, 200, and 128, respectively. The number of the input to the network is the number of real parts and that of imaginary parts of an OFDM frame containing both a pilot block and a data block. In contrast, the number of outputs equals the number of real parts plus the number of the imaginary parts of the data symbols derived from the data block.

In our performance study, the exponential linear unit (ELU) function (Goodfellow et al., 2016) is used as the activation function in all the layers other than the last layer. The ELU function is

$$f(\alpha, x) = f_{ELU}(\alpha, x) = \begin{cases} \alpha(e^x - 1), & x < 0 \\ x, & x \geq 0 \end{cases} \tag{12.39}$$

where α is a constant. Furthermore, at the output layer, as shown in Figure 12.13, the tanh function is applied to map the output received from the last hidden layer to the interval $[-1, 1]$

Figure 12.13 DNN network structure for OFDM systems.

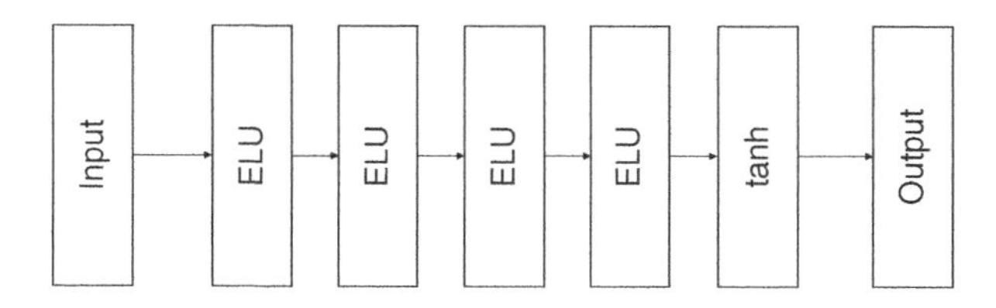

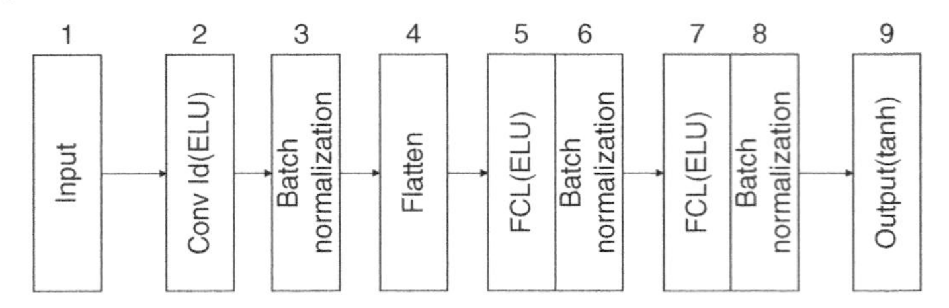

Figure 12.14 The CNN network structure for OFDM systems.

via the operation

$$g(x) = tanh(x) = \frac{e^x - e^{-x}}{e^x + e^{-x}}, \tag{12.40}$$

In the context of an OFDM system with a CNN-based equalizer, Figure 12.14 illustrates the structure of a nine-layer CNN. Thanks to the CNN's properties of facilitation of channel equalization, as mentioned in Section 12.4.3, a one-dimension (1D) convolution layer is introduced in the first hidden layer to extract the useful features of the received OFDM frames.

The inputs are OFDM frames, where every frame is constructed by the in-phase and quadrature-phase (IQ) parts. In order to improve the performance and stability of the CNN, batch normalization is always used by hidden layers to normalize the output. Then, a flattened layer is concatenated to flatten the input without affecting the batch size. After that, two identical sub-blocks are connected in tandem. Each sub-block consists of a fully connected layer (FCL) with ELU activation function and batch normalization. Finally, we use the tanh activation function to output data at the last layer.

To be more specific, the parameter settings of the CNN shown in Figure 12.14, which consists of one input layer, one output layer, and seven hidden layers, are given as follows. Note that the layer index has been labeled above each module. At the first layer, the inputs to the CNN are None $\times$ 2 $\times$ 128 3-dimension (3D) vectors, where "None" is the batch size, 2 is the width of the input, and 128 is the number of input channels. Subsequently, a 1D convolution operation with an ELU activation function is carried out in TensorFlow, where the number of kernels is set to 256. After the batch normalization and flattening operations, at the FCL layer, the number of neurons is set to 1024. Similarly, except for the batch normalization layers which do not require parameters, the neuron numbers of the following FCL and output tanh layers are 512 and 128, respectively.

12.5.3 Offline Training and Online Deployment

To obtain an effective NN for equalization, two stages are required: offline training and online deployment. In the offline training stage, we obtain training data by simulations based on a system with channel modeling. Specifically, for each simulation, a random data sequence is generated and modulated to form the transmitted symbols $\boldsymbol{s}$, which are then grouped into OFDM frames in combination with a sequence of pilot symbols. The pilot symbols are arranged in the first block of the OFDM frame, followed by the block of data symbols. We assume that the pilot symbols remain the same during both the training stage and the deployment stage. The channel used for training is obtained according to the channel models derived from practice measurements. We assume block channel fading, meaning the channel remains constant in spanning the pilot block and the data blocks of a frame, but changes from one frame to another independently. After undergoing channel distortions, the received signal $\boldsymbol{r}$ from the pilot block and data blocks of a frame as well as the transmitted symbols $\boldsymbol{s}$ are used, respectively, as the input and the expected output of the NN for its training.

As mentioned previously in Section 12.4, the NN is trained with the objective to minimize the difference between the actual output of the NN $\hat{\boldsymbol{s}}$ and the targeted output $\boldsymbol{s}$. Hence, in our experimental setting, the loss function to achieve this objective is

$$L_2(\hat{\boldsymbol{s}}, \boldsymbol{s}) = \| \hat{\boldsymbol{s}} - \boldsymbol{s} \|_2 \tag{12.41}$$

where $\hat{\boldsymbol{s}}$ is the actual output of the NN when $\boldsymbol{r}$ is the input of the NN, while $\boldsymbol{s}$ is the corresponding transmitted symbols, which also represent the targeted output of the NN.

12.5.4 Simulation Results and Analyses

In this subsection, we provide a range of BER performance results for the OFDM system with a DNN- or CNN-based equalizer. The BER performance of a DNN- or CNN-based equalizer is compared with that of the conventional least square (LS) method (Farzamnia et al., 2017) and that of the MMSE method (Edfors et al., 1998).

In our experiments, we assume an OFDM system with 64 subcarriers, 64 pilot symbols, and a CP length of 16. We assume 4-QAM for data modulation and a Rayleigh fading channel with a single path, whose real and imaginary parts are Gaussian distributed with zero mean and variance 0.5. We also consider the extended pedestrian A model (EPA) (3GPP, 2017) in which the number of paths is set to 7 and the maximum delay spread is 9 sampling periods. In addition, the extended typical urban model (ETU) (3GPP, 2017) with 9 paths and a maximum delay spread of 16 sampling periods is taken into account.

In Figure 12.15, we compare the BER performance of OFDM systems employing, respectively, our proposed DNN- and CNN-based equalizers with that of the OFDM systems with the traditional LS- and MMSE-based equalizers, when assuming communication over a single-path Rayleigh fading channel. Figure 12.15 shows that the LS method performs the worst among the four equalizers, with the result that it does not exploit the prior knowledge of the channel statistics, which is useful for channel equalization. In contrast, the MMSE method

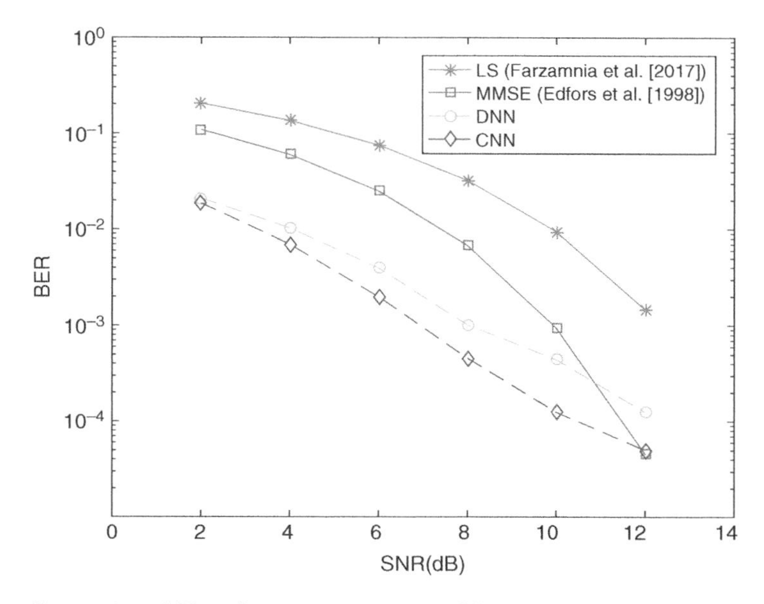

Figure 12.15 BER performance comparison of OFDM systems with DNN, CNN, LS, and MMSE-based equalizers, when communicating over single-path Rayleigh fading channels.

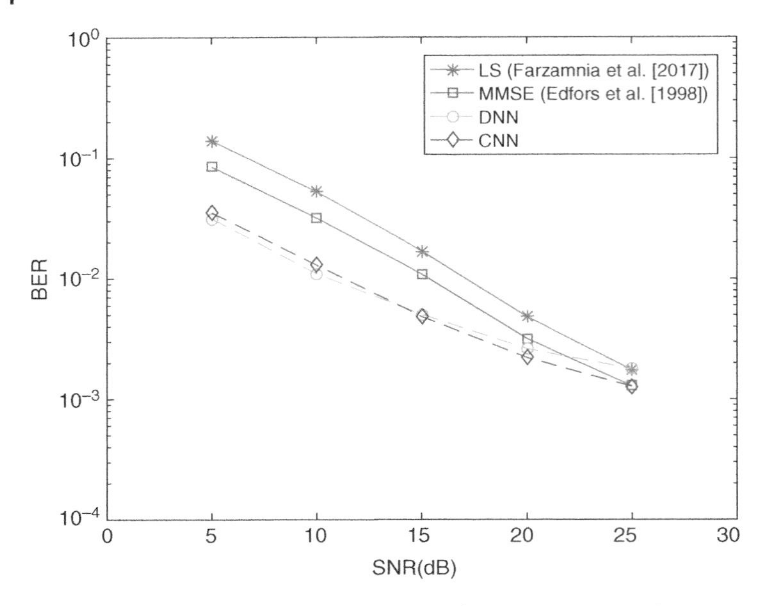

Figure 12.16 BER performance comparison of OFDM systems employing, respectively, DNN-, CNN-, LS-, and MMSE-based equalizers, when communicating over wireless channels following the EPA model (3GPP, 2017).

outperforms the LS method owing to utilizing the second-order statistics of the channel. It can be observed that both DNN- and CNN-based equalizers are capable of attaining better BER performance than the LS and MMSE methods from low to moderate SNRs. However, in the high SNR region, DNN- and CNN-based equalizers may be outperformed by the MMSE-based equalizer. Furthermore, the CNN-based equalizer achieves a lower BER than the DNN-based equalizer, thanks to the CNN's properties, which contribute more time-domain information of the signal and channel.

In Figure 12.15, the frequency non-selective fading channels were considered. In contrast, in Figures 12.16 and 12.17, we compare the BER performance of OFDM systems employing, respectively, the previously mentioned four equalizers, assuming communications over the frequency-selective fading channels modeled by the EPA model and the ETU model, respectively. As seen from the results, DNN- and CNN-based equalizers achieve better BER performance than the LS and MMSE methods over the main SNR regions considered in both the EPA and ETU channels. In the low SNR region, the BER of the DNN-based equalizer is either similar to or slightly lower than that of the CNN-based equalizer. In contrast, when the SNR is sufficiently high, the BER performance of the CNN-based equalizer is better than that of the DNN-based equalizer, when communicating over both the EPA and ETU channels. The observation implies that DNN-based equalizers are efficient for combating frequency-selective fading, but not as efficient as CNN-based equalizers for handling background noise. Furthermore, we should point out that CNN-based equalizers usually take much less time to converge than DNN-based equalizers.

12.6 Conclusions and Discussion

In this chapter, we first discussed various NN structures and learning methods in order to address the issues of channel equalization. Then, several representative NN algorithms were

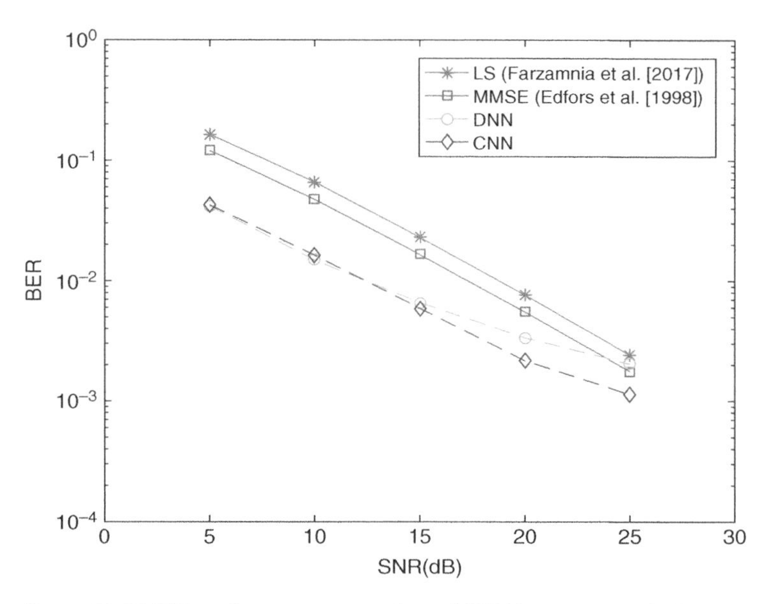

Figure 12.17 BER performance comparison of OFDM systems employing, respectively, DNN-, CNN-, LS-, and MMSE-based equalizers, when communicating over frequency-selective fading channels described by the ETU model (3GPP, 2017).

considered. Furthermore, two ML-based equalizers, DNN- and CNN-based equalizers, were presented, and their performance was investigated and compared with two conventional equalizers developed based on the LS and MMSE principles. Our study and performance results show that unlike traditional channel equalization and detection methods, no channel statistics are required to be separately computed in ML-based equalization methods. More importantly, the BER performance of OFDM systems employing ML-based equalizers is typically better than that of OFDM systems employing conventional LS- and MMSE-based equalizers. This implies that ML-based equalizers have a higher capability to learn and analyze complicated properties of wireless channels, and are also more effective to combat ISI. Meanwhile, ML-based equalization approaches exhibit better BER performance than LS- and MMSE-based equalization approaches, when communicating over channels with or without frequency-selective fading. Owing to their properties that no prior knowledge about the structure and characteristics of wireless channels are required, NN-based equalizers constitute promising candidates for channel equalization in future ultra-high-data-rate wireless communication systems.

Bibliography

3GPP. Evolved Universal Terrestrial Radio Access (E-UTRA); Base Station (BS) radio transmission and reception. Technical Specification (TS) 36.104, 3rd Generation Partnership Project (3GPP), 03 2017. URL www.3gpp.org. Version 14.3.0.

S. Arcens, J. Cid-Sueiro, and A.R. Figueiras-Vidal. Pao networks for data transmission equalization. In *International Joint Conference on Neural Networks (IJCNN)*, volume 2, pages 963–968, 1992.

K. Burse, R.N. Yadav, and S.C. Shrivastava. Channel equalization using neural networks: A review. *IEEE Transactions on Systems, Man and Cybernetics, Part C (Applications and Reviews)*, 40(3): 352–357, 2010.

I. Cha and S. Kassam. Channel equalization using adaptive complex radial basis function networks. *IEEE Journal on Selected Areas in Communications*, 13(1): 122–131, 1995.

P. Chang and B. Wang. Adaptive decision feedback equalization for digital satellite channels using multilayer neural networks. *IEEE Journal on Selected Areas in Communications*, 13(2): 316–324, 1995.

Y-J. Chang and C-L. Ho. Decision feedback equalizers using self-constructing fuzzy neural networks. In *2009 Fourth International Conference on Innovative Computing, Information and Control (ICICIC)*, pages 1483–1486. IEEE, 2009.

Y-J. Chang, S-S. Yang, and C-L. Ho. Fast self-constructing fuzzy neural network-based decision feedback equaliser in time-invariant and time-varying channels. *IET Communications*, 4(4): 463–471, 2010.

S. Chen, G.J. Gibson, C.F.N. Cowan, and P.M. Grant. Adaptive equalization of finite non-linear channels using multilayer perceptrons. *Signal Processing*, 20(2): 107–119, 1990.

S. Chen, B. Mulgrew, and S. McLaughlin. Adaptive Bayesian decision feedback equalizer based on a radial basis function network. In *SUPERCOMM/ICC '92 Discovering a New World of Communications*, pages 1267–1271, 1992.

S. Chen, B. Mulgrew, and S. McLaughlin. A clustering technique for digital communications channel equalization using radial basis function network. *IEEE Transactions on Neural Networks*, 4(4): 570–590, 1993a.

S. Chen, B. Mulgrew, and S. McLaughlin. Adaptive Bayesian equalizer with decision feedback. *Signal Processing*, 41(9): 2918–2927, 1993b.

S. Chen, S. Mclaughlin, and B. Mulgrew. Complex-valued radial basis function network, part II: Application to digital communications channel equalization. *Signal Processing*, 36: 165–188, 1994.

M. Chiu and C. Chao. Analysis of LMS-adaptive MLSE equalization on multipath fading channels. *IEEE Transactions on Communications*, 44(12): 1684–1692, 1996.

J. Cid-Sueiro and A.R. Figueiras-Vidal. Recurrent radial basis function networks for optimal blind equalization. In *Neural Networks for Signal Processing III - Proceedings of the 1993 IEEE-SP Workshop*, pages 562–571, 1993.

T.M. Cover. Geometrical and statistical properties of systems of linear inequalities with applications in pattern recognition. *IEEE transactions on electronic computers*, (3): 326–334, 1965.

S. Parkvall E. Dahiman and J. Skold. *4G: LTE/LTE-Advanced for mobile broadband*. Elsevier, 2014.

O. Edfors, M. Sandell, J. van de Beek, S.K. Wilson, and P.O. Borjesson. OFDM channel estimation by singular value decomposition. *IEEE Transactions on Communications*, 46(7): 931–939, 1998.

A. Farzamnia, N.W. Hlaing, M.K. Haldar, and J. Rahebi. Channel estimation for sparse channel OFDM systems using least square and minimum mean square error techniques. In *2017 International Conference on Engineering and Technology (ICET)*, pages 1–5, 2017.

M. Gao, Y. Guo, Z. Liu, and Y. Zhang. Feed-forward neural network blind equalization algorithm based on super-exponential iterative. In *2009 International Conference on Intelligent Human-Machine Systems and Cybernetics*, pages 335–338, 2009.

G.J. Gibson, S. Siu, and C.F. N. Cowan. Multilayer perceptron structures applied to adaptive equalisers for data communication. In *IEEE International Conference on Acoustic, Speech, and Signal Processing (ICASSP)*, pages 1183–1186, 1989a.

G.J. Gibson, S. Siu, and C.F.N. Cowan. Application of multilayer perceptrons as adaptive channel equalisers. In *Adaptive Systems in Control and Signal Processing*, pages 573–578, 1989b.

G.J. Gibson, S. Siu, and C.F.N. Cowan. The application of nonlinear structures to the reconstruction of binary signals. *IEEE Transactions on Signal Processing*, 39(8): 1877–1884, 1991.

A. Goldsmith. *Wireless Communications*. Cambridge University Press, 2005.

I. Goodfellow, Y. Bengio, and A. Courville. *Deep Learning*. MIT Press, 2016. www.deeplearningbook.org.
D.R. Guha and S.K. Patra. ISI and burst noise interference minimization using wilcoxon generalized radial basis function equalizer. In *2009 Fifth International Conference on MEMS NANO, and Smart Systems*, pages 89–92, 2009.
S. Haykin. *Adaptive Filter Theory*. Prentice Hall, 1986.
J-S.R. Jang, C-T. Sun, and E. Mizutani. *Neuro-Fuzzy and Soft Computing: A Computational Approach to Learning and Machine Intelligence*. Prentice Hall, 1997.
G. Kechriotis, E. Zervas, and E.S. Manolakos. Using recurrent neural networks for adaptive communication channel equalization. *IEEE Transactions on Neural Networks*, 5(2): 267–278, 1994.
E.A. Lee and D.G. Messerschmitt. *Digital Communication*. Kluwer Academic Publishing, 2nd Edition, 1994.
Y. Li, M. Chen, Y. Yang, M. Zhou, and C. Wang. Convolutional recurrent neural network-based channel equalization: An experimental study. In *2017 23rd Asia-Pacific Conference on Communications (APCC)*, pages 1–6, 2017.
R-C. Lin, W-D. Weng, and C-T. Hsueh. Design of an SCRFNN-based nonlinear channel equaliser. *IEEE Proceedings-Communications*, 152(6): 771–779, 2005.
B. Mulgrew. Applying radial basis functions. *IEEE Signal Processing Magazine*, 13: 50–65, 1996.
P.M. Olmos, J.J. Murillo-Fuentes, and F. Perez-Cruz. Joint nonlinear channel equalization and soft LDPC decoding with Gaussian processes. *IEEE Transactions on Signal Processing*, 58(3): 1183–1192, 2010.
T.J. O'Shea, J. Corgan, and T. C. Clancy. Convolutional radio modulation recognition networks. In *Engineering Applications of Neural Networks*, pages 213–226, 2016.
D.C. Park. Equalization for a wireless ATM channel with a recurrent neural network pruned by genetic algorithm. In *2008 Ninth ACIS International Conference on Software Engineering, Artificial Intelligence, Networking and Parallel/Distributed Computing, SNPD 2008 in conjunction with 2nd International Workshop on Advanced Internet Technology and Applications, AITA 2008*, pages 670–674, 2008.
J.C. Patra and A.C. Kot. Nonlinear dynamic system identification using Chebyshev functional link artificial neural networks. *IEEE Transactions on Systems, Man and Cybernetics, Part B (Cybernetics)*, 32(4): 505–511, 2002.
J.C. Patra and R.N. Pal. A functional link artificial neural network for adaptive channel equalization. *Signal Processing*, 43: 181–195, 1995.
J.C. Patra, R.N. Pal, R. Baliarsingh, and G. Panda. Nonlinear channel equalization for QAM signal constellation using artificial neural networks. *IEEE Transactions on Systems, Man and Cybernetics, Part B (Cybernetics)*, 29(2): 262–271, 1999.
J.C. Patra, W.B. Poh, N.S. Chaudhari, and A. Das. Nonlinear channel equalization with QAM signal using Chebyshev artificial neural network. In *IEEE International Joint Conference on Neural Networks*, pages 3214–3219, 2005.
J.C. Patra, W.C. Chin, P.K. Meher, and G. Chakraborty. Legendre-FLANN-based nonlinear channel equalization in wireless communication system. In *IEEE International Conference on Systems, Man and Cybernetics*, pages 1826–1831, 2008.
M.N. Patwary and P.B. Rapajic. Decision feedback per-survivor processing: New reduced complexity adaptive MLSE receiver. In *2005 Asia-Pacific Conference on Communications*, pages 802–806, 2005.
M. Peng, C.L. Nikias, and J.G. Proakis. Adaptive equalisers for PAM and QAM signals with neural network. In *Conference Record of the Twenty-Fifth Asliomar Conference on Signals, System & Computers*, pages 496–500, 1991.

M. Peng, C.L. Nikias, and J.G. Proakis. Adaptive equalization with neural networks: New multi-layer perceptron structures and their evaluation. In *IEEE International Conference on Acoustic, Speech, and Signal Processing (ICASSP)*, pages 301–304, 1992.

F. Perez-Cruz, A. Navia-Vazquez, P.L. Alarcon-Diana, and A. Artes-Rodriguez. SVC-based equalizer for burst TDMA transmissions. *Signal Processing*, 81(8): 1681–1693, 2001.

F. Perez-Cruz, J.J. Murillo-Fuentes, and S. Caro. Nonlinear channel equalization with Gaussian processes for regression. *IEEE Transactions on Signal Processing*, 56(10): 5283–5286, 2008.

S.S. Ranhotra, A. Kumar, M. Magarini, and A. Mishra. Performance comparison of blind and non-blind channel equalizers using artificial neural networks. In *2017 Ninth International Conference on Ubiquitous and Future Networks (ICUFN)*, pages 243–248, 2017.

T.S. Rappaport. *Wireless Communications: Principle and Practice*. Prentice Hall, 2nd Edition, 2002.

H. Rong, N. Sundararajan, G. Huang, and P. Saratchandran. Sequential adaptive fuzzy inference system (SAFIS) for non linear system identification and prediction. *Fuzzy Sets and Systems*, 157(9): 1260–1275, 2006.

H. Rong, G. Huang, N. Sundararajan, and P. Saratchandran. Online sequential fuzzy extreme learning machine for function approximation and classification problems. *IEEE Transactions on Systems, Man and Cybernetics, Part B (Cybernetics)*, 39(4): 1067–1072, 2009.

D.J. Sebald and J.A. Bucklew. Support vector machine techniques for nonlinear equalization. *IEEE Transactions on Signal Processing*, 48(11): 3217–3226, 2000.

E. Soria-Olivas, J.D. Martin-Guerrero, G. Camps-Valls, A.J. Serrano-Lopez, J. Calpe-Maravilla, and L. Gomez-Chova. A low-complexity fuzzy activation fuzzy activation function for artificial neural networks. *IEEE Transactions on Neural Networks*, 14(6): 1576–1579, 2003.

G.L. Stuber. *Principles of Mobile Communications*. Kluwer Academic Publishing, 1st Edition, 1999.

Y. Tan, J. Wang, and J.M. Zurada. Nonlinear blind source separation using a radial basis function network. *IEEE Transactions on Neural Networks*, 12(1): 124–134, 2001.

J. Tang, C. Deng, and G. Huang. Extreme learning machine for multilayer perceptron. *IEEE Transactions on Neural Networks And Learning Systems*, 27(4): 809–821, 2016.

M. Uesugi, K. Honma, and K. Tsubaki. Adaptive equalization in TDMA digital mobile radio. In *1989 IEEE Global Telecommunications Conference and Exhibition 'Communications Technology for the 1990s and Beyond'*, volume 1, pages 95–101, 1989.

A. Uncini and F. Piazza. Blind signal processing by complex domain adaptive spline neural networks. *IEEE Transactions on Neural Networks*, 14(2): 399–412, 2003.

Y. Wang, L. Yang, F. Wang, and L. Bai. Blind equalization using the support vector regression via PDF error function. In *2016 8th International Conference on Intelligent Human-Machine Systems and Cybernetics (IHMSC)*, volume 2, pages 212–216, 2016.

W.D. Weng and C.T. Yen. Reduced-decision feedback FLANN nonlinear channel equaliser for digital communication systems. *IEEE Proceedings-Communications*, 151(4): 305–311, 2004.

Y. Xiao, Y. Dong, and Z. Li. Blind equalization in underwater acoustic communication by recurrent neural network with bias unit. In *2008 Seventh World Congress on Intelligent Control and Automation (WCICA)*, pages 2407–2410, 2008.

N. Xie and H. Leung. Blind equalization using a predictive radial basis function neural network. *IEEE Transactions on Neural Networks*, 16(3): 709–720, 2005.

W. Xu, Z. Zhong, Y. Be'ery, X. You, and C. Zhang. Joint neural network equalizer and decoder. *arXiv:1807.02040v1*, 2018.

R. Yang, L. Yang, J. Zhang, C. Sun, W. Cong, and S. Zhu. Blind equalization of QAM signals via extreme learning machine. In *2018 Tenth International Conference on Advanced Computational Intelligence (ICACI)*, pages 34–39, 2018.

S. Yang, Y. Han, X. Wu, R. Wood, and R. Galbraith. A soft decodable concatenated LDPC code. *IEEE Transactions on Magnetics*, 51(11): 1–4, 2015.

H. Yao, S. Zhang, and Z. Wang. Blind equalization algorithm based on least squares support vector regressor. In *2011 International Conference on Electronic Mechanical Engineering and Information Technology*, volume 5, pages 2675–2678, 2011.

H. Ye and G.Y. Li. Initial results on deep learning for joint channel equalization and decoding. In *2017 IEEE 86th Vehicular Technology Conference (VTC-Fall)*, pages 1–5, 2017.

H. Ye, G.Y. Li, and B. Juang. Power of deep learning for channel estimation and signal detection in OFDM systems. *IEEE Wireless Communications Letters*, 7(1): 114–117, 2018.

C. You and D. Hong. Nonlinear blind equalization schemes using complex-valued multilayer feedforward neural networks. *IEEE Transactions on Neural Networks*, 9(6): 1442–1455, 1998.

X. Zhang, H. Wang, L. Zhang, and X. Zhang. A blind equalization algorithm based on bilinear recurrent neural network. In *2004 Fifth World Congress on Intelligent Control and Automation (WCICA) 2004*, pages 1982–1984, 2004.

13

Neural Networks for Signal Intelligence: Theory and Practice

Jithin Jagannath[1], Nicholas Polosky[1], Anu Jagannath[1], Francesco Restuccia[2], and Tommaso Melodia[2]

[1] *Marconi-Rosenblatt AI/ML Innovation Laboratory, ANDRO Computational Solutions LLC, New York, USA*
[2] *Department of Electrical and Computer Engineering, Northeastern University, Boston, USA*

The significance of robust wireless communication in both commercial and military applications is indisputable. The commercial sector struggles to balance limited spectral resources with the ever-growing bandwidth demand that includes multimedia support with specific quality of service (QoS) requirements. In tactical scenarios, it has always been challenging to operate in a hostile congested and contested environment. Both these scenarios can benefit from efficient spectrum-sensing and signal-classification capabilities. While this problem has been studied for decades, the recent rejuvenation of machine learning has made a significant footprint in this domain. Accordingly, this chapter aims to provide readers with a comprehensive account of how machine learning techniques, specifically artificial neural networks, have been applied to solve some of the key problems related to gathering signal intelligence. To accomplish this, we begin by presenting an overview of artificial neural networks. Next, we discuss the influence of machine learning on the physical layer in the context of signal intelligence. Thereafter, we discuss directions taken by the community towards hardware implementation. Finally, we identify the key hurdles associated with the applications of machine learning at the physical layer.

13.1 Introduction

According to the latest *Ericsson Mobility Report*, there are now 5.2 billion mobile broadband subscriptions worldwide, generating more than 130 exabytes per month of wireless traffic (Ericsson Incorporated, 2018). Moreover, it is expected that by 2020, over 50 billion devices will be absorbed into the Internet, generating a global network of “things” of dimensions never seen before (Cisco Systems, 2017). Given that only a few RF spectrum bands are available to wireless carriers (Federal Communications Commission [2016]), technologies such as radio frequency (RF) spectrum sharing through beamforming (Shokri-Ghadikolaei et al., 2016; Vázquez et al., 2018; Lv et al., 2018), dynamic spectrum access (DSA) (Jin et al., 2018; Chiwewe and Hancke, 2017; Jagannath et al., 2018a; Federated Wireless, 2018; Agarwal and De, 2016), and anti-jamming technologies (Zhang et al., 2017; Huang et al., 2017; Chang et al., 2017) will become essential in the near future.

Machine Learning for Future Wireless Communications, First Edition. Edited by Fa-Long Luo.
© 2020 John Wiley & Sons Ltd. Published 2020 by John Wiley & Sons Ltd.

Software-defined radios were introduced as a solution to the limitations associated with a rigid radio hardware design that prevents reconfigurability and operational flexibility. Equipping radios with the ability to learn and observe operational scenarios to make cognitive decisions can improve spectrum sharing and spectral situation awareness. Spectrum sharing will allow radios to sense and utilize unused/underutilized spectrums to avoid spectrum congestion caused by unintentional/intentional interference. Further, signal sensing and classification can bolster the spectral knowledge of the radios to reinforce and foster situational awareness. Commercial and tactical military operators can exploit this cognitive ability to maximize spectrum utility and provide more robust communications links.

The recent introduction of machine learning (ML) to wireless communications has in part to do with the newfound pervasiveness of ML throughout the scientific community and in part to do with the nature of the problems that arise in wireless communications. With the advent of advances in computing power and the ability to collect and store massive amounts of data, ML techniques have found their way into many different scientific domains in an attempt to put both of the aforementioned to good use. This concept is equally true in wireless communications. Additionally, problems that arise in wireless communication systems are frequently formulated as classification, detection, estimation, and optimization problems, all of which ML techniques can provide elegant and practical solutions to. In this context, the application of ML to wireless communications seems almost natural and presents a clear motivation (Bkassiny et al., 2013; Jiang et al., 2017; Chen et al.,).

The objective of this chapter is to provide detailed insight into the influence artificial neural networks (ANNs) have had on the physical layer. To begin, we provide an overview of ANNs in Section 13.2. In Section 13.3, we discuss the applications of ANNs to the physical layer specifically to acquire signal intelligence. Next, in Section 13.4, we discuss the implications of hardware implementations in the context of ML. Finally, in Section 13.5, we discuss the open problems that may be currently debilitating the application of ML in wireless systems.

13.2 Overview of Artificial Neural Networks

Before we begin, we would like to introduce some standard notations that will be used throughout this chapter. We use boldface uppercase and lowercase letters to denote matrices and column vectors, respectively. For a vector $\mathbf{x}$, x_i denotes the i-th element, $\|\mathbf{x}\|$ indicates the Euclidean norm, $\mathbf{x}^T$ its transpose, and $\mathbf{x} \cdot \mathbf{y}$ the Euclidean inner product of $\mathbf{x}$ and $\mathbf{y}$. For a matrix $\mathbf{H}$, H_{ij} will indicate the (i,j)-th element of $\mathbf{H}$. The notation $\mathcal{R}$ and $\mathcal{C}$ will indicate the set of real and complex numbers, respectively. The notation $\mathbb{E}_{x \sim p(x)}[f(x)]$ is used to denote the expected value, or average of the function $f(x)$ where the random variable x is drawn from the distribution $p(x)$. When a probability distribution of a random variable, x, is conditioned on a set of parameters, θ, we write $p(x; \theta)$ to emphasize the fact that θ parameterizes the distribution and reserve the typical conditional distribution notation, $p(x|y)$, for the distribution of the random variable x conditioned on the random variable y. We use the standard notation for operations on sets where $\cup$ and $\cap$ are the infix operators denoting the union and intersection of two sets, respectively. We use $S_k \subseteq S$ to say that S_k is either a strict subset of or equal to the set S and $x \in S$ to denote that x is an element of the set S. $\emptyset$ is used to denote the empty set and $|S|$ the cardinality of a set S. Lastly, the convolution operator is denoted as $*$.

13.2.1 Feedforward Neural Networks

The original formulation of feedforward neural networks was proposed by Rosenblatt (1962). It can be seen as an extension to the perceptron algorithm, originally developed by Rosenblatt

(1957), with an element-wise nonlinear transition function applied to the linear classifier. This nonlinear transition function allows the hyperplane decision boundary to take a nonlinear form, allowing the model to separate training data that is not linearly separable. The formulation for a single layer is as follows,

$$y = \sigma(\mathbf{w}^T\mathbf{x} + b) \tag{13.1}$$

where $\mathbf{x}$ is the training example input, y is the layer output, $\mathbf{w}$ are the layer weights, b is the bias. One common approach to handling the bias is to add an additional parameter to the weight vector and append a 1 to the input vector. When a bias term is omitted, this formulation can be assumed unless otherwise stated throughout the section.

The nonlinear transition function, σ, is also referred to the *activation function* throughout literature. This is often chosen from a handful of commonly used nonlinear functions for different applications. The most widely used activation functions are the following:

$$\sigma(z) = \frac{1}{1 + e^{-z}}, \tag{13.2}$$

$$ReLU = \max(0, z), \text{ and} \tag{13.3}$$

$$\tanh(z) = \frac{e^z - e^{-z}}{e^z + e^{-z}} \tag{13.4}$$

Additionally, the radial basis function (RBF) kernel function can be used as an activation function, and doing so gives rise to the radial basis function neural network (RBFNN), as introduced by Broomhead and Lowe (1988). To increase the complexity of the model, and thus its ability to learn more complex relationships between the input features, network layers can be subsequently added to the model that accept the previous layer's output as input. Doing so results in a deep neural network (DNN). The function of the network as a whole, $\phi(\mathbf{x})$, becomes,

$$\phi(\mathbf{x}) = \mathbf{W}^{(3)}\sigma(\mathbf{W}^{(2)}\sigma(\mathbf{W}^{(1)}\mathbf{x})) \tag{13.5}$$

where the weight matrices $\mathbf{W}^{(i)}$ are indexed according to the layer they belong to. Intuitively, this allows the first layer to learn linear functions between the input features, the second layer to learn nonlinear combinations of these functions, and the third layer to learn increasingly more complex nonlinear combinations of these functions. This formulation additionally gives rise to a nice graphical interpretation of the model, which is widely used in literature and given in Figure 13.1.

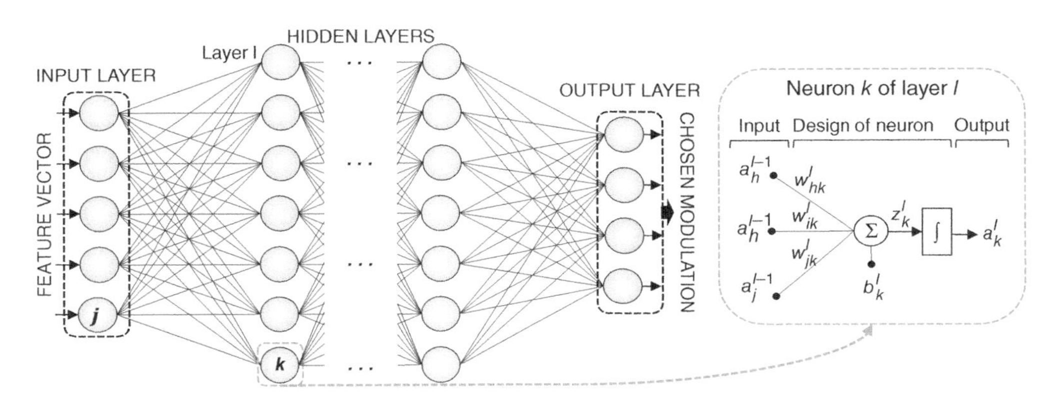

Figure 13.1 Standard framework of a feedforward neural network.

This graphical interpretation is also where the feedforward neural network gets its loose biological interpretation. Each solid line in Figure 13.1 denotes a weighted connection in the graph. The input, output, and hidden layers are denoted as such in the graph, and a close-up of one node in the graph is provided. This close-up calls the single node a neuron but it can equivalently be referred to simply as a *unit* in this text and throughout literature. The close-up also shows the inputs to the neuron, the weighted connections from the previous layer, the weighted sum of inputs, and the activation value, denoted as a_i^{l-1}, w_{ik}^l, z_k^l, and a_k^l, respectively. Occasionally, a neuron employing a given activation function may be referred to as that type of unit in this text and throughout literature, i.e. a unit with a *ReLU* activation function may be called a *ReLU unit*.

The most common way to train most types of neural networks is the optimization method called stochastic gradient descent (SGD). SGD is similar to well-known gradient descent methods with the exception that the true gradient of the loss function with respect to the model parameters is not used to update the parameters. Usually, the gradient is computed using the loss with respect to a single training example or some subset of the entire training set, which is typically referred to as a *mini-batch*, resulting in mini-batch SGD. This results in the updates of the network following a noisy gradient, which in fact often helps the learning process of the network by being able to avoid convergence on local minima that are prevalent in the non-convex loss landscapes of neural networks. The standard approach to applying SGD to the model parameters is through the repeated application of the chain rule of derivation using the famous back-propagation algorithm developed by Rumelhart et al. (1986).

The last layer in a given neural network is called the output layer. The output layer differs from the inner layers in that the choice of the activation function used in the output layer is tightly coupled with the selection of the loss function and the desired structure of the output of the network. Generally, the following discussion of output layers and loss functions applies to all neural networks, including the ones introduced later in this section.

Perhaps the simplest of output unit activation functions is that of the linear output function. It takes the following form,

$$\hat{y} = \mathbf{W}^T\mathbf{h} + b \tag{13.6}$$

where $\mathbf{W}$ is the output layer weight matrix, $\mathbf{h}$ are the latent features output from the previous layer, and $\hat{y}$ are the estimated output targets. Coupling a linear output activation function with a mean squared error loss function results in the maximizing the log-likelihood of the following conditional distribution,

$$p(y|\mathbf{x}) = N(y; \hat{y}, I) \tag{13.7}$$

Another task prominent among ML problems is that of binary classification. In a binary classification task, the output target assumes one of two values and thus can be characterized by a Bernoulli distribution, $p(y = 1|x)$. Since the output of a purely linear layer has a range over the entire real line, we motivate the use of a function that "squashes" the output to lie in the interval $[0, 1]$, thus obtaining a proper probability. The logistic *sigmoid* does exactly this and is, in fact, the preferred method to obtain a Bernoulli output distribution. Accordingly, the output layer becomes,

$$\hat{y} = \sigma(\mathbf{W}^T\mathbf{x} + b) \tag{13.8}$$

The negative log-likelihood loss function, used for maximum likelihood estimation, of this output layer is given as,

$$L(\theta) = -log(p(y|x)) = f((1 - 2y)z) \tag{13.9}$$

where $f(x) = log(1 + e^x)$ is called the *softplus* function and $z = \mathbf{W}^T\mathbf{x} + b$ is called the *activation value*. The derivation of Eq. (13.9) is not provided here but can be found in (Goodfellow et al., 2016) for the interested reader.

In the case when the task calls for a multi-class classification, we want a Multinoulli output distribution rather than a Bernoulli output distribution. The Multinoulli distribution assigns a probability that a particular example belongs to a particular class. Obviously, the sum over class probabilities for a single example should be equal to 1. The Multinoulli distribution is given as the conditional distribution: $\hat{y}_i = p(y = i|\mathbf{x})$. It is important to note that the output, $\hat{\mathbf{y}}$, is now an n-dimensional vector containing the probability that $\mathbf{x}$ belongs to class $i \in [0, n]$ at each index i in the output vector. The targets for such a classification task are often encoded as an n-dimensional vector containing $(n - 1)$ zeros and a single one, located at an index j denoting that the associated training example belongs to the class j. This type of target vector is commonly referred to as a *one-hot vector*. The output function that achieves the Multinoulli distribution in the maximum likelihood setting is called the *softmax* function and is given as,

$$softmax(\mathbf{z})_i = \frac{e^{\mathbf{z}}}{\sum_j e^{z_j}} \tag{13.10}$$

where z_j is the linear activation at an output unit j. Softmax output units are almost exclusively coupled with a negative log-likelihood loss function. Not only does this give rise to the maximum likelihood estimate for the Multinoulli output distribution, but the log in the loss function is able to undo the exponential in the softmax, which keeps the output units from saturating and allows the gradient to be well-behaved, allowing the learning to proceed (Goodfellow et al., 2016).

13.2.2 Convolutional Neural Networks

The convolutional neural network (CNN) was originally introduced by LeCun et al. (1989) as a means to handle grid-like input data more efficiently. Input of this type could be in the form of a time-series but is more typically found as image-based input. The formulation of CNNs additionally has biological underpinnings related to the human visual cortex.

CNNs are very similar to the feedforward networks introduced previously, with the exception that they use a convolution operation in place of a matrix multiplication in the computation of a unit's activation value. In this section, we assume that the reader is familiar with the concept of the convolution operation on two continuous functions, where one function, the input function, is convolved with the convolution kernel. The primary differences from the aforementioned notion of convolution and convolution in the CNN setting are that the convolution operation is discretized (for practical implementation purposes) and that it is often truly the cross-correlation operation that is performed in CNNs rather than true convolution. This means the kernel is not typically flipped before convolving it with the input function. This is primarily done for practical implementation purposes and does not typically affect the efficacy of the CNN in practice.

Convolution in the context of CNNs is thus defined as the following, for an input image I,

$$S(i,j) = (K * I)(i,j) = \sum_m \sum_n I(m,n)K(i - m, j - n) \tag{13.11}$$

where K is the convolution kernel, and the output, S, is often referred to as the *feature map* throughout literature. It is important to note that this formulation is for two-dimensional convolution but can be extended to input data of different dimensions. The entries of K can be seen as analogous to the weight parameters described previously (Section 13.2.1) and can be learned in a similar manner using SGD and the back-propagation (BP) algorithm Rumelhart et al. (1986). Intuitively, one can imagine having multiple K kernels in a single CNN layer being analogous to having multiple neurons in a single feedforward neural network layer. The output feature maps will be grid-like and subsequent convolutional layers can be applied to these feature maps after the element-wise application of one of the aforementioned nonlinear activation functions.

In addition to convolutional layers, CNNs often employ a separate kind of layer called *pooling layers*. The primary purpose of a pooling layer is to replace the output of the network at a certain location with a type of summarization of the outputs within a local neighborhood. Examples of pooling layers include max pooling (Zhou and Chellappa, 1988), average pooling, L^2 norm pooling, and distance weighted average pooling. A max pooling layer would summarize some rectangular region of the input image by selecting only the maximum activation value present in the region as output from the pooling layer. Pooling layers improve the efficacy of CNNs in a few different ways. First, they help make the learned representation of the input invariant to small translations, which is useful when aiming to determine the presence of a feature in the input rather than its location. Second, pooling layers help condense the size of the network since convolutional layers do not inherently do so. A binary classification task taking image data with size $256 \times 256 \times 3$ will need to reduce the size of the net to a single output neuron to make use of the output layer and cost function pairs described previously in Section 13.2.1. Lastly, pooling layers lead to infinitely strong prior distributions, making the CNN more statistically efficient (Goodfellow et al., 2016).

Some common adaptations applied to CNNs come in the form of allowing information flow to skip certain layers within the network. While the following adaptions were demonstrated on CNNs and LSTM (a type of RNN), the concepts can be applied to any of the networks presented in this chapter. A RN, or ResNet (He et al., 2015), is a neural network that contains a connection from the output of a layer, say L_{i-2}, to the input of the layer L_i. This connection allows the activation of the L_{i-2} layer to skip over the layer L_{i-1} such that a "residual function" is learned from layer L_{i-2} to layer L_i. The RN uses an identity operation on the activation of the L_{i-2} layer, meaning the values are unchanged, prior to adding them to the values input to layer L_i. Conversely, a highway neural network (Srivastava et al., 2015), allows a similar skip connection over layers but additionally applies weights and activation functions to these connections so a nonlinear relationship can be learned. Lastly, a dense neural network (Huang et al., 2016) is a network that employs such weighted connections between each layer and all of its subsequent layers. The motivation behind each of these techniques is similar in that they attempt to mitigate learning problems associated with vanishing gradients (Hochreiter et al., 2001). For each of these networks, the BP algorithm that Rumelhart et al. (1986) used must be augmented to incorporate the flow of error over these connections.

13.3 Neural Networks for Signal Intelligence

ML techniques for signal intelligence typically manifest themselves as solutions to discriminative tasks. That is, many applications focus on multi-class or binary classification tasks. Perhaps the most prevalent signal intelligence task solved using ML techniques is that of automatic modulation classification (AMC). In short, this task involves determining what scheme was used to modulate the transmitted signal, given the raw signal observed at the receiver. Other signal intelligence tasks that employ ML solutions include wireless interference classification. In this section, different state-of-the-art ML solutions to these signal intelligence tasks are discussed in further detail.

13.3.1 Modulation Classification

Deep learning (DL) solutions to modulation classification tasks have received significant attention in the last two years (O'Shea et al., 2018; O'Shea and Hoydis, 2017; Wang et al., 2017; West and O'Shea, 2017; Kulin et al., 2018; Karra et al., 2017). O'Shea et al. (2018) present several DL models to address the modulation recognition problem, while Karra et al. (2017) train hierarchical deep neural networks to identify data type, modulation class, and modulation order. Kulin et al. (2018) present a conceptual framework for end-to-end wireless DL, followed by a comprehensive overview of the methodology for collecting spectrum data, designing wireless signal representations, forming training data, and training deep neural networks for wireless signal classification tasks.

The task of AMC is pertinent in signal intelligence applications as the modulation scheme of the received signal can provide insight into what type of communication frameworks and emitters are present in the local RF environment. The problem at large can be formulated as estimating the conditional distribution, $p(y|x)$, where y represents the modulation structure of the signal and x is the received signal.

Traditionally, AMC techniques are broadly classified as maximum likelihood–based approaches (Ozdemir et al., 2013, 2015; Wimalajeewa et al., 2015; Foulke et al., 2014; Jagannath et al., 2015), feature-based approaches (Azzouz and Nandi, 1996; Hazza et al., 2012; Kubankova et al., 2010), and hybrid techniques (Jagannath et al., 2017). Prior to the introduction of ML, AMC tasks were often solved using complex hand-engineered features computed from the raw signal. While these features alone can provide insight about the modulation structure of the received signal, ML algorithms can often provide a better generalization to new unseen datasets, making their employment preferable over solely feature-based approaches. The logical remedy to the use of complex hand-engineered feature-based classifiers are models that aim to learn directly from received data. Recent work done by O'Shea and Corgan (2016) shows that deep convolutional neural networks (DCNNs) trained directly on complex time-domain signal data outperform traditional models using cyclic moment feature-based classifiers. In the work done by Shengliang Peng and Yao (2017), the authors propose a DCNN model trained on two-dimensional constellation plots generated from received signal data and show that their approach outperforms other approaches using cumulant based classifiers and SVMs (support vector machines).

While strictly feature-based approaches may become antiquated with the advent of the application of ML to signal intelligence, expert feature analysis can provide some useful inputs to ML algorithms. In (Jagannath et al., 2018b), we compute hand-engineered features directly from the raw received signal and apply a feedforward neural network classifier to the features to provide an AMC. The discrete time complex valued received signal can be represented as

$$y(n) = h(n)x(n) + w(n), \qquad n = 1, ..., N \tag{13.12}$$

where $x(n)$ is the discrete-time transmitted signal, $h(n)$ is the complex valued channel gain that follows a Gaussian distribution, and $w(n)$ is the additive complex zero-mean white Gaussian noise process at the receiver with two-sided power spectral density (PSD) $N_0/2$. The received signal is passed through an automatic gain control prior to the computation of feature values.

The first feature value computed from the received signal is the variance of the amplitude of the signal and is given by

$$Var(|y(n)|) = \frac{\sum_{N_s}(|y(n)| - \mathbb{E}(|y(n)|))^2}{N_s} \tag{13.13}$$

where $|y(n)|$ is the absolute value of the over-sampled signal and $\mathbb{E}(|y(n)|)$ represents the mean computed from N_s samples. This feature provides information that helps distinguish frequency shift keying (FSK) modulations from the phase shift keying (PSK) and quadrature amplitude modulation (QAM) modulation structures also considered in the classification task. The second and third features considered are the mean and variance of the maximum value of the power spectral density of the normalized centered-instantaneous amplitude, which is given as

$$\gamma_{max} = \frac{max|FFT(a_{cn}(n))|^2}{N_s}, \tag{13.14}$$

where $FFT(.)$ represents the fast Fourier transform (FFT) function, $a_{cn}(n) \triangleq \frac{a(n)}{m_a} - 1$, $m_a = \frac{1}{N_s}\sum_{n=1}^{N_s} a(n)$, and $a(n)$ is the absolute value of the complex-valued received signal. This feature provides a measure of the deviation of the PSD from its average value. The mean and variance of this feature computed over subsets of a given training example are used as two separate entries in the feature vector input into the classification algorithm, corresponding to the second and third features, respectively.

The fourth feature used in our work was computed using higher-order statistics of the received signal: cumulants, which are known to be invariant to the various distortions commonly seen in random signals and are computed as follows,

$$C_{lk} = \sum_{p}^{\text{No. of partitions in } l} (-1)^{p-1}(p-1)! \prod_{j=1}^{p} \mathbb{E}\{y^{l_j-k_j} y^{*k_j}\}, \tag{13.15}$$

where l denotes the order and k denotes the number of conjugations involved in the computation of the statistic. We use the ratio C_{40}/C_{42} as the fourth feature, which is computed using

$$C_{42} = \mathbb{E}(|y|^4) - |\mathbb{E}(y^2)|^2 - 2\mathbb{E}(|y|^2)^2, \tag{13.16}$$

$$C_{40} = \mathbb{E}(y^4) - 3\mathbb{E}(y^2)^2. \tag{13.17}$$

The fifth feature used in our work is called the in-band spectral variation as it allows discrimination between the FSK modulations considered in the task. We define $Var(f)$ as

$$Var(f) = Var\left(\mathcal{F}\left(y(t)\right)\right), \tag{13.18}$$

where $\mathcal{F}(y(t)) = \left\{Y(f) - Y(f - F_0)\right\}_{f=-f_i}^{+f_i} / F_0$, F_0 is the step size, $Y(f) = FFT(y(t))$, and $[-f_i, +f_i]$ is the frequency band of interest.

The final feature used in the classifier is the variance of the deviation of the normalized signal from the unit circle, which is denoted as $Var(\Delta_o)$. It is given as

$$\Delta_o = \frac{|y(t)|}{\mathbb{E}(|y|)} - 1. \tag{13.19}$$

This feature helps the classifier discriminate between PSK and QAM modulation schemes.

The modulations considered in the work are the following: binary phase shift keying (BPSK), quadrature phase shift keying (QPSK), 8-PSK, 16-QAM, continuous phase frequency shift keying (CPFSK), Gaussian frequency shift keying (GFSK), and Gaussian minimum shift keying (GMSK). This characterizes a seven-class classification task using the aforementioned six features computed from each training example. To generate the dataset, a total of 35,000 examples were collected: 1,000 examples for each modulation at each of the five SNR scenarios considered in the work. Three different feedforward neural network structures were trained at each SNR scenario using a training set consisting of 80% of the data collected at the given SNR and a test set consisting of the remaining 20%. The three feedforward nets differed in the number of hidden layers, ranging from one to three. Qualitatively, the feedforward network with one hidden layer outperformed the other models in all but the least favorable SNR scenario, achieving the highest classification accuracy of 98% in the most favorable SNR scenario. The seemingly paradoxical behavior is attributed to the over-fitting of the training data when using the higher-complexity models, leading to poorer generalization in the test set.

This work has been further extended to evaluate other ML techniques using the same features. Accordingly, we found that training a random forest classifier for the same AMC task yielded similar results to the feedforward network classifier. Additionally, the random forest classifier was found to outperform the DNN approach in scenarios when the exact center frequency of the transmitter was not known, which was assumed to be given in Jagannath et al. (2018b). The random forest classifier comprised 20 classification and regression trees (CARTs) constructed using the gini impurity function. At each split, a subset of the feature vectors with cardinality equal to 3 was considered.

An alternative approach to the previously described method is to learn the modulation of the received signal from different representations of the raw signal. Kulin et al. (2018) train DCNNs to learn the modulation of various signals using three separate representations of the raw received signal. In the work, the raw complex valued received signal training examples are denoted as $r_k \in C^N$, where k indexes the procured training dataset and N is the number of complex valued samples in each training example. We inherit this notation for presentation of their findings. The dataset in the work was collected by sampling a continuous transmission for a period of time and subsequently segmenting the received samples into N dimensional data vectors.

Kulin et al. (2018) train separate DCNNs on three different representations of the raw received signal and compare their results to evaluate which representation provides the best classification accuracy. The first of the three signal representations are given as a $2 \times N$ dimensional in-phase/quadrature (I/Q) matrix containing real-valued data vectors carrying the I/Q information of the raw signal, denoted x_i and x_q, respectively. Mathematically,

$$x_k^{IQ} = \begin{bmatrix} x_i^T \\ x_q^T \end{bmatrix} \tag{13.20}$$

where $x_k^{IQ} \in R^{2\times N}$. The second representation used is a mapping from the complex values of the raw received signal into two real-valued vectors representing the phase, Φ and the magnitude, A,

$$x_k^{A/\Phi} = \begin{bmatrix} x_A^T \\ x_\Phi^T \end{bmatrix} \tag{13.21}$$

where $x_k^{A/\Phi} \in R^{2\times N}$ and the phase vector $x_\Phi^T \in R^N$ and magnitude vector $x_A^T \in R^N$ have elements

$$x_{\Phi_n} = \arctan\left(\frac{r_{q_n}}{r_{i_n}}\right), x_{A_n} = (r_{q_n}^2 + r_{i_n}^2)^{\frac{1}{2}} \tag{13.22}$$

respectively. The third representation is a frequency domain representation of the raw time-domain complex signal. It is characterized by two real-valued data vectors, one containing the real components of the complex FFT, $\Re(X_k)$, and the other containing the imaginary components of the complex FFT, $\Im(X_k)$, giving

$$x_k^F = \begin{bmatrix} \Re(X_k)^T \\ \Im(X_k)^T \end{bmatrix} \tag{13.23}$$

Using these three representations of the raw signal, three DCNNs with identical structure are trained on each representation, and the accuracy of the resultant models is compared to determine which representation allows for learning the best mapping from raw signal to modulation structure.

Each training example comprised $N = 128$ samples of the raw signal sampled at 1 MS/s (mega-samples per seconds), and the following 11 modulation formats were considered in the classification task: BPSK, QPSK, 8-PSK, 16-QAM, 64-QAM, CPFSK, GFSK, 4-PAM, wideband frequency modulation (WBFM), amplitude modulation–double-sideband modulation (AM-DSB), and amplitude modulation–single-sideband modulation (AM-SSB). Thus, the training targets $y_k \in R^{11}$ are encoded as one-hot vectors where the index holding an i corresponds to the modulation of the signal. A total of 220,000 training examples $x_k \in R^{2\times128}$ were acquired uniformly over different SNR scenarios ranging from $-20dB$ to $+20dB$.

The DCNN structure used for each signal representation is the same and consists of two convolutional layers, a fully connected layer, and a softmax output layer trained using the negative log-likelihood loss function. The activation function used in each of the convolutional layers and the fully connected layer is the ReLU function. The DCNNs were trained using a training set comprising 67% of the total dataset, with the rest of the dataset being used as test and validation sets. An Adam optimizer (Kingma and Ba, 2014) was used to optimize the training processes for a total of 70 epochs. The metrics used to evaluate each of the models include the precision, recall, and F1 score of each model. In the work, a range of values is provided for the three aforementioned metrics for the CNN models trained on different data representations for three different SNR scenarios: high, medium, and low, corresponding to $18dB$, $0dB$, and $-8dB$, respectively. In the high SNR scenario, it is reported that the precision, recall, and F1 score of each of the three CNN models fall in the range of 0.67–0.86. For the medium and low SNR scenarios, the same metrics are reported in the ranges of 0.59–0.75 and 0.22–0.36, respectively. This relatively low performance can be attributed to the choice of the channel model used when generating the data, namely, a time-varying multipath fading channel.

Furthermore, the evaluation criteria also includes the classification accuracy of each of the three models trained using different data representations under similar SNR conditions. Qualitatively, each of the three DCNN models performs similarly at low SNR, while the DCNN trained on the I/Q representation of data yields a better accuracy at medium SNR, and the DCNN trained on the amplitude and phase representation yields a better accuracy at high SNR. Interestingly, the DCNN trained on the frequency domain representation of the data performs significantly worse than the I/Q and A/ϕ DCNNs at high SNR. This could potentially be due to similar characteristics exhibited in the frequency domain representation of the PSK and QAM modulations used in the classification problem. The primary takeaways from this work are that learning to classify modulation directly from different representations of the raw signal can be an effective means of developing a solution to the AMC task; however, the efficacy of the classifier is dependent on how the raw signal is represented to the learning algorithm.

13.3.2 Wireless Interference Classification

Wireless interference classification (WIC) is a classification task that is concerned with identifying what type of wireless emitter exists in the environment. The motivation behind such a task is that it can often be helpful to know what type of emitters are present (WiFi, Zigbee, Bluetooth, etc.) so that you can effectively attempt to avoid interference and coexist with other emitters sharing the resources. Recent work done by Selim et al. (2017) shows the use of DCNNs to classify radar signals using both spectrogram and amplitude-phase representations of the received signal. In the work presented by Akeret et al. (2017), DCNN models are proposed to accomplish interference classification on two-dimensional time-frequency representations of the received signal to mitigate the effects of radio interference in cosmological data. Additionally, the authors of Czech et al. (2018) employ DCNN and LSTM models to achieve a similar end.

Kulin et al. (2018) propose to employ DCNNs for the purpose of the wireless interference classification of three different wireless communication systems based on the WiFi, Zigbee, and Bluetooth standards. They look at 5 different channels for each of the 3 standards and construct a 15-class classification task for which they obtain 225,225 training vectors consisting of 128 samples each, where samples were collected at 10 MS/s. A flat fading channel with additive white Gaussian noise is assumed for this classification task.

Three DCNNs were trained and evaluated using the wireless interference classification dataset described. Each of the three DCNNs was trained on one of the representations of the data that were presented in the previous section that discussed AMC. The DCNN architectures were also the same as presented previously in Section 13.3.1.

Each of the three DCNNs trained using different data representations were evaluated in a similar fashion to the evaluation method described in Section 13.3.1, namely, using precision, recall, and F1 score under different SNR scenarios. For the wireless interference classification task, the precision, recall, and F1 score of each of the three DCNNs all fell in the interval from 0.98–0.99 under the high SNR scenario. For the medium and low SNR scenarios, the analogous intervals were from 0.94–0.99 and 0.81–0.90, respectively.

Additionally, Kulin et al. (2018) provide an analysis of classification accuracy for each of the three DCNN models at varying SNRs. For the task of wireless interference classification, the DCNN model trained on the frequency-domain representation of the data outperforms the other models at all SNRs and especially so in lower SNR scenarios. These findings are due to the fact that the wireless signals that were considered have more expressive features in the frequency domain as they have different bandwidth, modulation, and spreading characteristics.

Youssef et al. (2017) take a different approach to the wireless interference classification task and primarily compare different types of learning models rather than different types of data representation. The proposed models include deep feedforward networks, deep convolutional networks, support vector machines using two different kernels, and a MST (minimum-spanning-tree) algorithm using two different learning algorithms. In the work, 12 different transmitters are considered, and 1,000 packets from each transmitter are collected for a total of 12,000 packets forming the entire dataset. Each transmitter transmitted the same exact 1,000 packets, which were generated using pseudo-random values injected into the modem. All of the transmitters used a proprietary orthogonal frequency-division multiplexing (OFDM) protocol with a QPSK modulation scheme and a baseband transmitter sample rate of 1.92 MS/s. At the receiver, each packet is represented by 10,000 time-domain I/Q samples. Each of the models was trained on datasets consisting of training examples made up of 32, 64, 128, 256, 512, and 1024 samples from each packet, and their performance is compared across datasets. Given the complex-valued received signal,

$$f = (f_1, f_2, ..., f_N) \tag{13.24}$$

N samples were selected by skipping the first N_0 samples of a packet where $|\mathcal{R}(f_i)| < \tau$ for some $\tau > 0$ yielding the signal vector g,

$$g = (f_{N_0}, f_{N_0+1}, ..., f_{N_0+N-1}) \tag{13.25}$$

For the DNN, SVM, and MST models, each training example was constructed by concatenating the real and imaginary parts of the signal vector, yielding a vector of dimension $2N$. For the DCNN model, the real and imaginary parts of the signal vector were stacked to generate $2 \times N$ dimensional training vectors.

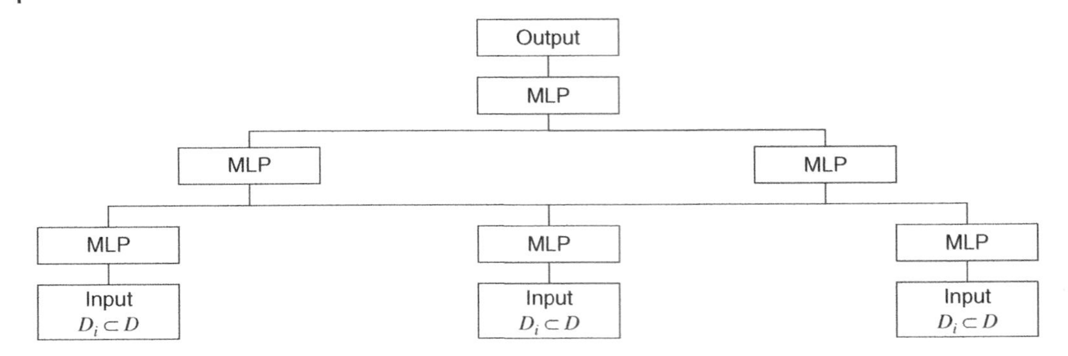

Figure 13.2 Adaptation of MST MLP used by Youssef et al. (2017).

The DNN architecture considered in the work consisted of two fully connected hidden layers, consisting of 128 ReLU units each and an output layer consisting of logistic sigmoid units. The network was trained using the Adam optimizer (Kingma and Ba, 2014) and a mini-batch size of 32.

The DCNN model used by the authors was composed of two convolutional layers using 64 (8×2) and 32 (16×1) filters, respectively. Each convolutional layer was input into a max-pool layer with a pool size of 2×2 and 2×1, respectively. The output of the second max-pool layer was fed into a fully connected layer consisting of 128 ReLU units. An output layer employing logistic sigmoid units was used on top of the fully connected layer.

The two SVM architectures analyzed in the work differ only in the kernel function used. The first architecture employed the polynomial kernel and the second employed the Pearson VII universal kernel (Üstün et al., 2005). Both architectures used Platt's minimization optimization algorithm to compute the maximum-margin hyperplanes.

Furthermore, an analysis of the performance of MST multi-layer perceptrons (MLPs) trained using first-order and second-order methods is provided. A high-level description of MST MLPs is presented here, and we refer the interested reader to Youssef et al. (2015) for a more rigorous derivation. The MST method to training neural networks, as presented in the work, is essentially a hierarchical way to solve an optimization problem by solving smaller constituent optimization problems. To this end, in what is called the first stage, a number of separate MLPs would be trained on different subsets of the training dataset. This can be seen in the lowest layer of the hierarchical representation adapted from Youssef et al. (2017) and provided in Figure 13.2.

Once the first stage is trained, a second stage is trained by taking the concatenation of the network outputs from the first stage as input. Training can continue in this fashion for subsequent stages. One of the advantages of training networks in this way is that the many smaller MLPs that form the larger classifier can be efficiently trained using second-order optimization methods. Second-order optimization methods such as Newton, Gauss-Newton, and Levenberg-Marquardt methods are usually intractable due to the size of typical networks but can provide better convergence when applicable. Two three-stage MST systems were trained in the work, one using the first-order method of SGD, and the other using the second-order accelerated Levenberg-Marquardt method (K. Youssef). Each MST system had an identical structure where stage 1 consisted of 60 MLPs with 2 hidden layers and 10 units in each layer. Stages 2 and 3 had the same architecture and consisted of 30 MLPs, with each MLP consisting of 2 hidden layers made up of 15 units each. All hidden units employed the tanh activation function, and all output layers contained linear units.

All of the models described were trained on 10 different iterations of the collected dataset, and their performance was compared. Five datasets were constructed using training examples

Table 13.1 Summary of ML solutions for signal intelligence

Classifiers	Task	Representation	Model
Jagannath et al. (2018b)	AMC	Feature-based	DNN
Kulin et al. (2018)	AMC	I/Q, A/Φ, FFT	DCNN
O'Shea and Corgan (2016)	AMC	I/Q	DCNN
Shengliang Peng and Yao (2017)	AMC	Constellation	DCNN
Kulin et al. (2018)	WIC	I/Q, A/Φ, FFT	DCNN
Selim et al. (2017)	WIC	2D time-frequency, A/Φ	DCNN
Akeret et al. (2017)	WIC	2D time-frequency	DCNN
Czech et al. (2018)	WIC	2D time-frequency	DCNN, LSTM
Youssef et al. (2017)	WIC	I/Q	DNN, DCNN, SVM, MST

made up of 32, 64, 128, 256, and 512 samples, and then each model was trained twice, using a training set consisting of 90% and 10% of the total dataset, for a total of 10 different datasets for each model. In general, the MST system trained using second-order methods on 90% of the training data performed best across all sizes of training examples, yielding a classification accuracy of 100% for each dataset. All of the models performed better when trained using 90% of the dataset as opposed to 10% of the training dataset. Generally, each model performed better when provided with training examples that contained more samples, with the exception of the deep feedforward network model, which could be attributed to the fact that longer sequences of samples may contain an increasing number of artifacts to which the DNN may not be robust.

A summarization of the different models presented in this section is provided in Table 13.1.

13.4 Neural Networks for Spectrum Sensing

One of the key challenges in enabling real-time inference from spectrum data is how to *effectively* and *efficiently* extract *meaningful* and *actionable* knowledge out of the tens of millions of I/Q samples received every second by wireless devices. Indeed, a single 20 MHz-wide WiFi channel generates an I/Q stream rate of about 1.28 Gbit/s, if I/Q samples are each stored in a 4-byte word. Moreover, the RF channel is significantly time-varying (i.e. in the order of milliseconds), which imposes strict timing constraints on the *validity* of the extracted RF knowledge. If (for example) the RF channel changes every 10 ms, a knowledge-extraction algorithm must run with latency (much) less than 10ms to both (i) offer an accurate RF prediction and (ii) drive an appropriate physical-layer response; for example, change in modulation/coding/beamforming vectors due to adverse channel conditions, LO frequency due to spectrum reuse, and so on.

As discussed earlier, DL has been a prominent technology of choice for solving classification problems for which no well-defined mathematical model exists. It enables the analysis of unprocessed I/Q samples without the need of application-specific and computational-expensive feature extraction and selection algorithms (O'Shea et al., 2018), thus going far beyond traditional low-dimensional ML techniques. Furthermore, DL architectures are application-insensitive, meaning the same architecture can be retrained for different learning problems.

Decision-making at the physical layer may leverage the spectrum knowledge provided by DL. On the other hand, RF DL algorithms must execute in *real time* (i.e. with static, known a priori latency) to achieve this goal. Traditional CPU-based knowledge-extraction algorithms

(Abadi et al., 2016) are unable to meet strict time constraints, as a general-purpose CPU can be interrupted at will by concurrent processes and thus introduce additional latency to the computation. Moreover, transferring data to the CPU from the radio interface introduces unacceptable latency for the RF domain. Finally, processing I/Q rates in the order of Gbit/s would require the CPU to run continuously at maximum speed, and thus consume enormous amounts of energy. For these reasons, RF DL algorithms must be closely integrated into the RF signal processing chain of the embedded device.

13.4.1 Existing Work

Most of the existing work is based on traditional low-dimensional machine learning (Wong and Nandi, 2001; Xu et al., 2010; Pawar and Doherty, 2011; Shi and Karasawa, 2012; Ghodeswar and Poonacha,), which requires (i) extraction and careful selection of complex features from the RF waveform (i.e. average, median, kurtosis, skewness, high-order cyclic moments, etc.); and (ii) the establishment of tight decision bounds between classes based on the current application, which are derived either from mathematical analysis or by learning a carefully crafted dataset (Shalev-Shwartz and Ben-David, 2014). In other words, since feature-based machine learning is (i) significantly application-specific in nature and (ii) introduces additional latency and computational burden due to feature extraction, its application to real-time hardware-based wireless spectrum analysis becomes impractical, as the wireless radio hardware should be changed according to the specific application under consideration.

Recent advances in DL (LeCun et al., 2015) have prompted researchers to investigate whether similar techniques can be used to analyze the sheer complexity of the wireless spectrum. For a compendium of existing research on the topic, the reader can refer to Mao et al. (2018). Among other advantages, DL is significantly amenable to be used for real-time hardware-based spectrum analysis, since different model architectures can be reused to different problems as long as weights and hyperparameters can be changed through software. Additionally, DL solutions to the physical-layer modulation recognition task have been given much attention over recent years, as previously discussed in this chapter. The core issue with existing approaches is that they leverage DL to perform offline spectrum analysis only. On the other hand, the opportunity of real-time hardware-based spectrum knowledge inference remains substantially uninvestigated.

13.4.2 Background on System-on-Chip Computer Architecture

Due to its several advantages, we contend that one of the most appropriate computing platform for RF DL is a system on chip (SoC). An SoC is an integrated circuit (also known as IC or chip) that integrates all the components of a computer, i.e. CPU, RAM, input/output (I/O) ports, and secondary storage (e.g. SD card) – all on a single substrate (Molanes et al., 2018). SoCs have low power consumption (Pete Bennett, 2004) and allow the design and implementation of *customized hardware* on the field-programmable gate array (FPGA) portion of the chip, also called programmable logic (PL). Furthermore, SoC brings unparalleled flexibility, as the PL can be reprogrammed at will according to the desired learning design. The PL portion of the SoC can be managed by the processing system (PS), i.e. the CPU, RAM, and associated buses.

SoCs use the advanced extensible interface (AXI) bus specification to exchange data (i) between functional blocks inside the PL and (ii) between the PS and PL. There are three main AXI sub-specifications: *AXI-Lite, AXI-Stream,* and *AXI-Full.* AXI-Lite is a lightweight, low-speed AXI protocol for register access, and it is used to configure the circuits inside the PL. AXI-Stream is used to transport data between circuits inside the PL. AXI-Stream is widely used, since it provides (i) standard inter-block interfaces and (ii) rate-insensitive design; since all the AXI-Stream interfaces share the same bus clock, the high-level synthesis (HLS) design tool will handle the handshake between DL layers and insert FIFO for buffering incoming/outgoing samples. AXI-Full is used to enable burst-based data transfer from PL to PS (and vice versa). Along with AXI-Full, direct memory access (DMA) is usually used to allow PL circuits to read/write data obtained through AXI-Stream to the RAM residing in the PS. The use of DMA is crucial since the CPU would be fully occupied for the entire duration of the read/write operation, and thus unavailable to perform other work.

13.4.3 A Design Framework for Real-Time RF Deep Learning

One of the fundamental challenges to be addressed is how to transition from a software-based DL implementation (e.g. developed with the TensorFlow Abadi et al. (2016) engine) to a hardware-based implementation on an SoC. Basic notions of high-level synthesis and a hardware design framework are presented in Sections 13.4.3.1 and 13.4.3.2, respectively.

13.4.3.1 High-Level Synthesis

HLS is an automated design process that interprets an algorithmic description of a desired behavior (e.g. C/C++) and creates a model written in hardware description language (HDL) that can be executed by the FPGA and implements the desired behavior (Winterstein et al., 2013). Designing digital circuits using HLS has several advantages over traditional approaches. First, HLS programming models can implement almost any algorithm written in C/C++. This allows the developer to spend less time on the HDL code and focus on the algorithmic portion of the design, and at the same time avoid bugs and increase efficiency, since HLS optimizes the circuit according to the system specifications. The clock speed of today's FPGA is a few orders of magnitude slower than CPUs (i.e. up to 200–300 MHz in the very best FPGAs). Thus, parallelizing the circuit's operations is crucial. In traditional HDL, transforming the signal-processing algorithms to fit FPGA's parallel architecture requires challenging programming efforts. On the other hand, an HLS toolchain can tell how many cycles are needed for a circuit to generate all the outputs for a given input size, given a target parallelization level. This helps to reach the best trade-off between hardware complexity and latency. In addition, as shown in the following, loop pipelining and loop unrolling could be used for a better silicon convergence in terms of performance, power consumption, and latency.

Loop Pipelining: In high-level languages (such as C/C++), the operations in a loop are executed sequentially, and the next iteration of the loop can only begin when the last operation in the current loop iteration is complete. Loop pipelining allows the operations in a loop to be implemented in a concurrent manner.

Figure 13.3 shows an example of loop pipelining, where a simple loop of three operations, i.e. read (RD), execute (EX), and write (WR), is executed twice. For simplicity, we assume that each operation takes one clock cycle to complete. Without loop pipelining, the loop would take six clock cycles to complete. Conversely, with loop pipelining, the next RD operation is executed concurrently to the EX operation in the first loop iteration. This brings the total loop latency to four clock cycles. If the loop length were to increase to 100, then the latency decrease would be even more evident: 300 versus 103 clock cycles, corresponding to a speedup of about 65%. An important term for loop pipelining is the initiation interval (II), which is the number of clock cycles between the start times of consecutive loop iterations. In the example of Figure 13.3, the II is equal to one, because there is only one clock cycle between the start times of consecutive loop iterations.

Loop Unrolling: Loop unrolling creates multiple copies of the loop body and adjusts the loop iteration counter accordingly. For example, if a loop is processed with an unrolling factor (UF)

```
for (int i=0; i<2;i++) {
    Op_Read;      /* RD */
    Op_Execute;   /* EX */
    Op_Write;     /* WR */
}
```

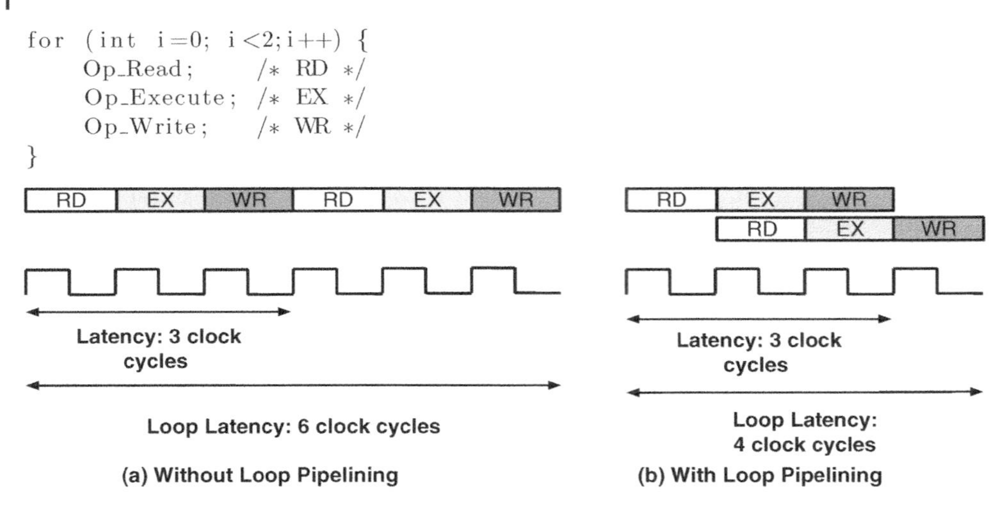

Figure 13.3 Loop pipelining.

```
for(int i = 0; i < 10; i++) {
    sum += a[i];
}

for(int i = 0; i < 10; i+=2) {
    sum += a[i];
    sum += a[i+1];
}
```

Figure 13.4 Loop unrolling.

equal to 2 (i.e. two subsequent operations in the same clock cycle, as shown in Figure 13.4), it may reduce a loop's latency by a factor of 50%, since a loop will execute in half the iterations usually needed. Higher UF and II may help achieve low latency but at the cost of higher hardware resource consumption. Thus, the trade-off between latency and hardware consumption should be thoroughly explored.

13.4.3.2 Design Steps

Our framework presents several design and development steps, which are illustrated in Figure 13.5. Steps that involve hardware, middleware (i.e. HDL), and software have been depicted with different shades of grey.

The first major step of the framework is to take an existing DL model and convert it into HLS language, so it can be optimized and later synthesized in hardware. Another critical challenge is how to make the hardware implementation fully reconfigurable, i.e. the weights of the DL model may need to be changed by the *Controller* according to the specific training. To address these issues, we distinguish between (i) the DL model architecture, which is the set of layers and hyperparameters that compose the model itself; and (ii) the parameters of each layer, i.e. the neurons' and filters' weights.

To generate the HLS code describing the software-based DL model, an HLS library provides a set of HLS functions that parse the software-based DL model architecture and generates the HLS design corresponding to the desired architecture. The HLS library supports the generation of a convolutional, fully connected, rectified linear unit, and pooling layers, and operates on fixed-point arithmetic for better latency and hardware resource consumption. The HLS code is subsequently translated to HDL code by an automated tool that takes into account optimization

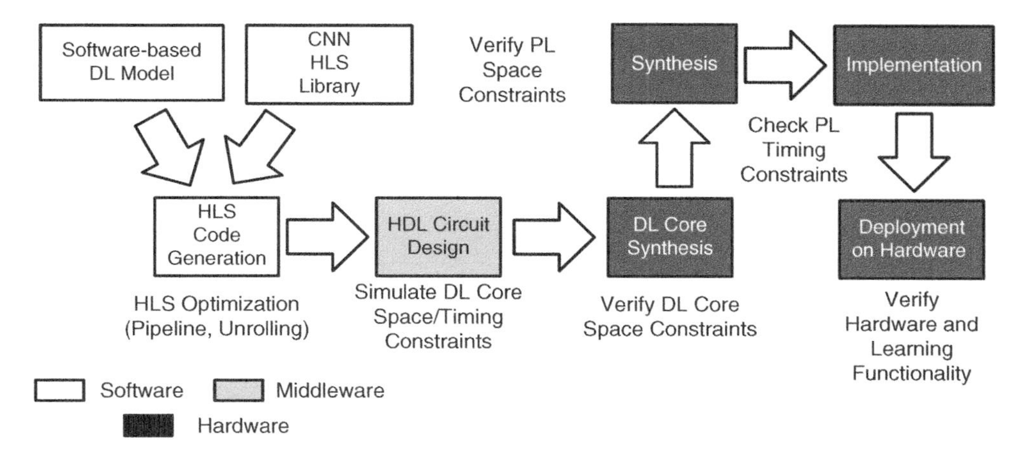

Figure 13.5 A hardware design framework for RF deep learning.

directives such as loop pipelining and loop unrolling. At this stage, the HDL describing the DL core can be simulated to (i) calculate the amount of PL resources consumed by the circuit (i.e. flip-flops, BRAM blocks, etc) and (ii) estimate the circuit latency in terms of clock cycles. After a compromise between space and latency as dictated by the application has been found, the DL core can be synthesized and integrated with the other PL components, and thus total space constraints can be verified. After implementation (i.e. placing/routing), the PL timing constraints can be verified, and finally the whole system can be deployed on the SoC and its functionality tested.

13.5 Open Problems

In this section, we discuss a set of open challenges, the overcoming of which will accelerate the induction of ML techniques into future wireless communications and networking.

13.5.1 Lack of Large-Scale Wireless Signal Datasets

It is well known that learning algorithms require a considerable amount of data to be able to effectively learn from a training dataset. Moreover, to compare the performance of different learning models and algorithms, it is imperative to use the same sets of data. More mature learning fields, such as computer vision and natural language processing (NLP), already have standardized datasets for these purposes (Deng, 2012; Deng et al., 2009). However, the still lacks large-scale datasets for RF ML.

This is not without a reason. Although the wireless domain allows the synthetic generation of signals having the desired characteristics (e.g. modulation, frequency content, and so on), problems such as RF fingerprinting and jamming detection require data that captures the unique characteristics of devices and wireless channels. Therefore, significant research effort must be put forth to build large-scale wireless signal datasets to be shared with the research community at large.

13.5.2 Choice of I/Q Data Representation Format

It is still a subject of debate within the research community what the best data representation is for RF DL applications. For example, an I/Q sample can be represented as a tuple of real numbers or a single complex number, while a set of I/Q samples can be represented as a matrix or a single set of numbers represented as a string. It is a common belief that there is no one-size-fits-all data representation solution for every learning problem and that the right format might depend, among others, on the learning objective, choice of the loss function, and learning problem considered (O'Shea et al., 2018).

13.5.3 Choice of Learning Model and Architecture

While there is a direct connection between images and tensors, the same cannot be concluded for wireless signals. For example, while 3-D tensors have been proven to effectively model images (i.e. red, green, and blue channels), and kernels in convolutional layers are demonstrably powerful tools to detect edges and contours in a given image, it is still unclear if and how these concepts can be applied to wireless signals. Another major difference is that, while images can be considered stationary data, RF signals are inherently stochastic, non-stationary, and time-varying. This peculiar aspect poses significant issues in determining the right learning strategy in the wireless RF domain. For example, while CNN seems to be able to effective at solving problems such as modulation recognition (West and O'Shea, 2017; Karra et al., 2017; O'Shea et al., 2018), it is still unclear if this is the case for complex problems such as RF fingerprinting. Moreover, DL has traditionally been used in static contexts (Krizhevsky et al., 2012; Hinton et al., 2012), where the model latency is usually not a concern. Another fundamental issue absent in traditional DL is the need to satisfy strict constraints on resource consumption. Indeed, models with high number of neurons/layers/parameters will necessarily require additional hardware and energy consumption, which are clearly scarce resources in embedded systems. Particular care must be devoted, therefore, when designing learning architectures to solve learning problems in the RF domain.

13.6 Conclusion

This chapter provides a comprehensive account of advancements in the physical layer rendered by the application of ANNs. To accomplish this, we first provided readers with an overview of the most prevalent ANNs employed in wireless communication networks. Next, we discussed the impact of ANNs on designing a physical layer for gathering signal intelligence. Realizing the importance of extending these techniques to hardware implementation, we discussed some steps that can be taken in those directions to ensure a rapid transition of these techniques to commercial hardware. Finally, we discussed some of the open problems that need to be tackled to further ease the adoption of ANNs for wireless networks. The overarching goal of this chapter has been to enable researchers with the fundamental tools to understand the applications of ANNs in the context of signal intelligence in wireless communication and apprise them of the latest advancements that will consequently motivate new and existing works.

Bibliography

Martín Abadi, Paul Barham, Jianmin Chen, Zhifeng Chen, Andy Davis, Jeffrey Dean, Matthieu Devin, Sanjay Ghemawat, Geoffrey Irving, Michael Isard, et al. TensorFlow: A system for large-scale machine learning. In *OSDI*, volume 16, pages 265–283, 2016.

Satyam Agarwal and Swades De. eDSA: energy-efficient dynamic spectrum access protocols for cognitive radio networks. *IEEE Transactions on Mobile Computing*, 15(12): 3057–3071, 2016.

J. Akeret, C. Chang, A. Lucchi, and A. Refregier. Radio frequency interference mitigation using deep convolutional neural networks. *Astronomy and Computing*, 18: 35–39, January 2017. doi: 10.1016/j.ascom.2017.01.002.

Elsayed Elsayed Azzouz and Asoke Kumar Nandi. *Automatic Modulation Recognition of Communication Signals*. Kluwer Academic Publishers, Norwell, MA, 1996. ISBN 0792397967.

M. Bkassiny, Y. Li, and S. K. Jayaweera. A survey on machine-learning techniques in cognitive radios. *IEEE Communications Surveys & Tutorials*, 15(3): 1136–1159, Third 2013. ISSN 1553-877X. doi: 10.1109/SURV.2012.100412.00017.

David Broomhead and David Lowe. Radial basis functions, multi-variable functional interpolation and adaptive networks. *Royal Signals and Radar Establishment Malvern (United Kingdom)*, RSRE-MEMO-4148, 03 1988.

Guey-Yun Chang, Szu-Yung Wang, and Yuen-Xin Liu. A jamming-resistant channel hopping scheme for cognitive radio networks. *IEEE Transactions on Wireless Communications*, 16(10): 6712–6725, 2017.

Mingzhe Chen, Ursula Challita, Walid Saad, Changchuan Yin, and Mérouane Debbah. Machine learning for wireless networks with artificial intelligence: A tutorial on neural networks. *CoRR*, abs/1710.02913, 2017. URL http://arxiv.org/abs/1710.02913.

Tapiwa Moses Chiwewe and Gerhard Petrus Hancke. Fast convergence cooperative dynamic spectrum access for cognitive radio networks. *IEEE Transactions on Industrial Informatics*, 2017.

Cisco Systems. Cisco Visual Networking Index: Global Mobile Data Traffic Forecast Update, 2016-2021 white paper. http://tinyurl.com/zzo6766, 2017.

D. Czech, A. Mishra, and M. Inggs. A CNN and LSTM-based approach to classifying transient radio frequency interference. *Astronomy and Computing*, 25: 52–57, October 2018. doi: 10.1016/j.ascom.2018.07.002.

Jia Deng, Wei Dong, Richard Socher, Li-Jia Li, Kai Li, and Li Fei-Fei. Imagenet: A large-scale hierarchical image database. In *Computer Vision and Pattern Recognition, 2009. CVPR 2009. IEEE Conference on*, pages 248–255. IEEE, 2009.

Li Deng. The mnist database of handwritten digit images for machine learning research [best of the web]. *IEEE Signal Processing Magazine*, 29(6): 141–142, 2012.

Ericsson Incorporated. Ericsson mobility report, February 2018. https://www.ericsson.com/assets/local/mobility-report/documents/2018/emr-interim-feb-2018.pdf, 2018.

Federal Communications Commission [2016]. Spectrum crunch. https://www.fcc.gov/general/spectrum-crunch.

Federated Wireless. Citizens broadband radio service (CBRS) shared spectrum: An overview. https://www.federatedwireless.com/wp-content/uploads/2017/09/CBRS-Spectrum-Sharing-Overview.pdf, 2018.

S. Foulke, J. Jagannath, A.L. Drozd, T. Wimalajeewa, P. K. Varshney, and W. Su. Multisensor Modulation Classification (MMC) Implementation considerations - USRP case study. In *Proc. of IEEE Conf. on Military Communications (MILCOM)*, Baltimore, MD, USA, October 2014.

S. Ghodeswar and P.G. Poonacha. An SNR estimation based adaptive hierarchical modulation classification method to recognize M-ary QAM and M-ary PSK signals. In *Proc. of International Conference on Signal Processing, Communication and Networking (ICSCN)*, pages 1–6, Chennai, India, March 2015. doi: 10.1109/ICSCN.2015.7219867.

Ian Goodfellow, Yoshua Bengio, Aaron Courville, and Yoshua Bengio. *Deep learning*, volume 1. MIT Press Cambridge, 2016b.

Alharbi Hazza, Mobien Shoaib, Saleh AlShebeili, and Alturki Fahd. Automatic modulation classification of digital modulations in presence of HF noise. *EURASIP Journal on Adv. in Signal Processing*, 2012: 238, 2012.

Kaiming He, Xiangyu Zhang, Shaoqing Ren, and Jian Sun. Deep residual learning for image recognition. *CoRR*, abs/1512.03385, 2015. URL http://arxiv.org/abs/1512.03385.

Geoffrey Hinton, Li Deng, Dong Yu, George E. Dahl, Abdel-rahman Mohamed, Navdeep Jaitly, Andrew Senior, Vincent Vanhoucke, Patrick Nguyen, Tara N. Sainath, et al. Deep neural networks for acoustic modeling in speech recognition: The shared views of four research groups. *IEEE Signal processing magazine*, 29(6): 82–97, 2012.

Sepp Hochreiter, Yoshua Bengio, Paolo Frasconi, and Jürgen Schmidhuber. Gradient flow in recurrent nets: the difficulty of learning long-term dependencies. In *A Field Guide to Dynamical Recurrent Neural Networks*, 2001.

Gao Huang, Zhuang Liu, and Kilian Q. Weinberger. Densely connected convolutional networks. *CoRR*, abs/1608.06993, 2016. URL http://arxiv.org/abs/1608.06993.

Jen-Feng Huang, Guey-Yun Chang, and Jian-Xun Huang. Anti-jamming rendezvous scheme for cognitive radio networks. *IEEE Transactions on Mobile Computing*, 16(3): 648–661, 2017.

J. Jagannath, H.M. Saarinen, and A.L. Drozd. Framework for automatic signal classification techniques (FACT) for software defined radios. In *Proc. of IEEE Symposium on Computational Intelligence in Security and Defense Applications (CISDA)*, Verona, NY, USA, May 2015.

J. Jagannath, D. O'Connor, N. Polosky, B. Sheaffer, L.N. Theagarajan, S. Foulke, P.K. Varshney, and S.P. Reichhart. Design and evaluation of hierarchical hybrid automatic modulation classifier using software defined radios. In *Proc. of IEEE Annual Computing and Communication Workshop and Conference (CCWC)*, Las Vegas, NV, USA, January 2017.

J. Jagannath, S. Furman, T. Melodia, and A. Drozd. Design and experimental evaluation of a cross-layer deadline-based joint routing and spectrum allocation algorithm. *IEEE Transactions on Mobile Computing*, pages 1–1, 2018a. ISSN 1536-1233. doi: 10.1109/TMC.2018.2866093.

J. Jagannath, N. Polosky, D. O'Connor, L. Theagarajan, B. Sheaffer, S. Foulke, and P. Varshney. Artificial neural network based automatic modulation classifier for software defined radios. In *Proc. of IEEE International Conference on Communications (ICC)*, Kansas City, MO, USA, May 2018b.

C. Jiang, H. Zhang, Y. Ren, Z. Han, K.C. Chen, and L. Hanzo. Machine learning paradigms for next-generation wireless networks. *IEEE Wireless Communications*, 24(2): 98–105, April 2017. ISSN 1536-1284. doi: 10.1109/MWC.2016.1500356WC.

Xiaocong Jin, Jingchao Sun, Rui Zhang, Yanchao Zhang, and Chi Zhang. SpecGuard: spectrum misuse detection in dynamic spectrum access systems. *to appear, IEEE Transactions on Mobile Computing*, 2018.

L-S. Bouchard K. Youssef. Training artificial neural networks with reduced computational complexity. URL https://gtp.autm.net/public/project/34861/.

K. Karra, S. Kuzdeba, and J. Petersen. Modulation recognition using hierarchical deep neural networks. In *Proc. of IEEE International Symposium on Dynamic Spectrum Access Networks (DySPAN)*, pages 1–3, Baltimore, MD, USA, March 2017. doi: 10.1109/DySPAN.2017.7920746.

Diederik P. Kingma and Jimmy Ba. Adam: A method for stochastic optimization. *CoRR*, abs/1412.6980, 2014. URL http://arxiv.org/abs/1412.6980.

Alex Krizhevsky, Ilya Sutskever, and Geoffrey E. Hinton. Imagenet classification with deep convolutional neural networks. In *Advances in Neural Information Processing Systems*, pages 1097–1105, 2012.

A. Kubankova, J. Prinosil, and D. Kubanek. Recognition of digital modulations based on mathematical classifier. In *Proc. of the European Conference of Systems (ECCS)*, Stevens Point, WI, 2010.

M. Kulin, T. Kazaz, I. Moerman, and E. De Poorter. End-to-end learning from spectrum data: A deep learning approach for wireless signal identification in spectrum monitoring applications. *IEEE Access*, 6: 18484–18501, 2018. doi: 10.1109/ACCESS.2018.2818794.

Yann LeCun, Yoshua Bengio, and Geoffrey Hinton. Deep learning. *Nature*, 521(7553): 436, 2015.

Yann LeCun et al. Generalization and network design strategies. *Connectionism in Perspective*, pages 143–155, 1989.

Lu Lv, Jian Chen, Qiang Ni, Zhiguo Ding, and Hai Jiang. Cognitive non-orthogonal multiple access with cooperative relaying: A new wireless frontier for 5G spectrum sharing. *IEEE Communications Magazine*, 56(4): 188–195, 2018.

Q. Mao, F. Hu, and Q. Hao. Deep learning for intelligent wireless networks: A comprehensive survey. *IEEE Communications Surveys & Tutorials*, 2018. doi: 10.1109/COMST.2018.2846401.

R.F. Molanes, J.J. Rodríguez-Andina, and J. Faria. Performance characterization and design guidelines for efficient processor - FPGA communication in Cyclone V FPSoCs. *IEEE Transactions on Industrial Electronics*, 65(5): 4368–4377, May 2018. ISSN 0278-0046. doi: 10.1109/TIE.2017.2766581.

Timothy J. O'Shea and Johnathan Corgan. Convolutional radio modulation recognition networks. *CoRR*, abs/1602.04105, 2016. URL http://arxiv.org/abs/1602.04105.

Timothy James O'Shea and Jakob Hoydis. An introduction to deep learning for the physical layer. *IEEE Transactions on Cognitive Communications and Networking*, 3(4): 563–575, 2017.

T.J. O'Shea, T. Roy, and T.C. Clancy. Over-the-air deep learning based radio signal classification. *IEEE Journal of Selected Topics in Signal Processing*, 12(1): 168–179, Feb 2018. ISSN 1932-4553. doi: 10.1109/JSTSP.2018.2797022.

O. Ozdemir, Ruoyu Li, and P.K. Varshney. Hybrid maximum likelihood modulation classification using multiple radios. *IEEE Communications Letters*, 17(10): 1889–1892, October 2013.

O. Ozdemir, T. Wimalajeewa, B. Dulek, P. K. Varshney, and W Su. Asynchronous linear modulation classification with multiple sensors via generalized EM algorithm. *IEEE Transactions on Wireless Communications*, 14(11): 6389–6400, November 2015.

S.U. Pawar and J.F. Doherty. Modulation recognition in continuous phase modulation using approximate entropy. *IEEE Transactions on Information Forensics and Security*, 6(3): 843–852, Sept 2011. ISSN 1556-6013. doi: 10.1109/TIFS.2011.2159000.

Pete Bennett. The why, where and what of low-power SoC design. https://www.eetimes.com/document.asp?doc_id=1276973, 2004.

F. Rosenblatt. *The Perceptron, a Perceiving and Recognizing Automaton Project Para*. Report: Cornell Aeronautical Laboratory. Cornell Aeronautical Laboratory, 1957. URL https://books.google.com/books?id=P_XGPgAACAAJ.

F. Rosenblatt. *Principles of neurodynamics: perceptrons and the theory of brain mechanisms*. Report (Cornell Aeronautical Laboratory). Spartan Books, 1962. URL https://books.google.com/books?id=7FhRAAAAMAAJ.

D.E. Rumelhart, G.E. Hinton, and R.J. Williams. Learning representations by back-propagating errors. *Nature*, 323: 533–536, 1986.

Ahmed Selim, Francisco Paisana, Jerome A. Arokkiam, Yi Zhang, Linda Doyle, and Luiz A. DaSilva. Spectrum monitoring for radar bands using deep convolutional neural networks. *CoRR*, abs/1705.00462, 2017. URL http://arxiv.org/abs/1705.00462.

Shai Shalev-Shwartz and Shai Ben-David. *Understanding machine learning: From theory to algorithms*. Cambridge University Press, 2014.

Huaxia Wang Hathal Alwageed Shengliang Peng, Hanyu Jiang and Yu-Dong Yao. Modulation classification using convolutional neural network based deep learning model. *WOCC*, 2017.

Q. Shi and Y. Karasawa. Automatic modulation identification based on the probability density function of signal phase. *IEEE Transactions on Communications*, 60(4): 1033–1044, April 2012. ISSN 0090-6778. doi: 10.1109/TCOMM.2012.021712.100638.

Hossein Shokri-Ghadikolaei, Federico Boccardi, Carlo Fischione, Gabor Fodor, and Michele Zorzi. Spectrum sharing in mmwave cellular networks via cell association, coordination, and beamforming. *IEEE Journal on Selected Areas in Communications*, 34(11): 2902–2917, 2016.

Rupesh Kumar Srivastava, Klaus Greff, and Jürgen Schmidhuber. Highway networks. *CoRR*, abs/1505.00387, 2015. URL http://arxiv.org/abs/1505.00387.

Bülent Üstün, Willem J. Melssen, and Lutgarde M.C. Buydens. Facilitating the application of support vector regression by using a universal Pearson VII function based kernel. 2005.

Miguel Angel Vázquez, Luis Blanco, and Ana I Pérez-Neira. Hybrid analog–digital transmit beamforming for spectrum sharing backhaul networks. *IEEE transactions on Signal Processing*, 66(9): 2273, 2018.

Tianqi Wang, Chao-Kai Wen, Hanqing Wang, Feifei Gao, Tao Jiang, and Shi Jin. Deep learning for wireless physical layer: Opportunities and challenges. *China Communications*, 14(11): 92–111, 2017.

N. E. West and T. O'Shea. Deep architectures for modulation recognition. In *Proc. of IEEE International Symposium on Dynamic Spectrum Access Networks (DySPAN)*, pages 1–6, Baltimore, MD, USA, March 2017. doi: 10.1109/DySPAN.2017.7920754.

T. Wimalajeewa, J. Jagannath, P.K. Varshney, A.L. Drozd, and W. Su. Distributed asynchronous modulation classification based on hybrid maximum likelihood approach. In *Proc. of IEEE Conf. on Military Communications (MILCOM)*, Tampa, FL, USA, October 2015.

Felix Winterstein, Samuel Bayliss, and George A. Constantinides. High-level synthesis of dynamic data structures: A case study using vivado hls. In *Proc. of International Conference on Field-Programmable Technology (FPT)*, pages 362–365, Kyoto, Japan, 2013.

M.L.D. Wong and A.K. Nandi. Automatic digital modulation recognition using spectral and statistical features with multi-layer perceptrons. In *Proc. of the Sixth International Symposium on Signal Processing and its Applications (Cat.No.01EX467)*, volume 2, pages 390–393, 2001. doi: 10.1109/ISSPA.2001.950162.

J.L. Xu, W. Su, and M. Zhou. Software-defined radio equipped with rapid modulation recognition. *IEEE Transactions on Vehicular Technology*, 59(4): 1659–1667, May 2010. ISSN 0018-9545. doi: 10.1109/TVT.2010.2041805.

K. Youssef, N.N. Jarenwattananon, and L. Bouchard. Feature-preserving noise removal. *IEEE Transactions on Medical Imaging*, 34(9): 1822–1829, Sept 2015. ISSN 0278-0062. doi: 10.1109/TMI.2015.2409265.

K. Youssef, L.S. Bouchard, K.Z. Haigh, H. Krovi, J. Silovsky, and C.P. Vander Valk. Machine learning approach to RF transmitter identification. *ArXiv e-prints*, November 2017.

Liyang Zhang, Francesco Restuccia, Tommaso Melodia, and Scott Pudlewski. Learning to detect and mitigate cross-layer attacks in wireless networks: Framework and applications. In *Proc. of IEEE Conf. on Communications and Network Security*, Las Vegas, NV, USA, October 2017.

Yi Ting Zhou and Rama Chellappa. Computation of optical flow using a neural network. *IEEE 1988 International Conference on Neural Networks*, pages 71–78 vol. 2, 1988.

14

Channel Coding with Deep Learning: An Overview

Shugong Xu

SICS, Shanghai University, Shanghai, China

This chapter is devoted to the use of various neural networks and related learning algorithms in the channel coding (encoder and decoder) of wireless communications and networking. Due to its powerful nonlinear mapping and distributed processing capability, neural network–based machine learning technology could offer a more powerful channel coding solution than conventional approaches in many aspects including coding performance, computational complexity, power consumption, and processing latency. The neural networks discussed in this chapter mainly include deep neural networks (DNNs), convolutional neural networks (CNNs), and recurrent neural networks (RNNs).

This chapter is organized into the following five sections. Section 14.1 focuses on the background information of channel coding and deep learning, together with the motivation for the use of machine learning in channel coding. Sections 14.2–14.4 introduce the channel coding schemes with DNN, CNN, and RNN networks, respectively, and then discuss potential coding/decoding performance, computational complexity, power consumption, and processing delay. Section 14.5 offers some further discussion and conclusions.

14.1 Overview of Channel Coding and Deep Learning

14.1.1 Channel Coding

Due to the channel fading environments in wireless communications, the channel coding scheme is widely adopted in modern wireless communication systems; a simple system diagram can be found in Figure 14.1. Although channel coding schemes can be classified as the error-correction code and the error-detection code, most research efforts nowadays focus on dealing with channel variations via advanced channel-correction codes.

In the early 1950s, Hamming code, a kind of linear block code, was proposed by R. Hamming in (Hamming, 1950), where the generation matrix and parity check matrix are often applied at the transmitter and the receiver sides. A similar structure was used to develop Reed-Muller code in 1954 and low-density parity-check (LDPC) code in 1969 (Gallager, 1962). Linear block codes have been widely used in our daily lives, such as Hamming code for flash memory and Reed-Muller code for optical fiber communication. However, LDPC code was not used in wireless communication systems until 1996 (MacKay and Neal, 1996), when the WiFi standard was commercially deployed. This is partially because of the large decoding delay of linear block codes, as they have to receive all the codewords before decoding.

Machine Learning for Future Wireless Communications, First Edition. Edited by Fa-Long Luo.
© 2020 John Wiley & Sons Ltd. Published 2020 by John Wiley & Sons Ltd.

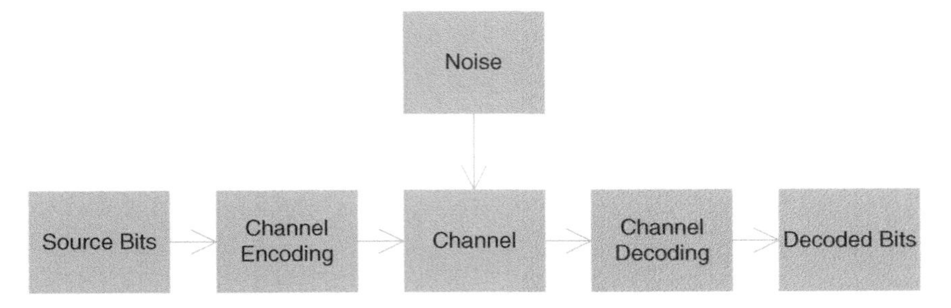

Figure 14.1 Channel coding model.

Table 14.1 Summary of modern channel coding schemes.

	Year	Application
Gray code	1940	Space exploration
Hamming code	1950	Flash error correction
Reed-Muller code	1954	Space exploration
Convolutional code	1955	GSM, 3G, Wi-Fi
LDPC code	1969	DVB-S.2, WiMax, 5G
Turbo code	1993	4G
Polar code	2008	5G

To reduce the decoding delay, another type of channel coding scheme was invented in 1955, where the convolution operation rather than linear matrix multiplication was adopted in the code generation. In 1993, a special Turbo-like structure was proposed by (Berrou et al., 1993), where two or more convolutional operations are cascaded together with random scramblers. To decode the convolutional-based channel coding scheme, the decoder takes full advantage of the corrections among different information blocks, which greatly reduces the decoding delay in practical wireless systems. As a result, these convolutional codes have been widely deployed in cellular systems, such as GSM, UMTS, and LTE.

From the theoretical point of view, both Turbo code and LDPC code can approach to the Shannon limit with a certain gap. However, the first channel code to reach the Shannon bound was not invented until 2009, where the channel polarization effects were discovered by Arikan in (Arıkan, 2009). In Table 14.1, we summarize the history of different channel coding schemes and their typical applications in various communication tasks.

14.1.2 Deep Learning

Together with the development of channel coding, machine learning technology has been widely explored to deal with challenging tasks including voice recognition (Hinton et al., 2012), natural language processing (Collobert and Weston, 2008), and computer vision problems (Rosten and Drummond, 2006). In particular, the deep learning (DL) technique (Ian Goodfellow, 2016) has been widely deployed since its development. By increasing the number of neural processing layers, the DL technique is able to describe highly nonlinear relations between input and output vectors by feeding a sufficient amount of training data. Typical neural networks used in the DL area include DNNs, CNNs, and RNNs, as explained in the following.

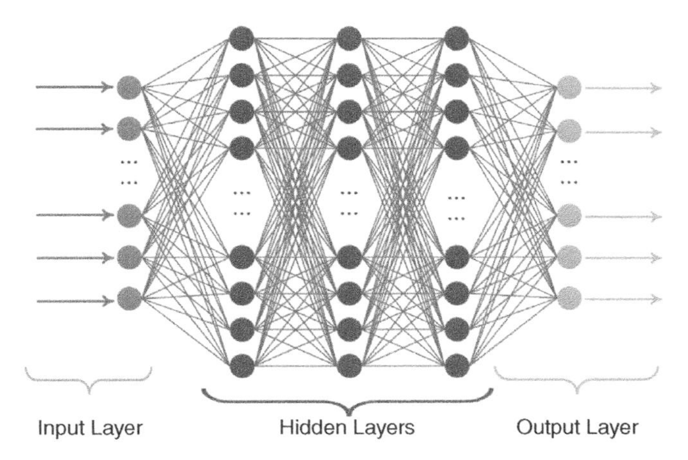

Figure 14.2 An illustration of a DNN network architecture, including input layer, output layer, and hidden layers.

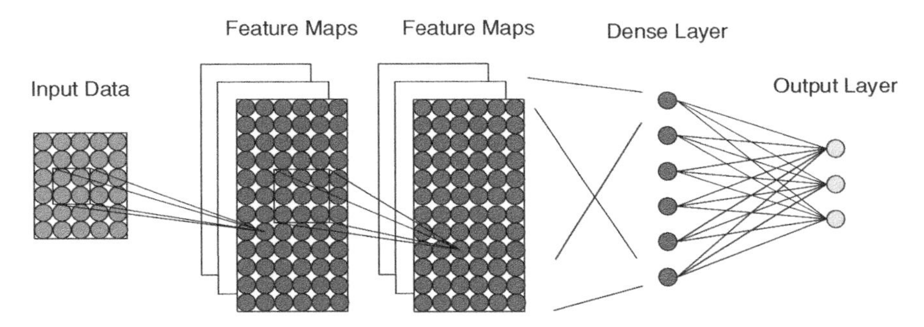

Figure 14.3 An illustration of a CNN network architecture including convolutional layers.

A DNN is a standard multi-layer neural network architecture with fully connected neurons, as shown in Figure 14.2, which generally contains a single input layer, a single output layer, and several hidden layers. In each hidden layer, a series of neurons are deployed to connect input and output nodes via linear (weighted sum) and nonlinear (activation function) mapping relations; the generalization ability usually scales with the number of hidden layers, which is also known as the *depth* of a neural network. With some supervision information during the training stage, DNNs can be adjusted to deal with some computer vision tasks, such as (Nguyen et al., 2015).

As a DNN needs to fully connect all the nodes in hidden layers, the number of parameters usually grows exponentially. To address this issue, CNNs were proposed, where several convolutional layers were utilized to extract high-dimensional features of input data. As shown in Figure 14.3, CNN exploits the correlations of neighboring pixels to reduce the number of parameters as well as the potential computational complexity in neural networks, and it has been widely adopted in some image-processing tasks. Typical examples of CNNs include AlexNet (Krizhevsky et al., 2012), VGG (Simonyan and Zisserman, 2014), and ResNet (He et al., 2016).

Another issue for DNNs is their limited capability of handling time-correlated objective detection and tracking tasks with video streaming. To incorporate the time-domain correlation, recurrent architecture has been utilized on top of the standard DNN architecture, as shown in Figure 14.4; the time-domain information is modeled through recurrent propagations among different neurons in each layer. Although the RNN structure can extract the

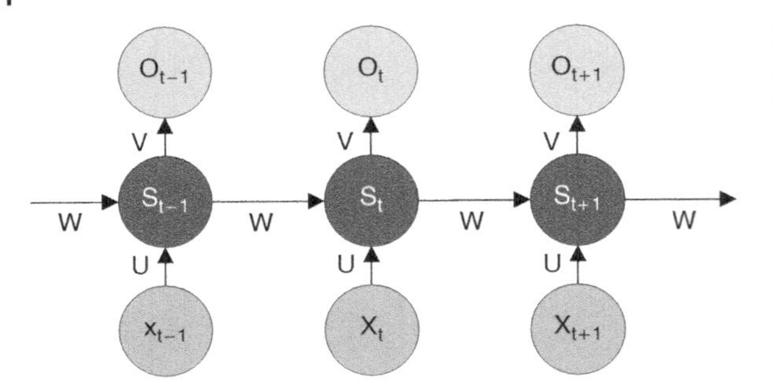

Figure 14.4 An illustration of a RNN network architecture that connects the output of the neuron back to the input of the neuron.

time-correlation features as mentioned, it often suffers from the gradient-diminishing problem (Ian Goodfellow, 2016); several improved network architecture have been invented recently, such as long short-term memory (Gers et al., 1999) and bidirectional RNNs (BRNNs) (Graves and Schmidhuber, 2005).

Among all these neural networks, activation functions and loss functions are usually considered the most important components: the former introduces nonlinearity, and the latter proposes the evaluation metric of neural networks. Typical activation functions are tanh (Glorot et al., 2011), sigmoid, and rectified linear unit (ReLU) (Hao Ye, 2017), while cross-entropy (CE) and mean square error (MSE) are often classified as common loss functions.

As the evaluation of massive layers of activation functions and loss functions require significant computational power, DL technology was not widely deployed until 2012, when high-performance graphics processing units (GPUs) and tensor processing units (TPUs) (Jouppi et al., 2017) become popular. In the meantime, in addition to traditional computer vision and speech-processing tasks, DL technology has been used to solve many challenging problems in other areas, such as medical image diagnosis (Litjens et al., 2017), chess games (Silver et al., 2016), and wireless communications (O'Shea and Hoydis, 2017). For example, in wireless communications, the optimal multiple-input and multiple-output (MIMO) detection algorithm can be well-approximated by autoencoder networks, as shown in (Yan et al., 2017), and a joint channel-estimation and signal-detection structure using DL has been proposed in (Hao Ye, 2017).

As a key element in modern wireless communication systems, channel coding is often considered a computing-intensive task, where an application-specific integrated circuit (ASIC) is usually regarded as the only solution. With the rising demand to support different types of channel codes in wireless communications, the design challenge is ever increasing. Thanks to the generalization capability of neural networks, DL technology has been merged with different channel coding schemes, and several pioneer works have been proposed. In the rest of this chapter, we classify them according to different types of neural networks, which may pave the way for future universal encoders and decoders.

14.2 DNNs for Channel Coding

A DNN is a relatively simple neural network that enhances learning performance by increasing the number of layers in the network. The input and output of a DNN are consistent with those of channel coding, which makes it feasible to use DNN for decoding.

In digital communication systems, the information sequence is transformed into an encoded sequence by channel encoding. Usually these sequences are binary sequences, so this mapping can be seen as a classification problem that can be learned by neural networks. This inspires us to use DNNs instead of traditional decoders to decode the received sequence.

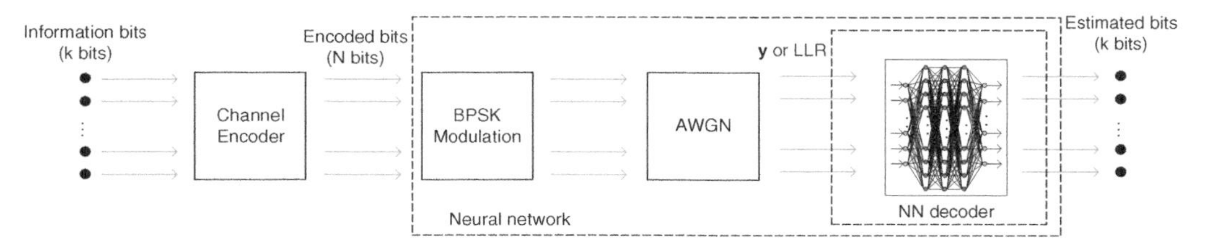

Figure 14.5 DNN method for direct polar decoding.

Representative works about this issue include using DNNs to decode directly (Gruber et al., 2017), combining DNNs with traditional decoding methods such as belief propagation (BP) and the successive cancellation (SC) scaling method to decode medium-length code (Cammerer et al., 2017) (Doan et al., 2018), using DNNs to jointly equalize and decode (Hao Ye, 2017)(Xu et al., 2018), and using DNNs to decode multicodes(Wang et al., 2018).

14.2.1 Using DNNs to Decode Directly

The authors in (Gruber et al., 2017) proposed a neural network decoder (NND) for polar codes with length (16, 8). A NND is a DNN with layers (16, 128, 64, 32, 8), whose structure is shown in Figure 14.5.

As we can see from Figure 14.5, k information bits are encoded into N-bit polar codes, which are then modulated by BPSK and sent by the transmitter. At the receiver side, the signals transmitted through the channel are received and decoded. To replace the whole traditional decoder with the DNN, the input of the network is either the actual value $\mathbf{y}$ or the log-likelihood ratio (LLR), where the output is the estimated codeword. Here the LLR is defined as:

$$\text{LLR} = 2 * \mathbf{y}/\sigma^2, \tag{14.1}$$

where $\mathbf{y}$ is the received signals and σ^2 is the noise variance.

The activation function for the hidden layers is ReLU. A sigmoid activation function is used after the output layer, which forces the output neurons to be between zero and one. This is determined by the shape of the sigmoid function. If the result of the network output is closer to zero, then this estimated bit can be considered to be zero. If the result of the network output is closer to one, then this estimated bit can be considered to be one.

DNNs inherently describe a highly parallelizable structure, enabling one-shot decoding. There are two types of codes in this experiment: random codes and polar codes. They represent different coding schemes: unstructured coding and structured coding. The method is training all 2^k possible sequences at one specific SNR point and testing random 0-1 sequences at different SNR points to show the generalization ability.

The simulation results are shown in Figures 14.6 and 14.7. It can be seen that for both code families, the larger the number of training epochs, the closer the gap between maximum a posteriori (MAP) and NND performance. And for polar codes, close to MAP performance is already achieved at $M_{ep} = 2^{20}$. The simulation results show that polar codes that have a deterministic coding structure can be learned better by the neural network than random codes.

In order to measure the generalization ability of DNNs, e.g how well the trained model can adapt to a new environment, a new performance metric, called the normalized validation error (NVE), is defined as

$$\text{NVE}(\rho_t) = \frac{1}{S} \sum_{s=1}^{S} \frac{\text{BER}_{\text{NND}}(\rho_t, \rho_{v,s})}{\text{BER}_{\text{MAP}}(\rho_{v,s})} \tag{14.2}$$

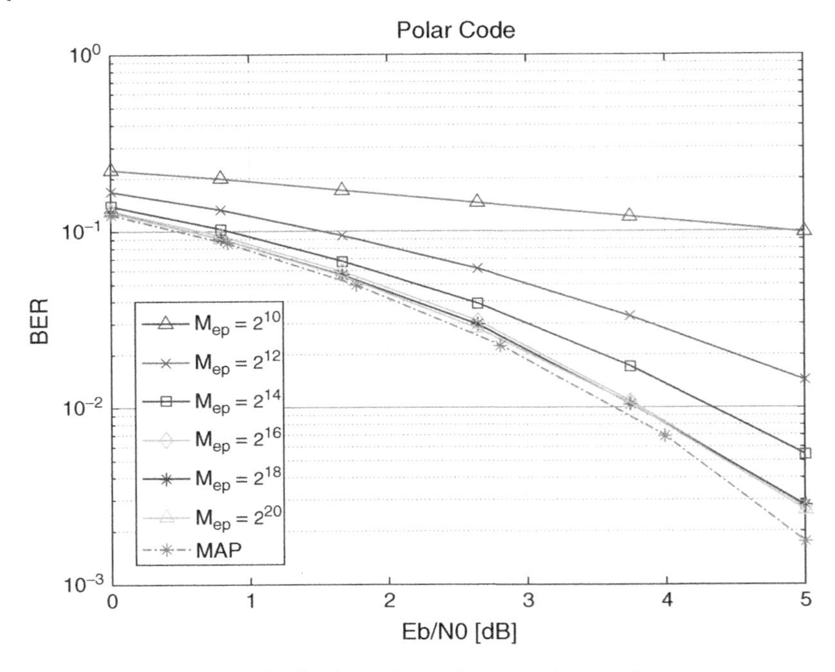

Figure 14.6 The BER result of polar codes under a neural network.

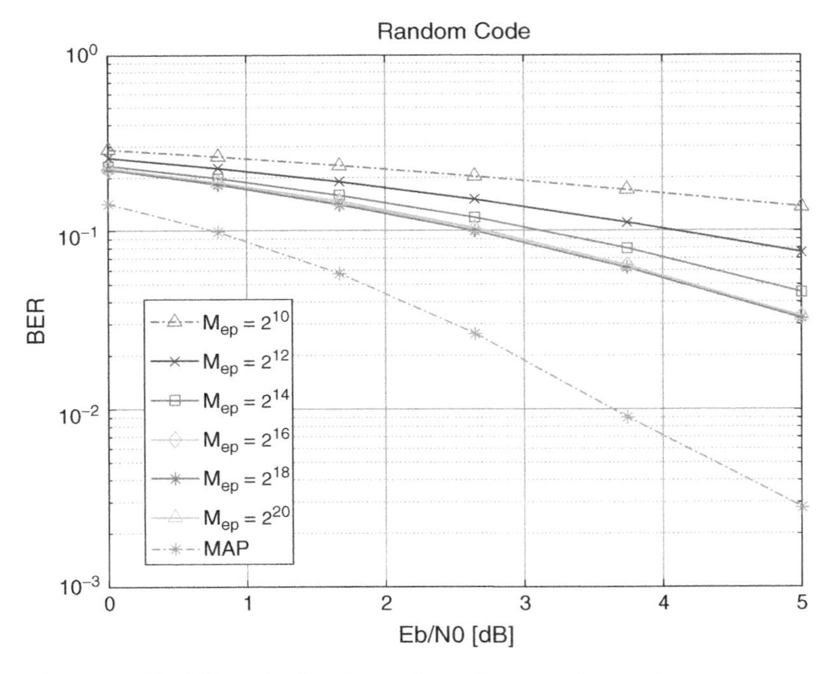

Figure 14.7 The BER result of random codes under a neural network.

A more detailed definition can be found in (Gruber et al., 2017). This definition tells us the performance of training at a single signal-to-noise ratio (SNR) and then testing at different SNRs. It is obvious that the lower the NVE is, the closer the performance is between NND and MAP.

As mentioned in Figure 14.5, the input value of the NND can be either direct channel values **y** or LLR. The authors also show the performance of using **y** or LLR as the input of the NND, and using MSE or binary CE(BCE) as a loss function. It seems that it doesn't matter if **y** or LLR is used as input and which loss function is employed, because with an increase in training epochs, similar NVE values are obtained using these input values and loss functions. The reason is the both channel values and LLRs can represent the features of polar codes. And the two loss functions can both measure the gap between the label and the output of the network precisely.

14.2.2 Scaling DL Method

Unfortunately, the limitation of the previous method lies in its training complexity: for a short code of length N and rate r, 2^{N*r} different codewords exist; that is to say, the training complexity grows exponentially with increasing code length. So in practice training, the neural network may not be trained fully due to the computer's power. Therefore, the code length used in the previous NND is limited to 16.

To solve this problem, the authors in (Cammerer et al., 2017) proposed a scalable method to decode medium-length polar code. The method is based on a combination of NND and traditional BP decoding. They define partitionable code in a sense that each sub-block can be decoded independently. Then, each sub-graph is coupled with the other sub-blocks via the remaining BP results. The whole partitioned neural network (PNN) decoding structure is depicted in Figure 14.8.

This structure consists of many independent NND networks in the same way as the earlier NNDs. The input of the NND structure is the LLR value, which has been propagated in several stages. Then the first NND block is decoded first. The decoded bits are treated as fixed bits, e.g. the LLRs are set to be infinite. After decoding the first NND block, the remaining block can also be decoded using this method. With this structure, the entire decoding process can be presented like a pipeline.

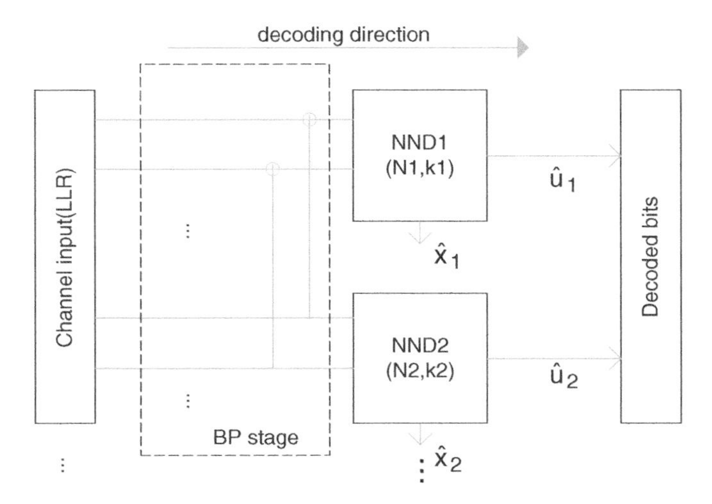

Figure 14.8 Partitioned neural network polar decoding structure.

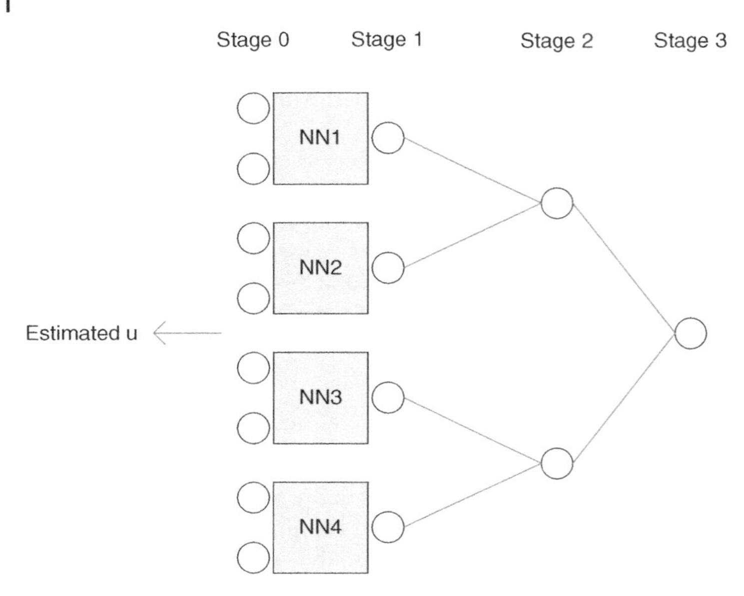

Figure 14.9 NSC decoding with NNDs at Stage 1.

By using this neural network decoder, decoding with a medium code length (128, 64) is proposed in (Cammerer et al., 2017), where the authors also report the simulation results and corresponding comparisons of this proposed solution using eight partitions. Each block has a code length of 16 and a different number of information bits k. The simulation results show that decoders with PNN can realize BER performance similar to other methods such as successive cancellation (SC) and successive cancellation list (SCL). However, due to the performance loss caused by partitioning, the length of decoding by this method is still short. Readers can refer to (Cammerer et al., 2017) to see more simulation details.

Another similar work is shown in (Doan et al., 2018). In this paper, the authors propose NSC (neural SC) decoding, which is constructed of multiple NNDs connected with SC decoding in a way similar to the previous paper (Cammerer et al., 2017). The result shows that the NSC decoder has similar BER and FER performance in comparison with PNN in (Cammerer et al., 2017), SC, and BP decoders. However, from the perspective of complexity, the NSC decoder can reduce the number of time steps required for the PNN decoder with the same error-correction performance. The decoding architecture (N=8) is shown in Figure 14.9. For more detailed information, readers can refer to (Doan et al., 2018).

14.2.3 DNNs for Joint Equalization and Channel Decoding

In addition to channel decoding, training multiple modules (such as equalization and decoding) simultaneously in a communication system has also become a subject of research. This is because neural networks (such as DNNs) are essentially black boxes that can simultaneously learn the features of these modules.

In practical communication systems, inter-symbol interference (ISI) exists because the channel has memory and nonlinear distortions that are introduced by amplifiers and converters. Channel equalization is used to deal with ISI and recover the transmitted signals. But the noise will no longer be Gaussian white noise after the equalization step. As a result, it is hard to make sure performance is still good when considering equalization and decoding. Instead, a DL-based approach for joint consideration of these two steps may deal with this problem.

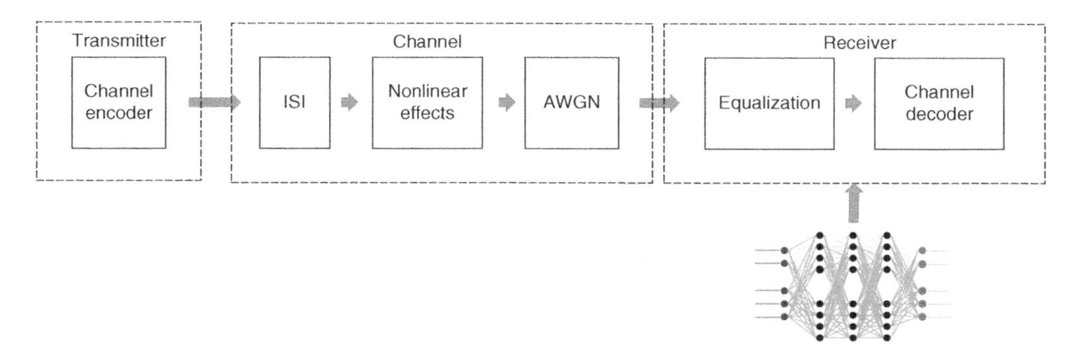

Figure 14.10 The architecture for jointly equalizing and decoding using neural networks.

The initial results on this subject are shown in (Hao Ye, 2017). Figure 14.10 is the schematic architecture of this proposed solution for NN-based joint equalization and decoding. We can see that all the equalization and decoding parts are replaced by neural networks, which can realize a one-shot equalization and decoding approach. Under this circumstance, the authors consider two scenarios: dispersive channels with nonlinear distortion, the frequency-selective time-varying channels.

The input data of the network is the received signal **y** suffering from ISI, and the original messages **b** serve as the true labels. The loss function is MSE to measure performance.

In the first scenario, e.g. dispersive channels with nonlinear distortion, let us use g, h to denote the nonlinear function caused by amplifiers and the channel impulse response, respectively. As an illustration example, the channel impulse h can be:

$$h(z) = 0.3482 + 0.8704z^{-1} + 0.3482z^{-2} \tag{14.3}$$

while the nonlinear distortion function of the channel is:

$$|g(v)| = |v| + 0.2|v|^2 - 0.1|v|^3 + 0.5cos(\pi|v|) \tag{14.4}$$

A six-layer DNN with (16, 256, 128, 64, 32, 8) is used in the previous reference to jointly equalize and decode polar codes with size (16, 8). In terms of BER performance, this joint solution is compared with a baseline system that uses a Gaussian process for classification (GPC) algorithm for equalization and a SC algorithm for decoding. Under this channel model, the performance of the DNN is better than the baseline GPC+SC system, with a gain of about 1-2 dB; this shows the advantages of the DNN approach.

In the second scenario, frequency-selective time-varying channels are considered, where channel responses are constantly changing. In this scenario, a DNN model is trained to decode information under an OFDM system, where channel estimation is achieved simultaneously. An OFDM system with 64 subcarriers and 16 cyclic-prefixes is used. In this reference work, QAM serves as the the modulation method, and 16 pilots are used for channel estimation. The DNN model is the same as the former one, other than the number of inputs being changed to fit the number of OFDM symbols.

The authors of this reference work compare the proposed DL-based detection method with a traditional MMSE estimator with identity matrix (MMSE-I) and MMSE estimator with perfect channel statistics information (MMSE-S). The DL method outperforms the MMSE-I method, but there is still a performance gap compared to MMSE-S. As for decoding, the DL-based method has better performance compared to MMSE with the SC decoder, which shows the advantage of using a neural network to take both detection and decoding into consideration. Readers can refer to (Hao Ye, 2017) to see detailed simulation results.

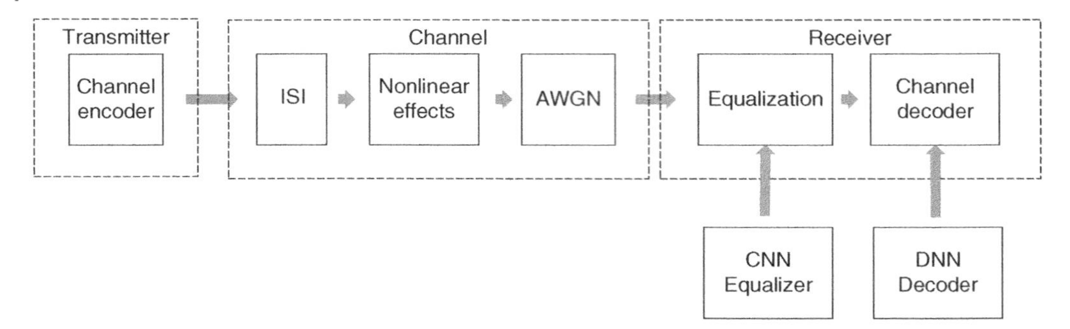

Figure 14.11 System architecture for equalizing and decoding, respectively.

To further improve performance, another paper (Xu et al., 2018) proposes a joint channel equalizer and decoder. Unlike the previous architecture, they use a CNN for equalization and a DNN for decoding, respectively. The architecture is shown in Figure 14.11.

The CNN equalizer is first used to compensate for the signal distortion. Then the other polar DNN decoder is concatenated after the CNN to decode the recovered sequence. The reason for using CNN for equalization is that a CNN has strong feature-extraction capabilities. CNNs can deal with the nonlinear channel model, since ISI exists only between consecutive bits of the transmitted sequence, and the effects of nonlinear distortion are independent for each bit.

Under the circumstances of channel equalization, the input format of CNN is a 1-D vector instead of a typical 2-D image in the field of image processing. Then a multiple-layer DNN can be used to decode the output of the CNN network equalizer. In other words, the input of the DNN is a 1-D vector from the CNN equalizer.

In the simulation part, a five-layer DNN with structure (16, 128, 64, 32, 8) is introduced as the neural network decoder to decode polar codes with length (16, 8). The NND receives the soft output of the CNN equalizer to recover the originally transmitted bits. The nonlinear function of the channel and the impulse response of the dispersive channel with ISI and AWGN can be found in (Xu et al., 2018).

The CNN output may not follow the same distribution of AWGN channel output, which will potentially cause degradation of performance. Hence, joint training NND using the soft output of the CNN equalizer (CNN+NND–Joint) can compensate for the performance loss. Polar codes (16, 8) are tested, and the simulations compared with (Hao Ye, 2017) is also given. The initial result shows that the traditional GPC+SC method, which consists of the GPC algorithm for equalization and the SC algorithm for decoding, is worse than CNN+NND by 1 dB. Joint training of CNN and NND (CNN+NND–joint) has about a 0.5 dB gain over CNN+DNN. Performance of the CNN+DNN–joint scheme is very close to that of the method in (Hao Ye, 2017), which uses a DNN for jointly equalizing and decoding.

These cases show us that a neural network can be used for more than operating a single communication part like channel coding. Learning multiple modules in communication systems using DL is also a promising direction.

14.2.4 A Unified Method to Decode Multiple Codes

In the 5G eMBB scenario, polar codes and LDPC codes are used as the channel coding method for the control channel and data channel, respectively. Under this circumstance, we require reliable control signaling and high-throughput data transmission simultaneously, so a hybrid solution is needed. From the perspective of hardware, the challenge is that the hardware resource

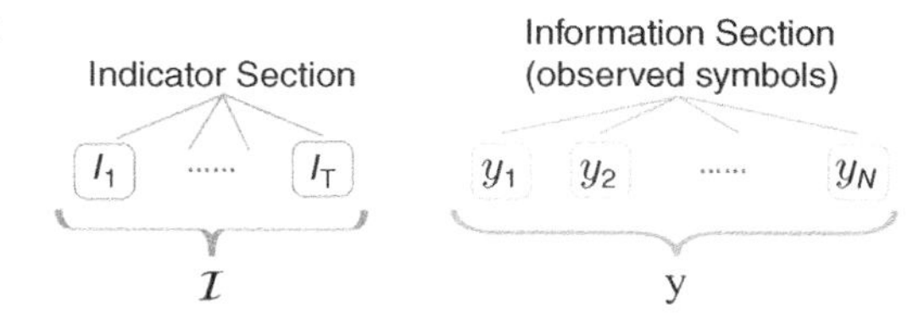

Figure 14.12 The proposed indicator section was put before the received sequence (Wang et al., 2018).

needs careful design of the decoding structure, and generalization to different coding rates will be challenging. From the perspective of DL, whether a DL-based decoding structure can be applied to support multiple coding schemes is still open.

In (Wang et al., 2018), the authors tackle this problem by inserting an indicator section to identify different coding types. The proposed unified approach exploits the similarity between neural networks and merged BP decoding algorithms. The proposed architecture is shown in Figure 14.12, including the indicator section and information section.

In the proposed design, the indicator section $\mathcal{I}$ and the information section (observed symbols) **y** are concatenated as the input layer for the network. The indicator section $\mathcal{I}$ varies when added before different coding types: for example, 1 for polar codes and -1 for LDPC codes. The function of the indicator section is to provide the neural network with the difference between the two coding types so that we can use one neural network to decode two types of codes.

Based on this idea, the entire architecture is given in Figure 14.13. The architecture puts both polar and LDPC encoded code bits into one neural network by adding the indicator section and tries to find a unified mapping algorithm. The neural network they used is a three-hidden-layer (512, 256, 128) DNN architecture.

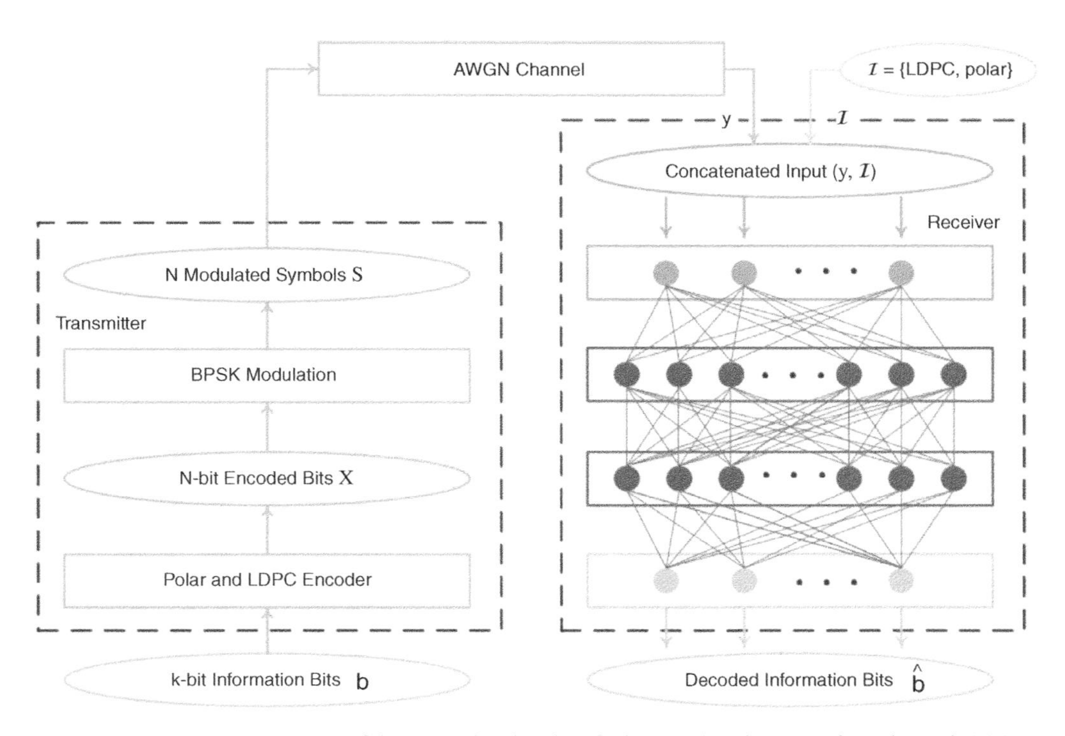

Figure 14.13 A system overview of the generalized and unified network architecture for polar and LDPC decoders (Wang et al., 2018).

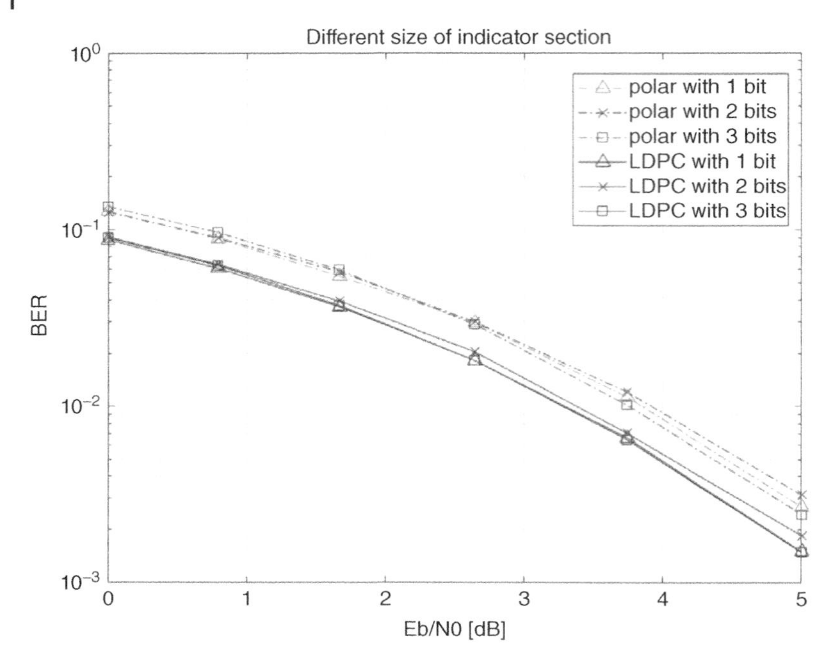

Figure 14.14 Performance comparison of different sizes of indicator sections. (Wang et al., 2018)

In the simulation part, the authors test both codes with length (16, 8) with different number of indicator sections and find that the indicator section with only one bit provides sufficient BER versus SNR performance for both the LDPC and polar cases. The result is shown in Figure 14.14.

In addition, the proposed DNN-based unified polar-LDPC decoder is compared with traditional BP decoders and isolated schemes. The comparison results of BER versus SNR performance are shown in Figure 14.15. In the polar decoding case, the proposed DNN-based unified polar-LDPC decoder achieves more or less the same decoding performance (less than 0.2 dB gap at BER equal to 10^{-2}) as conventional polar BP detection. This small gap may be due to the size of the network or the introduction of the indicator section. In the LDPC decoding case, the proposed solution outperforms the traditional LDPC BP decoding scheme by 0.8 dB at BER equal to 10^{-2}. From these results, we can see that the proposed unified decoder can approach or surpass traditional BP decoding performance.

The decoding efficiency of different decoding schemes is compared in Table 14.2. It can be seen that from the throughput point of view, the DNN-based approaches provide three orders of magnitude improvement if compared with traditional BP-based decoding schemes. Meanwhile, compared with the individual DNN-based polar or LDPC decoders, the unified approach provides nearly the same throughput (less than 0.4% loss) with marginal network overhead (e.g. less than 0.3% overhead in terms of total parameters) from the indicator section.

14.2.5 Summary

Despite a lot of research effort on using DNNs for decoding linear codes, there are still some limitations at present. The first is the problem of code length, because the complexity of the calculations increases exponentially as the code length increases. So, there is currently no good way to apply DL decoding to the standard code length of 5G. The second limitation is the problem of joint optimization. On the other hand, the combination of decoding and other

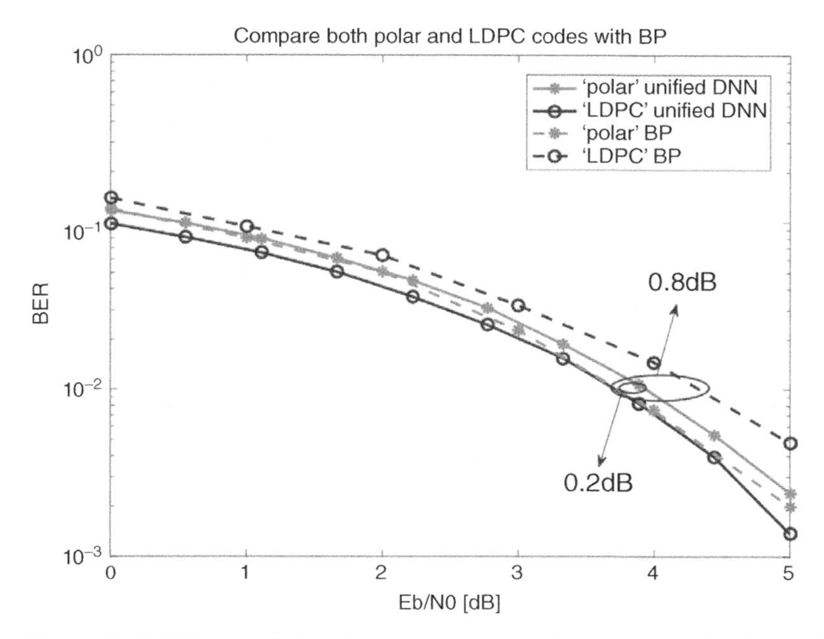

Figure 14.15 BER versus SNR performance comparison for the proposed DNN-based unified polar-LDPC decoder and conventional BP-based polar and LDPC decoders. (Wang et al., 2018)

Table 14.2 Throughput and network size comparison among different implementation schemes.

	Polar (DNN)	Polar (BP)	LDPC (DNN)	LDPC (BP)	Unified(polar/LDPC)
Input	$\mathbf{y}$	$\mathbf{y}$	$\mathbf{y}$	$\mathbf{y}$	$(\mathbf{y},\mathcal{I})$
Throughput (Kbps)	1.1485×10^3	4.59	1.1372×10^3	0.38	$1.14657/1.13379 \times 10^3$
Total parameters	173960	–	173960	–	174472

communication modules is another research direction for DL. These directions are likely to become new choices in the future by replacing traditional time-consuming algorithms with DNN-based solutions.

14.3 CNNs for Decoding

CNNs are commonly used to extract correlations for high-dimensional data in image processing, natural language processing, and computer vision tasks, but applications dealing with one-dimensional tasks are still under investigation. Instead of treating the specific channel code itself, most of the state-of-art results focus on eliminating correlated channel noise, as follows.

14.3.1 Decoding by Eliminating Correlated Channel Noise

In practical communication systems, there exists a type of noise called *correlated noise*. This is due to filtering, oversampling, channel fading, and multi-user interference. The difficulty in handling this issue mainly comes from the high complexity introduced by the colored noise.

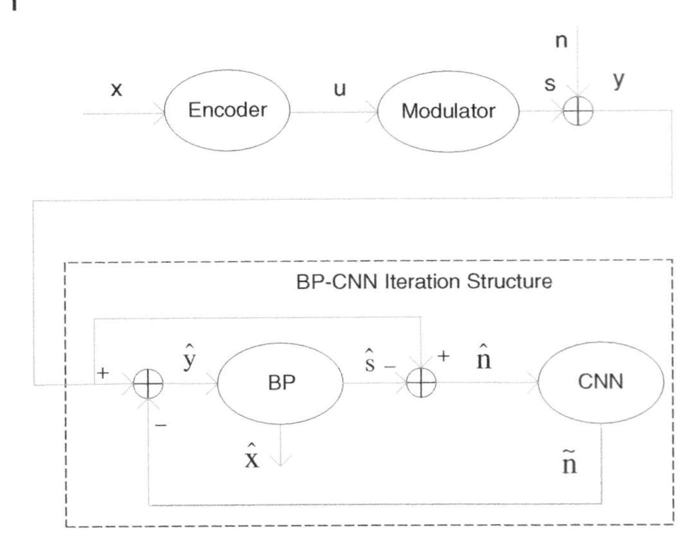

Figure 14.16 The proposed iterative decoding architecture, which consists of a BP decoder and a feed-forward CNN.

Traditional methods cannot have a balance between computational complexity and performance, but a DL-based method may solve this issue. Fortunately, some researchers have focused on this issue and have gotten some initial results.

The authors in (Liang et al., 2018) propose a novel iterative BP-CNN architecture for channel decoding under correlated noise. To train a well-behaved CNN model, the authors define a new loss function that involves not only the accuracy of the noise estimation but also the normality test for the estimation errors, i.e. to measure how likely it is that the estimation follows a Gaussian distribution.

Including a CNN network and a standard BP decoder, the proposed receiver architecture is shown in Figure 14.16. It should be noted that the function of the CNN is not only to estimate the channel noise by extracting the noise correlation but also to generate an output to the BP decoder through the feedback path.

We can see from this structure that after being encoded and modulated, the signals are added by correlated noise $\mathbf{n}$ of length N. Essentially, we find that the correlation in channel noise $\mathbf{n}$ can be considered a "feature" that may be exploited in channel decoding. Using $\tilde{\mathbf{n}}$ to denote the CNN output and subtracting it from the received vector $\mathbf{y}$ results in $\hat{\mathbf{y}} = \mathbf{y} - \tilde{\mathbf{n}} = \mathbf{s} + \mathbf{n} - \tilde{\mathbf{n}} = \mathbf{s} + \mathbf{r}$, where $\mathbf{r} = \mathbf{n} - \tilde{\mathbf{n}}$, which is the residual noise that denotes the difference between the output of the CNN and the true channel noise. The output of the CNN is fed back into the input of the BP decoder, which results in the structure of iterative decoding.

Based on this architecture, two types of loss functions are proposed. One is the traditional MSE, which is called *baseline BP-CNN*. This measures how close the output of the CNN and the true channel noise are. The expression is

$$\mathbf{Loss}_A = \frac{\| \mathbf{r} \|^2}{N} \tag{14.5}$$

The other loss function introduces a normality test so we can measure how likely it is that the residual noise samples follow a Gaussian distribution. The second loss function is called *enhanced BP-CNN*. The expression is

$$\mathbf{Loss}_B = \frac{\| \mathbf{r} \|^2}{N} + \lambda(S^2 + \frac{1}{4}(C - 3)^2) \tag{14.6}$$

where the second term, adopted from the Jarque-Bera test, represents a normality test to determine how much a dataset is modeled by a Gaussian distribution.

14.3.1.1 BP-CNN Reduces Decoding BER

One of the simulation results is reported as follows. Three different correlation factors η: 0.8, 0.5, 0 are used. η=0.8 represents a relatively strong correlation model: that is to say, the correlation in channel noise is strong. η=0.5 represents a moderate model. When η=0, the noise in the channel equals Gaussian noise. For the BER performance using these different factors, it is reported that in the strong (η=0.8) or moderate (η=0.5) correlation model, both baseline and enhanced BP-CNN can achieve performance gains in comparison with the traditional BP decoding method. Enhanced BP-CNN further outperforms the baseline strategy. This is because the enhanced method can reduce the residual noise and also reshape the distribution of the output, and thus is better suited for concatenation with the BP decoder. And with decreasing correlation factors, the performance of the neural networks declines. This is because as the correlation factors decrease, the information, i.e. the correlation in the noise that the neural network can learn, is also reduced. On the other hand, the reported results show that without correlation noise, the proposed NN method can also achieve BER performance comparable to that of the traditional BP decoder.

14.3.1.2 Multiple Iterations Between CNN and BP Further Improve Performance

Using K{BP(n)-CNN}-BP(n) to denote the iterative BP-CNN decoder structure with K iterations between BP and CNN, and n iterations inside BP, different K values and different n values can affect the entire decoding performance. Taking K=1,2,3,4, n=5, and the correlation η=0.8, the simulations and related analyses show that larger numbers of iterations can achieve greater improvement. More specifically, two BP-CNN iterations can improve decoding performance by 0.7 dB at BER = 10^{-4} compared to BP(5)-CNN-BP(5). In addition, after four BP-CNN iterations, the performance improvement becomes less obvious. This is because the CNN has reached its maximum capacity and cannot further reduce the residual noise power. Readers can refer to (Liang et al., 2018) to see more detailed simulation results.

14.3.2 Summary

Although CNNs cannot be directly applied to channel encoding or decoding at the current stage, they can be used to extract features of correlated noise and exploit the nonlinear relations among different BP iterations. Based on this understanding, we believe other iterative detection algorithms can be merged with CNNs to provide more reliable decoding performance.

14.4 RNNs for Decoding

As RNNs are able to extract time-domain correlations in traditional video streaming applications, a natural extension is to apply them to classical sequential codes. Meanwhile, RNNs can also be adopted to traditional BP or other types of decoding algorithms.

14.4.1 Using RNNs to Decode Sequential Codes

Sequential codes such as convolutional codes or Turbo codes are particularly attractive because they have a natural recurrent structure that is aligned with RNNs. In addition, sequential coding

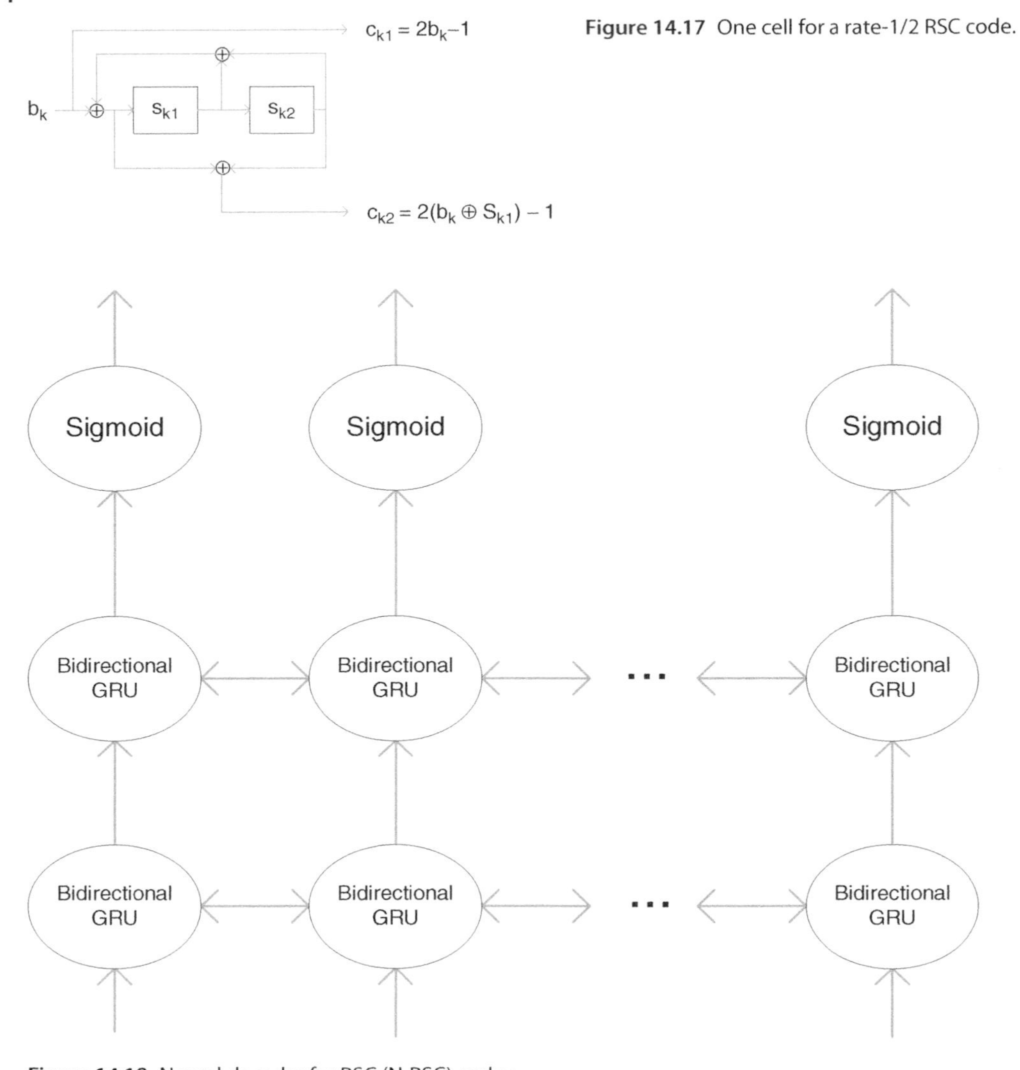

Figure 14.17 One cell for a rate-1/2 RSC code.

Figure 14.18 Neural decoder for RSC (N-RSC) codes.

schemes have many advantages such as arbitrary coding length and the arbitrary coding rate that can be achieved.

Based on this, in (Kim et al., 2018), the authors find that creatively designed and trained RNN architectures can decode well-known sequential codes such as recursive systematic convolutional (RSC) and Turbo codes.

What is considered in this RNN-based decoding scheme is a traditional convolutional code, which is rate-1/2 RSC code. The encoder structure is depicted in Figure 14.17, where we can see the encoded code is a time sequence.

This inspires us to use a RNN to form a decoder, called a neural decoder for RSC (N-RSC), by selecting bidirectional gated recurrent units (GRUs) as a building block, as shown in Figure 14.18. A two-layer architecture and batch normalization are used in the proposed solution, as well.

In the training stage, a mismatch between training SNR and testing SNR, i.e. $SNR_{train} = \min\{SNR_{test,0}\}$, can be used. To compare different traditional methods, such as MAP and Viterbi decoders, with the proposed N-RSC decoder, the authors of (Kim et al., 2018) considered two decoded bit lengths: 100 and 10,000. Their simulation result shows that the N-RSC can learn to decode and achieve BER performance comparable to that of the MAP and Viterbi decoders. In terms of generalization, the N-RSC trained on length 100 can be directly applied to codes of length 10,000 and still achieve optimal performance. That is to say, the N-RSC can generalize to unseen codewords.

Turbo codes are essentially concatenated convolutional codes. In the previous reference work, the authors also report that corresponding stacking of multiple layers of the convolutional neural decoders leads to a neural Turbo decoder that is able to achieve performance close to that of standard state-of-the-art Turbo decoders on the AWGN channel.

The proposed neural decoder for Turbo codes is called N-Turbo, which contains several N-BCJRs. The N-BCJR is an algorithm defined on a trellis diagram to maximize the posterior probability of error-correction code that is stacked by several N-RSC decoders. The N-BCJR architecture is a new type of N-RSC that can take flexible bitwise prior distribution as input.

The simulation compared with the traditional Turbo decoder shows that the proposed N-Turbo matches the performance of the Turbo decoder for block length 100. When the code length increases to 1000, the N-Turbo decoder also has performance similar to the Turbo decoder. There is also some research on how robust and adaptive the neural networks are, and a new coding scheme is proposed. Readers can refer to (Kim et al., 2018) to see more details.

14.4.2 Improving the Standard BP Algorithm with RNNs

The BP algorithm is a commonly used decoding method, which can be used to decode LDPC codes or polar codes. The BP decoding algorithm can be constructed from the Tanner graph, which can present the parity matrix that describes the code. Messages are transmitted over edges. Each node calculates its outgoing message based on all the incoming messages received over all the other edges.

The entire BP decoding process can be denoted as the following equations:

$$x_{i,e} = (v, c) = l_v + \sum_{e'=(v,c'),c'\neq c} x_{i-1,e'} \tag{14.7}$$

$$x_{i,e} = 2\tan\text{h}^{-1}\left(\prod_{e'=(v,c'),v'\neq v} \tanh\left(\frac{x_{i-1,e'}}{2}\right)\right) \tag{14.8}$$

The final vth output of the network is

$$o_v = l_v + \sum_{e'=(v,c')} x_{2\text{L},e'} \tag{14.9}$$

where L denotes the iteration number with the ith hidden layer, i=1,2,…,2L. Denote e=(v,c) as the index of the processing element in ith hidden layer. x is the output message of the processing element.

Although the performance of BP decoding is good enough for many decoding processes, its complexity is very high, and it is not user-friendly in hardware. In order to solve this problem, researchers have proposed a variety of improved BP decoders, including min-sum decoders, normalized min-sum (NMS), offset min-sum (OMS), neural normalized min-sum (NNMS), neural offset min-sum (NOMS), etc. (Nachmani et al., 2018).

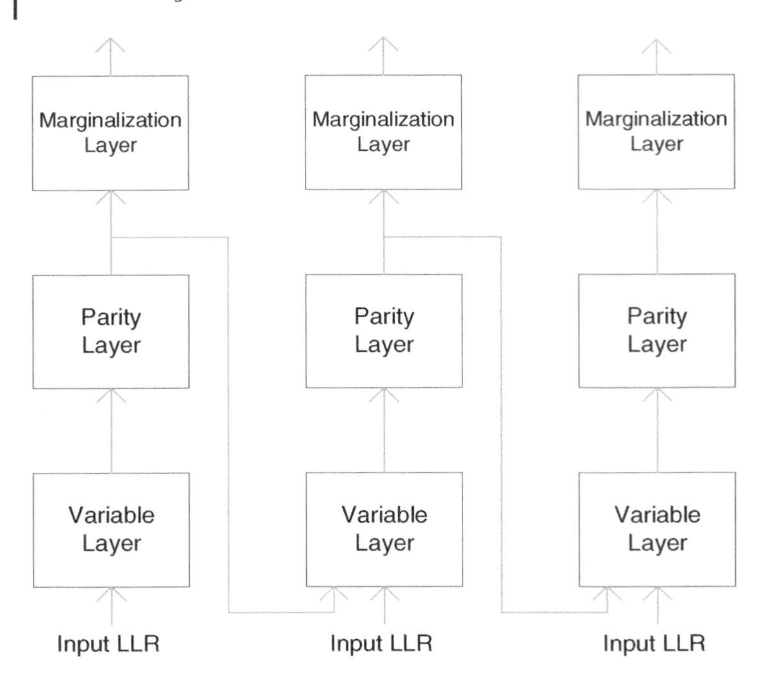

Figure 14.19 Recurrent neural network architecture with unfold three, which corresponds to three full BP iterations.

More importantly, (Nachmani et al., 2018) initiate some related work about RNN with BP decoding. They use the tied weights of the edges in the Tanner graph to represent a RNN called a BP-RNN. The modified equations are

$$x_{t,e=(v,c)} = \tanh\left(\frac{1}{2}\left(w_v l_v + \sum_{e'=(c',v),c'\neq c} w_{e,e'} x_{t-1,e'}\right)\right) \tag{14.10}$$

$$x_{t,e=(c,v)} = 2\tanh^{-1}\left(\prod_{e'=(v,c'),v'\neq v} x_{t,e'}\right) \tag{14.11}$$

where t is the iteration number, $x_{t,e=(v,c)}$ $(x_{e,t=(c,v)})$ denotes, in different iteration t, message from variable node v (check node c) to check node c (variable node v).

CE is used as the loss function. The proposed BP-RNN has a property that there exists a final marginalization layer after every time step. We can compute the loss after this layer. Furthermore, a multi-loss function illustrated in Eq.(14.12) can be used to increase the gradient update:

$$L(o,y) = -\frac{1}{N}\sum_{t=1}^{T}\sum_{v=1}^{N} y_v \log(o_{v,t}) + (1-y_v)\log(1-o_{v,t}) \tag{14.12}$$

The entire network architecture is illustrated in Figure 14.19. Nodes in different layers implement different equations: nodes in the variable layer implement Eq.(14.10), while nodes in the parity layer implement Eq.(14.11). Nodes in the marginalization layer implement a sigmoid function. The goal of this BP-RNN structure is to train the parameters w to minimize the multi-loss function.

Table 14.3 BER results for BCH (63, 45) code trained with a right-regular parity check matrix.

Decoding types	BP	BP-FF	BP-RNN	BP-RNN (multiloss)	BP-FF (multiloss)
Performance gain (dB)	0	0.8	1	1.3	1.3

Table 14.4 State-of-the-art results of utilizing DL in channel coding design.

Neural types	Applications in channel coding
DNN	Decoding linear codes. Joint optimization. Decoding multicodes.
CNN	Processing correlated noise in channel.
RNN	Decoding sequential codes. Optimizing traditional algorithms.

Based on this theoretical basis, several simulation results are presented in Table 14.3. Taking BCH (63, 45) as an example, these simulations compare different decoding types including BP, BP feed-forward (BP-FF), BP-RNN, BP-RNN (multiloss), and BP-FF (multiloss). It should be noted that in Table 14.3, BP at BER=10^{-4} is considered the baseline to measure the performance gain.

The previous reference work also introduces a modified random redundant iterative algorithm (mRRD) with a neural BP decoder and a relaxed method for BP-RNN using a relaxation γ. Readers can refer to (Nachmani et al., 2018) to see more details.

14.4.3 Summary

As wireless communications are performed in a time-correlated environment, RNNs are promising to improve general detection performance by carefully learning time-domain information. However, RNNs rely on the recurrent structure to improve detection performance, usually with a significant processing delay. Thus, a low-latency RNN-based decoding scheme will be more desirable.

14.5 Conclusions

In this chapter, we have summarized the state-of-the-art research progresses utilizing DL technology in channel coding design. As concluded in Table 14.4, more research is needed on different types of neural networks suitable for different tasks in channel coding areas, and a unified framework to understand the fundamental relations is yet to be explored. Moreover, whether DL technology can be applied to design a universal encoder/decoder has not been answered yet. As a result, jointly considering DL technology and channel code schemes is still a promising research direction in the near future.

Bibliography

E. Arıkan. Channel polarization: A method for constructing capacity-achieving codes for symmetric binary-input memoryless channels. *IEEE Transactions on Information Theory*, 55 (7): 3051–3073, Jul. 2009.

Claude Berrou, Alain Glavieux, and Punya Thitimajshima. Near Shannon limit error-correcting coding and decoding: Turbo-codes. 1. In *Communications, 1993. ICC'93 Geneva. Technical Program, Conference Record, IEEE International Conference on*, volume 2, pages 1064–1070. IEEE, 1993.

Sebastian Cammerer, Tobias Gruber, Jakob Hoydis, and Stephan Ten Brink. Scaling deep learning-based decoding of polar codes via partitioning. *arXiv preprint arXiv:1702.06901*, 2017.

Ronan Collobert and Jason Weston. A unified architecture for natural language processing: Deep neural networks with multitask learning. In *Proceedings of the 25th International Conference on Machine learning*, pages 160–167. ACM, 2008.

Nghia Doan, Seyyed Ali Hashemi, and Warren J. Gross. Neural successive cancellation decoding of polar codes. In *2018 IEEE 19th International Workshop on Signal Processing Advances in Wireless Communications (SPAWC)*, pages 1–5. IEEE, 2018.

Robert Gallager. Low-density parity-check codes. *IRE Transactions on Information Theory*, 8 (1): 21–28, 1962.

Felix A. Gers, Jürgen Schmidhuber, and Fred Cummins. Learning to forget: Continual prediction with LSTM. IEE, 1999.

Xavier Glorot, Antoine Bordes, and Yoshua Bengio. Deep sparse rectifier neural networks. In *Proceedings of the 14th International Conference on Artificial Intelligence and Statistics*, pages 315–323, 2011.

Alex Graves and Jürgen Schmidhuber. Framewise phoneme classification with bidirectional LSTM and other neural network architectures. *Neural Networks*, 18 (5–6): 602–610, 2005.

Tobias Gruber, Sebastian Cammerer, Jakob Hoydis, and Stephan ten Brink. On deep learning-based channel decoding. In *IEEE Annual Conference on Information Sciences and Systems (CISS)*, pages 1–6, 2017.

Richard W. Hamming. Error detecting and error correcting codes. *Bell System technical journal*, 29 (2): 147–160, 1950.

Geoffrey Ye Li Hao Ye. Initial results on deep learning for joint channel equalization and decoding. pages 1–5, Toronto, ON, Canada, 2017.

Kaiming He, Xiangyu Zhang, Shaoqing Ren, and Jian Sun. Deep residual learning for image recognition. In *Proceedings of the IEEE Conference on Computer Vision and Pattern Recognition*, pages 770–778, 2016.

Geoffrey Hinton, Li Deng, Dong Yu, George E. Dahl, Abdel-rahman Mohamed, Navdeep Jaitly, Andrew Senior, Vincent Vanhoucke, Patrick Nguyen, Tara N. Sainath, et al. Deep neural networks for acoustic modeling in speech recognition: The shared views of four research groups. *Signal Processing*, 29 (6): 82–97, 2012.

Aaron Courville Ian Goodfellow, Yoshua Bengio. *Deep Learning*. MIT Press, 2016. www .deeplearningbook.org.

Norman P. Jouppi, Cliff Young, Nishant Patil, David Patterson, Gaurav Agrawal, Raminder Bajwa, Sarah Bates, Suresh Bhatia, Nan Boden, Al Borchers, et al. In-datacenter performance analysis of a tensor processing unit. In *Computer Architecture (ISCA), 2017 ACM/IEEE 44th Annual International Symposium on*, pages 1–12. IEEE, 2017.

Hyeji Kim, Yihan Jiang, Ranvir Rana, Sreeram Kannan, Sewoong Oh, and Pramod Viswanath. Communication algorithms via deep learning. *arXiv preprint arXiv:1805.09317*, 2018.

Alex Krizhevsky, Ilya Sutskever, and Geoffrey E. Hinton. Imagenet classification with deep convolutional neural networks. In *Advances in neural information processing systems*, pages 1097–1105, 2012.

Fei Liang, Cong Shen, and Feng Wu. An iterative BP-CNN architecture for channel decoding. *IEEE Journal of Selected Topics in Signal Processing*, 12 (1): 144–159, 2018.

Geert Litjens, Thijs Kooi, Babak Ehteshami Bejnordi, Arnaud Arindra Adiyoso Setio, Francesco Ciompi, Mohsen Ghafoorian, Jeroen Awm Van Der Laak, Bram Van Ginneken, and Clara I. Sánchez. A survey on deep learning in medical image analysis. *Medical Image Analysis*, 42: 60–88, 2017.

David J.C. MacKay and Radford M. Neal. Near Shannon limit performance of low density parity check codes. *Electronics Letters*, 32 (18): 1645–1646, 1996.

Eliya Nachmani, Elad Marciano, Loren Lugosch, Warren J. Gross, David Burshtein, and Yair Be'ery. Deep learning methods for improved decoding of linear codes. *IEEE Journal of Selected Topics in Signal Processing*, 12 (1): 119–131, 2018.

Anh Nguyen, Jason Yosinski, and Jeff Clune. Deep neural networks are easily fooled: High confidence predictions for unrecognizable images. In *Proceedings of the IEEE Conference on Computer Vision and Pattern Recognition*, pages 427–436, 2015.

Timothy O'Shea and Jakob Hoydis. An introduction to deep learning for the physical layer. *IEEE Transactions on Cognitive Communications and Networking*, 3 (4): 563–575, 2017.

Edward Rosten and Tom Drummond. *Machine learning for high-speed corner detection*. In *European Conference on Computer Vision*, pages 430–443. Springer, 2006.

David Silver, Aja Huang, Chris J. Maddison, Arthur Guez, Laurent Sifre, George Van Den Driessche, Julian Schrittwieser, Ioannis Antonoglou, Veda Panneershelvam, Marc Lanctot, et al. Mastering the game of Go with deep neural networks and tree search. *Nature*, 529 (7587): 484, 2016.

Karen Simonyan and Andrew Zisserman. Very deep convolutional networks for large-scale image recognition. *arXiv preprint arXiv:1409.1556*, 2014.

Yaohan Wang, Zhizhao Zhang, and Xu Shugong Zhang Shunqing, Cao Shan. A unified deep learning based polar-LDPC decoder for 5G communication systems. In *2018 The Tenth International Conference on Wireless Communications and Signal Processing (WCSP)*, pages 1–6, Hangzhou, China, 2018. IEEE.

Weihong Xu, Zhiwei Zhong, Xiaohu You, and Chuan Zhang. Joint neural network equalizer and decoder. In *2018 15th International Symposium on Wireless Communication Systems (ISWCS)*, pages 1–5. IEEE, 2018.

Xin Yan, Fei Long, Jingshuai Wang, Na Fu, Weihua Ou, and Bin Liu. Signal detection of MIMO-OFDM system based on auto encoder and extreme learning machine. In *Neural Networks (IJCNN), 2017 International Joint Conference on*, pages 1602–1606. IEEE, 2017.

15

Deep Learning Techniques for Decoding Polar Codes

Warren J. Gross, Nghia Doan, Elie Ngomseu Mambou, and Seyyed Ali Hashemi

Department of Electrical and Computer Engineering, McGill University, Montreal, Quebec, Canada

In recent decades, the challenge of designing capacity-approaching codes has been one of the main focuses in digital communications. In this regard, polar codes were introduced as the first class of error-correcting codes that provably achieve the capacity for any channel at infinite code length. Recently, deep learning (DL) has shown great potential in a wide range of applications in digital communications including channel coding for forward error correction (FEC) codes. Therefore, we believe that a literature review on the intersection between DL and FEC codes, especially polar codes, can contribute to the coding community. Organized into four sections, this chapter first provides background and motivation for the use of DL in various FEC schemes used for wireless communication systems. Section 15.2 introduces polar codes and their traditional decoding algorithms. In Section 15.3, three major DL-based approaches for decoding polar codes are presented in terms of decoding performance, algorithm complexity, and decoding latency. The last section of this chapter, Section 15.4, offers further discussion and conclusions.

15.1 Motivation and Background

Deep learning (DL) (LeCun et al., 2015) has been widely used in digital communications through applications such as channel decoding, prediction, equalization, modulation/demodulation, detection, quantization, compression, and spectrum sensing (Ibnkahla, 2000). In (O'Shea and Hoydis, 2017), a new approach in communication systems based on DL was introduced, where the entire channel transmission was abstracted as an autoencoder: that is, an end-to-end pipeline designed by joint concatenation of transmitters and receivers in a single process. It has been established that the block error rate (BLER) of traditional communication systems can be improved through this scheme. Furthermore, this idea of end-to-end channel modeling was extended to multiple-input and multiple-output (MIMO) systems. MIMO suffers from interference between channels in a way that obtaining the optimal signal is very difficult. It has been shown that such systems can be represented as multiple-input and -output generative adversarial nets (Goodfellow et al., 2014), which can be optimized using a customized loss function for the joint model. The challenges and opportunities related to DL for the wireless physical layer were presented in (Wang et al., 2017).

In (Farsad et al., 2018), a DL-based approach for joint source-channel coding of a text was proposed. This approach is quite similar to the idea presented in (O'Shea and Hoydis, 2017), where the source is jointly trained with the channel to reduce transmission distortion. This

Machine Learning for Future Wireless Communications, First Edition. Edited by Fa-Long Luo.
© 2020 John Wiley & Sons Ltd. Published 2020 by John Wiley & Sons Ltd.

architecture makes use of a long short-term memory (LSTM)-based model, as proposed in (Graves and Schmidhuber, 2005, Hasim et al., 2014), over a binary erasure channel (BEC) to ensure a gain in word error rate (WER) when compared to separate source-channel word processing. However, the fixed-length word-processing scheme does not allow flexibility at the incoming source. Similarly, an end-to-end neural network (NN) system was proposed for communication over the air between two software-defined radios (SDRs) in (Dörner et al., 2018).

The concept of an end-to-end NN was also extended for MIMO relays in (Sun and Jing, 2012). In this work, channel training and coherent decoding under estimated channel error were studied for relay networks, and the setup of a single-antenna as transmitter, R distinct antenna relays, and a single R antenna receiver is considered. Note that many coherent cooperative decoding methods assume perfect and global channel state information (CSI), which is in reality imperfect, especially in cooperative relay networks. Two types of approaches were considered: decoding where CSI is assumed noise-free, and matched decoding where the noise estimation is evaluated. Simulations demonstrate that at least $3R$ symbol intervals of training are needed for mismatched decoding versus $R + 2$ for matched decoding to achieve full diversity. In addition, adaptive decoding was presented to compensate for the complexity in the matched decoding. Several works concerning NNs applied to MIMO systems can be found in (Wen et al., 2018, O'Shea and Hoydis, 2017, Samuel et al., 2017).

It was established that DL decoding for linear block codes is equivalent to deriving the maximum energy function of a NN (Bruck and Blaum, 1989). In order to maximize the energy function of a NN, (Bruck and Blaum, 1989) suggested that decoding of FEC codes can be done through maximizing polynomials over the N-cube for a (N, K) block code, where N is the length and K is the dimension of the code. With similar logic, (Tallini and Cull, 1995) predicted that the NP-complete problem of receiving an error-free message through a channel can be solved through NN. In (Wu et al., 2002), a neural structure was described as a perceptron with a higher-order polynomial as a discriminant function. A $(2^m - 1, 2^m - 1 - m)$ Hamming code was decoded through only $m + 1$ assigned weights on each perceptron, m being a positive integer. This architecture combines two layers: the first is made of a set of parity bits, and the second is a linear classifier. This proves that high-order codes such as Bose Chaudhuri Hocquenghem (BCH) codes can be learned successfully through a multilayer perceptron (MLP).

In channel coding, DL-based decoders can provide improvements in error probabilities over conventional decoders. In (Nachmani et al., 2018), the conventional belief propagation (BP) decoding algorithm is formalized as a partially connected NN. In addition, by assigning trainable weights to the BP-based NN, neural BP decoders can achieve the same performance as the conventional sum-product BP decoding algorithm with a significantly smaller number of iterations. In (Kim et al., 2018), the sequential decoding of turbo and convolutional codes (Berrou et al., 1993, Viterbi, 1971) was performed through recurrent neural network (RNN)-based models. It has been demonstrated that a trained RNN architecture can decode these codes over additive white Gaussian noise (AWGN) channels with performance near that of the maximum a posteriori (MAP) decoding given by the Bahl-Cocke-Jelinek-Raviv (BCJR) algorithm and the maximum likelihood (ML) decoding given by Viterbi algorithm for sequential codes. This result was confirmed in (Kim et al., 2018) through the same architecture and extended to codes with larger block lengths. In (Bennatan et al., 2018), a syndrome-based DL technique was proposed to decode linear block codes. For a BCH code of length 127, an 11-layer vanilla MLP and a RNN of 4 stacks were considered separately to estimate the channel noise rather than the transmitted codeword. The syndrome decoding and soft channel reliabilities were helpful to eliminate the problem of overfitting of the training codeword set. BCH decoding performance was compared under traditional BP decoding, syndrome-based vanilla MLP, and syndrome-based stacked RNN. Results showed that syndrome-based stacked

RNN performance approached that of the ordered statistics decoding (OSD) of order 2 while outperforming that of the syndrome-based vanilla MLP at the cost of high complexity and latency.

In the following section, we will focus on DL techniques used for polar codes as a case study of DL for FEC codes.

15.2 Decoding of Polar Codes: An Overview

Polar codes are a recent breakthrough in the field of channel coding, as they were proven to achieve channel capacity with efficient encoding and decoding algorithms (Arıkan, 2009). Successive cancellation (SC) and BP decoding algorithms are first introduced to decode polar codes (Arıkan, 2009). Although SC decoding can provide a low-complexity implementation, its serial nature prevents the decoder from reaching a high decoding throughput. In addition, the error-correction performance of SC decoding for short to moderate polar codes does not satisfy the requirements of the fifth generation of cellular mobile communications (5G). SC list (SCL) decoding was introduced in (Tal and Vardy, 2015) to improve the performance of SC decoding. SCL can provide a significant improvement in terms of error probability if the decoder is aided by a cyclic redundancy check (CRC). With these appealing properties, polar codes have been selected to be used in the enhanced mobile broadband (eMBB) control channel of 5G, together with a CRC (3GPP, 2018).

Recently, it has been shown that polar codes can also be decoded using off-the-shelf DL decoders, which may lead to high decoding throughput thanks to their one-shot-decoding property (Cammerer et al., 2017). In addition, it was observed that by assigning trainable weights to the unrolled factor graph of polar codes, neural BP decoders can provide a significant error-correction performance gain over conventional BP decoding (Nachmani et al., 2018). Other approaches such as using DL models for channel noise estimation have also demonstrated great potential for DL techniques when applied to well-established problems in channel coding (Bennatan et al., 2018, Liang et al., 2018).

In this section, we first provide some basic knowledge about polar codes and conventional polar decoders. In the next section, several DL-based decoding algorithms and their variants for polar codes are discussed, followed by a detailed evaluation concerning error-correction performance and decoding latency of state-of-the-art DL-aided decoders for a 5G polar code.

15.2.1 Problem Formulation of Polar Codes

A polar code $\mathcal{P}(N, K)$ of length N with K information bits is constructed by applying a linear transformation to the message word $\boldsymbol{u} = \{u_0, u_1, \ldots, u_{N-1}\}$ as $\boldsymbol{x} = \boldsymbol{u}\boldsymbol{G}^{\otimes n}$ where $\boldsymbol{x} = \{x_0, x_1, \ldots, x_{N-1}\}$ is the codeword, $\boldsymbol{G}^{\otimes n}$ is the n-th Kronecker power of the polarizing matrix $\boldsymbol{G} = \begin{bmatrix} 1 & 0 \\ 1 & 1 \end{bmatrix}$, and $n = \log_2 N$. The vector $\boldsymbol{u}$ contains a set $\mathcal{A}$ of K information bits and a set $\mathcal{A}^c$ of $N - K$ frozen bits. The locations of the frozen bits are known to both the encoder and the decoder, and their values are set to 0. The codeword $\boldsymbol{x}$ is then modulated and sent through the channel where binary phase-shift keying (BPSK) modulation and AWGN channel model are considered. The soft vector of the transmitted codeword received by the decoder in this setting can be written as

$$\boldsymbol{y} = (\mathbf{1} - 2\boldsymbol{x}) + \boldsymbol{z}, \tag{15.1}$$

where $\mathbf{1}$ is an all-one vector of size N, and $\boldsymbol{z} \in \mathbb{R}^N$ is the AWGN vector with variance σ^2 and zero mean. In the log-likelihood ratio (LLR) domain, the LLR vector of the transmitted codeword is

$$\mathbf{LLR}_{\boldsymbol{x}} = \ln \frac{\Pr(\boldsymbol{x} = 0|\boldsymbol{y})}{\Pr(\boldsymbol{x} = 1|\boldsymbol{y})} = \frac{2\boldsymbol{y}}{\sigma^2} . \tag{15.2}$$

15.2.2 Successive-Cancellation Decoding

The SC decoding algorithm can be illustrated on a polar-code factor graph as shown in Figure 15.1a, where $N = 8$ and $K = 4$. The messages are propagated through the SC processing elements (SCPEs) as shown in Figure 15.1b, where $\beta_{t,s}$ denotes a left-to-right message and $\alpha_{t,s}$ denotes a right-to-left message of the t-th bit index at stage s of the factor graph.

In SC decoding, the right-to-left messages $\alpha_{t,s}$ are soft LLR values, and the left-to-right messages $\beta_{t,s}$ are hard decision bits. The messages in the factor graph are updated as

$$\alpha_{t,s} = f(\alpha_{t,s+1}, \alpha_{t+2^s,s+1}) , \tag{15.3}$$

$$\alpha_{t+2^s,s} = g(\alpha_{t,s+1}, \alpha_{t+2^s,s+1}, \beta_{t,s}) , \tag{15.4}$$

$$\beta_{t,s+1} = \beta_{t,s} \oplus \beta_{t+2^s,s} , \tag{15.5}$$

$$\beta_{t+2^s,s+1} = \beta_{t+2^s,s} , \tag{15.6}$$

where

$$f(a, b) = 2\text{arctanh}\left(\tanh\left(\frac{a}{2}\right)\tanh\left(\frac{b}{2}\right)\right) \approx \text{sgn}\,(a)\ \text{sgn}\,(b)\min\,(|a|, |b|) , \tag{15.7}$$

$$g(a, b, c) = b + (1 - 2c)a, \tag{15.8}$$

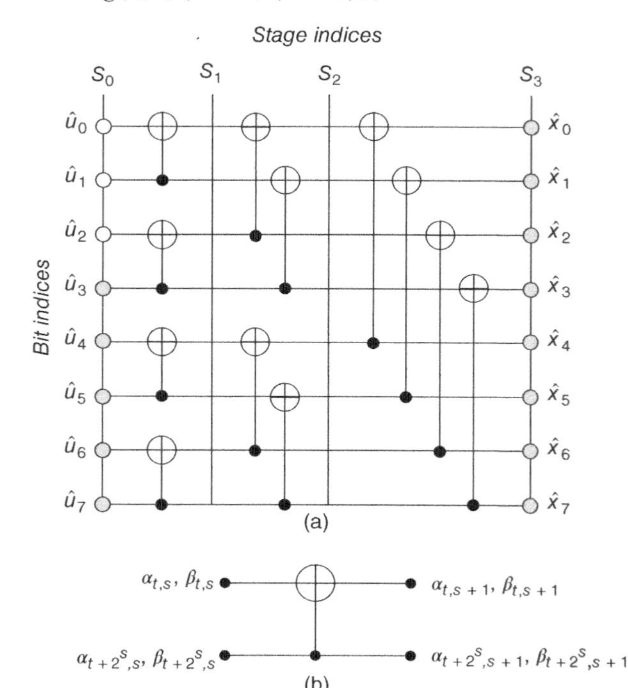

Figure 15.1 (a) Factor graph representation of a polar code with $N = 8$, $K = 4$, and $\{u_0, u_1, u_2\} \in \mathcal{A}^c$; (b) an SC processing element (SCPE).

and $\oplus$ is the bitwise XOR operation. SC decoding is initialized by setting $\alpha_{t,n} = y_t$, and the decoding schedule is such that the bits are decoded one by one from u_0 to u_{N-1}. At layer 0, the elements of $\boldsymbol{u}$ are estimated as

$$\hat{u}_t = \begin{cases} 0, & \text{if } u_t \in \mathcal{A}^c \text{ or } \alpha_{t,0} \geq 0, \\ 1, & \text{otherwise.} \end{cases} \tag{15.9}$$

15.2.3 Successive-Cancellation List Decoding

SCL decoding improves the error-correction performance of SC decoding by running multiple SC decoders in parallel. Instead of using Eq. (15.9) to estimate $\boldsymbol{u}$ as in SC decoding, each bit is estimated considering both its possible values 0 and 1. Therefore, at each bit estimation, the number of candidates doubles. In order to constrain the high complexity of SCL decoding, at each bit estimation, a set of only L candidates are allowed to survive, based on a path metric that is calculated as (Balatsoukas-Stimming et al., 2015, Hashemi et al., 2016)

$$\mathrm{PM}_{t_\ell} = \sum_{j=0}^{t} \ln(1 + \mathrm{e}^{-(1-2\hat{u}_{j_\ell})\alpha_{j,0_\ell}}), \tag{15.10}$$

$$\approx \frac{1}{2} \sum_{j=0}^{t} \mathrm{sgn}(\alpha_{j,0_\ell})\alpha_{j,0_\ell} - (1 - 2\hat{u}_{j_\ell})\alpha_{j,0_\ell}, \tag{15.11}$$

where ℓ is the path index and $\hat{u}_{j_\ell}$ is the estimate of bit j at path ℓ.

15.2.4 Belief Propagation Decoding

Figure 15.2a demonstrates BP decoding on the factor graph representation of $\mathcal{P}(8, 5)$. The messages are iteratively propagated through the BP processing elements (BPPEs) (Arıkan,

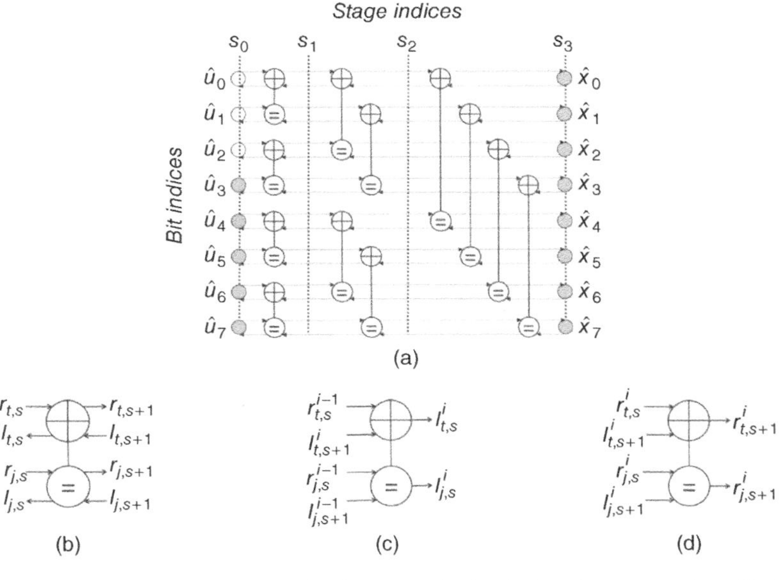

Figure 15.2 (a) BP decoding on the factor graph of $\mathcal{P}(8, 5)$ with $\{u_0, u_1, u_2\} \in \mathcal{A}^c$; (b) a BPPE; (c) a right-to-left message update of a BPPE on an unrolled factor graph; (d) a left-to-right message update of a BPPE on an unrolled factor graph.

2010) located in each stage. Each update iteration starts with a right-to-left message pass that propagates the LLR values from the channel (rightmost) stage to the information bit (leftmost) stage, and ends with the left-to-right message pass, which occurs in the opposite order. Figure 15.2b illustrates a BPPE with its corresponding soft messages, where $r_{t,s}$ denotes a left-to-right message and $l_{t,s}$ denotes a right-to-left message of the t-th bit index at stage s. One can also apply BP decoding on the unrolled polar-code factor graph (Doan et al., 2018b); thus the BP iterations in this setup are performed sequentially. Figures 15.2c and 15.2d illustrate the input and output messages of a BPPE for the right-to-left and left-to-right message updates on an unrolled factor graph, where the superscript i denotes the iteration number. The update rule (Arıkan, 2010) for the right-to-left messages of a BPPE is

$$\begin{cases} l_{t,s}^{i} &= f(l_{t,k}^{i}, r_{j,s}^{i-1} + l_{j,k}^{i}), \\ l_{j,s}^{i} &= f(l_{t,k}^{i}, r_{t,s}^{i-1}) + l_{j,k}^{i}, \end{cases} \tag{15.12}$$

and for the left-to-right messages is

$$\begin{cases} r_{t,k}^{i} &= f(r_{t,s}^{i}, l_{j,k}^{i} + r_{j,s}^{i}), \\ r_{j,k}^{i} &= f(r_{t,s}^{i}, l_{t,k}^{i}) + r_{j,s}^{i}, \end{cases} \tag{15.13}$$

where $j = t + 2^s$, $k = s + 1$.

The BP decoding performs a predetermined $I_{\max}$ update iterations, where the messages are propagated through all BPPEs in accordance with Eqs. (15.12) and (15.13). Initially, for $0 \leq t < N$ and $\forall i \leq I_{\max}$, $l_{t,n}^{i}$ are set to the received channel LLR values $\mathbf{LLR}_{x}$, and $r_{t,0}^{i}$ are set to the LLR values of the information and frozen bits as

$$\mathbf{LLR}_{\mathcal{A} \cup \mathcal{A}^c} = \begin{cases} 0, & \text{if } u_t \in \mathcal{A}, \\ +\infty, & \text{if } u_t \in \mathcal{A}^c \, . \end{cases} \tag{15.14}$$

All the other left-to-right and right-to-left messages of the BPPEs at the first iteration are set to 0. After running $I_{\max}$ iterations, the decoder makes a hard decision on the LLR values of the t-th bit at the information bit stage to obtain the estimated message word as

$$\hat{u}_t = \begin{cases} 0, & \text{if } r_{t,0}^{I_{\max}} + l_{t,0}^{I_{\max}} \geq 0, \\ 1, & \text{otherwise.} \end{cases} \tag{15.15}$$

15.3 DL-Based Decoding for Polar Codes

As mentioned earlier, this section is devoted to the use of DL in decoding polar codes with emphasis on off-the-shelf DL decoders and DL-aided decoders by addressing their working principles, algorithm details, and performance evaluations.

15.3.1 Off-the-Shelf DL Decoders for Polar Codes

In (Gruber et al., 2017), it was shown that a MLP decoder can generalize structured codes, e.g. polar codes, more effectively than random codes, where the MAP performance was obtained for structured codes but not for random codes. However, the considered code length in (Gruber et al., 2017) is limited to 16. A more detailed investigation was carried out in (Lyu et al., 2018), where a comparison was performed between different off-the-shelf network models including MLP, a convolutional neural network (CNN), and a RNN for polar codes. The RNN model outperforms the MLP and CNN models in terms of error-correction performance, at the cost of the highest decoding complexity. On the other hand, the CNN model provided a performance gain over the MLP model, with higher computational time. It was also observed in (Lyu et al., 2018) that the code length impacts the fitting (underfitting versus overfitting) of the deep NN, and each type of NN has a saturation code length, which is related to the learning capabilities of the model.

For all the aforementioned off-the-shelf decoders, the networks are formalized to solve a multi-category classification problem where the correct codewords are used as the training labels and the corresponding values of $\mathbf{LLR}_x$ are used as the network's input. Normally, the size of the DL decoder scales with the size of the codeword and the natural architecture of the DL models in use. Finding the network parameters or weights is done by backpropagation (LeCun et al., 2015) with various optimization methods such as ADAM (Kingma and Ba, 2014) or RMSPROP (Hinton et al.). Polar decoding under off-the-shelf DL decoders is carried out by performing the inference phase of the trained DL models, given the channel LLR values.

The main problem associated with off-the-shelf DL decoders when applied to polar codes or other linear block codes is *the curse of dimensionality* (Gruber et al., 2017), which states that the number of required training samples scales exponentially with code length. To overcome this issue, a scaling approach was introduced in (Cammerer et al., 2017) constraining DL decoders to only work with sub-codes with small code sizes. Specifically, a partitioned NN (PNN) decoder for a polar code of size 128 was proposed. The considered polar code was divided into smaller sub-blocks, and the partitioned DL decoders are trained individually so that the performance obtained for each sub-block was close to that of MAP decoding. However, the bit-error-rate (BER) performance of the integrated system is only similar to that of SC decoding. It is worth mentioning that the latency of the proposed decoder can be reduced as parallel computations can be exploited for the DL decoders thanks to their one-shot-decoding property.

In (Doan et al., 2018a), a neural decoder was introduced on the basis of the partitioning idea of (Cammerer et al., 2017). The proposed neural SC (NSC) in (Doan et al., 2018a) preserves the same decoding performance in terms of BER and frame error rate (FER) as that of PNN, with a decoding latency improved by 42.5% for a polar code of length 128 and rate 0.5.

15.3.2 DL-Aided Decoders for Polar Codes

15.3.2.1 Neural Belief Propagation Decoders

In contrast to off-the-shelf DL decoders, another approach is to exploit *domain knowledge* to design DL-aided decoders. In (Nachmani et al., 2018), a deep neural BP decoder is proposed to improve conventional BP decoding where trainable weights are assigned to the edges of the unrolled factor graph. The deep network in this case is constrained to be a partially connected NN, and its inference functions resemble the operations of conventional BP decoding. This idea was first evaluated on high-density parity check (HDPC) codes in (Nachmani et al., 2018) and then on polar codes in (Doan et al., 2018b). It was observed that the trainable weights help the conventional BP decoding to mitigate the detrimental effects of the code's short cycles, which are often found in practical linear codes.

Another problem associated with BP decoding is the costly sum-product (SP) algorithm (Ryan and Lin, 2009). Instead, a low complexity min-sum (MS) algorithm is used in practical applications (Ryan and Lin, 2009). However, the MS algorithm also introduces decoding errors due to its poor estimation compared to the SP algorithm. With the objective of tackling this challenge, neural offset min-sum (NOMS) decoding was proposed in (Lugosch and Gross, 2017). The NOMS algorithm trains offset parameters and uses them to correct the MS approximation. It should be noted that NOMS decoding only requires additions, which makes this decoder particularly attractive for hardware implementation.

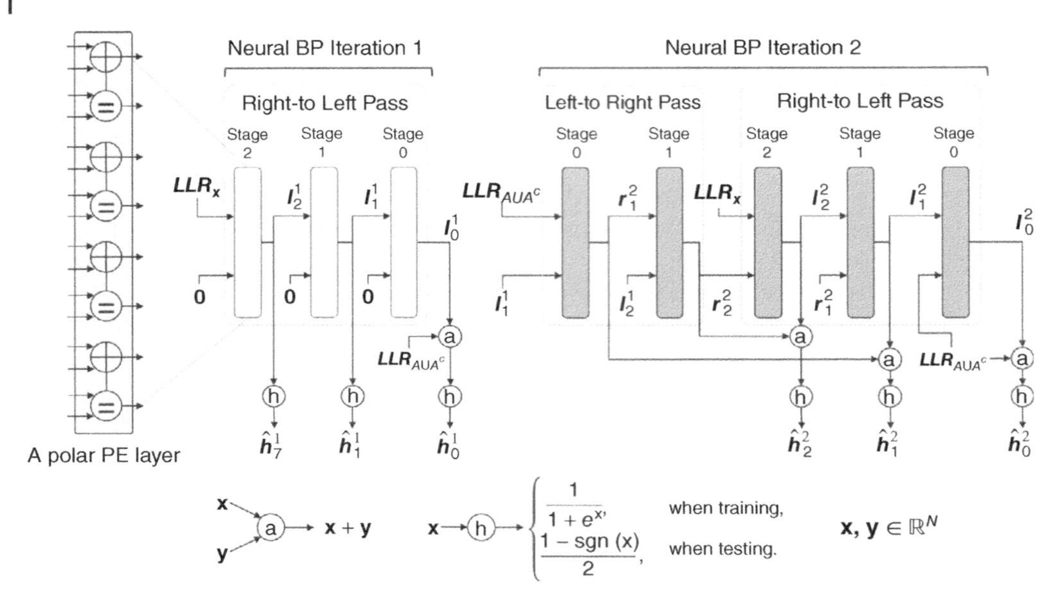

Figure 15.3 A neural BP decoder architecture with two iterations for $\mathcal{P}(8,5)$.

In (Nachmani et al., 2018), the architectures from (Nachmani et al., 2018) and (Lugosch and Gross, 2017) were changed to resemble an RNN by reusing the weights at each iteration; this reduces their complexity, as fewer parameters are needed. Furthermore, an RNN architecture based on a successive relaxation technique was constructed, which further improved the proposed RNN-like neural BP decoders.

In (Xu et al., 2017), a neural normalized min-sum (NNMS) decoder was developed, which adapts the idea in (Nachmani et al., 2018) for the case of polar codes. NNMS also uses a multiplicative weight to correct the min-sum approximation. This setup can be scaled to large size polar codes while still maintaining low decoding latency and complexity.

It was shown in (Ren et al., 2015) that the CRC capability is only used as an early stopping criterion with incremental error-correction performance improvement for BP decoding of polar codes. In (Doan et al., 2018b), by assigning trainable weights to the CRC-Polar concatenated graph, the proposed decoder has shown a performance gain of 0.5 dB over the conventional CRC-aided BP at the FER of 10^{-5}, for a 5G polar code of length 128. The authors in (Doan et al., 2018b) also derived a general neural BP decoder architecture specified for polar codes. Figure 15.3 illustrates an example of this architecture where the weights are shared between each neural BP decoding iteration. The weight-assignment schemes of state-of-the-art neural BP decoders when applied to the BPPE update functions in Eqs. (15.12) and (15.13) are summarized as follows:

- NNMS-RNN (Nachmani et al., 2018)

$$\begin{cases} l^i_{t,s} &= w_0 f(l^i_{t,k}, w_1 r^{i-1}_{j,s} + w_2 l^i_{j,k}), \\ l^i_{j,s} &= w_4(w_3 f(l^i_{t,k}, r^{i-1}_{t,s})) + w_5 l^i_{j,k}, \end{cases} \tag{15.16}$$

$$\begin{cases} r^i_{t,k} &= w_6 f(r^i_{t,s}, w_7 l^i_{j,k} + w_8 r^i_{j,s}), \\ r^i_{j,k} &= w_{10}(w_9 f(r^i_{t,s}, l^i_{t,k})) + w_{11} r^i_{j,s}, \end{cases} \tag{15.17}$$

- NOMS (Lugosch and Gross, 2017)

$$\begin{cases} l^i_{t,s} &= \text{sgn}(l^i_{t,k})\text{sgn}(r^{i-1}_{j,s} + l^i_{j,k}) \max\left(0, \min\left(|l^i_{t,k}|, |r^{i-1}_{j,s} + l^i_{j,k}|\right) - w_0\right), \\ l^i_{j,s} &= \text{sgn}(l^i_{t,k})\text{sgn}(r^{i-1}_{t,s}) \max\left(0, \min\left(|l^i_{t,k}|, |r^{i-1}_{t,s}|\right) - w_3\right), \end{cases} \tag{15.18}$$

$$\begin{cases} r^i_{t,k} &= \text{sgn}(r^i_{t,s})\text{sgn}(l^i_{j,k} + r^i_{j,s}) \max\left(0, \min\left(|r^i_{t,s}|, |l^i_{j,k} + r^i_{j,s}|\right) - w_6\right), \\ r^i_{j,k} &= \text{sgn}(r^i_{t,s})\text{sgn}(l^i_{t,k}) \max\left(0, \min\left(|r^i_{t,s}|, l^i_{t,k}|\right) - w_9\right) + r^i_{j,s}, \end{cases} \tag{15.19}$$

- NNMS (Xu et al., 2017)

$$\begin{cases} l^i_{t,s} &= w_0 f(l^i_{t,k}, r^{i-1}_{j,s} + l^i_{j,k}), \\ l^i_{j,s} &= w_3 f(l^i_{t,k}, r^{i-1}_{t,s}) + l^i_{j,k}, \end{cases} \tag{15.20}$$

$$\begin{cases} r^i_{t,k} &= w_6 f(r^i_{t,s}, l^i_{j,k} + r^i_{j,s}), \\ r^i_{j,k} &= w_9 f(r^i_{t,s}, l^i_{t,k}) + r^i_{j,s}, \end{cases} \tag{15.21}$$

where $w_m \in \mathbb{R}$ $(0 \leq m \leq 11)$ are the trainable weights.

Optimizing the weights of the neural BP decoders, as depicted in Figure 15.3, can be done through backpropagation in order to minimize the following objective function

$$Loss = \sum_{i=1}^{I_{\max}} \sum_{s=0}^{n-1} H_{\text{CE}}(\hat{\boldsymbol{h}}^i_s, \boldsymbol{h}_s), \tag{15.22}$$

where H_{CE} is the cross-entropy function, and $\boldsymbol{h}_s$ is the correct hard value vector at stage s of the polar code factor graph that is obtained from the training samples. In the decoding phase, only the hard estimated values at stage 0 of the polar code factor graph, i.e. $\hat{\boldsymbol{h}}^i_0$ $(1 \leq i \leq I_{\max})$, is required to obtain the decoded message bits.

15.3.2.2 Joint Decoder and Noise Estimator

In (Liang et al., 2018) a CNN-based noise estimator is used to remove interference noise between channels. The noise estimator is then coupled with a conventional BP decoder. Although the proposed iterative denoising-decoding approach in (Liang et al., 2018) showed a significant performance gain in the case of strong correlations between channels, when there is no correlation, this approach only achieved a negligible error probability gain over the conventional BP decoder.

A multiplicative noise model was introduced in (Bennatan et al., 2018) based on the decoding syndrome. This noise model has two advantages compared to that of the additive noise-estimation model used in (Liang et al., 2018). First, it reduces the regression problem of the noise model in (Liang et al., 2018) to a classification problem, i.e. only learn to estimate the sign of the noise instead of the actual noise value, which is more feasible given the high dimensional state of the model inputs. Second, the multiplicative noise-estimation model also preserves the code symmetry conditions (Richardson and Urbanke, 2008), allowing for the use of all-zero codewords during training.

Figure 15.4 depicts a joint decoder-noise estimator approach derived from (Liang et al., 2018, Bennatan et al., 2018), where the noise estimator is a DL model such as NN, CNN, or RNN. Unlike off-the-shelf DL decoders, which try to directly predict the message words given the channel LLRs, the joint decoder and noise estimator approach only utilizes off-the-shelf DL models to denoise the channel LLR values, while the main decoding algorithm is still carried out by conventional BP decoding.

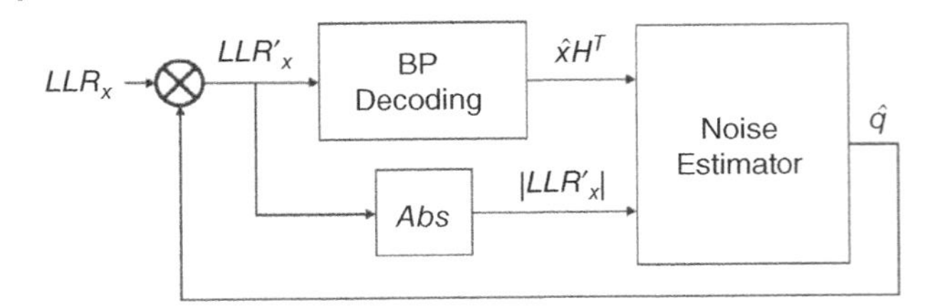

Figure 15.4 A joint BP decoder-DL noise estimator as proposed in (Liang et al., 2018). The input of the noise estimator is the syndrome of the estimated codeword and the magnitude of the estimated channel LLRs (Bennatan et al., 2018).

The iterative decoding algorithm in Figure 15.4 starts with the first decoding attempt by running the conventional BP decoding given the channel input $\mathbf{LLR}_x$. If the estimated codeword $\hat{\boldsymbol{x}}$ and the estimated message word $\hat{\mathrm{u}}$ do not satisfy the G-matrix-based termination condition (Yuan and Parhi, 2013), the channel LLR values will be denoised and followed by another BP decoding attempt. Given the syndrome of the conventional BP decoding algorithm, $\hat{\mathbf{x}}\mathbf{H}^{\mathrm{T}}$, where $\boldsymbol{H}$ is the parity-check matrix of polar codes, and the absolute values of the channel LLR values, $|\mathbf{LLR}_x|$, the DL-based noise estimator predicts the channel noise by estimating its sign values, $\hat{\boldsymbol{q}}$. The channel LLR values are then updated by flipping the signs at certain positions predicted by the noise estimator, which results in the denoised channel LLR values, $\mathbf{LLR'}_x$. Another BP decoding attempt is then carried out given the denoised LLR values. Finally, the decoding is terminated if the mentioned termination condition is satisfied or a predetermined maximum number of decoding attempts is reached.

It is worth noticing that the training samples of the noise estimator depicted in Figure 15.4 only include the erroneous syndromes after the first BP decoding, i.e. when $\hat{\boldsymbol{x}}\boldsymbol{H}^T$ is a nonzero vector, and the corresponding absolute values of the channel LLRs. The label $\boldsymbol{q}, \boldsymbol{q} \subset \{-1, 1\}^N$, is used as the correct output label, where $q_j = -1$ indicates a flip at the j-th element of the channel LLRs, while $q_j = 1$ indicates there is no change required for the j-th element.

15.3.3 Evaluation

In this section, we provide a performance comparison in terms of FER for various state-of-the-art DL-aided BP decoders when applied to a polar code. In addition, the FER performance of conventional decoders including BP (Arıkan, 2010) and SCL (Balatsoukas-Stimming et al., 2015, Hashemi et al., 2016) decoding is also plotted. The examined polar code has a code length of 64, with 32 information bits, and is selected for the eMBB control channel of 5G (3GPP, 2018). We denote BP$I_{\max}$, where $I_{\max} \in \{5, 30\}$, as the conventional min-sum BP decoder with $I_{\max}$ iterations, and SCL32 (Tal and Vardy, 2015) as the SCL decoder with a list size of 32. All the neural BP decoders considered in this section contain five unrolled BP iterations.

We also examine a joint decoder and channel equalizer decoding system by exploiting the idea proposed in (Liang et al., 2018, Bennatan et al., 2018). We denote the joint decoding systems as BP5-MLP-BP5 and BP5-LSTM-BP5, where MLP and LSTM refer to the noise-estimator models using fully connected NNs and stacked LSTM networks, respectively. Note that the network architectures and parameters of the DL-based noise estimators are adopted from (Bennatan et al., 2018).

As all the DL-aided decoders considered in this section satisfy the code symmetry conditions (Richardson and Urbanke, 2008), only all-zero codewords are required for training. The training data is obtained for various E_b/N_0 values where $E_b/N_0 \in \{1, 2, 3, 4, 5, 6, 7, 8\}$. At each E_b/N_0 value, 10^5 all-zero codewords are obtained using BPSK modulation and the AWGN

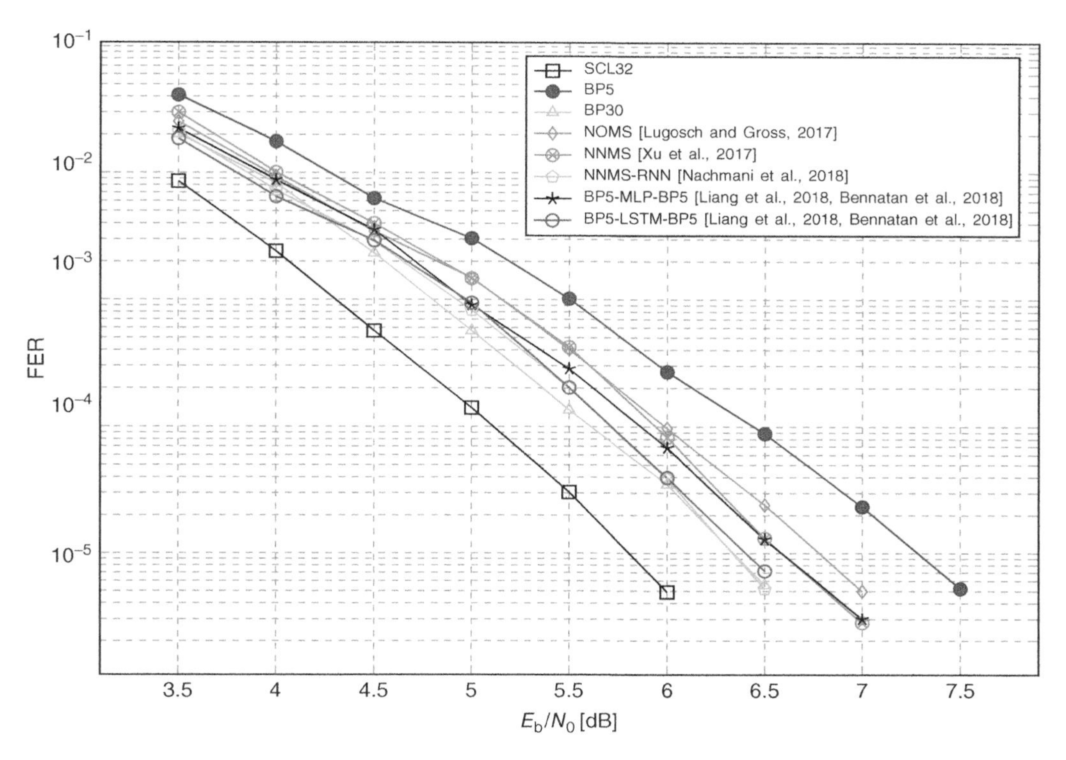

Figure 15.5 FER performance of various decoders for $\mathcal{P}(64, 32)$ selected for 5G.

channel model. All neural BP-based decoders are trained for 100 epochs, while BP5-MLP-BP5 and BP5-LSTM-BP5 are trained for 1000 epochs. The mini-batch size is set to 320, and RMSPROP (Hinton et al.) is used as the optimization algorithm for training. Keras (Chollet et al., 2015) and TensorFlow (Abadi et al., 2016) are used as our DL frameworks. During testing, each decoder decodes at least 10^4 random codewords to obtain at least 50 frames in errors at each E_b/N_0 value.

Figure 15.5 illustrates the FER performance of the mentioned decoders. Table 15.1 summarizes the error-correction performance gains of all the decoders in Figure 15.5 with respect to the baseline BP5 decoder at a target FER of 10^{-5}. As observed from Table 15.1, NOMS provides a gain of 0.5 dB compared to the baseline BP5 decoder, while NNMS and BP5-MLP-BP5 both have a gain of 0.7 dB. On the other hand, NNMS-RNN and BP30 have the same error-correction performance, which is around 1.0 dB better than that of the baseline BP5, while BP5-LSTM-BP5 is slightly worse than NNMS-RNN. It is worth mentioning that the best neural BP decoder, NNMS-RNN, is still 0.5 dB away from SCL32.

Table 15.1 Performance gain in dB when compared with BP5 at FER=10^{-5}.

SCL32	BP30	NOMS	NNMS	NNMS-RNN	BP5-MLP-BP5	BP5-LSTM-BP5
1.5	1.0	0.5	0.7	1.0	0.7	0.9

The decoding latency in terms of time steps for a polar code of size N under BP decoding with $I_{\max}$ iterations can be calculated as (Arıkan, 2010)

$$\mathcal{T}_{\mathrm{BP}\,I_{\max}} = 2I_{\max}\log_2(N). \tag{15.23}$$

As the unrolled factor graphs of the NOMS, NNMS, and NNMS-RNN decoders are equivalent to that of a traditional BP decoder with five iterations, their decoding latency can also be calculated by Eq. (15.23). For BP5-MLP-BP5 and BP5-LSTM-BP5 decoders, their decoding latency in time steps is the sum of the time steps consumed by two successive BP decoders with five iterations, and a deep NN with a depth of 5. Therefore, their decoding latency can be calculated as (Liang et al., 2018):

$$\mathcal{T}_{\mathrm{BP}\,I_{\max}\text{-MLP/LSTM-BP}\,I_{\max}} = 4I_{\max}\log_2(N) + Depth_{\mathrm{MLP/LSTM}}. \tag{15.24}$$

On the other hand, the SCL32 decoder of (Balatsoukas-Stimming et al., 2015) requires $(2N + K - 2)$ time steps.

Figure 15.6 illustrates the decoding latency in time steps for all the neural decoders considered in Figure 15.5. It should be noted that by assigning trainable weights to the factor graph of polar codes, NNMS-RNN with 5 iterations is able to achieve the same error-correction performance of BP30, which also results in a saving of 300 time steps. In addition, the decoding latency of BP5-MLP-BP5 and BP5-LSTM-BP5 is 65 time steps greater than that of NNMS-RNN.

Table 15.2 gives a detailed comparison in the number of weights required by different DL-aided BP decoders in Figure 15.5. Although it is demonstrated in Figure 15.5 that off-the-shelf deep networks are able to estimate channel noise, this approach shows inefficiency since relatively large DL models are required for the task. On the contrary, by incorporating the conventional BP decoding algorithm to define a constrained network model, NNMS-RNN can provide a reasonable error probability while only requiring a small number of weights compared to those of BP5-MLP-BP5 and BP5-LSTM-BP5. Furthermore, although NOMS and NNMS only

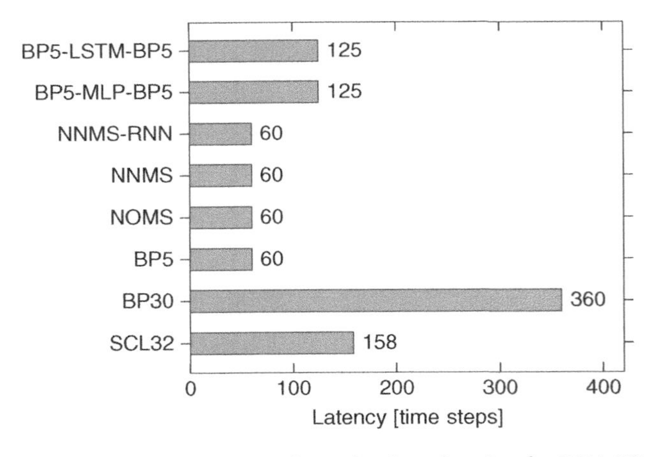

Figure 15.6 Latency comparison of various decoders for $\mathcal{P}(64, 32)$ selected for 5G.

Table 15.2 Number of trainable parameters required for different DL-aided BP decoders.

NOMS	NNMS	NNMS-RNN	BP5-MLP-BP5	BP5-LSTM-BP5
288	288	864	5403712	3446976

require 33.33% of the number of weights consumed by NNMS-RNN, the smaller number of weights results in a considerable error-correction performance loss, as observed in Figure 15.5.

15.4 Conclusions

In this chapter, we have discussed a wide range of fruitful applications of digital communications where deep learning can play a vital role. We provided an overview of DL techniques with a focus on FEC codes and examined state-of-the-art DL-aided decoders for polar codes as our case study. It was demonstrated that off-the-shelf DL decoders can reach MAP decoding performance for short code lengths and that they enable parallel execution thanks to their one-shot-decoding property. However, for longer code lengths, off-the-shelf DL decoders require a training dataset that scales exponentially with the code length. This issue becomes the main challenge for those decoders in practical applications. On the other hand, domain knowledge can be exploited to design DL-aided decoders, as demonstrated by various neural BP decoders. It was shown that neural BP decoders can obtain significant decoding performance gains over their conventional counterpart, while maintaining the same decoding latency.

Future research on applying DL techniques to FEC can be carried out in various directions, such as designing jointly trained systems of customized DL-aided decoders and neural channel-noise estimators for various nonlinear communication channels. In addition, the sequential decoding of linear block codes such as polar codes can be suitably formalized as a RL problem, thus greatly enabling the applications of state-of-the-art RL algorithms to FEC. Other approaches may include the use of DL techniques as optimization methods for well-defined problems of conventional decoding algorithms, whose solutions are obtained based on an approximation or a massive Monte Carlo simulation.

Bibliography

3GPP. Multiplexing and channel coding (Release 10) 3GPP TS 21.101 v10.4.0. Oct. 2018. URL http://www.3gpp.org/ftp/Specs/2018-09/Rel-10/21_series/21101-a40.zip.

M. Abadi, P. Barham, J. Chen, Z. Chen, A. Davis, et al. TensorFlow: A system for large-scale machine learning. *Proceedings of the 12th USENIX Conference on Operating Systems Design and Implementation*, pages 265–283, 2016.

E. Arıkan. Channel polarization: A method for constructing capacity-achieving codes for symmetric binary-input memoryless channels. *IEEE Transactions on Information Theory*, 55 (7): 3051–3073, July 2009. ISSN 0018-9448. doi: 10.1109/TIT.2009.2021379.

E. Arıkan. Polar codes: A pipelined implementation. *Proceedings of the 4th International Symposium on Broadband Communications*, pages 11–14, 2010.

A. Balatsoukas-Stimming, M.B. Parizi, and A. Burg. LLR-based successive cancellation list decoding of polar codes. *IEEE Transactions on Signal Processing*, 63 (19): 5165–5179, Oct 2015. ISSN 1053-587X. doi: 10.1109/TSP.2015.2439211.

A. Bennatan, Y. Choukroun, and P. Kisilev. Deep learning for decoding of linear codes - a syndrome-based approach. *2018 IEEE International Symposium on Information Theory (ISIT)*, pages 1595–1599, June 2018. ISSN 2157-8117. doi: 10.1109/ISIT.2018.8437530.

C. Berrou, A. Glavieux, and P. Thitimajshima. Near Shannon limit error-correcting coding and decoding: Turbo-codes. *Proceedings of ICC '93 - IEEE International Conference on Communications*, 2: 1064–1070 vol.2, May 1993. doi: 10.1109/ICC.1993.397441.

J. Bruck and M. Blaum. Neural networks, error-correcting codes, and polynomials over the binary n-cube. *IEEE Transactions on Information Theory*, 35 (5): 976–987, Sept 1989. ISSN 0018-9448. doi: 10.1109/18.42215.

S. Cammerer, T. Gruber, J. Hoydis, and S.T. Brink. Scaling deep learning-based decoding of polar codes via partitioning. *IEEE Global Communication Conference*, pages 1–6, December 2017. doi: 10.1109/GLOCOM.2017.8254811.

F. Chollet et al. Keras. https://keras.io, 2015.

N. Doan, S.A. Hashemi, and W.J. Gross. Neural successive cancellation decoding of polar codes. *2018 IEEE 19th International Workshop on Signal Processing Advances in Wireless Communications (SPAWC)*, pages 1–5, June 2018a. ISSN 1948-3252. doi: 10.1109/SPAWC.2018.8445986.

N. Doan, S.A. Hashemi, E.N. Mambou, T. Tonnellier, and W.J. Gross. Neural belief propagation decoding of CRC-Polar concatenated codes. *arXiv preprint arXiv:1811.00124*, 2018b.

S. Dörner, S. Cammerer, J. Hoydis, and S.T. Brink. Deep learning based communication over the air. *IEEE Journal of Selected Topics in Signal Processing*, 12 (1): 132–143, Feb 2018. ISSN 1932-4553. doi: 10.1109/JSTSP.2017.2784180.

N. Farsad, M. Rao, and A. Goldsmith. Deep learning for joint source-channel coding of text. *2018 IEEE International Conference on Acoustics, Speech and Signal Processing (ICASSP)*, pages 2326–2330, April 2018. ISSN 2379-190X. doi: 10.1109/ICASSP.2018.8461983.

I. Goodfellow, J. Pouget-Abadie, M. Mirza, B. Xu, D. Warde-Farley, S. Ozair, A. Courville, and Y. Bengio. Generative adversarial nets. *Advances in Neural Information Processing Systems 27*, pages 2672–2680, 2014.

A. Graves and J. Schmidhuber. Framewise phoneme classification with bidirectional lstm and other neural network architectures. *Neural Networks*, 18 (5): 602–610, 2005. ISSN 0893-6080. doi: https://doi.org/10.1016/j.neunet.2005.06.042. IJCNN 2005.

T. Gruber, S. Cammerer, J. Hoydis, and S.T. Brink. On deep learning-based channel decoding. *2017 51st Annual Conference on Information Sciences and Systems (CISS)*, pages 1–6, March 2017. doi: 10.1109/CISS.2017.7926071.

S.A. Hashemi, C. Condo, and W.J. Gross. A fast polar code list decoder architecture based on sphere decoding. *IEEE Transactions on Circuits and Systems I: Regular Papers*, 63 (12): 2368–2380, Dec 2016. ISSN 1549-8328. doi: 10.1109/TCSI.2016.2619324.

S. Hasim, W.S. Andrew, and F. Beaufays. Long short-term memory recurrent neural network architectures for large scale acoustic modeling. *Fifteenth annual conference of the international speech communication association (INTERSPEECH)*, 2014.

G. Hinton, N. Srivastava, and K. Swersky. Neural networks for machine learning lecture 6a overview of mini-batch gradient descent. URL https://cs.toronto.edu/~tijmen/csc321/slides/lecture_slides_lec6.pdf.

M. Ibnkahla. Applications of neural networks to digital communications–a survey. *Signal processing*, 80 (7): 1185–1215, 2000. doi: 10.1016/s0165-1684(00)00030-x.

H. Kim, Y. Jiang, R. Rana, S. Kannan, S. Oh, and P. Viswanath. Communication algorithms via deep learning. *International Conference on Learning Representations (ICLR)*, 2018.

D.P. Kingma and J. Ba. Adam: A method for stochastic optimization. *arXiv preprint arXiv:1412.6980*, 2014.

Y. LeCun, Y. Bengio, and G. Hinton. Deep learning. *Nature*, 521 (7553): 436, May 2015.

F. Liang, C. Shen, and F. Wu. An iterative BP-CNN architecture for channel decoding. *IEEE Journal of Selected Topics in Signal Processing*, 12 (1): 144–159, Feb 2018. ISSN 1932-4553. doi: 10.1109/JSTSP.2018.2794062.

L. Lugosch and W.J. Gross. Neural offset min-sum decoding. *2017 IEEE International Symposium on Information Theory (ISIT)*, pages 1361–1365, June 2017. ISSN 2157-8117. doi: 10.1109/ISIT.2017.8006751.

W. Lyu, Z. Zhang, C. Jiao, K. Qin, and H. Zhang. Performance evaluation of channel decoding with deep neural networks. *2018 IEEE International Conference on Communications (ICC)*, pages 1–6, May 2018. ISSN 1938-1883. doi: 10.1109/ICC.2018.8422289.

E. Nachmani, E. Marciano, L. Lugosch, W.J. Gross, D. Burshtein, and Y. Be'ery. Deep learning methods for improved decoding of linear codes. *IEEE Journal of Selected Topics in Signal Processing*, 12 (1): 119–131, February 2018. ISSN 1932-4553. doi: 10.1109/JSTSP.2017.2788405.

T.J. O'Shea and J. Hoydis. An introduction to deep learning for the physical layer. *IEEE Transactions on Cognitive Communications and Networking*, 3 (4): 563–575, Dec 2017. ISSN 2332-7731. doi: 10.1109/TCCN.2017.2758370.

Y. Ren, C. Zhang, X. Liu, and X. You. Efficient early termination schemes for belief-propagation decoding of polar codes. *IEEE 11th International Conference on ASIC*, pages 1–4, Nov 2015. doi: 10.1109/ASICON.2015.7517046.

T. Richardson and R. Urbanke. *Modern coding theory*. Cambridge university press, 2008.

W. Ryan and S. Lin. *Channel codes: classical and modern*. Cambridge University Press, 2009.

N. Samuel, T. Diskin, and A. Wiesel. Deep MIMO detection. *2017 IEEE 18th International Workshop on Signal Processing Advances in Wireless Communications (SPAWC)*, pages 1–5, July 2017. ISSN 1948-3252. doi: 10.1109/SPAWC.2017.8227772.

S. Sun and Y. Jing. Training and decodings for cooperative network with multiple relays and receive antennas. *IEEE Transactions on Communications*, 60 (6): 1534–1544, June 2012. ISSN 0090-6778. doi: 10.1109/TCOMM.2012.050912.110380.

I. Tal and A. Vardy. List decoding of polar codes. *IEEE Transactions on Information Theory*, 61 (5): 2213–2226, May 2015. ISSN 0018-9448. doi: 10.1109/TIT.2015.2410251.

L.G. Tallini and P. Cull. Neural nets for decoding error-correcting codes. *IEEE Technical Applications Conference and Workshops. Northcon/95. Conference Record*, pages 89–, Oct 1995. doi: 10.1109/NORTHC.1995.485019.

A. Viterbi. Convolutional codes and their performance in communication systems. *IEEE Transactions on Communication Technology*, 19 (5): 751–772, October 1971. ISSN 0018-9332. doi: 10.1109/TCOM.1971.1090700.

T. Wang, C. Wen, H. Wang, F. Gao, T. Jiang, and S. Jin. Deep learning for wireless physical layer: Opportunities and challenges. *China Communications*, 14 (11): 92–111, Nov 2017. ISSN 1673-5447. doi: 10.1109/CC.2017.8233654.

C. Wen, W. Shih, and S. Jin. Deep learning for massive MIMO CSI feedback. *IEEE Wireless Communications Letters*, 7 (5): 748–751, Oct 2018. ISSN 2162-2337. doi: 10.1109/LWC.2018.2818160.

J.L. Wu, Y.H. Tseng, and Y.M. Huang. Neural network decoders for linear block codes. *International Journal of Computational Engineering Science*, 3 (3): 235–256, 2002. doi: 10.1142/S1465876302000629.

W. Xu, Z. Wu, Y. Ueng, X. You, and C. Zhang. Improved polar decoder based on deep learning. *2017 IEEE International Workshop on Signal Processing Systems (SiPS)*, pages 1–6, Oct 2017. ISSN 2374-7390. doi: 10.1109/SiPS.2017.8109997.

B. Yuan and K.K. Parhi. Architecture optimizations for bp polar decoders. *2013 IEEE International Conference on Acoustics, Speech and Signal Processing*, pages 2654–2658, May 2013. ISSN 1520-6149. doi: 10.1109/ICASSP.2013.6638137.

16

Neural Network–Based Wireless Channel Prediction

Wei Jiang[1]*, Hans Dieter Schotten*[2]*, and Ji-ying Xiang*[3]

[1] *Intelligent Networking Group, German Research Center for Artificial Intelligence, Kaiserslautern, Germany*
[2] *Institute for Wireless Communication and Navigation, University of Kaiserslautern, Kaiserslautern, Germany*
[3] *ZTE Ltd., Shenzhen, China*

16.1 Introduction

The advantages of wireless communication over wired are its flexibility, scalability, mobility, convenience, and economical efficiency, thanks to the free propagation of electromagnetic waves through a wireless channel from the transmitter to the receiver. Due to the reflection, diffraction, and scattering of electromagnetic waves traveling along different paths, as well as the mobility of surrounding objects or mobile stations, a wireless channel exhibits an extremely challenging condition for the design and implementation of wireless systems. By adapting transmission parameters such as the constellation size, coding rate, transmit power, time or frequency resources, transmit or receive antennas, and relaying nodes to instantaneous channel conditions, adaptive transmission systems can potentially aid the achievement of great performance. To fully realize this potential, the transmitter needs to know accurate channel state information (CSI). In a frequency-division duplex (FDD) system, the CSI is estimated at the receiver and fed back to the transmitter through a limited feedback channel. Owing to time delays in the process of channel estimation, signal processing, and feedback, the available CSI at the transmitter may become outdated before its actual usage. In a time-division duplex (TDD) system, although the feedback delay can be avoided by taking advantage of channel reciprocity, it is still possible that the CSI is outdated, especially in high-mobility environments.

It has been extensively recognized that outdated CSI severely deteriorates the performance of a wide variety of adaptive transmission systems, including precoded multiple-input and multiple-output (MIMO) (Zheng and Rao, 2008), multi-user MIMO (Wang et al., 2014a), massive MIMO (Truong and Heath, 2013), beamforming (Kim et al., 2014), interference alignment (Aquilina and Ratnarajah, 2015), closed-loop transmit diversity (Onggosanusi et al., 2001), transmit antenna selection (Yu et al., 2017), orthogonal frequency-division multiple access (Wang et al., 2014b), opportunistic relaying (Vicario et al., 2009), coordinated multi-point (CoMP) transmission (Ramirez et al., 2014), physical layer security (Hyadi et al., 2016), mobility management (Teng et al., 2017), etc. In the fifth generation (5G) system, new applications and services such as Industry 4.0, the Internet of Things, the Tactile Internet, virtual and augmented reality, and autonomous driving impose a great demand for high-data-rate, ultra-reliable, low-latency, ubiquitous, high-mobility, secure wireless connections, where adaptive transmission systems are required to play more important roles. However, the fluctuation of time-varying channels speeds up if the velocity of moving objects increases or the wavelength

Machine Learning for Future Wireless Communications, First Edition. Edited by Fa-Long Luo.
© 2020 John Wiley & Sons Ltd. Published 2020 by John Wiley & Sons Ltd.

of radio signals decreases according to the Doppler effect of electromagnetic radiation. Some 5G deployment scenarios, e.g. millimeter wave-enabled networks (having shorter wavelength), unmanned aerial vehicles, and high-speed trains (with higher velocity), experience wireless channels changing more rapidly, leading to difficult or impossible availability of accurate CSI.

To cope with outdated CSI, a large number of mitigation algorithms and protocols have been proposed in the literature. These methods compensate for the performance loss passively at the cost of scarce wireless resources (Jiang et al., 2016) or aim to achieve only a portion of the full performance potential by recognizing the assumption of imperfect CSI (Love et al., 2008). In contrast, an alternative technique referred to as *channel prediction* provides an efficient and effective approach to improve the accuracy of CSI directly without spending extra wireless resources, and therefore has attracted much attention from researchers. Through statistical modeling of wireless channels, two classical model-based prediction schemes – parametric models (Adeogun et al., 2014) and autoregressive (AR) models (Baddour and Beaulieu, 2005) – were developed. The former assumes that a fading channel is a superposition of a finite number of complex sinusoids, and its parameters, e.g. amplitude, angles of arrival and departure, Doppler shift, and number of scattering sources, vary slowly relative to the channels' fluctuation rate and can be estimated accurately. But the estimation process is tedious, and the estimated parameters will quickly expire with the fluctuation of the fading channel and therefore need to be re-estimated iteratively, leading to high computational complexity. In contrast, the AR model approximates the fading channel as an AR process and extrapolates future CSI using a weighted linear combination of past and current CSI. Although the AR model is simple, it is sensitive to noise, and the problem of error propagation makes it unattractive in multi-step predictions.

In March 2016, when AlphaGo, a computer program developed by Google DeepMind (Silver et al., 2016), achieved an overwhelming victory versus a human champion in the game of Go, the passion for exploring artificial intelligence (AI) technology was sparked in almost all scientific and engineering branches (Jiang et al., 2017). Actually, the wireless research community started to apply AI techniques to solve communication problems long ago. Making use of its capability of time-series prediction (Connor et al., 1994), a recurrent neural network (RNN) was first proposed in (Liu et al., 2006) to build a narrow-band single-antenna predictor and was further extended to MIMO channels in (Potter et al., 2008, Ding and Hirose, 2014). The feasibility of applying a deep neural network to predict fading channels was also studied in (Liao et al., 2018). In (Jiang and Schotten, 2018a), the authors proposed to employ a real-valued RNN to implement a multi-step predictor and further verified its effectiveness in a MIMO system (Jiang and Schotten, 2018b). In (Jiang and Schotten, 2019b), a frequency-domain predictor was designed and validated for frequency-selective multi-antenna channels in wideband communications.

This chapter provides a comprehensive introduction to channel prediction methods with an emphasis on neural network (NN)-based prediction. The rest of this chapter is organized as follows: Section 16.2 briefly describes adaptive transmission systems using transmit antenna selection and opportunistic relaying as examples, followed by the impact of outdated CSI on the performance of adaptive transmission systems in Section 16.3. Then, Section 16.4 reviews two kinds of classical prediction methods: parametric and AR models. Section 16.5 details the principles of RNN-based predictors applied from flat-fading single-antenna channels to frequency-selective multi-antenna channels, as well as their achievable performance and computational complexity. Finally, conclusive remarks are made in Section 16.6.

16.2 Adaptive Transmission Systems

The temporal fluctuation and frequency selectivity of channels impose a fundamental barrier on wireless communications to achieve large capacity and high reliability. By adapting transmission parameters such as scheduled users, modulation and coding schemes, transmit power, relaying nodes, time slots, sub-carriers, and transmit or receive antennas to the instantaneous channel condition, adaptive transmission systems can remarkably boost system performance. In this section, we briefly review two representative examples, i.e. transmit antenna selection (TAS) in a MIMO system and opportunistic relaying in a cooperative network, in order to make clear their working mechanisms.

16.2.1 Transmit Antenna Selection

Time and frequency resources in a wireless system are extremely constrained, so a particularly appealing approach is to exploit the spatial domain by the application of antenna arrays, achieving attractive multiplexing and diversity gains by simply installing additional antennas. With the capability of remarkably improving the capacity and reliability of wireless communications, multiple-antenna systems, also known as MIMO, have been widely adopted in prevalent commercial systems such as 4G and WiFi. However, the design and implementation of the radio frequency (RF) part of a transceiver become challenging because the complexity, size, and cost scale with the number of antennas. As a low-cost, low-complexity alternative, a technique known as *antenna selection* that is able to alleviate this limitation and at the same time capture many of the benefits from MIMO systems has been exploited (Yu et al., 2017).

Figure 16.1 shows the principle of TAS in a MIMO system with N_t transmit and N_r receive antennas. Relying on antenna-specific pilot symbols inserted in transmitted signals, instantaneous channel conditions can be estimated at the receiver. Without loss of generality, consider a frequency-flat fading channel; an $N_r \times N_t$ channel matrix at time t denoted by $\mathbf{H}(t) = [h_{n_r n_t}(t)]_{N_r \times N_t}$ is available, where $h_{n_r n_t} \in \mathbb{C}^{1\times1}$ represents the complex channel gain between transmit antenna n_t and receive antenna n_r. Assuming L out of N_t transmit antennas are selected, the total number of possible choices is a combination of *n choose k* notated by $\binom{N_t}{L}$. Regarding the jth choice, where $1 \leq j \leq \binom{N_t}{L}$, use $\mathbf{H}_j(t)$ with a dimension of $N_r \times L$ to indicate the channel matrix from L selected transmit antennas to N_r receive antennas, which

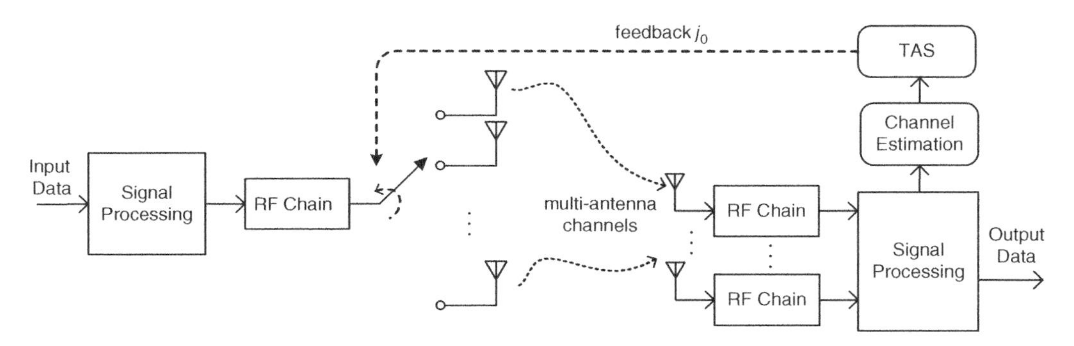

Figure 16.1 Block diagram of transmit antenna selection in a MIMO system.

is a subset of $\mathbf{H}(t)$. With the knowledge of CSI, namely $\mathbf{H}(t)$ in this case, the receiver finds the best choice that has the largest total channel gain:

$$j_0 = \arg\max_{1\leq j\leq \binom{N_t}{L}} \|\mathbf{H}_j(t)\|^2, \tag{16.1}$$

where $\|\cdot\|$ denotes the Frobenius norm of a matrix. The receiver feeds the index of the selected choice j_0 back to the transmitter through a feedback channel. Once it receives the feedback, the transmitter activates the antennas belonging to choice j_0 to transmit signals.

16.2.2 Opportunistic Relaying

Because of the constraints on power supply, cost, and hardware size of antenna arrays at sub-6GHz frequency bands, it is difficult for a mobile terminal to obtain the benefits of MIMO technology. Alternatively, taking advantage of the broadcast nature of radio signals, a technique called *cooperative relaying* (Sendonaris et al., 2003), which can alleviate these constraints by forming a virtual antenna array using multiple single-antenna terminals, draws much attention from researchers.

Figure 16.2 illustrates a typical dual-hop cooperative network consisting of a single source s, a single destination d, and K decode-and-forward relays. To avoid harmful self-interference between the transmitter and receiver, the relays operate in a half-duplex mode. Without loss of generality, signal transmission can be divided into two phases, i.e. broadcasting and forwarding. In the broadcasting phase, as shown in Figure 16.2, the source sends out a transmit signal x_t. Those relays that can correctly decode this signal constitute a *decoding subset*, which is mathematically defined as

$$\begin{aligned} \mathcal{DS} &\triangleq \{k : \log_2(1+\gamma_{s,k}) \geq 2R\} \\ &= \{k : \gamma_{s,k} \geq 2^{2R}-1\}, \end{aligned} \tag{16.2}$$

where $\gamma_{s,k} = \frac{|h_{s,k}|^2 P_t}{\sigma^2}$ denotes the instantaneous received signal-to-noise ratio (SNR) at relay k, $h_{s,k}$ is the channel gain experienced in the channel between the source and relay k, $P_t = \mathbb{E}[|x_t|^2]$ stands for the transmit power, σ^2 is the variance of additive white noise, and R is an end-to-end target data rate. Note that a relay can be regarded as correctly decoding the transmit signal only if the instantaneous capacity for the corresponding source-relay channel is higher than $2R$, recalling that the relay works in a half-duplex mode.

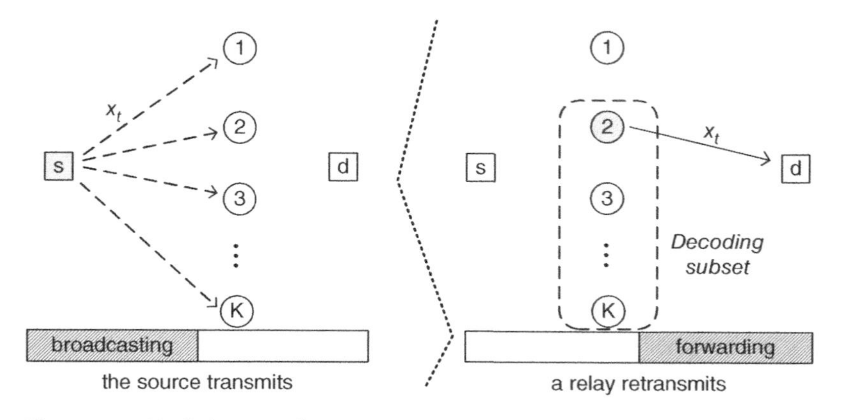

Figure 16.2 Block diagram of an opportunistic relay system.

In the forwarding phase, several *multi-relay* selection methods have been studied in the literature. Choosing a number of N relays to orthogonally forward the original signal, a scheme called *generalized selection combining* (Xiao and Dong, 2006) suffers from substantially degraded spectral efficiency to $1/N$ compared to the full potential. To avoid this loss, a simultaneous transmission method referred to as *distributed beamforming* has been proposed in (Jing and Jafarkhani, 2009). However, beamforming is very sensitive to phase noise, and *a priori* knowledge of forward channels is mandatory. The authors of (Laneman and Wornell, 2003) presented an approach called *distributed space-time coding* (DSTC) to achieve full diversity. But designing such a code is infeasible since the number of distributed antennas is unknown and randomly varies. In addition, the problem of synchronization among simultaneously transmitting relays for multi-relay methods is difficult to solve, especially when the number of relays is large. In contrast, a *single-relay* approach referred to as *opportunistic relaying* or *opportunistic relay selection* (ORS) (Bletsas et al., 2006) has been recognized as a simple but efficient solution to achieve a full cooperative diversity. Despite using only a single node with the best channel condition serving as the relay, it can achieve full diversity the same as that of DSTC. From the viewpoint of a multiplexing-diversity trade-off, it is optimal, while avoiding the implementation complexity of multi-relay methods.

As shown in Figure 16.2, a relay with the largest SNR in relay-destination channels is selected from $\mathcal{DS}$ to serve as the best relay k_0, i.e.

$$k_0 = \arg\max_{k\in\mathcal{DS}} \gamma_{k,d}, \tag{16.3}$$

where $\gamma_{k,d} = \frac{|h_{k,d}|^2 P_t}{\sigma^2}$ denotes instantaneous received SNR at the destination, $h_{k,d}$ is the channel gain experienced in the channel between relay k and the destination, and the transmit power of the best relay is assumed to be identical to that of the source.

16.3 The Impact of Outdated CSI

From a practical point of view, the CSI at the time of selecting transmission parameters may substantially differ from the CSI at the instant of using the selected parameters to transmit. Utilizing an outdated version of the CSI rather than the actual CSI may severely deteriorate system performance. This section first provides a mathematical model to quantify the inaccuracy of outdated CSI and then uses the ORS system as an example to illustrate the impact of outdated CSI on the performance of adaptive transmission systems.

16.3.1 Modeling Outdated CSI

For simplicity, let us ignore time indices and denote the actual CSI by $\mathbf{H} = [h_{n_r n_t}]_{N_r\times N_t}$ and the outdated CSI $\mathbf{H}' = [h'_{n_r n_t}]_{N_r\times N_t}$. To quantify the inaccuracy of the outdated CSI, the correlation coefficient between $\mathbf{H}$ and $\mathbf{H}'$ in the condition of independent and identically distributed (i.i.d.) channels is introduced, as follows:

$$\rho = \frac{|cov(h_{n_r n_t}, h'_{n_r n_t})|}{\mu_h \mu_{h'}}, \tag{16.4}$$

where $cov(\cdot)$ stands for the covariance of two random variables, μ is the standard deviation, and $h_{n_r n_t}$ is the gain of the channel between transmit antenna n_t and receive antenna n_r. Due to i.i.d. elements in $\mathbf{H}$ and $\mathbf{H}'$, ρ is independent of n_r and n_t. Thus, Eq. (16.4) can be simplified into

$$\rho = \frac{|cov(h, h')|}{\mu_h \mu_{h'}}. \tag{16.5}$$

Since the elements of $\mathbf{H}$ and $\mathbf{H}'$ are both zero mean circularly symmetric Gaussian distributed, according to (Ramya and Bhashyam, 2009), their relationship can be given by

$$\mathbf{H}' = \rho\mathbf{H} + \sqrt{1-\rho^2}\mathbf{E}, \tag{16.6}$$

where $\mathbf{E} = [\varepsilon_{n_r n_t}]_{N_r \times N_t}$ is a matrix consisting of normalized Gaussian random variables, i.e. $\varepsilon_{n_r n_t} \sim \mathcal{CN}(0,1)$. Assuming the Jakes' scattering model, $\mathbf{H}$ and $\mathbf{H}'$ follow joint complex Gaussian distribution with correlation coefficient $\rho = J_0(2\pi f_d \tau)$, where f_d denotes the maximal Doppler frequency, τ is the delay, and $J_0(\cdot)$ represents the zeroth order Bessel function of the first kind. Thus, $\mathbf{H}$ conditioned on $\mathbf{H}'$ is also Gaussian distributed:

$$\mathbf{H}|\mathbf{H}' \sim \mathcal{CN}(\rho\mathbf{H}', 1-\rho^2).$$

Due to the assumption of a normalized channel gain $\mathbb{E}[|h|^2] = 1$, the long-term average SNR $\bar{\gamma} = \mathbb{E}\left[\frac{|h|^2 P_t}{\sigma^2}\right]$ is simplified to $\bar{\gamma} = P_t/\sigma^2$. Thus, an instantaneous SNR $\gamma = \frac{||\mathbf{H}||^2 P_t}{\sigma^2}$ can be rewritten as $\gamma = ||\mathbf{H}||^2\bar{\gamma}$. Conditioned on its outdated version $\gamma' = ||\mathbf{H}'||^2\bar{\gamma}$, γ follows a noncentral chi-square distribution with two degrees of freedom. As given in Eq. (3) of (Jiang et al., 2016), its probability density function (PDF) is expressed as

$$f_{\gamma|\gamma'}(\gamma|\gamma') = \frac{1}{\bar{\gamma}(1-\rho^2)} e^{-\frac{\gamma+\rho^2\gamma'}{\bar{\gamma}(1-\rho^2)}} J_0\left(\frac{2\sqrt{\rho^2\gamma\gamma'}}{\bar{\gamma}(1-\rho^2)}\right). \tag{16.7}$$

16.3.2 Performance Impact

Outdated CSI causes an inaccurate or even wrong selection of transmission parameters, leading to severe performance degradation of adaptive transmission systems. As an example, let us use the ORS system in this section to illustrate the severity of this impact. In contrast to Eq. (16.3), the best relay in the presence of outdated CSI is decided as follows

$$k_0 = \arg\max_{k\in DS} \gamma'_{k,d}, \tag{16.8}$$

where $\gamma'_{k,d}$ is the outdated version of the instantaneous received SNR for the relay-destination channel. The instantaneous channel capacity below a target rate of R, i.e. $C < R$, denotes an *outage* event, in which reliable communication cannot be realized no matter what coding is used. The probability measuring such an outage event is defined as the outage probability, notated by $P_{out}(R)$ and expressed as the following definition (Tse and Viswanath, 2005):

$$P_{out}(R) = Pr\{\log_2(1+\gamma) < R\}, \tag{16.9}$$

where Pr is the notation of mathematical probability. As the closed-form expression provided in Eq. (2) of (Vicario et al., 2009), the outage probability of opportunistic relaying in the presence of outdated CSI can be expressed as

$$P_{ors}(\gamma_o) = \left(1-e^{-\frac{\gamma_o}{\bar{\gamma}}}\right)^K + \sum_{l=1}^{K} l \sum_{m=0}^{l-1} \binom{l-1}{m} \frac{(-1)^m}{m+1}\left(1-e^{-\frac{(m+1)\gamma_o}{\bar{\gamma}(1+m(1-\rho^2))}}\right)$$
$$\cdot \binom{K}{l}\left(1-e^{-\frac{\gamma_o}{\bar{\gamma}}}\right)^{K-l} e^{-\frac{\gamma_o l}{\bar{\gamma}}}, \tag{16.10}$$

where $\gamma_o = 2^{2R}-1$ is the threshold SNR corresponding to the target rate R, and $\bar{\gamma} = P/\sigma^2$ denotes the average transmit SNR.

From Eq. (16.10), it is still difficult to offer insightful thoughts about the impact of outdated CSI. It is worth providing an asymptotic analysis in a high SNR regime to clarify their achievable

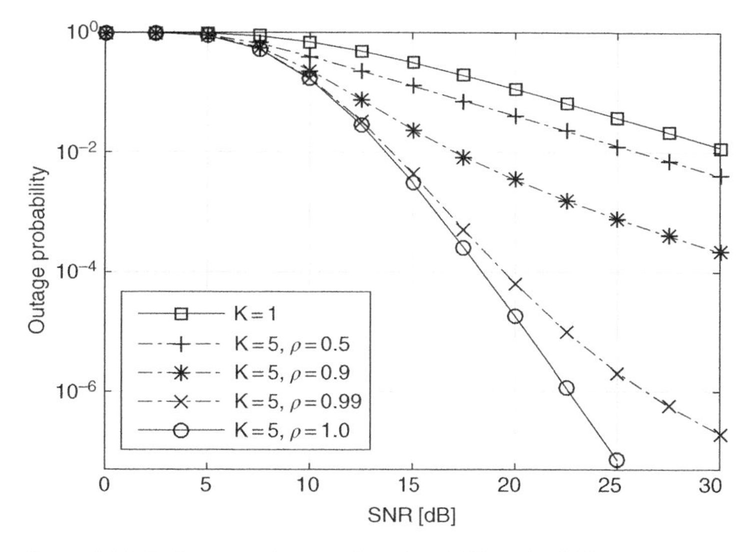

Figure 16.3 Performance impact of outdated CSI on the ORS system.

diversity, like the definition of $d = -\lim_{\bar{\gamma}\to\infty} \log(P_{out})/\log(\bar{\gamma})$ in (Vicario et al., 2009). The ORS system can achieve a full diversity of K with the prefect CSI. However, its diversity degrades to 1 in the presence of outdated CSI no matter how close the outdated CSI is to its actual value, even if the former arbitrarily tends to the latter ($\rho \to 1$). As proved in (Vicario et al., 2009), the achievable diversity order of the ORS system is

$$d = \begin{cases} 1, & \rho < 1 \\ K, & \rho = 1 \end{cases}. \tag{16.11}$$

Figure 16.3 illustrates the impact of outdated CSI on a cooperative network with $K = 5$ decode-and-forward relays, among which the best relay is selected. As a benchmark, the performance curve of another cooperative network having only $K = 1$ relay in between is also given, representing no diversity gain, namely $d = 1$. As indicated by the curve of $K = 5, \rho = 1$ in Figure 16.3, the outage probability of the ORS system decays at a rate of $1/\bar{\gamma}^5$ in high SNR, and a full diversity gain of $d = 5$ is achieved. The curves marked by $\rho = 0.5$, 0.9, and 0.99 are all parallel with that of $K = 1$ in high SNR. It is implied that the diversity order achieved by the ORS system in the presence of outdated CSI is only $d = 1$. The performance degradation is substantial, e.g. a performance loss of nearly 15 dB is observed in the case of $\rho = 0.9$ in comparison with that of the perfect CSI at a given outage probability of 10^{-4}. In other words, the performance gain originated from adaptive transmission systems might be thoroughly overwhelmed by the loss brought about by outdated CSI, highlighting the importance of channel prediction.

16.4 Classical Channel Prediction

Relying on traditional statistical methodology, a prediction model with a number of parameters can be formed to approximate the dynamics of a fading channel. Given the knowledge of current and past CSI, these parameters can be estimated, and then future CSI is extrapolated through this model. Existing model-based channel prediction is mainly differentiated into two categories, i.e. AR and parametric. Their principles, modelings, parameter estimation methods, and constraints are briefly discussed in this section.

16.4.1 Autoregressive Models

By exploiting temporal correlation, this scheme models the impulse response of a time-varying channel as an autoregressive process and employs a Kalman filter (KF) to estimate AR coefficients so as to build a linear predictor, which extrapolates future CSI by combining weighted current and past CSI (Eyceoz et al., 1998, Duel-Hallen et al., 2000, Peng et al., 2017, Wu and Lee, 2013). According to (Baddour and Beaulieu, 2005), a complex AR process of order p denoted by AR(p) can be generated via a time domain recursion

$$x[n] = \sum_{k=1}^{p} a_k x[n-k] + w[n], \tag{16.12}$$

where $w[n]$ is zero mean complex Gaussian noise with variance σ_p^2, and $\{a_1, a_2, \dots, a_p\}$ denote the AR model coefficients. The corresponding power spectral density (PSD) of the AR(p) process has a rational form as follows:

$$S_{xx}(f) = \frac{\sigma_p^2}{\left|1 + \sum_{k=1}^{p} a_k e^{-2\pi jfk}\right|^2}. \tag{16.13}$$

For a Rayleigh channel, the theoretical PSD associated with either in-phase or quadrature part of a fading signal has a well-know U-shaped band-limited form, i.e.

$$S(f) = \begin{cases} \frac{1}{\pi f_d \sqrt{1-\left(\frac{f}{f_d}\right)^2}}, & |f| \leq f_d \\ 0, & f > f_d \end{cases}, \tag{16.14}$$

where f_d is the maximum Doppler shift in Hertz. The corresponding discrete-time autocorrelation function is

$$R[n] = J_0(2\pi f_m |n|), \tag{16.15}$$

where $f_m = f_d T_s$ indicates the maximal Doppler shift normalized by the signal sampling rate $f_s = 1/T_s$. An arbitrary spectrum can be closely approximated by an AR model with sufficiently large order. The basic relationship between a desired autocorrelation function $R[n]$ and an AR(p) model parameters can be given in matrix form by

$$\mathbf{v} = \mathbf{R}\mathbf{a}, \tag{16.16}$$

where

$$\mathbf{R} = \begin{bmatrix} R[0] & R[-1] & \cdots & R[-p+1] \\ R[1] & R[0] & \cdots & R[-p+2] \\ \vdots & \vdots & \ddots & \vdots \\ R[p-1] & R[p-2] & \cdots & R[0] \end{bmatrix}, \tag{16.17}$$

$$\mathbf{a} = \begin{bmatrix} a_1 & a_2 & \cdots & a_p \end{bmatrix}^T, \tag{16.18}$$

$$\mathbf{v} = \begin{bmatrix} R[1] & R[2] & \cdots & R[p] \end{bmatrix}^T, \tag{16.19}$$

and

$$\sigma_p^2 = R[0] + \Sigma_{k=1}^{p} a_k R[k]. \tag{16.20}$$

Substituting Eqs. (16.17)–(16.19) into Eq. (16.16), $\{a_1, a_2, \dots, a_p\}$ are determined. Thus, we can get a KF predictor for a single-input and single-output (SISO) system in a frequency-flat channel:

$$\hat{h}[t+1] = \sum_{k=1}^{p} a_k h[t-k+1]. \tag{16.21}$$

By processing a MIMO channel as a set of parallel, independent SISO channels, a KF predictor for a multi-antenna system can also be derived:

$$\hat{\mathbf{H}}[t+1] = \sum_{k=1}^{p} a_k \mathbf{H}[t-k+1]. \tag{16.22}$$

This scheme is not optimal since it only takes advantage of temporal correlation of individual SISO channels, while ignoring spatial correlation among multiple antennas in a MIMO channel. Due to the fact that the KF predictor can only provide one-step prediction, multi-step prediction is achieved by reusing the extrapolated CSI at previous time instants, which causes the problem of error propagation. Moreover, this scheme is vulnerable to noise (Jiang and Schotten, 2019a, Fung and Chan, 2002), making it unattractive in practice.

16.4.2 Parametric Models

This scheme models a fading channel as a superposition of a finite number of complex sinusoids, each of which has its respective amplitude, Doppler shift, and phase (Adeogun et al., 2014). The rationale is based on an observation that multi-path parameters change slowly in comparison with the fading rate of channels, and future CSI within a certain range can be extrapolated if these parameters are known.

Following a commonly used sum of sinusoids model, a MIMO channel is expressed as the superposition of P scattering sources

$$\mathbf{H}(t) = \sum_{p=1}^{P} \alpha_p \mathbf{a}_r(\theta_p) \mathbf{a}_t^T(\phi_p) e^{j\omega_p t}, \tag{16.23}$$

where α_p is the amplitude of the pth scattering source; ω_p denotes its Doppler shift; θ_p and ϕ_p stand for the angles of arrival and departure, respectively; $\mathbf{a}_r$ represents the response vector of the receive antenna array; and $\mathbf{a}_t$ represents the response vecor for the transmit antenna array. Using a uniform linear array (ULA) with M equally spaced elements as an example, its steering vector can be formulated as

$$\mathbf{a}(\psi) = \left[1, e^{-j\frac{2\pi}{\lambda}d\sin(\psi)}, \dots, e^{-j\frac{2\pi}{\lambda}(M-1)d\sin(\psi)}\right]^T, \tag{16.24}$$

where ψ stands for the angle of arrival or departure, d is the antenna spacing, and λ denotes the wavelength of the carrier frequency. Prediction of a MIMO channel in terms of the model of Eq. (16.23) is essentially a problem of parameter estimation, in which the number of scattering sources, the amplitude and Doppler shift for each path, as well as its angles of arrival and departure need to be estimated. In other words, the main work of building a parametric model is to figure out $\hat{P}$ and $\{\hat{\alpha}_p, \hat{\theta}_p, \hat{\phi}_p, \hat{\omega}_p\}_{p=1}^{\hat{P}}$ with the knowledge of a number of discrete-time channel gain samples $\{\mathbf{H}[k] | k = 1, \dots, K\}$.

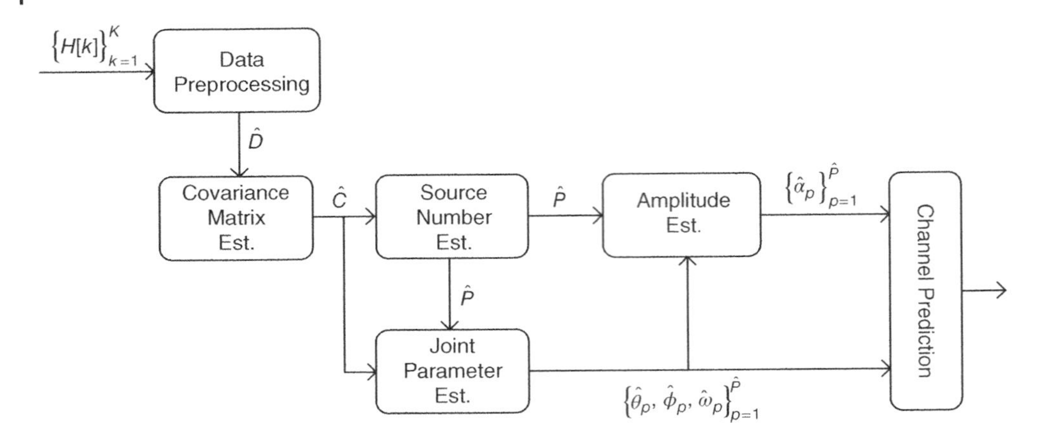

Figure 16.4 The procedure of parameter estimation in the parametric model. Source: Figure 4.1 in (Adeogun, 2014).

As shown in Figure 16.4, the procedure of parameter estimation for the parametric model is divided into the following stages:

i. Use the K available channel matrices to form a sufficiently large matrix exhibiting the required translational invariance structure in all dimensions. According to (Adeogun et al., 2013), therefore, form an $N_r Q \times N_t L$ block-Hankel matrix, which can be written as

$$\hat{\mathbf{D}} = \begin{bmatrix} \mathbf{H}[1] & \mathbf{H}[2] & \cdots & \mathbf{H}[S] \\ \mathbf{H}[2] & \mathbf{H}[3] & \cdots & \mathbf{H}[S+1] \\ \vdots & \vdots & \ddots & \vdots \\ \mathbf{H}[Q] & \mathbf{H}[Q+1] & \cdots & \mathbf{H}[K] \end{bmatrix}, \tag{16.25}$$

where Q is the size of Hankel matrix and $S = K - Q + 1$.

ii. From the transformed data, calculate a covariance matrix containing the temporal and spatial correlation. The spatio-temporal covariance matrix $\hat{\mathbf{C}}$ is then derived as $\hat{\mathbf{C}} = \hat{\mathbf{D}}\hat{\mathbf{D}}^H/(N_t S)$, where $(\cdot)^H$ denotes the Hermitian conjugate transpose.

iii. Then, the number of dominant scattering sources can be estimated using the minimum description length (MDL) criterion as

$$\hat{P} = \arg \min_{p=1,\dots,N_r Q-1} \left[S \log(\lambda_p) + \frac{1}{2}(p^2 + p) \log S \right], \tag{16.26}$$

where λ_p is the pth eigenvalue of $\hat{\mathbf{C}}$.

iv. The invariance structure in $\hat{\mathbf{C}}$ is exploited to jointly estimate the structural parameters. Making full use of classical estimation algorithms, such as multiple signal classification (MUSIC) and estimation of signal parameters by rotational invariance techniques (ESPRIT), the angles of arrival and departure, as well as the Doppler shifts, i.e. $\{\hat{\theta}_p, \hat{\phi}_p, \hat{\omega}_p\}_{p=1}^{\hat{P}}$, can be computed. For simplicity, the details of the calculation process and algorithms are omitted in this chapter.

v. By obtaining the estimated structural parameters $\{\hat{\theta}_p, \hat{\phi}_p, \hat{\omega}_p\}_{p=1}^{\hat{P}}$, together with $\hat{P}$, the complex amplitudes $\{\hat{\alpha}_p\}_{p=1}^{\hat{P}}$ then can be calculated.

vi. Once all parameters have been determined, the channel prediction is conducted as follows

$$\hat{\mathbf{H}}(\tau) = \sum_{p=1}^{\hat{P}} \hat{\alpha}_p \mathbf{a}_r(\hat{\theta}_p) \mathbf{a}_t^T(\hat{\phi}_p) e^{j\hat{\omega}_p \tau}, \tag{16.27}$$

where τ denotes a time range for which the CSI is to be predicted.

As we can see, the process of estimating parameters is tedious, leading to high computational complexity. More importantly, the estimated parameters become invalid quickly with the change of mobile propagation environments, especially in a fast-fading channel. That means these parameters need to be periodically estimated, which is unattractive from a practical viewpoint.

16.5 NN-Based Prediction Schemes

This section highlights novel channel prediction approaches making full use of the capability of time-series prediction enabled by NNs. First, the internal structure of a recurrent NN is provided, followed by an application of a RNN to implement a multi-step predictor for frequency-flat single-antenna channels, which is further extended to multi-antenna channels. Then, a frequency-domain predictor suited to frequency-selective multi-antenna channels is described, as well as its integration into a MIMO-OFDM system. Finally, the performance and computational complexity of the RNN predictor are analyzed and compared with those of the KF predictor.

16.5.1 The RNN Architecture

RNNs are an effective AI technique that has shown great potential in the field of time-series prediction (Connor et al., 1994). The internal structure of a RNN used for constructing a multi-step multi-antenna channel predictor is provided in Figure 16.5. Basically, it consists of three layers: an input layer with N_i neurons consisting of N_e external input and N_f feedback input, where $N_i = N_e + N_f$; a hidden layer with N_h neurons; and an output layer with N_o neurons. Using a feedback function $F(\cdot)$ to represent the transformation from an output vector denoted by $\mathbf{y} = [y_1, ..., y_{N_o}]$ to a desired feedback $\mathbf{f} = [f_1, ..., f_{N_f}]$, we have $\mathbf{f} = F(\mathbf{y})$. Denoting the external input by $\mathbf{x}_e = [x_1, ..., x_{N_e}]$, together with the feedback, the whole input can thus be written as an N_i-dimensional input vector $\mathbf{x} = [x_1, ..., x_{N_e}, f_1 ..., f_{N_f}]$.

The behavior of a RNN is decided by connection weights and transfer functions. Each connection between the output of a neuron in the predecessor layer and the input of a neuron in the successor layer is assigned a weight. As shown in Figure 16.5, $w_{l,n}$ denotes the weight connecting the nth input and the lth hidden neuron, while $c_{o,l}$ is the weight for connecting hidden neuron l and output o, where $1 \leq n \leq N_i$, $1 \leq l \leq N_h$, and $1 \leq o \leq N_o$. Transfer functions typically fall into one of three categories: linear, threshold, and sigmoid. For example, a sigmoid function generally used in a hidden neuron is defined as

$$S(x) = \frac{1}{1 + e^{-x}}. \tag{16.28}$$

The upper-right diagram in Figure 16.5 illustrates the internal architecture of a hidden neuron driven by a sigmoid function. The output of the lth hidden neuron can be expressed by

$$\begin{aligned} z_l &= S(\mathbf{w}_l \cdot \mathbf{x} + b_l) \\ &= S\left(\sum_{n=1}^{N_e} w_{l,n} x_n + \sum_{n=N_e+1}^{N_i} w_{l,n} f_{(n-N_e)} + b_l\right), \end{aligned} \tag{16.29}$$

where b_l denotes the bias, and $\mathbf{w}_l \cdot \mathbf{x}$ stands for the dot product of the input vector $\mathbf{x}$ and the lth weight vector that is defined as the collection of all weights upon connections pointing to the lth hidden neuron, i.e. $\mathbf{w}_l = [w_{l,1}, w_{l,2}, ..., w_{l,N_i}]$. Suppose the output neurons adopt a linear

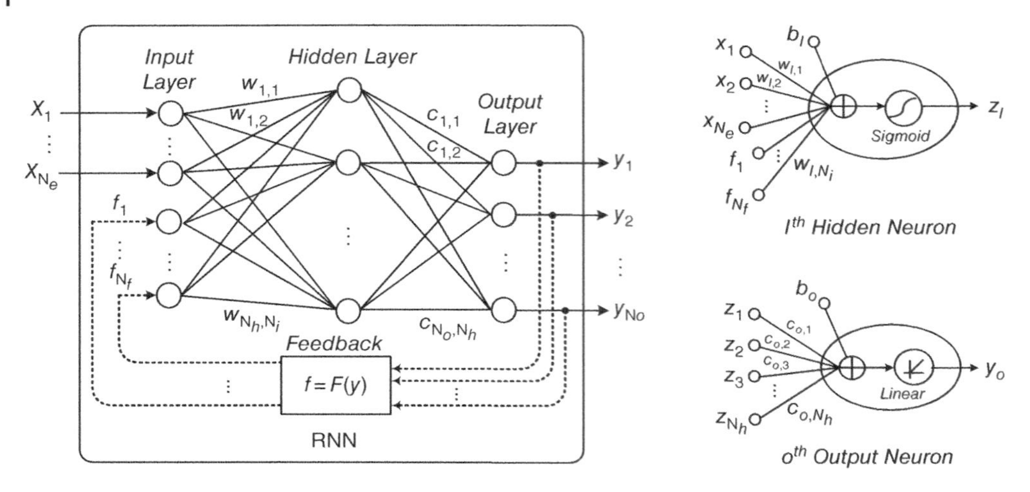

Figure 16.5 The internal structure of a recurrent neural network.

transfer function; the output of the oth output neuron can be expressed as

$$y_o = \sum_{l=1}^{N_h} c_{o,l} z_l + b_o. \tag{16.30}$$

Like other data-driven AI techniques, the operation of a RNN predictor has two phases: training and prediction. Once the parameters of a network have been determined, it is ready to be trained. Provided with a training dataset, the RNN processes each input data and compares its resulting output against the desired value. Errors are then propagated back through the network, causing the network to adjust its weights iteratively. Once the training process completed, the trained RNN can be used to predict upcoming samples based on the current state. The training of a NN has already been well studied and can be found in the literature such as (Nerrand et al., 1994, Fu et al., 2015).

16.5.2 Flat-Fading SISO Prediction

To begin with, consider a frequency-flat fading channel with one transmit antenna and one receive antenna:

$$r[t] = h[t]s[t] + n[t], \tag{16.31}$$

where $r[t]$ represents the received signal at time t, $s[t]$ is the transmitted symbol, n stands for additive white Gaussian noise (AWGN), and $h[t]$ denotes the channel gain. Due to feedback and processing delays, the channel condition at the time of selecting adaptive transmission parameters may be outdated before its actual usage, namely $h[t] \neq h[t+\tau]$, where τ denotes the delay. Outdated CSI imposes a severely negative impact on a wide variety of adaptive transmission systems. The aim of channel prediction is to get a predicted gain $\hat{h}[t+\tau]$ that is as close as possible to its actual value $h[t+\tau]$ at the instant $t+\tau$ when the actual transmission using the selected parameters happens.

16.5.2.1 Channel Gain Prediction with a Complex-Valued RNN

Since channel gains are complex-valued, a RNN with complex-valued weights (called a complex-valued RNN hereinafter) needs to be applied (Liu et al., 2006, Potter et al., 2008,

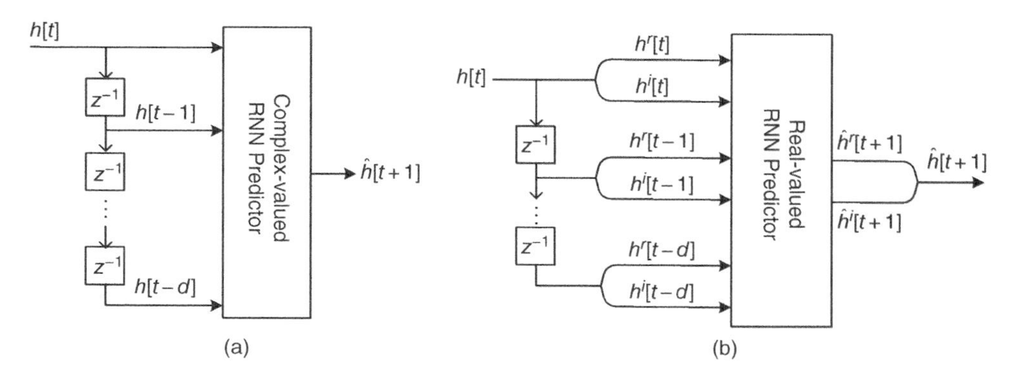

Figure 16.6 (a) Schematics of applying a complex-valued recurrent neural network to predict one-step-ahead channel state information $\hat{h}[t+1]$; (b) applying a real-valued RNN to predict a channel gain by means of processing the real and imaginary parts separately.

Ding and Hirose, 2014). At time t, $h[t]$ is obtained through channel estimation, for example, while a number of d past values $h[t-1], h[t-2], ..., h[t-d]$ can be stored through a tapped delay line, as illustrated in Figure 16.6. These $d+1$ channel gains are fed into the RNN as the external input, i.e.

$$\mathbf{x}_e[t] = [h[t], h[t-1], \ldots, h[t-d]]. \tag{16.32}$$

With the aid of the delayed feedback, the prediction of a future channel gain $\hat{h}[t+1]$ at the next time instant $t+1$ is obtained, as shown in Figure 16.6.

16.5.2.2 Channel Gain Prediction with a Real-Valued RNN

In comparison with a complex-valued RNN, a RNN with real-valued weights (called a real-valued RNN) has the advantages of lower computational complexity and higher prediction accuracy, but it can only deal with real-valued input. Fortunately, a complex-valued channel gain can be decomposed into two real values, namely the real and imaginary parts, i.e. $h = h^r + jh^i$, where $j^2 = -1$ is the imaginary unit. Therefore, a real-valued RNN was proposed in (Jiang and Schotten, 2018a) to build a simpler predictor with higher accuracy by means of separately predicting the real and imaginary parts and then combining them together. As shown in Figure 16.6, the external input of the RNN is then

$$\mathbf{x}_e[t] = [h^r[t], h^i[t], \ldots, h^r[t-d], h^i[t-d]]. \tag{16.33}$$

In this case, the output of the RNN is $\hat{h}^r[t+1]$ and $\hat{h}^i[t+1]$. By combining the predicted real and imaginary parts, the prediction of the channel gain at the next time instant can be obtained, i.e. $\hat{h}[t+1] = \hat{h}^r[t+1] + j\hat{h}^i[t+1]$.

16.5.2.3 Channel Envelope Prediction

Many adaptive transmission systems only need to know the envelope of the channel response, $|h|$, rather than a complex-valued coefficient h itself. Therefore, a real-valued RNN was proposed in (Jiang and Schotten, 2018a) to predict $|h|$ directly, which in turn can lower computational complexity, speed up the training process, and improve prediction accuracy, in comparison with predicting channel gains. The channel envelope at time t denoted by $|h[t]|$ is known, while a number of d past values $|h[t-1]|, |h[t-2]|, \ldots, |h[t-d]|$ can be kept through a tapped delay line. The CSI for the next time instant $|\hat{h}[t+1]|$ can be predicted by feeding these $d+1$ channel values into the RNN as the external input, which in this case is written as

$$\mathbf{x}_e[t] = [|h[t]|, |h[t-1]|, \ldots, |h[t-d]|]. \tag{16.34}$$

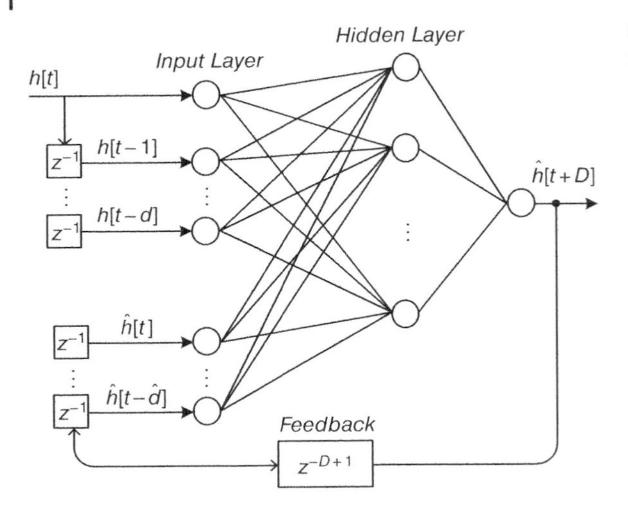

Figure 16.7 The architecture of a multi-step predictor.

16.5.2.4 Multi-Step Prediction

So far, the predictor is only set to forecast one step ahead, extrapolating $\hat{h}[t+1]$ from the current and past CSI $h[t], h[t-1], \dots, h[t-d]$. In practice, an adaptive transmission system probably needs to know a long-range prediction of the CSI, which is enabled only by a multi-step predictor. Luckily, the structure of a RNN is quite flexible, as shown in Figure 16.7, where a multi-step predictor is constructed by tuning the tapped delay lines for input and feedback. Using the response of a single-antenna channel as an example for simplicity, the external input contains the current CSI $h[t]$ and its delays $h[t-1], \dots, h[t-d]$. Meanwhile, a number of $\hat{d}+1$ channel gains $\hat{h}[t], ..., \hat{h}[t-\hat{d}]$ are input as the feedback. The output is $\hat{h}[t+D]$, where D stands for the number of steps being predicted ahead. The parameter D is an positive integer $D = 1, 2, 3, ...$ and the predictor returns back to the previous one-step-ahead prediction if $D = 1$. From the perspective of training, there is no intrinsic distinction between one-step and multi-step prediction. The only difference is that the desired value for calculating the prediction error in the training process is shifted from $h[t+1]$ to $h[t+D]$, resulting in different weights. With the increase of D, the prediction accuracy will degrade due to the intrinsic characteristics of the predicted channel itself.

16.5.3 Flat-Fading MIMO Prediction

A multi-antenna wireless system with N_t transmit and N_r receive antennas in a flat fading channel is modeled as

$$\mathbf{r}[t] = \mathbf{H}[t]\mathbf{s}[t] + \mathbf{n}[t], \tag{16.35}$$

where $\mathbf{r}[t]$ denotes the $N_r \times 1$ received symbol vector at time t, $\mathbf{s}[t]$ is the $N_t \times 1$ transmitted symbol vector, $\mathbf{n}$ stands for the vector of additive white noise, $\mathbf{H}[t] = [h_{n_r n_t}[t]]_{N_r \times N_t}$ is the channel matrix, and $h_{n_r n_t} \in \mathbb{C}^{1\times 1}$ represents the gain of the channel between transmit antenna n_t and receive antenna n_r, where $1 \leq n_r \leq N_r$ and $1 \leq n_t \leq N_t$. Due to feedback and processing delays, the obtained CSI may be outdated before its actual usage, namely $\mathbf{H}[t] \neq \mathbf{H}[t+\tau]$, which probably will degrade the performance of adaptive transmission systems severely. The task of MIMO channel prediction is to get a predicted value $\hat{\mathbf{H}}[t+\tau]$ that approximates $\mathbf{H}[t+\tau]$ as closely as possible.

16.5.3.1 Channel Gain Prediction

Analogous to a single-antenna system, a complex-valued RNN needs to be employed to deal with complex channel gains of a MIMO system. At time t, $\mathbf{H}[t]$ is obtained through channel estimation. To adapt to the input layer of a RNN, the channel matrix is required to be vectorized into a $1 \times N_r N_t$ vector, as follows:

$$\mathbf{h}[t] = \vec{\mathbf{H}}[t] = [h_{11}[t], h_{12}[t], ..., h_{N_r N_t}[t]]. \tag{16.36}$$

Together with a number of d past values $\mathbf{H}[t-1], \mathbf{H}[t-2], \ldots, \mathbf{H}[t-d]$, the external input of RNN this case is $\mathbf{x}_e[t] = [\mathbf{h}[t], \mathbf{h}[t-1], \ldots, \mathbf{h}[t-d]]$, resulting in a multi-step predictive value $\hat{\mathbf{h}}[t+D]$, which can be transformed to a predicted channel matrix $\hat{\mathbf{H}}[t+D]$.

Similar to a SISO channel, a real-valued RNN can be applied to predict the real and imaginary parts of a channel matrix separately so as to lower complexity and improve accuracy. Accordingly, a channel matrix $\mathbf{H}$ (time index dropped for brevity) can be decomposed into

$$\mathbf{H} = \mathbf{H}_R + j\mathbf{H}_I, \tag{16.37}$$

where $\mathbf{H}_R = \Re(\mathbf{H}) = [h^r_{n_r n_t}]_{N_r \times N_t}$ denotes a matrix composed of the real parts of channel gains and $\mathbf{H}_I = \Im(\mathbf{H}) = [h^i_{n_r n_t}]_{N_r \times N_t}$ is its imaginary counterpart. Also, these matrices are required to be vectorized, e.g.

$$\mathbf{h}_r = \vec{\mathbf{H}}_R = [h^r_{11}, h^r_{12}, ..., h^r_{N_r N_t}]. \tag{16.38}$$

Without the necessity of using two RNNs, the real and imaginary parts can be processed jointly in a single predictor, enabled by an external input as $\mathbf{x}_e[t] = [\mathbf{h}_r[t], \mathbf{h}_i[t], ..., \mathbf{h}_r[t-d], \mathbf{h}_i[t-d]]$. Together with the feedback, the predictive output $\mathbf{y} = [\hat{\mathbf{h}}_r[t+D], \hat{\mathbf{h}}_i[t+D]]$ is available and transformed into the predicted real and imaginary matrices $\hat{\mathbf{H}}_R[t+D]$ and $\hat{\mathbf{H}}_I[t+D]$, respectively. Then, a multi-step predicted channel matrix for time $t+D$ is reached simply by applying $\hat{\mathbf{H}}[t+D] = \hat{\mathbf{H}}_R[t+D] + j\hat{\mathbf{H}}_I[t+D]$.

16.5.3.2 Channel Envelope Prediction

As mentioned previously, only the envelope of a channel gain, rather than the gain itself, is required by some adaptive transmission systems, where the prediction of channel gains can be avoided so as to lower complexity, speed up the training process, and improve accuracy. Let $\mathbf{Q} = [|h_{n_r n_t}|]_{N_r \times N_t}$ denote a matrix in which the $(n_r, n_t)^{th}$ entry is the envelope of the channel between transmit antenna n_t and receive antenna n_r, denoted by $|h_{n_r n_t}|$. In order to adapt the input layer of a RNN, this matrix needs to be vectorized:

$$\mathbf{q} = \vec{\mathbf{Q}} = [|h_{11}|, |h_{12}|, ..., |h_{N_r N_t}|]. \tag{16.39}$$

The matrix $\mathbf{Q}[t]$ at time t, as well as its delays $\mathbf{Q}[t-1], ..., \mathbf{Q}[t-d]$, are fed into the RNN predictor as the external input, which thus is rewritten as $\mathbf{x}_e[t] = [\mathbf{q}[t], \mathbf{q}[t-1], ..., \mathbf{q}[t-d]]$. The RNN output is then a predicted channel vector at D steps ahead, i.e. $\hat{\mathbf{q}}[t+D]$, which can be transformed into $\hat{\mathbf{Q}}[t+D]$.

16.5.4 Frequency-Selective MIMO Prediction

The discrete-time baseband equivalent model for a single-antenna system in a frequency-selective channel is given by

$$r[t] = \sum_{l=0}^{L-1} h_l[t]s[t-l] + n[t], \tag{16.40}$$

where $h_l[t]$ denotes the lth tap for a time-varying channel filter, $s[t]$ and $r[t]$ represent the transmitted and received signals at time t, respectively, and $n[t]$ is additive noise. Dropping the time index for simplicity, a frequency-selective channel is modeled as a linear channel filter $\mathbf{h} = [h_0, h_1, \dots, h_{L-1}]^T$, where L is the filter length. This channel can be converted into N orthogonal flat-fading sub-carriers by means of the OFDM modulation (Jiang and Kaiser, 2016). The signal transmission over the nth sub-carrier at time t can be modeled as

$$\tilde{r}_n[t] = \tilde{h}_n[t]\tilde{s}_n[t] + \tilde{n}_n[t], \quad n = 0, 1, \dots, N-1, \tag{16.41}$$

where $\tilde{s}_n[t]$, $\tilde{r}_n[t]$, and $\tilde{n}_n[t]$ stand for the transmitted signal, received signal, and noise, respectively, in the frequency domain. According to the picket fence effect in discrete Fourier transform (DFT) (Oppenheim and Schafer, 1975), the frequency response of the channel filter denoted by $\tilde{\mathbf{h}} = [\tilde{h}_0, \tilde{h}_1, \dots, \tilde{h}_{N-1}]^T$ is the DFT of $\mathbf{h}' = [h_0, h_1, \dots, h_{L-1}, 0, \dots, 0]^T$, which is the filter $\mathbf{h}$ padding with $N - L$ zeros at the tail.

The extension of Eq. (16.41) to a multi-antenna system with N_t transmit and N_r receive antennas is straightforward by applying the same OFDM modulation into MIMO channels. Thus, on the nth sub-carrier, the signal transmission is represented by

$$\tilde{\mathbf{r}}_n[t] = \tilde{\mathbf{H}}_n[t]\tilde{\mathbf{s}}_n[t] + \tilde{\mathbf{n}}_n[t], \quad n = 0, 1, \dots, N-1, \tag{16.42}$$

where $\tilde{\mathbf{r}}_n[t]$ represents N_r received symbols for sub-carrier n at time t, $\tilde{\mathbf{s}}_n[t]$ corresponding to N_t transmit symbols, and $\tilde{\mathbf{n}}[t]$ is a vector of additive noise. The matrix $\tilde{\mathbf{H}}_n[t] = [\tilde{h}_n^{n_r n_t}[t]]_{N_r \times N_t}$ consists of the frequency responses of all subchannels on sub-carrier n at time t denoted by $\tilde{h}_n^{n_r n_t} \in \mathbb{C}^{1\times1}$, where $1 \le n_r \le N_r$ and $1 \le n_t \le N_t$. The frequency response for the subchannel between transmit antenna n_t and receive antenna n_r, i.e. $\tilde{\mathbf{h}}^{n_r n_t} = [\tilde{h}_0^{n_r n_t}, \tilde{h}_1^{n_r n_t}, \dots, \tilde{h}_{N-1}^{n_r n_t}]^T$, can be derived by conducting DFT on its channel filter denoted by $\mathbf{h}^{n_r n_t} = [h_0^{n_r n_t}, h_1^{n_r n_t}, \dots, h_{L-1}^{n_r n_t}]^T$.

Figure 16.8 illustrates the block diagram of a frequency-domain predictor proposed in (Jiang and Schotten, 2019b). The main idea is to convert a frequency-selective channel into a set of orthogonal flat-fading sub-carriers and then utilize a frequency-domain predictor to forecast the frequency response on each sub-carrier. Using a series of channel samples over an arbitrary sub-carrier $\{\tilde{\mathbf{H}}_n[t] | t = 1, 2, \dots\}$ to train a network, the trained RNN can be applied to predict unknown samples. At time t over sub-carrier n, as shown in Figure 16.8, the current CSI $\tilde{\mathbf{H}}_n[t]$, as well as its d-step delays $\tilde{\mathbf{H}}_n[t-1]$, ..., $\tilde{\mathbf{H}}_n[t-d]$, are fed into the RNN. To adapt the input

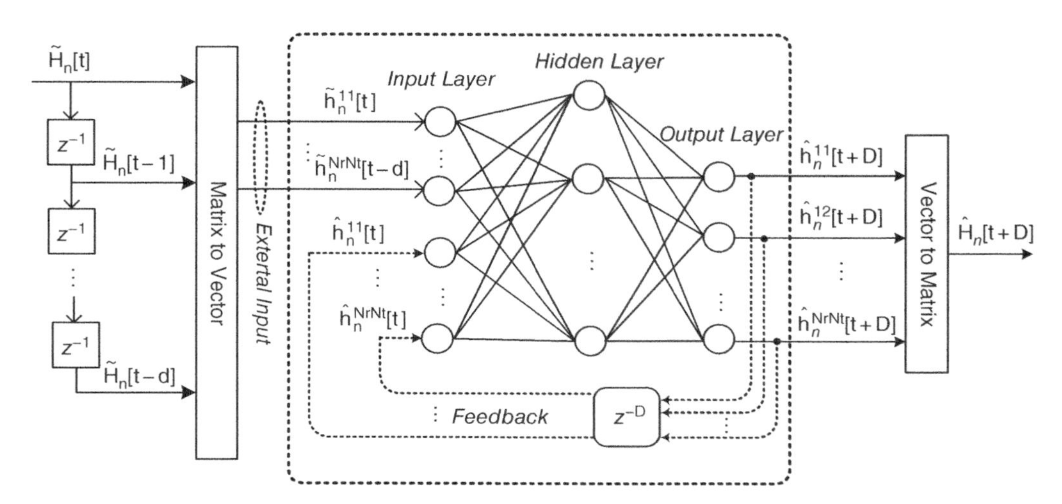

Figure 16.8 Illustration of a frequency-domain predictor for frequency-selective multi-antenna channels.

layer, these matrices need to be vectorized as:

$$\tilde{\mathbf{h}}_n = \text{vec}(\tilde{\mathbf{H}}_n) = [\tilde{h}_n^{11}, \tilde{h}_n^{12}, ..., \tilde{h}_n^{N_r N_t}]^T, \tag{16.43}$$

which is implemented through a matrix-to-vector module as shown in the figure. Together with the feedback from the output denoted by $\hat{\mathbf{h}}_n[t] = [\hat{h}_n^{11}[t], ..., \hat{h}_n^{N_r N_t}[t]]^T$, the whole input is thus $\tilde{\mathbf{h}}_n[t], \tilde{\mathbf{h}}_n[t-1], ..., \tilde{\mathbf{h}}_n[t-d]$ and $\hat{\mathbf{h}}_n[t]$. The RNN outputs a D-step prediction, i.e. $\hat{\mathbf{h}}_n[t+D] = [\hat{h}_n^{11}[t+D], ..., \hat{h}_n^{N_r N_t}[t+D]]^T$, which can be recovered to a predicted matrix $\hat{\mathbf{H}}_n[t+D]$ by a vector-to-matrix module.

From the perspective of a pilot-assisted system, only a subset of sub-carriers instead of all N sub-carriers needs to be predicted if the frequency correlation of channels is utilized. Suppose one pilot is inserted uniformly every N_P sub-carriers, which amounts to a total of $P = \left\lceil \frac{N}{N_P} \right\rceil$ pilot sub-carriers, where $\lceil \cdot \rceil$ denotes a ceiling function. Given their predicted CSI $\hat{\mathbf{H}}_p[t+D]$, $i = 1, \dots, P$, and supposing the indices of pilot sub-carriers are $p = (i-1)N_P$, for example, the prediction for all sub-carriers $\hat{\mathbf{H}}_n[t+D]$, $n = 0, \dots, N-1$ can be obtained by interpolating $\hat{\mathbf{H}}_0[t+D], \hat{\mathbf{H}}_{N_P}[t+D], \dots, \hat{\mathbf{H}}_{(P-1)N_P}[t+D]$.

16.5.5 Prediction-Assisted MIMO-OFDM

To further shed light on the mechanism of channel prediction, transmit antenna selection in a MIMO system with N_t transmit and N_r receive antennas in a frequency-selective fading channel is depicted as a representative application example. A frequency-selective channel can be converted into N orthogonal flat-fading sub-carriers by means of a fast Fourier transform (FFT) demodulator at the receiver and an inverse FFT (IFFT) modulator at the transmitter, in combination with the utilization of cyclic prefix (CP), as shown in Figure 16.9. There exist two selection strategies for TAS: bulk or per-tone, as mentioned in (Zhang and Nabar, 2008). Without loss of generality, the latter is used for a clear illustration, i.e. each sub-carrier decides its best antenna individually instead of the same selection for all sub-carriers.

An OFDM symbol carries a payload of M data symbols denoted by $\mathbf{d} = [d_1, d_2, \dots, d_M]^T$, while the remaining $P = N - M$ sub-carriers are reserved for comb-type pilot symbols that are uniformly inserted in sub-carriers $p = (i-1)N_P$, where $i = 1, \dots, P$ and N_P is the interval of pilots. Through estimating the pth pilot, the frequency response on pilot sub-carrier p denoted by $\tilde{\mathbf{H}}_p[t]$ can be known at the receiver. Taking advantage of the channel's frequency correlation, frequency-domain interpolation is conducted to recover the CSI on all sub-carriers, i.e. $\tilde{\mathbf{H}}_n[t]$, $n = 0, 1, \dots, N-1$. Following the per-tone selection scheme (Zhang and Nabar, 2008), each data sub-carrier chooses its own transmit antenna(s). Analogous to Eq. (16.1), the traditional TAS system directly applies outdated CSI $\tilde{\mathbf{H}}_n[t]$ to select a single antenna with the largest channel gain in sub-carrier n, following

$$\eta_n[t] = \arg\max_{1 \le n_t \le N_t} \|\tilde{\mathbf{h}}_n^{n_t}[t]\|^2, \tag{16.44}$$

where $\eta_n[t]$ represents the index of the best antenna at time t for sub-carrier n, $\tilde{\mathbf{h}}_n^{n_t}[t]$ is the n_t^{th} column vector of $\tilde{\mathbf{H}}_n[t]$, and $\| \cdot \|$ stands for the Euclidean norm of a vector. The receiver feeds the set of selected indices for all data sub-carriers $\{\eta_n[t] \mid 0 \le n \le N-1, n \ne p\}$ back to the transmitter through a feedback channel. In the nth sub-carrier of OFDM symbol $t + D$, the TAS precoder allocates a data symbol to the best antenna $\eta_n[t]$, while other antennas keep silent at this sub-carrier.

Due to the channel fading, outdated CSI $\tilde{\mathbf{H}}_n[t]$ may differ substantially from the actual value $\tilde{\mathbf{H}}_n[t+D]$, leading to remarkable performance degradation (Yu et al., 2017). With the aid of

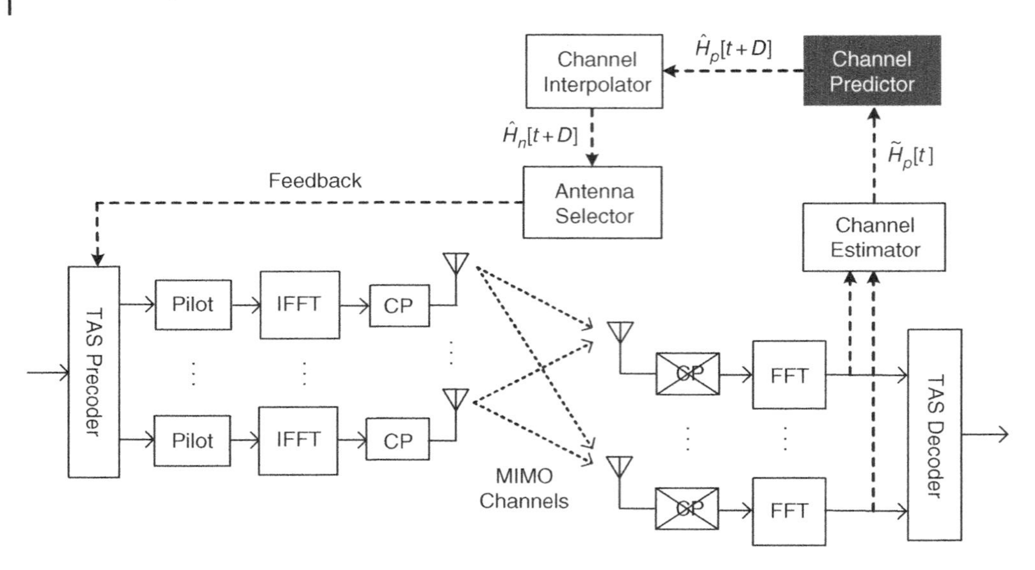

Figure 16.9 Block diagram of prediction-assisted transmit antenna selection in a multiple-input and multiple-output orthogonal frequency-division multiplexing system.

channel prediction, a selection decision can be made in terms of the predicted CSI that is possible to closely approximate the actual value. At time t, as depicted in Figure 16.9, estimating the pilots at the tth OFDM symbol can get the CSI of pilot sub-carriers $\tilde{\mathbf{H}}_p[t]$, fed into the channel predictor to extrapolate the predicted CSI $\hat{\mathbf{H}}_p[t+D]$. A frequency-domain interpolator is applied to obtain the CSI on all sub-carriers denoted by $\hat{\mathbf{H}}_n[t+D]$, $n=0,1,\ldots,N-1$ so as to replace $\tilde{\mathbf{H}}_n[t]$ to make decisions in a TAS system. Thus, the best antenna over sub-carrier n can be selected as

$$\hat{\eta}_n[t] = \arg\max_{1\leq n_t\leq N_t} \|\hat{\mathbf{h}}_n^{n_t}[t+D]\|^2. \tag{16.45}$$

16.5.6 Performance and Complexity

The computational complexity of RNNs is a concern for their application in practical systems. This section analyzes the complexity in terms of the number of complex multiplications required in the process of channel prediction. Meanwhile, the achievable performance of the RNN predictor, in comparison with the KF predictor, is illustrated through numerical results in terms of outage probability in a MIMO-OFDM system.

16.5.6.1 Computational Complexity

In general, the number of complex multiplications is used as a measure for computational complexity. As can be derived from Eqs. (16.29) and (16.30), the hidden and output layer need N_iN_h and N_oN_h times multiplication operations to conduct one time prediction, respectively, amounting to a total of $\Omega_{rnn}=N_h(N_i+N_o)$. The number of required input neurons is proportional to the number of MIMO subchannels N_rN_t: we have $N_i=(d+2)N_rN_t$ if the feedback contains only one channel matrix, as illustrated in Figure 16.8, and similarly the number of output neurons is $N_o=N_rN_t$. Then, the complexity of the RNN predictor can be computed by $\Omega_{rnn}=(d+3)N_hN_rN_t$. In contrast, derived from Eq. (16.22), the KF predictor requires $\Omega_{kf}=pN_rN_t$ times multiplication operations per prediction. Since a small filter order such as $p=4$ is generally optimal and thus $(d + 3)N_h > p$, it is concluded that the KF predictor is simpler than the RNN predictor.

Further, it is meaningful to make clear how many computing resources are required. The number of pilot sub-carriers per OFDM symbol is around N/N_P, and there are f_s/N OFDM symbols per second, from which the number of predictions per second can be figured out, i.e. $\psi = f_s/N_P$. The required multiplications per second by the KF predictor are exactly the product of ψ and Ω_{kf}, i.e. $\Omega_{kf}^{(s)} = f_s\Omega_{kf}/N_P$, and $\Omega_{rnn}^{(s)} = f_s\Omega_{rnn}/N_P$ in the case of the RNN predictor. Assuming a RNN with $N_h = 10$ and $d = 3$ is applied for a 4×1 MIMO system with a signal sampling rate of $f_s = 10^6$ Hz and $N_P = 4$, we have $\Omega_{kf}^{(s)} = 4 \times 10^6$ and $\Omega_{rnn}^{(s)} = 60 \times 10^6$. Compared with off-the-shelf digital signal processors (DSPs), e.g. TI 66AK2x, which provides a capability of nearly 2×10^4 million instructions executed per second (MIPS), the required resource of the RNN predictor is 0.3%. Even in a massive MIMO system with a dimension of 32×4, it consumes only 10% of the computing power of a single DSP. In summary, the computing resources required by a channel predictor are affordable, which is promising from a practical perspective.

16.5.6.2 Performance

The numerical results of the performance achieved by prediction-assisted TAS in a MIMO-OFDM system in a frequency-selective channel are illustrated. The signal bandwidth is 1 MHz, which is converted into $N = 64$ parallel sub-carriers by the OFDM modulation, resulting in a sub-carrier spacing around $\triangle f = 15$ KHz. The number of hidden neurons is $N_h = 10$, and the length of the tapped delay line is $d = 3$. The details of setting up the Monte Carlo simulation can be found in (Jiang and Schotten, 2019b).

To train a RNN, a training dataset containing a series of consecutive CSI $\{\tilde{\mathbf{H}}_{tr}[t] | t = 1, 2, \dots\}$ extracted from an arbitrary sub-carrier during 10 periods of fluctuation (i.e. channel's coherence time) is built. A training process starts from an initial state where all weights can be randomly selected. At iteration t, feeding $\tilde{\mathbf{H}}_{tr}[t]$ into the RNN, the resultant output is compared with the desired value, and the prediction error $\hat{\mathbf{H}}_{tr}[t + D] - \tilde{\mathbf{H}}_{tr}[t + D]$ is propagated back through the network so as to update the weights by means of training algorithms such as Levenberg-Marquardt (Fu et al., 2015). This process is iteratively carried out until the RNN reaches a certain convergence condition. In contrast, the KF predictor does not need training. Its filter coefficients required in Eq. (16.22) can be figured out if f_d and f_s are known. Once the training process of a RNN is completed and the coefficients of a Kalman filter are determined, channel prediction can be conducted. Figure 16.10 provides a direct view of prediction accuracy, where the amplitudes and phases of a series of predicted channel gains are compared with their actual values.

Suppose a multi-antenna system with a uniform linear array has $N_t = 4$ transmit and $N_r = 1$ receive antennas, and a single transmit antenna with the largest instantaneous channel gain is selected. Outage probability is investigated: it is an important performance metric over fading channels, defined as $P(R) = Pr\{\log_2(1 + SNR) < R\}$, where Pr is the notation of mathematical probability and R means a target end-to-end data rate that is set to 1bps/Hz in general. Three different CSI modes are compared:

- The perfect mode where a transmit antenna for subcarrier n at OFDM symbol $t + D$ is chosen in terms of the actual CSI $\tilde{\mathbf{H}}_n[t + D]$, despite it never existing in practice owing to delay and noise.
- As in a traditional TAS system, only outdated CSI $\tilde{\mathbf{H}}_n[t]$ is available.
- With the aid of channel prediction, the predicted CSI $\hat{\mathbf{H}}_n[t + D]$ that is possible to closely approximate the actual CSI is used.

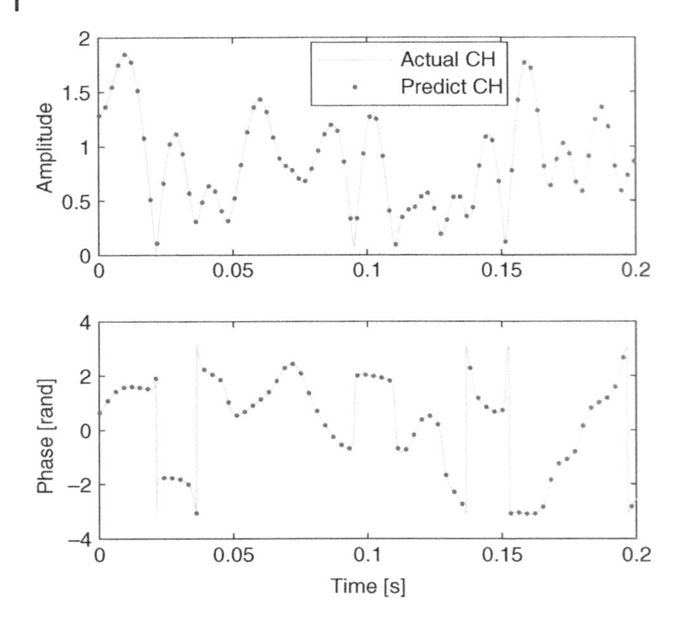

Figure 16.10 Predicted amplitudes and phases (dotted lines) versus actual values (solid lines).

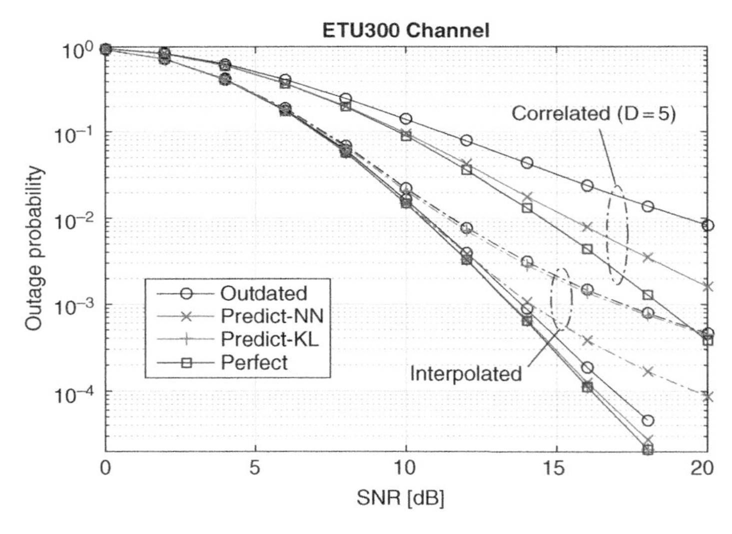

Figure 16.11 Performance comparison of outdated, predicted, and perfect channel state information in a multiple-input and multiple-output orthogonal frequency-division multiplexing system.

Figure 16.11 illustrates the performance achieved by a MIMO-OFDM system in a frequency-selective channel specified by the 3GPP extended typical urban (ETU) model with a maximal Doppler shift of $f_d = 300$ Hz. First, the RNN predictor is tuned to one-step prediction mode ($D = 1$) in order to make a direct comparison with the KF predictor. The prediction step $D = 1$ is equivalent to a time range of 64 us in contrast with a coherence time of $T_c \approx 1/f_d = 3.3$ ms. Using the curve of the perfect mode as a benchmark, the KF predictor achieves optimal performance. Although the RNN predictor is slightly inferior to the KF predictor, it is still quite close to optimal performance and clearly outperforms the outdated mode. Channel interpolation error, which is defined as the difference between the perfect CSI and the interpolated CSI, is also a concern in the process of channel estimation, so the impact of interpolation errors on performance is evaluated. For the purpose of a better illustration, the results for a pilot interval of $N_P = 3$ are selected to show in the figure. Even in the mode of one-step prediction $D = 1$, outdated CSI has a remarkable loss of 3.6 dB in comparison with the perfect mode at the outage probability of 10^{-3}. The KF predictor is vulnerable to interpolation errors, corresponding to a worse result that is very comparable to the outdated mode. In contrast, the RNN predictor is robust and outperforms the KF predictor with an SNR gain of 3.1 dB. To look at the effect of channel correlation, correlated channels generated by the correlation matrix recommended by 3GPP LTE standard for ETU channels are applied. Under the medium correlation indicated by $\alpha = 0.3$ and a multi-step mode of $D = 5$, outdated CSI has a performance loss of approximately 5 dB given $P(R) = 10^{-2}$ in comparison with the perfect mode, while the channel prediction can take back nearly 4 dB. As mentioned previously, the KF predictor can merely conduct one-step prediction and therefore is not applicable to this case.

16.6 Summary

This chapter provided a comprehensive view of channel prediction techniques with an emphasis on NN-based prediction. At the beginning, the principles of two representative adaptive transmission systems –: transmit antenna selection and opportunistic relaying – were briefly introduced, followed by the performance impact of outdated CSI. Then, classical channel prediction methods based on statistical modeling, i.e. parametric models and autoregressive models, were reviewed. After an explanation of the internal structure of a recurrent NN, the RNN-based multi-step prediction were detailed, which can be applied for either SISO or MIMO systems in both flat-fading and frequency-selective channels. To further shed light on the mechanism of RNN predictors, the integration of a predictor into a MIMO-OFDM system to improve the correctness of selecting antennas at the transmitter was illustrated. Performance results in multi-path fading environment specified by the 3GPP ETU channel model, taking into account a number of influential factors including the spatial correlation, Doppler shift, as well as interpolation error, were shown. It is verified that applying the RNN predictors to combat the problem of outdated CSI is effective and efficient. Although its computational complexity is higher than the Kalman filter, the required computing resources are still affordable relative to off-the-shelf hardware. In summary, the RNN exhibits great flexibility, generality, scalability, and applicability in the application of wireless fading-channel prediction, and can therefore be regarded as a very promising machine learning technique for future wireless communications.

Bibliography

R.O. Adeogun. Channel prediction for mobile MIMO wireless communication systems. *Ph.D Thesis*, 2014.

R.O. Adeogun et al. Parametric channel prediction for narrowband mobile MIMO systems using spatio-temporal correlation analysis. *Proceedings of IEEE Vehicular Tech. Conf.*, 2013.

R.O. Adeogun et al. Extrapolation of MIMO mobile-to-mobile wireless channels using parametric-model-based prediction. *IEEE Transactions on Vehicular Technology*, 64: 4487–4498, 2014.

P. Aquilina and T. Ratnarajah. Performance analysis of IA techniques in the MIMO IBC with imperfect CSI. *IEEE Transactions on Communications*, 63: 1259–1270, 2015.

K.E. Baddour and N.C. Beaulieu. Autoregressive modeling for fading channel simulation. *IEEE Transactions on Wireless Communications*, 4: 1650–1662, 2005.

A. Bletsas et al. A simple cooperative diversity method based on network path selection. *IEEE Journal on Selected Areas in Communications*, 24: 659–672, 2006.

J.T. Connor et al. Recurrent neural networks and robust time series prediction. *IEEE Transactions on Neural Networks*, 5: 240–254, 1994.

T. Ding and A. Hirose. Fading channel prediction based on combination of complex-valued neural networks and chirp Z-transform. *IEEE Transactions on Neural Networks and Learning Systems*, 25: 1686–1695, 2014.

A. Duel-Hallen et al. Long-range prediction of fading signals. *IEEE Signal Processing Magazine*, 17: 62–75, 2000.

T. Eyceoz et al. Deterministic channel modeling and long range prediction of fast fading mobile radio channels. *IEEE Communications Letters*, 2: 254–256, 1998.

X. Fu et al. Training recurrent neural networks with the Levenberg-Marquardt algorithm for optimal control of a grid-connected converter. *IEEE Transactions on Neural Networks*, 26: 1900–1912, 2015.

C.Y. Fung and S.C. Chan. Estimation of fast fading channel in impulse noise environment. *Proceedings of IEEE International Symposium on Circuits and Systems*, pages 497–500, 2002.

A. Hyadi et al. An overview of physical layer security in wireless communication systems with CSIT uncertainty. *IEEE Access*, 4: 6121–6132, 2016.

W. Jiang and T. Kaiser. From OFDM to FBMC: Principles and Comparisons. In F.L. Luo and C. Zhang, editors, *Signal Processing for 5G: Algorithms and Implementations*, chapter 3. John Wiley & Sons, United Kindom, 2016.

W. Jiang and H.D. Schotten. Multi-antenna fading channel prediction empowered by artificial intelligence. *Proceedings of IEEE Vehicular Tech. Conf. (VTC)*, 2018a.

W. Jiang and H.D. Schotten. Neural network-based channel prediction and its performance in multi-antenna systems. *Proceedings of IEEE Vehicular Tech. Conf. (VTC)*, 2018b.

W. Jiang and H.D. Schotten. A comparison of wireless channel predictors: Artificial intelligence versus Kalman filter. *Proceedings of IEEE Intl. Commu. Conf. (ICC)*, 2019a.

W. Jiang and H.D. Schotten. Recurrent neural network-based frequency-domain channel prediction for wideband communications. *Proceedings of IEEE Vehicular Tech. Conf. (VTC)*, 2019b.

W. Jiang, T. Kaiser, and A.J.H. Vinck. A robust opportunistic relaying strategy for co-operative wireless communications. *IEEE Transactions on Wireless Communications*, 15: 2642–2655, 2016.

W. Jiang, M. Strufe, and H.D. Schotten. Experimental results for artificial intelligence-based self-organized 5G networks. *Proceedings of IEEE PIMRC*, 2017.

Y. Jing and H. Jafarkhani. Network beamforming using relays with perfect channel information. *IEEE Transactions on Information Theory*, 55: 2499–2517, 2009.

J. Kim et al. Cooperative distributed beamforming with outdated CSI and channel estimation errors. *IEEE Transactions on Communications*, 62: 4269–4280, 2014.

J.N. Laneman and G.W. Wornell. Distributed space-time-coded protocols for exploiting cooperative diversity in wireless networks. *IEEE Transactions on Information Theory*, 49: 2415–2425, 2003.

R. Liao et al. The Rayleigh fading channel prediction via deep learning. *Wireless Communications and Mobile Computing*, 2018, 2018.

W. Liu et al. Recurrent neural network based narrowband channel prediction. *Proceedings of IEEE VTC*, 2006.

D.J. Love et al. An overview of limited feedback in wireless communication systems. *IEEE Journal on Selected Areas in Communications*, 26: 1341–1365, 2008.

O. Nerrand et al. Training recurrent neural networks: why and how? an illustration in dynamical process modeling. *IEEE Transactions on Neural Networks*, 5: 178–184, 1994.

E.N. Onggosanusi et al. Performance analysis of closed-loop transmit diversity in the presence of feedback delay. *IEEE Transactions on Communications*, 49: 1618–1630, 2001.

A.V. Oppenheim and R.W. Schafer. *Digital Signal Processing*. Prentice-Hall, first edition, 1975.

W. Peng et al. Channel prediction in time-varying Massive MIMO environments. *IEEE Access*, 5: 23938–23946, 2017.

C. Potter et al. MIMO beam-forming with neural network channel prediction trained by a novel PSO-EA-DEPSO algorithm. *Proceedings of IEEE IJCNN*, 2008.

D.J. Ramirez et al. Coordinated multi-point transmission with imperfect CSI and other-cell interference. *IEEE Transactions on Wireless Communications*, 14: 1882–1896, 2014.

T.R. Ramya and S. Bhashyam. Using delayed feedback for antenna selection in MIMO systems. *IEEE Transactions on Wireless Communications*, 8: 6059–6067, 2009.

A. Sendonaris et al. User cooperation diversity-Part I and II. *IEEE Transactions on Communications*, 51: 1927–1948, 2003.

D. Silver et al. Mastering the game of Go with deep neural networks and tree search. *Nature*, 529: 484–489, 2016.

Y. Teng et al. Effect of outdated CSI on handover decisions in dense networks. *IEEE Communications Letters*, 21: 2238–2241, 2017.

K.T. Truong and R. W. Heath. Effects of channel aging in massive MIMO systems. *Journal of Communications and Networks*, 15: 338–351, 2013.

D. Tse and P. Viswanath. *Fundamentals of Wireless Communication*. Cambridge Univ. Press, Cambridge, UK, 2005.

J.L. Vicario et al. Opportunistic relay selection with outdated CSI: outage probability and diversity analysis. *IEEE Transactions on Wireless Communications*, 8: 2872–2876, 2009.

Q. Wang et al. Multi-user and single-user throughputs for downlink MIMO channels with outdated channel state information. *IEEE Wireless Communications Letters*, 3: 321–324, 2014a.

Z. Wang et al. Resource allocation in OFDMA networks with imperfect channel state information. *IEEE Communications Letters*, 18: 1611–1614, 2014b.

J. Wu and W. Lee. Optimal linear channel prediction for LTE-A uplink under channel estimation errors. *IEEE Transactions on Vehicular Technology*, 62: 4135–4142, 2013.

L. Xiao and X. Dong. Unified analysis of generalized selection combining with normalized threshold test per branch. *IEEE Transactions on Wireless Communications*, 5: 2153–2163, 2006.

X. Yu et al. Unified performance analysis of transmit antenna selection with OSTBC and imperfect CSI over Nakagami-m fading channels. *IEEE Transactions on Vehicular Technology*, 67: 494–508, 2017.

H. Zhang and R.U. Nabar. Transmit antenna selection in MIMO-OFDM systems: Bulk versus per-tone selection. *Proceedings of IEEE Intl. Conf. on Commu. (ICC)*, 2008.

J. Zheng and B.D. Rao. Capacity analysis of MIMO systems using limited feedback transmit precoding schemes. *IEEE Transactions on Signal Processing*, 56: 2886–2901, 2008.

Part III

Network Intelligence and Adaptive System Optimization

17

Machine Learning for Digital Front-End: a Comprehensive Overview

Pere L. Gilabert[1]*, David López-Bueno*[2]*, Thi Quynh Anh Pham*[1]*, and Gabriel Montoro*[1]

[1] *Dept. of Signal Theory and Communications, UPC-Barcelona Tech., Castelldefels, Barcelona, Spain*
[2] *CTTC/CERCA & Dept. of Signal Theory and Communications, UPC-Barcelona Tech., Castelldefels, Barcelona, Spain*

17.1 Motivation and Background

In wireless and wired communications, the power amplifier (PA) is a critical subsystem in the transmitter chain – not only because it is one of the most power-hungry devices and accounts for most of the direct current (DC) power consumed in macro base stations, but also because it is the main source of nonlinear distortion in the transmitter. Amplitude- and phase-modulated communication signals presenting a high peak-to-average power ratio (PAPR) have a negative impact on the transmitter's power efficiency, because the PA has to be operated at high power back-off levels to avoid introducing nonlinear distortion. As shown in Figure 17.1, to prevent the peaks of the signal from going into compression, it is necessary to operate far from saturation, where the PA is more efficient. Consequently, the mean power added efficiency (PAE) is low, mainly in linear but inefficient class-A or class-AB PAs.

Power amplifier system level linearizers, such as digital predistortion (DPD), as shown in Figure 17.2, extend the linear range of PAs. Properly combined with crest factor reduction (CFR) techniques, DPD allows PAs to be driven harder into compression while meeting linearity requirements (López et al. (2014)). DPD linearization can overcome or at least mitigate the efficiency versus linearity trade-off in PAs. However, the resulting power efficiency achieved with linearization techniques applied to PAs operating as controlled current sources (e.g. class A, B, AB) is limited. To avoid wasting excessive power resources when handling high PAPR signals, either the operating conditions of a current source mode PA can be forced to follow its envelope, or switched-mode amplifying classes can be properly introduced. Among the set of techniques aimed at dynamic bias or load adaptation, envelope tracking (ET) PAs (Wang (2015), Popovic (2017), Watkins and Mimis (2018)), Doherty PAs (Pengelly et al. (2016), Darraji et al. (2016)), and LINC or outphasing PAs (Barton (2016), Popovic and García (2018)) are the most widely proposed in literature. In either case, these highly efficient topologies demand linearization techniques to guarantee the linearity levels specified in the communications standards.

In 5G-NR (Shafi et al. (2017)), the same network infrastructure will be able to efficiently serve different types of traffic with a very wide range of requirements, such as a huge number of users for the Internet of Things, ultra-low latency and high reliability for mission-critical systems, and enhanced transmission rates for broadband mobile communications. 5G-NR intends to provide very high data rates everywhere. To achieve this goal, bandwidths up to GHz will be allocated at mmWave bands, while at sub-6 GHz, bandwidths of hundreds of MHz will be required.

Achieving these new capabilities requires coping with multiple demanding challenges that, particularizing for the design of radio transceivers, are related to: (i) ensuring the linearity of

Machine Learning for Future Wireless Communications, First Edition. Edited by Fa-Long Luo.
© 2020 John Wiley & Sons Ltd. Published 2020 by John Wiley & Sons Ltd.

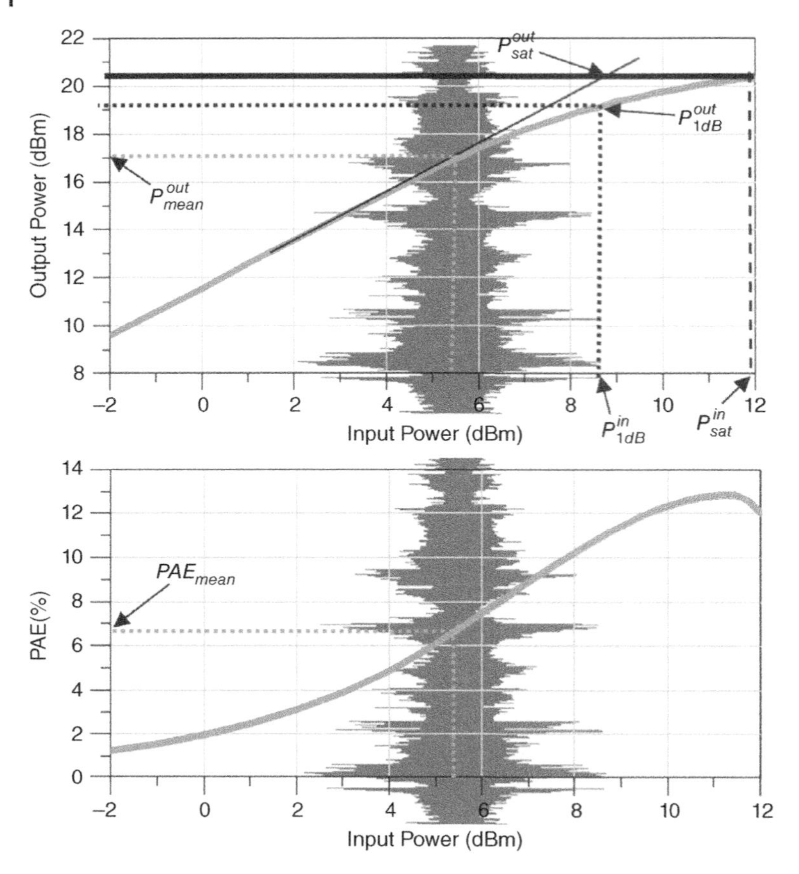

Figure 17.1 Linearity versus power efficiency trade-off.

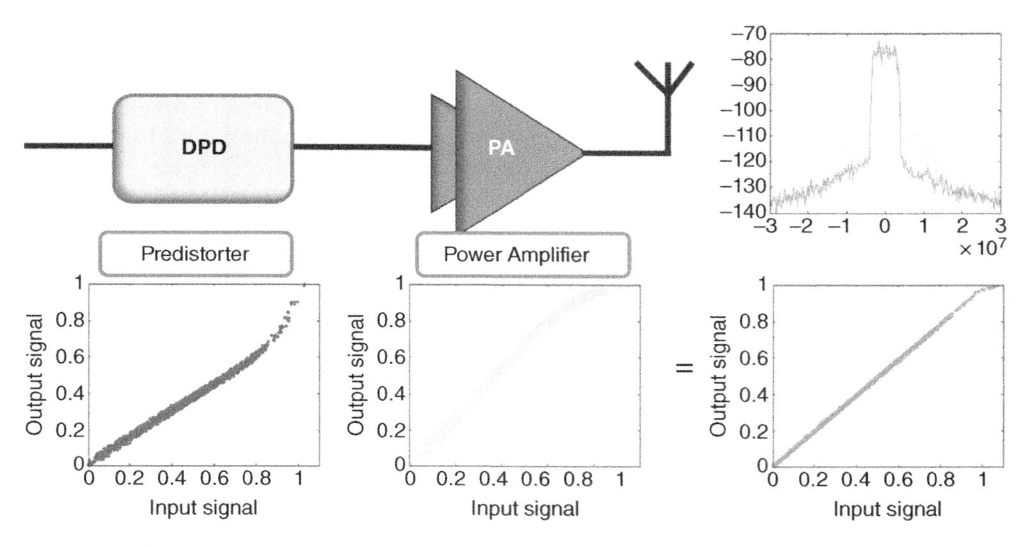

Figure 17.2 Block diagram of DPD linearization.

signals having bandwidths of several hundreds of MHz and peak factors exceeding 10 dB in order to ensure high transmission rates; (ii) improving energy and computational efficiency, as more dense deployments of base stations is expected to scale down the need for transmitted power; (iii) transmitting architectures with multiple antennas (massive multiple-input and multiple-output [MIMO] in millimeter bands) and multiple power amplifiers to apply beamforming techniques that allow increasing capacity and focusing energy where needed to minimize interference; and (iv) simultaneous transmission and reception (full-duplex FDD in sub-6 GHz bands).

The use of 5G spectrally efficient waveforms featuring high PAPR and occupying wider bandwidths in multiple-antenna transmitters (Suryasarman and Springer (2015)) only aggravates the inherent PA linearity versus efficiency trade-off. When considering wide bandwidth signals, carrier aggregation, or multi-band configurations (Jaraut et al. (2018)) in highly efficient transmitter architectures, such as Doherty PAs, envelope tracking PAs, or outphasing transmitters, the number of parameters required in the DPD model to compensate for both static nonlinearities and dynamic memory effects can be unacceptably high. This has a negative impact on the DPD model extraction/adaptation process, because it increases the computational complexity, which may provoke overfitting and uncertainty in the DPD estimation stages (Chani-Cahuana et al. (2017)). However, by applying regularization or dimensionality reduction techniques (Braithwaite (2017)), we can both avoid the numerical ill-conditioning of the estimation and reduce the number of coefficients of the DPD function in the forward path, which ultimately impacts the baseband processing computational complexity and power consumption.

This chapter is devoted to the use of machine learning (ML) algorithms in digital front-end with the emphasis on CFR techniques, DPD linearization, and in-phase/quadrature (I/Q) imbalance mitigation. Due to their powerful nonlinear mapping and distributed processing capability, neural network (NN)-based ML technology can offer a more powerful digital front-end (DFE) solution than conventional approaches in many aspects including system performance, computational complexity, power consumption, and processing latency. The rest of this chapter is organized into the following five sections. Section 17.2 focuses on the problem formulations and fundamental principles of the use of ML and artificial NNs (ANNs) in the DFE, by providing an overview of the need for CFR and DPD techniques and the importance of regularization. Section 17.3 addresses feature-selection and feature-extraction techniques used to reduce the number of parameters of the DPD linearization system as well as to ensure proper, well-conditioned estimation for related variables. Sections 17.4 and 17.5 discuss some advanced solutions for ANNs and support vector regression (SVR) approaches to model and compensate for unwanted nonlinear effects in the transmitter chain as well as to reduce the PAPR of the signals. Finally, Section 17.6 will further discuss the use of ML techniques in DFE linearization and provides some conclusions.

17.2 Overview of CFR and DPD

17.2.1 Crest Factor Reduction Techniques

When dealing with signals presenting high PAPR, the digital-to-analog converter (DAC) and PA of the transmitter require large dynamic ranges to avoid amplitude clipping (and thus avoid introducing nonlinear distortion), which implies increasing both the power consumption and cost of the transceiver. In addition, signals with large dynamic range lead to increased power dissipation in DACs, as well as to a shrinking of both the signal-to-noise ratio (SNR) and spurious-free dynamic range. Therefore, high PAPR makes the converter behave as if it were of a lower number of bits than it actually is (Giannopoulos and Paliouras). In addition, as discussed in the previous subsection, operating the PA with significant back-off levels to prevent the signal peaks from going into compression degrades the power efficiency of the overall amplification system. Combining CFR with linearization techniques can enhance overall PA power efficiency while preserving required linearity levels at the output of the PA.

In OFDM-based multi-carrier systems (e.g. LTE, LTE-A, WiMax), when some of the subcarriers are added with the same phase, what is produced is a peak power that increases the PAPR. In the literature, it is possible to find several published CFR techniques aimed at reducing the PAPR, mainly for OFDM-based signals (Han and Lee (2005), Jiang and Wu (2008), Kaur and Saini (2017)), such as the following examples:

- *Coding*: The idea of the coding schemes is to reduce the occurrence probability of the same phase of the signals by selecting codewords that minimize the PAPR (avoiding in-phase addition of signals) in the transmission. Several coding techniques have been published in literature, such as simple block coding, (Fragiacomo et al. (1998)), complement block coding (Jiang and Zhu (2005)), and modified complement block coding (Jiang and Zhu (2004)), among others.
- *Partial transmit sequence (PTS)*: In the PTS technique presented in Müller and Huber (1997), an input data block of N frequency-domain symbols is partitioned into disjoint sub-blocks. Then, the sub-carriers in each sub-block are IFFT transformed into time-domain partial transmit sequences and independently rotated (weighted) by phase factors. These phase factors are selected in such a manner as to minimize the PAPR of the output signal that results from the combination of each of the sub-blocks. The phase information vector needs to be transmitted to the receiver for the correct decoding of the transmitted bit sequence.
- *Selected mapping technique (SLM)*: Similarly to PTS, in the SLM technique the input data (consisting of N frequency-domain symbols) are multiplied by a vector of phase-shifts to generate an alternative (rotated) input. This operation is done in parallel R times. Each of these R alternative input data sequences is IFFT processed and then the PAPR is evaluated for each of these possible candidates. Finally, the data sequence with the lowest PAPR is selected for transmission (Bäuml et al. (1996)).
- *Interleaving technique*; Similarly to the SLM technique, a set of interleavers (instead of phase sequences) is used to generate new data blocks targeting the PAPR reduction of the OFDM-based signals. The interleaver takes a block of N symbols and reorders or permutes them (Hill et al. (2000), Han and Lee (2005)). From the original data block, $R - 1$ new data blocks can be obtained by permuting the original data using interleavers. After the R IFFT operations, the data block with the lowest PAPR is chosen for transmission.
- *Tone reservation (TR) and tone injection (TI)*: Both TR and TI are the methods based on adding a time-domain signal to the original multi-carrier signal in such a way that its contribution reduces the overall PAPR (Tellado (1999)). The time-domain signal is computed at the transmitter side and simply removed at the receiver side. In the case of TR, the additional subcarriers (or time-domain signal) used to reduce the PAPR are reserved for this purpose (i.e. not used for data transmission) and known by the transmitter and receiver. In TI, instead of using reserved tones, the subcarriers carrying data information are used to reduce the PAPR by extending the original constellation size (Tellado (1999)). Consequently, each of the points in the original constellation can be mapped into several equivalent points in the expanded constellation, providing several extra degrees of freedom to be used for minimizing the PAPR.
- *Active constellation extension (ACE)*: Similarly to TI, in ACE (Krongold and Jones (2003)) some of the outer signal constellation points in the data block are dynamically extended

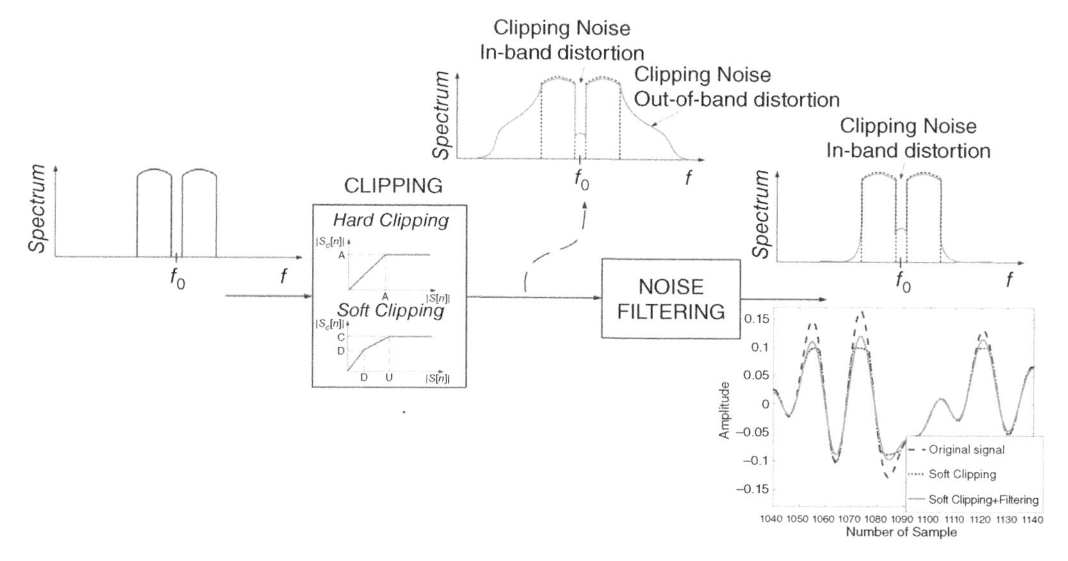

Figure 17.3 Block diagram of the clipping and filtering CFR technique.

toward the outside of the original constellation to reduce the PAPR of the signal. The constellation symbols (in M-QAM modulations) allocated in corners may be allocated within the quarter-plane outside of the nominal constellation point. By properly changing the modulus and phase of these symbols, the signal PAPR can be reduced at the expense of increasing the mean power of the transmit signal.

- *Clipping and filtering*: Clipping techniques may be classified as hard-clipping, soft-clipping, and companding. In hard-clipping, the output signal is strictly limited at the established threshold; while in soft-clipping, the output signal follows a piecewise law where several threshold levels are defined. In the companding technique, the dynamic range of the signal is compressed at the transmitter side by means of a memoryless transformation (i.e. companding function). As observed in Figure 17.3, these clipping techniques require some kind of spectral shaping procedure to mitigate the clipping noise that appears as spectral regrowth in the adjacent channels. To cope with the spectral shaping of the clipping noise, several techniques have been proposed, such as *clipping pulses* (Kim et al. (2007)), *pulse windowing* (Vaananen et al. (2005)), or *noise shaping* (Saul (2004)).

The amount of PAPR reduction that can be achieved depends on the chosen CFR technique and always comes at a price. The harmful effects that appear when reducing the signal's PAPR depend on the specific CFR technique; and, in general, the more PAPR reduction, the more critical are these side effects. Some of the collateral factors that need to be taken into account are (Han and Lee (2005)): (i) power increase in the transmit signal (e.g. TR, TI, or ACE); (ii) BER increase at the receiver (e.g. clipping and filtering techniques: TR, TI, and ACE if the transmit signal power is fixed; SLM, PTS, and interleaving if the side information is not properly received); (iii) loss in data rate (e.g. block coding technique, SLM, PTS, and interleaving due to the side information sent to inform the receiver of what has been done in the transmitter); and (iv) computational complexity increase (e.g. SLM, PTS, interleaving, TI, and ACE, have to run search for the best configurations).

As will be addressed in subsection 17.4.3, several of the aforementioned CFR techniques make use of ML strategies. For example, ANNs are employed to reduce the computational complexity of CFR techniques such as SLM, PTS, TI, and ACE that require running intensive search (that

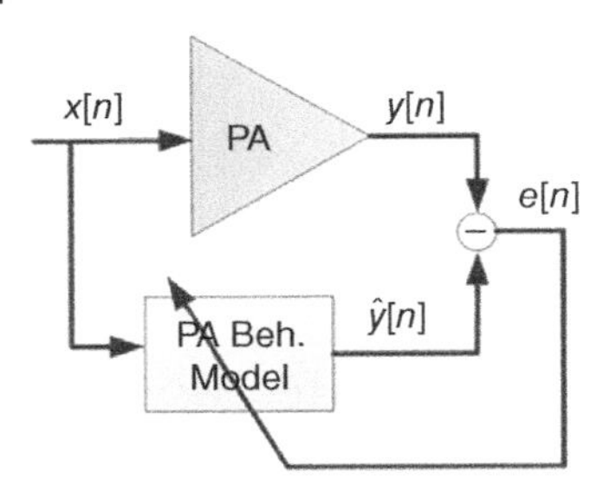

Figure 17.4 Identification of the power amplifier behavior.

may take several iterations) to find the best configuration of their parameters; or, in the case of iterative clipping and filtering algorithms, to avoid several FFT/IFFT complex operations.

17.2.2 Power Amplifier Behavioral Modeling

Power amplifier behavioral models, or black-box models, are mathematical descriptors of the non-ideal behavior of a power amplifier, mainly describing its nonlinear behavior and memory effects (i.e. PA dynamic behavior). Unlike physical models, where it is necessary to know the electronic elements that form the PA, their constitutive relations, and the theoretical rules describing their interactions, the extraction of PA behavioral models relies only on a set of input-output observations. Consequently, their accuracy is highly sensitive to the adopted model structure and the parameter extraction procedure. In general, the same model used for approximating the response of the PA is also used to estimate its inverse response. For this reason, the behavioral models listed in this subsection are valid approximations of the inverse behavior of the PA and will be used in the following subsection when describing the digital predistortion linearizer.

It is possible to find in literature an enormous number of publications on PA behavioral modeling to address not only single-input and single-output (SISO) systems but also multiple-input and single-output (MISO) systems, for example when having to characterize concurrent multi-band transmissions or dynamic supply modulation strategies for the PA. Some of the most commonly used polynomial-based behavioral models can be seen as a simplified approximations of the general Volterra series. Volterra series are aimed at describing time-invariant nonlinear systems with fading memory. The discrete-time low-pass equivalent Volterra series formulation is described in the following. Considering the general input-output notation in Figure 17.4, the estimated output $\hat{y}[n]$ of the Volterra series is

$$\hat{y}[n] = \sum_{p=1}^{P} \sum_{q_p=0}^{Q_p-1} \cdots \sum_{q_1=0}^{Q_1-1} h_p(q_1, \cdots, q_p) \prod_{i=1}^{p} x[n - q_i]. \tag{17.1}$$

The series is composed by P kernels of increasing dimensional order. The main drawback of using the full Volterra series is that the number of parameters grows exponentially when considering higher-order kernels, and typical communication signals do not present enough richness to fully excite these kernels, which ultimately may lead to an ill-conditioned problem.

One of the most widely used models in literature due to its simplicity is the memory polynomial (MP), presented in Kim and Konstantinou (2001). Another widely used model for SISO systems is the generalized memory polynomial (GMP) behavioral model, proposed in D.R. Morgan, Z. Ma et al. (2006). Unlike the MP, the GMP has bi-dimensional kernels (considering cross-term products between the complex signal and the lagging and leading envelope terms), which increases the accuracy of the modeling at the price of increasing the number of parameters. There are plenty of other behavioral models in literature used for DPD purposes in SISO systems: just to mention a couple of examples, the NARMA model proposed in Montoro et al. (2007) and the dynamic deviation reduction Volterra series in Zhu et al. (2006). Further information on PA behavioral models for SISO systems can be found in Scheurs et al. (2009). In addition, when considering concurrent multi-band transmissions such as in Roblin et al. (2013), or combined with PA dynamic supply modulation strategies as in Gilabert and Montoro (2015), or also in multi-antenna systems where each transmit path has its own PA and antenna element as in Hausmair et al. (2018), MISO behavioral models are required to characterize the different sources of nonlinear behavior. In addition, as an alternative to polynomial-based behavioral models, ANN and SVR approaches have been used in literature for PA behavioral modeling and DPD linearization purposes. As will be presented in Sections 17.4 and 17.5, ANNs and SVR can outperform the modeling capabilities of classical polynomial-based solutions (inherently local approximations) by providing global approximation and better extrapolation capabilities.

In general, the estimated PA behavioral model output $\hat{y}[n]$ (for $n = 0, 1, \cdots, N-1$) can be defined following a matrix notation as

$$\hat{\mathbf{y}} = \mathbf{X}\mathbf{w} \tag{17.2}$$

where $\mathbf{w} = (w_1, \cdots, w_i, \cdots, w_M)^T$ is the $M \times 1$ vector of parameters and $\mathbf{X}$ is the $N \times M$ data matrix (with $N \gg M$) containing the basis functions or components. The data matrix can be defined as

$$\mathbf{X} = (\boldsymbol{\varphi}_x[0], \boldsymbol{\varphi}_x[1], \cdots, \boldsymbol{\varphi}_x[n], \cdots, \boldsymbol{\varphi}_x[N-1])^T \tag{17.3}$$

where $\boldsymbol{\varphi}_x[n] = (\phi_1^x[n], \cdots, \phi_i^x[n], \cdots, \phi_M^x[n])^T$ is the $M \times 1$ vector of basis functions $\phi_i^x[n]$ (with $i = 1, \cdots M$) at time n. This general equation can be particularized for any behavioral model.

Generally, the problem in Eq. (17.2) has no exact solution since it is *over-determined* (i.e. more equations than unknowns). To identify the vector of coefficients $\mathbf{w}$, we define a cost function that takes into account the identification error $\mathbf{e}$, expressed, as depicted in Figure 17.4, as

$$\mathbf{e} = \mathbf{y} - \hat{\mathbf{y}} = \mathbf{y} - \mathbf{X}\mathbf{w}. \tag{17.4}$$

Taking the ℓ_2-norm squared of the identification error, the least squares (LS) minimization problem can be defined as follows:

$$\min_{\mathbf{w}} \|\mathbf{e}\|_2^2 = \min_{\mathbf{w}} \|\mathbf{y} - \mathbf{X}\mathbf{w}\|_2^2. \tag{17.5}$$

Taking the derivative of the cost function $J(\mathbf{w}) = \|\mathbf{e}\|_2^2$ and setting it to zero, it can be proved that the solution to the LS minimization problem in Eq. (17.5) is given by

$$\mathbf{w} = (\mathbf{X}^H\mathbf{X})^{-1}\mathbf{X}^H\mathbf{y}. \tag{17.6}$$

The most common numerical methods (Trefethen and Bau (1997)) used to solve the LS problem are Cholesky factorization, QR factorization, and singular value decomposition (SVD).

17.2.3 Closed-Loop Digital Predistortion Linearization

In the forward path, the input-output relationship at the DPD block can be described as

$$x[n] = u[n] - d[n] \tag{17.7}$$

where $x[n]$ is the signal at the output of the DPD block, $u[n]$ is the input signal, and $d[n]$ is the distortion signal that can be described using the aforementioned PA behavioral models that can be found in the literature. Therefore, in general,

$$x[n] = u[n] - \boldsymbol{\varphi}_u{}^T[n]\mathbf{w}[n] \tag{17.8}$$

where $\boldsymbol{w}[n] = (w_1[n], \cdots, w_i[n], \cdots, w_M[n])^T$ is a vector of coefficients at time n with dimensions $M \times 1$, with M being the order of the behavioral model; $\boldsymbol{\varphi}_u{}^T[n] = (\phi_1^u[n], \cdots, \phi_i^u[n], \cdots, \phi_M^u[n])$ is the vector containing the basis functions $\phi_i^u[n]$ (with $i = 1, \cdots, M$) at time n. As explained in the previous section, the same behavioral model or basis functions used for approximating the response of the PA can be also used for DPD purposes to estimate the inverse response of the PA. Now, considering a matrix notation, Eq. (17.8) can be rewritten as

$$\boldsymbol{x} = \boldsymbol{u} - \boldsymbol{U}\boldsymbol{w} \tag{17.9}$$

where $\boldsymbol{x} = (x[0], \cdots, x[n], \cdots, x[N-1])^T$ and $\boldsymbol{u} = (u[0], \cdots, u[n], \cdots, u[N-1])^T$, with $n = 0, \cdots, N-1$, are the predistorted and input $N \times 1$ vectors, respectively. The $N \times M$ data matrix is defined as

$$\boldsymbol{U} = (\boldsymbol{\varphi}_u[0], \cdots, \boldsymbol{\varphi}_u[n], \cdots, \boldsymbol{\varphi}_u[N-1])^T. \tag{17.10}$$

The DPD function in the forward path described in Eq. (17.8) can be implemented to operate in real time in a programmable logic (PL) device following different approaches, such as look-up tables (LUTs) (e.g. in Molina et al. (2017), Gilabert et al. (2007)), complex multipliers following a polynomial approach using the Horner's rule as in Mrabet et al. (2012), or some combination of complex multipliers and memory blocks as in Cao et al. (2017).

Unlike the DPD in the forward path, the identification/adaptation of the parameters does not need to be carried out in real time, and thus it can be implemented in a processing system (PS). Therefore, the DPD coefficients can be extracted and adapted iteratively in a slower time scale than real time. The extraction of the DPD coefficients can be carried out following either a direct learning or an indirect learning approach.

The block diagram of a closed-loop adaptive DPD architecture following an indirect learning approach is shown in Figure 17.5. The DPD function in the forward path is described in Eq. (17.8) or, considering a matrix notation, in Eq. (17.9). However, with the indirect learning approach, the inverse PA model is estimated as a postdistortion. That is, the assumption made is that the coefficients of the postdistortion are equal to the coefficients of the predistortion; as mentioned in Braithwaite (2015), the approximation can be considered a valid approximation as long as there is no saturation in either the transmit path or observation path.

The block diagram of a closed-loop adaptive DPD architecture following a direct learning approach is shown in Figure 17.6. As explained in Braithwaite (2015), in comparison to the indirect learning approach, with the direct learning estimation we gain robustness against noisy PA output observations and avoid the offset of the coefficient vector from its optimal value.

In both direct and indirect learning approaches, the coefficients can be extracted iteratively:

$$\boldsymbol{w}^{j+1} = \boldsymbol{w}^j + \mu\boldsymbol{\Delta}w. \tag{17.11}$$

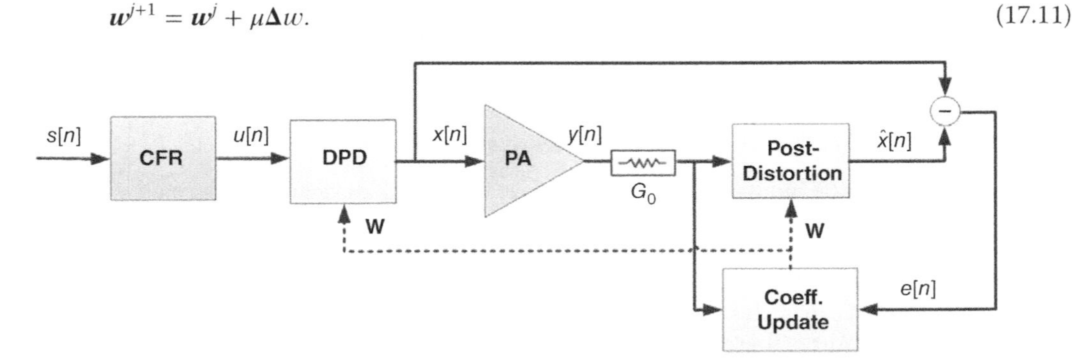

Figure 17.5 Closed-loop digital predistortion linearization: indirect learning approach.

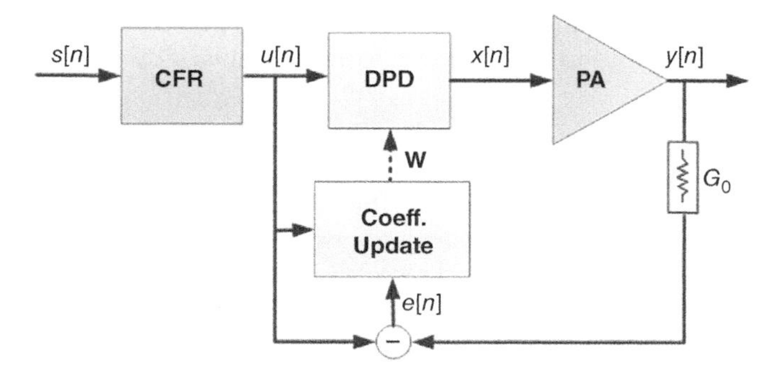

Figure 17.6 Closed-loop digital predistortion linearization: direct learning approach.

However, considering a direct learning approach, Δw is obtained by finding the following LS solution,

$$\Delta w = (U^{H}U)^{-1}U^{H}e \tag{17.12}$$

where e is the $N \times 1$ vector of the identification error defined as

$$e = \frac{y}{G_0} - u \tag{17.13}$$

where G_0 determines the desired linear gain of the PA, and where y and u are the $N \times 1$ vectors of the PA output and the transmitted input, respectively.

17.2.4 Regularization

The method of LS performs well to approximate the solution of overdetermined systems when considering big datasets. However, it may face the risk of *underfitting* or *overfitting* (see Figure 17.7). An underfitted model lacks essential coefficients in the model description. On the contrary, an overfitted model contains more parameters than the model really needs. Both underfitted and overfitted models tend to misrepresent the training data and will therefore have poor predictive performance.

Coefficient estimates for the PA models described in Eq. (17.6) assume the independence of the model basis functions. When the basis functions are correlated and the columns of the data matrix X have an approximate linear dependence, the inverse of the covariance matrix $(X^{H}X)^{-1}$ becomes close to singular. Consequently, the LS estimate becomes highly sensitive to random errors in the observed response y (e.g. random noise, quantization noise of the measurement setup, etc.).

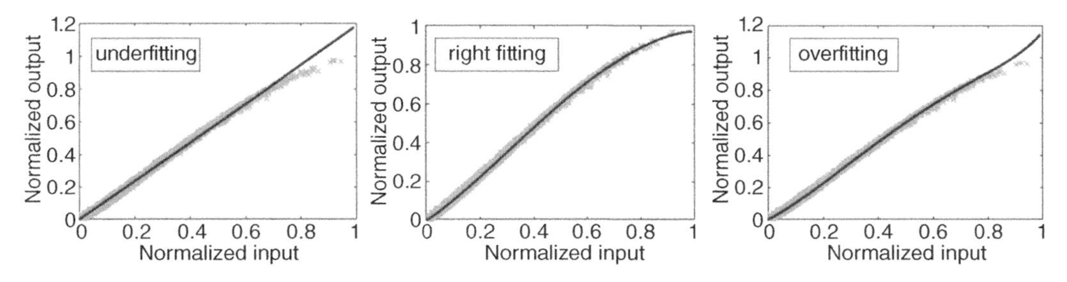

Figure 17.7 Underfitting and overfitting in the least square identification of the power amplifier behavior.

Regularization is a process of introducing additional information in order to prevent overfitting. In general, the main idea of the regularization techniques is to add a regularization term $\boldsymbol{R}(\boldsymbol{w})$ to the cost function:

$$J(\boldsymbol{w}) = \|\boldsymbol{y} - \boldsymbol{X}\boldsymbol{w}\|_2^2 + \lambda \boldsymbol{R}(\boldsymbol{w}) \tag{17.14}$$

In the following, the regularization term will be particularized. Therefore, the cost functions for the Ridge regression, the least absolute shrinkage and selection operator (LASSO), and the elastic net will be presented as constrained versions of the ordinary least squares (OLS) regression cost function. In the case of Ridge regression or Tikhonov regularization (Tikhonov and Arsensin (1977)), it will be subject to a constraint on the squared ℓ_2-norm (Euclidean norm) of the vector of coefficients, while in the case of LASSO (Tibshirani (1994)), it will be subject to a constraint on the ℓ_1-norm of the vector of coefficients. Finally, the elastic net (Zou and Hastie (2005)) combines both the Ridge regression and the Lasso constraints.

17.2.4.1 Ridge Regression or Tikhonov ℓ_2 Regularization

In ℓ_2 regularization, the goal is to minimize the residual sum of squares subject to a constraint on the sum of squares of the coefficients:

$$\begin{aligned} \min_{\boldsymbol{w}} \quad & \sum_{n=0}^{N-1} (y[n] - \boldsymbol{\varphi}_x^H[n]\boldsymbol{w}[n])^2 \\ \text{subject to} \quad & \sum_{i=1}^{M} |w_i[n]|^2 \leq t_2. \end{aligned} \tag{17.15}$$

This constraint forces the coefficients to stay within a sphere of radius t_2. As depicted in Figure 17.8, the contours represent the values of coefficients estimated by the least squares regression. The solution to the Ridge regression or Tikhonov regularization is the coefficients on the contours that meet the constraint. The constrained cost function can also be written as

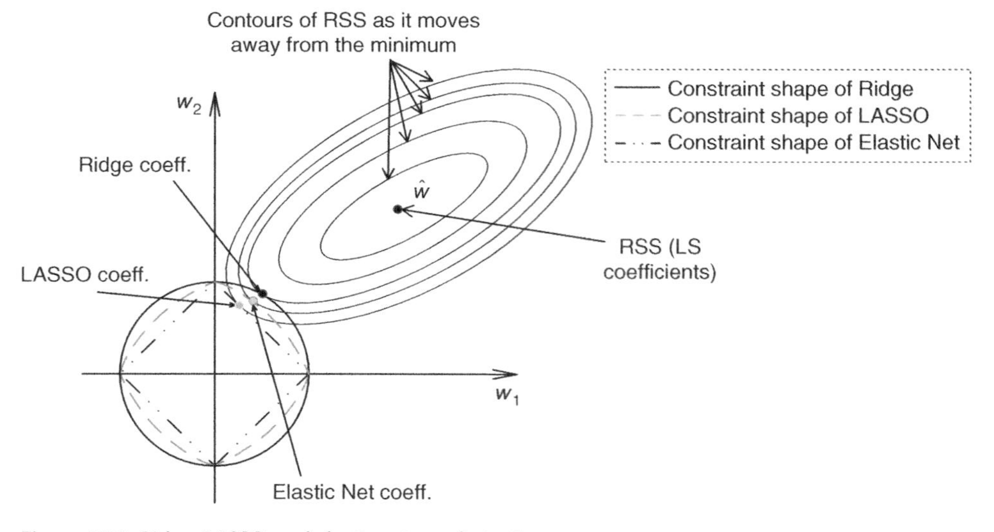

Figure 17.8 Ridge, LASSO, and elastic net regularization.

a penalized residual sum of squares

$$\begin{aligned} J(\boldsymbol{w}) &= \sum_{n=0}^{N-1} (y[n] - \boldsymbol{\varphi}_x^H[n]\boldsymbol{w}[n])^2 + \lambda_2 \sum_{i=1}^{M} |w_i[n]|^2 \\ &= (\boldsymbol{y} - \boldsymbol{X}\boldsymbol{w})^H(\boldsymbol{y} - \boldsymbol{X}\boldsymbol{w}) + \lambda_2 \|\boldsymbol{w}\|_2^2 \\ &= \|\boldsymbol{y} - \boldsymbol{X}\boldsymbol{w}\|_2^2 + \lambda_2 \|\boldsymbol{w}\|_2^2 \end{aligned} \tag{17.16}$$

where λ_2 ($\lambda_2 > 0$) is the shrinkage parameter. Taking the derivative of the cost function and setting it to zero, we obtain the following solution,

$$\boldsymbol{w}_{Ridge} = (\boldsymbol{X}^H\boldsymbol{X} + \lambda_2\boldsymbol{I})^{-1}\boldsymbol{X}^H\boldsymbol{y}, \tag{17.17}$$

with $\boldsymbol{I}$ being the identity matrix. This approach also avoids the problem of rank deficiency because $(\boldsymbol{X}^H\boldsymbol{X} + \lambda_2\boldsymbol{I})$ is invertible even if $(\boldsymbol{X}^H\boldsymbol{X})$ is not (Hoerl and Kennard (1970)). As shown in Eq. (17.17), the coefficients' solution $\boldsymbol{w}_{Ridge}$ depends on the shrinkage parameter λ_2. It controls the size of the coefficients and thus the amount of regularization (i.e. as $\lambda_2 \rightarrow 0$, $\boldsymbol{w}_{Ridge}$ tends to the OLS solution; while as $\lambda_2 \rightarrow \infty$, $\boldsymbol{w}_{Ridge}$ tends to 0). A common approach to properly tune the λ_2 parameter is to use K-fold cross validation.

As an example, Figure 17.9-left shows the normalized mean square error (NMSE) of a PA nonlinear behavior identification when considering a memory polynomial model and different configurations of nonlinear order and memory. It can be observed that adding more terms does not guarantee better NMSE: on the contrary, the NMSE starts degrading when the parameter identification is ill-conditioned due to overparametrization, and consequently the estimated parameters take high power values, as depicted Figure 17.10-left. After Ridge or Tikhonov regularization, the possible LS solutions are the ones that meet the constraint on the power of the coefficients (bounded coefficients values in Figure 17.10-right), and acceptable NMSE values are maintained even when the system is clearly overfitted, as depicted in Figure 17.9-right.

17.2.4.2 LASSO or ℓ_1 Regularization

The least absolute shrinkage and selection operator (LASSO) regression analysis method was introduced in Tibshirani (1994). Similarly to Ridge regression, LASSO can be used for both regularization and to generate a sparse model (i.e. reducing the number of parameters or components of the model). Whereas the constraint of the Ridge regression is the sum of square of the

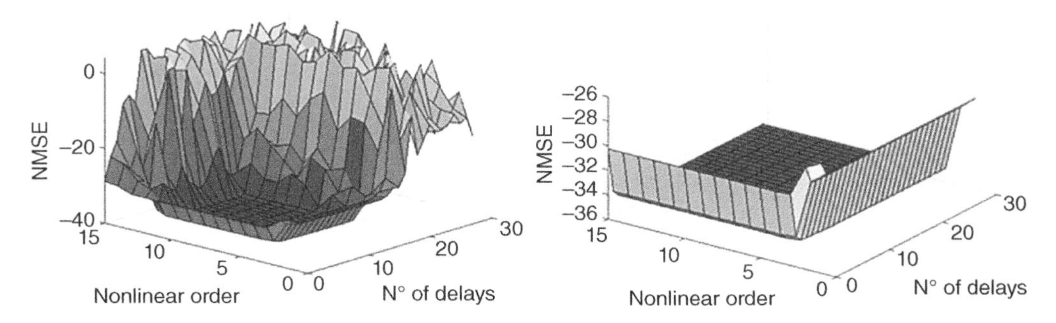

Figure 17.9 NMSE for different values of nonlinear order and memory taps when considering a memory polynomial model, without (left) and with (right) Tikhonov regularization.

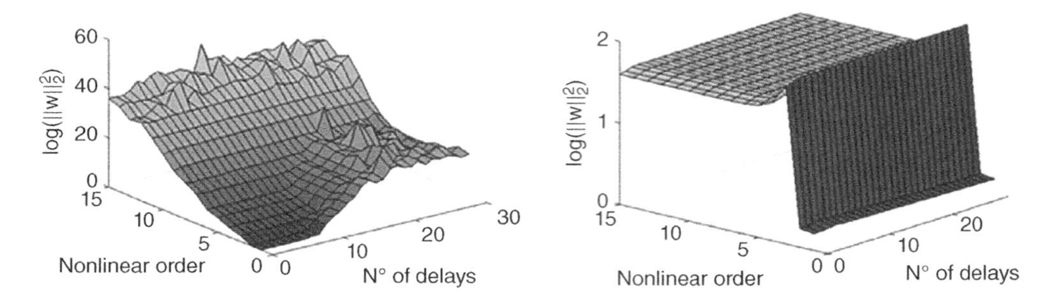

Figure 17.10 Squared norm of the vector of coefficients of the PA behavior identification for different values of nonlinear order and memory taps using a memory polynomial model, without (left) and with (right) Tikhonov regularization.

coefficients, the LASSO constraint consists of the sum of the absolute value of the coefficients. Thus, the solution of LASSO regression satisfies the following ℓ_1 optimization problem:

$$\begin{aligned} \min_{w} \quad & \sum_{n=0}^{N-1} (y[n] - w^T[n]\varphi_x[n])^2 \\ \text{subject to} \quad & \sum_{i=1}^{M} |w_i[n]| \le t_1. \end{aligned} \tag{17.18}$$

As depicted in Figure 17.8, this constraint forces the coefficients to stay within the diamond shape. The constrained cost function can also be written as a penalized residual sum of squares

$$\begin{aligned} J(w) &= \sum_{n=0}^{N-1} (y[n] - \varphi_x^H[n]w[n])^2 + \lambda_1 \sum_{i=1}^{M} |w_i[n]| \\ &= (y - Xw)^H(y - Xw) + \lambda_1 \|w\|_1 \\ &= \|y - Xw\|_2^2 + \lambda_1 \|w\|_1 \end{aligned} \tag{17.19}$$

where ($\lambda_1 > 0$) is the shrinkage parameter.

Unlike Ridge regression, LASSO has no closed form. The regression coefficients are estimated as

$$w_{LASSO} = (X^H X)^{-1}(X^H y - \frac{\lambda_1}{2} b), \tag{17.20}$$

where the elements b_i of b are either $+1$ or -1, depending on the sign of the corresponding regression coefficient $w_i[n]$. Despite the fact that the original implementation involves quadratic programming techniques from convex optimization, Efron et al. in Tibshirani et al. (2004) proposed the least angle regression (LARS) algorithm that can be used for computing the LASSO path efficiently.

17.2.4.3 Elastic Net

The elastic net was proposed in Zou and Hastie (2005) to overcome the LASSO limitations of selecting at most N components (or basis functions) when the number of components M is bigger than the number of observations N (i.e. $M > N$), and of selecting only one component from a group of highly correlated components.

The elastic net combines both the Ridge regression and the LASSO constraints (see Figure 17.8),

$$\min_{w} \sum_{n=0}^{N-1} (y[n] - \boldsymbol{w}^T[n]\boldsymbol{\varphi}_x[n])^2$$
$$\text{subject to} \quad \sum_{i=1}^{M} |w_i[n]|^2 \leq t_2 \quad \text{and} \sum_{i=1}^{M} |w_i[n]| \leq t_1. \tag{17.21}$$

The constrained cost function can also be written as a penalized residual sum of squares

$$\begin{aligned} J(\boldsymbol{w}) &= \sum_{n=0}^{N-1} (y[n] - \boldsymbol{\varphi}_x^H[n]\boldsymbol{w}[n])^2 + \lambda_2 \sum_{i=1}^{M} |w_i[n]|^2 + \lambda_1 \sum_{i=1}^{M} |w_i[n]| \\ &= (\boldsymbol{y} - \boldsymbol{X}\boldsymbol{w})^H(\boldsymbol{y} - \boldsymbol{X}\boldsymbol{w}) + \lambda_2 \|\boldsymbol{w}\|_2^2 + \lambda_1 \|\boldsymbol{w}\|_1 \\ &= \|\boldsymbol{y} - \boldsymbol{X}\boldsymbol{w}\|_2^2 + \lambda_2 \|\boldsymbol{w}\|_2^2 + \lambda_1 \|\boldsymbol{w}\|_1 \end{aligned} \tag{17.22}$$

where $\lambda_2 > 0$ and $\lambda_1 > 0$ are the shrinkage parameters. For the elastic net, the regression coefficients are estimated as

$$\boldsymbol{w}_{E-net} = (\boldsymbol{X}^H\boldsymbol{X} + \lambda_2\boldsymbol{I})^{-1}(\boldsymbol{X}^H\boldsymbol{y} - \frac{\lambda_1}{2}\boldsymbol{b}). \tag{17.23}$$

The minimization of the elastic net cost function in Eq. (17.22) is similar to minimizing the LASSO cost function, and all the elastic net regularization paths can be estimated almost as efficiently as the LASSO paths with the LARS-EN algorithm proposed in Zou and Hastie (2005).

17.3 Dimensionality Reduction and ML

17.3.1 Introduction

The objective of dimensionality-reduction techniques is to reduce the number of features (dimensions, variables, basis functions, components) under consideration in a given dataset by obtaining a set of the principal features (i.e. eliminating redundant or irrelevant variables), which can allow keeping or even improving the model's performance. These techniques can be sorted as follows:

- *Feature selection*, selecting the most relevant variables from a random set of original variables.
- *Feature extraction*, creating a reduced set of new variables that are linear or nonlinear combinations of the original variables.

Figure 17.11 illustrates the hierarchical structure of dimensionality-reduction techniques.

Feature-selection techniques are typically presented in three categories: filter methods, wrapper methods, and embedded methods.

Filter methods are relatively fast pre-processing (as in Figure 17.12) algorithms that do not assume the use of a specific model (which makes them less accurate than, for example, wrapper methods). These methods suppress the least-interesting variables based only on general features. Filter approaches select features according to their scores evaluated via statistical measures. First, a feature-ranking technique is used to evaluate the features and rank them. Then, low-score features are suppressed, leaving only the features with higher scores. Cross-validation (Kohavi (1995)) can be used to decide the cut-off point in the ranked list of features. The score of the features can be considered by: the distance from them to their class and the class nearby (as

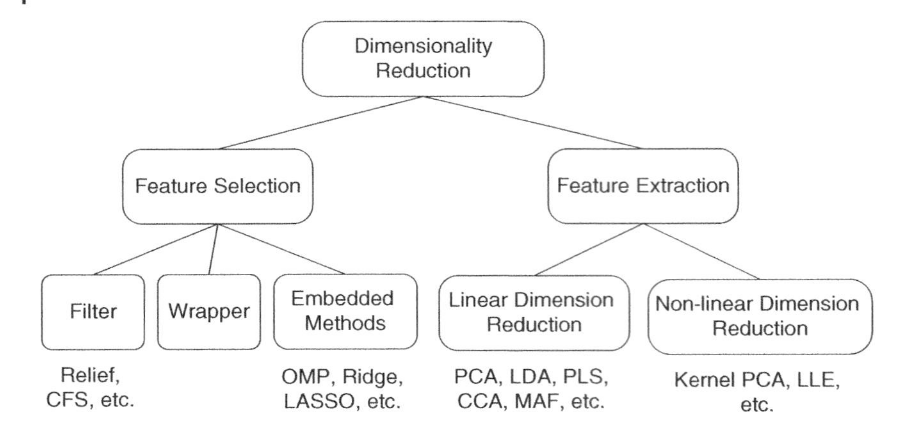

Figure 17.11 Hierarchical structure of dimensionality-reduction techniques.

in Relief), the correlation between the features and their class (as in correlation-based feature selection [CFS] (Li et al. (2011)), and fast correlated based filter [FCBF] (Yu and Liu (2003))), mutual information (as presented in Guyon and Elisseeff (2003)), pointwise mutual information (as in Yang and Pedersen (1997)), etc. Obviously, different ranking techniques can lead to different rankings and thus to different selected subsets.

Wrapper methods, unlike filter methods, employ specific learning algorithms to score the candidate feature subsets. The learning algorithm is chosen depending on the target of the problem Jovic et al. (2015): for example, for regression problems, the wrapper approach evaluates subsets based on the performance of a regression algorithm (e.g. support vector machines [SVMs] Li et al. (2011), LASSO regression Tibshirani (1994), or Ridge regression Hoerl and Kennard (1970)); for clustering, the wrapper approach rates subsets based on the performance of a clustering algorithm (e.g. K-means Khan and Ahmad (2004) or mean-shift clustering Comaniciu and Meer (2002)); for classification tasks, the wrapper technique evaluates subsets based on the performance of a classification algorithm (e.g. naive Bayes Buzic and Dobsa (2018) or decision trees Quinlan (1986)). The evaluation is repeated for each subset; therefore, with wrapper methods, it is possible to obtain better results than with filter methods at the price of being less general and more computationally expensive.

Embedded methods try to combine the advantages of both previous methods. As schematically depicted in Figure 17.12, embedded methods differ from other feature-selection methods in the way feature selection and learning interact. Filter methods do not incorporate learning. Wrapper methods use a learning machine to measure the quality of subsets of features without incorporating knowledge about the specific structure of the classification or regression function, and can therefore be combined with any learning machine. In contrast to filter and wrapper approaches, in embedded methods the learning part and the feature selection part cannot be separated – the structure of the class of functions under consideration plays a

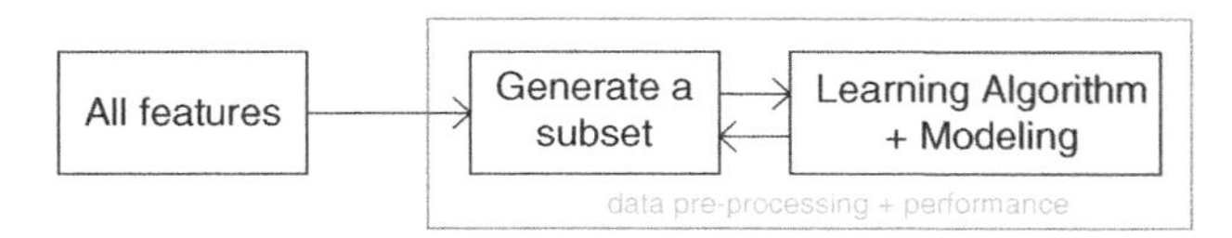

Figure 17.12 Feature selection embedded approach.

crucial role. The embedded methods can be divided into three groups: (i) forward-backward methods, (ii) optimization of scaling factors, and (iii) sparsity term. For example, one of the greedy-search algorithms used for dimensionality reduction in DPD linearization applications is orthogonal matching pursuit (OMP), which belongs to the family of forward-backward methods – in particular, to the family of sequential forward-selection algorithms (Marcano-Cedeno et al. (2010)).

Feature-extraction techniques are aimed at finding a reduced set of features that are a combination of the original ones. Feature extraction can be classed in two subgroups: linear dimension reduction and nonlinear dimension reduction. As will be presented in the following subsections, in the field of DPD linearization, some popular linear dimensionality reduction methods such as principal component analysis (PCA) Gilabert et al. (2013b) and partial least squares (PLS) P. L. Gilabert, G. Montoro, et al. (2016) have been used. Linear dimensionality reduction methods perform the reduction by first generating the new components, which are the linear combinations of the original basis, and then retaining the most significant components and suppressing the irrelevant ones. These methods consider many data features of interest, such as covariance, correlation between datasets, and input-output relationships.

17.3.2 Dimensionality Reduction Applied to DPD Linearization

As depicted in the block diagram in Figure 17.13, the DPD linearization system can be divided into two subsystems: a forward-path subsystem operating in real time, where the input signal is conveniently predistorted; and a feedback or observation path subsystem, where the coefficients characterizing the nonlinear DPD function in the forward path are estimated and updated in a more relaxed time scale. When targeting an implementation in a signal-processing platform – for example, in a system on chip (SoC) FPGA device – the DPD function in the forward path can be implemented in a programmable logic (PL) unit: for example, by following a LUT approach, as in Gilabert et al. (2007), Molina et al. (2017); or by considering a polynomial approach using Horner's rule, as in Mrabet et al. (2012); or by combining both complex multipliers/adders and memory, as in Cao et al. (2017). Therefore, the DPD function in the forward

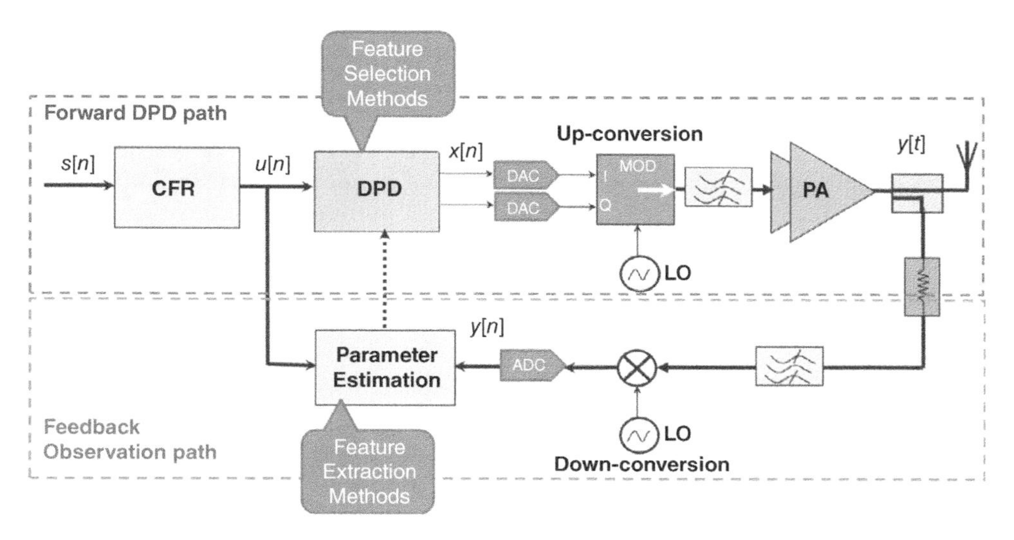

Figure 17.13 Block diagram of the digital predistortion linearization forward and feedback/observation paths.

path should be designed as simply as possible (i.e. including the minimum and most relevant basis functions) to save as many hardware logic resources and memory as possible. On the other hand, the adaptation of the DPD coefficients can be carried out in a processing system (PS) in a much slower time scale than in the forward path (i.e. not in real time).

In the field of DPD linearization, dimensionality-reduction techniques are used with a double objective: on the one hand, to ensure proper, well-conditioned parameter identification; and on the other hand, to reduce the number of coefficients to be estimated and thus relax the computational complexity and memory requirements of a hardware implementation.

Some of the proposed solutions for dimensionality reduction of DPD linearizers are based on feature-selection techniques. The objective of these techniques is to enforce the sparsity constraint on the vector of parameters by minimizing the number of active components (i.e. ℓ_0-norm) subject to a constraint on the ℓ_2-norm squared of the identification error. For example, particularizing for the identification of the PA behavioral model coefficients described in Eqs. (17.2)-(17.6), the optimization problem can be described as

$$\begin{aligned} &\min_{w} \|\boldsymbol{w}\|_0 \\ &\text{subject to} \quad \|\boldsymbol{y} - \boldsymbol{X}\boldsymbol{w}\|_2^2 \leq \varepsilon . \end{aligned} \tag{17.24}$$

Unfortunately, this is a non-deterministic polynomial-time hard (NP-hard) combinatorial search problem. Therefore, in the field of DPD linearization, several sub-optimal approaches have been proposed, targeting both robust identification and model order reduction, such as LASSO, used for example by Wisell et al. in (Wisell et al. (2008)) and consisting of a ℓ_1-norm regularization; the Ridge regression, used for example by Guan et al. in (Guan and Zhu (2012)) and consisting of a ℓ_2-norm regularization; the sparse Bayesian learning (SBL) algorithm, used by Peng et al. in (Peng et al. (2016)); and the orthogonal matching pursuit (OMP), a greedy algorithm for sparse approximation used in (J. Reina-Tosina, M. Allegue et al. (2015)) by Reina et al. to select the most relevant basis functions of the DPD function.

Another approach to address the dimensionality reduction in DPD linearization consists of applying feature-extraction techniques. In commercial products and in publications addressing DPD implementation, one of the most common solutions used to solve the least squares regression problem consists of extracting the parameters through QR factorization combined with recursive least squares (QR-RLS) (Muruganathan and Sesay (2006)). However, by considering feature-extraction techniques such as PCA (Gilabert et al. (2013a)) and PLS (Pham et al. (2018b)), it is possible to ensure both a proper, well-conditioned estimation and a reduction in the number of parameters in the identification process. The DPD dimensionality reduction is carried out by calculating a new, reduced set of orthogonal components that are linear combinations of the original basis functions. However, unlike feature-selection techniques, with feature-extraction techniques the number of coefficients of the DPD function in the forward path are not reduced.

Alternatively, both feature-selection and feature-extraction techniques can be properly combined as in Pham et al. (2018c), by:

- Doing an a priori offline search (e.g. OMP, LASSO) to reduce the number of basis functions of the DPD function in the forward path.
- Using PCA or PLS techniques for the parameter extraction in the adaptation path.

In the following, examples of feature-selection and feature-extraction techniques that have been used for DPD dimensionality reduction will be further described. In particular, further details of the OMP greedy algorithm and the PCA and PLS techniques applied to DPD linearization will be given.

17.3.3 Greedy Feature-Selection Algorithm: OMP

The OMP is a greedy algorithm, also referred to as forward greedy selection in the ML literature (Mallat and Zhang (1993)). The OMP algorithm can be used to perform an a priori offline study, to properly select the best basis functions that will contribute to linearize the PA. Therefore, this study is carried out once and then applied to both reduce the number of coefficients of the forward path behavioral model and improve the conditioning and robustness of the adaptation subsystem.

In order to minimize the number of coefficients being required by the DPD function in the forward path, we assume that the optimal subset of selected basis functions of the DPD function will be the same as that used for PA behavioral modeling. Therefore, the OMP algorithm is the sub-optimal approach considered to solve Eq. (17.24).

The support set containing the indices of the basis functions describing the PA behavioral model is defined as $\mathbf{S}^{(m)}$. Considering that m_{max} is the number of basis functions under study (i.e. $m_{max} = M$), the OMP algorithm is defined in Algorithm 1. At every iteration of the OMP search, the basis function that better contributes to minimize the residual error is selected and added to the support set $\boldsymbol{S}^{(m)}$. The elements of $\boldsymbol{X}_{S^{(m)}}$ have been normalized in power to simplify the index $i^{(m)}$ calculation in line 6 of the algorithm, which can be obtained by maximizing the absolute value of the correlation between the basis function $\boldsymbol{X}_{\{i\}}$ and the residual error $\boldsymbol{e}^{(m-1)}$ of the previous iteration. After a complete OMP search, we obtain a vector $\boldsymbol{S}^{(m_{max})}$ with the indices of all the original basis functions (active components) sorted according to their relevance. Then, by using some information criterion, such as the Akaike (AIC) or Bayesian (BIC) (J. Reina-Tosina, M. Allegue et al. (2015)), it is possible to determine the optimum number of coefficients (m_{opt}), where $m_{opt} < m_{max}$. Finally, the subset of selected basis functions, $\boldsymbol{X}_{S^{(m_{opt})}}$, is used in Eqs. (17.8)–(17.9) as $\boldsymbol{U}_{S^{(m_{opt})}}$ to carry out the DPD.

Algorithm 1 Orthogonal matching pursuit algorithm

1: **procedure** OMP($\boldsymbol{y}, \boldsymbol{X}$)
2: initialization:
3: $\boldsymbol{e}^{(0)} = \boldsymbol{y} - \hat{\boldsymbol{y}}^{(0)}$; with $\hat{\boldsymbol{y}}^{(0)} = 0$
4: $\boldsymbol{S}^{(0)} = \{\}$
5: **for** $m = 1$ **to** m_{max} **do**
6: $i^{(m)} = \arg\min_{i} \min_{w_i} \|\boldsymbol{e}^{(m-1)} - \boldsymbol{X}_{\{i\}} w_i\|_2^2 \approx \arg\max_{i} \left|\boldsymbol{X}_{\{i\}}^H \boldsymbol{e}^{(m-1)}\right|$
7: $\boldsymbol{S}^{(m)} \leftarrow \boldsymbol{S}^{(m-1)} \bigcup i^{(m)}$
8: $\boldsymbol{w}^{(m)} = \left(\boldsymbol{X}_{S^{(m)}}^H \boldsymbol{X}_{S^{(m)}}\right)^{-1} \boldsymbol{X}_{S^{(m)}}^H \boldsymbol{y}$
9: $\hat{\boldsymbol{y}}^{(m)} = \boldsymbol{X}_{S^{(m)}} \boldsymbol{w}^{(m)}$
10: $\boldsymbol{e}^{(m)} = \boldsymbol{y} - \hat{\boldsymbol{y}}^{(m)}$
11: **end for**
12: return $\boldsymbol{S}^{(m_{max})}$
13: **end procedure**

17.3.4 Principal Component Analysis

PCA is a statistical learning technique suitable for converting an original basis of eventually correlated features or components into a new, uncorrelated orthogonal basis set called *principal components*. The principal components are linear combinations of the original variables oriented to capture the maximum variance in the data.

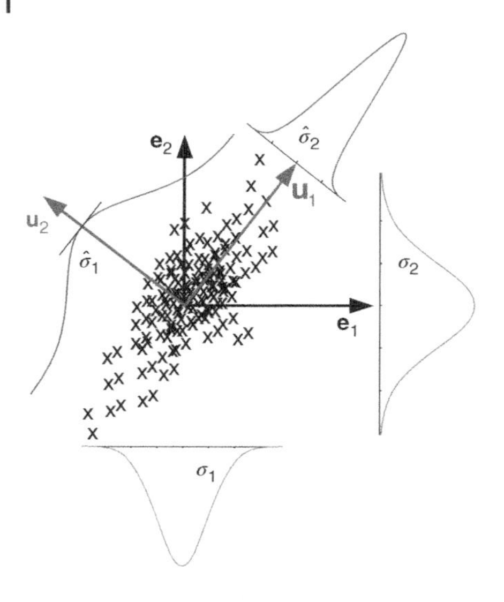

Figure 17.14 Principal component analysis transformation considering two-dimensional data.

Figure 17.14 presents an example of PCA transformation considering two-dimensional data. The original coordinate axes are $\boldsymbol{e}_1$ and $\boldsymbol{e}_2$. The new coordinate axes are $\boldsymbol{u}_1$ and $\boldsymbol{u}_2$, corresponding to the two eigenvectors of the original data. The first eigenvector (corresponding to axis $\boldsymbol{u}_1$) has much larger variance than the second eigenvector (corresponding to axis $\boldsymbol{u}_2$) (i.e. $\hat{\sigma}_1 > \hat{\sigma}_2$). In the original coordinate axes, the difference of the variances of the data on the two axes is not significantly different. In this example, $\boldsymbol{u}_1$ is the principal component of the considered data. Therefore, if we discard the unimportant component $\boldsymbol{u}_2$ and project the data on the dimension $\boldsymbol{u}_1$, the information loss will not be very significant, whereas, in the original coordinate axes, if we simply remove one dimension and retain the other, we will encounter relevant information loss.

Following the notation of the DPD linearization in Eqs. (17.9)–(17.10) in Section 17.2.3, the PCA theory is used to generate a new basis set of orthogonal components, as explained in Gilabert et al. (2013a). The new orthogonal basis is obtained through a change of basis using a transformation matrix $\boldsymbol{V}$ that contains the eigenvectors of the covariance matrix of $\boldsymbol{U}$,

$$cov(\boldsymbol{U}) = \frac{1}{N-1}((\boldsymbol{U} - E\{\boldsymbol{U}\})^H(\boldsymbol{U} - E\{\boldsymbol{U}\})) \approx \boldsymbol{U}^H\boldsymbol{U} \tag{17.25}$$

where $E\{\cdot\}$ is the expected value. The principal components of the basis functions (i.e. columns of $\boldsymbol{U}$) are the eigenvectors of $\boldsymbol{U}\boldsymbol{U}^H$. However, as it will be proved, $\boldsymbol{U}^H\boldsymbol{U}$ and $\boldsymbol{U}\boldsymbol{U}^H$ have the same eigenvalues, and, moreover, their eigenvectors are related as described in the following,

$$(\boldsymbol{U}^H\boldsymbol{U})\boldsymbol{v}_i = \lambda_i\boldsymbol{v}_i \rightarrow (\boldsymbol{U}\boldsymbol{U}^H)\boldsymbol{U}\boldsymbol{v}_i = \lambda_i\boldsymbol{U}\boldsymbol{v}_i \tag{17.26}$$

with $\boldsymbol{v}_i$ being the ith eigenvector of $\boldsymbol{U}^H\boldsymbol{U}$. For each i,

$$(\boldsymbol{U}^H\boldsymbol{U})\boldsymbol{V} = \boldsymbol{\lambda}\boldsymbol{V} \rightarrow (\boldsymbol{U}\boldsymbol{U}^H)\boldsymbol{U}\boldsymbol{V} = \boldsymbol{\lambda}\boldsymbol{U}\boldsymbol{V} \tag{17.27}$$

where $\boldsymbol{V} = (\boldsymbol{v}_1, \cdots, \boldsymbol{v}_i, \cdots, \boldsymbol{v}_L)$ is the $M \times L$ transformation matrix with $L \leq M$. The linear combination $\boldsymbol{U}\boldsymbol{V}$ corresponds to the eigenvectors of the matrix $\boldsymbol{U}\boldsymbol{U}^H$, which are the desired principal components of the basis functions (i.e. columns) of $\boldsymbol{U}$. Moreover, $\boldsymbol{\lambda}$ is the diagonal matrix containing the eigenvalues of both the $\boldsymbol{U}\boldsymbol{U}^H$ and the $\boldsymbol{U}^H\boldsymbol{U}$ matrices. Therefore, the new transformed matrix is found as

$$\hat{\boldsymbol{U}} = \boldsymbol{U}\boldsymbol{V} \tag{17.28}$$

with $\hat{\boldsymbol{U}} = (\boldsymbol{\psi}_{\boldsymbol{u}}[0], \cdots, \boldsymbol{\psi}_{\boldsymbol{u}}[n], \cdots, \boldsymbol{\psi}_{\boldsymbol{u}}[N-1])^T$ being the $N \times L$ data matrix and where $\boldsymbol{\psi}_{\boldsymbol{u}}{}^T[n] = (\vartheta_1^u[n], \cdots, \vartheta_j^u[n], \cdots, \vartheta_L^u[n])$ is the $1 \times L$ data vector containing the new orthogonal basis functions (or components) $\vartheta_j^u[n]$ (with $j = 1, \cdots, L$) at time n.

An independent DPD parameter estimation based on the adaptive PCA (APCA) algorithm for dimensionality reduction was proposed in López-Bueno et al. (2018). The proposed solution addresses the DPD model parameter extraction in order to enable implementation in an FPGA containing a programmable logic device and a processing system.

Taking into account the transformation matrix $\hat{\boldsymbol{U}}$ in Eq. (17.28) with orthogonal basis functions, the direct learning coefficients extraction in Eqs. (17.11) and (17.12) can be rewritten as

$$\hat{\boldsymbol{w}}^{n+1} = \hat{\boldsymbol{w}}^n + \mu(\hat{\boldsymbol{U}}^H \hat{\boldsymbol{U}})^{-1} \hat{\boldsymbol{U}}^H \boldsymbol{e} \tag{17.29}$$

where

$$\boldsymbol{w} = \boldsymbol{V}\hat{\boldsymbol{w}} \tag{17.30}$$

and then by taking into account the orthogonal basis functions in $\hat{\boldsymbol{U}}$, we have

$$(\hat{\boldsymbol{U}}^H \hat{\boldsymbol{U}})^{-1} = diag(\lambda_1^{-1}, \cdots, \lambda_j^{-1} \cdots, \lambda_L^{-1}) \tag{17.31}$$

with λ_j being the eigenvalues of $\boldsymbol{U}^H\boldsymbol{U}$ and $\boldsymbol{U}\boldsymbol{U}^H$ (with $j = 1, \cdots L$). The coefficients can be now estimated independently in a least mean square (LMS) fashion at every sample iteration n, and thus Eq. (17.29) becomes

$$\begin{pmatrix} \hat{w}_1[n+1] \\ \vdots \\ \hat{w}_j[n+1] \\ \vdots \\ \hat{w}_L[n+1] \end{pmatrix} = \begin{pmatrix} \hat{w}_1[n] \\ \vdots \\ \hat{w}_j[n] \\ \vdots \\ \hat{w}_L[n] \end{pmatrix} + \mu \begin{pmatrix} \lambda_1^{-1}\vartheta_1^u[n] \\ \vdots \\ \lambda_j^{-1}\vartheta_j^u[n] \\ \vdots \\ \lambda_L^{-1}\vartheta_L^u[n] \end{pmatrix} e[n]. \tag{17.32}$$

By exploiting the orthogonality of the resulting transformed basis functions, the coefficient adaptation can therefore be carried out independently as follows,

$$\hat{w}_j[n+1] = \hat{w}_j[n] + \mu\lambda_j^{-1}\vartheta_j^u[n]e[n] \tag{17.33}$$

with $j = 1, \cdots, L$ and where $\vartheta_j^u[n]$ is the jth transformed basis function at time n. A schematic flowchart describing the independent DPD extraction is depicted in Figure 17.15. The goal is to estimate the minimum necessary number of transformed coefficients $\hat{w}_j$ to meet the target linearity levels, specified in terms of adjacent channel power ratio (ACPR) and NMSE. As explained in Pham et al. (2018a), with the proposed block deflated adaptive principal component (BD-APCA) algorithm, the columns $\boldsymbol{r}_j$ ($j = 1, 2, \cdots, L$) of the transformation matrix $\boldsymbol{R}$ are iteratively found one by one. Therefore, the next column is estimated by using the values of the previously extracted components. Eventually, the $M \times L$ transformation matrix $\boldsymbol{R} = (\boldsymbol{r}_1, \cdots, \boldsymbol{r}_j, \cdots, \boldsymbol{r}_L)$ will converge to $\boldsymbol{V}$. Therefore, as shown in Figure 17.15, until the desired linearity levels are met, the algorithm increases the number of transformed coefficients to be estimated. More specifically, after several APCA iterations, a new transformation vector $\boldsymbol{r}_j$ is obtained. This vector, together with the other $j-1$ previously calculated vectors (i.e. $j-1, j-2, \cdots, 1$), defines the j components of the transformation matrix $\boldsymbol{R}^{(j)}$. Then, each one of the j transformed coefficients $\hat{w}_j[n]$ can be estimated/updated independently (e.g. in parallel) by following an LMS approach,

$$\hat{w}_j[n+1] = \hat{w}_j[n] + \mu\lambda_j^{-1}\boldsymbol{\varphi}_{\boldsymbol{u}}{}^T[n]\boldsymbol{r}_j[n]e[n]. \tag{17.34}$$

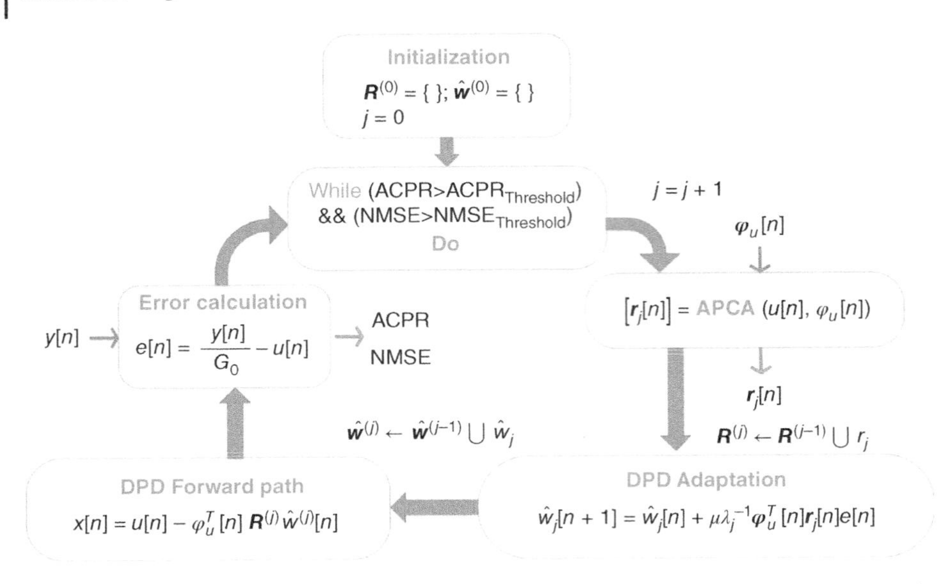

Figure 17.15 Flowchart of the independent digital predistortion identification process using adaptive principal components analysis.

Consequently, the input-output relationship of the DPD function in the forward path described in Eq. (17.8)- Eq. (17.9) can be rewritten as

$$x[n] = u[n] - \boldsymbol{\psi}_u{}^T[n]\hat{\boldsymbol{w}}[n] \tag{17.35}$$

or, alternatively, be expressed in terms of the original basis functions $\boldsymbol{\varphi}_u[n]$ and the transformation matrix $\boldsymbol{R}$

$$x[n] = u[n] - \boldsymbol{\varphi}_u{}^T[n]\boldsymbol{R}\hat{\boldsymbol{w}}[n]. \tag{17.36}$$

Experimental results showing the viability and robustness of the independent DPD algorithm described in Figure 17.15 can be found in López-Bueno et al. (2018).

17.3.5 Partial Least Squares

Similarly to PCA, PLS is a statistical technique used to construct a new basis of components that are linear combinations of the original basis functions. However, while PCA obtains new components that maximize their own variance, PLS finds linear combinations of the original basis functions that maximize the covariance between the new components and the reference signal. This enables PLS to outperform PCA in applications such as dimensionality reduction for PA behavioral modeling and DPD linearization P. L. Gilabert, G. Montoro, et al. (2016), Pham et al. (2018c).

In Pham et al. (2018c), for example, with the PLS technique, a set of new components is generated from the original basis functions. By properly selecting the most relevant components from the set, it is possible to guarantee a well-conditioned identification while reducing the number of estimated parameters without loss of accuracy. In addition, thanks to the orthonormality among the components of the new basis, the matrix inversion operation of the LS Moore-Penrose inverse is significantly simplified.

To obtain with PLS a basis of orthonormal components, the iterative SIMPLS algorithm was proposed in de Jong (1993). In a similar manner as it was done for PCA in Eq. (17.28), the new basis $\hat{\boldsymbol{U}}$ is defined by means of the $M \times L$ (with $L \leq M$) transformation matrix $\boldsymbol{P}$,

$$\hat{\boldsymbol{U}} = \boldsymbol{U}\boldsymbol{P} \tag{17.37}$$

with $\boldsymbol{U}$ being the $N \times M$ matrix of basis functions defining the DPD linearizer in Eq. (17.9). The new orthonormal components of the transformed matrix $\hat{\boldsymbol{U}}$ are sorted according to their contribution to maximize the covariance between the new components and the error signal **e**. Taking into account the transformation matrix $\hat{\boldsymbol{U}}$ in Eq. (17.37) with orthonormal basis functions, the direct learning coefficients extraction in Eqs. (17.11) and (17.12) can be rewritten as

$$\hat{\boldsymbol{w}}^{j+1} = \hat{\boldsymbol{w}}^{j} + \mu(\hat{\boldsymbol{U}}^{H}\hat{\boldsymbol{U}})^{-1}\hat{\boldsymbol{U}}^{H}\boldsymbol{e}. \tag{17.38}$$

Now, considering the orthonormal property of the transformed matrix $\hat{\boldsymbol{U}}$ (i.e. $\hat{\boldsymbol{U}}^{H}\hat{\boldsymbol{U}} = \mathbf{I}$), the update of the transformed coefficients is significantly simplified:

$$\hat{\boldsymbol{w}}^{j+1} = \hat{\boldsymbol{w}}^{j} + \mu\hat{\boldsymbol{U}}^{H}\boldsymbol{e}. \tag{17.39}$$

Finally, the original coefficients are obtained through the following anti-transformation:

$$\boldsymbol{w} = \boldsymbol{P}\hat{\boldsymbol{w}}. \tag{17.40}$$

When comparing the accuracy versus coefficient reduction between the PLS and PCA techniques, such as in P. L. Gilabert, G. Montoro, et al. (2016), Pham et al. (2018c), we observe that the PLS technique is more robust than PCA in terms of performance degradation (e.g. in linearization or modeling accuracy) when reducing the number of parameters of the estimation (i.e. coefficients of the transformed basis). This is because PLS, unlike PCA, also considers the information of the PA output signal for creating the transformation matrix.

Another approach presented in Pham et al. (2019) consists of using the PLS technique for estimating and adapting the DPD coefficients with a dynamic basis matrix. Therefore, PLS is employed inside the DPD adaptation loop to actively adjust the basis matrix in the DPD identification subsystem. The dynamic basis reduction is carried out at every iteration according to the *residual linearization error*, defined as the difference between the actual and the desired linear PA output signals. In comparison to the QR decomposition (commonly used in conventional DPD estimation/adaptation to avoid the costly mathematical computation of the LS Moore-Penrose inverse of the covariance matrix), the proposed technique allows dynamic adjustment of the number of coefficients to meet the targeted linearity level.

Figure 17.16 depicts the DPD estimation/adaptation employing the dynamic basis matrix approach presented in Pham et al. (2019). The proposed dynamic orthonormal transformation matrix (DOTM) algorithm, which is a modification of the iterative SIMPLS algorithm in de Jong (1993), calculates the linear combinations of the original basis with maximum covariance between the new basis and the signal to be estimated. Whereas the size of the transformation matrix $\boldsymbol{P}$ is predetermined and given as an input information in SIMPLS, in DOTM, the number of columns of $\boldsymbol{P}$ are iteratively added and calculated until the power of the estimated error is close enough to a desired threshold E_{th}, defined as a percentage δ of the power of the error signal $\boldsymbol{e}$. Therefore, thanks to the DOTM algorithm, at each iteration j of the DPD adaptation, the number of columns L (with $L \leq M$) of the transformation matrix $\boldsymbol{P}$ varies, and only the minimum necessary number of columns that meet the E_{th} threshold requirements are selected. By taking into account the orthonormality among the components, the transformed coefficients increment $\boldsymbol{\Delta}\hat{w}$ is calculated as described in Eq. (17.39), i.e. $\boldsymbol{\Delta}\hat{w} = \mu\hat{\boldsymbol{U}}^{H}\boldsymbol{e}$. Then, as shown in Figure 17.16, after the antitransformation, the estimated DPD coefficients are used in the forward path to predistort the input signal, and the DPD adaptation will continue until it achieves the desired ACPR level.

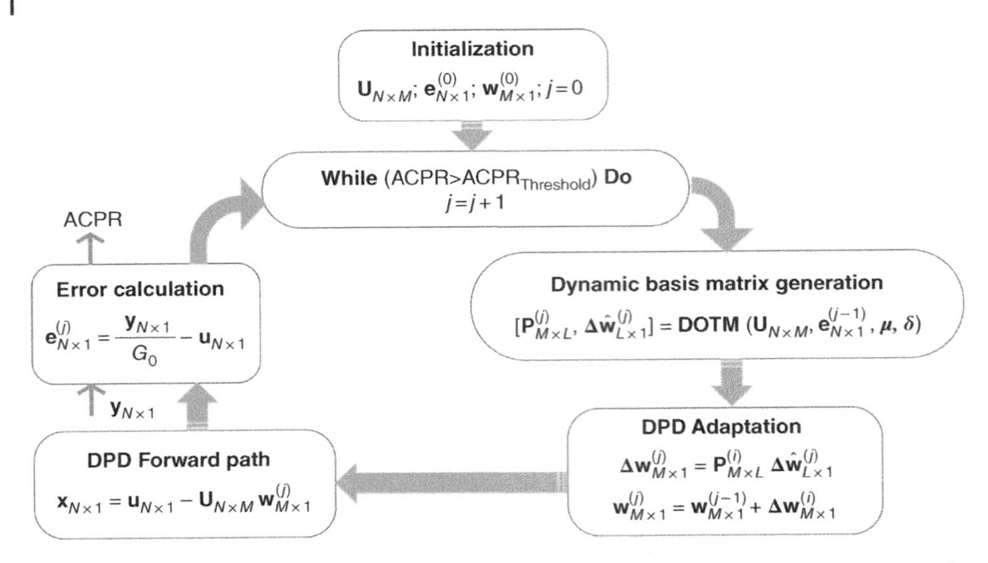

Figure 17.16 Flowchart of digital predistortion estimation/adaptation using a dynamic basis matrix.

17.4 Nonlinear Neural Network Approaches

17.4.1 Introduction to ANN Topologies

An artificial NN (ANN) is a modeling technique, originally inspired by some partial knowledge on the behavior of the neurons in the human brain, that can be trained to learn the structure of the data and model complex nonlinear functions. Essentially, the neurons are distributed between different layers and communicate with each other through neuron output-to-input weighted interconnections (or synapses). Based on the interconnection pattern or architecture, we can distinguish between feedforward networks (FNNs) and recurrent (or feedback) networks (RNNs).

FNNs, which are among the most used ANN, have unidirectional interconnections between the neurons of every layer since the flow of data is from input to outputs, without feedback (one input pattern produces one output). The most common FNN is called a multilayer perceptron (MLP), which is composed of fully connected layers where all the output activations are composed of a weighted sum of input activations (the neurons of a specific layer are fed by the outputs of all the neurons of the preceding layer). The larger the weight, the more influential the corresponding input will be. Enabling full connection in a densely populated NN may require significant hardware resources, but in many applications the weight of some interconnections can be set to zero without loss of accuracy, which results in sparsely connected layers. In a RNN, the inputs of the neurons of a specific layer may be fed by the output of the neurons either in the same layer or at any of the following layers, which senses time and memory of previous states. These concepts are shown in Figure 17.17a, where a modified MLP-based FNN is displayed (a classic MLP would have full connection between at all layers).

In more detail, Figure 17.17b shows a single-layer perceptron model and the operation of this fundamental building block of an MLP NN. The jth neuron of the kth layer receives as input each x_i from the previous layer. Each x_i, with $i = 1, 2, \cdots, N$, is then multiplied by a weight w_{ji} and the resulting values are all summed together. A single bias or offset value θ_j is added to the summation and, finally, an activation or transfer function $\varphi^k(\cdot)$ (different activation functions can be applied to different layers) is applied to provide the output of the jth neuron found in the

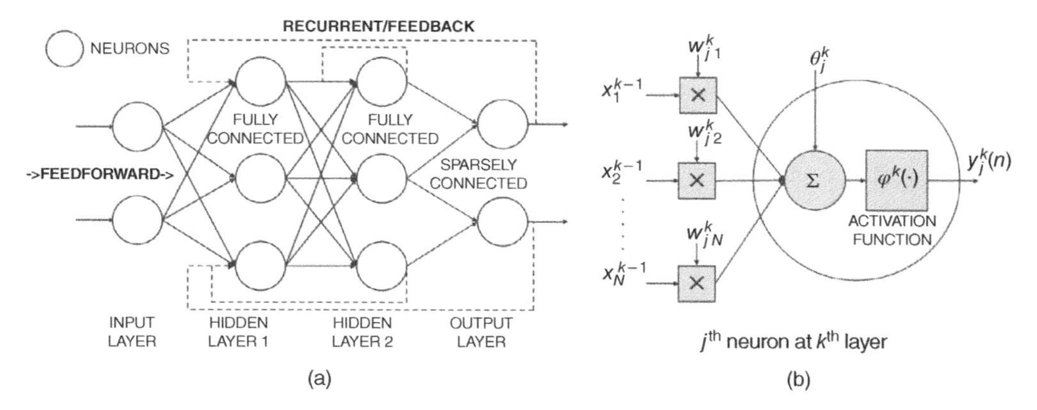

Figure 17.17 (a) Feedforward network and recurrent network architectures; (b) single-layer perceptron (SLP) model.

k^{th} layer, as shown in Eq. (17.41):

$$y_j^k(n) = \varphi^k\left(\sum_{n=1}^{N} w_{ji}^k x_i^{k-1}(n) + \theta_j^k\right). \tag{17.41}$$

A brief historical review follows. When the SLP was defined in the late 1950s by Rosenblatt (1958), the activation function being used was the step or threshold function, and the concept of hidden layers was not yet exploited. This model was a first practical implementation that could be used for simple linearly separable binary classification problems but was not valid for more complicated modeling requiring nonlinear outputs. The solution to this limitation came in the mid and late 1980s thanks to a few works, such as Rumelhart and McClelland (1986), that considered a MLP with hidden layers to enable the NN to learn more complicated features, proposed backpropagation algorithms to adjust the weights and minimize the difference between the actual output and the desired output, and employed nonlinear activation functions such as the sigmoid function that could enable gradually changing the weights of the NN and introduce nonlinearity. The universal approximation theorem in Cybenko (1989) proved that a feedforward ANN with a single hidden layer (a three-layer network considering the input and output layers) and non-constant, bounded, and monotone-increasing continuous activation function can approximate any nonlinear function with any desired error. Figure 17.18 shows a summary of the main activation functions.

In the past, and generally speaking, FNNs were considered static and memoryless in the sense that the response of an input was independent of the previous network state, while RNNs were considered dynamic systems because of the feedback connections. Nowadays, and given the highest complexity of RNNs versus FNNs, RNN architectures are frequently unrolled in a way that they are redrawn and reformulated similarly to a FNN to simplify the processing complexity. In addition, the need for modeling nonlinear system dynamics considering memory effects has grown over the last decades in multiple applications. For instance, the ever-increasing signal bandwidth at each wireless communication standard generation makes the modeling of PA memory effects, which are more evident when this component is excited by higher bandwidth signals, a relevant topic for enhancing the performance of the physical layer.

In order to solve time-series prediction and thus enable dynamic nonlinear system identification, focused time-delayed NNs (FTDNNs), which include tapped delay lines to generate delayed samples of the input variables, have been proposed. The FTDNN can be seen as combining a linear time invariant (LTI) system such as a finite impulse response (FIR) filter, which

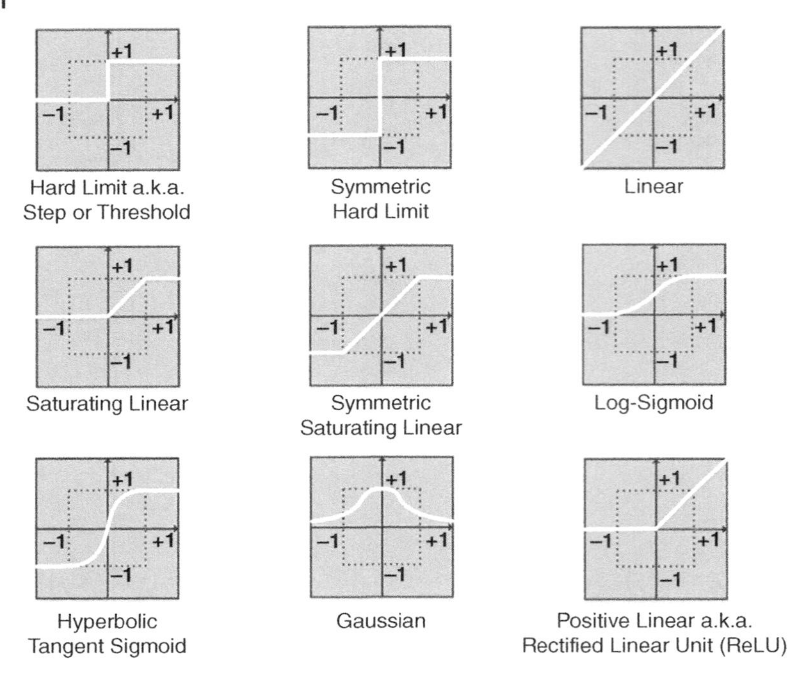

Figure 17.18 Activation or transfer functions.

enables performing dynamic mappings depending on past input values, and a nonlinear memoryless MLP network that can be trained using static backpropagation algorithms. The memory depth of the system being modeled will be reflected in the length of the taps imposed by the required bandwidth accuracy. A FTDNN structure could be seen as a special case of the Wiener model (i.e. a linear time-invariant system followed by a memoryless nonlinear system). Figure 17.19 shows a four-layer architecture (with two hidden layers) of a fully connected FTDNN whose input-output relation is defined in Eq. (17.42) according to the notation of the aforementioned SLP concept.

In this example, the input layer contains $N+1$ neurons (including the input signal and all the delayed versions, z^{-1} is the unit delay operator), the first hidden layer has M neurons, the second hidden layer has L neurons, and there is a final output layer with a single neuron. For the sake of simplicity, the output layer in this example is considered a unitary weighted summation ($w^3_{1l}=1$ for $l=1,2,\cdots,L$) of the signals coming from the previous layer, just followed by a pure linear activation function. The total number of coefficients to be tuned in backpropagation would be the sum of the number of weights ($M(N+1+L)$) and biases ($M+L$). For example, if we considered four memory taps (5 neurons in the input layer), 8 neurons in the first hidden layer, and 6 neurons in the second hidden layer, we would need to tune 88 weights and 14 biases totaling 102 parameters:

$$
\begin{aligned}
y(n) &= \sum_{k=1}^{L} \varphi^2\left(\sum_{j=1}^{M} w^2_{kj} y^1_j(n) + \theta^2_k\right) \\
&= \sum_{k=1}^{L} \varphi^2\left(\sum_{j=1}^{M} w^2_{kj}\varphi^1\left(\sum_{i=0}^{N} w^1_{ji} x_i(n-\tau_i) + \theta^1_j\right) + \theta^2_k\right).
\end{aligned}
\tag{17.42}
$$

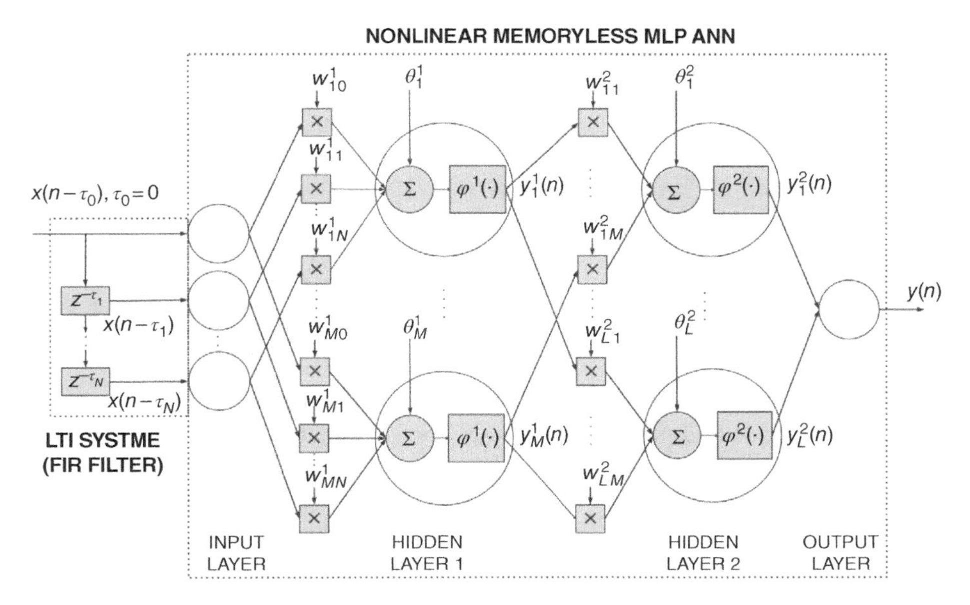

Figure 17.19 Four-layer focused time-delayed neural network architecture.

Generally speaking, and when comparing the ANN approach with the polynomial one (discussed in previous sections), the polynomials have inherent local approximating properties in contrast to the global approximation capability of ANNs, when modeling strongly nonlinear systems. In addition, when compared to classical models, the ANN may adapt better to extrapolating beyond the zone exploited for parameter extraction (Gilabert (2007)).

Based on this introduction and illustration, we now present more details of the use of ANN in digital front-end with the emphasis on addressing undesired effects such as PA nonlinearity, in-phase/quadrature (I/Q) modulator imbalances, DC offsets, and multi-antenna cross-couplings that have a negative impact on today's complex 5G communication systems. There is not a universal recipe to set up the best ANN architecture, learning algorithm, or activation function given a specific problem. Trial and error is frequently employed, but some physical knowledge of the phenomena to be modeled can be important when optimizing the resources and aiming to reach the best modeling performance. However, some design considerations (architectures, activation functions, backpropagation detail and learning algorithms, metrics, etc.) found in literature and being experimentally validated and benchmarked will be given to assist in the modeling and compensation of the previously mentioned RF transceiver impairments.

17.4.2 Design Considerations for Digital Linearization and RF Impairment Correction

ANNs are considered an alternative to complex Volterra-based nonlinear models that require an unaffordable complexity to characterize the RF impairments in highly demanding transceiver architectures such as massive MIMO. The FTDNN architecture, combined with a back propagation learning algorithm (BPLA), over the last 10 years has been one of the most attractive approaches for dynamic nonlinear modeling. Another frequent type of FNN being widely exploited to predict the behavior of the PA is the radial basis function NN (RBFNN), which can progressively keep increasing the number of neurons in the hidden layer until the desired performance is met, as shown in Isaksson et al. (2005). RNNs, which have been preferred to model dynamic nonlinear systems with feedback paths or frequency-dependent phenomena, allow for better characterization of the interaction between input and output samples and the cross terms. However, these networks have traditionally employed lengthy training algorithms, making real-time implementation difficult. While MLPNN variations are predominantly employed for PA modeling and DPD, RNNs are found in works dealing with I/Q modulator gain and phase imbalances or to apply CFR. When all the impairments, including strong PA nonlinearities, have to be solved with a common architecture, the RNN may be hardly implementable or outperformed by the MLPNN-based approaches, as will be shown in the next subsections. The better the dynamic nonlinearities are modeled, the better the linearization performance will be.

17.4.2.1 ANN Architectures for Single-Antenna DPD

In order to extract amplitude and phase information from modulated complex waveforms, ANNs need to consider operating with either complex-valued (CV) input signals, weights and activation outputs, or real-valued (RV) double-inputs double-outputs (and real weights and activation outputs), i.e. in the form of multiple I and Q components. Complex-valued operation leads to heavy calculations and a longer training phase. In addition, the architectures that employ independent NNs to separately model the AM/AM and AM/PM behavior may fail in the synchronous convergence of the two NNs and thus tend to overtrain the fastest-converging one (Rawat et al. (2010)). RV FTDNNs, which combine I/Q RV processing with input time-delay lines (TDLs) to handle memory effects (but not output-to-input TDLs, as would happen in a RNN), can offer superior performance and easy baseband implementation when used for inverse modeling of PAs with strong nonlinearities and memory effects. As seen in Rawat et al. (2010), these ANNs utilize a similar structure to that shown in Figure 17.19; but instead of a single-input, single output NN, now we have double-input, double-output (I/Q inputs and outputs). In this case, the weighted summation at each first hidden layer neuron will include a sequence of input samples both for the I and Q components (this information will propagate throughout the NN according to the activation functions), and the output layer may have a non-unitary weighted summation at each I and Q output accounting for the contributions of each neuron output in the second hidden layer. This ANN has therefore a maximum of $2M(N+1)+LM+2L$ weights and $M+L+2$ biases (considering the notation in Figure 17.19) that will be adjusted using feedforward backpropagation.

As previously introduced, RNNs can be modified or unrolled in most cases in such a way as to emulate a FNN scheme where consolidated BPLAs are applied. A relevant design consideration is that choosing an ANN architecture without taking into account which sources generate the nonlinearities may impact negatively on performance or be highly resource inefficient. For instance, when modeling PA nonlinear dynamic effects, it can be worth paying attention to the PA physical model to reflect output-to-input interactions or account for memory effects given a signal bandwidth. One example of the previous design considerations is found in Mkadem and Boumaiza (2011) and is displayed in Figure 17.20.

In this example, nonlinear activation functions are used to model static PA nonlinearities (typically, the stronger the nonlinearities, the more neurons are required) while linear activation functions will be used to model the feedback mechanism FIR filter (with memory depth K). The input FIR (input signal) models the memory effects that result from combining a wideband modulated signal with a non-flat response input matching network (N is the memory depth of the input signal). For FNN deployment, the connection between the initial input samples and the neurons is 1 (at the first hidden layer). To train this NN, only measured past data from the PA output will be used at the input (not during validation, since the delayed output data will be fed back once the necessary outputs are produced).

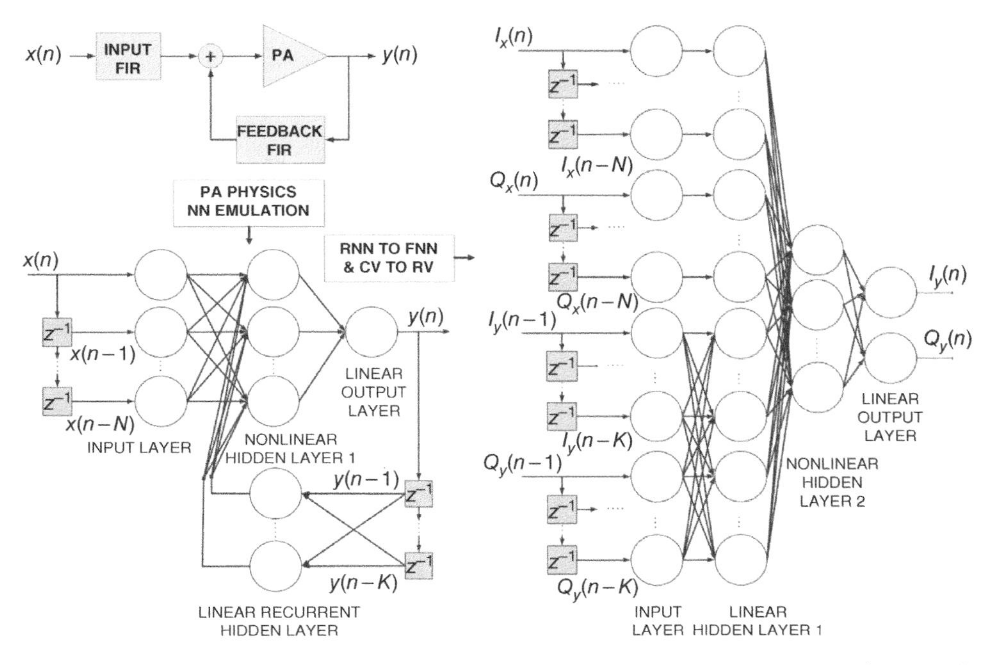

Figure 17.20 Power amplifier physics-aware complex-valued recurrent neural network, reformulated as a modified real-valued focused time-delayed neural network.

Moreover, some recent works have shown the benefits brought by adding envelope-dependent terms as inputs to the ANN. Jueschke and Fischer (2017) propose a two-hidden layer RV FTDNN that includes one additional input based on the calculation of the modulus of the I and Q samples. This input is fed not only to the first hidden layer but also directly to the output layer, all of which helps to improve numerical stability and training convergence. Wang et al. (2019) inject additional envelope-dependent term combinations (i.e. between the modulus raised to the power of two, three, four, and five) as inputs of a single-hidden-layer NN (no direct connection between envelope-related inputs and the output layer is enabled in this case). This obtains better modeling performance since these new terms are able to generate some desired even-order intermodulation terms throughout the NN that cannot be obtained if only I and Q components are used as inputs.

17.4.2.2 ANN Architectures for MIMO DPD, I/Q Imbalances, and DC Offset Correction

Over the last 10 years, the need for wireless communication technologies fulfilling user mobile broadband capacity requirements has been coupled with the need to reduce costs and CO_2 emissions. This has boosted the research, development, and industrial release of MIMO transceiver solutions with a high number of elements both for macro base stations and for the next generation of small cells to be used in ultra-dense deployments. The high number of RF transceiver chains in these solutions makes integration, power consumption, and cost-effectiveness prominent design constraints that may play against employing the best-performing solution. High-channel density radio frequency integrated circuits (RFICs) integrating the data-conversion stages, I/Q modulators and demodulators, and LO signal synthesis and distribution are employed together with typically moderate-cost PAs from high-volume markets to benefit from economies of scale. Figure 17.21a shows a generic architecture of a base station highly integrated MIMO transmitter (the Rx blocks have been omitted). At the Tx side, crosstalk may occur in the multi-channel I/Q modulation stages

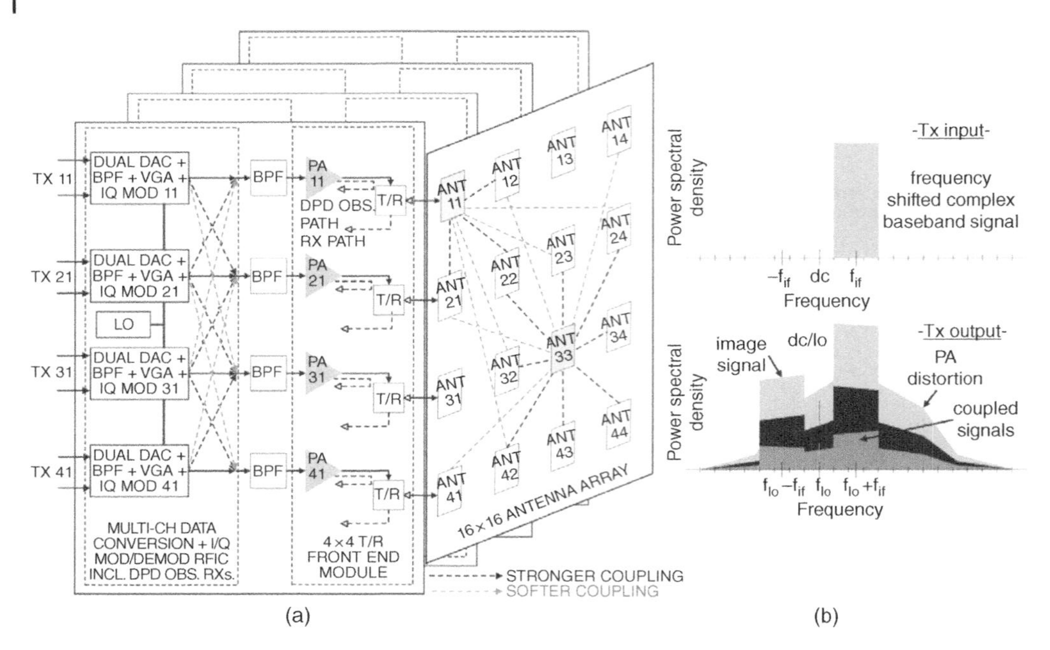

Figure 17.21 (a) Multi-antenna transmitter architecture; (b) spectra of the complex baseband signal and the radio frequency signal at power amplifier output (with RF impairments).

sharing a common LO in the same integrated circuit or at the PA output (for instance, due to the antenna cross-couplings). These effects invalidate the classical single-antenna DPD approaches since they cannot fulfill the desired performance under strong or medium-level couplings (i.e. typically −15—30 dB). The use of direct-conversion Tx/Rx chains and thus I/Q modulators/demodulators is clearly advantageous with respect to the use of superheterodyne architectures in terms of integration and cost; however, this architecture suffers from I/Q gain and phase imbalances due to mismatches between the I and Q branches, DC offsets, and in-band LO couplings. The image rejection ratio (IRR) characterizes the I/Q imbalances through measuring the ratio between the image signal generated by the imperfections and the signal of interest at the output of the modulator (i.e. typical values are −20—40 dB). The joint effect of all these impairments is depicted in Figure 17.21b.

Many of the existing MIMO DPD models that account for cross-couplings and that consider nonlinear crosstalk at the PA input and linear crosstalk at the PA output are mainly based on the crossover memory polynomial model (COMPM) found in Bassam et al. (2009). The parallel Hammerstein (PH) model in Amin et al. (2014) shows better linearization performance and includes terms compensating for nonlinear crosstalk at the PA output, but requires more coefficients. These models are insufficient to mitigate the I/Q modulator imbalances and DC offsets. By adding a complex conjugate function and a DC term, the PH model will be able to handle not only crosstalk but the modulator imbalances, as found in Khan et al. (2017). However, this model requires extraction of a high number of coefficients and inverse modeling for each Tx path, which is unaffordable for systems employing a large number of antennas. Hausmair et al. (2018) present a model to deal with the PA nonlinearities, any crosstalk between Tx channels, and the PA mismatch effects due to antenna couplings (i.e. seeking to enable circulator-less MIMO transceiver operation). The authors in the previous reference propose employing dual-input DPD blocks at each PA branch and a single common linear crosstalk and mismatch model that feeds them all to reduce the complexity of previous developments. Regarding the use of ANNs

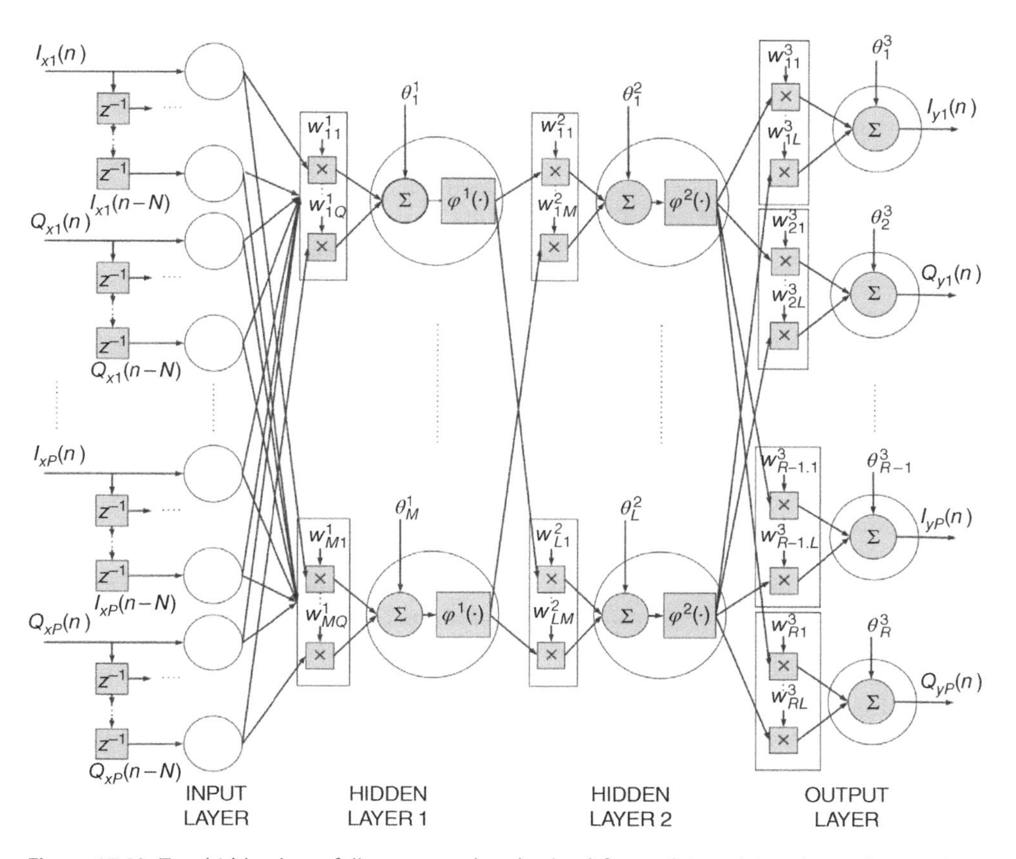

Figure 17.22 Two-hidden-layer, fully connected, real-valued, focused time-delayed neural network to model nonlinear power amplifier distortion for multiple-input and multiple-output architectures with cross-couplings, in-phase/quadrature imbalances, and direct current offsets.

in MIMO DPD applications, the authors in Zayani et al. (2010) presented one of the first works aiming at using a MLP-based NN to compensate both for cross-couplings and PA nonlinearities for a STBC OFDM MIMO system. However, the reach of this work is limited since only system-level simulation is conducted with a Saleh PA model. In Jaraut et al. (2018), the authors propose and experimentally validate an ANN for inverse nonlinear modeling of a number of transmitters with a single DPD block and thus overcome some previous implementation burdens. This modified RV FTDNN architecture, shown in Figure 17.22, is able to deal with all the mentioned RF impairments more efficiently than some previous works. In this ANN, we have four fully connected layers with Q inputs and R outputs, where $Q = (N + 1)R$, $R = 2P$, P is the number of Tx antennas, I_{xh} and Q_{xh} with $h = 1, 2, \cdots, P$ are the input signal I/Q pairs, and I_{yh} and Q_{yh} are the output signal I/Q pairs. Each input I/Q pair has $N + 1$ terms, with N being the memory depth, to account for memory effects. In the first and second hidden layers, the number of neurons is M and L, respectively. This ANN therefore has a maximum of $MQ + LM + RL$ weights and $M + L + R$ biases that will be adjusted using feedforward backpropagation.

17.4.2.3 ANN Training and Parameter Extraction Procedure

The ANN structure is typically trained with relevant I and Q baseband signals with expanded bandwidth and sample rate to allow for the DPD out-of-band compensation to fulfill ACPR requirements, and some additional envelope-dependent terms that can contribute to enhance performance and the training speed and numerical stability. These ANN excitations lead to output signals that are typically compared with those taken from measurements at the PA output after RF-to-baseband down-conversion, time alignment, and gain compensation. In this batch-trained supervised learning environment, the BPLA is applied to tune the weights and biases given the selected parameter subset of layers, neurons, and activation functions.

As introduced when comparing the CV- and RV-data operation in ANNs, using I and Q components features significantly faster training by using real weights instead of complex weights. Selecting convenient initial weight and bias values for the ANN can be crucial to avoid training divergence or long training periods that do not learn significantly and thus deliver under-performing models. A general rule to be followed is to avoid extreme values (either the smallest or the largest) and symmetrical distribution of weights, which make the neurons perform similarly and thus provokes unnecessary redundancy and lower performance. If no initial knowledge is considered, the weights are chosen in a way that the input to the next activation function typically lies in the region between linear and saturated (see Figure 17.18). Random initialization of −0.8–+0.8 leads to a reasonably good starting point, while values below −1 and above +1 are avoided since the neuron learning will be very slow or will be stopped.

At every training epoch (or iteration), there is both a forward pass, where the error cost function is calculated with the outputs of the ANN and the desired outputs, and a backward pass that calculates the increment to be applied to the NN weights and biases in order to minimize that cost function. Having an adaptive digitally assisted linearization or RF impairment compensation technique in the baseband modem puts some constraints on the number of epochs used for learning to reach the desired modeling performance. A categorization of the fast BPLA techniques is found in Rawat et al. (2010). In the first category, we find the heuristic techniques (more detail is provided in Haykin (2009)), which are derivations from the analysis of the standard steepest-descent algorithm. Here we can include the gradient descent with momentum (GDM), which prevents from falling into bad local minima; and variants such as variable learning rate (GDA), momentum and adaptive learning rule (GDX), and resilient back-propagation (RP). The standard numerical optimization techniques are in the second category (the information is expanded in Hagan et al. (1996)). This category includes (i) conjugate gradient-based techniques such as Polak-Ribiere (CGP), Fletcher-Powell (CGF), Powell-Beale (CGB) and scaled conjugate gradient (SGC); (ii) quasi-Newton algorithms such as Broyden-Fletcher-Goldfarb-Shanno (BFG) and one-step secant (OSS); and (iii) the Levenberg-Marquardt (LM) algorithm. The LM combines the gradient descent and Gauss-Newton methods and is vastly used in ANNs to minimize the cost function in DPD-related applications given its fast convergence properties, which are paired with good modeling performance and fair implementation complexity (Haykin (2009)).

The forward-backward pass process is repeated until the desired modeling performance is met or the ANN fails in the validation procedure or generalization. In the example shown in Figure 17.22, the performance index (or cost function) can be formulated as

$$\begin{aligned} E &= \frac{1}{2N}\sum_{h=1}^{P}\sum_{n=1}^{K}\left[(I_{yh}(n)-\hat{I}_{yh}(n))^2+(Q_{yh}(n)-\hat{Q}_{yh}(n))^2\right] \\ &= \frac{1}{2K}\sum_{n=1}^{K}\{\boldsymbol{e}^T(n)\boldsymbol{e}(n)\} \end{aligned} \tag{17.43}$$

where K is the data batch length, I_{yh} and Q_{yh} are the expected output signal I/Q pairs, $\hat{I}_{yh}$ and $\hat{Q}_{yh}$ are the output signal I/Q pairs produced at the output layer of the ANN, and $\boldsymbol{e}$ is the data batch error vector. E is minimized according to the LM algorithm and with respect to a parameter $\boldsymbol{g}$ depending on the overall weights and biases of the ANN. When going backward, $\boldsymbol{g}$ is updated at every epoch f as

$$\boldsymbol{g}^{f+1} = \boldsymbol{g}^{f} - [\boldsymbol{J}^T\boldsymbol{J} + \mu\boldsymbol{I}]^{-1}\boldsymbol{J}^T\boldsymbol{e} \tag{17.44}$$

where

$$\boldsymbol{g} = [w_{11}^1 \cdots w_{MQ}^1 \theta_1^1 \cdots \theta_M^1 w_{11}^2 \cdots w_{LM}^2 \theta_1^2 \cdots \theta_L^2 w_{11}^3 \cdots w_{RL}^3 \theta_1^3 \cdots \theta_R^3]^T \tag{17.45}$$

where $\boldsymbol{I}$ is the identity matrix, μ is a learning rate parameter, and $\boldsymbol{J}$ is the Jacobian matrix being calculated over the error vector $\boldsymbol{e}$ with respect to $\boldsymbol{g}$ as

$$\begin{pmatrix} \frac{\partial e(1)}{\partial w_{11}^1} & \frac{\partial e(1)}{\partial w_{12}^1} & \cdots & \frac{\partial e(1)}{\partial \theta_{R-1}^3} & \frac{\partial e(1)}{\partial \theta_R^3} \\ \frac{\partial e(2)}{\partial w_{11}^1} & \frac{\partial e(2)}{\partial w_{12}^1} & \cdots & \frac{\partial e(2)}{\partial \theta_{R-1}^3} & \frac{\partial e(2)}{\partial \theta_R^3} \\ & & \vdots & & \\ \frac{\partial e(K)}{\partial w_{11}^1} & \frac{\partial e(K)}{\partial w_{12}^1} & \cdots & \frac{\partial e(K)}{\partial \theta_{R-1}^3} & \frac{\partial e(K)}{\partial \theta_R^3} \end{pmatrix} \tag{17.46}$$

whose elements can be computed as shown in Jaraut et al. (2018).

In order to guarantee the convergence of the BPLA, the learning rate and momentum terms are introduced in the algorithms minimizing the estimation error. As found in classical DPD learning schemes, the learning rate controls the convergence speed. If it is too small, convergence is very slow and reaching the desired modeling performance requires more epochs, while if it is too high it can make the algorithm diverge. In Bertsekas and Tsitsiklis (1996), the best learning rate is found from the Hessian matrix of the input signal that, however, changes significantly with time and is computationally complex to track. In Mkadem and Boumaiza (2011), the authors propose applying to the learning factor either an increasing or decreasing rate at every epoch depending on whether the error between the network outputs and the desired output is, respectively, meeting the desired performance or not. Given the fact that even with an appropriate choice of the learning rate the BPLA may suffer from convergence to a local optimum, in order to better approach a global optimum, the authors follow the procedure by Plaut et al. (1986) and include the momentum term into to the BPLA. This factor adds the relative contribution of the current and past errors to the current change of the estimated parameters in the shape of an oscillatory descent solution. Therefore, ANN modeling performance can be benchmarked choosing first between a static or a dynamic learning rate and then between a static or a dynamic momentum term. By considering Eq. (17.44), the authors in Jaraut et al. (2018) employed a learning rate that started at a low value of 0.01. Depending on whether performance index E had increased or decreased, the learning rate was either multiplied or divided, respectively, by an additional factor β set to 10.

To achieve a better trade-off between modeling performance and processing complexity, the following procedures should be followed in finding an ANN-based solution, although no universal rule exists:

- *Input data memory depth*: The memory depth of the input signals is chosen typically by benchmarking different depth values in terms of modeling performance or NMSE (characterizing the error between the expected output and that obtained by the ANN) and complexity. For instance, a setting of memory taps that is 2 dB below the best NMSE attained could be the optimal one if the number of taps could be significantly reduced and the NMSE obtained was sufficient according to the application requirements. The knowledge of the PA physics and the designer expertise evaluating the PA response under a wideband modulated signal can help reduce the number of cases to be evaluated.

- *Number of hidden layers and neurons*: The universal approximation theorem by Cybenko (1989) has been previously introduced in this chapter to justify the capacity of a single-hidden-layer ANN to approximate any nonlinear function with any desired error with a convenient activation function. However, the theorem does not specify the best solution in terms of learning time (or epochs), implementation complexity, number of hidden neurons, or generalization capability with non-trained data, and assumes noise-free training data that is not always met in practice when the data is taken from measurements. Several works analyzing the performance of two-hidden-layer versus single-hidden-layer ANN schemes concluded that the two-hidden-layer ANNs provide better generalization and stability against training data noise. Chester (1990) proved that adding the second hidden layer filters out the measurement noise that the single-hidden-layer operation does not (since it models the noise instead of filtering it out). Hush and Horne (1993) proved that a two-hidden-layer network may require a lower overall number of neurons than a single-hidden-layer scheme to approximate a modeling function. In general, it is hard to find generalized deterministic approaches to choose the number of hidden layers since they would need to be validated under a massive number of different datasets. Therefore, the final empirical selection of hidden layers may be driven in the end by trading off the overall size of the ANN or the complexity, learning time, and modeling accuracy (Thomas et al. (2016)).
  There is not a specific rule in selecting the optimal number of neurons at each hidden layer, despite the fact that the stronger the PA nonlinearities, the greater the number of neurons in hidden layers (with nonlinear activation functions) will be. However, the complexity of the ANN can be set by evaluating the generalization error obtained when combining the bias-variance dilemma (Geman et al. (1992)) and the cross-validation technique (Stone (1978), Haykin (2009)). The bias error can be seen as how far from the expected data the output data of the ANN model is when using the training or estimation dataset. A high bias error is indicative of underfitting. The variance error comes from the sensitivity to small variations over the training dataset when the output of the ANN is evaluated with the validation dataset. A high variance error is an indicator showing that the ANN is modeling the random noise in the training data instead of the intended outputs and thus is overfitting. When the number of hidden neurons increases, the bias error typically decreases and the variance error increases. These parameters are taken into account in a backpropagation algorithm that learns in stages, moving from the realization of simpler to more complex mapping functions as the training session progresses and the iteration or epoch number increases. By using this procedure, the training session is stopped periodically (i.e. every five epochs) and, given the obtained ANN weight and bias values, the model is tested on the validation subset at each of these periods. The MSE of the estimation during training decreases monotonically for an increasing number of epochs while in validation the MSE curve first decreases to a minimum and then increases (the learning algorithm starts modeling the noise given the training dataset). An early MSE minimum could define the stopping point at which the ANN parameters are selected; however, the number of epochs is typically increased beyond this stopping point to check whether the early MSE minimum is local or not and then choose the most convenient stopping point. A few improved versions of the cross-validation method and pruning procedures are found in Haykin (2009).
- *Activation functions*: Regarding the type of activation functions, there is not a systematic approach to set the suitable function in the hidden layers. As shown in previous examples, for the output layer, a pure linear function is typically used to sum up the outputs of hidden

neurons and linearly map them at the output. According to Karray and Silva (2006), the linear activation functions are used typically for regression while nonlinear activation functions are used for input-output modeling. It is well-known that faster training can be achieved by using antisymmetrical activation functions such as the antisymmetric hyperbolic tangent. Again, the benchmarking of different activation functions for the hidden layers can help to determine which is the best option. Some information in this regard is provided in the next subsection.

17.4.2.4 Validation Methodologies and Key Performance Index

In this section, we will further provide the quantitative analyses and evaluation results about the usability and performance of ANNs in behavioral modeling, DPD, and RF impairment compensation applications. Not only will the impact of the ANN architecture parameter selection be assessed, but a comparison between ANN architectures with respect to classical polynomial-based approaches will also be targeted.

Comparison between ANN architectures and classical polynomial-based approaches: Let us first focus on RV FTDNN and RV RNN for single-DPD applications. The RV FTDNN architecture would be equivalent to that shown in Figure 17.22 but considering just a single input I/Q pair (which includes all the necessary delay taps) and an output I/Q pair. This architecture was presented in Rawat et al. (2010), where it is compared with the RV RNN architecture when modeling a highly nonlinear Doherty PA at 2.14 GHz with multiple wideband code division multiple access (WCDMA) signals aggregation of up to 15 MHz bandwidth. Once the optimal number of neurons is found for the two ANNs, the RV FTDNN has a PA modeling NMSE about 10 dB better by employing 30% fewer coefficients than the RV RNN. The performance of the modeling, apart from being quantified, can be also visually evaluated when comparing the estimated and the expected output amplitude and phase test signals for each networks and the spectra being generated. The time-domain measurements show that the RV RNN does not model well during fast transition states of the waveform either in amplitude or in phase, and the spectra plots denote that the out-of-band distortion modeling is not good (the adjacent channel error power ratio [ACEPR] could also be used as a quantitative indicator for out-of-band modeling performance). This could be provoked by (i) the recursive nature of the RV RNN since the input is dependent on the model itself and this impacts initial convergence and uncertainty, and by (ii) the fact that the PA is part of an open-loop transmitter without any feedback between the output and the input. Therefore, the RV FTDNN is in this case closer to the physical analogy of the PA. Further refinement can be produced to the RV RNN to raise the performance to the same level, but this would lead to higher training periods and hardware complexity. In contrast with the previous experiments, in Mkadem and Boumaiza (2011) the PA under test shows output to input interactions in the physical model. For this reason, the proposed ANN architecture results from unrolling a RV RNN, which maps these physical interactions, to turn it into a RV FNN. This architecture is compared with the RV FTDNN structure used in Rawat et al. (2010). After an extensive parameter benchmarking and sensitivity analysis, the physics-aware RV FNN performs better than both the memory polynomial approach and the RV FTDNN (i.e. showing 3.5 dB and 2.5 dB better NMSE, respectively).

If we now focus not only on the topology but also on the complexity of the network. For multiple-layer ANNs, the combination in number of neurons at the first hidden layer and the second hidden layer can be benchmarked in terms of NMSE (once the optimal number of neurons at the first hidden layer is set at a fixed number, the number of neurons in the second is then evaluated). The memory depth is also another parameter to be benchmarked. Different number of memory taps can be assessed where, typically, for RV inputs the same configuration will be employed, while in RV RNN different taps configurations between input and feedback

Table 17.1 Complexity comparison between polynomial-based models and the MIMO RV FTDNN

Model	Coefficients/parameters for $P{\times}P$ antennas	Coefficients/parameters[a] for 2×2 — 3×3 — 4×4 ant.
COMPPM	$P^2(N+1)K$	20 — 270 — 480
PH	$P^2(N+1)(K+P-1)!/P!$	420 — 2520 — 10800
ACC-PH	$\sim 2P^2(N+1)(K+P-1)!/P!$	840 — 5040 — 21600
MIMO RV FTDNN	$2P(N+1)M+LM+2PL$ weights and $M+L+2P$ biases	431 — 587 — 743

a) with $N=4$, $K=6$, $M=14$ and $L=7$.

signals can be used depending on the physics to be modeled. Having an indication of how the ANN paradigm compares with classical polynomial-based dynamic nonlinear modeling approaches in terms of complexity can be of interest. In order to choose a currently relevant application environment to do the comparison, the MIMO DPD architecture compensating I/Q impairments and DC offsets shown in Figure 17.22 is considered. Jaraut et al. (2018) provide an interesting coefficient number and performance comparison among the MIMO RV FTDNN, the COMPM, the PH, and the augmented complex conjugate PH (ACC-PH) compensating I/Q imbalance and DC offsets, for 2×2 and 3×3 MIMO DPD models. This comparison is expanded in Table 17.1, where P is the number of antennas (or Tx channels), N is the memory depth, K is the polynomial order (for COMPM, PH and ACC-PH), and M and L are, respectively, the number of neurons in the first and second hidden layers (MIMO RV FTDNN).

Table 17.1 shows that the MIMO RV FTDNN modeling requires fewer parameters when compared to PH and ACC-PH, whose number of coefficients increases exponentially with the number of antennas. The authors in this work provide rich experimental results comparing the COMPM, PH, and ANN schemes for 2×2 and 3×3 with a power amplifier operating at 2.14 GHz and being excited with LTE carrier-aggregated signals of 5 MHz and 10 MHz bandwidth (i.e. LTE 101 totaling 15 MHz bandwidth and 30 MHz bandwidth, respectively). The test setup includes several couplers at the input and output of the PA to provoke equal and unequal cross-couplings of −15 dB and/or −20 dB between the two or three Tx antennas (i.e. under unequal conditions, both strong and soft cross-couplings will be combined to demonstrate ANN inverse modeling generalization or validity). Instrumentation-based Tx hardware is used to have control over the I and Q imbalances and set any desired IRR value (i.e. 4-degree phase imbalance, and −20 dB IRR and DC offset with respect to the main signal). Several tests are conducted for 2×2 and 3×3 configurations with and without I/Q modulator imperfections and considering −15 dB couplings. When considering no I/Q impairments and using the COMPM, the PH, and the MIMO RV FTDNN model, each of these models outperforms the previous one by 2 dB NMSE and 3 dB ACPR (they are listed from worse to best performance), and thus there is a 4 dB NMSE and 6 dB ACPR difference between COMPM and MIMO RV FTDNN. When having I/Q modulator imperfections, both the COMPM and the PH feature unacceptable performance at all levels. The MIMO RV FTDNN outperforms the ACC-PH by around 3–4 dB in terms of both NMSE and ACPR, and 1–2 dB in terms of IRR, employing a significantly lower number of HW resources (i.e. half the coefficients for 2×2 and one-tenth for 3×3).

Comparison of BPLA algorithms: Considering the previously presented classification of BPLA techniques and the single-antenna RV FTDNN, the authors in Rawat et al. (2010) also evaluate the modeling performance of the algorithms after 50 epochs in terms of MSE. The results show that the standard numerical optimization techniques (LM, BFG, and CGB) significantly outperform the conventional gradient descent methods (RP, GDX, GDA, and GDM). The LM is without doubt the best option and the one reaching top modeling performance values (i.e. below −65 dB MSE) in the fewest epochs (i.e. in about 15). At 50 epochs, the LM is followed by the BFG (beyond 10 dB worse MSE) and by the CGB (beyond 20 dB worse MSE). The conventional gradient descent methods cannot provide sufficiently good modeling performance (i.e. 30 to near 50 dB worse MSE when compared to LM when trained at 50 epochs) unless the number of epochs is increased very significantly and thus consumes unaffordable time. Two additional parameters that can be evaluated, as shown in Mkadem and Boumaiza (2011), are the learning and momentum rates. First, different constant learning rates are applied to get the modeling performance in terms of NMSE. This is followed by evaluation of dynamic learning rates combining the initial value with an increasing rate and a decreasing rate. Once the best learning rate strategy is chosen, similar tests are applied to the momentum term to select the best option again. The results being reported show differences of about 1–2 dB of NMSE, but do not state the impact on the training period length when choosing the different options which is another factor to be taken into account.

Training signals: In Rawat et al. (2010), different WCDMA channel aggregation combinations are considered. In scenarios where the bandwidth of the signal or the spectra occupation can be different over time, it is important to train the ANN with adequate training signals. For instance, when operating with 5 MHz bandwidth channels and aggregations of up to four channels in WCDMA (i.e. 1111 is equivalent to aggregating four 5 MHz bandwidth signals totaling an overall 20 MHz bandwidth signal), training the network with WCDMA signals with lower-bandwidth channel aggregations will result in worse modeling of the reverse model (i.e. in DPD applications). As can be expected, the higher the overall bandwidth is, the higher the in-band noise and the lower the MSE metric; however, depending on the frequency or bandwidth occupation of the training signals, a specific validation signal configuration can lead to inverse modeling performances differing up to 5 dB in MSE. Training with WCDMA 1111 and validating with WCDMA 1 will lead to good results, but training with WCDMA 1 and validating with WCDMA1111 will lead to modeling performance degradation (which means that it will be worse but not necessarily insufficient). Jaraut et al. (2018) also demonstrate how the length of the training signal versus that employed for validation may have an impact in the inverse modeling results given the fact that larger training signals will be more representative of the physical phenomena to be modeled. In this reference, the inverse modeling performance when training with 40000 sample waveforms is compared for different validation signals lengths larger than the training one. In this case, experiments with the validation signals having the same length as that used for training, and those with the validation signals three times larger, just brings 1 dB NMSE difference. This is due to the fact that the training signal already captures sufficiently well the dynamic nonlinear phenomena. If the comparison was done employing shorter-length training signals, the result could differ more significantly.

Activation functions: The authors in Mkadem and Boumaiza (2011) show some NMSE results for the ANN under study when combining different activation functions for the first and the second hidden layers. Linear and symmetric saturating linear functions are used for the first hidden layer (note again that in this case the RV RNN is reformulated as a FNN) while the log-sigmoid and the hyperbolic tangent sigmoid are used for the second layer. Combining the symmetric saturating linear function with the hyperbolic tangent sigmoid brings the best NMSE (3.5 dB above other combinations).

Having in mind the aforementioned procedures and benchmarkings to find the ANN structure and parameters can lead to achieve better results than random selection of the parameters. Non-optimized parameters will lead to finding many constant response saturated nonlinear neurons that do not contribute to the modeling of the dynamic nonlinear behavior, while the contrary happens in optimized designs. Some previous results show that the ANNs can outperform classical Volterra-based approaches when modeling complex multi-antenna and multi-impairment problems at lower complexity. However, these challenging scenarios still require techniques to prune the ANN parameters to make these ANNs implementable for beyond 4×4 RF transceiver subsystems.

17.4.3 ANN for CFR: Design and Key Performance Index

ANNs have been proposed to overcome several limitations of the classical PAPR reduction schemes presented in Subsection 17.2.1 of this chapter. This section provides an overview of ANN CFR applications that provide a more convenient alternative to both signal scrambling-based probabilistic schemes requiring explicit side information such as SLM, PTS, tone injection, or ACE, and signal-distortion techniques such as peak cancellation or clipping and filtering.

17.4.3.1 SLM and PTS

Many works that envisaged the use of ANN for PAPR reduction started considering the Hopfield NN (HNN), which is a type of RNN. The HNN is based on a set of neurons and unit-time delays in the shape of a multiple-loop feedback system from one neuron to all the rest of neurons in the same layer (i.e. no self-feedback). RNNs can be hard to analyze and may either reach a stable state, oscillate, or behave chaotically. However, if the connections between binary output neurons are symmetric, by setting the right energy function (depending on the connection weights and the binary states of the neurons), one can find that the binary threshold decision rule makes the energy function output decrease and iteratively reach an energy minimum. One of the key contributions of Hopfield was the application of the Lyapunov stability theory to the analysis of RNN (Hagan et al. (1996)) to find the attractor points or energy function minima.

The authors in Ohta and Yamashita formulate the PAPR reduction problem as a combinatorial optimization solved by using a chaotic NN (CNN) suitable for real-time implementation. This scheme, now introducing chaos to the HNN, outperforms previous HNN-based approaches that employ the gradient descent method and typically fall into local minima. A generic and simplified ODFM transmitter and receiver block diagram including an ANN-assisted SLM scheme is shown in Figure 17.23 (note that for PTS, the IFFT would appear before the phase factors multiplications). To target the PAPR reduction by means of phase rotation, the complex symbols that will be mapped into a specific subcarrier are multiplied by a set of unit modulus complex

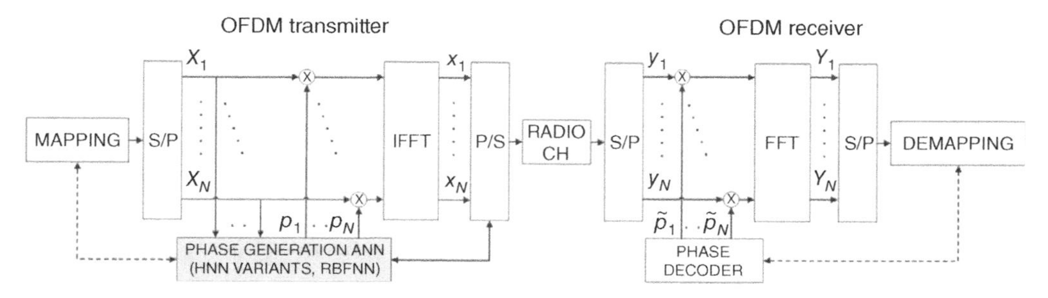

Figure 17.23 Simplified orthogonal frequency-division multiplexing transmit/receive block diagram with an artificial neural network–assisted selected mapping technique scheme.

values that are fed into an objective function that gives the lowest value when the resulting combination of phase rotated symbols features the lowest PAPR.

In order to reduce the search complexity, the unit modulus complex values can be reduced to +1 or −1, which turns the process into a simple combinatorial optimization that is first solved by a HNN and later by means of a CNN. These two ANN schemes are compared with SLM for a given number of phase rotation patterns M, considering 128 QPSK symbols. The HNN, with a maximum iteration step of 200, can suppress about 5 dB the PAPR as SLM does for $M = 128$ at $Prob\{PAPR > PAPRth\} = 10^{-4}$. A CNN with up to 64 maximum iteration step provides near 1.7 dB additional PAPR reduction on top of the previous results. In this system, the information of the phase-rotation factors needs to be transmitted, thus reducing system efficiency. In Ohta et al. (2006), the same authors combine the neural phase rotator with a biased polynomial cancellation coded OFDM (BPCC-OFDM) to reduce the PAPR of these waveforms and at the same time avoid the side information by sending the phase-rotation factors through the pilot symbols. When comparing BPCC-OFDM(SLM) and BPCC-OFDM(ANN), the results show both 1.5 dB higher PAPR reduction and about 3 dB BER versus E_b/N_o improvement at $Prob\{PAPR > PAPRth\} = 10^{-4}$ in the former. Another work by Wang (2006) proposes a similarly performing HNN-modified scheme by using a phase generator now based on a stochastic HNN that avoids local minima by changing the neuron output functions by adding a random disturbance into the neural state (for instance, a logistic distribution with zero mean value). Sohn and Shin (2006) present an RBFNN mapper before the IFFT to reduce the PAPR. The three-layer RBFNN employs the nonlinear Gaussian activation function in the hidden layer and the linear activation function in the output layer. The centers of the hidden layer can be determined by using K-means (clustered) competitive learning algorithms (Haykin (2009)) where the hidden neurons compete with each other and where the center vector of the winner is updated with a decreasing learning rate. The possible transmit data symbols can be represented in the hidden layers by N dimensional statistically independent rotation sequences whose weights can be adapted at every iteration by using LMS and are a function of the output RBF mapper, the RBF basis function, the desired optimum rotation pattern index, and the input data stream. With $N = 128$ QPSK modulated subcarriers and considering 1024 hidden layer centers with N dimensional statistically independent rotation sequences, the RBFNN mapper is compared in terms of performance with the HNN and the SLM. The results show that the RBFNN is between the SLM and the HNN in terms of PAPR reduction, but at much lower complexity. Wang et al. (2008) propose combining the HNN with the immune clonal selection algorithm (ICSA) to provide better performance at a lower number of iterations for a multi-carrier MC-CDMA system. The authors implement a feedback-control mechanism to minimize the number of generation of the ICSA algorithm from 30 to 20 to reach good performance. When comparing the ICSA-HNN and HNN at 20 iterations in terms of PAPR reduction at $Prob\{PAPR > PAPR_{th}\} = 10^{-3}$, the difference is around 0.8 dB (1 dB if we compare ICSA-HNN at 20 iterations and HNN at 10 iterations).

17.4.3.2 Tone Injection

As mentioned at the beginning of this chapter, this technique employs data subcarriers to reduce the PAPR by expanding the original constellation size, which can be seen as injecting a tone into the original multi-carrier signal. In order to compute the optimum tone selection, avoiding greedy algorithms, which easily get trapped into local minimum points, the authors in Mizutani et al. (2007) propose now combining the HNN and the CNN with an architecture that aims at pruning IFFTs for neuron state updating to reduce the complexity. An oversampled OFDM signal is used to capture all the continuous-time peaks (as it happens in most of the herein described PAPR reduction techniques). Whenever an M-QAM modulated data symbols contributes significantly to having a large peak power, the tone to be injected is multiplied by factors that will lead the data symbol to shift to an equivalent constellation point at a distance that is chosen conveniently to avoid overlap with the original constellation points and that can be easily demodulated at the receiver side without need of side information. This process is repeated until the PAPR falls below the desired maximum value. Considering 128 subcarriers, an oversampling factor of 4, 32 iterations for the ANN, and 30 for the conventional TI process, the PAPR is further reduced by 0.7 dB with the ANN $Prob\{PAPR > PAPR_{th}\} = 10^{-4}$. The interlaced ANN scheme is then proposed to reduce by half the complexity of the network by rewriting the motion equation in a way that either the even or odd sampling points of the oversampled OFDM signal are alternatively used for updating the neuron state at each iteration. With this technique, the PAPR reduction is decreased only by about 0.2 dB but the computational time is reduced by 35%.

17.4.3.3 ACE

In order to reduce the envelope fluctuations and thus the PAPR values, in ACE some of the outer signal constellation symbols are dynamically expanded to a region that does not affect the demodulation decision. ACE does not require side information; however, obtaining the reduced PAPR signal can take considerable time due to its typically slow convergence. With the goal of reducing the complexity and enhance the performance of some classical schemes such as SLM, PTS, and smart gradient-project (SGP) ACE, the authors in Jabrane et al. (2010) propose combining the approximate gradient-project (AGP) ACE and a RV MLP ANN. The AGP algorithm (prioritizing convergence over speed) in the ACE module will provide the training and validation signals to the ANN that will learn on the characteristics of a low envelope-fluctuation signal with the backpropagation LM algorithm. Since the time-domain RV ANN can hardly learn which constellation regions are allowed and which ones are not, a second ANN working on the frequency domain is concatenated to the previous ANN (once the time-domain ANN model has been validated with the AGP time-domain selected test data). In the second ANN, the DFT is applied both to the input real and imaginary training data coming from the previous ANN, and to the values provided by the AGP. After that, the training data is separated into four constellation regions to train eight ANNs with the LM algorithm (there are four ANNs for each of the transformed real and imaginary components); once the ANNs are finally trained the AGP frequency domain, selected test data is used for validation. This information is visually represented in Figure 17.24 (i.e. when all the switches are in an "off" state, only the time-domain ANN is operative; when the switches are in an "on" state, both the time- and frequency-domain ANNs are operative). This procedure is understood as an initial offline training to set the overall ANN parameters. The inner ANNs have one hidden layer with two neurons making use of the triangular activation function, all of which results in a number of integer multiplications and integer additions proportional to the subcarrier number. When comparing with SLM and PTS, the AGP-ACE post-trained RV MLP ANN scheme becomes simpler (it totals about one order of magnitude fewer operations) since it does not need as many IFFT operations (only two are needed) and the operations are simpler (i.e. predominantly integer operations) and more parallelizable. The authors in the previous reference also provide simulation results for 50,000 randomly generated QPSK and 16-QAM symbols modulating 512, 1024, and 2048 subcarriers. About 70% of the data is used for training and 30% for validation. For 512 subcarriers, AGP-ACE performs 0.9 dB better than the proposed ANN approaches (either the time domain or the time and frequency ANN) in terms of the cubic metric (CM), characterizing more precisely the PA back-off than the PAPR metric, and only about 0.35 dB better for 2048 subcarriers. Compared to SLM, the ANNs are 0.3 dB and 0.9 dB better, considering 512 and 1024 subcarriers, respectively, and perform similarly to PTS at 512 subcarriers but about 0.75 dB better for

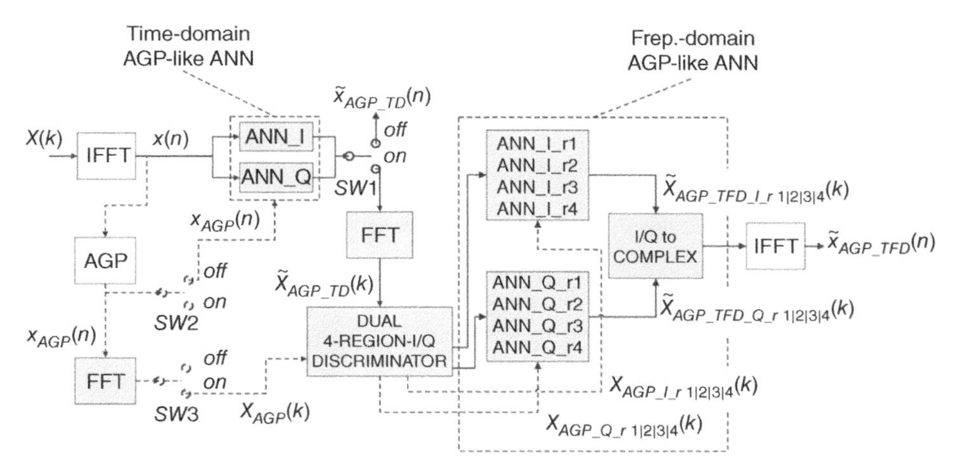

Figure 17.24 Real-value multilayer perceptron time and frequency artificial neural networks for approximate gradient project–active constellation extension emulation.

2048 dB subcarriers. Therefore, the RV MLP time and frequency ANNs perform better than SLM and PTS and slightly worse than AGP-ACE, but at a reduced computational complexity. Regarding the BER versus SNR characteristic, the time- and frequency-domain ANN performs better than the rest of the approaches (more evidently when the modulation order increases). This is due to the fact that, in the ANN approach, the symbols mapped in the outer constellation points are less expanded from the original position when compared to ACE and thus the constellation energy is kept more concentrated. Moreover, since the number of symbols being expanded to the allowed region is higher (note that in this region the symbols are less affected by the system noise), only a few symbols will experience an effective lower SNR. An adaptive neural fuzzy interference system (ANFIS) is built by the authors on top of the same time- and frequency-domain ANN architecture in Jiménez et al. (2011). The ANFIS is used to synthesize unknown behaviors by applying fuzzy heuristic rules as an adaptive network that can compute backpropagation gradient vectors in combination with the LS method. The time-domain ANFIS learns which time-domain signals feature lower envelope fluctuations, while the time- and frequency-domain ANFIS learns which constellation regions are allowed and which are not. A detailed complexity benchmarking is also provided for SLM and PTS variants, SGP-ACE, and both the time-domain ANFIS and the time- and frequency-domain ANFIS. One modified version of these approaches is to remove the frequency-domain ANN as presented in Sohn (2014). This architecture, however, builds another time-domain ANN at the receive side (right after channel equalization). The Rx ANN is trained with the Tx ANN outputs and the original OFDM signal is used as the desired signal (at the Tx side, the ANN desired signal is that given by ACE as in previous example works). Again, all these ANN schemes have a two-neuron single hidden layer. In terms of PAPR reduction, this latter approach performs similarly to previous works and near 0.7 dB worse than the ACE scheme at $Prob\{PAPR > PAPR_{th}\} = 10^{-3}$. When testing the BER versus Eb/No for QPSK and 16-QAM, assuming a quasi-static frequency-selective channel and perfect channel estimation, the ANN-based scheme shows significant BER improvement when compared to the ACE scheme since the Rx ANN improves the M-QAM demodulator performance (i.e. at $E_b/N_o = 20$dB, the BER using the ACE scheme is 1×10^{-2} and the BER using the ANN-based scheme is 3.5×10^{-4}). At a shorter extent, the Tx-Rx ANN approach also performs better than the previous Tx time- and frequency-domain works, but with the advantage of featuring a complexity reduction factor of around two.

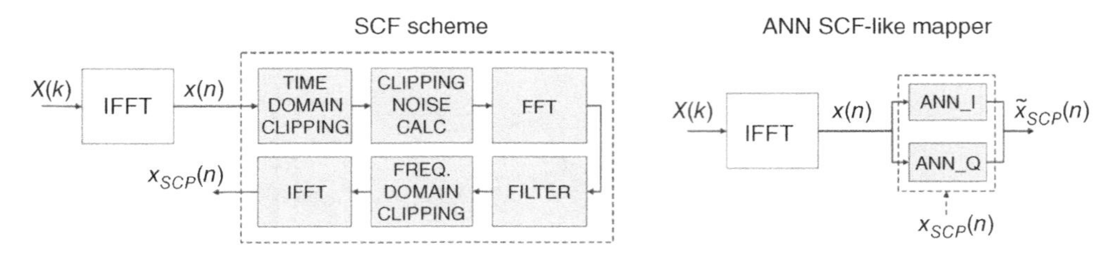

Figure 17.25 Simplified clipping and filtering block diagram, and proposed artificial neural network–based architecture emulating SCF.

17.4.3.4 Clipping and Filtering

As described before, these techniques feature a trade-off between PAPR reduction and the amount of in-band and out-of-band distortion induced by the clipping process. Since the filtering process to suppress out-of-band distortion may provoke some spectral regrowth, this technique requires a certain amount of iterations. The iterative clipping and filtering (ICF) algorithm in Armstrong (2002) is a well-known approach but requires a large number of repetitions involving complex FFT operations. The simplified CF (SCF) in Wang and Tellambura (2005) can obtain the same PAPR reduction as ICF in a single iteration but also requires several FFT/IFFT complex operations (increasing with the number of subcarriers being employed). In this context, the authors in Sohn and Kim (2015) propose a RV MLPNN mapper that emulates the SCF featuring similar CM performance but at much lower complexity, as can be seen in Figure 17.25. Two independent NN modules do the processing of the I and the Q signal components separately. These modules have again two neurons in the hidden layer and a triangular activation function, and use the backpropagation LM learning algorithm to tune the parameters, minimizing the error between the ANN output and the expected signals. The I and Q components of the original OFDM signals are used as training data, while the I and Q components of the time-domain SCF signal are used as the desired output data. The computational complexity analysis shows that the proposed ANN remarkably reduces the computational complexity in terms of complex operations by one order of magnitude when compared with SLM, one-seventh when compared with ICF, and one-third when compared with SCF. In terms of the CM, the ANN performs in simulation similarly to the ICF and the SCF schemes and about 0.8 dB better than the SLM at $Prob\{CM > CM_{th}\} = 10^{-3}$. Regarding the BER versus E_b/N_o characteristic (considering AWGN channel), and as expected, all the clipping and filtering techniques do worse than the SLM. Considering QPSK subcarrier modulation (considering 256 subcarriers), the ICF, SCF, and proposed ANN perform similarly. For 16-QAM, the proposed ANN has a performance loss of about 1.5 dB when compared with SCF, about 2.4 dB when compared with ICF, and 2.8 dB when compared with SLM. Increasing the constellation size makes it harder for the ANN to emulate the SCF mapping, which results in partial frequency-domain information loss.

All of this shows that much more investigation is needed on the use of ANNs in digital front-end, as will be further discussed in Section 17.6.

17.5 Support Vector Regression Approaches

Support vector machine (SVM) is a well-established supervised learning algorithm in the field of ML that can be used either for classification or for predicting and modeling. In the later

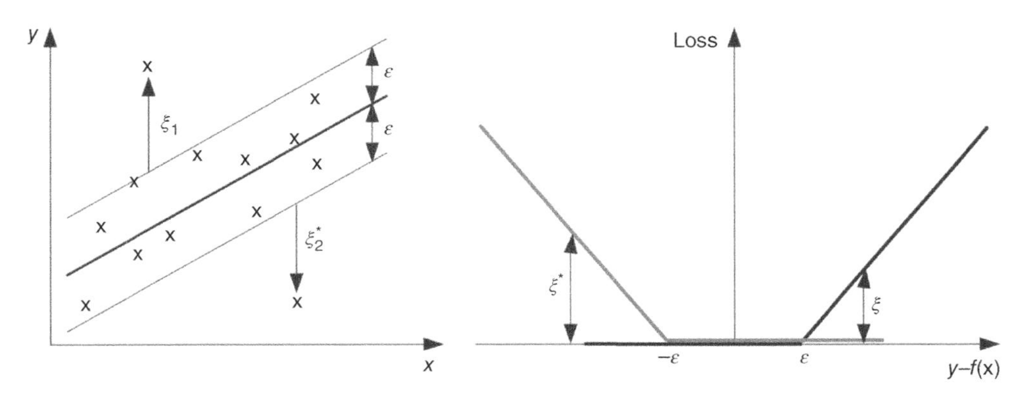

Figure 17.26 The soft margin loss setting for a linear support vector regressor (Smola and Schölkopf (2004)).

case, when the SVM is used for regression, it is known as support vector regressor (SVR). Both use very similar algorithms but predict different types of variables. SVR has been successfully extended to nonlinear regression problems (Vapnik (1998)) by converting it into a linear regression one. In particular, SVR is oriented at optimizing the generalization bounds of a nonlinear function through a linear method in a high-dimensional feature space, by emphasizing the data on the margins of the search space, known as *support vectors* (Chen and Brazil (2018)). In order to reduce the computational cost, SVR uses kernel functions to perform a nonlinear mapping to a high-dimensional feature space. In comparison with ANN techniques, the SVR method can obtain the optimal model in a short time with performance quality similar to that obtained with ANN. Due to these good properties, SVR has been employed in different applications in the DFE: for example, for the modeling and design of filters, antennas, and PAs, and even for nonlinear equalization and DPD linearization.

The basic idea of support vector regression theory is presented in Smola and Schölkopf (2004). Given training data with m dimensional variables and 1 target-variable $\{(\boldsymbol{x}_1, y_1), (\boldsymbol{x}_2, y_2), \cdots, (\boldsymbol{x}_n, y_n)\}$, where $\boldsymbol{x} \in \Re^m$ and $y \in \Re$, the objective is to find a function $f(\boldsymbol{x})$ that returns the best fit, taking into account certain margin ε, and that at the same time is as flat as possible. As shown in Figure 17.26, certain deviation (at most ε) from the actually obtained targets y_i is tolerated for all the training data.

Assuming that the relationship between $\boldsymbol{x}$ and y is approximately linear, $f(\boldsymbol{x})$ takes the form

$$f(\boldsymbol{x}) = y = \boldsymbol{w}^T\boldsymbol{x} + b \tag{17.47}$$

where $\boldsymbol{w}$ is the vector of coefficients and b is an intercept or bias term. In the case of Eq. (17.47), flatness means that the solution has to consider a small $\boldsymbol{w}$, i.e. a regularization condition (as in Ridge regression) consisting of minimizing the Euclidean norm $\|\boldsymbol{w}\|^2$ (also know as ℓ_2-norm $\|\boldsymbol{w}\|_2^2$). Formally, we can write this problem as a convex optimization problem such as

$$\begin{aligned} \text{minimize} \quad & \frac{1}{2} \parallel \boldsymbol{w}\|^2 \\ \text{subject to:} \quad & y_i - \boldsymbol{w}^T\boldsymbol{x}_i - b \leq \varepsilon \\ & \boldsymbol{w}^T\boldsymbol{x}_i + b - y_i \leq \varepsilon. \end{aligned} \tag{17.48}$$

In Eq. (17.48), it is assumed that the convex optimization problem is feasible. When this is not the case, one can introduce slack variables ξ_i, ξ_i^* (i.e. that allow for some errors) to cope with otherwise infeasible constraints of the optimization problem in Eq. (17.48). This leads to the formulation stated in Vapnik (1995),

$$\begin{aligned} &\text{minimize} \quad \frac{1}{2} \parallel \boldsymbol{w} \|^2 + C \sum_{i=1}^{n} \xi_i + \xi_i^* \\ &\text{subject to:} \quad y_i - \boldsymbol{w}^T \boldsymbol{x}_i - b \leq \varepsilon + \xi_i \\ &\qquad \boldsymbol{w}^T \boldsymbol{x}_i + b - y_i \leq \varepsilon + \xi_i^* \\ &\qquad \xi_i, \xi_i^* \geq 0 \end{aligned} \tag{17.49}$$

where C is a regularization term that determines the trade-off between the flatness of the model and the constraint violation error. As shown in Figure 17.26, only the points outside the shaded region contribute to the cost, as the deviations are penalized in a linear fashion (Smola and Schölkopf (2004)).

This constrained optimization problem is more easily solved in its dual formulation, since it allows extending SVR to nonlinear functions. Therefore, using the method of Lagrange multipliers (Fletcher (1987)), follows

$$\begin{aligned} L = \frac{1}{2} \|\boldsymbol{w}\|^2 + C \sum_{i=1}^{n} \xi_i + \xi_i^* - \sum_{i=1}^{n} (\eta_i \xi_i + \eta_i^* \xi_i^*) \\ - \sum_{i=1}^{n} \alpha_i (\varepsilon + \xi_i - y_i + \boldsymbol{w}^T \boldsymbol{x}_i + b) \\ - \sum_{i=1}^{n} \alpha_i^* (\varepsilon + \xi_i^* + y_i - \boldsymbol{w}^T \boldsymbol{x}_i - b) \end{aligned} \tag{17.50}$$

where L is the Lagrangian and $\eta_i, \eta_i^*, \alpha_i, \alpha_i^* \geq 0$ are Lagrange multipliers. The partial derivatives of L with respect to the primal variables $(\boldsymbol{w}, b, \xi_i, \xi_i^*)$ have to vanish for optimality,

$$\begin{aligned} \frac{\partial L}{\partial \boldsymbol{w}} &= \boldsymbol{w} - \sum_{i=1}^{n} (\alpha_i - \alpha_i^*) x_i = 0 \\ \frac{\partial L}{\partial b} &= \sum_{i=1}^{n} (\alpha_i - \alpha_i^*) = 0 \\ \frac{\partial L}{\partial \xi_i} &= C - \alpha_i - \eta_i = 0 \\ \frac{\partial L}{\partial \xi_i^*} &= C - \alpha_i^* - \eta_i^* = 0. \end{aligned} \tag{17.51}$$

Substituting Eq. (17.51) into Eq. (17.50) yields the dual optimization problem

$$\begin{aligned} &\text{maximize} \quad -\frac{1}{2} \sum_{i,j=1}^{n} (\alpha_i - \alpha_i^*)(\alpha_j - \alpha_j^*) < x_i, x_j > \\ &\qquad - \varepsilon \sum_{i=1}^{n} (\alpha_i - \alpha_i^*) + \sum_{i=1}^{n} y_i (\alpha_i - \alpha_i^*) \\ &\text{subject to:} \quad \sum_{i=1}^{n} (\alpha_i - \alpha_i^*) x_i = 0 \\ &\qquad 0 \leq \alpha_i, \alpha_i^* \leq C. \end{aligned} \tag{17.52}$$

From Eq. (17.51), $\boldsymbol{w}$ can be defined as $\boldsymbol{w} = \sum_{i=1}^{n}(\alpha_i - \alpha_i^*)x_i$, and thus the SVR fitting equation can be described as

$$f(\boldsymbol{x}) = y = \boldsymbol{w}^T\boldsymbol{x} + b = \sum_{i=1}^{n}(\alpha_i - \alpha_i^*) < x_i, x > +b. \tag{17.53}$$

This is the so-called support vector expansion, where $\boldsymbol{w}$ can be described as a linear combination of the training patterns x_i. As explained in Smola and Schölkopf (2004), note that the complexity of a function's representation by support is independent of the dimensionality of the input space and depends only on the number of support vectors. In addition, the complete algorithm can be described in terms of dot products between the data and, even for evaluating $f(\boldsymbol{x})$, it is not necessary to explicitly compute $\boldsymbol{w}$.

For applications such as PA modeling, nonlinear equalization, and DPD linearization, the next step is to make the SV algorithm nonlinear. For a nonlinear regression problem, SVR converts it into a linear regression one. At first, we convert training sets to a high-dimension feature space by utilizing a nonlinear function $\Phi(\cdot)$, and then get the linear regression function in this space:

$$f(\boldsymbol{x}) = y = \boldsymbol{w}^T\Phi(\boldsymbol{x}) + b. \tag{17.54}$$

Instead of computing a mapping function, $\Phi(\cdot)$, explicitly, the inner product in the feature space can be expressed by a kernel function, $K(x_i, x)$, which has to satisfy the Mercer theorem (Mercer (1909)). Thus, the support vector expansion in Eq. (17.53) can be written as

$$f(\boldsymbol{x}) = y = \boldsymbol{w}^T\Phi(\boldsymbol{x}) + b = \sum_{i=1}^{n}(\alpha_i - \alpha_i^*)K(x_i, x) + b. \tag{17.55}$$

Some examples of kernel functions (Smola and Schölkopf (2004), Cai et al. (2018)) are:

- Homogeneous polynomial kernels

$$K(x, x') =< x, x' >^p \tag{17.56}$$

- Inhomogeneous polynomial kernels

$$K(x, x') = (< x, x' > +d)^p \tag{17.57}$$

- Radial basis function (RBF) kernel

$$K(x, x') = e^{-\frac{\|x-x'\|^2}{2\sigma^2}} \tag{17.58}$$

- Hyperbolic tangent (or sigmoidal) kernel

$$K(x, x') = tanh(\gamma < x, x' > +d) \tag{17.59}$$

As discussed in Section 17.2, for PA behavioral modeling and DPD linearization we use input-output baseband signals, i.e. datasets of complex data ($\boldsymbol{x}, \boldsymbol{y} \in \mathbb{C}$). However, the nonlinear function in Eq. (17.55) is defined as a real-valued function of one real variable. Thus, in order to use SVR for PA modeling and linearization, both the input and the output values are separated into real part and imaginary part, $x = Re\{x\} + jIm\{x\}$, as in Cai et al. (2018); or magnitude and phase components, $x = |x|e^{j\angle x}$, as in Chen et al. (2005). For example, Figure 17.27 shows the block diagram of a PA behavioral model that includes two different SVR machines giving the real and imaginary parts of the output signal as distinct outputs. The real part and the imaginary part of the input signal are separated and taken as the inputs of

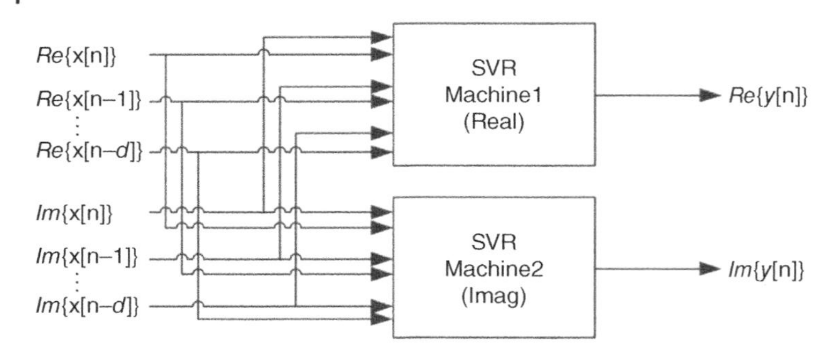

Figure 17.27 Block diagram of a power amplifier behavioral model based on time-delay support vector regression (Cai et al. (2018)).

the two SVR machines. In Cai et al. (2018), the authors provide a comparison of the modeling capabilities among different PA behavioral models, including their proposed SVR solution. In particular, they compared the modeling performance in terms of NMSE between the actual PA output and the modeled one, number of parameters used by the model, and simulation time. The PA behavioral models under comparison were: generalized memory polynomial (GMP) model (D.R. Morgan, Z. Ma et al. (2006)), decomposed vector rotation (DVR) model (Cao and Zhu (2017)), a two-layer real-valued time-delay NN (RVTDNN) model (Liu et al. (2004)), a RBF NN model (Isaksson et al.) and the SVR model proposed by the authors. In terms of NMSE (or modeling accuracy), the SVR model outperformed the models based on analytic functions and ANNs, but at the price of using more coefficients and simulation time. The ANN prediction accuracy could have been improved through careful modification of the optimization algorithm and NN complexity. This process, however, would be time-consuming.

SVRs have been used not only for PA behavioral modeling, but also (or mainly) for circuit-level modeling and PA design. For example, in Chen and Brazil (2018), the authors use SVR to design an optimization-oriented method to design a broadband Doherty power amplifier (DPA); in Chen and Brazil (2017), a method to classify load-pull contours for the design of a broadband high-efficiency; and in Chen et al. (2015), to design an automatic optimization method to control the harmonic impedances of continuous Class-F PAs. In addition, in Guo et al. (2007), a SVR method is presented for electrothermal modeling of power FETs; and in Maoliu and Xiaojie (2008), SVRs are used to obtain large signal behavior models for RF PAs. In Zhou and Huang (2013), instead of modeling PAs, the authors present an intelligent alignment method for an automatic tuning device of microwave filters based on SVR that is particularly suited to computer-aided tuning devices or an automatic tuning robot of volume-producing filters. Similarly, in Koziel and Bandler (2008), the authors propose an SVR method for microwave device modeling. Another field of application for SVR is nonlinear equalization (Nguyen et al. (2016), Mallouki et al. (2016)). For example, in Sebald and Bucklew (2000), SVRs are reported to perform as well as conventional decision feedback equalizers, with the advantage that a smaller number of parameters for the model can be identified in a manner that does not require the extent of prior information or heuristic assumptions being required by some previous techniques. Regarding DPD linearization, some works have been published (e.g. Eda et al. (2001) and Chen et al. (2005)) providing simulation results while lacking the detailed performance comparison of their solution with that achieved by ANNs. In the field of DPD linearization, ANN is the most used ML approach reported in literature, as discussed in Section 17.4.

In summary, for the aforementioned fields of device design or modeling, SVR methods generally require fewer samples in statistical learning and are free of local minima in optimization in comparison with ANN-based methods.

17.6 Further Discussion and Conclusions

In this chapter, we have discussed some ML algorithms and strategies to be applied in the DFE. These techniques are mainly aimed at coping with the unwanted nonlinear distortion effects that appear when dealing with current high PAPR and wide-bandwidth communications signals, while targeting high power-efficient operation to save energy consumption.

The principles of CFR techniques and DPD for nonlinear distortion mitigation were discussed in Section 17.2. In order to deal with current highly efficient power amplifier architectures, MIMO, or concurrent multi-band systems, the number of parameters required for DPD or PA behavioral modeling based on parametric approaches grows exponentially, which increases the computational complexity and may provoke overfitting and uncertainty in estimation. Feature-selection and feature-extraction dimensionality-reduction techniques oriented at reducing the number of required parameters and guaranteeing a well-conditioned extraction were discussed and particularized for DPD linearization in Section 17.3.

As an alternative to polynomial-based models, in Sections 17.4 and 17.5, we presented some design guidelines for ANNs and SVR approaches to model and compensate for the nonlinear distortion introduced by the PA or other unwanted distortion effects such as I/Q imbalances and DC offset. In addition, CFR strategies based on ANNs were also discussed, pointing out their advantages in terms of performance and computational complexity reduction.

Some pros and cons that need to be taken into account in order to decide whether to use the solutions based on ANNs or SVRs are discussed in the following. While ANN-based models permit a compact representation of a multidimensional function, they suffer from two main weaknesses: (i) ANNs often converge on local minima rather than global minima, and (ii) they can suffer from the overfitting problem due to the traditional empirical risk-minimization principle employed by ANNs. SVR techniques instead use the structural risk-minimization principle that avoids the model becoming too strongly tailored to the particularities of the training set, thus preventing it from overfitting. Additionally, the solution to SVR is global and unique. Regarding the dimensions, ANNs are parametric models whose size is fixed and depends on their specific architecture, while SVRs (based on kernels) consist of a set of support vectors, selected from the training set, with a weight for each. In the worst case, the number of support vectors is exactly the number of training samples, and in general its model size scales linearly. Thus, if the number of training samples is high, so is the computational complexity of the SVR. Finally, while ANNs may have any number of outputs, SVRs have only one. The most direct way to create a MIMO system with SVR is to create multiple support vector machines and train each of them one by one.

Under the current practical circumstances, ANNs have been successfully employed in several applications in the DFE for their capability to approximate any continuous function. ANNs have been reported in the literature to be used for modeling at the transistor, circuit, and system level. The SVR technique has been successfully applied to circuit modeling for characterizing and designing antennas, filters, and power amplifiers. At the system level, however, solutions based on SVR methods to model or linearize PAs through DPD are not as extended as those based on ANNs. It is strongly believed that ANN- or SVR-based solutions will play a very important role in future DFE implementation and deployment, including intelligent vector processors and functional IP blocks.

Bibliography

Shoaib Amin, Per N. Landin, Peter Handel, and Daniel Ronnow. Behavioral modeling and linearization of crosstalk and memory effects in RF MIMO transmitters. *IEEE Transactions on Microwave Theory and Techniques*, 62 (4): 810–823, apr 2014. doi: 10.1109/tmtt.2014.2309932.

J. Armstrong. Peak-to-average power reduction for OFDM by repeated clipping and frequency domain filtering. *Electronics Letters*, 38 (5): 246, 2002. doi: 10.1049/el:20020175.

Taylor Barton. Not just a phase: Outphasing power amplifiers. *IEEE Microwave Magazine*, 17 (2): 18–31, feb 2016. doi: 10.1109/mmm.2015.2498078.

S.A. Bassam, M. Helaoui, and F.M. Ghannouchi. Crossover digital predistorter for the compensation of crosstalk and nonlinearity in MIMO transmitters. *IEEE Transactions on Microwave Theory and Techniques*, 57 (5): 1119–1128, may 2009. doi: 10.1109/tmtt.2009.2017258.

R.W. Bäuml, R.F.H. Fischer, and J.B. Huber. Reducing the peak-to-average power ratio of multicarrier modulation by selected mapping. *Electronics Letters*, 32 (22): 2056, 1996. doi: 10.1049/el:19961384.

Dimitri P. Bertsekas and J.N. Tsitsiklis. Neuro-dynamic programming. In *Encyclopedia of Optimization*, pages 2555–2560. Springer US, 1996. doi: 10.1007/978-0-387-74759-0_440.

R. Neil Braithwaite. A comparison of indirect learning and closed loop estimators used in digital predistortion of power amplifiers. In *2015 IEEE MTT-S International Microwave Symposium*. IEEE, may 2015. doi: 10.1109/mwsym.2015.7166826.

R.N. Braithwaite. Digital predistortion of an RF power amplifier using a reduced volterra series model with a memory polynomial estimator. *IEEE Trans. on Microw. Theory and Tech.*, 65 (10): 3613–3623, Oct. 2017. doi: 10.1109/tmtt.2017.2729513.

Dalibor Buzic and Jasminka Dobsa. Lyrics classification using naive bayes. In *2018 41st International Convention on Information and Communication Technology, Electronics and Microelectronics (MIPRO)*. IEEE, may 2018. doi: 10.23919/mipro.2018.8400185.

Jialin Cai, Chao Yu, Lingling Sun, Shichang Chen, and Justin B. King. Dynamic behavioral modeling of RF power amplifier based on time-delay support vector regression. *IEEE Transactions on Microwave Theory and Techniques*, pages 1–11, 2018. doi: 10.1109/tmtt.2018.2884414.

Wenhui Cao and Anding Zhu. A modified decomposed vector rotation-based behavioral model with efficient hardware implementation for digital predistortion of RF power amplifiers. *IEEE Transactions on Microwave Theory and Techniques*, 65 (7): 2443–2452, jul 2017. doi: 10.1109/tmtt.2016.2640318.

Wenhui Cao, Yue Li, and Anding Zhu. Magnitude-selective affine function based digital predistorter for RF power amplifiers in 5G small-cell transmitters. In *2017 IEEE MTT-S International Microwave Symposium (IMS)*. IEEE, jun 2017. doi: 10.1109/mwsym.2017.8058921.

Jessica Chani-Cahuana, Mustafa Ozen, Christian Fager, and Thomas Eriksson. Digital predistortion parameter identification for RF power amplifiers using real-valued output data. *IEEE Trans. on Circ. and Sys. II: Express Briefs*, 64 (10): 1227–1231, Oct. 2017. doi: 10.1109/tcsii.2017.2686004.

Kaiya Chen, Minxi Wang, and Xusheng Li. Amplifier predistortion method based on support vector machine. In *2005 Asia-Pacific Microwave Conference Proceedings*. IEEE, 2005. doi: 10.1109/apmc.2005.1607109.

Peng Chen and Thomas J. Brazil. Classifying load-pull contours of a broadband high-efficiency power amplifier using a support vector machine. In *2017 Integrated Nonlinear Microwave and Millimetre-wave Circuits Workshop (INMMiC)*. IEEE, apr 2017. doi: 10.1109/inmmic.2017.7927305.

Peng Chen and Thomas J. Brazil. Optimization-oriented method for broadband doherty power amplifier designs using support vector regression. In *2018 IEEE MTT-S International*

Microwave Workshop Series on 5G Hardware and System Technologies (IMWS-5G). IEEE, aug 2018. doi: 10.1109/imws-5G.2018.8484335.

Peng Chen, Brian M. Merrick, and Thomas J. Brazil. Support vector regression for harmonic optimization in continuous class-F power amplifier design. In *2015 Integrated Nonlinear Microwave and Millimetre-wave Circuits Workshop (INMMiC)*. IEEE, oct 2015. doi: 10.1109/inmmic.2015.7330376.

D.L. Chester. Why two hidden layers are better than one. *Proc. IJCNN, Washington, D.C.*, 1: 265–268, 1990. URL https://ci.nii.ac.jp/naid/10000055979/en/.

D. Comaniciu and P. Meer. Mean shift: a robust approach toward feature space analysis. *IEEE Transactions on Pattern Analysis and Machine Intelligence*, 24 (5): 603–619, may 2002. doi: 10.1109/34.1000236.

G. Cybenko. Approximation by superpositions of a sigmoidal function. *Mathematics of Control, Signals, and Systems*, 2 (4): 303–314, dec 1989. doi: 10.1007/bf02551274.

Ramzi Darraji, Pedram Mousavi, and Fadhel M. Ghannouchi. Doherty goes digital: Digitally enhanced doherty power amplifiers. *IEEE Microwave Magazine*, 17 (8): 41–51, aug 2016. doi: 10.1109/mmm.2016.2561478.

Sijmen de Jong. SIMPLS: An alternative approach to partial least squares regression. *Chemometrics and Intelligent Laboratory Systems*, 18 (3): 251–263, Mar. 1993. doi: 10.1016/0169-7439(93)85002-x.

D.R. Morgan, Z. Ma et al. A generalized memory polynomial model for digital predistortion of RF power amplifiers. *IEEE Trans. on Signal Processing*, 54 (10): 3852–3860, Oct. 2006. doi: 10.1109/TSP.2006.879264.

Toshiyuki Eda, Takanori Ito, Hiromitsu Ohmori, and Akira Sano. Adaptive compensation of nonlinearity in high power amplifier by support vector machine. *IFAC Proceedings Volumes*, 34 (14): 243–248, aug 2001. doi: 10.1016/s1474-6670(17)41629-6.

Roger Fletcher. *Practical Methods of Optimization*. John Wiley & Sons, New York, NY, USA, 2nd edition, 1987.

S. Fragiacomo, C. Matrakidis, and J.J. O'Reilly. Multicarrier transmission peak-to-average power reduction using simple block code. *Electronics Letters*, 34 (10): 953, 1998. doi: 10.1049/el:19980699.

Stuart Geman, Elie Bienenstock, and René Doursat. Neural networks and the bias/variance dilemma. *Neural Computation*, 4 (1): 1–58, jan 1992. doi: 10.1162/neco.1992.4.1.1.

Th. Giannopoulos and V. Paliouras. A novel technique for low-power D/A conversion based on PAPR reduction. In *2006 IEEE International Symposium on Circuits and Systems*. IEEE. doi: 10.1109/iscas.2006.1693754.

Pere L. Gilabert. *Multi lookup table digital predistortion for RF power amplifier linearization*. PhD thesis, Department of Signal Theory and Communications, Universitat Politècnica de Catalunya, 2007.

Pere L. Gilabert and Gabriel Montoro. 3-D distributed memory polynomial behavioral model for concurrent dual-band envelope tracking power amplifier linearization. *IEEE Transactions on Microwave Theory and Techniques*, 63 (2): 638–648, Feb. 2015. doi: 10.1109/tmtt.2014.2387825.

Pere L. Gilabert, Gabriel Montoro, David López,Nikolaos Bartzoudis, Eduard Bertran, Miquel Payaro, and Alain Hourtane. Order reduction of wideband digital predistorters using principal component analysis. In *Microwave Symposium Digest (IMS), 2013 IEEE MTT-S International*, pages 1–4, 2013a. doi: 10.1109/MWSYM.2013.6697687.

Pere L. Gilabert, Gabriel Montoro, David López, and Jose A. García. 3D Digital predistortion for dual-band envelope tracking power amplifiers. In *Microwave Conference Proceedings (APMC), 2013 Asia-Pacific*, pages 734–736, 2013b. doi: 10.1109/APMC.2013.6694913.

P.L. Gilabert, A. Cesari, G. Montoro, E. Bertran, and J.M. Dilhac. Multi look-up table FPGA implementation of a digital adaptive predistorter for linearizing RF power amplifiers with memory effects. *IEEE Trans. on Microwave Theory and Techniques*, 2007.

Lei Guan and Anding Zhu. Optimized low-complexity implementation of least squares based model extraction for digital predistortion of RF power amplifiers. *IEEE Trans. Microw. Theory Techn.*, 60 (3): 594–603, Mar. 2012. ISSN 0018-9480. doi: 10.1109/TMTT.2011.2182656.

Yunchuan Guo, Yuehang Xu, Lei Wang, and Ruimin Xu. A support vector machine method for electrothermal modeling of power FETs. In *2007 International Symposium on Microwave, Antenna, Propagation and EMC Technologies for Wireless Communications*. IEEE, aug 2007. doi: 10.1109/mape.2007.4393537.

Isabelle Guyon and André Elisseeff. An introduction to variable and feature selection. *J. Mach. Learn. Res.*, 3: 1157–1182, March 2003. ISSN 1532-4435. URL http://dl.acm.org/citation.cfm?id=944919.944968.

M. T. Hagan, H. B. Demuth, and M. H. Beale. *Neural Network Design*. PWS Publishing, Boston, 1996.

Seung Hee Han and Jae Hong Lee. Modulation, coding and signal processing for wireless communications – An overview of peak-to-average power ratio reduction techniques for multicarrier transmission. *IEEE Wireless Communications*, 12 (2): 56–65, apr 2005. doi: 10.1109/mwc.2005.1421929.

Katharina Hausmair, Per N. Landin, Ulf Gustavsson, Christian Fager, and Thomas Eriksson. Digital predistortion for multi-antenna transmitters affected by antenna crosstalk. *IEEE Transactions on Microwave Theory and Techniques*, 66 (3): 1524–1535, mar 2018. doi: 10.1109/tmtt.2017.2748948.

S.S. Haykin. *Neural Networks and Learning Machines*. Neural networks and learning machines. Prentice Hall, 2009. ISBN 9780131471399.

G.R. Hill, M. Faulkner, and J. Singh. Reducing the peak-to-average power ratio in OFDM by cyclically shifting partial transmit sequences. *Electronics Letters*, 36 (6): 560, 2000. doi: 10.1049/el:20000366.

Arthur E. Hoerl and Robert W. Kennard. Ridge regression: Biased estimation for nonorthogonal problems. *Technometrics*, 12 (1): 55–67, feb 1970. doi: 10.1080/00401706.1970.10488634.

D.R. Hush and B.G. Horne. Progress in supervised neural networks. *IEEE Signal Processing Magazine*, 10 (1): 8–39, jan 1993. doi: 10.1109/79.180705.

M. Isaksson, D. Wisell, and D. Ronnow. Nonlinear behavioral modeling of power amplifiers using radial-basis function neural networks. In *IEEE MTT-S International Microwave Symposium Digest*, 2005. IEEE. doi: 10.1109/mwsym.2005.1517128.

M. Isaksson, D. Wisell, and D. Ronnow. Wide-band dynamic modeling of power amplifiers using radial-basis function neural networks. *IEEE Transactions on Microwave Theory and Techniques*, 53 (11): 3422–3428, nov 2005. doi: 10.1109/tmtt.2005.855742.

J. Reina-Tosina, M. Allegue et al. Behavioral modeling and predistortion of power amplifiers under sparsity hypothesis. *IEEE Trans. on Microw. Theory and Tech.*, 63 (2): 745–753, Feb. 2015. ISSN 0018-9480. doi: 10.1109/TMTT.2014.2387852.

Younes Jabrane, Victor P. Gil Jiménez, Ana García Armada, Brahim Ait Es Said, and Abdellah Ait Ouahman. Reduction of power envelope fluctuations in OFDM signals by using neural networks. *IEEE Communications Letters*, 14 (7): 599–601, jul 2010. doi: 10.1109/lcomm.2010.07.100385.

P. Jaraut, M. Rawat, and F. M. Ghannouchi. Harmonically related concurrent tri-band behavioral modeling and digital predistortion. *IEEE Trans. on Circ. and Sys. II: Express Briefs*, pages 1–5, Oct. 2018. ISSN 1549-7747. doi: 10.1109/TCSII.2018.2873251.

Praveen Jaraut, Meenakshi Rawat, and Fadhel M. Ghannouchi. Composite neural network digital predistortion model for joint mitigation of crosstalk, I/Q imbalance, nonlinearity in MIMO transmitters. *IEEE Transactions on Microwave Theory and Techniques*, pages 1–10, 2018. doi: 10.1109/tmtt.2018.2869602.

Tao Jiang and Yiyan Wu. An overview: Peak-to-average power ratio reduction techniques for OFDM signals. *IEEE Transactions on Broadcasting*, 54 (2): 257–268, jun 2008. doi: 10.1109/tbc.2008.915770.

Tao Jiang and Guangxi Zhu. OFDM peak-to-average power ratio reduction by complement block coding scheme and its modified version. In *IEEE 60th Vehicular Technology Conference, 2004. VTC2004-Fall. 2004*. IEEE, 2004. doi: 10.1109/vetecf.2004.1400043.

Tao Jiang and Guangxi Zhu. Complement block coding for reduction in peak-to-average power ratio of OFDM signals. *IEEE Communications Magazine*, 43 (9): S17–S22, sep 2005. doi: 10.1109/mcom.2005.1509967.

Víctor P. Gil Jiménez, Younes Jabrane, Ana García Armada, Brahim Ait Es Said, and Abdellah Ait Ouahman. Reduction of the envelope fluctuations of multi-carrier modulations using adaptive neural fuzzy inference systems. *IEEE Transactions on Communications*, 59 (1): 19–25, jan 2011. doi: 10.1109/tcomm.2010.102910.100079.

A. Jovic, K. Brkic, and N. Bogunovic. A review of feature selection methods with applications. In *2015 38th International Convention on Information and Communication Technology, Electronics and Microelectronics (MIPRO)*. IEEE, may 2015. doi: 10.1109/mipro.2015.7160458.

Patrick Jueschke and Georg Fischer. Machine learning using neural networks in digital signal processing for RF transceivers. In *2017 IEEE AFRICON*. IEEE, sep 2017. doi: 10.1109/afrcon.2017.8095513.

O. Karray and C. De Silva. Soft computing and intelligent systems design, theory, tools and applications. *IEEE Transactions on Neural Networks*, 17 (3): 825–825, may 2006. doi: 10.1109/tnn.2006.875966.

Sukhraj Kaur and Gurpreet Singh Saini. Review paper on PAPR reduction techniques in OFDM system. *Indian Journal of Science and Technology*, 9 (48), jan 2017. doi: 10.17485/ijst/2016/v9i48/106893.

Shehroz S. Khan and Amir Ahmad. Cluster center initialization algorithm for k-means clustering. *Pattern Recognition Letters*, 25 (11): 1293–1302, aug 2004. doi: 10.1016/j.patrec.2004.04.007.

Zain Ahmed Khan, Efrain Zenteno, Peter Handel, and Magnus Isaksson. Digital predistortion for joint mitigation of i/q imbalance and MIMO power amplifier distortion. *IEEE Transactions on Microwave Theory and Techniques*, 65 (1): 322–333, jan 2017. doi: 10.1109/tmtt.2016.2614933.

J. Kim and K. Konstantinou. Digital Predistortion of Wideband Signals Based on Power Amplifier Model with Memory. In *Electronics Letters*, volume 37, pages 1417–1418, Nov. 2001. doi: 10.1049/el:20010940.

Wan-Jong Kim, Kyoung-Joon Cho, Shawn P. Stapleton, and Jong-Heon Kim. Doherty feed-forward amplifier performance using a novel crest factor reduction technique. *IEEE Microwave and Wireless Components Letters*, 17 (1): 82–84, jan 2007. doi: 10.1109/lmwc.2006.887287.

Ron Kohavi. A study of cross-validation and bootstrap for accuracy estimation and model selection. pages 1137–1143. Morgan Kaufmann, 1995.

Slawomir Koziel and John W. Bandler. Support-vector-regression-based output space-mapping for microwave device modeling. In *2008 IEEE MTT-S International Microwave Symposium Digest*. IEEE, jun 2008. doi: 10.1109/mwsym.2008.4633241.

B.S. Krongold and D.L. Jones. PAR reduction in OFDM via active constellation extension. *IEEE Transactions on Broadcasting*, 49 (3): 258–268, sep 2003. doi: 10.1109/tbc.2003.817088.

Boyang Li, Qiangwei Wang, and Jinglu Hu. Feature subset selection: a correlation-based SVM filter approach. *IEEJ Transactions on Electrical and Electronic Engineering*, 6 (2): 173–179, jan 2011. doi: 10.1002/tee.20641.

T. Liu, S. Boumaiza, and F. M. Ghannouchi. Dynamic Behavioral Modeling of 3G Power Amplifiers Using Real-Valued Time-Delay Neural Networks. *IEEE Trans. on Microwave Theory and Techniques*, 52 (3): 1025–1033, March 2004. doi: 10.1109/TMTT.2004.823583.

David López, Pere L. Gilabert, Gabriel Montoro, and Nikolaos Bartzoudis. Peak cancellation and digital predistortion of high-order QAM wideband signals for next generation wireless backhaul equipment. In *2014 International Workshop on Integrated Nonlinear Microwave and Millimetre-wave Circuits (INMMiC)*. IEEE, apr 2014. doi: 10.1109/inmmic.2014.6815111.

David López-Bueno, Quynh Anh Pham, Gabriel Montoro, and Pere L. Gilabert. Independent digital predistortion parameters estimation using adaptive principal component analysis. *IEEE Transactions on Microwave Theory and Techniques*, 66 (12): 5771–5779, dec 2018. doi: 10.1109/tmtt.2018.2870420.

S.G. Mallat and Zhifeng Zhang. Matching pursuits with time-frequency dictionaries. *IEEE Transactions on Signal Processing*, 41 (12): 3397–3415, 1993. doi: 10.1109/78.258082.

Nasreddine Mallouki, Bechir Nsiri, Mohammad Ghanbarisabagh, Walid Hakimi, and Mahmoud Ammar. Improvement of downlink LTE system performances using nonlinear equalization methods based on SVM and wiener–hammerstein. *Wireless Networks*, 23 (8): 2447–2454, may 2016. doi: 10.1007/s11276-016-1290-3.

Lin Maoliu and Hua Xiaojie. Research on SVM-based large signal behavior model method for RF power amplifier. In *2008 Conference on Precision Electromagnetic Measurements Digest*. IEEE, jun 2008. doi: 10.1109/cpem.2008.4574837.

A. Marcano-Cedeno, J. Quintanilla-Dominguez, M. G. Cortina-Januchs, and D. Andina. Feature selection using sequential forward selection and classification applying artificial metaplasticity neural network. In *IECON 2010 – 36th Annual Conference on IEEE Industrial Electronics Society*. IEEE, nov 2010. doi: 10.1109/iecon.2010.5675075.

J. Mercer. Functions of positive and negative type, and their connection with the theory of integral equations. *Philosophical Transactions of the Royal Society, London*, 209: 415–446, 1909.

Keiichi Mizutani, Masaya Ohta, Yasuo Ueda, and Katsumi Yamashita. A PAPR reduction of OFDM signal using neural networks with tone injection scheme. In *2007 6th International Conference on Information, Communications & Signal Processing*. IEEE, 2007. doi: 10.1109/icics.2007.4449855.

Farouk Mkadem and Slim Boumaiza. Physically inspired neural network model for RF power amplifier behavioral modeling and digital predistortion. *IEEE Transactions on Microwave Theory and Techniques*, 59 (4): 913–923, apr 2011. doi: 10.1109/tmtt.2010.2098041.

Albert Molina, Kannan Rajamani, and Kamran Azadet. Concurrent dual-band digital predistortion using 2-D lookup tables with bilinear interpolation and extrapolation: Direct least squares coefficient adaptation. *IEEE Transactions on Microwave Theory and Techniques*, 65 (4): 1381–1393, Apr. 2017. doi: 10.1109/tmtt.2016.2634001.

G. Montoro, P.L. Gilabert, E. Bertran, A. Cesari, and D.D. Silveira. A new digital predictive predistorter for behavioral power amplifier linearization. *IEEE Microwave and Wireless Components Letters*, 17 (6): 448–450, June 2007. doi: 10.1109/LMWC.2007.897797.

Nizar Mrabet, Imaduddin Mohammad, Farouk Mkadem, Chiheb Rebai, and Slim Boumaiza. Optimized hardware for polynomial digital predistortion system implementation. In *2012 IEEE Topical Conf. on Power Amplifiers for Wireless and Radio Appl. (PAWR)*, pages 83–84. IEEE, Jan. 2012. doi: 10.1109/pawr.2012.6174914.

S.H. Müller and J.B. Huber. OFDM with reduced peak-to-average power ratio by optimum combination of partial transmit sequences. *Electronics Letters*, 33 (5): 368, 1997. doi: 10.1049/el:19970266.

S.D. Muruganathan and A.B. Sesay. A QRD-RLS-based predistortion scheme for high-power amplifier linearization. *IEEE Trans. Circuits Syst. II, Express Briefs*, 53 (10): 1108–1112, Oct. 2006. doi: 10.1109/tcsii.2006.882182.

Tu Nguyen, Sofien Mhatli, Elias Giacoumidis, Ludo Van Compernolle, Marc Wuilpart, and Patrice Megret. Fiber nonlinearity equalizer based on support vector classification for coherent optical OFDM. *IEEE Photonics Journal*, 8 (2): 1–9, apr 2016. doi: 10.1109/jphot.2016.2528886.

M. Ohta and K. Yamashita. A chaotic neural network for reducing the peak-to-average power ratio of multicarrier modulation. In *Proceedings of the International Joint Conference on Neural Networks*, 2003. IEEE. doi: 10.1109/ijcnn.2003.1223803.

M. Ohta, H. Yamada, and K. Yamashita. BER performance improvement of biased PCC-OFDM with neural phase rotator by suppressing ICI. In *The 2006 IEEE International Joint Conference on Neural Network Proceedings*. IEEE, 2006. doi: 10.1109/ijcnn.2006.246645.

P. L. Gilabert, G. Montoro, et al. Comparison of model order reduction techniques for digital predistortion of power amplifiers. In *46th European Microw. Conf. (EuMC)*, pages 182–185, Oct. 2016. doi: 10.1109/EuMC.2016.7824308.

Jun Peng, Songbai He, Bingwen Wang, Zhijiang Dai, and Jingzhou Pang. Digital predistortion for power amplifier based on sparse bayesian learning. *IEEE Trans. on Circ. and Sys. II: Express Briefs*, 63 (9): 828–832, Sep. 2016. doi: 10.1109/tcsii.2016.2534718.

Raymond Pengelly, Christian Fager, and Mustafa Ozen. Doherty's legacy: A history of the doherty power amplifier from 1936 to the present day. *IEEE Microwave Magazine*, 17 (2): 41–58, feb 2016. doi: 10.1109/mmm.2015.2498081.

Quynh Anh Pham, David López-Bueno, Gabriel Montoro, and Pere L. Gilabert. Adaptive principal component analysis for online reduced order parameter extraction in PA behavioral modeling and DPD linearization. In *2018 IEEE MTT-S Int. Microw. Symp. (IMS)*, pages 160–163, Jun. 2018a.

Quynh Anh Pham, David López-Bueno, Teng Wang, Gabriel Montoro, and Pere L. Gilabert. Multi-dimensional LUT-based digital predistorter for concurrent dual-band envelope tracking power amplifier linearization. In *Proc. 2018 IEEE Topical Conf. on RF/Microw. Power Amplifiers for Radio and Wireless Appl. (PAWR)*, pages 47–50, Jan. 2018b.

Quynh Anh Pham, David López-Bueno, Teng Wang, Gabriel Montoro, and Pere L. Gilabert. Partial least squares identification of multi look-up table digital predistorters for concurrent dual-band envelope tracking power amplifiers. *IEEE Transactions on Microwave Theory and Techniques*, 66 (12): 5143–5150, dec 2018c. doi: 10.1109/tmtt.2018.2857819.

Quynh Anh Pham, David López-Bueno, Gabriel Montoro, and Pere L. Gilabert. Dynamic selection and update of digital predistorter coefficients for power amplifier linearization. In *Proc. 2019 IEEE Topical Conf. on RF/Microw. Power Amplifiers for Radio and Wireless Appl. (PAWR)*, pages 1–4, Jan. 2019.

D.C. Plaut, S.J. Nowlan, and G.E. Hinton. Experiments on learning by back propagation. In *Technical Report CMU-CS-86-126*. Carnegie-Mellon University, 1986.

Zoya Popovic. Amping up the PA for 5G: Efficient GaN power amplifiers with dynamic supplies. *IEEE Microwave Magazine*, 18 (3): 137–149, may 2017. doi: 10.1109/mmm.2017.2664018.

Zoya Popovic and Jose A. García. Microwave class-E power amplifiers: A brief review of essential concepts in high-frequency class-E PAs and related circuits. *IEEE Microwave Magazine*, 19 (5): 54–66, jul 2018. doi: 10.1109/mmm.2018.2822202.

J.R. Quinlan. Introduction of decision trees. *Machine Learning*, 1 (1): 81–106, 1986. doi: 10.1023/a:1022643204877.

M. Rawat, K. Rawat, and F.M. Ghannouchi. Adaptive digital predistortion of wireless power amplifiers/transmitters using dynamic real-valued focused time-delay line neural networks. *IEEE Transactions on Microwave Theory and Techniques*, 58 (1): 95–104, jan 2010. doi: 10.1109/tmtt.2009.2036334.

Patrick Roblin, Christophe Quindroit, Naveen Naraharisetti, Shahin Gheitanchi, and Mike Fitton. Concurrent linearization: The state of the art for modeling and linearization of multiband power amplifiers. *IEEE Microwave Magazine*, 14 (7): 75–91, nov 2013. doi: 10.1109/mmm.2013.2281297.

F. Rosenblatt. The perceptron: A probabilistic model for information storage and organization in the brain. *Psychological Review*, 65 (6): 386–408, 1958. doi: 10.1037/h0042519.

David E. Rumelhart and James L. McClelland. *Parallel Distributed Processing: Explorations in the Microstructure of Cognition (2 Volume Set) (Vol.1)*. MIT Press, 1986. ISBN 0262181231.

A. Saul. Peak reduction for OFDM by shaping the clipping noise. In *IEEE 60th Vehicular Technology Conference, 2004. VTC2004-Fall. 2004*. IEEE, 2004. doi: 10.1109/vetecf.2004.1400042.

D. Scheurs, M. O'Droma, A.A. Goacher, and M. Gadringer, editors. *RF Power Amplifier Behavioural Modeling*. Cambridge University Press, 2009.

D.J. Sebald and J.A. Bucklew. Support vector machine techniques for nonlinear equalization. *IEEE Transactions on Signal Processing*, 48 (11): 3217–3226, 2000. doi: 10.1109/78.875477.

Mansoor Shafi, Andreas F. Molisch, Peter J. Smith, Thomas Haustein, Peiying Zhu, Prasan De Silva, Fredrik Tufvesson, Anass Benjebbour, and Gerhard Wunder. 5G: A tutorial overview of standards, trials, challenges, deployment, and practice. *IEEE Journal on Selected Areas in Communications*, 35 (6): 1201–1221, jun 2017. doi: 10.1109/jsac.2017.2692307.

Alex J. Smola and Bernhard Schölkopf. A tutorial on support vector regression. *Statistics and Computing*, 14 (3): 199–222, aug 2004. doi: 10.1023/b:stco.0000035301.49549.88.

Insoo Sohn. A low complexity PAPR reduction scheme for OFDM systems via neural networks. *IEEE Communications Letters*, 18 (2): 225–228, feb 2014. doi: 10.1109/lcomm.2013.123113.131888.

Insoo Sohn and Sung Chul Kim. Neural network based simplified clipping and filtering technique for PAPR reduction of OFDM signals. *IEEE Communications Letters*, 19 (8): 1438–1441, aug 2015. doi: 10.1109/lcomm.2015.2441065.

Insoo Sohn and Jaeho Shin. PAPR reduction of OFDM signals using radial basis function neural. In *2006 International Conference on Communication Technology*. IEEE, nov 2006. doi: 10.1109/icct.2006.341659.

M. Stone. Cross-validation:a review. *Series Statistics*, 9 (1): 127–139, jan 1978. doi: 10.1080/02331887808801414.

Padmanabhan Madampu Suryasarman and Andreas Springer. A comparative analysis of adaptive digital predistortion algorithms for multiple antenna transmitters. *IEEE Trans. on Circ. and Sys. I*, 62 (5): 1412–1420, May 2015. doi: 10.1109/tcsi.2015.2403034.

J. Tellado. *Peak to Average Power Ratio Reduction for Multicarrier Modulation*. PhD thesis, University of Stanford, 1999.

Alan J. Thomas,, Simon D. Walters, Saeed Malekshahi Gheytassi, Robert E. Morgan, and Miltos Petridis. On the optimal node ratio between hidden layers: A probabilistic study. *International Journal of Machine Learning and Computing*, 6 (5): 241–247, oct 2016. doi: 10.18178/ijmlc.2016.6.5.605.

Robert Tibshirani. Regression shrinkage and selection via the LASSO. *Journal of the Royal Statistical Society, Series B*, 58: 267–288, 1994.

Robert Tibshirani, Iain Johnstone, Trevor Hastie, and Bradley Efron. Least angle regression. *The Annals of Statistics*, 32 (2): 407–499, apr 2004. doi: 10.1214/009053604000000067.

A N Tikhonov and V Y Arsensin. *Solution of ill-posed problems*. V H Winston, Washington DC, 1977.

Lloyd N. Trefethen and David Bau. *Numerical Linear Algebra*. SIAM, 1997. ISBN 0898713617.

O. Vaananen, J. Vankka, and K. Halonen. Simple algorithm for peak windowing and its application in GSM, EDGE and WCDMA systems. *IEE Proceedings – Communications*, 152 (3): 357, 2005. doi: 10.1049/ip-com:20059014.

Vladimir N. Vapnik. *The Nature of Statistical Learning Theory*. Springer-Verlag, Berlin, Heidelberg, 1995. ISBN 0-387-94559-8.

Vladimir N. Vapnik. *Statistical Learning Theory*. Wiley-Interscience, 1998.

A. Wang, J. An, and Z. He. PAPR reduction for MC-CDMA system based on ICSA and hopfield neural network. In *2008 IEEE International Conference on Communications*. IEEE, 2008. doi: 10.1109/icc.2008.951.

Dongming Wang, Mohsin Aziz, Mohamed Helaoui, and Fadhel M. Ghannouchi. Augmented real-valued time-delay neural network for compensation of distortions and impairments in wireless transmitters. *IEEE Transactions on Neural Networks and Learning Systems*, 30 (1): 242–254, jan 2019. doi: 10.1109/tnnls.2018.2838039.

Haiming Wang. PAPR reduction for OFDM system with a class of HNN. In *2006 International Symposium on Communications and Information Technologies*. IEEE, oct 2006. doi: 10.1109/iscit.2006.339995.

Luqing Wang and C. Tellambura. A simplified clipping and filtering technique for PAR reduction in OFDM systems. *IEEE Signal Processing Letters*, 12 (6): 453–456, jun 2005. doi: 10.1109/lsp.2005.847886.

Zhancang Wang. Demystifying envelope tracking: Use for high-efficiency power amplifiers for 4G and beyond. *IEEE Microwave Magazine*, 16 (3): 106–129, apr 2015. doi: 10.1109/mmm.2014.2385351.

Gavin T. Watkins and Konstantinos Mimis. How not to rely on Moore's law alone: Low-complexity envelope-tracking amplifiers. *IEEE Microwave Magazine*, 19 (4): 84–94, jun 2018. doi: 10.1109/mmm.2018.2813840.

D. Wisell, J. Jalden, and P. Handel. Behavioral power amplifier modeling using the LASSO. In *2008 IEEE Inst. and Meas. Tech. Conf.*, pages 1864–1867, May 2008. doi: 10.1109/IMTC.2008.4547349.

Yiming Yang and Jan O. Pedersen. A comparative study on feature selection in text categorization. In *Proceedings of the Fourteenth International Conference on Machine Learning*, ICML '97, pages 412–420, San Francisco, CA, USA, 1997. Morgan Kaufmann Publishers Inc. ISBN 1-55860-486-3. URL http://dl.acm.org/citation.cfm?id=645526.657137.

Lei Yu and Huan Liu. Feature selection for high-dimensional data: A fast correlation-based filter solution. In T. Fawcett and N. Mishra, editors, *Proceedings, Twentieth International Conference on Machine Learning*, volume 2, pages 856–863, 2003. ISBN 1577351894.

Rafik Zayani, Ridha Bouallegue, and Daniel Roviras. Crossover neural network predistorter for the compensation of crosstalk and nonlinearity in MIMO OFDM systems. In *21st Annual IEEE International Symposium on Personal, Indoor and Mobile Radio Communications*. IEEE, sep 2010. doi: 10.1109/pimrc.2010.5671770.

Jinzhu Zhou and Jin Huang. Intelligent tuning for microwave filters based on multi-kernel machine learning model. In *2013 5th IEEE International Symposium on Microwave, Antenna, Propagation and EMC Technologies for Wireless Communications*. IEEE, oct 2013. doi: 10.1109/mape.2013.6689881.

Anding Zhu, Jos C. Pedro, and Thomas J. Brazil. Dynamic deviation reduction-based volterra behavioral modeling of RF power amplifiers. *IEEE Transactions on Microwave Theory and Techniques*, 54 (12): 4323–4332, dec 2006. doi: 10.1109/tmtt.2006.883243.

Hui Zou and Trevor Hastie. Regularization and variable selection via the elastic net. *Journal of the Royal Statistical Society: Series B (Statistical Methodology)*, 67 (2): 301–320, apr 2005. doi: 10.1111/j.1467-9868.2005.00503.x.

18

Neural Networks for Full-Duplex Radios: Self-Interference Cancellation

Alexios Balatsoukas-Stimming

Department of Electrical Engineering, Eindhoven University of Technology, Eindhoven, The Netherlands

Bi-directional wireless communications are usually achieved by separating the uplink and downlink signals using either time-division duplexing (TDD) or frequency-division duplexing (FDD). In-band full-duplex (FD) is a promising method to increase the spectral efficiency of current communications systems by transmitting and receiving data simultaneously in the same frequency band Jain et al. (2011), Duarte et al. (2012), Bharadia et al. (2013). A fundamental challenge in FD communications is that the transmitter of a node induces a self-interference (SI) signal at the receiver of the same node. This SI signal is several orders of magnitude stronger than the signal that the node is trying to receive. Thus, in order for an FD node to operate correctly, the SI signal needs to be canceled, ideally to the level of the receiver noise floor.

A combination of SI cancellation in both the radio frequency (RF) domain and the digital domain is usually required in order to cancel the SI signal to the level of the receiver noise floor. RF cancellation can be achieved either through physical isolation between the transmitter and the receiver (*passive RF cancellation*) or through the injection of a cancellation signal (*active RF cancellation*), and it is necessary in order to avoid saturating the analog front-end of the receiver. Passive RF cancellation can be obtained through several passive devices and techniques, such as circulators, directional antennas, beamforming, polarization, or shielding. Active RF cancellation is commonly implemented by coupling into the transmitted RF signal; adding an appropriate time delay, phase rotation, and attenuation; and adding the resulting SI cancellation signal to the received SI signal Jain et al. (2011), Bharadia et al. (2013). Alternatively, a second transmitter chain can be used to generate the SI cancellation signal Duarte et al. (2012).

Perfect RF cancellation is challenging and costly to achieve, meaning that a residual SI signal is usually still present at the receiver after the RF cancellation stage. In principle, this residual SI signal can be easily canceled in the digital domain, since it is caused by a known transmitted baseband signal. Unfortunately, in practice this is not the case, as several transceiver nonlinearities distort the SI signal. Some examples of nonlinearities include baseband nonlinearities (e.g. digital-to-analog converter [DAC] and analog-to-digital converter [ADC]) Balatsoukas-Stimming et al. (2015), in-phase/quadrature (I/Q) imbalance Balatsoukas-Stimming et al. (2015), Korpi et al. (2014), phase-noise Sahai et al. (2013), Syrjala et al. (2014), and power amplifier (PA) nonlinearities Balatsoukas-Stimming et al. (2015), Korpi et al. (2014), Anttila et al. (2014), Korpi et al. (2017). Complicated nonlinear cancellation methods, which are usually based on polynomial expansions, are required in order to fully cancel the SI to the level of the receiver noise floor. A commonly used nonlinear

Machine Learning for Future Wireless Communications, First Edition. Edited by Fa-Long Luo.
© 2020 John Wiley & Sons Ltd. Published 2020 by John Wiley & Sons Ltd.

SI cancellation method employs a parallel Hammerstein model that incorporates both PA nonlinearities and IQ imbalance Korpi et al. (2017).

Polynomial models have been shown to work well in practice, but they generally have a high implementation complexity as the number of estimated parameters grows rapidly with the maximum considered nonlinearity order and because a large number of nonlinear basis functions have to be computed. Principal component analysis (PCA) is an effective complexity-reduction technique that can identify the most significant nonlinearity terms in a parallel Hammerstein model Korpi et al. (2017). However, with PCA-based methods, the transmitted digital baseband samples need to be multiplied with a transformation matrix to generate the SI cancellation signal, thus introducing additional complexity. Moreover, whenever the SI channel changes significantly, the high-complexity PCA operation needs to be re-run. Neural networks (NNs) have been widely used in the literature to model and compensate for nonlinear effects in communications systems (see e.g. Ibnkahla (2000), Naskas and Papananos (2004), Rawat et al. (2010), Mkadem and Boumaiza (2011) and references therein) and, recently, they have also been used for SI cancellation in FD radios Balatsoukas-Stimming (2018), Guo et al. (2018). Due to their powerful nonlinear modeling capabilities, NN-based solutions for digital SI cancellation have been shown to provide a good trade-off between computational complexity and SI cancellation performance in FD radios. This chapter covers technical aspects of digital SI cancellation in FD radios in a self-contained manner, using both conventional polynomial models and NNs.

The rest of this chapter is organized as follows. Section 18.1 describes how the nonlinear effects of various transceiver components can be modeled and derives a comprehensive nonlinear SI model. Section 18.2 describes various linear and nonlinear cancellation methods, with emphasis on a NN canceler. In this section, the computational complexity of each approach is also analyzed and compared. For further illustration and comparison, Section 18.3 provides experimental results using measured samples from a hardware testbed, which demonstrate that a simple NN-based nonlinear canceler can match the performance of a state-of-the-art polynomial model for nonlinear cancellation with a significantly lower computational complexity. Finally, Section 18.4 concludes this chapter by discussing a number of interesting future research directions.

18.1 Nonlinear Self-Interference Models

Each active component in the transceiver chain shown in Figure 18.1 is essentially a dynamic nonlinear system, which can be modeled in a variety of ways. The Volterra series is one of the most accurate models, but it has a very large number of parameters and is thus rarely used in practice. Instead, a simplification of the Volterra series, called a *parallel Hammerstein model*, is often used. In the parallel Hammerstein model, the input-output relation of a dynamic nonlinear system with input $x[n] \in \mathbb{C}$ and output $y[n] \in \mathbb{C}$ is modeled as:

$$y[n] = \sum_{p=1}^{P} \sum_{l=L_1}^{L_2} h_p[l] x[n-l] |x[n-l]|^{p-1}, \tag{18.1}$$

where P is the maximum considered nonlinearity order, L_1 and L_2 are the numbers of considered pre-cursor and post-cursor memory taps, respectively, and $h_p[l] \in \mathbb{C}$ are the model parameters. To simplify the notation, in this section it is assumed that $x[n]$ is already pre-shifted by L_1 samples, so that Eq. (18.1) can be equivalently rewritten using a single parameter L as:

$$y[n] = \sum_{p=1}^{P} \sum_{l=0}^{L-1} h_p[l] x[n-l] |x[n-l]|^{p-1}. \tag{18.2}$$

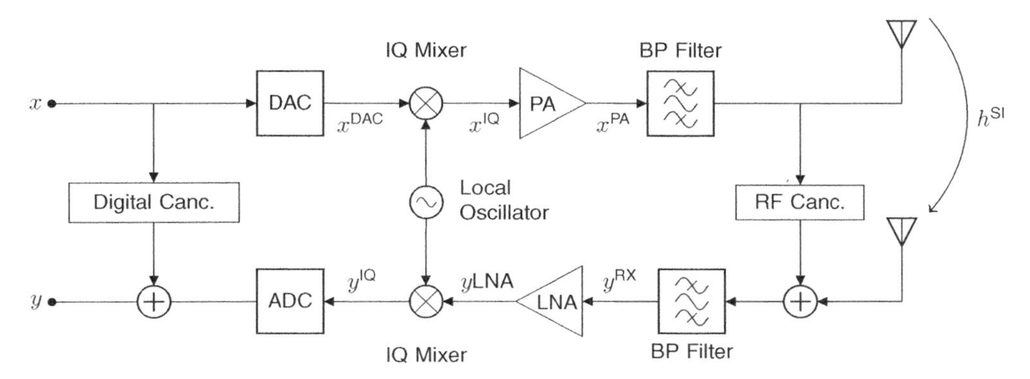

Figure 18.1 Model of a full-duplex transceiver with active radio frequency cancellation and active digital cancellation. A few components have been omitted for simplicity; a more detailed diagram can be found in Korpi et al. (2017).

18.1.1 Nonlinear Self-Interference Model

Let the complex-valued digital signal that enters the digital-to-analog converter (DAC) in Figure 18.1 at time instant n be denoted by $x[n]$. The transmitter uses two distinct DACs for the real and the imaginary parts of $x[n]$, respectively. As these are distinct components, they generally have different nonlinear characteristics. As such, each DAC has to be modeled separately using a parallel Hammerstein model, and the complex-valued output of the two DACs is given by:

$$x^{\mathrm{DAC}}[n] = \sum_{p=1}^{P^{\mathrm{DAC}}} \sum_{l=0}^{L^{\mathrm{DAC}}-1} (h_{p,\Re}^{\mathrm{DAC}}[l]\Re\{x[n-l]\}^p + jh_{p,\Im}^{\mathrm{DAC}}[l]\Im\{x[n-l]\}^p), \tag{18.3}$$

where $h_{p,\Re}^{\mathrm{DAC}}[l] \in \mathbb{R}$ and $h_{p,\Im}^{\mathrm{DAC}}[l] \in \mathbb{R}$ are the model parameters for the real and imaginary parts of $x[n]$, respectively. The IQ mixer introduces IQ imbalance and phase noise, which can be modeled as:

$$x^{\mathrm{IQ,TX}}[n] = (K_1^{\mathrm{TX}} x^{\mathrm{DAC}}[n] + K_2^{\mathrm{TX}} (x^{\mathrm{DAC}}[n])^*) e^{j\phi^{\mathrm{TX}}[n]}, \tag{18.4}$$

where $K_1^{\mathrm{TX}} \in \mathbb{C}$ and $K_2^{\mathrm{TX}} \in \mathbb{C}$ are parameters, and $\phi^{\mathrm{TX}}[n]$ is the baseband equivalent of the transmitter phase noise process at discrete time-instant n. The upconverted signal $x^{\mathrm{IQ,TX}}[n]$ is amplified by a power amplifier (PA), which introduces further nonlinearities. The even-powered nonlinearity terms are filtered out by the transmitter bandpass filter, so the PA nonlinearities can be modeled as:

$$x^{\mathrm{PA}}[n] = \sum_{\substack{p=1,\\ p \text{ odd}}}^{P^{\mathrm{PA}}} \sum_{l=0}^{L^{\mathrm{PA}}-1} h_p^{\mathrm{PA}}[l] x^{\mathrm{IQ,TX}}[n-l] |x^{\mathrm{IQ,TX}}[n-l]|^{p-l}. \tag{18.5}$$

The output signal of the PA is transmitted over the air and received at the receiving antenna through a linear SI channel, which can be modeled as:

$$y^{\mathrm{RX}}[n] = \sum_{l=0}^{L^{\mathrm{SI}}-1} h^{\mathrm{SI}}[l] x^{\mathrm{PA}}[n-l]. \tag{18.6}$$

Note that the effect of active RF cancellation is captured by the linear SI channel, in the sense that if active RF cancellation is present, the average power of $h^{\mathrm{SI}}[l]$ is significantly lower. The received signal is amplified using a low-noise amplifier (LNA), whose nonlinear effects can be modeled as:

$$y^{\text{LNA}}[n] = \sum_{\substack{p=1,\\ p\text{ odd}}}^{P^{\text{LNA}}} \sum_{l=0}^{L^{\text{LNA}}-1} h_p^{\text{LNA}}[l] y^{\text{RX}}[n-l] |y^{\text{RX}}[n-l]|^{p-l}. \tag{18.7}$$

After the LNA, the signal $y^{\text{LNA}}[n]$ is downconverted using an I/Q mixer, which introduces I/Q imbalance and phase noise:

$$y^{\text{IQ,RX}}[n] = (K_1^{\text{RX}} y^{\text{LNA}}[n] + K_2^{\text{RX}} (y^{\text{LNA}}[n])^*) e^{-j\phi^{\text{RX}}[n]}, \tag{18.8}$$

where $K_1^{\text{RX}} \in \mathbb{C}$ and $K_2^{\text{RX}} \in \mathbb{C}$ are parameters, and $\phi^{\text{RX}}[n]$ is the baseband equivalent of the receiver phase noise process at discrete time-instant n. Finally, $y^{\text{IQ,RX}}[n]$ passes through a potentially nonlinear analog-to-digital converter (ADC), which introduces a nonlinearity of the form:

$$y_{\text{SI}}[n] = \sum_{p=1}^{P^{\text{ADC}}} \sum_{l=0}^{L^{\text{ADC}}-1} (h_{p,\Re}^{\text{ADC}}[l] \Re\{y^{\text{IQ,RX}}[n-l]\}^p + j h_{p,\Im}^{\text{ADC}}[l] \Im\{y^{\text{IQ,RX}}[n-l]\}^p), \tag{18.9}$$

where $h_{p,\Re}^{\text{ADC}}[l] \in \mathbb{R}$ and $h_{p,\Im}^{\text{ADC}}[l] \in \mathbb{R}$ are the model parameters for the real and imaginary parts of $y^{\text{IQ,RX}}[n]$, respectively.

In these derivations, the thermal noise and the desired signal received from a remote node were omitted for simplicity. The actual received signal $y[n]$ at time-instant n is given by:

$$y[n] = d[n] + y_{\text{SI}}[n] + z[n], \tag{18.10}$$

where $d[n]$ is the desired signal, $y_{\text{SI}}[n]$ is given in Eq. (18.9), and $z[n] \sim \mathcal{CN}(0, \sigma^2)$ is complex-valued additive white Gaussian noise.

18.2 Digital Self-Interference Cancellation

The goal of digital SI cancellation is to produce an approximation of $y_{\text{SI}}[n]$, denoted by $\hat{y}_{\text{SI}}[n]$, based on $x[n]$ as accurately as possible using digital signal processing (DSP) techniques and to subtract $\hat{y}_{\text{SI}}[n]$ from $y[n]$ in order to cancel the SI. The achieved SI cancellation over a sample window of length N, measured in dB, is commonly calculated as:

$$C_{\text{dB}} = 10\log_{10}\left(\frac{1}{N}\sum_{n=0}^{N-1} \frac{|y[n]|^2}{|y[n] - \hat{y}_{\text{SI}}[n]|^2}\right). \tag{18.11}$$

The SI cancellation is usually calculated when no desired signal $d[n]$ is present. If all transceiver nonlinearities described in Section 18.1 are taken into account, then the overall system model is the composition of the seven functions given in Eqs. (18.3)–(18.9). Such a model, while potentially very accurate, has a prohibitively large number of parameters to be practically useful. As such, significantly simpler models that attempt to focus on the dominant nonlinearities are used in practice.

18.2.1 Linear Cancellation

The simplest form of digital cancellation ignores all nonlinear effects described in Eqs. (18.3)–(18.5) and Eqs. (18.7)–(18.9) and only considers the effect of the (linear) SI channel of length L^{SI} given in Eq. (18.6). Let $L = L^{\text{SI}}$ for simplicity. Then, the SI cancellation signal $\hat{y}_{\text{SI}}[n]$ is constructed as:

$$\hat{y}_{\text{SI}}[n] = \sum_{l=0}^{L-1} \hat{h}^{\text{SI}}[l]x[n-l], \tag{18.12}$$

where the SI channel $\hat{h}^{\text{SI}}[l]$ can be estimated on the basis of training samples using e.g. least-squares (LS) estimation.

18.2.2 Polynomial Nonlinear Cancellation

Linear cancellation alone is, in most scenarios, not powerful enough to cancel a sufficient portion of the SI signal. As such, polynomial models that consider a subset of the nonlinearities described in Eqs. (18.3)–(18.9) are often used in order to achieve better SI cancellation. It has been shown that, in most cases, the transmitter IQ imbalance and the PA nonlinearities, given in Eq. (18.4) and Eq. (18.5), respectively, dominate all remaining nonlinearities. This is true in particular when the transmitter and receiver chains use the same local oscillator signal for upconversion, as shown in Figure 18.1, so that $\phi^{\text{TX}}[n] = \phi^{\text{RX}}[n]$ and the effect of phase noise becomes negligible. As such, the SI cancellation signal $\hat{y}[n]$ can be constructed as Anttila et al. (2014), Korpi et al. (2017):

$$\hat{y}_{\text{SI}}[n] = \sum_{\substack{p=1,\\ p\text{ odd}}}^{P^{\text{PA}}} \sum_{l=0}^{L^{\text{PA}}+L^{\text{SI}}-1} \hat{h}_p[l](K_1^{\text{TX}}x[n-l] + K_2^{\text{TX}}x^*[n-l])|(K_1^{\text{TX}}x[n-l] + K_2^{\text{TX}}x^*[n-l])|^{p-l}. \tag{18.13}$$

where $\hat{h}_p[l] \in \mathbb{C}$ is the convolution of $\hat{h}_p^{\text{PA}}[l]$ and $\hat{h}^{\text{SI}}[l]$. Let $L = L^{\text{PA}} + L^{\text{SI}}$ and $P = P^{\text{PA}}$ for simplicity. Then, with some arithmetic manipulations $\hat{y}_{\text{SI}}[n]$ can be rewritten as Anttila et al. (2014), Korpi et al. (2017):

$$\hat{y}_{\text{SI}}[n] = \sum_{\substack{p=1,\\ p\text{ odd}}}^{P} \sum_{q=0}^{p} \sum_{m=0}^{L-1} \hat{h}_{p,q}[l]x[n-l]^q x^*[n-l]^{p-q}, \tag{18.14}$$

where $\hat{h}_{p,q}[l] \in \mathbb{C}$ captures the joint effect of $\hat{h}_p[l]$ and IQ imbalance parameters K_1^{TX} and K_2^{TX}. The parameters $\hat{h}_{p,q}[l]$ can be estimated based on training samples using e.g. LS estimation. The *basis functions* of the polynomial model in Eq. (18.14) are defined as:

$$BF_{p,q}(x[n]) = x[n]^q x^*[n]^{p-q}. \tag{18.15}$$

Using Eq. (18.15), the expression for $\hat{y}_{\text{SI}}[n]$ in Eq. (18.14) can be rewritten in a more compact form:

$$\hat{y}[n] = \sum_{\substack{p=1,\\ p\text{ odd}}}^{P} \sum_{q=0}^{p} \sum_{m=0}^{L-1} \hat{h}_{p,q}[l]BF_{p,q}(x[n-l]), \tag{18.16}$$

Linear cancellation is actually a special case of the previous model for $p = 1$ and $q = 1$.

18.2.3 Neural Network Nonlinear Cancellation

NN SI cancelers are an attractive alternative to polynomial-based SI cancelers, as they can extract the essential structure of the SI signal from training data, thus significantly reducing the complexity of the SI model Balatsoukas-Stimming (2018), Guo et al. (2018). The main challenge with NN cancelers is that the training process is inherently noisy due to the use of mini-batches for gradient estimation. Thus, it is difficult to achieve very accurate reconstruction of the SI signal, which is essential to achieve high levels of SI cancellation. One way to overcome this challenge is to use a NN to reconstruct only a particular part of the SI signal, while using conventional (e.g. linear) cancellation for the remaining part Balatsoukas-Stimming (2018). More specifically, the SI signal can be conceptually decomposed into a linear component and a nonlinear component:

$$y_{\mathrm{SI}}[n] = y_{\mathrm{linear}}[n] + y_{\mathrm{nonlinear}}[n]. \tag{18.17}$$

Then, the SI cancellation can be carried out in two steps. First, standard linear cancellation is used in order to reconstruct $\hat{y}_{\mathrm{linear}}[n]$ as:

$$\hat{y}_{\mathrm{linear}}[n] = \sum_{l=0}^{L^{\mathrm{SI}}-1} \hat{h}^{\mathrm{SI}}[l]x[n-l]. \tag{18.18}$$

The parameters $\hat{h}^{\mathrm{SI}}[l]$ are obtained using LS estimation while considering the (significantly weaker) signal $\hat{y}_{\mathrm{nonlinear}}[n]$ as noise. The linear SI cancellation signal is subtracted from the SI signal in order to obtain:

$$y_{\mathrm{nonlinear}}[n] \approx \hat{y}_{\mathrm{SI}}[n] - \hat{y}_{\mathrm{linear}}[n]. \tag{18.19}$$

The goal of the NN is to reconstruct each $y_{\mathrm{nonlinear}}[n]$ sample based on the subset of $x[n]$ that this $y_{\mathrm{nonlinear}}[n]$ sample depends on (see Eq. (18.9)). Since NNs generally operate on real numbers, all complex-valued baseband signals are split into their real and imaginary parts.

Due to the universal approximation theorem Hornik (1991), a feedforward NN with one hidden layer, as depicted in Figure 18.2, can be used in order to reconstruct the nonlinear SI signal Balatsoukas-Stimming (2018). The NN has $2L$ input nodes, which correspond to the real and imaginary parts of the L delayed versions of $x[n]$ in Eq. (18.9), and two output nodes, which correspond to the real and imaginary parts of the target $\hat{y}_{\mathrm{nonlinear}}[n]$ sample. The number of hidden nodes is denoted by N_h and is a parameter that can be chosen freely.

More specifically, let the vector $\mathbf{l}_i$ contain the $2L$ inputs to the NN:

$$\mathbf{l}_i = \left[\Re\{x[n]\} \;\; \Im\{x[n]\} \;\; \dots \;\; \Re\{x[n-L+1]\} \;\; \Im\{x[n-L+1]\}\right]^T. \tag{18.20}$$

Then, the outputs of the hidden layer neurons are given by:

$$\mathbf{l}_h = f_h(\mathbf{W}_h\mathbf{l}_i + \mathbf{b}_h), \tag{18.21}$$

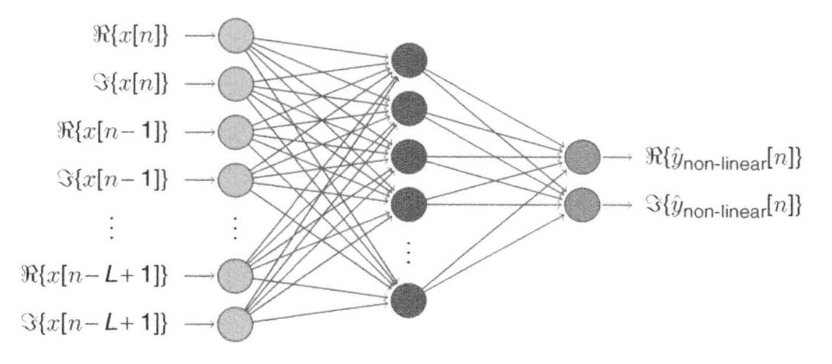

Figure 18.2 Example of a neural network with one hidden layer used for the reconstruction of the nonlinear self-interference cancellation signal $\hat{y}_{\mathrm{nonlinear}}[n]$.

where $\mathbf{W}_h$ is a $N_h \times 2L$ matrix containing the hidden layer weights, $\mathbf{b}_h$ is an $N_h \times 1$ vector containing the hidden layer biases, and f_h is the (vectorized) activation function used in the hidden layer. The outputs of the output layer neurons are in turn given by:

$$\mathbf{l}_o = f_o(\mathbf{W}_o \mathbf{l}_h + \mathbf{b}_o) \quad (18.22)$$
$$= f_o(\mathbf{W}_o f_h(\mathbf{W}_h \mathbf{l}_i + \mathbf{b}_h) + \mathbf{b}_o), \quad (18.23)$$

where $\mathbf{W}_o$ is a $2 \times N_h$ matrix containing the output layer weights, $\mathbf{b}_o$ is a 2×1 vector containing the output layer biases, and f_o is the activation function used in the output layer. As can be seen in Figure 18.2, for $\mathbf{l}_o$ we have:

$$\mathbf{l}_o = \left[\Re\{\hat{y}_{\text{nonlinear}}[n]\} \;\; \Im\{\hat{y}_{\text{nonlinear}}[n]\}\right]^T. \quad (18.24)$$

The goal of the NN is to minimize the following mean squared error between the expected NN output and the actual NN output:

$$\text{MSE} = \frac{1}{2N}\sum_{n=0}^{N-1}(\Re\{y_{\text{non-linear}}[n]\} - \Re\{\hat{y}_{\text{non-linear}}[n]\})^2$$
$$+ \frac{1}{2N}\sum_{n=0}^{N-1}(\Im\{y_{\text{non-linear}}[n]\} - \Im\{\hat{y}_{\text{non-linear}}[n]\})^2, \quad (18.25)$$

where N is the total number of training samples. In practice, the expected NN output can be easily obtained by transmitting a frame of known $x[n]$ samples using an experimental testbed and recording the corresponding SI at the receiver.

The MSE can be minimized by choosing appropriate values for $\mathbf{W}_h$, $\mathbf{b}_h$, $\mathbf{W}_o$, and $\mathbf{b}_o$. These values can be computed iteratively by starting from randomly initialized values, calculating the derivative of MSE with respect to each weight and each bias through backpropagation Rumelhart et al. (1986), and using a weight-adaptation algorithm (e.g. Adam Kingma and Ba (2015)).

18.2.4 Computational Complexity

In this section, the computational complexity of the polynomial canceler and the NN canceler is examined. The polynomial canceler operates on complex numbers, while the NN canceler operates on real numbers. Thus, in order to perform a fair comparison, the number of real additions and real multiplications that are required to perform the computations of each method are counted. Let $a, b \in \mathbb{C}$ and let $a_R = \Re\{a\}$ and $a_I = \Im\{a\}$ for simplicity. Then, complex addition can be written as:

$$a + b = (a_R + b_R) + j(a_I + b_I), \quad (18.26)$$

and complex multiplication can be written as:

$$ab = (a_R + ja_I)(b_R + jb_I) \quad (18.27)$$
$$= (a_R b_R - a_I b_I) + j((a_R + a_I)(b_R + b_I) - a_R b_R - a_I b_I). \quad (18.28)$$

Thus, one complex addition is equivalent to two real additions, while one complex multiplication is equivalent to three real multiplications and five real additions.

18.2.4.1 Linear Cancellation

The evaluation of the linear cancellation expression in Eq. (18.18) requires L complex multiplications and $L - 1$ complex additions. Equivalently, it requires $N_{\text{MUL,linear}} = 3L$ real multiplications and $N_{\text{MUL,linear}} = 7L - 2$ real additions.

18.2.4.2 Polynomial Nonlinear Cancellation

The total number of complex-valued parameters $\hat{h}_{p,q}[l]$ in Eq. (18.14) can be calculated as Korpi et al. (2017):

$$N_{\text{poly}} = \frac{L}{4}(P+1)(P+3), \tag{18.29}$$

which grows quadratically with the PA nonlinearity order P. In order to perform a best-case complexity analysis for the polynomial canceler, it is assumed that the calculation of the basis functions $BF_{p,q}(x[n])$ in Eq. (18.14) comes at no computational cost Balatsoukas-Stimming (2018). One complex multiplication is performed for each of the N_{poly} complex-valued parameters $\hat{h}_{p,q}[l]$ in Eq. (18.14), requiring a total of $3N_{\text{poly}}$ real multiplications and $5N_{\text{poly}}$ real additions. Moreover, the results of the N_{poly} complex multiplications need to be summed up in order to calculate $\hat{y}[n]$, which requires a total of $2(N_{\text{poly}} - 1)$ real additions. Thus, the polynomial canceler requires a total of:

$$N_{\text{MUL,poly}} = 3N_{\text{poly}} = \frac{3}{4}L(P+1)(P+3), \tag{18.30}$$

$$N_{\text{ADD,poly}} = 5N_{\text{poly}} + 2(N_{\text{poly}} - 1) = \frac{7}{4}L(P+1)(P+3) - 2, \tag{18.31}$$

real multiplications and real additions, respectively.

18.2.4.3 Neural Network Nonlinear Cancellation

Apart from the connections that are visible in Figure 18.2, each node also has a *bias* input, which is omitted from the figure for simplicity. Thus, the total number of real-valued weights in the NN is:

$$N_w = (2L+1)N_h + 2(N_h + 1). \tag{18.32}$$

Excluding the biases that are not involved in multiplications, there are $2LN_h$ real weights in the hidden layer that are multiplied with the real input values, and $2N_h$ real weights in the output layer that are multiplied with the real output values from the hidden nodes. Moreover, the linear cancellation stage that precedes the NN requires L complex multiplications, which correspond to $3L$ real multiplications and $5L$ real additions. Thus, the total number of real multiplications required by the NN canceler is:

$$N_{\text{MUL,NN}} = (2L+2)N_h + 3L. \tag{18.33}$$

For each of the N_h hidden neurons, $2L + 1$ incoming real values need to be summed, which requires a total of $2LN_h$ real additions. Moreover, at each of the 2 output neurons, $N_h + 1$ real values need to be summed, which requires a total of $2N_h$ real additions. The computation of each of the N_h ReLU activation functions requires one multiplexer (and one comparator with zero, which can be trivially implemented by looking at the most significant bit of the input). Moreover, the linear cancellation stage that precedes the NN requires summing up L complex values, which requires $2(L - 1)$ real additions. Thus, assuming a worst case where a multiplexer has the same complexity as an addition, the total number of real additions required by the NN canceler is:

$$N_{\text{ADD,NN}} = (2L+3)N_h + 7L - 2. \tag{18.34}$$

The complexity expressions for the polynomial and the NN-based canceler are summarized in Table 18.1. It is important to note that the complexity expressions for the two methods cannot be compared directly because they contain different sets of parameters. Thus, in order to perform a fair comparison, appropriate values for L, P, and N_h are selected in Section 18.3 so that the two methods have the same SI cancellation performance.

Table 18.1 Computational complexity of polynomial and neural network cancelers.

	Polynomial	Neural network
Real additions	$\frac{7}{4}L(P+1)(P+3)-2$	$(2L+3)N_h+7L-2$
Real multiplications	$\frac{3}{4}L(P+1)(P+3)$	$(2L+2)N_h+3L$

18.3 Experimental Results

In this section, the performance and complexity of the standard polynomial nonlinear canceler described in Section 18.2.2 is compared with the NN canceler described in Section 18.2.3 using experimental results from a FD testbed.

18.3.1 Experimental Setup

The FD hardware testbed that is used to obtain the results in this section uses a National Instruments FlexRIO device and two FlexRIO 5791R RF transceiver modules as described in more detail in Balatsoukas-Stimming et al. (2013), Belanovic et al. (2013), Balatsoukas-Stimming et al. (2015). The transmitted signal is a QPSK-modulated OFDM signal with a passband bandwidth of 10 MHz and $N_c = 1024$ carriers. At the receiver, the signal is sampled with a sampling frequency of 20 MHz so that the signal side-lobes can be observed. Each transmitted OFDM frame consists of 20,480 baseband samples, out of which 90% are used for training and the remaining 10% are used to calculate the achieved SI cancellation, both for the polynomial canceler and for the NN canceler. The average transmit power is 10 dBm and the two-antenna FD testbed setup provides a passive analog cancellation of 53 dB. No active RF cancellation is performed as the achieved passive cancellation is sufficient for the results presented in this section.

For both cancelers, it was found through trial and error that $L = 13$ memory taps are sufficient to model the equivalent SI channel. Moreover, for the polynomial canceler, a maximum nonlinearity order of $P = 7$ is used, since further increasing this parameter results in very limited gains in the achieved SI suppression, and after some point even decreased performance due to overfitting. The NN was implemented using the Keras framework with a TensorFlow back-end. Moreover, the Adam optimization algorithm is used for training with a mean squared error cost function, a learning rate of $\lambda = 0.004$, and a mini-batch size of $B = 32$. All remaining parameters have their default values. The NN has $2L = 26$ input units and $N_h = 18$ hidden units. The neurons in the hidden layer use a rectified linear unit (ReLU) activation function, defined as $f_h(x) = \max(0, x)$, while the output neurons use the identity activation function, defined as $f_o(x) = x$.

18.3.2 Self-Interference Cancellation Results

Figure 18.3 shows SI cancellation results using the polynomial canceler of Section 18.2.2 and the NN canceler of Section 18.2.2. It can be observed that digital linear cancellation provides approximately 37.9 dB of cancellation, reducing the residual SI power to −80.6 dBm. Both the polynomial canceler and the NN canceler reduce the SI by an additional 6.9 dB, leading to a residual SI power of −87.5 dBm, which is only 3.3 dB away from the receiver noise floor. In Figure 18.5 it can be observed that after only four training epochs, the NN can already achieve a

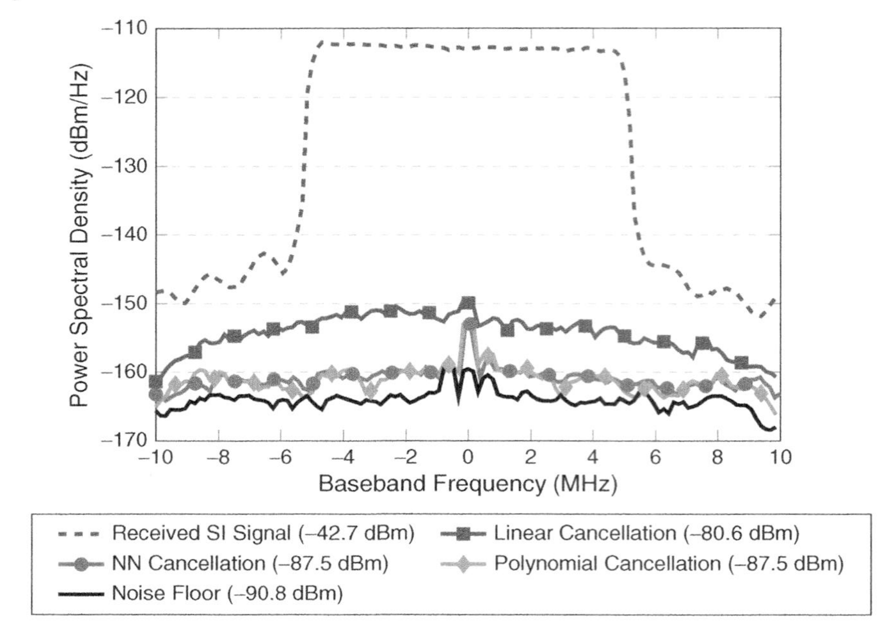

Figure 18.3 Power spectral densities of the self-interference signal, the SI signal after linear cancellation, as well as the SI signal after nonlinear cancellation using both a polynomial canceler and a neural network canceler.

nonlinear SI cancellation of over 6 dB on both the training and the test frames. After 20 training epochs, the nonlinear SI cancellation reaches approximately 7 dB and there is no obvious indication of overfitting since the SI cancellation on the training and on the test data is very similar.

In Figure 18.4, it can be seen that, if no linear cancellation is performed before the NN canceler, then the achieved SI cancellation is significantly worse. Specifically, in this case even a large NN canceler with $N_h = 100$ hidden neurons trained for a total of 1000 epochs achieves only 37.1 dB of cancellation (care was taken to ensure that no overfitting occurred). This amount of SI cancellation is similar to the cancellation achieved by simply using the (much lower complexity) linear canceler. In principle the NN should be able to learn to jointly cancel both the linear and the nonlinear part of the signal. However, because the nonlinear part of the SI signal is significantly weaker than the linear part, it seems that the noise in the gradient computation due to the use of mini-batches essentially completely hides the nonlinear structure of the SI signal from the learning algorithm.

18.3.3 Computational Complexity

Having found the set of parameters $P = 7$, $L = 13$, and $N_h = 18$ that lead to the same SI cancellation performance, it is possible to fairly compare the complexity of the polynomial canceler with the complexity of the NN canceler by evaluating the complexity expressions derived in Section 18.2.4. More specifically, the polynomial canceler requires $N_{\text{ADD,poly}} = 1818$ real additions and $N_{\text{MUL,poly}} = 780$ real multiplications, while the NN canceler requires $N_{\text{ADD,NN}} = 611$ real additions and $N_{\text{MUL,NN}} = 543$ real multiplications. In other words, the NN canceler requires 66% fewer real additions and 30% fewer real multiplications than the polynomial canceler. This complexity reduction is also reflected well in hardware implementations of the two types of cancelers. Specifically, a NN canceler was shown to be 11% smaller and 60% faster than a

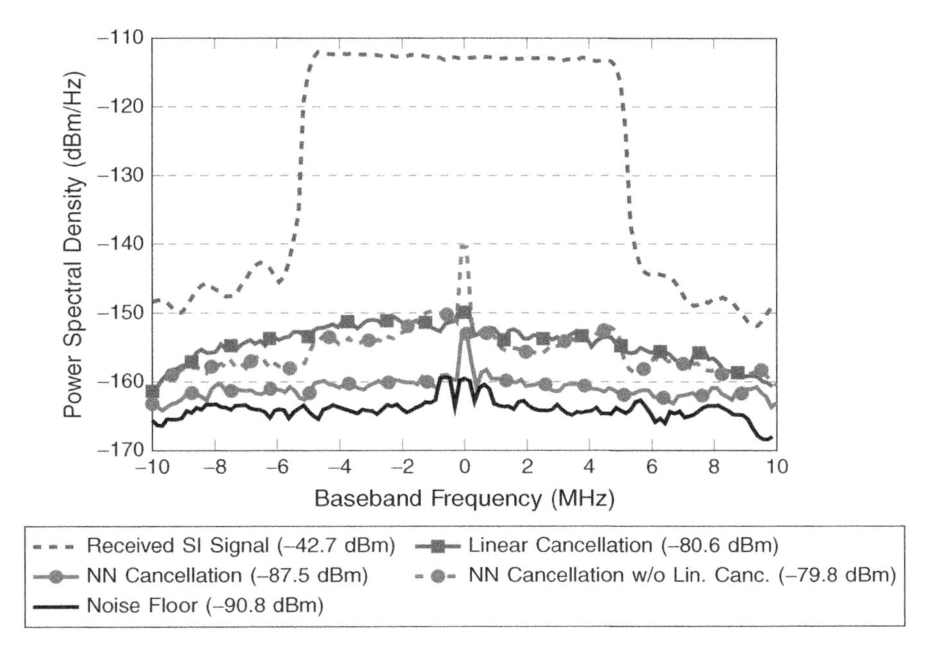

Figure 18.4 Power spectral densities of the self-interference signal, the SI signal after linear cancellation, and the SI signal after neural network–based nonlinear cancellation with and without a linear cancellation stage.

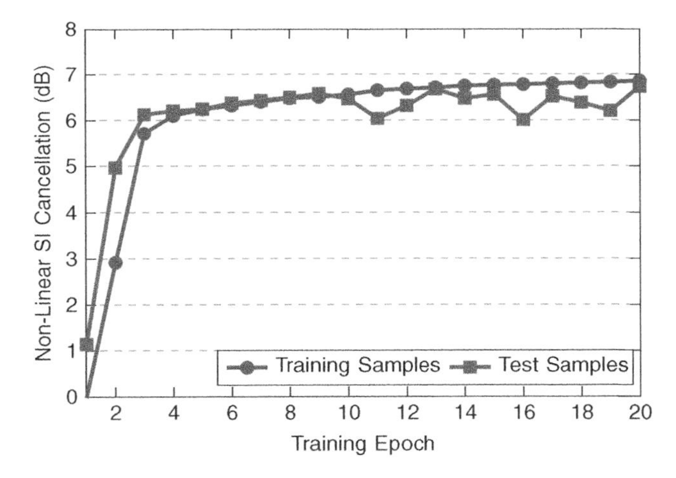

Figure 18.5 Achieved nonlinear self-interference cancellation on the training frames and the test frames as a function of the number of training epochs.

polynomial canceler when implemented as an ASIC using a 28 nm FD-SOI technology Kurzo et al. (2018). All implementation results are summarized in Table 18.2.

18.4 Conclusions

This chapter provided a self-contained and comprehensive overview of FD communications and digital SI cancellation methods. In particular, a detailed FD transceiver model was

Table 18.2 Comparison of polynomial and neural network canceler implementations.

	Polynomial	Neural network	Gain
Real additions	1818	611	66%
Real multiplications	780	543	30%
ASIC area (mm^2)	0.36	0.32	11%
ASIC throughput (Msamples/s)	17.4	27.8	60%

presented and the main transceiver non-idealities that make digital SI cancellation challenging were described and modeled. Based on the aforementioned models, a polynomial SI cancellation method was presented, which is commonly used in the literature for SI cancellation. Then, an alternative NN-based SI cancellation method was presented and the computational complexity of the two methods was examined and compared. Finally, experimental results showed that the NN-based SI canceller achieves identical digital SI cancellation performance to the polynomial SI canceller, but with significantly lower computational and hardware implementation complexity.

18.4.1 Open Problems

The NN-based cancellation method described in this chapter uses a simple single-layer feedforward NN. Even though the method works well, significant improvements in performance and/or complexity could be achieved by considering other NN architectures. For example, deep feedforward NNs have been shown to generally outperform single-layer feedforward NNs for a wide variety of applications. As such, a deep and narrow (i.e. with few neurons per layer) NN may outperform a shallow and wide NN for the same number of trainable parameters, or it may provide similar performance with fewer trainable parameters. Moreover, since the SI signal has memory, a natural choice would be to use a recurrent NN (RNN) that can reproduce the SI memory with only two inputs (i.e. the real and imaginary parts of $x[n]$), instead of the $2L$ inputs that are required by the feedforward NN.

The SI channel changes over time and needs to be tracked. For the polynomial canceler, the SI channel can be tracked by re-estimating the parameters $\hat{h}_{p,q}[l]$ in Eq. (18.16), using either standard least-squares estimation or an adaptive version of the least-squares estimation algorithm, such as least mean squares or recursive least squares. For the NN-based canceler, the SI channel can be tracked by re-running the backpropagation training algorithm. It is essential to examine and compare the SI channel-tracking methods for the polynomial and the NN-based canceler. It particular, it would be interesting to compare the computational complexity, the convergence speed, and the required number of training samples for each tracking method.

Finally, existing NN-based methods for SI cancellation do not take any expert knowledge about the problem into account. Both the SI transceiver system diagram in Figure 18.1 and Eq. (18.16) provide valuable information that could be used in order to derive physically inspired NN architectures for digital SI cancellation, similarly to the physically inspired RF power amplifier modeling of Mkadem and Boumaiza (2011).

Bibliography

L. Anttila, D. Korpi, E. Antonio-Rodríguez, R. Wichman, and M. Valkama. Modeling and efficient cancellation of nonlinear self-interference in MIMO full-duplex transceivers. In *Globecom Workshops*, pages 777–783, 2014.

A. Balatsoukas-Stimming. Non-linear digital self-interference cancellation for in-band full-duplex radios using neural networks. In *IEEE International Workshop on Signal Processing Advances in Wireless Communications (SPAWC)*, pages 1–5, June 2018.

A. Balatsoukas-Stimming, P. Belanovic, K. Alexandris, and A. Burg. On self-interference suppression methods for low-complexity full-duplex MIMO. In *Asilomar Conference on Signals, Systems and Computers*, pages 992–997, November 2013. doi: 10.1109/ACSSC.2013.6810439.

A. Balatsoukas-Stimming, A.C.M. Austin, P. Belanovic, and A. Burg. Baseband and RF hardware impairments in full-duplex wireless systems: experimental characterisation and suppression. *EURASIP Journal on Wireless Communications and Networking*, 2015 (142), 2015.

P. Belanovic, A. Balatsoukas-Stimming, and A. Burg. A multipurpose testbed for full-duplex wireless communications. In *IEEE International Conference on Electronics, Circuits, and Systems (ICECS)*, pages 70–71, December 2013. doi: 10.1109/ICECS.2013.6815349.

D. Bharadia, E. McMilin, and S. Katti. Full duplex radios. In *ACM SIGCOMM*, pages 375–386, 2013.

Melissa Duarte, Chris Dick, and Ashutosh Sabharwal. Experiment-driven characterization of full-duplex wireless systems. *IEEE Transactions on Wireless Communications*, 11 (12): 4296–4307, Dec. 2012.

H. Guo, J. Xu, S. Zhu, and S. Wu. Realtime software defined self-interference cancellation based on machine learning for in-band full duplex wireless communications. In *International Conference on Computing, Networking and Communications (ICNC)*, pages 779–783, March 2018.

Kurt Hornik. Approximation capabilities of multilayer feedforward networks. *Neural Networks*, 4 (2): 251–257, 1991.

Mohamed Ibnkahla. Applications of neural networks to digital communications—a survey. *Elsevier Signal Processing*, 80 (7): 1185–1215, July 2000.

M. Jain, J.I. Choi, T. Kim, D. Bharadia, S. Seth, K. Srinivasan, P. Levis, S. Katti, and P. Sinha. Practical, real-time, full duplex wireless. In *Proc. 17th International Conference on Mobile Computing and Networking*, pages 301–312. ACM, 2011.

Diederik P. Kingma and Jimmy Ba. Adam: A method for stochastic optimization. In *International Conference for Learning Representations (ICLR)*, May 2015.

D. Korpi, L. Anttila, V. Syrjala, and M. Valkama. Widely linear digital self-interference cancellation in direct-conversion full-duplex transceiver. *IEEE J. Sel. Areas Commun.*, 32 (9): 1674–1687, Sep. 2014.

D. Korpi, L. Anttila, and M. Valkama. Nonlinear self-interference cancellation in MIMO full-duplex transceivers under crosstalk. *EURASIP Journal on Wireless Comm. and Netw.*, 2017 (1): 24, February 2017.

Y. Kurzo, A. Burg, and A. Balatsoukas-Stimming. Design and implementation of a neural network aided self-interference cancellation scheme for full-duplex radios. In *Asilomar Conference on Signals, Systems, and Computers*, pages 1–5, October 2018.

F. Mkadem and S. Boumaiza. Physically inspired neural network model for RF power amplifier behavioral modeling and digital predistortion. *IEEE Transactions on Microwave Theory and Techniques*, 59 (4): 913–923, April 2011.

N. Naskas and Y. Papananos. Neural-network-based adaptive baseband predistortion method for RF power amplifiers. *IEEE Transactions on Circuits and Systems II: Express Briefs*, 51 (11): 619–623, November 2004.

M. Rawat, K. Rawat, and F.M. Ghannouchi. Adaptive digital predistortion of wireless power amplifiers/transmitters using dynamic real-valued focused time-delay line neural networks. *IEEE Transactions on Microwave Theory and Techniques*, 58 (1): 95–104, January 2010.

David E. Rumelhart, Geoffrey E. Hinton, and Ronald J. Williams. Learning representations by back-propagating errors. *Nature*, 323: 533–536, October 1986.

A. Sahai, G. Patel, C. Dick, and A. Sabharwal. On the impact of phase noise on active cancelation in wireless full-duplex. *IEEE Transactions on Vehicular Technology*, 62 (9): 4494–4510, Nov. 2013.

V. Syrjala, M. Valkama, L. Anttila, T. Riihonen, and D. Korpi. Analysis of oscillator phase-noise effects on self-interference cancellation in full-duplex OFDM radio transceivers. *IEEE Transactions on Wireless Communications*, 13 (6): 2977–2990, June 2014.

19

Machine Learning for Context-Aware Cross-Layer Optimization

Yang Yang[1], Zening Liu[1], Shuang Zhao[2], Ziyu Shao[1], and Kunlun Wang[1]

[1] *SHIFT, School of Information Science and Technology, ShanghaiTech University, Shanghai, China*
[2] *Interactive Entertainment Group, Tencent Inc., Shanghai, China*

19.1 Introduction

In recent years, global mobile data traffic has experienced an explosive growth. It is expected to grow to 49 exabytes per month by 2021, a sevenfold increase over 2016 Cisco (2016). Current wireless technologies, such as 4G and WiFi, do not have localized data analysis and processing capabilities so that they cannot handle such a bursty traffic increase Chen et al. (2014), Yang et al. (2018b). As machine-type communications (MTC) have been adopted in 5G networks Ge et al. (2016), Shi et al. (2014), Wang et al. (2016), Chen et al. (2016a), new flexible network architectures and service strategies are desperately needed to support more and more data-centric and delay-sensitive Internet-of-Things (IoT) applications Yang (2019), Chen et al. (2018a), such as smart city, environment surveillance, intelligent manufacturing, and autonomous driving. If only centralized cloud computing architecture is applied to those various IoT applications, it is envisaged that the underlayer communication networks, especially backhaul connections, will face heavy bursty traffic burdens and experience dramatic performance degradation. On the other hand, Moore's Law has significantly driven down the prices of computing and storage devices, and more and more smart network nodes and user terminals are deployed and connected into modern communication networks. They provide a rich collection of ubiquitous local computing, communication, and storage resources. In view of this technological trend, the concept of fog computing is proposed to enable computing anywhere along the cloud-to-thing continuum Bonomi et al. (2012), Vaquero and Rodero-Merino (2014), Ouyang et al. (2018), Chen et al. (2018). In other words, fog-enabled network architecture and services can effectively leverage those local resources to support fast-growing data-centric and delay-sensitive IoT-applications in regional environments, thus reducing backhual traffic transmissions and centralized computing needs, and at the same time improving the overall network throughput performance and users' quality of experience (QoE) Chiang and Zhang (2016), Yang et al. (2017b), Chen et al. (2017).

Without loss of generality, let us consider a multi-tier content delivery wireless network consisting of user terminals (UTs), access tier, and control tier, as shown in Figure 19.1, where (i) a node in access tier is typically located close to the UTs and is called a *fog access node*(FAN); (ii) a node in the control tier, which is usually far away from UTs, manages a group of FANs through reliable but expensive backhaul connections and is called a *fog control node*(FCN). As UTs are moving around and can make requests of any contents at anytime anywhere, it is obvious that popular contents should be placed in multiple FANs in advance, according to their resources

Machine Learning for Future Wireless Communications, First Edition. Edited by Fa-Long Luo.
© 2020 John Wiley & Sons Ltd. Published 2020 by John Wiley & Sons Ltd.

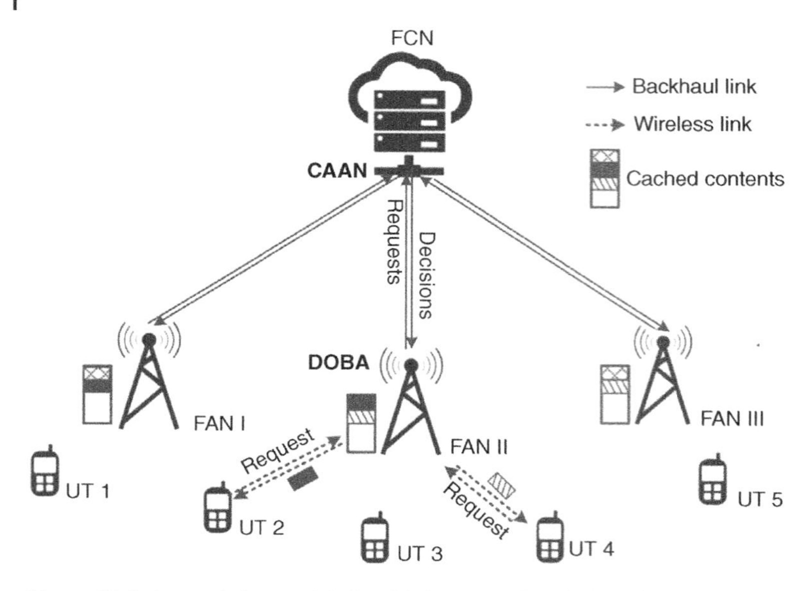

Figure 19.1 A sample fog-enabled multi-tier network with three fog access nodes and five user terminals.

and capabilities. In doing so, most content delivery requests are handled in the network edge, and thus service delay and backhaul traffic transmission can be greatly reduced.

In such a complex multi-tier network with heterogeneous node capabilities and dynamic network resources, in terms of computing power, storage capacity, transmission power, and communication bandwidth, how to conduct effective multi-tier operations scheduling is a key challenge to be resolved Zhang et al. (2018b,a), Yang et al. (2018e). To deal with the node capacity and dynamic network resources management in fog computing networks, many techniques have been explored. Recently, applying AI to solve the resource management in such complex networks has gained considerable attention Yang et al. (2019), Huang et al. (2018), Yang et al. (2018c). Luong et al. (2018) proposed a deep learning (DL)-based algorithm for edge resource management. Li et al. (2018) introduced DL for IoT into the edge computing environment to improve learning performance as well as to reduce network traffic. He et al. (2017) presented a novel big data deep reinforcement learning (RL) approach. Chen et al. (2018) proposed an efficient RL-based resource management algorithm, which learned on-the-fly the optimal policy of dynamic workload offloading. However, the shortcomings of these existing machine learning techniques cannot be overlooked:

(1) The training data for both DL and deep RL is essential for training the models, while it is difficult to collect, especially in large-scale fog networks.
(2) The number of optimized parameters is typically very large for such a large-scale fog network, and thus the computing resources and time required for training the models are usually rather high.
(3) The network is dynamic, and learning from the models is time-consuming, as well as resource-consuming, so it is hard for DL or deep RL to effectively realize online or real-time resource management.
(4) The performance of DL and deep RL cannot be guaranteed in theory.

Different from these machine learning techniques for resource management, in this chapter, we employ Lyapunov optimization-based learning techniques and propose an online (real-time) low-complexity fog-enabled multi-tier operations scheduling (FEMOS) algorithm, which is data-free and performance-guaranteed. The FEMOS algorithm simultaneously addresses the following challenging problems:

(1) When popular contents are randomly cached at different FANs, how to identify the most feasible FAN for every UT's request in order to maximize network throughput and global fairness.
(2) Under dynamic wireless network conditions and fading channel characteristics, how to effectively allocate communication bandwidth for associated FAN-UT pairs in order to minimize service delay.

It is worth noting that, since the file placement has been widely investigated in the design of content delivery networks Baştug et al. (2015), Li et al. (2015), Liu et al. (2017), the dynamic association between UTs and FANs is the focus of this chapter, which has rarely been investigated.

Although the most commonly used AI techniques, such as DL and deep RL, may be not best suited to be immediately applied to conduct all the resource management in fog networks, is it possible to deploy AI techniques to realize partial functions, and as a result all resource management can benefit from it? Motivated by this question and the recent advancement in data mining for learning user behaviors Kumar and Tomkins (2010), Huang et al. (2016b), Yang et al. (2017a), Zhang et al. (2014), Wang et al. (2015), this chapter presents a predictive scheduling model and develops the predictive multi-tier operations scheduling (PMOS) algorithm, where the FCN is assumed to be aware of users' future request information within a limited future time window.

In addition, this chapter further addresses a cost model and the resulting cost-minimization user-scheduling problem in multi-tier fog computing networks. Generally, the FCN is operated by a telecom operator that signs a service contract with UTs, while the FANs belong to different individuals. To better motivate the FANs to share resources, the cost model, especially for FANs, should be taken into consideration. For this purpose, this chapter presents a unified multi-tier cost model, including the service delay and a linear inverse demand dynamic payment scheme. Correspondingly, a cost-oriented user scheduling (COUS) algorithm, based on a potential game, is reported in this chapter.

The rest of this chapter is organized as follows. The system model is presented in Section 19.2. Under the fog-enabled network architecture, the problem is formulated in Section 19.3. The online FEMOS algorithm is proposed in Section 19.3.1 and corresponding performance analysis is conducted in Section 19.3.2. Section 19.4 further develops the PMOS algorithm based on the proposed FEMOS algorithm and predicted users' information. Furthermore, Section 19.5 proposes a unified multi-tier cost model to motivate the FANs for resources sharing, and develops the COUS algorithm to effectively solve the resulted cost-minimization user scheduling problem. Section 19.6 concludes this chapter.

19.2 System Model

We consider a fog-enabled multi-tier network with heterogeneous nodes as shown in Figure 19.1, which involves a FCN tier, a FAN tier, and a set of multiple stationary or low-mobility UTs in the region under consideration. Each UT possesses a small storage capacity, minor communication ability, and little or no computation ability. UTs request files to be downloaded from the FAN tier through wireless links. Due to the restricted transmission power and dynamic wireless environment, the communication channels between a FAN and its neighboring UTs are unreliable and time-varying. Each FAN in the FAN tier is equipped with limited storage capacity, medium computation ability, but strong communication ability. All of them cached a subset of popular files. Through reliable backhaul links, FANs are connected with a FCN, which is next to the cloud and core network, has the global information about the network, and is the server of the file library. The FCN has both sufficient storage capacity and powerful computation ability. We emphasize here that in our model, the backhaul links transmit control information from the FCN to FANs, and files cached at each FANs are only refreshed at off-peak times. To be clear, if the UT requested file is not cached on the FANs, the FAN will not fetch the file from the operation center through the backhaul link. Under these conditions, we can see more clearly the benefit of dynamic assignment of FAN and resource scheduling.

Denote the set of FANs as $\mathcal{H}$, the set of UTs as $\mathcal{U}$, and the file library as $\mathcal{F}$. The large-scale fading and small-scale fading coefficients seen by each FAN are assumed to be mutually independent. We assume that the network operates in a slotted system, indexed by $t \in \{0, 1, 2, \dots\}$, and the time slot length is $\mathcal{T}$. The FCN determines the dynamic FAN assignment for UTs at the beginning of each slot, to optimize the network throughput in a memoryless pattern, based on the network states, request queue length and disregarding all such previous decisions. Each UT requested file will be then transmitted by its associated FAN through the wireless link. The queue length of current unserved request buffers will in turn influence the FCN's decision about FAN assignment in the next slot. Each FAN then independently implements its per-slot scheduling policy including the bandwidth and service rate allocation over the UTs associated with it. Consistent with this setting, the requested file f can only be downloaded from the FAN that has cached it. The service rate would be zero if the requested file is not cached on the FAN that the UT is associated with.

FAN Assignment and File Placement Define the FAN assignment (UT-FAN association) as a bipartite graph $\mathcal{G} = (\mathcal{U}, \mathcal{H}, \mathcal{E})$, where $\mathcal{E}$ contains edges for the pairs (u, h) such that there exists a potential transmission link between FAN $h \in \mathcal{H}$ and UT $u \in \mathcal{U}$. We assume $\mathcal{G}$ varies in different time slots. Let $\mathbf{X}(t)$ denote a $|\mathcal{U}| \times |\mathcal{H}|$ association matrix of $\mathcal{G}$ between UTs and FANs in time slot t, where $|\mathcal{U}|(|\mathcal{H}|)$ denotes the cardinality of the set $\mathcal{U}(\mathcal{H})$, $\mathbf{X}(t) \triangleq [x_{uh}(t)]_{u,h}$. Here $x_{uh}(t) = 1$ if $(u, h) \in \mathcal{E}$, and 0 otherwise.

Define the file placement (FAN-File association) as a bipartite graph $\tilde{\mathcal{G}} = (\mathcal{H}, \mathcal{F}, \tilde{\mathcal{E}})$, where edges $(h,f) \in \tilde{\mathcal{E}}$ indicates that files with type f are cached in FAN h. The file set cached at each FAN is $\mathcal{N}$, $\mathcal{N} \subseteq \mathcal{F}$, with $|\mathcal{N}|$ different file types.

Let us first focus on dynamic FAN assignment and the resource scheduling problem. As said before, the backhaul updates the storages at a time scale much larger than the time scale of UTs placing file requests. Therefore, assuming fixed file placement (FAN-File Association) is justified. Let $\mathbf{Y}$ denote a $|\mathcal{H}| \times |\mathcal{F}|$ file placement matrix of $\tilde{\mathcal{G}}$ and $\mathbf{Y} \triangleq [y_{hf}]_{h,f}$. Here, $y_{hf} = 1$ if $(h,f) \in \tilde{\mathcal{E}}$, and 0 otherwise.

We assume that each UT can be associated with at most one FAN and each FAN can associate with at most M UTs in one time slot. Thus $\mathbf{X}(t)$ should be chosen from the feasible set $\mathcal{A}$,

$$\mathcal{A} = \left\{ \mathbf{X}(t) \in \{0,1\}^{|\mathcal{U}|\times|\mathcal{H}|} \;\middle|\; \begin{array}{ll} \sum_{u\in\mathcal{U}} x_{uh}(t) \le M, & \forall h \in \mathcal{H}; \\ \sum_{h\in\mathcal{H}} x_{uh}(t) \le 1, & \forall u \in \mathcal{U}. \end{array} \right\}. \tag{19.1}$$

UT Traffic Model All UTs are assumed to generate file-request traffic randomly in each time slot, and this traffic generation is independent of the FCN's operation.

Let $\mathbf{A}(t)$ denote the request arrival vector in time slot t and $\mathbf{A}^T(t) \triangleq [A_1(t), \dots, A_{|\mathcal{U}|}(t)]$, where random variable $A_u(t)$ (with the unit kbits) denotes the requested amount in time slot t and the operation $(\cdot)^T$ denotes vector transposition. Here we assume that $A_u(t)$ is i.i.d. with $\mathbb{E}\{A_u(t)\} = \lambda_u$, and there exists a positive constant $A_{\max}$ such that $0 \le A_u(t) \le A_{\max}$.

Let $\mathbf{I}(t)$ denote the $|\mathcal{U}| \times |\mathcal{F}|$ requested file type matrix in time slot t and $\mathbf{I}(t) \triangleq [I_{uf}(t)]_{u,f}$. Here $I_{uf}(t) = 1$ if the requested file type by UT u is f in time slot t, and 0 otherwise. We assume that each UT can request at most only one type of file in one time slot, which means that the row weight of $\mathbf{I}(t)$ is at most 1. The requested probability of each file $f \in \mathcal{F}$ is subject to Zipf distribution Zink et al. (2009).

The Transmission Model The wireless channels between UTs and FANs are assumed to be flat-fading channels Tse and Viswanath (2005), and all FANs transmit at constant power. We assume that the additive white Gaussian noise (AWGN) at the UTs follows Gaussian distribution with $\mathcal{N}(0, \sigma^2)$. Note that the maximum service rate of UT u can be obtained if it has been allocated the total bandwidth by its associated FAN. Then the maximum backlog that can be served in time slot t over link $(u, h) \in \mathcal{E}$ is given by

$$C_{uh}(t) = \mathcal{T} B_h(t) \cdot \mathbb{E}\left[\log_2\left(1 + \frac{P_h g_{hu}(t)|s_{hu}|^2}{\sigma^2 + \sum_{h' \in \mathcal{H}\backslash h} P_{h'} g_{h'u}(t)|s_{h'u}|^2}\right)\right], \tag{19.2}$$

where $B_h(t)$ is the total bandwidth of FAN h in time slot t; P_h is the transmit power of FAN h; $g_{hu}(t)$ is the large-scale fading from FAN h to UT u, which contains pathloss and shadow; and s_{hu} is the small-scale fading, which follows the Rayleigh distribution. For simplicity, currently implemented rate adaption schemes Biglieri et al. (1998) Ong et al. (2011) are consistent in assuming slowly varying pathloss coefficients $g_{hu}(t)$ change across slots in an i.i.d. manner, and each FAN h being aware of $g_{hu}(t)$ for all $u \in \mathcal{U}$ at the beginning of each time slot t.

We also assume that each FAN h serves its associated UTs by using orthogonal FDMA or TDMA, which is consistent with most current wireless standards. Let $\nu_{uh}(t)$ be the proportion of bandwidth allocated to UT u by FAN h. Then $\nu_{uh}(t)$ satisfies $0 < \nu_{uh}(t) \le 1$ when $x_{uh}(t) = 1$, otherwise $\nu_{uh}(t) = 0$. Denote $\boldsymbol{\nu}(t) \triangleq [\nu_{uh}(t)]_{u,h}$ as the bandwidth allocation matrix, which is chosen from the feasible set $\mathcal{B}$,

$$\mathcal{B} = \left\{ \boldsymbol{\nu}(t) \in \mathbb{R}_+^{|\mathcal{U}|\times|\mathcal{H}|} \;\middle|\; \begin{array}{l} \sum_{u \in \mathcal{U}} \nu_{uh}(t) x_{uh}(t) \le 1; \\ \nu_{uh} = 0 \quad \text{if} \quad x_{uh} = 0, \quad \forall h \in \mathcal{H}. \end{array} \right\}. \tag{19.3}$$

Let $\mu_u(t)$ denote the amount of backlog that can be served for UT u in time slot t with maximum value $\mu_{\max}$, which is called the *service rate* hereafter. Define $\boldsymbol{\mu}^{\mathrm{T}}(t) \triangleq [\mu_1(t), \dots, \mu_{|\mathcal{U}|}(t)]$. Note that each UT can associate with at most one FAN in a time slot, thus $\mu_u(t)$ can be expressed as follows:

$$\mu_u(t) = \sum_{h \in \mathcal{H}} C_{uh}(t) \nu_{uh}(t) x_{uh}(t), \forall u \in \mathcal{U}. \tag{19.4}$$

Queuing In each time slot, the arrived requests of all UTs will be queued in the request buffers at the FCN. We assume that FCN has $|\mathcal{F}|$ request buffers for each UT $u \in \mathcal{U}$. Denote the queue length of the amount of request for file with type f at the beginning of the tth time slot as $Q_{uf}(t)$. Define $Q_u^{\mathrm{sum}}(t) \triangleq \sum_{f \in \mathcal{F}} Q_{uf}(t)$ and denote $\boldsymbol{Q}^T(t) = [Q_u^{\mathrm{sum}}(t), \cdots, Q_{|\mathcal{U}|}^{\mathrm{sum}}(t)]$ as the queue length vector. We assume that all queues are initially empty, i.e. $Q_{uf}(0) = 0, \forall u \in \mathcal{U}, f \in \mathcal{F}$.

Let $\mu_{uf}(t)$ denote the service rate for the requested file f scheduled by the FCN according to a certain queuing discipline Huang et al. (2016a), such as FIFO, LIFO, or random discipline. We adopt the fully efficient scheduling policy given in Huang et al. (2016a) for queues, which means:

$$\sum_{f \in \mathcal{F}} \mu_{uf}(t) = \mu_u(t), \tag{19.5}$$

where $\mu_u(t)$ is defined in Eq. (19.4) and $\mu_{uf}(t) = 0$ if $y_{hf} = 0$.

The queue length $Q_{uf}(t)$ is updated in every time slot t according to the following rules:

$$Q_{uf}(t+1) = [Q_{uf}(t) - \mu_{uf}(t)]^{+} + A_u(t) \cdot I_{uf}(t), \tag{19.6}$$

where $[x]^{+} = \max\{x, 0\}$.

The queuing process $Q_{uf}(t)$ is stable if the following condition holds Neely (2010):

$$\overline{Q}_u^{\text{sum}} = \lim_{t\to\infty} \frac{1}{t} \sum_{\tau=0}^{t-1} \mathbb{E}[Q_u^{\text{sum}}(\tau)] < \infty. \tag{19.7}$$

19.3 Problem Formulation and Analytical Framework

Performance Metrics

We focus on the total throughput of the network. Therefore, we adopt the time-averaged sum service rate of different UTs in the network as the performance metric, which is defined as follows:

$$\phi_{av} \triangleq \overline{\phi(\boldsymbol{\mu})} = \sum_{u\in\mathcal{U}} \overline{\mu}_u, \tag{19.8}$$

where $\overline{\mu}_u(t)$ is the averaged expected service rate of UT u.

To support for transmission latency sensitive applications, i.e. online video, service delay is a key metric that needs to be considered Ahlehagh and Dey (2012). According to Little's law Ross (2014), the average service delay experienced by each UT is proportional to the averaged amount of its unserved requests waiting at the FCN, which is the sum of the queue length for different files. Thus, the average delay per UT can be computed as the ratio between the average queue length and the mean traffic arrival rate, which is shown as follows:

$$\Lambda_{av} \triangleq \frac{\sum_{u\in\mathcal{U}} \overline{Q}_u^{\text{sum}}}{\sum_{u\in\mathcal{U}} \lambda_u}. \tag{19.9}$$

Average Network Throughput Maximization Problem

The system objective is to find a feasible FAN assignment $\mathbf{X}(t)$ and bandwidth allocation $\boldsymbol{\nu}(t)$ to maximize the average network throughput while maintaining the stability of all the queues in the network. The average network throughput maximization problem can be formulated in $\mathcal{P}1$:

$$\begin{aligned} \mathcal{P}1: \quad & \max_{\mathbf{X}(t),\boldsymbol{\nu}(t)} \ \overline{\phi(\boldsymbol{\mu})} \\ & \text{s.t.}\ \overline{Q}_u^{\text{sum}} < \infty, \quad \forall u \in \mathcal{U}, \\ & \qquad \mathbf{X}(t) \in \mathcal{A}, \quad \boldsymbol{\nu}(t) \in \mathcal{B} \quad \forall t, \end{aligned} \tag{19.10}$$

where the requirement of finite $\overline{Q}_u^{\text{sum}}$ corresponds to the strong stability condition for all the queues Neely (2010). Queuing stability implies that buffered file requests are processed with finite delay. We will show that our proposed algorithms guarantee upper bounds for $\overline{Q}_{uf}$ and thus achieve the bounded service delay.

It is not difficult to identify that $\mathcal{P}1$ is a highly challenging stochastic optimization problem with a large amount of stochastic information to be handled (including channel conditions and request buffer state information) and two optimization variables to be determined, which requires the design of an online operation and scheduling scheme for such a network. In addition, to maximize network throughput, it is essential to jointly optimize the FAN-UT association and the resource allocation, which is always a complicated mixed integer programming problem. Further, the optimal decisions are temporally correlated due to the random-arrival traffic demands. Furthermore, the FCN needs to reduce the delay per UT while maintaining the average network throughput, which requires the FCN to maintain a good balance between network throughput and average delay.

Lyapunov Optimization-Based Analytical Framework

In the following, we focus on solving this challenging problem $\mathcal{P}1$ by using the Lyapunov optimization technique Neely (2010), with which we can transfer the challenging stochastic optimization problem $\mathcal{P}1$ to be a deterministic per-slot problem in each time slot.

We first define a quadratic Lyapunov function as follows:

$$L(\boldsymbol{Q}(t)) \triangleq \frac{1}{2}\boldsymbol{Q}^{\mathrm{T}}(t)\boldsymbol{Q}(t) = \frac{1}{2}\sum_{u\in\mathcal{U}}(Q_u^{\mathrm{sum}}(t))^2. \tag{19.11}$$

We then define a one-slot conditional Lyapunov drift $\Delta(\boldsymbol{Q}(t))$ as follows:

$$\Delta(\boldsymbol{Q}(t)) \triangleq \mathbb{E}\{L(\boldsymbol{Q}(t+1)) - L(\boldsymbol{Q}(t))|\boldsymbol{Q}(t)\}. \tag{19.12}$$

Accordingly, the one-slot conditional Lyapunov *drift-plus-penalty* function is shown as follows:

$$\Delta_V(\boldsymbol{Q}(t)) = \Delta(\boldsymbol{Q}(t)) - V\mathbb{E}\{\phi(\boldsymbol{\mu}(t))|\boldsymbol{Q}(t)\}, \tag{19.13}$$

where $V > 0$ is the policy control parameter.

19.3.1 Fog-Enabled Multi-Tier Operations Scheduling (FEMOS) Algorithm

Dynamic Online Bandwidth Allocation (DOBA)

To solve Problem $\mathcal{P}1$ based on the Lyapunov optimization method Neely (2010), it needs to design an algorithm to minimize the upper bound of the *Lyapunov drift-plus-penalty* term in each time slot. Ignoring the constant components in the upper bound of $\Delta_V(\boldsymbol{Q}(t))$ and rearranging them, the upper-bound minimization problem is converted to:

$$\min_{\mathbf{X}(t)\in\mathcal{A},\mathbf{v}(t)\in\mathcal{B}} -\sum_{u\in\mathcal{U}}\left[V + Q_u^{\mathrm{sum}}(t)\right]\mu_u(t). \tag{19.14}$$

Note that $Q_u^{\mathrm{sum}}(t)$ is observed at the beginning of each time slot, which can be viewed as constant per time slot. Therefore the upper-bound minimization only depends on $\mu_u(t)$, which involves the FAN assignment and bandwidth allocation. For convenience, we define $W_{uh}(t) \triangleq [V + Q_u^{\mathrm{sum}}(t)]C_{uh}(t)$, which is constant per time slot. From the definition of $\mu_u(t)$ in Eq. (19.4), the upper-bound minimization of $\Delta_V(\boldsymbol{Q}(t))$ is transferred to the following equivalent problem:

$$\mathcal{P}_{\mathrm{AR}}: \max_{\mathbf{X}(t)\in\mathcal{A},\mathbf{v}(t)\in\mathcal{B}} \sum_{h\in\mathcal{H}}\sum_{u\in\mathcal{U}} W_{uh}(t)v_{uh}(t)x_{uh}(t). \tag{19.15}$$

$\mathcal{P}_{\mathrm{AR}}$ is a joint optimization problem of access node assignment and bandwidth allocation. The main idea of the proposed FEMOS algorithm is to solve the deterministic optimization problem $\mathcal{P}_{\mathrm{AR}}$ in each time slot. By doing so, the number of requests waiting in the queues can be maintained at a small level, and network throughput can be maximized at the same time.

Note that $\mathcal{P}_{\mathrm{AR}}$ is a nonlinear integer programming problem, for which the computational complexity of the brute-force search is prohibitive. By exploiting the structure information of

$\mathcal{P}_{\text{AR}}$, we transfer problem $\mathcal{P}_{\text{AR}}$ to the following problem $\mathcal{P}'_{\text{AR}}$, which is proven equivalent with $\mathcal{P}_{\text{AR}}$:

$$\mathcal{P}'_{\text{AR}} : \max_{\mathbf{X}(t)} \sum_{h \in \mathcal{H}} \sum_{u \in \mathcal{U}} W_{uh}(t) x_{uh}(t) \tag{19.16}$$
$$\text{s.t.} \sum_{u \in \mathcal{U}} x_{uh}(t) \leq 1, \quad \forall h \in \mathcal{H},$$
$$\sum_{h \in \mathcal{H}} x_{uh}(t) \leq 1, \forall u \in \mathcal{H}, \qquad x_{uh}(t) \in \{0, 1\},$$

and the corresponding bandwidth allocation for each UT $u \in \mathcal{U}$ can be expressed as follows:

$$\nu_{uh}(t) = \begin{cases} 1 & x_{uh} = 1, \forall h \in \mathcal{H}, \\ 0 & \text{otherwise}. \end{cases} \tag{19.17}$$

Centralized Assignment of Access Node (CAAN)

The dynamic assignment of FANs in each time slot can be obtained by solving the problem $\mathcal{P}'_{\text{AR}}$. Note that the constraint of maximum connectable UTs for each FAN in $\mathcal{P}_{\text{AR}}$ is M, while it reduces to 1 in $\mathcal{P}'_{\text{AR}}$.

To solve $\mathcal{P}'_{\text{AR}}$ efficiently, we demonstrate that the problem $\mathcal{P}'_{\text{AR}}$ can be formulated as the maximization of a normalized modular function subject to the intersection of two partition matroid constraints in the following. This structure can be exploited to design computationally efficient algorithms for Problem $\mathcal{P}'_{\text{AR}}$ with provable approximation gaps.

FEMOS Algorithm

The proposed FEMOS algorithm is summarized in Algorithm 1, which will be implemented at both the FCN tier and the FAN tier in practice. Specifically, in each time slot, within the global network information and queue length of each request buffer, the FCN will execute CAAN by run the greedy FAN assignment algorithm. The assignment decision is then transmitted to the FAN tier. Next, each FAN performs the DOBA independently and schedule the service rate for the associated UT under full efficient scheduling policy. Finally, the FCN updates the request buffers for each UT, the length of which will influence operations scheduling in the next slot.

Algorithm 1 FEMOS algorithm.

1: Set $t = 0$, $\mathbf{Q}(0) = \mathbf{0}$;
2: **While** $t < t_{\text{end}}$, **do**
3: At beginning of the tth time slot, observe $A_u(t)$, $g_{hu}(t)$ and $Q_{uf}(t)$;
4: CAAN:$\mathbf{X}^\star(t)$ is obtained by run Algorithm 1;
5: DOBA:

$$\nu^\star_{uh}(t) = \begin{cases} 1 & x^\star_{uh}(t) = 1 \\ 0 & \text{otherwise}; \end{cases}$$

6: Schedule the service rates $\mu^\star_{uf}(t)$ to the queues $Q_{uf}(t)$ according to Eq. (19.5) with any pre-specified queuing discipline;
7: Update $\{Q_{uf}(t)\}$ according to Eq. (19.6) for each UT based on $\mathbf{X}^\star(t)$, $\boldsymbol{\nu}^\star(t)$ and $\mu^\star_{uf}(t)$;
8: $t \leftarrow t + 1$.
9: **end While**

Interestingly, a closer inspection of $\mathcal{P}'_{\text{AR}}$ reveals that in each time slot, each FAN can associate with at most only one UT, and the number of UTs that can dynamically associate with FANs depends on the number of FANs, which is $|\mathcal{H}|$. What makes our greedy FAN assignment algorithm outstanding from the common assignment method, which directly select FAN with the best channel condition for each UT or each FAN associates with the UT with the longest queue length, is that we provide a quantitative analysis and selection method for UT-FAN association. In each time slot, the association between UTs and FANs depends on corresponding UT-FAN pair gain $W_{uh}(t) = [V + Q_u^{\text{sum}}(t)]C_{uh}(t)$. The greedy FAN assignment algorithm greedily selects the UT-FAN pair with the largest $W_{uh}(t)$ within the feasible set. For each FAN h, it will associate with one UT with either large queue backlog or good channel condition. If V is small, i.e. $V \ll Q_u^{\text{sum}}(t)$, both the queue backlog and the channel condition will determine the decision about UT-FAN association. The FAN h will associate with UT with largest $Q_u^{\text{sum}}(t)C_{uh}(t)$. Conversely, if V is large enough, i.e. $V \gg Q_u^{\text{sum}}(t)$, the FAN h will more effectively invoke the willingness to associate with a UT with a good channel condition $C_{uh}(t)$. Under large V, the UT with weak channel conditions cannot access the network for a long time, leading to large accumulated queue length and influence on the UT-FAN decision in turn. Therefore, the parameter V actually controls the FANs' willingness to serve UTs, i.e. performing UT-FAN association. In other words, it controls the trade-off between network throughput and transmission delay.

19.3.2 Theoretical and Numerical Analysis

19.3.2.1 Theoretical Analysis

We provide the main theoretical results for FEMOS, which characterize the lower bounds for average network throughput as well as the upper bounds for the average sum queue length of the requests of all the UTs. Additionally, the trade-off between the network average throughput and average delay will also be revealed.

Theorem 19.1 For the network defined before, the centralized assignment of access node and dynamic online bandwidth-allocation policy obtained through FEMOS algorithm achieves the following performance:

$$\begin{aligned}\phi_{av}^{\text{FEMOS}} &\triangleq \lim_{t\to\infty} \inf \phi\left(\frac{1}{t}\sum_{\tau=0}^{t-1}\mathbb{E}\{\phi(\boldsymbol{\mu}(\tau))\}\right)\\ &\geq \beta\phi^{\text{opt}} - \frac{\mathcal{K}}{V},\end{aligned} \tag{19.18}$$

$$\begin{aligned}\Lambda_{av}^{\text{FEMOS}} &\triangleq \frac{1}{\sum_u \lambda_u} \lim_{t\to\infty} \sup \frac{1}{t}\sum_{\tau=0}^{t-1}\sum_u \mathbb{E}\{Q_u^{\text{sum}}(t)\}\\ &\leq \frac{\mathcal{K} + V(\phi^{\text{opt}} - \phi_\epsilon)}{\beta\epsilon\sum_u \lambda_u}.\end{aligned} \tag{19.19}$$

More analysis and proof can be found in Zhao et al. (2018). Theorem 1 shows that under the proposed fog-enabled multi-tier operation scheduling algorithm, the lower bound of average network throughput increases inversely proportional to V, while the upper bound of average service delay per UT experienced increases linearly with V. If a larger V is used to pursue the better network throughput performance, it will introduce severe service delay. Hence, there exists an $[O(1/V), O(V)]$ trade-off between these two objects. Through adjusting V, we can balance the network throughput and service delay.

19.3.2.2 Numerical Analysis

We consider a network with $|\mathcal{H}| = 9$ fixed FANs and $|\mathcal{U}| = U$ randomly deployed UTs. Each FAN can associate with at most $M = 12$ UTs. The size of the file types in the file library is set to $|\mathcal{F}| = 1000$ and the cached file types of each FAN is set to $|\mathcal{N}| = 500$. The system region has a size of $20 \times 20\ m^2$. Each UT requests files with type $f \in \mathcal{F}$ according to the Zipf distribution with parameter η_r, i.e. $p_f = \frac{f^{-\eta_r}}{\sum_{i \in \mathcal{F}} i^{-\eta_r}}$ Golrezaei et al. (2014) and $\eta_r = 0.56$ in our simulations. A FIFO queuing discipline is applied in the simulations. The simulation results are averaged over 3000 constant time slots with $\mathcal{T} = 100$ milliseconds intervals.

We assume that each FAN operates on a $B_h = 18$MHz bandwidth (100 resource blocks with 180 kHz for each resource block) and transmits at a fixed power level $P = 20\quad W$. In addition, the noise power σ^2 is assumed to be $2 \times 10^{-7}\ W$. Based on the WINNER II channel model in small-cell scenarios Khan and Oestges (2011), the pathloss coefficients between the FAN h and UT u is defined by $g_{hu}(t) = 10^{-\frac{PL(d_{hu}(t))}{10}}$, where $d_{hu}(t)$ is the distance from FAN h to UT u in time slot t, and $PL(d) = A\log_{10}(d) + B + C\log_{10}(f_0/5) + \mathcal{X}_{dB}$, where f_0 is the carrier frequency, $\mathcal{X}_{dB}$ is a shadowing log-normal variable with variance σ^2_{dB}; $A = 18.7$, $B = 46.8$, $C = 20$, and $\sigma^2_{dB} = 9$ in line-of-sight (LOS) condition; $A = 36.8$, $B = 43.8$, $C = 20$, and $\sigma^2_{dB} = 16$ in non-line-of-sight (NLOS) condition. Each link is in LOS or NLOS independently and randomly, with probabilities of $p_l(d)$ and $1 - p_l(d)$, respectively, where

$$p_l(d) = \begin{cases} 1 & d \leq 2.5\quad m \\ 1 - 0.9(1 - (1.24 - 0.6\log(d))^3)^{1/3} & \text{otherwise.} \end{cases} \tag{19.20}$$

In this subsection, we evaluate the performance of the proposed FEMOS algorithm and compare its greedy FAN assignment with the other two dynamic schemes: *select best channel* (SBC) and *selecq longest queue* (SLQ). In the SBC scheme, each FAN associates with the UT with the best channel condition between them in each time slot. In the SLQ scheme, in each time slot, the UTs with top-$|\mathcal{H}|$ queue length access the network, and each of them associates with the FAN with the best transmission condition.

Figure 19.2 first validates the theoretical results for the proposed FEMOS algorithm derived in Theorem 1. The average network throughput performance is shown in Figure 19.2a, while the average service delay per UT is shown in Figure 19.2b. It can be observed from Figure 19.2a that the average network throughput obtained by the greedy FAN assignment in FEMOS increases as V increases and converges to the maximum value when V is sufficiently large. Meanwhile, as shown in Figure 19.2b, the average service delay experienced by per UT of FEMOS increases linearly with the control parameter V. This is in accordance with the analysis that with V increasing, the importance of request queue backlogs decreases, which makes the dynamic assignment bias toward good channel conditions. Those observations verify the $[O(1/V), O(V)]$ trade-off between average network throughput and average queue backlog as demonstrated in Theorem 1.

Figure 19.2 also compares the greedy FAN assignment in FEMOS with SBC and SLQ assignment schemes. We observe that the average network throughput performance of SBC stabilizes around 1.7×10^4 kbits/slot, which approaches the maximum value of FEMOS. However, the average network throughput performance deteriorates severely under SLQ assignment scheme for any V. The average service delay in both SBC and SLQ assignment schemes is much larger than that of greedy assignment in FEMOS when $V < 2000$. Specifically, when $V = 1$, the average delay in FEMOS approaches zero, but it is $5s$ and $45s$ for the SLQ and SBC schemes, respectively. These comparisons demonstrate the advantages of the greedy assignment in FEMOS.

In Figure 19.3, we further explore the relationship between network throughput and average service delay of the FEMOS algorithm regarding different workloads ($A_{\max}$) and different UT amounts (U). Specifically, in Figure 19.3a we explore the average network throughput versus

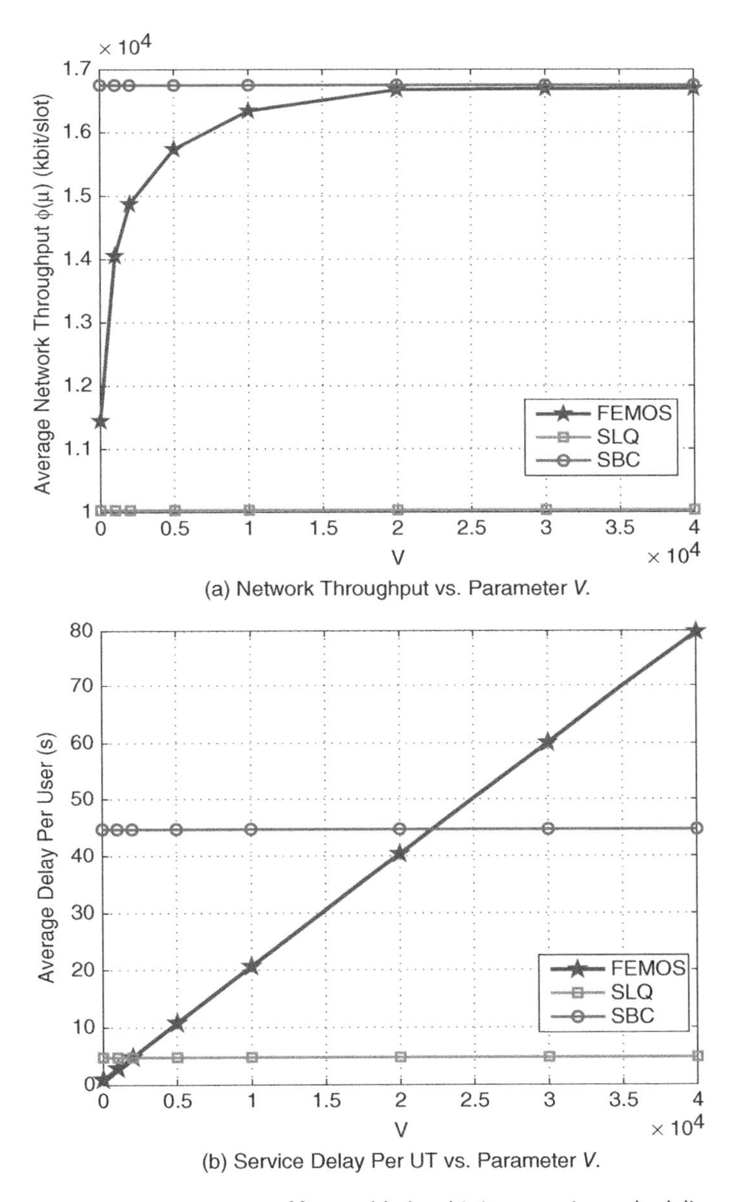

(a) Network Throughput vs. Parameter V.

(b) Service Delay Per UT vs. Parameter V.

Figure 19.2 Comparison of fog-enabled multi-tier operations scheduling and other fog access node assignment schemes, $A_{max} = 100$ kbits, $U = 100$.

average service delay in terms of different numbers of workloads ($A_{max} = 60$ kbits, 80 kbits, 100 kbits, and 120 kbits, respectively). The random caching strategy is adopted here. It can be observed that for a given control parameter V, the proposed FEMOS algorithm with a lower workload obtains better network throughput performance and experiences shorter service delay. For example, when $V = 2000$, the average delay per UT experienced in FEMOS with $A_{max} = 120$ kbits is 3.2 s, which is about twice as large a that with $A_{max} = 60$ kbits. The average network throughput obtained by FEMOS with $A_{max} = 60$ kbits is 1.512×10^4 kbits/slot when $V = 2000$, while it is 1.317×10^4 kbits/slot obtained by FEMOS with $A_{max} = 120$ kbits.

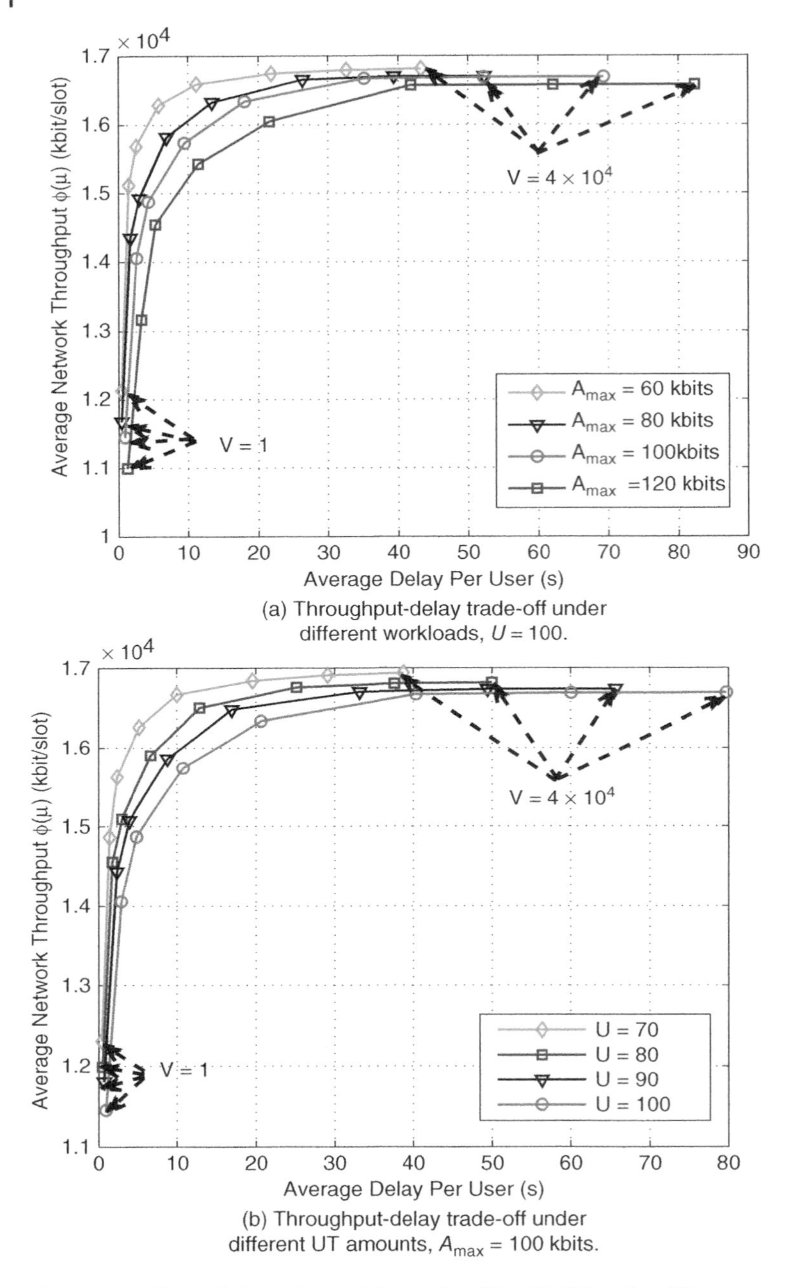

Figure 19.3 Network throughput-delay trade-off for FEMOS under different parameters.

The reason behind these observations is, according to Eqs. (19.18) and (19.19) in *Theorem 1*, both the lower bound of the average network throughput ϕ_{av}^{FEMOS} and the upper bound of average service delay $\Lambda_{av}^{\text{FEMOS}}$ are determined by parameters V and $\mathcal{K}$, where $\mathcal{K} = \frac{|\mathcal{U}|}{2}(\mu_{\max}^2 + A_{\max}^2)$. Larger workload leads to larger $\mathcal{K}$, and thus a smaller average network throughput and a longer average service delay are obtained.

In Figure 19.3b, we show the impacts of UT amount U on the network throughput-delay trade-off performance of the FEMOS algorithm. A similar phenomenon exists for the curves with different workloads. A network throughput-delay trade-off also exists under different UT amounts. In addition, it can be observed that by increasing U, the average service delay increases and the average network throughput decreases for a given V. This is also consistent with Theorem 1 that the UT amount U influences the value of parameter $\mathcal{K}$, and thus impacts both throughput and service-delay performance. Intuitively, in this case along with different workload scenarios, the FCN has to fully utilize the network resources to serve UTs' traffic demands, and either high workload or large UT amount certainly will deteriorate both network throughput and delay performance.

19.4 Predictive Multi-tier Operations Scheduling (PMOS) Algorithm

Motivated by recent advancements in data mining for learning user behaviors Kumar and Tomkins (2010), the FCN is assumed to be aware of users' future request information within a limited future time window. With this predictive information, the FCN can control the FANs serving the upcoming requests and pre-push files to UTs beforehand when the link condition is good, instead of waiting for them to submit their requests, which may lead to a large service latency.

19.4.1 System Model

Similar to the system model in Zhao et al. (2018), a general fog-enabled multi-tier network architecture is shown in Figure 19.4. The network involves a powerful FCN tier with both strong computation and sufficient storage capacities, a FAN tier with limited storage capacity and computation capacities but strong communication capacity, and a set of multiple stationary or low-mobility UTs in the region under consideration. Each UT possesses small storage

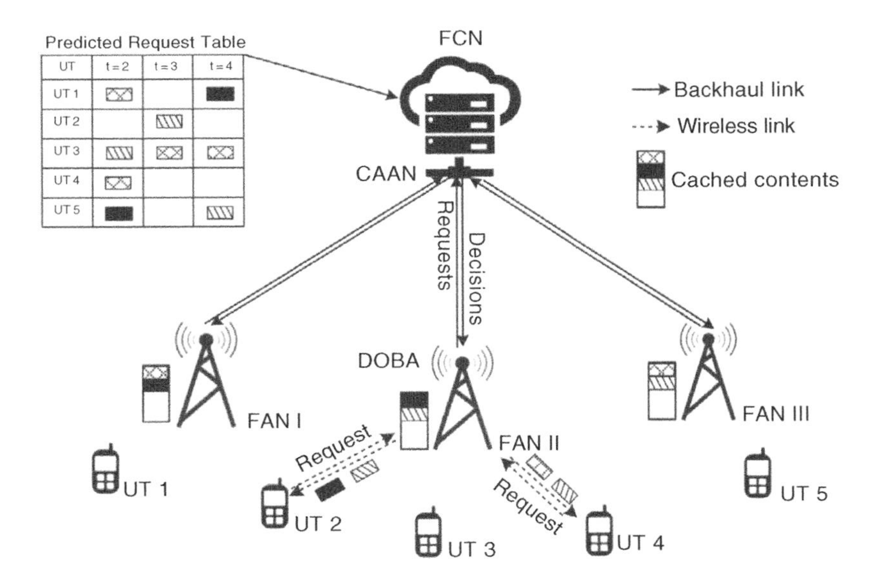

Figure 19.4 A sample fog-enabled multi-tier network with three FANs and five UTs, operating predictive scheduling in slot $t = 1$.

capacity, minor communication ability, and little or no computation ability. Each FAN in the FAN tier caches a subset of popular files. FANs are connected to a FCN through backhaul links, which is next to the cloud and core network and acts as the server of the file library. UTs request files to be downloaded from the FAN tier through wireless links. Due to the restricted transmit power and dynamic wireless environment, the communication channels between a FAN and its neighboring UTs are unreliable and time varying.

Different from the model in Zhao et al. (2018), the FCN here is powerful enough to predict the UTs' request information within a limited future time window. With the global information about the network and the limited future information, the FCN performs the centralized assignment of access node for UTs and controls the FANs to preserve the future requests in advance when link condition is good. Then each FAN in the FAN tier executes a predictive online bandwidth allocation (POBA) scheme and proactive online service rate-scheduling scheme locally, to guarantee the transmission delay and ensure the QoE for UTs.

We introduce a *prediction window* in our predictive scheduling model that was developed by Huang et al. (2016a). Specifically, the FCN is assumed to have access to future arrival information $\{A_u(t), \dots, A_u(t + D_u - 1)\}$ and $\{I_{uf}(t), \dots, I_{uf}(t + D_u - 1)\}, \forall u \in \mathcal{U}$ within the prediction window. Here D_u is the prediction window size of UT u with $D_u \geq 1$. Without loss of generality, the imperfect prediction is considered in such a predictive scheduling model.

The expression $\{\tilde{\mu}^d_{uf}(t)\}_{d=0}^{D_u-1}$ is introduced to denote the service rate scheduled by the FCN in time slot t serving for arrival requests in time slot $t + d$. Here, $d = \{0, 1, \dots D_u - 1\}$ is the predictive phase. Let $\tilde{\mu}^{-1}_{uf}(t)$ denote the service rate allocated for the file requests already in the system queues in time slot t.

In each time slot, the arrived but not yet served requests will be queued in the request buffers at the FCN with sufficiently large capacity. Next, we introduce prediction queues, which record the residual requests for different type of files in the prediction window $[t, t + D_u - 1]$. Specifically, $\tilde{Q}^{-1}_{uf}(t)$ denotes the number of file request queues already in the system at the beginning of time slot t. The expression $\tilde{Q}^d_{uf}(t)$ denotes the number of file request queues in future slot $t + d$. Note that $\tilde{Q}^{-1}_{uf}(t)$ is the only actual backlog in the network and the network is stable if and only if $\tilde{Q}^{-1}_{uf}(t)$ is stable. The expression $\{\tilde{Q}^d_{uf}(t)\}_{d=0}^{D_u-1}$ represents virtual queues that simply record the residual arrivals in the prediction window.

Under the scenario of imperfect prediction, the false predicted arrivals will appear in prediction queues $\tilde{Q}^d_{uf}(t)$ $(0 \leq d \leq D_u - 1)$, and the fraction of false predicted arrivals is e on average. Thus, the effective queue length of $\tilde{Q}^d_{uf}(t)$ is $(1 - e)\tilde{Q}^d_{uf}(t)$. Define $\tilde{\mathbf{Q}}(t) \triangleq [\tilde{Q}^{\text{sum}}_1(t), \dots, \tilde{Q}^{\text{sum}}_{|\mathcal{U}|}(t)]$ as the vector of total effective queue length of request buffers, specifically,

$$\tilde{Q}^{\text{sum}}_u(t) = \sum_{f \in \mathcal{F}} \tilde{Q}^{-1}_{uf}(t) + (1 - e) \sum_{f \in \mathcal{F}} \sum_{d=0}^{D_u-1} \tilde{Q}^d_{uf}(t). \tag{19.21}$$

The prediction queues $\{\tilde{Q}^d_{uf}\}_{d=-1}^{D_u-1} (\forall\, u \in \mathcal{U}, f \in \mathcal{F})$ are updated according to the following rules Huang et al. (2016a):

1. If $d = D_u - 1$, then:

$$\tilde{Q}^d_{uf}(t + 1) = A_u(t + D_u) \cdot I_{uf}(t + D_u). \tag{19.22}$$

2. If $0 \leq d \leq D_u - 2$, then:

$$\tilde{Q}^d_{uf}(t + 1) = \left[\tilde{Q}^{d+1}_{uf}(t) - \tilde{\mu}^{d+1}_{uf}(t)\right]^+. \tag{19.23}$$

3. If $d = -1$, then:

$$\begin{aligned}\tilde{Q}_{uf}^{-1}(t+1) = {} & \left[\tilde{Q}_{uf}^{-1}(t) - \tilde{\mu}_{uf}^{-1}(t)\right]^{+} \\ & + (1-e)\left[\tilde{Q}_{uf}^{0}(t) - \tilde{\mu}_{uf}^{0}(t)\right]^{+},\end{aligned} \tag{19.24}$$

where $[x]^{+} = \max\{x, 0\}$, $\tilde{Q}_{uf}^{-1}(0) = 0$ and $\tilde{Q}_{uf}^{d}(0) = A_u(0+d) \cdot I_{uf}(0+d)$.

The queuing process $\tilde{Q}_{u}^{\text{sum}}(t)$ is stable if the following condition holds Neely (2010):

$$\overline{Q}_{u}^{\text{sum}} = \lim_{t\to\infty} \frac{1}{t} \sum_{\tau=0}^{t-1} \mathbb{E}[\tilde{Q}_{u}^{\text{sum}}(\tau)] < \infty. \tag{19.25}$$

As opposed to the system without prediction, in predictive scheduling, the system schedules the service rate $\{\tilde{\mu}_{uf}^{d}(t)\}_{d=-1}^{D_u-1}$ serving the request that has already or will arrive in the system according to a certain rate-allocation discipline Huang et al. (2016a), such as FIFO, LIFO, or random discipline. We adopt a fully efficient scheduling policy for queues, which means:

$$\sum_{f\in\mathcal{F}} \sum_{d=-1}^{D_u-1} \tilde{\mu}_{uf}^{d}(t) = \mu_u(t), \tag{19.26}$$

and where $\mu_u(t)$ is defined in (19.4) and $\tilde{\mu}_{uf}^{d}(t) = 0$ if $y_{hf} = 0$.

The problem formulation and analysis framework are the same as those with FEMOS in the previous sections. Based on this, we will show the theoretical and numerical analysis for PMOS in the following.

19.4.2 Theoretical Analysis

The term ϕ_{av}^{PMOS} is defined as the long-term expected average network throughput of PMOS, and Q_{av}^{PMOS} is defined as the long-term expected average queue length of users. The performance of PMOS is described in the following theorem.

Theorem 19.2 PMOS achieves the following average network throughput:

$$\phi_{av}^{\text{PMOS}} \triangleq \liminf_{t\to\infty} \phi\left(\frac{1}{t}\sum_{\tau=0}^{t-1}\mathbb{E}\{\boldsymbol{\mu}(\tau)\}\right) \geq \phi_{\beta}^{\text{opt}} - \frac{\mathcal{K}}{V}, \tag{19.27}$$

with bounded queue backlog:

$$\begin{aligned}Q_{av}^{\text{PMOS}} & \triangleq \limsup_{t\to\infty} \frac{1}{t}\sum_{\tau=0}^{t-1}\sum_{u}\mathbb{E}\{\tilde{Q}_{u}^{\text{sum}}(t)\} \\ & \leq \frac{\mathcal{K} + V(\phi^{\text{opt}} - \phi_{\epsilon})}{\beta\epsilon},\end{aligned} \tag{19.28}$$

where $\phi_{\beta}^{\text{opt}}$ is the optimal expected average network throughput for the β-reduced problem defined in Zhao et al. (2018), $\beta = \frac{1}{2}$, and $\mathcal{K}$ is a constant.

The theorem shows that under the proposed PMOS algorithm, the lower bound of average network throughput increases inversely proportional to V, which performs the optimality of the proposed algorithm. It also demonstrates that the upper bound of average queue length Q_{av}^{PMOS}, including the true backlog $\tilde{Q}_{uf}^{-1}(t)$ and prediction queues $\{\tilde{Q}_{uf}^{d}(t)\}_{d=0}^{D_u-1}$, increases linearly with V. Hence, there exists an $[O(1/V), O(V)]$ trade-off between these two objects. Through adjusting V, we can balance the network throughput and average delay.

In the following, we will show the average backlog reduction due to predictive scheduling in PMOS. To do so, we employ one theorem from Huang and Neely (2011), which shows that the queue vector of the network is within distance $O(\log(V))$ from a fixed point. First, we define the following optimization problem:

$$\max: \quad g(\boldsymbol{\ell}), \qquad \text{s.t.} \quad \boldsymbol{\ell} \succeq 0, \tag{19.29}$$

where $g(\boldsymbol{\ell})$ is called the dual function with scaled objective (by V) of the original problem that maximizes the average network throughput. The expression $\boldsymbol{\ell} = [l_1, \dots, l_{|\mathcal{U}|}]$ is the Lagrange multiplier. The term $g(\boldsymbol{\ell})$ is defined as follows:

$$g(\boldsymbol{\ell}) = \inf_{\boldsymbol{\mu}(t)} \mathbb{E}\left\{ V\phi(\boldsymbol{\mu}(t)) + \sum_{u\in\mathcal{U}} l_u[\lambda_u - \mathbb{E}\{\mu_u(t)\}] \right\}.$$

Let $\boldsymbol{\ell}^*$ denote the optimal solution of problem (19.29) and $\boldsymbol{\ell}^*$ be either $O(V)$ or 0 according to Huang and Neely (2011). Now we have the following Theorem 3, which is listed as Theorem 1 in Huang and Neely (2011).

Theorem 19.3 Suppose that (i) $\boldsymbol{\ell}^*$ is unique and dual function $g(\boldsymbol{\ell})$ satisfies:

$$g(\boldsymbol{\ell}^*) \geq g(\boldsymbol{\ell}) + L \parallel \boldsymbol{\ell}^* - \boldsymbol{\ell} \parallel, \qquad \forall \boldsymbol{\ell} \succeq 0, \tag{19.30}$$

for some constant $L > 0$ independent of V, (ii) the θ-slack condition is satisfied with $\theta > 0$ Zhao et al. (2018). Then, there exists constants G, K, and c, such that for any $m \in \mathbb{R}_+$,

$$\mathcal{P}_r(G, Km) \leq ce^{-m}, \tag{19.31}$$

where $\mathcal{P}_r(G, Km)$ is defined as $\mathcal{P}_r(G, Km) \triangleq \limsup_{t\to\infty} \frac{1}{t} \sum_{\tau=0}^{t-1} Pr\{\exists u | \tilde{Q}_u^{\text{sum}}(\tau) - l_u^*| > G + Km\}$.

The proof can be obtained in Huang and Neely (2011), which is omitted here for the sake of space.

Next, we state Theorem 4 regarding the average backlog reduction due to predictive scheduling.

Theorem 19.4 Suppose that (i) the assumption in Theorem 2 holds, (ii) there exists a steady-state distribution of $\tilde{\boldsymbol{Q}}(t)$ under PMOS, (iii) $D_u = O\left(\frac{1}{A_{\max}}[l_u^* - G - K(\log(V))^2 - \mu_{\max}]^+\right)$ for all $u \in \mathcal{U}$, and (iv) FIFO is used in PMOS. Then PMOS achieves the following result with a sufficiently large V:

$$\tilde{Q}_{av}^{-1} \leq Q_{av}^{\text{FEMOS}} - \sum_{u\in\mathcal{U}} D_u\left[\lambda_u - O\left(\frac{1}{V^{\log(V)}}\right)\right]^+, \tag{19.32}$$

where Q_{av}^{FEMOS} denotes the queue backlog of fog-enabled multi-tier operations scheduling without prediction.

According to the proof given in Huang et al. Huang et al. (2016a), Theorem 4 implies that compared with multi-tier operations scheduling without prediction, the average true queue length (average delay) performance of the proposed PMOS is roughly reduced by $\sum_{u\in\mathcal{U}} \lambda_u D_u$.

19.4.3 Numerical Analysis

The detailed simulation parameters are the same as those for the earlier FEMOS algorithm. We first evaluate the performance of PMOS under different prediction window sizes and compare it with the FEMOS algorithm in Zhao et al. (2018) without prediction ($D = 0$). The FIFO rate-allocation discipline is adopted and the perfect prediction is considered, $e = 0$. The maximum network traffic arrival is set to $A_{\max} = 100$ kbits.

Figure 19.5a shows the average network throughput against V for PMOS with different prediction window sizes $D \in \{5, 10, 15\}$. It can be observed from Figure 19.5a that the average network throughput increases and converges to the maximum value when V is sufficient large. In addition, Figure 19.5a also shows that PMOS with different prediction window sizes can achieve the same average network throughput.

Figure 19.5b presents the average delay per UT against V for PMOS. It can be observed that the average delay experienced by per UT increases linearly with the control parameter V, which along with the observations in Figure 19.5a verifies the $[O(1/V), O(V)]$ trade-off between average network throughput and average queue backlog derived in Theorem 1. This indicates that a proper V should be chosen to balance these two objects. We also observe that PMOS always generates a smaller average delay than the multi-tier operations scheduling without prediction ($D = 0$). As the predictive window size D increases, the average delay per UT decreases almost linearly in D, which is in accordance with the theoretical analysis in Theorem 4. The reason behind those observations is that the predictive information helps the operator design a better FAN assignment and bandwidth allocation strategy, which utilizes the network more efficiently.

19.5 A Multi-tier Cost Model for User Scheduling in Fog Computing Networks

Although different aspects of user scheduling in multi-tier fog computing networks have been widely discussed in the literature, such as Zhao et al. (2018), Shah-Mansouri and Wong (2018), Liu et al. (2018b,a), the effective user scheduling scheme still faces challenges, especially when the cost model is considered. Generally, the FCN is operated by a telecom operator that signs a service contract with UTs, while the FANs belong to different individuals. To better motivate FANs to share resources and anticipate in caching, the cost model, especially for FANs, should be taken into consideration.

19.5.1 System Model and Problem Formulation

System Model

Just like Figure 19.1, a multi-tier fog computing network consisting of one FCN, M FANs, and N UTs is considered. As mentioned, the FCN is operated by a telecom service provider, which provides services to N UTs, i.e. service subscribers, while the FANs belong to different individual owners. To reduce service delay and improve QoS, the FCN is willing to pay money to FANs if they provide services to UTs. For ease of expression, we take caching as an example in the following context. [1] In the fog-enabled caching network, the FCN can allocate files to FANs during off-peak time, i.e. file placement, and thus the UTs can be associated with proper FANs or FCN to download files during peak time, i.e. user scheduling.

1 Our model and algorithm also apply to other services, such as computing.

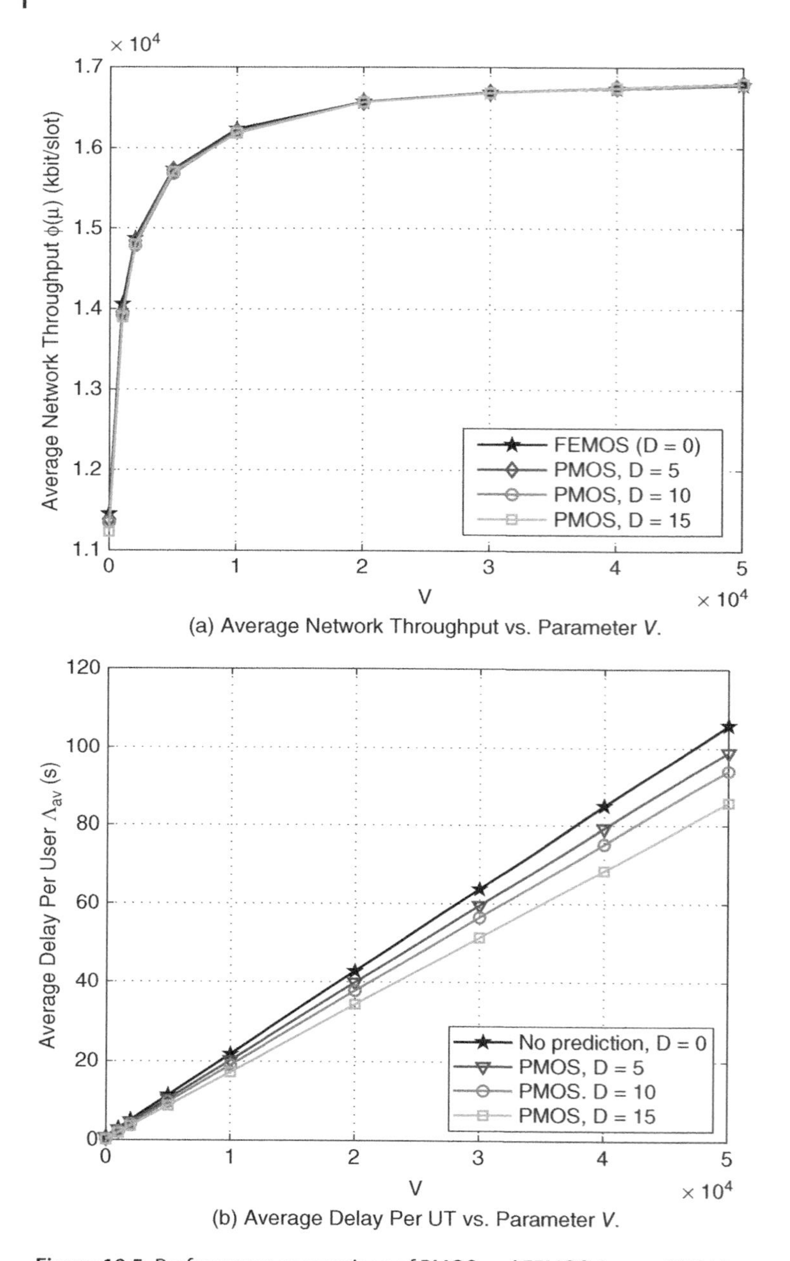

Figure 19.5 Performance comparison of PMOS and FEMOS,$A_{\max}$ = 100kbits, $e = 0$.

Denote the set of N UTs and the set of M FANs by $\mathcal{N} = \{1, ..., N\}$ and $\mathcal{M} = \{1, ..., M\}$, respectively. We further denote the library of F files as $\mathcal{F} = \{1, ..., F\}$. Without loss of generality, all files are assumed to have a uniform size with L bits. Define the association vector of UT n as $\mathbf{a}_n = (a_{n,0}, ..., a_{n,M})$, where $a_{n,x} \in \{0, 1\}$, $x \in \{0\} \cup \mathcal{M}$, with $\sum_{x \in \{0\} \cup \mathcal{M}} a_{n,x} = 1$, is an association indicator between UT n and FCN, FANs. To be specific, $a_{n,m} = 1$ indicates that UT n is associated with FAN m; otherwise, $a_{n,m} = 0$. Especially, $a_{n,0}$ is the association indicator between

UT n and FCN. We further define the association profile as $\mathbf{A} = (\mathbf{a}_1^T, \mathbf{a}_2^T, ..., \mathbf{a}_N^T)^T$. The FANs usually have constrained storage size and communication capability, and thus can cache limited files and serve limited UTs. We further introduce a matrix $\mathbf{B} = (\mathbf{b}_1^T, \mathbf{b}_2^T, ..., \mathbf{b}_N^T)^T$, with $\mathbf{b}_n = \{b_{n,1}, b_{n,2}, ..., b_{n,M}\}$, $b_{n,m} \in \{0, 1\}, \forall n \in \mathcal{N},\ m \in \mathcal{M}$, to denote the availability of UTs' requested files at FANs, and matrix $\mathbf{C} = (\mathbf{c}_1^T, \mathbf{c}_2^T, ..., \mathbf{c}_N^T)^T$ to denote the connectivity between UTs and FANs, where $\mathbf{c}_n = \{c_{n,1}, c_{n,2}, ..., c_{n,M}\}$, $c_{n,m} \in \{0, 1\}$, $\forall n \in \mathcal{N},\ m \in \mathcal{M}$. Specifically, if $b_{n,m} = 1$, the requested file of UT n is available at FAN m; otherwise, $b_{n,m} = 0$. If $c_{n,m} = 1$, UT n can connect to FAN m; otherwise, $c_{n,m} = 0$.

Similar to many previous works, such as Shah-Mansouri and Wong (2018), Liu et al. (2018b, 2018), Yang et al. (2018d), a quasi-static scenario, wherein the UTs remain unchanged during a user scheduling interval, is assumed.

Cost Model

Service Delay If UT n is associated with FAN m, the downloading delay of a file can be expressed as

$$t_{n,m} = \frac{L}{R_{n,m}}, \tag{19.33}$$

where $R_{n,m}$ is the transmission rate from FAN m to UT n.

If UT n cannot obtain the requested file from neighboring FANs or it is more cost-effective to get the file from FCN, UT n will be associated with FCN. Similarly, the downloading delay of a file can be written as

$$t_{n,0} = \frac{L}{R_{n,0}}, \tag{19.34}$$

where $R_{n,0}$ is the transmission rate from FCN to UT n.

Payment Just like the payment scheme for online advertisement, i.e. cost per click Moon and Kwon (2011), the FANs are assumed to charge by usage amounts or downloads. To motivate more UTs to download files from them, and thus earn more revenue, the FANs set their prices as an inverse demand function Lã et al. (2016). Assume a linear inverse demand function, and the price for single download or the payment for downloading a file from FANs is given by

$$\alpha_m - \beta_m \sum_{n=1}^{N} a_{n,m}, \forall m \in \mathcal{M}, \tag{19.35}$$

where α_m and β_m are two price-related constants set by FAN m. It is worth noting that the FCN pays money to FANs for UTs downloading files from them.

If the UTs download files from the FCN, the FCN will pay extra for electric power consumption. Without loss of generality, a constant payment or cost γ per download is considered here.

Cost Function The overall cost function for the FCN is defined as a combination of service delay and payment Shah-Mansouri and Wong (2018), Yang et al. (2018a), Chen et al. (2016b), which is given by

$$\begin{aligned} O_n(\mathbf{a}_n, \mathbf{A}_{-n}) = {} & \lambda_n^T a_{n,0} \frac{L}{R_{n,0}} + \lambda_n^C a_{n,0} \gamma + \lambda_n^T \sum_{m=1}^{M} a_{n,m} \frac{L}{R_{n,m}} \\ & + \lambda_n^C \sum_{m=1}^{M} a_{n,m} \left(\alpha_m - \beta_m \sum_{n=1}^{N} a_{n,m} \right), \end{aligned} \tag{19.36}$$

where $\lambda_n^T, \lambda_n^C \in [0, 1]$ denote the weighting parameters of service delay and payment set by the FCN, respectively, and $\mathbf{A}_{-n}$ is the association vectors of all UTs except n. If a UT is more sensitive to delay, or a UT subscribes to better services, λ_n^T is higher; otherwise, λ_n^T may be lower.

Problem Formulation

Focusing on the user scheduling problem, the solution becomes to minimize the overall cost of the FCN, i.e.

$$\min_{\mathbf{A}} \sum_{n=1}^{N} O_n(\mathbf{a}_n, \mathbf{A}_{-n}) \tag{19.37a}$$

$$s.t.\ a_{n,0}, a_{n,m} \in \{0, 1\}, \forall n \in \mathcal{N}, m \in \mathcal{M}, \tag{19.37b}$$

$$a_{n,0} + \sum_{m \in \mathcal{M}} a_{n,m} = 1, \tag{19.37c}$$

$$a_{n,m} \leq b_{n,m}, \forall m \in \mathcal{M}, \tag{19.37d}$$

$$a_{n,m} \leq c_{n,m}, \forall m \in \mathcal{M}. \tag{19.37e}$$

Constraints (19.37b) and (19.37c) ensure that each UT is associated with only one FAN or FCN. Constraint (19.37d) guarantees that each UT is associated with the FANs that cache its requested file. Constraint (19.37e) assures that each UT is associated with the FANs to which it can connect.

The optimization problem (19.37a) is an NP-hard combinatorial problem, which has a high computational complexity. In the following section, we will reformulate the problem into a user scheduling game, which can be proven to be a potential game, and thus can be effectively solved by a distributed algorithm called COUS.

19.5.2 COUS Algorithm

Algorithm Design

Define our user scheduling game as $G = \{\mathcal{N}, \{\mathcal{A}_n\}_{n \in \mathcal{N}}, \{O_n\}_{n \in \mathcal{N}}\}$, where $\mathcal{A}_n = \{\mathbf{a}_n | a_{n,0}, a_{n,m} \in \{0, 1\}, a_{n,0} + \sum_{m \in \mathcal{M}} a_{n,m} = 1, a_{n,m} \leq b_{n,m}, a_{n,m} \leq c_{n,m}, \forall m \in \mathcal{M}\}$is the association strategy space of UT n.

Definition 19.1 The *best-response function* $b_n(\mathbf{A}_{-n})$ of UT n to the given $\mathbf{A}_{-n}$ is a set of strategies for UT n such that

$$b_n(\mathbf{A}_{-n}) = \{\mathbf{a}_n | O_n(\mathbf{a}_n, \mathbf{A}_{-n}) \leq O_n(\mathbf{a}'_n, \mathbf{A}_{-n}), \forall\ \mathbf{a}'_n \in \mathcal{A}_n\}. \tag{19.38}$$

Definition 19.2 An association profile $\overline{\mathbf{A}} = \{\overline{\mathbf{a}}_1^T, \overline{\mathbf{a}}_2^T, \ldots, \overline{\mathbf{a}}_N^T\}^T$ is a *pure-strategy Nash equilibrium* of the user scheduling game G if and only if

$$\overline{\mathbf{a}}_n \in b_n(\overline{\mathbf{A}}_{-n}),\ \forall n \in \mathcal{N}. \tag{19.39}$$

At the Nash equilibrium (NE) point $\overline{\mathbf{A}}$, no UT can change its association strategy to further reduce its cost, while keeping other UTs' association strategies fixed.

Theorem 19.5 The user scheduling game G possesses at least one pure-strategy NE and guarantees the *finite improvement property*.

The critical step of proof is to prove that the game G is a *weighted potential game* Monderer and Shapley (1996) with potential function

$$\begin{aligned}\Phi(\mathbf{A}) = &\sum_{n=1}^{N}\frac{\lambda_n^T}{\lambda_n^C}a_{n,0}\frac{L}{R_{n,0}} + \sum_{n=1}^{N}a_{n,0}\gamma + \sum_{n=1}^{N}\frac{\lambda_n^T}{\lambda_n^C}\sum_{m=1}^{M}a_{n,m}\frac{L}{R_{n,m}}\\ &+\sum_{n=1}^{N}\sum_{m=1}^{M}a_{n,m}\alpha_m - \frac{1}{2}\sum_{n=1}^{N}\sum_{k=1,k\neq n}^{N}\sum_{m=1}^{M}\beta_m a_{n,m}a_{k,m},\\ &-\sum_{n=1}^{N}\sum_{m=1}^{M}\beta_m a_{n,m}^2\end{aligned} \tag{19.40}$$

such that

$$\begin{aligned}&O_n(\mathbf{a}_n, \mathbf{A}_{-n}) - O_n(\mathbf{a}'_n, \mathbf{A}_{-n}) = w_n(\Phi(\mathbf{a}_n, \mathbf{A}_{-n}) - \Phi(\mathbf{a}'_n, \mathbf{A}_{-n})),\\ &\forall\, \mathbf{a}_n, \mathbf{a}'_n \in \mathcal{A}_n, \mathbf{A}_{-n} \in \prod_{m\neq n}\mathcal{A}_m,\end{aligned} \tag{19.41}$$

where $(w_n)_{n\in\mathcal{N}}$ is a vector of positive numbers, i.e. weights.

As stated in Theorem 5, any asynchronous better or best response update process is guaranteed to reach a pure-strategy NE within a finite number of iterations. By employing such a property, a distributed user scheduling algorithm called the COUS algorithm can be obtained and shown in Algorithm 2 according to Chen et al. (2016b).

Price of Anarchy

In game theory, price of anarchy (PoA) is most often used to evaluate the efficiency of an NE solution. It answers the question of how far the overall performance of an NE is from the socially optimal solution. To be specific, let Γ be the set of NEs of the user scheduling game G and

Algorithm 2 COUS algorithm.

1: **initialization:**
2: each UT n chooses to be associated with FCN, i.e. $\mathbf{a}_n(0) = [1, 0, \dots, 0]$.
3: **end initialization**
4: **repeat** for each UT n and each iteration in parallel:
5: send the pilot signal to FCN and available FANs.
6: receive the necessary information from FCN and available FANs.
7: compute the best response $b_n(\mathbf{A}_{-n}(t))$.
8: **if** $\mathbf{a}_n(t) \notin b_n(\mathbf{A}_{-n}(t))$ **then**
9: send RTU message to FCN for contending for the association strategy update opportunity.
10: **if** receive the UP message from FCN **then**
11: update the association strategy $\mathbf{a}_n(t+1) \in b_n(\mathbf{A}_{-n}(t))$ for next iteration.
12: **else**
13: maintain the current association strategy $\mathbf{a}_n(t+1) = \mathbf{a}_n(t)$ for next iteration.
14: **end if**
15: **else**
16: maintain the current association strategy $\mathbf{a}_n(t+1) = \mathbf{a}_n(t)$ for next iteration.
17: **end if**
18: **until** END message is received from FCN.

$\mathbf{A}^* = \{\mathbf{a}_1^{*T}, \mathbf{a}_2^{*T}, \cdots, \mathbf{a}_N^{*T}\}^T$ be the centralized optimal solution that minimizes the system cost. Then, the PoA is defined as

$$PoA = \frac{\max_{\mathbf{A}\in\Gamma} \sum_{n\in\mathcal{N}} O_n(\mathbf{A})}{\sum_{n\in\mathcal{N}} O_n(\mathbf{A}^*)}. \quad (19.42)$$

For the user scheduling game G, we have the following theorem.

Theorem 19.6 For the user scheduling game G, the PoA of the overall cost satisfies that

$$1 \leq PoA \leq \frac{\sum_{n=1}^{N} \min\{O_{n,0}, O_{n,m}^{\min}\}}{\sum_{n=1}^{N} \min\{O_{n,0}, O_n^{\min}\}}, \quad (19.43)$$

where $O_{n,0} \triangleq \lambda_n^T L/R_{n,0} + \lambda_n^T \gamma$, $\min_{m\in\mathcal{M}}(\lambda_n^T L/R_{n,m} + \lambda_n^C(\alpha_m - \beta_m)) \triangleq O_{n,m}^{\min}$, and $\min_{\mathbf{A}'\in\prod_{n\in\mathcal{N}}\mathcal{A}_n} O_n(\mathbf{A}') \triangleq O_n^{\min}$.

The detailed proof of Theorems 5 and 6 can be found in Liu et al. (2019), which is omitted here due to space.

19.5.3 Performance Evaluation

Simulation Setup

There are a total of $F = 10$ files, each of which is 10 Mbits, and $M = 10$ FANs, each of which can cache 4 random files. Assume that the UTs communicate with FCN and FANs via long-term evolution (LTE). As measured in Kwak et al. (2015), the average data rate of LTE is 5.85 Mbps, and thus the data rate between FANs and UTs, i.e. $R_{n,m}$, is randomly distributed in [5.35, 6.35] Mbps. Furthermore, generally, we have $R_{n,0} < R_{n,m}$ Liu et al. (2018a), and thus the data rate between FCN and UTs, i.e. $R_{n,0}$, is randomly selected from [4.35, 5.35] Mbps. Set $\gamma = 4$, while α_m is uniformly distributed over [5.5, 6.5]. To guarantee that the revenues of FANs can increase with increasing downloads, β_m is randomly chosen from [0.05, 0.1] in our simulation. In addition, λ_n^T and λ_n^C are uniformly and randomly selected from [0.5, 1] and [0.1, 0.2], respectively. All numerical results are averaged over 500 simulation trials. In each simulation trial, the requested file of each UT is randomly determined.

Overall Cost

Figure 19.6 compares our proposed COUS algorithm with the following baseline solutions in terms of the overall cost:

- *Optimal scheduling (Optimal)*: The near-optimal solution to the overall cost minimization is obtained, utilizing the Cross Entropy method Rubinstein and Kroese (2004).
- *Random scheduling (Random)*: Each UT is randomly associated with FCN or one FAN.
- *FCN scheduling (FCN)*: Each UT is associated with FCN.

As demonstrated in Figure 19.6, the overall cost increases as the number of UTs increases, and the COUS algorithm can always offer near-optimal performance. The COUS algorithm shows better performance than the random scheduling scheme and the FCN scheduling scheme, especially when the number of UTs is large.

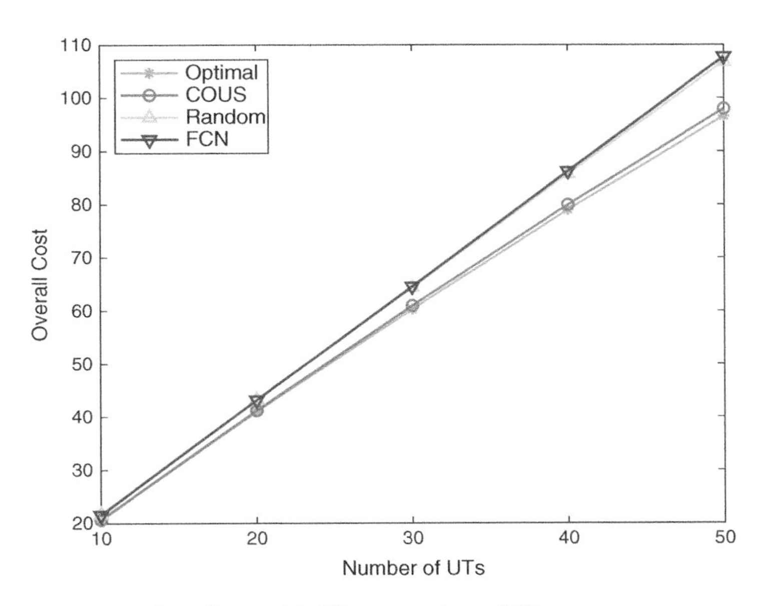

Figure 19.6 Overall cost with different numbers of UTs.

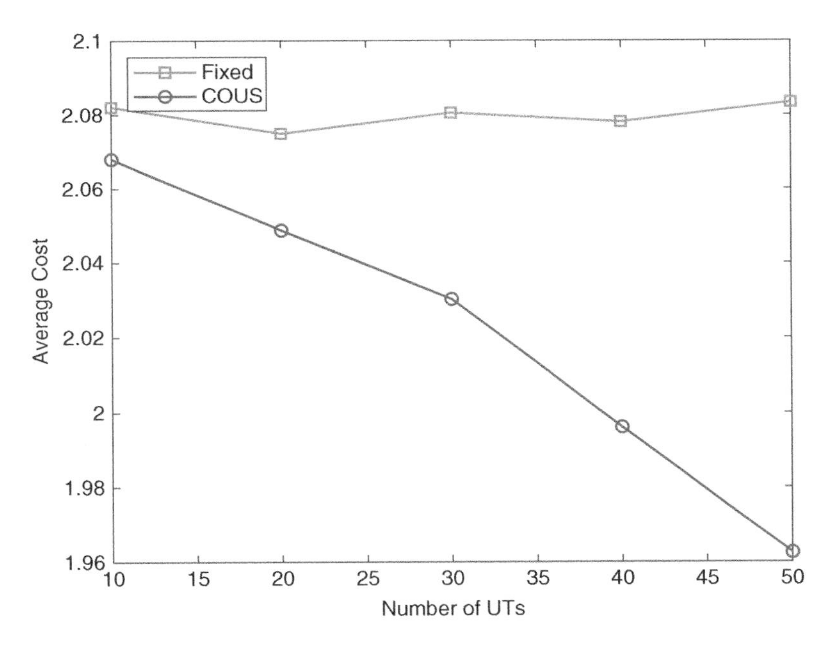

Figure 19.7 Average cost with different numbers of UTs.

Average Cost

Figure 19.7 shows the average cost with different number of UTs, under different payment schemes. We compare the solution achieved by our payment scheme and algorithm with the optimal solution (Fixed) under the fixed payment scheme, where the FANs set a fixed price for a single download, regardless of downloads.

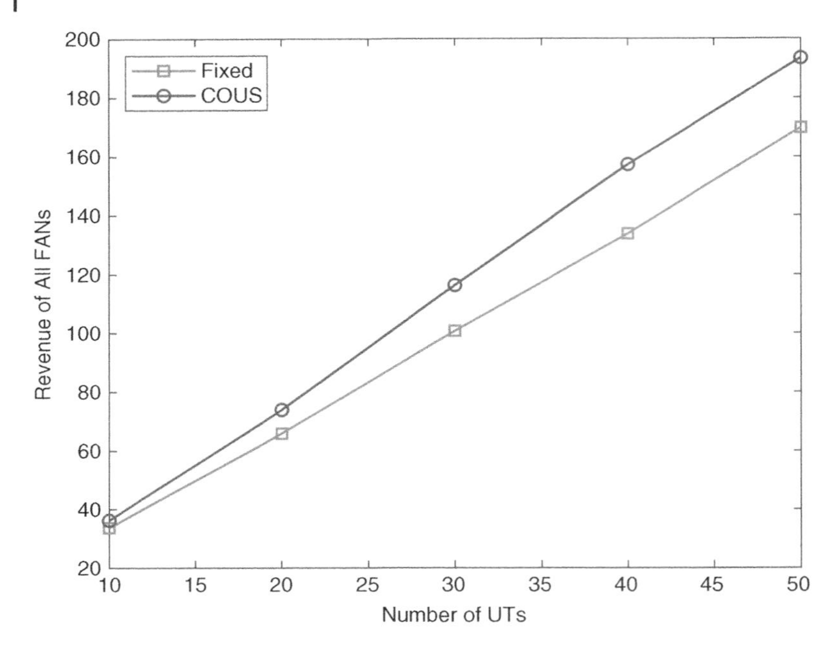

Figure 19.8 Revenue of all FANs with different UTs.

As shown in Figure 19.7, the average cost of our dynamic scheme is smaller than that of the fixed scheme. And the average cost offered by the dynamic scheme decreases with the increase in UTs, while the average cost achieved by fixed scheme almost remains the same, in despite of the number of UTs.

Service Revenue

Figure 19.8 demonstrates the revenue of all FANs with different numbers of UTs under the COUS algorithm and the fixed scheme, respectively. As demonstrated in Figure 19.8, the revenue of all FANs offered by the dynamic scheme is larger than that offered by the fixed scheme. Moreover, the revenues of all FANs offered by the dynamic scheme and the fixed scheme increase as the number of UTs increases, while the revenue of all FANs offered by our proposed dynamic scheme shows a faster growth trend.

Figures 19.7 and 19.8 show a win-win outcome for the proposed cost model, or dynamic payment scheme. To be specific, the proposed dynamic payment scheme can not only offer lower average cost for FCN, but also provide higher revenues for FANs, and this advantage becomes increasingly prominent as the number of UTs increases.

Workload Distribution

Figure 19.9 illustrates the number of UTs served by different FANs under different payment schemes when $N = 50$. It can be seen that our proposed dynamic payment scheme will incur unfair workload distribution among different FANs, compared with the fixed payment scheme.

19.6 Conclusion

In this chapter, we investigated online multi-tier operations scheduling in fog-enabled network architecture with heterogeneous node capabilities and dynamic wireless network conditions. A

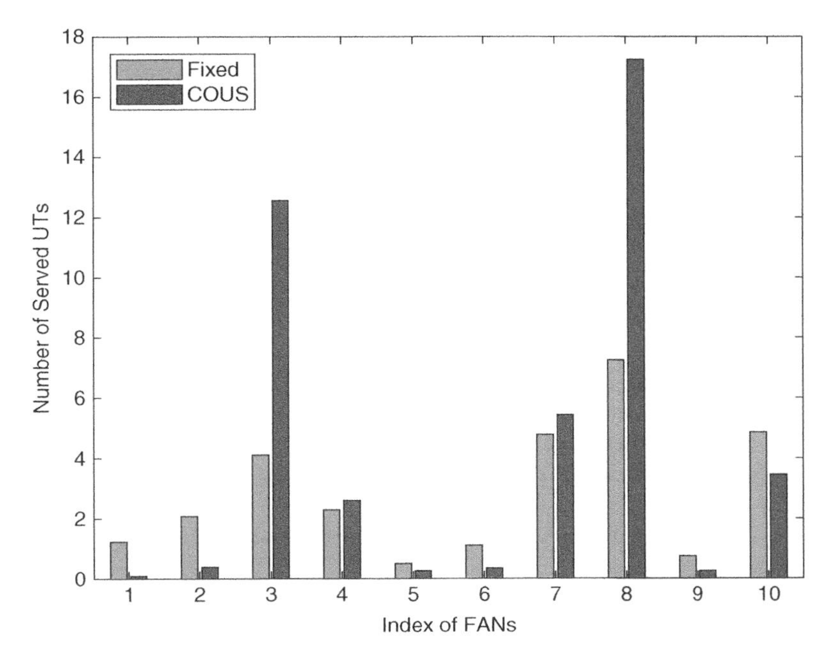

Figure 19.9 Number of UTs served by different FANs ($N = 50$).

low-complexity online algorithm based on Lyapunov optimization was proposed. Performance analysis, as well as simulations, explicitly characterized the trade-offs between average network throughput and service delay, and affirmed the benefit of centralized assignment of access node and dynamic online bandwidth allocation. Following this, we investigated the proactive FAN assignment and resource management problem, given the availability of predictive information, and proposed a predictive scheduling-based online algorithm to reduce service delay. In the end, a unified multi-tier cost model, including the service delay and a linear inverse demand dynamic payment scheme, was proposed to motivate the FANs for resources sharing. Furthermore, the COUS algorithm, based on a potential game, was provided to show how to minimize the overall cost for the FCN.

Bibliography

Hasti Ahlehagh and Sujit Dey. Video caching in radio access network: impact on delay and capacity. In *proceeding of 2012 IEEE Wireless Communications and Networking Conference (WCNC)*, pages 2276–2281, 2012.

Ejder Baştug, Mehdi Bennis, Marios Kountouris, and Mérouane Debbah. Cache-enabled small cell networks: Modeling and tradeoffs. *EURASIP Journal on Wireless Communications and Networking*, 2015 (1): 41, 2015.

Ezio Biglieri, John Proakis, and Shlomo Shamai. Fading channels: Information-theoretic and communications aspects. *IEEE Transactions on Information Theory*, 44 (6): 2619–2692, 1998.

Flavio Bonomi, Rodolfo Milito, Jiang Zhu, and Sateesh Addepalli. Fog computing and its role in the internet of things. In *Proceedings of the 1st Edition of the MCC Workshop on Mobile Cloud Computing*, pages 13–16, 2012.

Nanxi Chen, Yang Yang, Tao Zhang, Ming-Tuo Zhou, Xiliang Luo, and John K Zao. Fog as a service technology. *IEEE Communications Magazine*, 56 (11): 95–101, 2018a.

Shanzhi Chen, Hui Xu, Dake Liu, Bo Hu, and Hucheng Wang. A vision of IoT: Applications, challenges, and opportunities with China perspective. *IEEE Internet of Things Journal*, 1 (4): 349–359, 2014.

Shanzhi Chen, Fei Qin, Bo Hu, Xi Li, and Zhonglin Chen. User-centric ultra-dense networks for 5G: Challenges, methodologies, and directions. *IEEE Wireless Communications*, 23 (2): 78–85, 2016a.

Xu Chen, Lei Jiao, Wenzhong Li, and Xiaoming Fu. Efficient multi-user computation offloading for mobile-edge cloud computing. *IEEE/ACM Transactions on Networking*, 24 (5): 2795–2808, 2016b.

Xu Chen, Lingjun Pu, Lin Gao, Weigang Wu, and Di Wu. Exploiting massive D2D collaboration for energy-efficient mobile edge computing. *IEEE Wireless Communications*, 24 (4): 64–71, 2017.

Xu Chen, Wenzhong Li, Sanglu Lu, Zhi Zhou, and Xiaoming Fu. Efficient resource allocation for on-demand mobile-edge cloud computing. *IEEE Transactions on Vehicular Technology*, 67 (9): 8769–8780, 2018.

M. Chiang and T. Zhang. Fog and IoT: An overview of research opportunities. *IEEE Internet of Things Journal*, 3 (6): 854–864, 2016.

Cisco. Cisco Visual Networking Index: Global Mobile Data Traffic Forecast Update, 2015–2020, White Paper. 2016.

Xiaohu Ge, Song Tu, Guoqiang Mao, Cheng-Xiang Wang, and Tao Han. 5G ultra-dense cellular networks. *IEEE Wireless Communications*, 23 (1): 72–79, 2016.

Negin Golrezaei, Alexandros G. Dimakis, and Andreas F. Molisch. Scaling behavior for device-to-device communications with distributed caching. *IEEE Trans. Information Theory*, 60 (7): 4286–4298, 2014.

Ying He, F. Richard Yu, Nan Zhao, Victor C.M. Leung, and Hongxi Yin. Software-defined networks with mobile edge computing and caching for smart cities: A big data deep reinforcement learning approach. *IEEE Communications Magazine*, 55 (12): 31–37, 2017.

Jie Huang, Cheng-Xiang Wang, Lu Bai, Jian Sun, Yang Yang, Jie Li, Olav Tirkkonen, and Mingtuo Zhou. A big data enabled channel model for 5G wireless communication systems. *IEEE Transactions on Big Data*, Early Access, 2018.

L. Huang and M.J. Neely. Delay reduction via Lagrange multipliers in stochastic network optimization. *IEEE Transactions on Automatic Control*, 56 (4): 842–857, 2011.

L. Huang, S. Zhang, M. Chen, and X. Liu. When backpressure meets predictive scheduling. *IEEE/ACM Transactions on Networking*, 24 (4): 2237–2250, 2016a.

Longbo Huang, Shaoquan Zhang, Minghua Chen, Xin Liu, Longbo Huang, Shaoquan Zhang, Minghua Chen, and Xin Liu. When backpressure meets predictive scheduling. *IEEE/ACM Transactions on Networking*, 24 (4): 2237–2250, 2016b.

Nizabat Khan and Claude Oestges. Impact of transmit antenna beamwidth for fixed relay links using ray-tracing and WINNER II channel models. In *Antennas and Propagation (EUCAP). Proceedings of the 5th European Conference on*, pages 2938–2941. IEEE, 2011.

Ravi Kumar and Andrew Tomkins. A characterization of online browsing behavior. In *Proceedings of the 19th International Conference on Worldwide Web*, pages 561–570. ACM, 2010.

Jeongho Kwak, Yeongjin Kim, Joohyun Lee, and Song Chong. Dream: Dynamic resource and task allocation for energy minimization in mobile cloud systems. *IEEE Journal on Selected Areas in Communications*, 33 (12): 2510–2523, 2015.

Quang Duy Lã, Yong Huat Chew, and Boon-Hee Soong. *Potential Game Theory: Applications in Radio Resource Allocation*. Springer, 2016.

He Li, Kaoru Ota, and Mianxiong Dong. Learning iot in edge: Deep learning for the internet of things with edge computing. *IEEE Network*, 32 (1): 96–101, 2018.

Jun Li, Youjia Chen, Zihuai Lin, Wen Chen, Branka Vucetic, and Lajos Hanzo. Distributed caching for data dissemination in the downlink of heterogeneous networks. *IEEE Transactions on Communications*, 63 (10): 3553–3568, 2015.

Juan Liu, Bo Bai, Jun Zhang, and Khaled B Letaief. Cache placement in fog-RANs: From centralized to distributed algorithms. *IEEE Transactions on Wireless Communications*, 16 (11): 7039–7051, 2017.

Tingting Liu, Jun Li, BaekGyu Kim, Chung-Wei Lin, Shinichi Shiraishi, Jiang Xie, and Zhu Han. Distributed file allocation using matching game in mobile fog-caching service network. In *Proceedings of IEEE Conference on Computer Communications Workshops (INFOCOM WKSHPS)*, pages 499–504. IEEE, 2018a.

Yiming Liu, F. Richard Yu, Xi Li, Hong Ji, and Victor C.M. Leung. Hybrid computation offloading in fog and cloud networks with non-orthogonal multiple access. In *Proceedings of IEEE Conference on Computer Communications Workshops (INFOCOM WKSHPS)*, pages 154–159. IEEE, 2018b.

Zening Liu, Xiumei Yang, Yang Yang, Kunlun Wang, and Guoqiang Mao. DATS: Dispersive stable task scheduling in heterogeneous fog networks. *IEEE Internet of Things Journal*, 6 (2): 3423–3436, 2018.

Zening Liu, Yang Yang, Yu Chen, Kai Li, Ziqin Li, and Xiliang Luo. A multi-tier cost model for effective user scheduling in fog computing networks. In *Proceedings of IEEE Conference on Computer Communications Workshops (INFOCOM WKSHPS)*. IEEE, to appear, 2019.

Nguyen Cong Luong, Zehui Xiong, Ping Wang, and Dusit Niyato. Optimal auction for edge computing resource management in mobile blockchain networks: A deep learning approach. In *Proceedings of IEEE International Conference on Communications (ICC)*, pages 1–6. IEEE, 2018.

Dov Monderer and Lloyd S. Shapley. Potential games. *Games and Economic Behavior*, 14 (1): 124–143, 1996.

Yongma Moon and Changhyun Kwon. Online advertisement service pricing and an option contract. *Electronic Commerce Research and Applications*, 10 (1): 38–48, 2011.

Michael J. Neely. Stochastic network optimization with application to communication and queueing systems. *Synthesis Lectures on Communication Networks*, 3 (1): 1–211, 2010.

Eng Hwee Ong, Jarkko Kneckt, Olli Alanen, Zheng Chang, Toni Huovinen, and Timo Nihtilä. IEEE 802.11 ac: Enhancements for very high throughput WLANs. In *Proceeding of the 22nd IEEE International Symposium on Personal Indoor and Mobile Radio Communications (PIMRC)*, pages 849–853, 2011.

Tao Ouyang, Zhi Zhou, and Xu Chen. Follow me at the edge: Mobility-aware dynamic service placement for mobile edge computing. *IEEE Journal on Selected Areas in Communications*, 36 (10): 2333–2345, 2018.

Sheldon M. Ross. *Introduction to probability models*. Academic Press, 2014.

Reuven Y. Rubinstein and Dirk P. Kroese. *The Cross Entropy Method: A Unified Approach To Combinatorial Optimization, Monte-carlo Simulation (Information Science and Statistics)*. Springer-Verlag New York, Inc., 2004.

Hamed Shah-Mansouri and Vincent W.S. Wong. Hierarchical fog-cloud computing for IoT systems: A computation offloading game. *IEEE Internet of Things Journal*, 5 (4): 3246–3257, 2018.

Yuanming Shi, Jun Zhang, and Khaled B Letaief. Group sparse beamforming for green cloud-RAN. *IEEE Transactions on Wireless Communications*, 13 (5): 2809–2823, 2014.

David Tse and Pramod Viswanath. *Fundamentals of wireless communication*. Cambridge university press, 2005.

Luis M. Vaquero and Luis Rodero-Merino. Finding your way in the fog: Towards a comprehensive definition of fog computing. *ACM SIGCOMM Computer Communication Review*, 44 (5): 27–32, 2014.

Huandong Wang, Fengli Xu, Yong Li, Pengyu Zhang, and Depeng Jin. Understanding mobile traffic patterns of large scale cellular towers in urban environment. In *Proceedings of the 2015 Internet Measurement Conference*, pages 225–238. ACM, 2015.

R. Wang, X. Peng, J. Zhang, and K. B. Letaief. Mobility-aware caching for content-centric wireless networks: Modeling and methodology. *IEEE Communications Magazine*, 54 (8): 77–83, 2016.

Kan Yang, Qi Han, Hui Li, Kan Zheng, Zhou Su, and Xuemin Shen. An efficient and fine-grained big data access control scheme with privacy-preserving policy. *IEEE Internet of Things Journal*, 4 (2): 563–571, 2017a.

Peng Yang, Ning Zhang, Shan Zhang, Kan Yang, Li Yu, and Xuemin Shen. Identifying the most valuable workers in fog-assisted spatial crowdsourcing. *IEEE Internet of Things Journal*, 4 (5): 1193–1203, 2017b.

Xiumei Yang, Zening Liu, and Yang Yang. Minimization of weighted bandwidth and computation resources of fog servers under per-task delay constraint. In *Proceedings of IEEE International Conference on Communications (ICC)*, pages 1–6. IEEE, 2018a.

Yang Yang. Multi-tier computing networks for intelligent IoT. *Nature Electronics*, 2: 4–5, 2019.

Yang Yang, Yunsong Gui, Haowen Wang, Wuxiong Zhang, Yang Li, Xuefeng Yin, and Cheng-Xiang Wang. Parallel channel sounder for MIMO channel measurements. *IEEE Wireless Communications*, 25 (5): 16–22, 2018b.

Yang Yang, Yang Li, Kai Li, Shuang Zhao, Rui Chen, Jun Wang, and Song Ci. DECCO: Deep-learning enabled coverage and capacity optimization for massive MIMO systems. *IEEE Access*, 6: 23361–23371, 2018c.

Yang Yang, Kunlun Wang, Guowei Zhang, Xu Chen, Xiliang Luo, and Ming-Tuo Zhou. MEETS: Maximal energy efficient task scheduling in homogeneous fog networks. *IEEE Internet of Things Journal*, 5 (5): 4076–4087, 2018d.

Yang Yang, Yecheng Wu, Nanxi Chen, Kunlun Wang, Shanzhi Chen, and Sha Yao. LOCASS: Local optimal caching algorithm with social selfishness for mixed cooperative and selfish devices. *IEEE Access*, 6: 30060–30072, 2018e.

Yang Yang, Yang Li, Wuxiong Zhang, Qin Fei, Pengcheng Zhu, and Chengxiang Wang. Generative adversarial network-based wireless channel modeling: Challenges and opportunities. *IEEE Communications Magazine*, 57 (3): 22–27, 2019.

Daqiang Zhang, Min Chen, Mohsen Guizani, Haoyi Xiong, and Daqing Zhang. Mobility prediction in telecom cloud using mobile calls. *IEEE Wireless Communications*, 21 (1): 26–32, 2014.

Guowei Zhang, Fei Shen, Nanxi Chen, Pengcheng Zhu, Xuewu Dai, and Yang Yang. DOTS: Delay-optimal task scheduling among voluntary nodes in fog networks. *IEEE Internet of Things Journal*, 6 (2): 3533–3544, 2019.

Guowei Zhang, Fei Shen, Zening Liu, Yang Yang, Kunlun Wang, and Ming-Tuo Zhou. FEMTO: Fair and energy-minimized task offloading for fog-enabled IoT networks. *IEEE Internet of Things Journal*, 2018b.

Shuang Zhao, Yang Yang, Ziyu Shao, Xiumei Yang, Hua Qian, and Cheng-Xiang Wang. FEMOS: Fog-enabled multitier operations scheduling in dynamic wireless networks. *IEEE Internet of Things Journal*, 5 (2): 1169–1183, 2018.

Michael Zink, Kyoungwon Suh, Yu Gu, and Jim Kurose. Characteristics of YouTube network traffic at a campus network–measurements, models, and implications. *Computer Networks*, 53 (4): 501–514, 2009.

20

Physical-Layer Location Verification by Machine Learning

Stefano Tomasin, Alessandro Brighente, Francesco Formaggio, and Gabriele Ruvoletto

Department of Information Engineering, University of Padova, Padova, Italy

Information on the position of wireless communication devices is nowadays used for a variety of services, e.g. social networking, gaming, content distribution systems, and navigation. Position-related applications relying only on information reported by the device itself can be easily modified by a malicious user tampering with either the software or the global navigation satellite system (GNSS) receiver.

In order to be effective against these attacks, location verification systems verify the user position independently of the reported one. In particular, *in region location verification* (IRLV) aims at assessing if the user is in a specific authorized area or *region of interest* (ROI). To this end, the network examines the *features* of the physical-layer channel over which communications occur. While in general any channel feature related to user position – such as components related to line-of-sight reception, reflections, scattering, and Doppler phenomena – could be exploited, this chapter focuses on the attenuation at various frequencies of a broadband channel, thus taking into account multi-path, path loss, shadowing, and fading. Indeed, some of these characteristics (such as fading) actually act as a disturbance of the channel path-loss, which most appropriately identifies the position.

IRLV can be seen as a hypothesis-testing problem between the two hypotheses of the device being inside or outside the ROI. Due to the mentioned disturbances, the test outcome is affected by two types of error: false alarms (FAs), occurring when a user inside the ROI is classified as being outside; and misdetections (MDs), occurring when a user outside the ROI is classified as being inside. The optimal[1] test is given by the Neyman-Pearson (NP) theorem by Neyman et al. (1933).

In order to perform the NP test, the probability density functions (PDFs) of the channel features under both hypotheses are needed. However, closed-form expressions of the PDFs may not be immediately available, e.g. upon deployment in a new environment or when the environment is changing over time. In this context, machine learning (ML) approaches provide a suitable solution, as discussed by Brighente et al. (2018): during the learning phase, the channel features for trusted devices positioned both inside and outside the ROI are estimated and suitable machines are trained in order to classify the features into the two classes. Among ML

1 By *optimal* test we refer to the most powerful test that minimizes the FA probability for a given MD probability.

Machine Learning for Future Wireless Communications, First Edition. Edited by Fa-Long Luo.
© 2020 John Wiley & Sons Ltd. Published 2020 by John Wiley & Sons Ltd.

classification solutions, this chapter focuses on those using neural networks (NNs) and support vector machines (SVMs) (see Bishop (2006) for an introduction to these topics).

Literature Survey

A number of works on IRLV assume that feature measurements are not affected by errors and are related to the device position in a deterministic way, thus giving neither FAs nor MDs. Sastry et al. (2003) first studied the IRLV problem, where the access points (APs) of the verification network measure the round-trip time (related to AP–device distance) of ultrasound links with the device under test. Using an empirical approach, the location is verified if the round-trip time is below a threshold computed according to the claimed user position. Vora and Nesterenko (2006) exploit radio waves, as the device sends a sequence of messages with increasing power: a user is classified as inside the ROI if APs inside the ROI detect messages earlier than APs outside it. Singelee and Preneel (2005) use distance bounding protocols (DBPs) run in cascade by each AP to upper-bound the distance between the APs and the tested device, showing that with at least three APs and triangulation techniques, accurate IRLV can be obtained. Song et al. (2008) use DBPs in the context of vehicular ad hoc networks where IRLV protects from GNSS spoofing.

The possibility of test errors is considered by Wei and Guan (2013), who adopt a strategy similar to that of Vora and Nesterenko (2006) in a sensor network context. In this case, the user is classified inside the ROI if the probability of this event, given the observed features, is higher than a threshold, thus again the PDF of the channel features must be available. Yan et al. (2016) use the NP test under both a more complex model (the received power includes Gaussian spatially correlated shadowing) and more sophisticated attacks (the attacker can adjust the transmission power). In Brighente et al. (2018) the application of ML strategies to the IRLV problem is first introduced, and the connections of both NN and SVM classifiers to the NP test are established.

The ML approach has been already pursued in a problem closely related to IRLV, i.e. *user authentication*, aiming at verifying if a message is coming from the claimed sender by checking whether it goes through the same channel as previously authenticated messages. In this context, Pei et al. (2014) use three channel features (the channel impulse response, the received signal strength, and the time of arrival), and SVM or the linear Fisher's discriminant analysis algorithm (see Fukunaga (2013)) are adopted. Abyaneh et al. (2018), instead, apply a NN classifier on the received signal.

Chapter Objective and Outline

This chapter analyzes ML algorithms for IRLV with emphasis on multiple-layer NN and SVM approaches. As these solutions efficiently match the performance of the NP hypothesis test (see Brighente et al. (2018)) at convergence, they turn out to be both effective and efficient in terms of computational complexity, resource consumption, and processing latency.

The rest of the chapter is organized as follows. Section 20.1 describes the IRLV system model and revises the optimal hypothesis testing framework. Section 20.2 presents the ML IRLV procedure and quickly reviews the fundamentals of NNs and SVMs. We also discuss the optimality of ML-based classification. Section 20.3 reviews theoretical results on the complexity-performance trade-off of ML solutions, which will be used to assess performance of methods based on NNs. Section 20.4 presents results obtained over a dataset of experimental data with attenuation values of a cellular system; and in the last section we make offer further discussion and conclusions.

20.1 IRLV by Wireless Channel Features

The IRLV reference scenario is shown in Figure 20.1, where a set of N_{AP} APs measure the features of AP channels to user equipment (UE), whose position must be verified by the IRLV system.

The broadband channel is described by Q attenuation values at different frequencies,[2] and $a_n^{(f_i)}$ denotes the measured attenuation at the nth AP on frequency f_i. Vector $\boldsymbol{a} = [a_1^{(f_1)}, \dots, a_{N_{\text{AP}}}^{(f_1)}, \dots, a_1^{(f_Q)}, \dots, a_{N_{\text{AP}}}^{(f_Q)}]$ collects the attenuation values at all APs in all frequencies. In a Rayleigh fading-channel environment comprising path loss, shadowing, and fading, the channel gain (including its uniformly distributed phase $\xi_n^{(f_i)}$) $\sqrt{a_n^{(f_i)}}e^{j\xi_n^{(f_i)}}$ is a complex Gaussian-distributed random variable. Its mean is zero, while its variance is $\sigma_{a,n}^{(f_i)2} = P_{\text{PL},n}^{(f_i)}e^s$, where $P_{\text{PL},n}^{(f_i)}$ is the path loss coefficient; s is the zero-mean Gaussian distributed shadowing component, having variance σ_s^2, where we omitted the dependence on position and frequency for the sake of simpler notation. We first observe that various components of the attenuation are specifically related to the UE's (and APs') position. This is the case of path loss and shadowing, which are slowly time-varying and can thus be used for IRLV of users moving in and out of the ROI over a long time.

The path loss of a wireless link is directly related to the distance between the transmitter and the receiver. In particular, let $\boldsymbol{x}_{\text{AP}}^{(n)}$ be the position of AP $n = 1, \dots, N_{\text{AP}}$, and similarly let $\boldsymbol{x}_{\text{UE}}$ be the true position of the UE under verification. Finally, let $L(\boldsymbol{x}_{\text{UE}}, \boldsymbol{x}_{\text{AP}}^{(n)})$ be the distance between

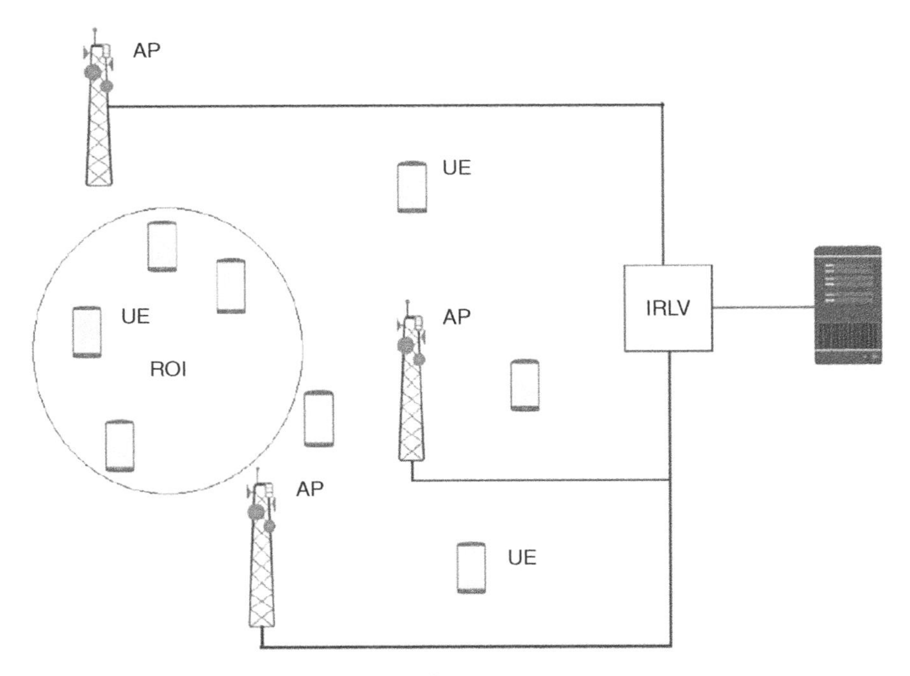

Figure 20.1 Typical in-region location verification scenario.

2 The channel phase is related to specific synchronization procedures and may vary significantly at each transmission, and thus is not considered for IRLV.

the UE and AP n. With reference to the scenario described by 3GPP (2018), two types of links are considered: line of sight (LOS) and non-LOS. For a LOS link, the path loss in dB can be modeled as

$$P_{\text{PL-LOS},n}^{(f_i)} = 10\nu \log_{10}\left(\frac{f_i 4\pi L(\boldsymbol{x}_{\text{UE}}, \boldsymbol{x}_{\text{AP}}^{(n)})}{c}\right), \tag{20.1}$$

where ν is the path loss coefficient and c is the speed of light. For a non-LOS link we have instead

$$P_{\text{PL-NLOS},n}^{(f_i)} = 40\log_{10}\left(\frac{L(\boldsymbol{x}_{\text{UE}}, \boldsymbol{x}_{\text{AP}}^{(n)})}{10^3}\right) + 21\log_{10}\left(\frac{f_i}{10^6}\right) + 80. \tag{20.2}$$

Note that shadowing is only statistically related to the distance, and thus it does not provide a direct information on the position. Instead, shadowing can also be seen as noise over the path loss component of the attenuation, thus complicating IRLV. Moreover, fading and measurement noise further prevent directly mapping the attenuation values into distances. Therefore, for given positions of both UE and APs, the measured attenuation vector $\boldsymbol{a}$ is a random vector and any IRLV solution will be in general affected by both FA and MD errors.

20.1.1 Optimal Test

The IRLV problem can be formulated as a hypothesis-testing problem between two hypotheses: $\mathcal{H}_0$, the UE is transmitting from inside the ROI; and $\mathcal{H}_1$, the UE is transmitting from outside the ROI.

Let $\hat{\mathcal{H}} \in \{\mathcal{H}_0, \mathcal{H}_1\}$ be the decision made by the IRLV system on the two hypotheses, whereas $\mathcal{H} \in \{\mathcal{H}_0, \mathcal{H}_1\}$ is the ground truth, i.e. the effective location of the UE. We also denote the FA probability as $P_{\text{FA}} = \mathbb{P}(\hat{\mathcal{H}} = \mathcal{H}_1 | \mathcal{H} = \mathcal{H}_0)$ and the MD probability as $P_{\text{MD}} = \mathbb{P}(\hat{\mathcal{H}} = \mathcal{H}_0 | \mathcal{H} = \mathcal{H}_1)$.

Let $p_{\boldsymbol{a}|\mathcal{H}}(\boldsymbol{a}|\mathcal{H}_i)$ be the PDF of attenuation vectors conditioned to the position of the UE. The log-likelihood ratio (LLR) is defined as the ratio between the two conditioned PDFs for a certain attenuation vector $\boldsymbol{a}$, i.e.

$$\mathcal{M}(\boldsymbol{a}) = \ln \frac{p_{\boldsymbol{a}|\mathcal{H}}(\boldsymbol{a}|\mathcal{H}_0)}{p_{\boldsymbol{a}|\mathcal{H}}(\boldsymbol{a}|\mathcal{H}_1)}. \tag{20.3}$$

According to the NP theorem, the most powerful test on an observed attenuation vector $\boldsymbol{a}$ is obtained by comparing $\mathcal{M}(\boldsymbol{a})$ with a threshold value Λ, obtaining the test function

$$\hat{\mathcal{H}} = \begin{cases} \mathcal{H}_0 & \text{if } \mathcal{M}(\boldsymbol{a}) \geq \Lambda, \\ \mathcal{H}_1 & \text{if } \mathcal{M}(\boldsymbol{a}) < \Lambda. \end{cases} \tag{20.4}$$

This procedure provides the minimum MD probability for a given FA probability.

20.2 ML Classification for IRLV

As the NP theorem requires the knowledge of the PDFs of the attenuation vectors conditioned on $\mathcal{H}_i$, $i = 0, 1$, which can be hard to obtain, a ML approach can be adopted, operating in two phases:

1. *Learning phase*: The APs collect attenuation vectors from a trusted UE moving both inside and outside the ROI. The UE reports its position to the APs, so that the network can learn how to classify the attenuations, by means of a suitable test function, according to the provided ground truth of the hypothesis.

2. *Exploitation phase*: In order to test the location of an untrusted UE, the IRLV system collects attenuation values from the APs and gives them to the test function designed in the learning phase.

During the learning phase, the APs collect k_t attenuation values. Since the channel is affected by fading, we consider k_f fading realizations for each fixed spatial position x_s, so that $k_t = x_s \cdot k_f$.

The details of the two phases are now described. When collecting training attenuation vectors $\mathbf{a}(\ell)$, $\ell = 1, \dots, k_t$, we associate them with *labels* t_ℓ, $\ell = 1, ..., k_t$, where $t_\ell = -1$ if the trusted UE is inside the ROI, and $t_\ell = 1$ if the trusted UE is outside the ROI. With these settings, the IRLV system designs the test function

$$\hat{t}(\mathbf{a}) \in \{-1, 1\} \tag{20.5}$$

that provides a decision $\hat{\mathcal{H}}$ ($\hat{\mathcal{H}} = \mathcal{H}_0$ for $\hat{t} = -1$ and $\hat{\mathcal{H}} = \mathcal{H}_1$ for $\hat{t} = 1$) for each attenuation vector $\boldsymbol{a}$. This function is then used to make a decision in the exploitation phase. Note that this solution does not explicitly evaluate the PDF and the LLR; rather it directly implements the test function with a ML algorithm.

20.2.1 Neural Networks

A NN is a function $g(\cdot)$ that maps an N-size input vector to an M-size output vector, i.e. $g : \mathbb{R}^N \to \mathbb{R}^M$. Function $g(\cdot)$ is implemented as the cascade of multiple functions that can be represented by a graph where nodes (*neurons*) are the functions whose input and output are represented by the connecting edges (*connections*).

Neurons are organized in L layers, the first layer being defined as the *input* layer, the last as the *output* layer, and all other layers as *hidden layers*. The neurons of layer ℓ are connected only to those of the two adjacent layers, and we focus on an implementation of the NN without loops, as shown in Figure 20.2.

All neurons of layer ℓ implement the same function. In particular, at neuron n of layer ℓ, the elements of output vector $\boldsymbol{y}^{(\ell-1)}$ of the previous layer are linearly combined with weights $\boldsymbol{w}_n^{(\ell-1)}$, and then a scalar bias $b_n^{(\ell-1)}$ is added. The result of these operations is the input of the *activation function* $\psi^{(\ell-1)}(\cdot)$ (the same for all neurons of the same layer). The output of the nth neuron at the ℓth layer is therefore

$$y_n^{(\ell)} = \psi^{(\ell-1)}(\boldsymbol{w}_n^{(\ell)}\boldsymbol{y}^{(\ell-1)} + b_n^{(\ell-1)}). \tag{20.6}$$

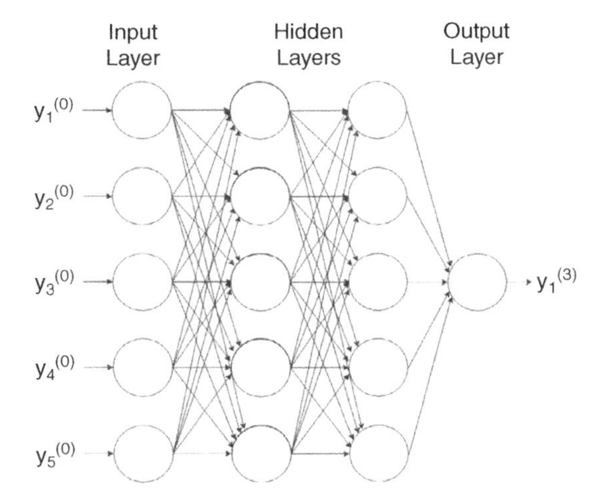

Figure 20.2 Example of a neural network architecture with five input values and $L = 4$ layers with $N_h = 5$ neurons in the hidden layers.

During training, weights $\boldsymbol{w}_n^{(\ell)}$ and biases $b_n^{(\ell-1)}$ are selected in order to minimize a suitably chosen *loss function* over the *training set*. The parameters optimization is typically based on gradient algorithms. Usually the *activation functions* are sigmoids, which generate a single output.

For the IRLV problem, the input to the NN is the $N_{\mathrm{AP}}Q$-size attenuation vector $\boldsymbol{a}$ and the output is a single ($M = 1$) value $\tilde{t}(\boldsymbol{a}) = y_1^{(L-1)}$, denoting the decision made. The NN output is then compared against a threshold λ, providing the decision[3]

$$\hat{t}(\boldsymbol{a}) = \begin{cases} 1 & \tilde{t}(\boldsymbol{a}) > \lambda, \\ -1 & \tilde{t}(\boldsymbol{a}) \leq \lambda. \end{cases} \tag{20.7}$$

Collecting into vectors $\boldsymbol{W}$ and $\boldsymbol{b}$ all the weights and bias factors, and using the mean square error (MSE) as loss function, the optimization problem for the NN training can be written as

$$(\boldsymbol{W}^*, \boldsymbol{b}^*) = \mathrm{argmin}_{(\boldsymbol{W},\boldsymbol{b})} \sum_{\ell=1}^{k_t} |\hat{t}(\boldsymbol{a}(\ell)) - t_\ell|^2 . \tag{20.8}$$

20.2.2 Support Vector Machines

A SVM used for classification purposes is again a function that maps an N-dimensional input into a scalar. In particular, the operations performed by an SVM can be described as

$$\tilde{t}(\boldsymbol{a}) = \boldsymbol{w}^T \phi(\boldsymbol{a}) + b, \tag{20.9}$$

where $\phi : \mathbb{R}^{N_{\mathrm{AP}}} \rightarrow \mathbb{R}^K$ is a fixed *feature-space transformation function*, $\boldsymbol{w} \in \mathbb{R}^K$ is the weight vector, and b is a bias parameter. As seen for NNs, in order to obtain a binary decision from the soft (continuous value) given by Eq. (20.9), we use Eq. (20.7), where now $y_1^{(L-1)}$ is replaced by $\tilde{t}(\boldsymbol{a})$ obtained by Eq. (20.9). In Eq. (20.9), function $\phi(\cdot)$ is fixed (see Goodfellow et al. (2016)), while both vector $\boldsymbol{w}$ and scalar b are properly chosen according to the specific modeled problem.

The design of the SVM parameters is obtained by minimizing various objective functions. *Maximum margin* classifiers aim at maximizing the margin between the two classes, and thus the design problem is

$$\min_{\boldsymbol{w},b} \quad \omega(\boldsymbol{w}, b) \triangleq \frac{1}{2}\boldsymbol{w}^T\boldsymbol{w} + C\frac{1}{2}\sum_{i=1}^{k_t} e_i \tag{20.10a}$$

$$e_i \geq 1 - t_i[\boldsymbol{w}^T\phi(\boldsymbol{a}(i)) + b] , \quad i = 1, \ldots, k_t , \tag{20.10b}$$

where C is a hyper-parameter. Inequalities in the constraints lead to a quadratic programming problem, typically solved in its dual version, where the feature function is implicitly defined by the *kernel*. In the least squares SVM (LS-SVM) by Suykens and Vandewalle (1999), instead of maximizing the margin, the LS error is minimized, providing the problem

$$\min_{\boldsymbol{w},b} \quad \omega(\boldsymbol{w}, b) \triangleq \frac{1}{2}\boldsymbol{w}^T\boldsymbol{w} + C\frac{1}{2}\sum_{i=1}^{k_t} e_i^2 \tag{20.11a}$$

$$e_i = t_i[\boldsymbol{w}^T\phi(\boldsymbol{a}(i)) + b] - 1, \quad i = 1, \ldots, k_t . \tag{20.11b}$$

Equality constraints in Eq. (20.11) yield a linear system of equations in the optimization values. Ye and Xiong (2007) have shown that both SVM and LS-SVM are equivalent under mild conditions.

3 Notice that this last step can be implemented as an additive layer with a single neuron that implements a comparator function and then provides as output the label associated with the input value.

20.2.3 ML Classification Optimality

A first question to address is whether the ML approaches are somehow equivalent to the optimal NP approach. Note that the performance of the ML approaches significantly depend on (i) the complexity/configuration of the model, e.g. the neurons of the NN or the feature-space transformation function, and (ii) the size/type, e.g. the balance between the attenuations in the two classes of the training set.

Thus, we expect to achieve the performance of NP test only for an infinite-size training set and for sufficiently (infinite) complex models. Indeed, Brighente et al. (2018) have shown that under these asymptotic conditions both the NN- and LS-SVM-based tests achieve the performance of the NP test.

In order to prove this, a MSE criterion is used for NN design, which Ruck et al. (1990) have shown to converge to the minimum MSE approximation of the Bayes optimal discriminant function

$$g_0(\boldsymbol{a}) = \mathbb{P}(\mathcal{H} = \mathcal{H}_0|\boldsymbol{a}) - \mathbb{P}(\mathcal{H} = \mathcal{H}_1|\boldsymbol{a}). \tag{20.12}$$

Then, by algebraic manipulations on the thresholding of $g_0(\boldsymbol{a}) > \lambda$, as from Eq. (20.7), we obtain a thresholding on the LLR of Eq. (20.3), as done in the NP test.

Similarly, starting from the LS optimization criterion of the SVM, it is possible to show first that it is equivalent to the MSE optimization of a parametric function similar to Eq. (20.8), from which, using the same reasoning for the MSE-designed NN, we conclude that LS-SVM is asymptotically equivalent to the optimal NP approach.

20.3 Learning Phase Convergence

A key issue for ML approaches is the convergence speed of the training phase: while having a larger training set yields better tuning of the machine parameters, the complexity (and time) entailed by training also increases. Here we first recall some theoretical bounds on the training set size as a function of the desired classification accuracy, and then present some numerical results for the considered IRLV scenario.

20.3.1 Fundamental Learning Theorem

The performance of a learning algorithm can be analyzed within the *probably approximately correct* (PAC) learning framework. To this end, we first recall two fundamental concepts: the PAC learnability and the Vapnik-Chervonensky (VC) dimension.

About the PAC learnability, let $\mathcal{E} = \{t(\boldsymbol{a}) \neq \hat{t}(\boldsymbol{a})\}$ be the error event of a wrong decision, where $t(\boldsymbol{a})$ is the ground truth associated with the UE giving the attenuation vector $\boldsymbol{a}$. For a parametric test function $\hat{t}(\boldsymbol{a})$, the error probability $\pi_{\hat{t}} = \mathbb{P}[\mathcal{E}]$ obtained in the exploitation phase depends on both the test function and the specific training set, and thus $\pi_{\hat{t}}$ is itself a random variable. Note also that this theory refers generically to *errors*, while the distinction between MDs and FAs is instead relevant for the IRLV problem. Let $\mathcal{A}$ be the set of possible attenuations and $\mathcal{T} = \{\hat{t}(\boldsymbol{a}) : \mathcal{A} \to \{-1, 1\}\}$ a set of test functions defined on $\mathcal{A}$.

We now recall the following definition:

Definition 20.1 ***($\epsilon - \delta$ agnostic PAC learnable problems):*** Consider a set $\mathcal{T}$ of test functions. The set $\mathcal{T}$ is $\epsilon - \delta$ *agnostic PAC learnable* if, for each $\epsilon \geq 0$ and $\delta \geq 0$, there exists an integer k_t^* such that, when running the learning algorithm on $k_t > k_t^*$ i.i.d. training samples, the resulting test function $\hat{t}^*(\boldsymbol{a}) \in \mathcal{T}$ ensures (see Shalev-Shwartz and Ben-David (2014))

$$\mathbb{P}\left[\pi_{\hat{t}^*} \leq \min_{\tau \in \mathcal{T}} \pi_\tau + \epsilon\right] = 1 - \delta , \tag{20.13}$$

where π_τ is the error probability computed for the test function τ over the training points, and the minimum is taken over all possible testing functions.

Let $\mathcal{A}^* \subset \mathcal{A}$ be a subset of $\mathcal{A}$, then the *restriction* of $\mathcal{T}$ to $\mathcal{A}^*$ is the set of all test functions of $\mathcal{T}$ defined on $\mathcal{A}^*$. Now, if the restriction of $\mathcal{T}$ on $\mathcal{A}^*$ is the set of all functions from $\mathcal{A}^*$ to $\{-1, 1\}$, we say that $\mathcal{T}$ *shatters* $\mathcal{A}^*$. We now recall the following definition:

Definition 20.2 ***(VC-dimension):*** The VC-dimension of a set of test functions $\mathcal{T}$ is the maximal size of a set $\mathcal{A}^* \subset \mathcal{A}$ that can be shattered by $\mathcal{T}$. If $\mathcal{T}$ shatters sets of arbitrary large size, we say that $\mathcal{T}$ has *infinite VC dimension*.

By the the *fundamental theorem of statistical learning* (see Shalev-Shwartz and Ben-David (2014)), it can be proven that $\mathcal{T}$ is $\epsilon - \delta$ agnostic PAC learnable if and only if it has a finite VC dimension.

Moreover, for $\epsilon - \delta$ agnostic PAC learnable problems, there exist two constants C_1 and C_2 satisfying

$$C_1 \frac{d + \log(1/\delta)}{\epsilon} \leq k_t^* \leq C_2 \frac{d + \log(1/\delta)}{\epsilon} \tag{20.14}$$

for all (related) values of ϵ, δ and k_t^*.

Now, Eq. (20.14) provides a bound on the training set size, according to the intrinsic characteristics of the considered class of test functions (through d) and the desired accuracy of the test (through δ and ϵ).

In particular, considering the class of test functions identified by a NN with E connections and a single output neuron, its VC dimension grows as $\mathcal{O}(E \log E)$, (see Shalev-Shwartz and Ben-David (2014)). Figure 20.3 shows the bounds of Eq. (20.14) for a scenario with $N_{\mathrm{AP}} = 5$ APs, each collecting attenuation values at a single carrier frequency $f_1 = 1800$ MHz and for channels with shadowing and without fading ($k_f = 1$). The NN comprises two hidden layers with $N_h = 5$ neurons each, and the hyperbolic tangent as activation function. The VC dimension is $d = 250$, and we set $\delta = 0.01$. The C_1 and C_2 values have been chosen such that their relative distance is the minimum guaranteeing that realization points are inside the bounds. Each dot represents the average value of ϵ obtained over 100 realizations of the attenuation maps for a fixed value of k_t. We notice that the error-performance saturation is well represented by the bound curves.

The results give a first intuition on the classification error probability, comprising both FA and MD errors, for a given size of the training set. We notice that, after a certain point, increasing the training set size k_t^* does not reduce the classification error. This is due to the size of the NN (in terms of number of neurons and layers), which is not sufficient to deal with the problem complexity.

20.3.2 Simulation Results

Due to the shortcomings of the PAC learnability theory when applied to classification problems where FA and MD probabilities must be clearly distinguished, either experiments or simulations are needed to assess performance.

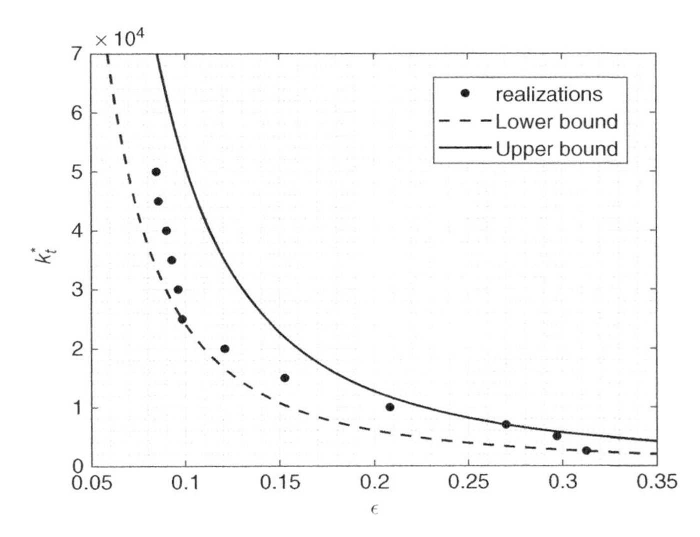

Figure 20.3 k_t^* vs. ϵ for a neural network with two hidden layers with $N_h = 5$.

Simulation Scenario Consider an IRLV network with $N_{\text{AP}} = 5$ APs, two carrier frequencies ($Q = 2$), namely $f_1 = 900$ MHz, and $f_2 = 1800$ MHz, and transmission power $P_{\text{tx}} = 0$ dBm. Moreover, for two UEs located at positions $\boldsymbol{x}_i$ and $\boldsymbol{x}_j$ and transmitting to the same AP, their shadowing parameters have correlation in space $\sigma_s^2 e^{-\frac{L(\boldsymbol{x}_i,\boldsymbol{x}_j)}{d_c}}$, where d_c is the shadowing decorrelation distance.

Path loss and shadowing are assumed to be time-invariant, while the fading component is independent at each attenuation estimate. Shadowing realizations at two different frequencies f_1 and f_2 have correlation $\rho(f_1,f_2) = \mathbb{E}[e^{s_1}e^{s_2}]$ as described in Van Laethem et al. (2012), where s_1 and s_2 are the shadowing parameters at frequencies f_1 and f_2, respectively. Channel measurements are assumed without noise.

In the following results, the channel is affected by path loss with $\nu = 3$, Rayleigh fading and shadowing with $\sigma_s^2 = 3.39$, decorrelation distance $d_c = 75$ m, and frequency correlation coefficient $\rho = 0.84$; see Van Laethem et al. (2012).

The NN is implemented with a single hidden layer of $N_h = 10$ neurons and trained with the MSE loss function. The activation functions of hidden layers are sigmoids, whereas the activation function of the output layer is a hyperbolic tangent. The SVM is designed according to the maximum margin criterion, with Gaussian kernel (see Bishop (2006)) and suitably optimized C parameter and kernel scale. We consider $k_f = 10$ fading realizations per spatial position.

Performance Results Figures 20.4 and 20.5 show the P_{MD} vs. k_t for different P_{FA} values for both NN- and SVM-based IRLV. Note that, as the FA probability increases, the training set size has a more significant impact on the MD probability. For both approaches note that the convergence of the MD probability is obtained for $k_t = 4 \cdot 10^4$ training points for all the FA probability values.

20.4 Experimental Results

The channel model considered for simulation is solid but still based on a number of assumptions, e.g. on the spatial correlation of the shadowing and its specific statistical distribution.

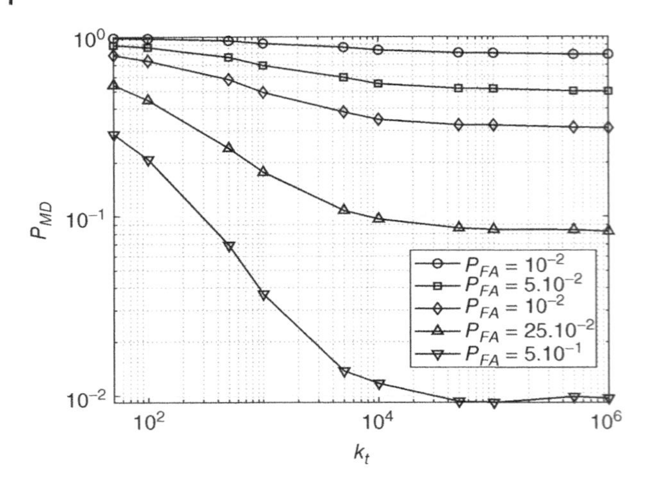

Figure 20.4 P_{MD} vs. k_t for different P_{FA} values and neural network–based IRLV.

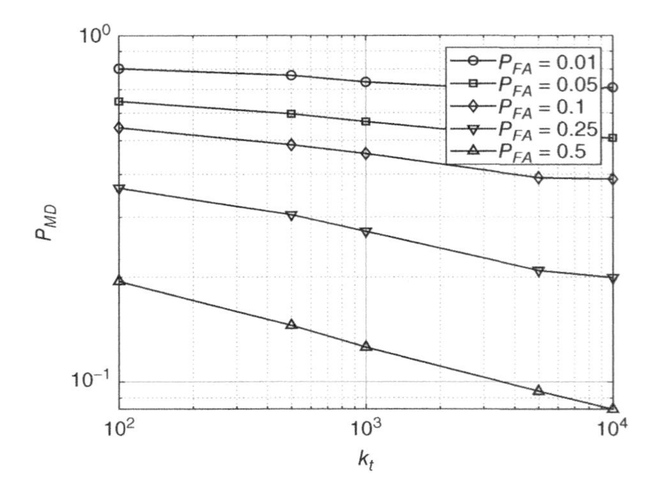

Figure 20.5 P_{MD} vs. k_t for different P_{FA} values and support vector machine–based IRLV.

In order to better assess the performance of ML-based IRLV, we can resort to experimental measurements.

An appropriate dataset is provided by the measurement campaign in Alexanderplatz, Berlin (Germany) of the MOMENTUM project described by Geerdes et al. (2013). Narrowband ($Q = 1$) attenuations at the frequency of the global system for mobile communications (GSM) have been measured for several APs in an area of 4500 m $\times$ 4500 m. Consider five attenuation maps corresponding to five APs located approximately at (2500, 2500) m (see Figure 20.6 for an example of the attenuation map), (500, 4000) m, (4000, 4000) m, (500, 500) m, and (4000, 500) m. The ROI has been positioned in the lower-right corner corresponding to a 20 m $\times$ 20 m square. The NN configuration is the same as the simulation results, with $L = 1$ and $N_h = 5$. The SVM configuration is the same as the simulation results.

First consider the average MD probability versus the FA probability, i.e. the receiver operating characteristic (ROC), which is the key performance measure of a binary classification system.

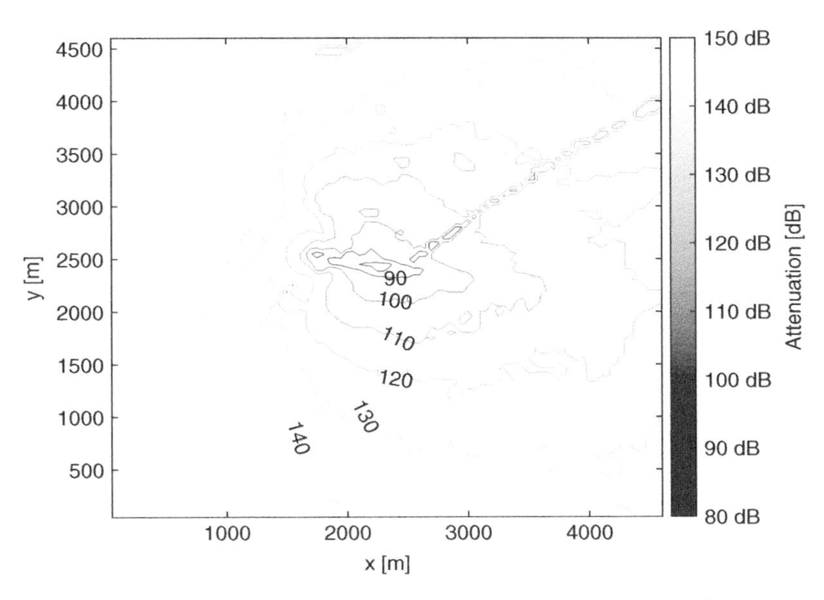

Figure 20.6 Example of an attenuation map of the MOMENTUM project for Alexanderplatz, Berlin.

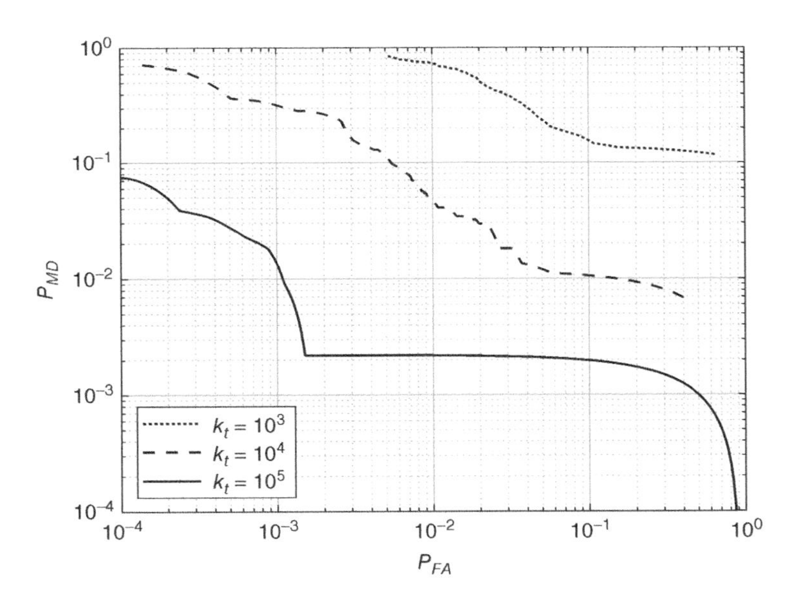

Figure 20.7 Receiver operating characteristics for different k_t values for neural network–based IRLV.

Figures 20.7 and 20.8 show the ROC for NN- and SVM-based IRLV, respectively, and various sizes of the training set k_t. Note that as the training set size increases, the ROC attains lower values of the $(P_{\text{MD}}, P_{\text{FA}})$ couples. Furthermore, note that the performance obtained by experimental data is significantly better than that obtained by simulation. When comparing the two ML approaches, we note that the NN-based IRLV achieves a much lower P_{MD} for a low P_{FA}, thus outperforming the SVM-based IRLV.

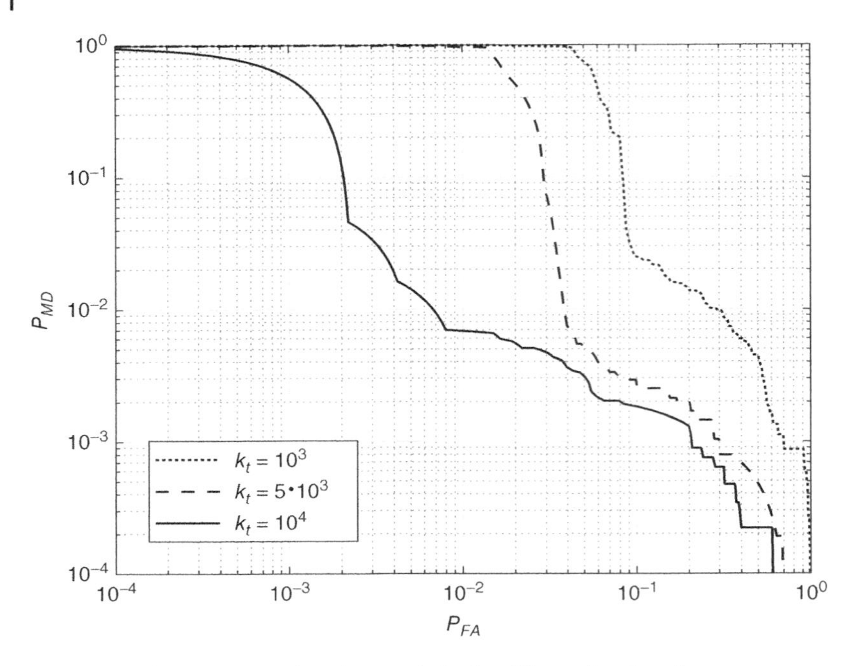

Figure 20.8 Receiver operating characteristics for different k_t values for support vector machine–based IRLV.

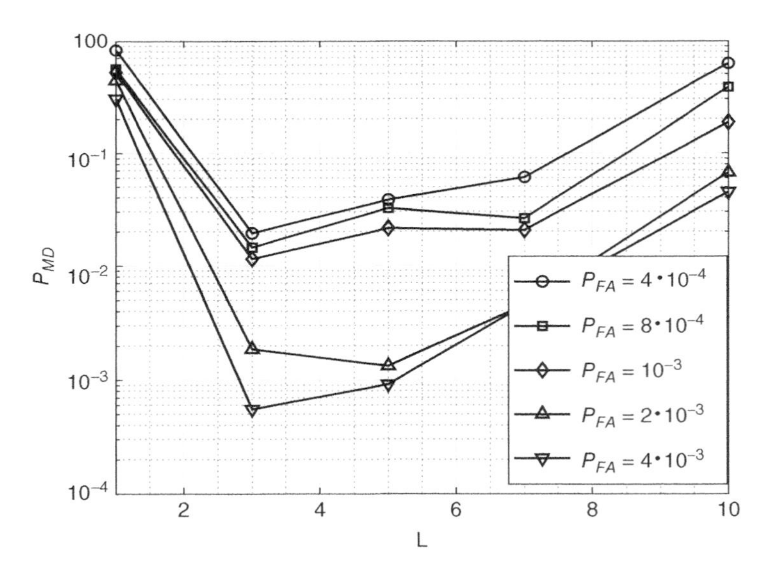

Figure 20.9 Average P_{MD} vs. L for different P_{FA} values and neural network–based IRLV.

We now assess the ML-based IRLV performance as a function of the complexity of the NN, i.e. the number of layers. Figure 20.9 shows the average MD probability as a function of the number of layers L, for fixed FA probabilities, $k_t = 10^5$ and $N_h = 10$ for each hidden layer. Note that the training set size is fixed, and it becomes insufficient for proper training as the number of parameters of the NN becomes large. For this particular setting, the smallest average MD probability is achieved for all the average FA probabilities with $L = 3$ layers.

20.5 Conclusions

In this chapter, we proposed a solution to IRLV formulated as a hypothesis-testing problem. Power attenuation values gathered by multiple APs were used as input the test function, implemented by a learning machine. A key design driver for ML algorithms is the number of training samples needed for parameter tuning, which provides a trade-off between complexity and classification accuracy. We recalled some key bounds on the training set size and applied them to ML-based IRLV.

We evaluated the performance of ML algorithms in terms of ROC for both a simulated propagation scenario and experimental data. We explored different NN architectures by varying the number of hidden layers and numerically evaluated the impact of training size on the MD probability.

The concept algorithm we considered opens new research opportunities in the ML framework, e.g. the NN architecture can be further optimized (in the number of neurons, layer shape, and activation function type) for a better adaptation to different IRLV scenarios.

Bibliography

3GPP. Evolved universal terrestrial radio access (E-UTRA); radio frequency (RF) system scenarios. *TR 36.942 version 15.0.0 Release 15*, 2018.

Abyaneh A.Y., Foumani A.H.G., and V. Pourahmadi. Deep neural networks meet CSI-based authentication. *arXiv preprint arXiv:1812.04715*, 2018.

Bishop C.M. *Pattern Recognition And Machine Learning*. Springer, 2006.

Brighente A., Formaggio F., Di Nunzio G.M., and Tomasin S. Machine learning for in-region location verification in wireless networks, *IEEE Journal on Selected Areas in Communications*. doi: 10.1109/JSAC.2019.2933970.

Fukunaga K. *Introduction to Statistical Pattern Recognition*. Elsevier, 2013.

Geerdes H.F., Lamers E., Lourenço P., Meijerink E., Türke U., Verwijmeren S., and Kürner T. Evaluation of reference and public scenarios. Technical Report D5.3, IST-2000-28088 MOMENTUM, 2013. URL http://www.zib.de/momentum/paper/momentum-d53.pdf.

Goodfellow I., Bengio Y., Courville A., and Bengio Y. *Deep Learning*. MIT Press, 2016.

Neyman J., Pearson E.S., and Pearson K. IX. On the problem of the most efficient tests of statistical hypotheses. *Philosophical Transactions of the Royal Society of London. Series A, Containing Papers of a Mathematical or Physical Character*, 231 (694-706): 289–337, 1933.

Pei C., Zhang N., Shen X.S., and Mark J.W. Channel-based physical layer authentication. In *Proc. Global Communications Conference (GLOBECOM), 2014 IEEE*, pages 4114–4119, 2014.

Ruck D.W., Rogers S.K., Kabrisky M., Oxley M.E., and Suter B.W. The multilayer perceptron as an approximation to a Bayes optimal discriminant function. *IEEE Trans on Neural Networks*, 1 (4): 296–298, Dec. 1990.

Sastry N., Shankar U., and Wagner D. Secure verification of location claims. In *Proc. of the 2nd ACM workshop on Wireless security*, pages 1–10, Sept. 2003.

Shalev-Shwartz S. and Ben-David S. *Understanding Machine Learning*. Cambridge University Press, 2014.

Singelee D. and Preneel B. Location verification using secure distance bounding protocols. In *Proc. Mobile Adhoc and Sensor Systems Conference, 2005. IEEE International Conference on*, pages 7–pp. 840, 2005.

Song J.H., Wong V.W.S., and Leung V.C.M. Secure location verification for vehicular ad-hoc networks. In *Proc. IEEE GLOBECOM 2008-2008 IEEE Global Communications Conference*, pages 1–5, 2008.

Suykens J.A.K. and Vandewalle J. Least squares support vector machine classifiers. *Neural processing letters*, 9 (3): 293–300, Jun. 1999.

Van Laethem B., Quitin F., Bellens F., Oestges C., and De Doncker P. Correlation for multi-frequency propagation in urban environments. *Progress In Electromagnetics Research*, 29: 151–156, 2012.

Vora A. and Nesterenko M. Secure location verification using radio broadcast. *IEEE Transactions on Dependable and Secure Computing*, 3 (4): 377–385, Oct. 2006.

Wei Y. and Guan Y. Lightweight location verification algorithms for wireless sensor networks. *IEEE Transactions on Parallel and Distributed Systems*, 24 (5): 938–950, May 2013.

Yan S., Nevat I., Peters G.W., and Malaney R. Location verification systems under spatially correlated shadowing. *IEEE Transactions on Wireless Comm.*, 15 (6): 4132–4144, Feb. 2016.

Ye J. and Xiong T. SVM versus least squares SVM. *Artificial Intelligence and Statistics*, pages 644–651, Apr. 2007.

21

Deep Multi-Agent Reinforcement Learning for Cooperative Edge Caching

M. Cenk Gursoy, Chen Zhong, and Senem Velipasalar

Department of Electrical Engineering and Computer Science, Syracuse University, Syracuse, New York, USA

Wireless edge caching is considered one of the key techniques to reduce data traffic congestion in backhaul links. This chapter is devoted to the application of deep reinforcement learning (DRL) strategies to edge caching at both small base stations and user equipment (UE). Organized into seven sections, this chapter first presents the motivation and background on wireless edge caching. Sections 21.2 and 21.3 address system modelling and problem formulation for different cellular deployment scenarios. In Section 21.4, we present a multi-agent actor-critic DRL framework for edge caching with the goal to increase the cache hit rate and reduce transmission delay. By providing extensive simulation results, Section 21.5 focuses on demonstrating the performance improvements with the proposed DRL policies and provides comparisons with the least recently used (LRU), least frequently used (LFU), and first-in-first-out (FIFO) caching strategies when caching is performed at small base stations. In terms of working principles and caching performance, Section 21.6 further compares the DRL with naive and probabilistic caching policies when caching is performed at UEs. Section 21.7 includes further discussion and conclusions.

21.1 Introduction

During the past decade, explosive growth in the number of smart devices has led to unprecedented increase in the demand on rich-media services, which generally require higher system capacity and data rates, making it even more challenging to satisfy the ever-growing data traffic. As summarized in Cisco Visual Networking Index, from 2012–2017, overall mobile data traffic has experienced 17-fold growth. This growth is primarily due to the increase in the mobile multimedia traffic, which is expected to grow further. For instance, it is predicted that video traffic will experience a ninefold increase by 2022 and will account for 79% of total mobile traffic. However, the increase in the mobile network connection speed, which is projected to grow threefold, will not be adequate to satisfy users' demands on high-quality streaming services. To reduce data traffic congestion in backhaul links, content caching has been proposed and is being envisioned as one of the key components of next-generation wireless networks (Jaber et al. (2016)).

With these motivations and to better serve the users, content-caching strategies have been studied recently. As noted before, content caching is considered a key technique to reduce data traffic by enabling content server nodes to store a part of popular content locally, so that when the cached content is requested, the server can deliver content directly to users or to the next content server node in the route to users and reduce the latency compared to requesting

Machine Learning for Future Wireless Communications, First Edition. Edited by Fa-Long Luo.
© 2020 John Wiley & Sons Ltd. Published 2020 by John Wiley & Sons Ltd.

content from upper-level servers. Based on this idea and considering different content server architectures, several caching strategies have been proposed and investigated. Considering the central content servers, such as the baseband unit in cloud radio access networks (C-RANs), centralized coded caching and delivery schemes are presented by Yan et al. (2017) and Yang et al. (2018). For decentralized caching, in-network caching at the content delivery network (CDN) server is widely adopted to reduce data congestion near content servers. The placement of the in-network cache usually involves a routing problem in CDNs. Dehghan et al. (2017) and Xu et al. (2018) proposed joint caching and routing polices to minimize the service delay. Even though CDNs have been shown to reduce data traffic, they can hardly handle the growing mobile data traffic because it is inevitable that content has to be transmitted through CDN nodes before arriving at the user.

More recently, proactive caching at the wireless network edge, such as at small base stations and UE, is proposed. This technique makes it possible to have popular content to be placed closer to end users and be directly transmitted, which can effectively reduce the time for routing in CDNs, and apparently save a considerable amount of waiting time for users and offload a portion of the data traffic at the CDN. The architectures and challenges of caching techniques are summarized in several comprehensive survey papers (see e.g. Lei et al. (2018), Wang et al. (2017) and Liu et al. (2016)). Zhang et al. (2018) and Li et al. (2018b) studied edge-caching policies aimed at minimizing transmission delay for the base station and device-to-device (D2D) users. Additionally, research on hierarchical caching has recently been conducted. For instance, in studies by Kwak et al. (2018), Chen et al. (2017b), Tran et al. (2017), hybrid content-caching schemes for joint content-caching control at the baseband unit and radio remote heads are presented. Li et al. (2018a) proposed an edge hierarchical caching policy for caching at the small base station and UE. Tandon and Simeone (2016) and Koh et al. (2017) considered delivery latency in fog radio access networks (F-RANs) and presented latency-centric caching policies. In the same context, Azimi et al. (2018) proposed an online caching policy. Furthermore, the edge-caching problem is also considered jointly with other problems. For example, Lee and Molisch (2018) jointly considered the caching policy and cooperation distance design in a base station–assisted wireless D2D caching network, and Chen et al. (2017a) addressed joint optimization of caching and scheduling policies.

In the literature, different methods have been applied to determine optimal caching policies. For the case of decentralized caching, Kvaternik et al. (2016) presented a decentralized optimization method for the design of caching strategies that aimed at minimizing energy consumption of the network, while Wang et al. (2018) proposed a decentralized framework for proactive caching according to blockchains from a game-theoretic point of view. In a study by Zhou et al. (2017), caching and multi-cast problems are jointly solved using dynamic programming. A belief propagation algorithm for caching in D2D networks is presented by Hao et al. (2018), and a clustering approach is studied by Zhang et al. (2016). Moreover, machine learning (ML) techniques are also applied in this field. Leconte et al. (2016) proposed an age-based threshold policy that caches all content that has been requested more than a threshold. Furthermore, popularity-based content-caching policies named StreamCache and PopCaching were studied by Li et al. (2016b) and Li et al. (2016a), respectively. Chang et al. (2018) investigated the application of ML methods to content replacement problems.

In this chapter, we focus on edge-caching policies based on DRL methods. For edge caching at both small base stations and UEs, caching policy is driven by content popularity. Therefore, knowing the content popularity is key to solve the caching problem. In previous works, content popularity is either assumed to be known to the content server as presented by Wang et al. (2018), ElBamby et al. (2014), or is estimated before caching actions as proposed by Zhu et al. (2018), Li et al. (2018b). The former assumption makes the framework less practical when content popularity is time-varying, and frequent estimation of content popularity or the arrival intensity of user requests will lead to the consumption of significant amount of resources and time. To avoid such drawbacks, ML methods are introduced to determine efficient caching policies. For example, Lei et al. (2017) trained a deep neural network in order to identify the optimal caching algorithm. And different DRL algorithms are used to find caching strategies that can better adapt to the changing environment. Sadeghi et al. (2018) implemented a Q-learning algorithm to find the optimal caching policy. In studies by Zhong et al. (2018) and Wei et al. (2018), the focus is the use of actor-critic DRL frameworks for caching. And for cooperative caching policies in decentralized caching networks, multi-agent Q-learning solutions were studied by Sung et al. (2016) and Jiang et al. (2018). Song et al. (2017) and Sengupta et al. (2014) presented two different multi-armed bandit–based caching schemes.

As seen in previous studies, content popularity distribution is always critical in solving the content-caching problem. In DRL algorithms, the agent needs to observe enough features of the environment to ensure the accuracy of its decisions. To better address this issue, this chapter presents a deep actor-critic reinforcement learning (RL) multi-agent framework for cooperative edge caching (see e.g. studies by Lowe et al. (2017), Foerster et al. (2016), Gupta et al. (2017) for multi-agent DRL). However, before the details on this multi-agent framework, we first provide the related system models and problem formulations.

21.2 System Model

We consider two different network models described in the following subsections.

21.2.1 Multi-Cell Network Model

As shown in Figure 21.1, in this subsection we introduce a communication network with a cloud data center and N base stations. It is assumed that the data center has sufficient storage

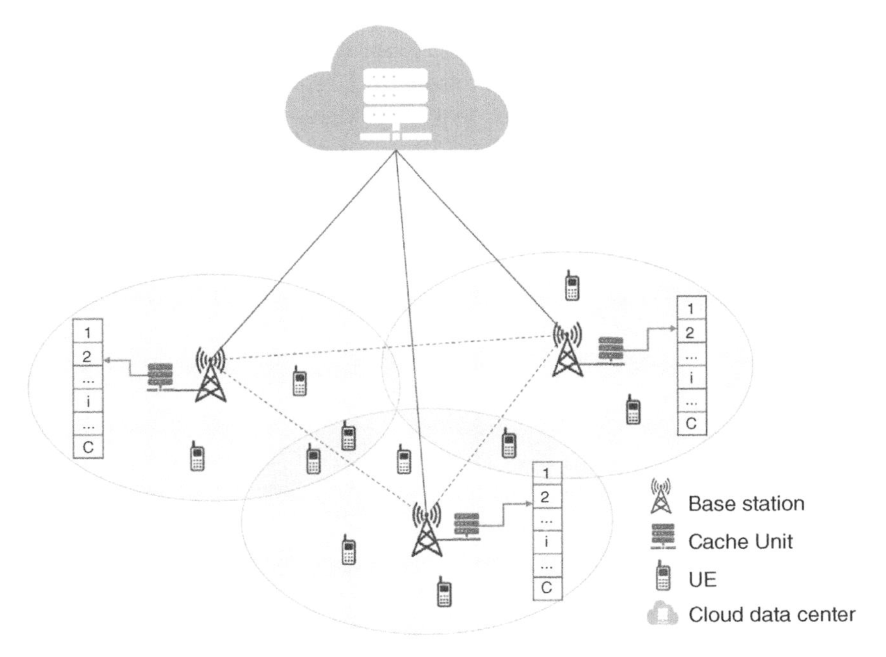

Figure 21.1 Multi-cell network model.

space to cache all content files, while each base station has a fixed cache capacity of C. All base stations can connect with the cloud data center and request files from it. And each base station decides whether to cache a file or not. A base station is assumed to cover a fixed circular cellular region with the corresponding base station at the center. We assume that the radii of the cells are fixed and all users in the cell can access the corresponding base station. There are U users randomly distributed in the system, and they are located in at least one cellular region covered by a base station to ensure service. We assume that in a given time slot, the users' locations do not change and those located at overlapped regions can be served by any one of the corresponding base stations. Users have their own preferences for content, and in each time slot each user can request only one content file. We denote the total number of content files as M, and use the content ID to denote requests for corresponding content. Also, in each operation cycle, users request a content based on their own preferences. The requests are sent to all base stations that can connect with the user, and the base station with the minimum transmission delay will finally transmit the content file to the user. In the meantime, all base stations will update their caches to minimize the average transmission delay based on users' requests. When making caching decisions, the base stations will compete with each other to get the chance to transmit and also cooperate with each other to reduce overall transmission delay.

21.2.2 Single-Cell Network Model with D2D Communication

In this subsection, we introduce the single-cell model with users having D2D communication capabilities. To offload data traffic from the base station, UE is assumed to be equipped with caches. In this setting, we consider a communication network with one base station and U users, and each user is equipped with one cache-enabled mobile device. To distinguish users and cache nodes, we define N as the number of local cache nodes, and in this case we have $N = U$. As shown in Figure 21.2, users are randomly distributed in the cellular region covered by the base station. The cellular region is described by a circle with the base station at the center, and the radius of the cell is d_b. In this system, all users are able to communicate with the base station, and each user can also communicate in D2D mode with other users within a distance d_u, and $d_u < d_b$. We assume that each UE has a fixed cache size C that can be updated depending on users' requests for content. Therefore, to minimize data traffic flows to the base station, when a user requests a content, it will first check the cache at its own equipment. If the

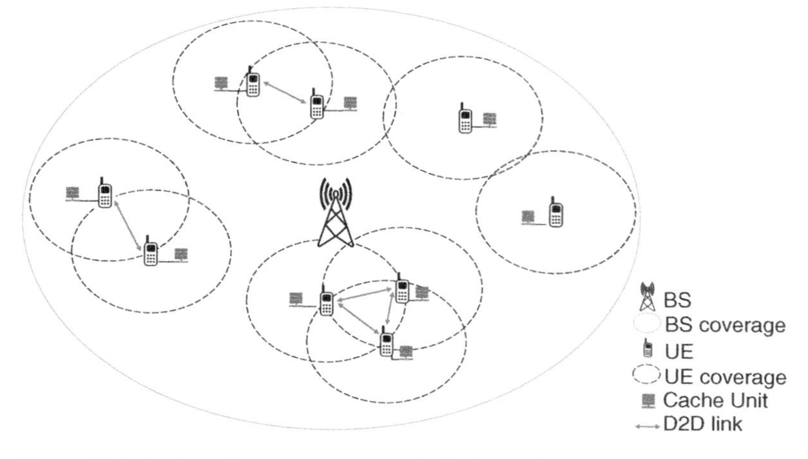

Figure 21.2 Single-cell network model with device-to-device communication.

requested content is cached, the user can serve itself directly and no traffic will be generated. If, on the other hand, the content is not cached by the user itself, caches of all other users that can communicate with the user in D2D mode will be checked, and if the content can be transmitted in D2D mode, the data traffic will be generated from the corresponding user. If none of the D2D transmitters has this content cached, the user will request the content from the base station. We assume that in a given time slot, users' locations do not change and those that can communicate in D2D mode with multiple users can be served by any one of the corresponding D2D transmitters. Users have their own preferences for content, and in each time slot a user can request only one content file. Let us again denote the total number of content files as M, and use the content ID to denote requests for corresponding content. In each operation cycle, users request a content based on their own preferences. The requests are sent to all D2D transmitters that can connect with the user, and all users update their caches. When making the caching decision, users need to consider their own preferences as well as potential collaborative file exchanges with other users in D2D mode in order to maximize the cache hit rates.

21.2.3 Action Space

Henceforth, in order to unify the descriptions and analysis of both cellular network models described, in the multi-cell network model we consider base stations as the local content servers and the cloud data center as an upper-level server. Correspondingly, in the single-cell network model with D2D communications, we consider cache-enabled UEs as local content servers and the single base station in the model as the upper-level server.

To find an edge-caching policy for these two network models, let us consider an actor-critic RL–based multi-agent framework. In this framework, there are N actor networks and one centralized critic network. We consider each local cache server an agent that adopts one of the actor networks to seek its own caching policy. And we assume there are control channels that allow local content servers to send the caching state and data traffic parameters to the upper-level server, so that the upper-level server can act as the centralized critic to evaluate the overall caching state. Similarly as in the study by Zhong et al. (2018), in each operation cycle, the agent can either keep the cache state the same or replace unpopular content files with the popular ones. However, since there can be more than one request arriving at a local content server at the same time, the agent needs to jointly decide which cached content will be deleted and which content requested by which user will be cached. We define the action space as $\mathcal{A}$, and let $\mathcal{A} = \{a_0, a_1, ..., a_D\}$, where a_v denotes a valid action. In our case, a_0 indicates that the current cache state is unchanged. For $v = \{1, 2, ..., D\}$, we define $D = \binom{C_i}{1}\binom{L_i}{1}$, where C_i is the number of files in the cache of local content server i, and L_i is the number of users that can connect with the local content server i. So each a_v stands for a possible combination to replace one of C_i cached content files with one of L_i currently requested content files. In each time slot, all agents must select their own action from the action space $\mathcal{A}$ and execute.

21.3 Problem Formulation

21.3.1 Cache Hit Rate

In this part, we evaluate the caching policy in terms of overall cache hit rate, which is defined as

$$P_{hit} = \frac{\sum_{i=1}^{U} \epsilon_i}{U}$$

where U is the total number of users, and ϵ_i is given by

$$\epsilon_i = \begin{cases} 1 & \text{if user } i \text{ is served by the local content server} \\ 0 & \text{if user } i \text{ is served by the upper-level server} \end{cases} .$$

Basically, the cache hit rate shows the percentage of requests served by the local content servers. The cache hit rate maximization problem can be formulated as follows:

$$\textbf{P1:} \quad \underset{\Upsilon}{\text{Maximize}} \quad P_{hit}$$
$$\text{Subject to} \quad \sum_{f=1}^{M} \upsilon_{i,f} \le C_i$$

where Υ is an $N \times M$ matrix that records the caching states of the N local content servers, and each element $\upsilon_{i,f}$ in the caching state matrix is an indicator to show if the file is cached:

$$\upsilon_{i,f} = \begin{cases} 1 & \text{if file } f \text{ is cached at the local content server } i \\ 0 & \text{if file } f \text{ is not cached at the local content server } i \end{cases} . \tag{21.1}$$

21.3.2 Transmission Delay

We also address the performance of the caching policy in terms of transmission delay. The transmission delay is defined as the number of time frames needed to transmit a content file, and can be expressed as

$$T = \min \left\{ \tilde{t} : F \le \sum_{\kappa=1}^{\tilde{t}} T_0 C[\kappa] \right\} \tag{21.2}$$

where F is the size of the content file to be transmitted. T_0 stands for the duration of each time frame, and $C[\kappa]$ is the instantaneous channel capacity in the κth time frame. And the channel capacity $C[\kappa]$ is expressed as

$$C[\kappa] = B \log_2 \left(1 + \frac{P_t}{B\sigma^2} z_\kappa \right) \quad \text{bits/s} \tag{21.3}$$

where P_t is the transmission power, B is the channel bandwidth, σ^2 is the noise variance, and z_κ is the magnitude square of the corresponding fading coefficient in the κth time frame. In the system, there are two types of transmitters, e.g. the cloud data center and the base stations in the multi-cell model (and the base station and D2D-capable users in the single-cell model). We assume that all transmitters transmit at their maximum power level to maximize the transmission rate. The transmission power is defined as

$$P_t = \begin{cases} P_b & \text{if the transmitter is the upper-level server} \\ P_i & \text{if the transmitter is the } i^{\text{th}} \text{ local content server} \end{cases} . \tag{21.4}$$

To be more concrete in the discussions and following descriptions, we specifically address the multi-cell network model. Note that the descriptions can easily be adapted to the single cell with D2D-capable users as well.

In the considered multi-cell model, if user j requests a content that is not cached at any base station that can connect with the user, the content file will be first transmitted from the cloud data center to base station $\hat{i}$, which is the closest base station to user j, and then from base station

$\hat{i}$ to user j. Thus, the minimum transmission delay $\hat{D}_j$ in the case of missing file in the cache can be expressed as

$$\hat{D}_j = T_{c,\hat{i}} + T_{\hat{i},j} \tag{21.5}$$

where $T_{c,\hat{i}}$ stands for the transmission delay from the cloud data center to base station $\hat{i}$, and $T_{\hat{i},j}$ is the transmission delay from base station $\hat{i}$ to user j.

However, if the requested file is cached at base station i, which can connect to user j, the transmission delay D_j for the case of hitting the cache can be expressed as

$$D_j = T_{i,j} \tag{21.6}$$

Now, the transmission delay for both cases of missing and hitting the cache is known, and we define the transmission delay reduction ΔD_j as

$$\Delta D_j = \hat{D}_j - D_j. \tag{21.7}$$

So, the average transmission delay reduction in an operation cycle is

$$\Delta D = \frac{1}{U}\sum_{j=1}^{U} \Delta D_j \tag{21.8}$$

$$= \frac{1}{U}\sum_{j=1}^{U} (\hat{D}_j - D_j) \tag{21.9}$$

$$= \frac{1}{U}\sum_{j=1}^{U} (T_{c,\hat{i}} + T_{\hat{i},j} - T_{i,j}) \tag{21.10}$$

where U is the total number of users. Now, our goal is to maximize the average transmission delay reduction, and the caching problem is formulated as follows:

$$\textbf{P2:} \quad \underset{\boldsymbol{\Phi}}{\text{Maximize}} \quad \Delta D \tag{21.11}$$

$$\text{Subject to} \quad \xi_{i,j} = 1 \tag{21.12}$$

$$\sum_{f=1}^{M} \phi_{i,f} F_f \leq C \tag{21.13}$$

where $\boldsymbol{\Phi}$ is an $N \times M$ matrix that records the caching states of the N local content servers, and each element $\phi_{i,f}$ in the caching state matrix is an indicator to show if the file is cached:

$$\phi_{i,f} = \begin{cases} 1 & \text{if the file } f \text{ is cached at the local content server } i \\ 0 & \text{if the file } f \text{ is not cached at the local content server } i \end{cases}. \tag{21.14}$$

F_f is the size of file f. If, without loss of generality, we assume all files have the same size, the condition in Eq. (21.13) can be rewritten as

$$\sum_{f=1}^{M} \phi_{i,f} \leq C \tag{21.15}$$

where C is the maximum number of files that can be stored at each base station and $\xi_{i,j}$ is an indicator describing if user j is in the area covered by base station i:

$$\xi_{i,j} = \begin{cases} 1 & \text{if user } j \text{ can connect to local content server } i \\ 0 & \text{if user } j \text{ cannot connect to local content server } i \end{cases}. \tag{21.16}$$

21.4 Deep Actor-Critic Framework for Content Caching

Having presented the previous system models and problem formulations, let us introduce a multi-agent actor-critic framework based on the partially observable Markov decision processes with N agents, where the critic network $V(\mathrm{x})$ and N actors $\pi_{\theta_i}(o_i)$, $i = i, 2, ..., N$, are parameterized by $\theta = \{\theta_c, \theta_1, \theta_2, ..., \theta_N\}$.

Actor: The actor is defined as a function to seek a caching policy $\pi = \{\pi_1, \pi_2, ..., \pi_N\}$, which can map the observation of the agent to a valid action chosen from the action space $\mathcal{A}$. In each time slot, agent i will select an action a_i based on its own observation o_i and policy π_i:

$$a_i = \pi_i(o_i).$$

Critic: The critic is employed to estimate the value function $V(\mathrm{x})$, where x stands for the observation of all agents, i.e. $\mathrm{x} = \{o_1, o_2, ..., o_N\}$. At time instant t, after actions $a_t = \{a_{1,t}, ..., a_{N,t}\}$ are chosen by the actor networks, the agents will execute the actions in the environment and send the current observation x_t along with feedback from the environment to the critic. The feedback includes reward r_t and the next time instant observation x_{t+1}. Then, the critic can calculate the temporal difference (TD) error:

$$\delta^{\pi_\theta} = r_t + \gamma V(\mathrm{x}_{t+1}) - V(\mathrm{x}_t)$$

where $\gamma \in (0, 1)$ is the discount factor.

Update: The critic is updated by minimizing the least squares temporal difference (LSTD):

$$V^* = \arg\min_V (\delta^{\pi_\theta})^2$$

where V^* denotes the optimal value function.

Actor i is updated by policy gradient. Here we use TD error to compute the policy gradient:

$$\Delta_{\theta_i} J(\theta_i) = E_{\pi_{\theta_i}}[\nabla_{\theta_i} \log \pi_{\theta_i}(o_i, a_i)\delta^{\pi_\theta}]$$

where $\pi_{\theta_i}(o_i, a_i)$ denotes the score of action a_i under the current policy. Then, the actor network i can be updated using the gradient decent method:

$$\theta_i \longleftarrow \theta_i + \alpha \nabla_{\theta_i} \log \pi_{\theta_i}(o_i, a_i)\delta^{\pi_\theta}.$$

The detailed steps of these processes are shown in Algorithm 1, and the structure is shown in Figure 21.3.

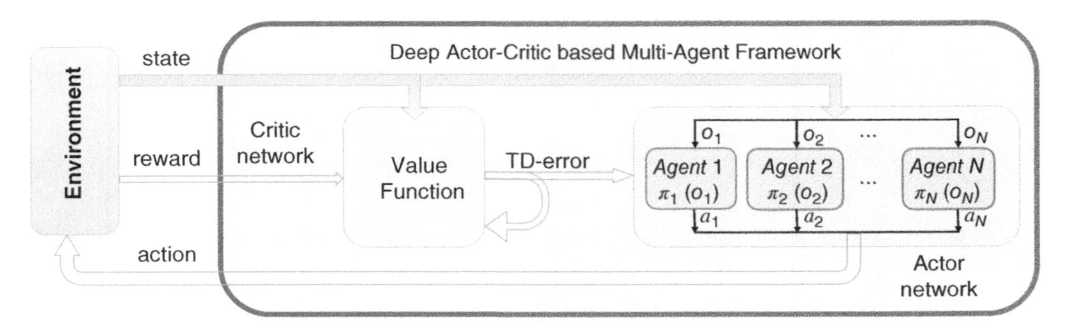

Figure 21.3 Deep actor-critic-based multi-agent structure.

Algorithm 1 Multi-agent actor-critic algorithm for edge caching.

Initialize critic network $V(\mathrm{x})$ and actor $\pi_{\theta_i}(o_i)$, parameterized by $\theta = \{\theta_c, \theta_1, \theta_2, \dots, \theta_N\}$.
Receive initial state $\mathrm{x} = \{o_1, o_2, \dots, o_N\}$.
for $t = 1, T$ **do**
  The base station receives users' requests $Req_t = \{req_{1,t}, req_{2,t}, \dots, req_{U,t}\}$.
  Extract observation at time t for each agent, and $\mathrm{x}_t = \{o_{1,t}, o_{2,t}, \dots, o_{N,t}\}$
  For each agent i, select action $a_i = \pi_{\theta_i}(o_{i,t})$ w.r.t. the current policy
  Execute actions $a_t = (a_{1,t}, a_{2,t}, \dots, a_{N,t})$ to update the cache state of each base station
  Observe reward r_t and new state x_{t+1}
  Critic calculates the TD error based on the current parameter: $\delta^{\pi_\theta} = r_t + \gamma V(\mathrm{x}_{t+1}) - V(\mathrm{x}_t)$
  Update the critic parameter θ_c by minimizing the loss: $\mathcal{L}(\theta) = (\delta^{\pi_\theta})^2$
  for agent $i = 1$ to N **do**
    Update the actor policy by maximizing the action value: $\Delta\theta_i = \nabla_{\theta_i} \log \pi_{\theta_i}(o_{i,t}, a_i)\delta^{\pi_\theta}$
  end for
  Update features space $\mathcal{F}$
end for

Environment: To perform the experiments, we consider wireless cellular networks with one upper-level server, N local content servers, and U users distributed in the servers' coverage regions. Allowing the agents to make their own caching decisions and cooperate with each other, the framework becomes a centralized critic network together with a decentralized actor network. Therefore, the agents will feed the actor network with their own observations and feed the critic network with the complete state space.

Agents' Observation and State Space: As noted before, the multi-agent actor-critic framework is based on a partially observable Markov decision process. Each agent i, $i = 1, 2, \dots, N$, can only observe the requests arriving at itself, and can select its own action only based on the observation o_i. In the environment, agent i can observe the content's features through its local request history. And for the centralized critic, the state space is defined as $\mathrm{x} = \{o_1, o_2, \dots, o_N\}$.

Feature Space: The feature space consists of three components: short-term feature $\mathcal{F}_s$, medium-term feature $\mathcal{F}_m$, and long-term feature $\mathcal{F}_l$, which represent the total number of requests for each content in a specific short-term, medium-term, or long-term, respectively. These features are updated as new requests arrive at agents. Then, we let f_{xj}, for $x \in \{s, m, l\}$ and $j \in \{1, \dots, M\}$ denote the feature of a specific content within a specific term, where M is the total number of content files. Thus, the observation for each agent i is defined as $o_i = \{\mathcal{F}_s; \mathcal{F}_m; \mathcal{F}_l\}$ where $\mathcal{F}_s = \{f_{s0}, f_{s1}, \dots, f_{sM}\}$, $\mathcal{F}_m = \{f_{m0}, f_{m1}, \dots, f_{mM}\}$, and $\mathcal{F}_l = \{f_{l0}, f_{l1}, \dots, f_{lM}\}$.

Reward: We consider the objective functions in problems **P1** and **P2** as the reward. When problem **P1** is targeted, for each operation cycle t, after the agents update their caches according to the selected actions, the cache hit rate for the requests in the next operation cycle $t + 1$ will be received as the reward within the multi-agent framework. Hence, we define the reward in the tth operation cycle as

$$r_t = P_{hit}^{t+1}. \tag{21.17}$$

When problem **P2** is set as the target of the agents, for each operation cycle t, after the agents update their caches according to the selected actions, the average delay reduction in transmitting the content files requested by the users in the next operation cycle $t + 1$ will be received as the reward within the multi-agent framework. Therefore, we define the reward in the tth operation cycle as

$$r_t = \Delta D_{t+1}. \tag{21.18}$$

In the following simulation results, we present the reduction in the transmission delay as a percentage, which is expressed as

$$\eta = \frac{\Delta D}{\frac{1}{U}\sum_{j=1}^{U}\hat{D}_j} \times 100\%. \tag{21.19}$$

Hence, η is the percentage of delay reduction per user in one operation cycle.

21.5 Application to the Multi-Cell Network

21.5.1 Experimental Settings

In this section, we provide simulation results for caching at small base stations. To better evaluate the proposed framework, we compare its performance with the following caching algorithms:

- *Least recently used (LRU)*: In this policy, the system keeps track of the most recent requests for every cached content. When the cache storage is full, the cached content requested the least recently will be replaced by new content.
- *Least frequently used (LFU)*: In this policy, the system keeps track of the number of requests for every cached content. When the cache storage is full, the cached content requested the fewest times will be replaced by new content.
- *First-in, first-out (FIFO)*: In this policy, the system, for each cached content, records the time when the content is cached. When the cache storage is full, the cached content stored earliest will be replaced by new content.

In their implementation, these three caching policies are executed at each base station independently.

21.5.2 Simulation Setup

Environment Settings: As shown in Figure 21.4, in the experiments, we consider a system with 5 base stations and 30 users randomly distributed in the area, each covered by at least one of the base stations. The cell radius is set as $R = 2.2$km, and the transmission power of all base stations is set as $P_i = 16.9$dB, $i = 1, 2, ..., 5$. The transmission power of the cloud data center is set as $P_c = 20$dB. As assumed, the content files are split into units of the same size, and the size of each unit is set as 100 bits. We assume Rayleigh fading with path loss $\mathbb{E}\{z\} = d^{-4}$, where d is the distance between the transmitter and receiver.

File/Content Request Generation: In our simulations, the raw data of users' requests is generated according to the Zipf distribution

$$f(k;\beta,M) = \frac{1/k^{\beta}}{\sum_{M}^{m=1}(1/n^{\beta})} \tag{21.20}$$

where the total number of files M is set as 500, and the Zipf exponent β is fixed at 1.3 in the study of the cache size. k is the rank of the file, and in the implementation, a user's preference for files is randomly generated. To encourage the base station to cache files that are popular for more users, the users are randomly divided into five groups. It is assumed that the users in the same group will have similar but not exactly the same rank for all files, and the group information will not influence the users' location. It is important to note that we generate the requests using the Zipf distribution, and group the users, but such information is totally unknown to the agents.

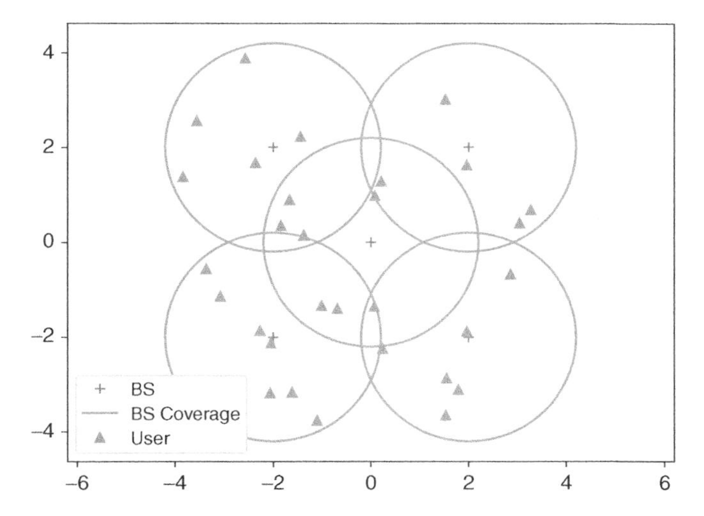

Figure 21.4 Coverage map of a system that contains 5 base stations and 30 users.

Feature Extraction: From the raw data of content requests we extract feature F and use it as the agents' observations of the network. As introduced in Section 21.4, the feature space consists of three components. Here, we extract the number of requests for each file from all the requests that have arrived at the agents in the most recent 10 time slots as the short-term feature, while the medium-term and long-term features are extracted from requests that have arrived in the most recent 100 and 1000 time slots, respectively.

21.5.3 Simulation Results

21.5.3.1 Cache Hit Rate

First, we investigate the relationship between cache hit rate and cache capacity. In this experiment, instead of directly using the cache capacity C, we consider the cache ratio $\sigma = \frac{C}{M}$ (where M is the total number of content files that can be requested by the users), so that we can analyze the impact of the cache capacity normalized by the potential data traffic flows into this system. Figure 21.5 shows the cache hit rates achieved by the proposed DRL agent and the

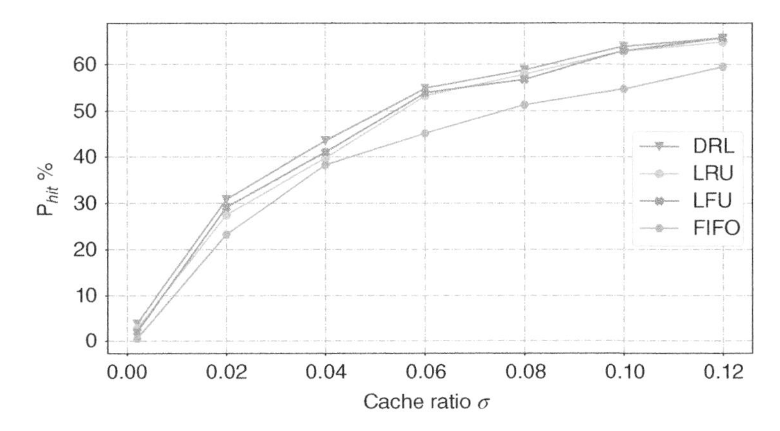

Figure 21.5 Percentage of cache hit rate P_{hit} vs. cache ratio σ, in cellular mode.

other three caching policies. As the cache ratio increases, the cache hit rates achieved by all the caching policies grow quickly at first, and then the rate of growth slows down. This is because popular files will be requested at higher probabilities, and initially caching the most popular files contributes more to improving the cache hit rate. And it is shown that the proposed DRL agent achieves higher overall average cache hit rates at all cache ratios in the experiment. LFR, LRU, and FIFO caching policies collect the files' popularity information directly when the requests arrive at the content servers, and make decisions based on the direct observations. However, with the centralized critic network, the multi-agent DRL framework is able to learn how the decision made by each agent influences the overall cache hit rate, and as a result the DRL agent is better at attaining a balance between the cache hit rate achieved by each agent and that of the whole system.

21.5.3.2 Transmission Delay

To determine the relationship of transmission delay and cache capacity, in Figure 21.6 we fix the Zipf exponent at $\beta = 1.3$ and plot the percentage of overall transmission delay reduction η as a function of the cache capacity ratio σ. It is shown that as the cache ratio σ increases, the reduction in transmission delay achieved by all four caching policies first rises quickly because the base stations can cache more files, and then the trend slows down after a certain value of σ. The upward trend starts to slow down because all these caching algorithms are encouraged to cache the most popular files following the statistics they learn. So when the cache ratio grows more and more, the caching agent will start caching less-popular content files. As more files are cached and transmission delay is further reduced, caching the less-popular files at the edge nodes leads to smaller improvements in reducing the transmission delay when compared with the contribution made by caching the most popular files. In other words, when the cache ratio is large enough to cache all of the most popular files, the system does not necessarily have to keep enlarging the cache capacity, considering the price to pay for storage and the relatively small reduction in transmission delay that will be achieved by storing the less-popular files. We also observe again that for all values of the cache ratio, the proposed framework achieves better performance for two reasons: (i) the proposed framework considers the reduction in the average transmission delay as the reward, so that the caching algorithm not only focuses on finding the most popular files, but also takes into account the users' locations and several less-popular files with potentially high delay penalties if not cached; and (ii) the critic network can facilitate

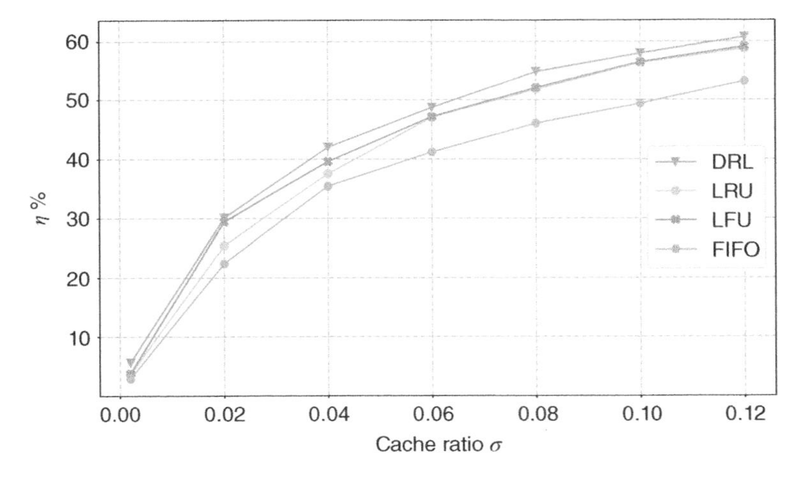

Figure 21.6 Percentage of transmission delay reduction η vs. cache ratio σ, in cellular mode.

the exchange of information among the base stations so that they can avoid caching the same files to serve users located in overlapped regions, and in this way, utilize the cache space more efficiently.

21.5.3.3 Time-Varying Scenario

As shown in Figures 21.5 and 21.6, though the proposed DRL agent can achieve better performance over the other policies, the improvement can be small in some cases. In this experiment, we present the simulation result in a time-varying scenario. In Figure 21.7, we demonstrate the ability of the caching policies to adapt to varying content popularity distributions. The experiment is conducted in time period $t = [0, 40000]$ (measured in per-unit runtime), where users' preferences for files change at every 10000 time slots. The users' requests are generated using Zipf distributions with their unique ranks of files and Zipf exponents. At every change point, these parameters vary randomly. The change points and Zipf parameters are all unknown to the caching agents. We only limit the Zipf exponent β to be in the range $[1.1, 1.5]$. Then we plot the average of the percentages of the average transmission delay reduction over time as $\overline{\eta}_T = \frac{1}{T}\sum_{t=1}^{T}\eta_t$, for $t = 1, 2, \ldots, 40000$. As shown in Figure 21.7, the proposed framework achieves lower performance at the beginning, because unlike the other three caching policies, the proposed framework does not directly collect statistics from the users' requests, but generally adjusts the parameters of the NNs and learns the popularity patterns of the files. After the NNs are trained well, the proposed framework achieves the best long-term performance. And each time the popularity distribution changes, even though performance slightly drops as the actor-critic framework updates the parameters to adapt to the new pattern, it is able to reach back to the previous level within a reasonable time frame because the previous experience has trained the network well. The LFU policy performs the best at the beginning, but due to frequency pollution, performance drops quickly at the first change point and goes all the way down. For the LRU and FIFO policies, performance is stable, because the cache size is limited and files that used to be popular and are less popular after the change can be replaced in a relatively short amount of time. However, as evidenced in this figure, the proposed framework is more suitable to be applied in scenarios that require long-term high performance and stability.

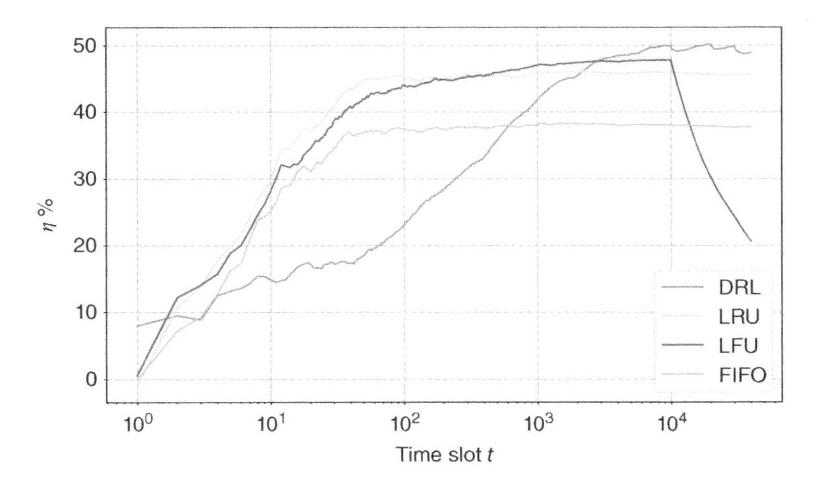

Figure 21.7 Percentage of transmission delay reduction η as the popularity distribution of content changes over time.

21.6 Application to the Single-Cell Network with D2D Communications

21.6.1 Experimental Settings

Now, we demonstrate the simulation results for caching at cache-enabled UEs in the single-cell network model with D2D communications enabled. In this section, we compare the performance of the DRL framework with the following caching algorithms:

- *Naive caching policy*: In this policy, it is assumed that the each user's preference for files is known by the cache-enabled mobile device, and the device caches the C most popular files.
- *Probabilistic caching policy*: In this policy, we assume that the probabilities that the user will request a specific file are known, and each time the corresponding caching agent makes a decision, it will take this popularity distribution into consideration, i.e, probabilistic caching of the files is performed.

21.6.2 Simulation Setup

Environment Settings: As shown in Figure 21.8, in the experiments, we consider a system with 1 base station and 30 users randomly distributed in the area. The cell radius is set as $R = 4$km, and the transmission power of the base station is set as $P_b = 16.9$dB. We assume that each user can connect with other users within distance $r = 1.5$km, and the transmission power of each UE is set as $P_i = 13.01$dB, $i = 1, 2, ..., 30$. The content files are split into units of the same size, and the size of each unit is set as 100 bits. We assume Rayleigh fading with path loss $\mathbb{E}\{z\} = d^{-4}$, where d is the distance between the transmitter and receiver.

File/Content Request Generation: In our simulations, the raw data of users' requests is generated according to the Zipf distribution given in Eq. (21.20) and rewritten here for ease of reference:

$$f(k; \beta, M) = \frac{1/k^{\beta}}{\sum_{M}^{m=1}(1/n^{\beta})} \tag{21.21}$$

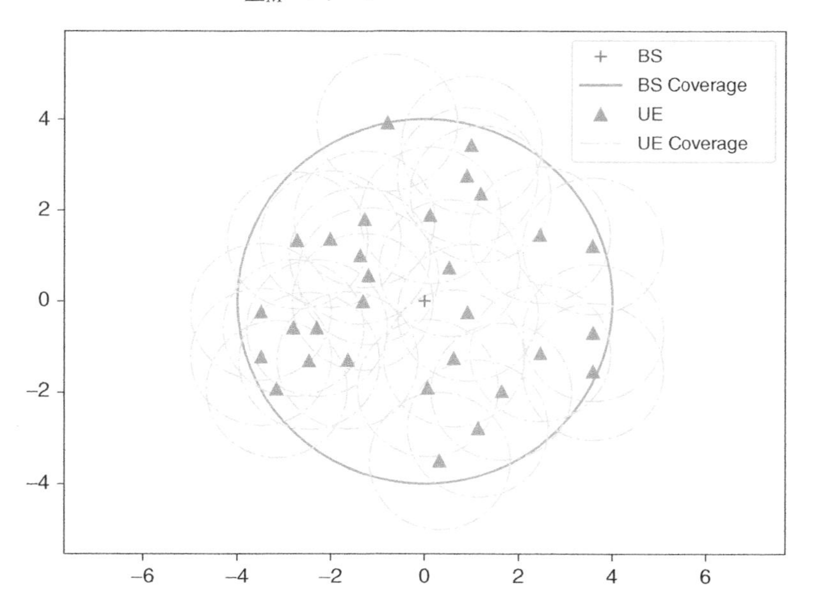

Figure 21.8 Coverage map of a system that contains 1 base station and 30 cache-enabled user equipments.

where the total number of files M is set as 100, and the Zipf exponent β is fixed at 1.3 in the study of the cache size. k is again the rank of the file, and in the implementation, a user's preference for files is randomly generated. To encourage UEs to cache files that are popular for multiple users, the users are randomly divided into five groups. The other assumptions for the single-cell case are the same as those for the multi-cell, as given in the previous section. More specifically, users in the same group will have similar but not exactly the same rank for all files, and the group information will not influence users' location. It is important to note that how we generate the requests using the Zipf distribution and how we group the users are totally unknown to the agents.

Feature Extraction: Similarly as in the case of a multi-cell network, the number of requests for each file is extracted from requests that have arrived in the most recent 10, 100, 1000 time slots to obtain the short-, medium-, and long-term features, respectively.

21.6.3 Simulation Results

21.6.3.1 Cache Hit Rate

Similarly as in the experiments for the multi-cell network model, we again consider first the relationship between the cache hit rate and the cache ratio σ. Figure 21.9 plots the overall average cache hit rates achieved by the proposed DRL framework, naive caching policy, and probabilistic caching policy as the cache capacity varies. The results show that the proposed DRL-based caching strategy provides significant improvements at all different cache ratios over the other two policies. The main reason for this improvement is that the centralized critic network enables information exchange between users, which in turn helps users better utilize D2D links in sharing locally cached content among each other. On the other hand, in the naive policy, when making a caching decision, a user only considers its own preference for files. As an example, let us assume that there are two users in each other's service coverage. If the two users have the same preference for files, they will cache exactly the same C files. Consequently, these two users will not get a chance to transmit to each other when a file other than the cached ones is requested. In the probabilistic caching policy, each user again considers only its own preference but the files are cached probabilistically in proportion to their probabilities. More specifically, the more popular a file is, the higher its probability of being cached. Note that in this case, with smaller probability, a less-popular file can be cached as well. And in this way, the chance that users can serve each other via the D2D links increases; consequently, performance improves over that of the naive caching policy.

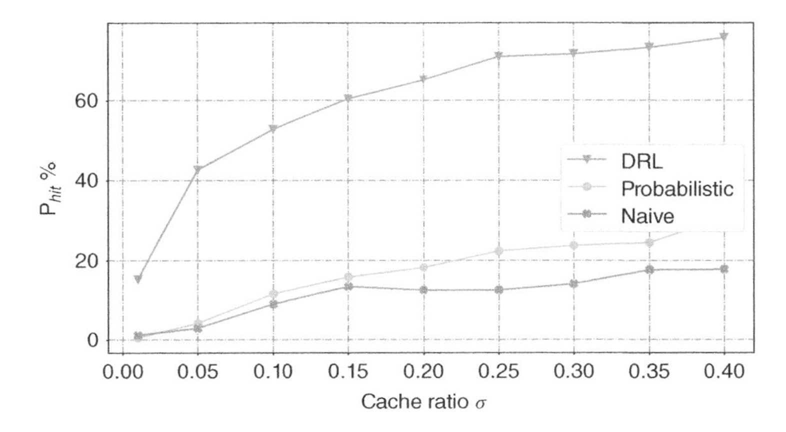

Figure 21.9 Percentage of cache hit rate P_{hit} vs. cache ratio σ, in device-to-device mode.

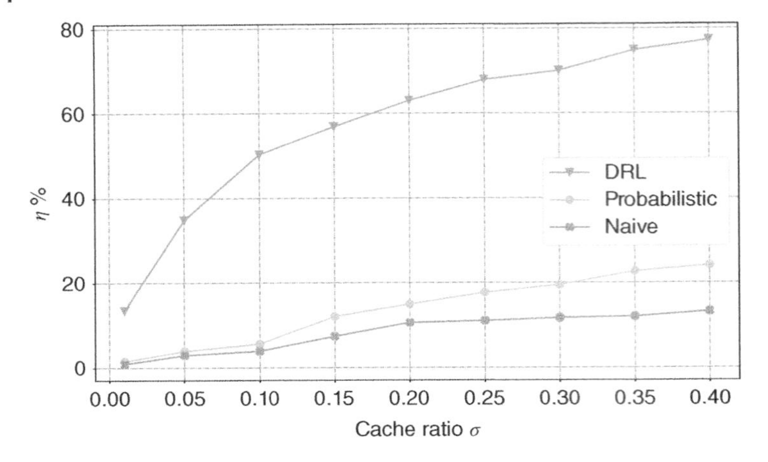

Figure 21.10 Percentage of transmission delay reduction η vs. cache ratio σ, in device-to-device mode.

21.6.3.2 Transmission Delay

In Figure 21.10, we plot the percentage of transmission-delay reduction as a function of the cache ratio. Similarly as in the study of how the cache hit rate varies as the cache ratio increases, we again observe that a substantial improvement is achieved with the proposed DRL caching framework when compared to the probabilistic and naive caching policies. Besides the analysis mentioned earlier, the reward is also an important reason to explain the advantage achieved by the DRL agent. As noted before, the probabilistic and naive caching policies are assumed to know users' preferences. However, in the study of transmission delay, it is not necessary for files that are requested most frequently to contribute the most to the reduction in transmission delay, because transmission delay also depends on the distance between transmitter and receiver. So it is possible that a less-frequently requested file causes more transmission delay than a more frequently requested one. Therefore, as their main drawback, the probabilistic and naive caching policies cannot adjust their strategies according to transmission delay. On the other hand, in the DRL framework, even though the observation is still the number of times that files were requested in the past, the reward is set as the transmission delay in this experiment, and therefore the DRL agent can update its policy to achieve a higher reduction in transmission delay.

21.7 Conclusion

In this chapter, we have investigated the application of deep reinforcement learning for edge caching. We have presented a deep actor-critic RL-based multi-agent framework for the edge-caching problem in both a multi-cell network and a single-cell network with D2D communication. To demonstrate the performance of the proposed DRL framework, we have provided simulation results for both network models in terms of the cache hit rate and reduction in transmission delay. In the multi-cell network, we have compared the proposed framework with LRU, LFU, and FIFO caching policies. In the single-cell network, we have provided comparisons with probabilistic and naive caching policies. We have verified that the proposed DRL framework attains better performances over the other caching policies.

Finally, we note that in addition to the proposed actor-critic framework, there are several other DRL structures that can be applied to identify efficient caching policies. For instance, the multi-agent deep Q-network can be used in a decentralized system with multiple caching nodes. Furthermore, as future research directions, multi-agent multi-task DRL frameworks can be studied to jointly solve the content-caching problem along with other optimization problems: for instance, related to power control and user scheduling.

Bibliography

Seyyed Mohammadreza Azimi, Osvaldo Simeone, Avik Sengupta, and Ravi Tandon. Online edge caching and wireless delivery in fog-aided networks with dynamic content popularity. *IEEE Journal on Selected Areas in Communications*, 36(6): 1189–1202, 2018.

Zheng Chang, Lei Lei, Zhenyu Zhou, Shiwen Mao, and Tapani Ristaniemi. Learn to cache: Machine learning for network edge caching in the big data era. *IEEE Wireless Communications*, 25(3): 28–35, 2018.

Binqiang Chen, Chenyang Yang, and Zixiang Xiong. Optimal caching and scheduling for cache-enabled D2D communications. *IEEE Communications Letters*, 21(5): 1155–1158, 2017a.

Mingzhe Chen, Walid Saad, Changchuan Yin, and Mérouane Debbah. Echo state networks for proactive caching in cloud-based radio access networks with mobile users. *IEEE Transactions on Wireless Communications*, 16(6): 3520–3535, 2017b.

Cisco Visual Networking Index. Global mobile data traffic forecast update 2017-2022 white paper. Feb. 2019. [Online]. Available: https://www.cisco.com/c/en/us/solutions/collateral/service-provider/visual-networking-index-vni/white-paper-c11-738429.html.

Mostafa Dehghan, Bo Jiang, Anand Seetharam, Ting He, Theodoros Salonidis, Jim Kurose, Don Towsley, and Ramesh Sitaraman. On the complexity of optimal request routing and content caching in heterogeneous cache networks. *IEEE/ACM Transactions on Networking (TON)*, 25(3): 1635–1648, 2017.

Mohammed S. ElBamby, Mehdi Bennis, Walid Saad, and Matti Latva-Aho. Content-aware user clustering and caching in wireless small cell networks. *arXiv preprint arXiv:1409.3413*, 2014.

Jakob Foerster, Ioannis Alexandros Assael, Nando de Freitas, and Shimon Whiteson. Learning to communicate with deep multi-agent reinforcement learning. In *Advances in Neural Information Processing Systems*, pages 2137–2145, 2016.

Jayesh K. Gupta, Maxim Egorov, and Mykel Kochenderfer. Cooperative multi-agent control using deep reinforcement learning. In *International Conference on Autonomous Agents and Multiagent Systems*, pages 66–83. Springer, 2017.

Shengqi Hao, Naifu Zhang, and Meixia Tao. A belief propagation approach for caching strategy design in D2D networks. In *2018 IEEE International Conference on Communications Workshops (ICC Workshops)*, pages 1–6. IEEE, 2018.

Mona Jaber, Muhammad Ali Imran, Rahim Tafazolli, and Anvar Tukmanov. 5G backhaul challenges and emerging research directions: A survey. *IEEE Access*, 4: 1743–1766, 2016.

Wei Jiang, Gang Feng, Shuang Qin, and Tak Shing Peter Yum. Efficient D2D content caching using multi-agent reinforcement learning. In *IEEE INFOCOM 2018-IEEE Conference on Computer Communications Workshops (INFOCOM WKSHPS)*, pages 511–516. IEEE,2018.

Jeongwan Koh, Osvaldo Simeone, Ravi Tandon, and Joonhyuk Kang. Cloud-aided edge caching with wireless multicast fronthauling in fog radio access networks. In *Wireless Communications and Networking Conference (WCNC), 2017 IEEE*, pages 1–6. IEEE,2017.

Karla Kvaternik, Jaime Llorca, Daniel Kilper, and Lacra Pavel. A methodology for the design of self-optimizing, decentralized content-caching strategies. *IEEE/ACM Transactions on Networking*, 24(5): 2634–2647, 2016.

Jeongho Kwak, Yeongjin Kim, Long Bao Le, and Song Chong. Hybrid content caching in 5G wireless networks: Cloud versus edge caching. *IEEE Transactions on Wireless Communications*, 17(5): 3030–3045, 2018.

Mathieu Leconte, Georgios Paschos, Lazaros Gkatzikis, Moez Draief, Spyridon Vassilaras, and Symeon Chouvardas. Placing dynamic content in caches with small population. In *Computer Communications, IEEE INFOCOM 2016-The 35th Annual IEEE International Conference on*, pages 1–9. IEEE, 2016.

Ming-Chun Lee and Andreas F. Molisch. Caching policy and cooperation distance design for base station-assisted wireless D2D caching networks: Throughput and energy efficiency optimization and tradeoff. *IEEE Transactions on Wireless Communications*, 17(11): 7500–7514, 2018.

Lei Lei, Lei You, Gaoyang Dai, Thang Xuan Vu, Di Yuan, and Symeon Chatzinotas. A deep learning approach for optimizing content delivering in cache-enabled hetnet. In *Wireless Communication Systems (ISWCS), 2017 International Symposium on*, pages 449–453. IEEE, 2017.

Lei Lei, Xiong Xiong, Lu Hou, and Kan Zheng. Collaborative edge caching through service function chaining: Architecture and challenges. *IEEE Wireless Communications*, 25(3): 94–102, 2018.

Suoheng Li, Jie Xu, Mihaela Van Der Schaar, and Weiping Li. Popularity-driven content caching. In *Computer Communications, IEEE INFOCOM 2016-The 35th Annual IEEE International Conference on*, pages 1–9. IEEE, 2016a.

Wenjie Li, Sharief M.A. Oteafy, and Hossam S. Hassanein. StreamCache: popularity-based caching for adaptive streaming over information-centric networks. In *Communications (ICC), 2016 IEEE International Conference on*, pages 1–6. IEEE,2016b.

Xiuhua Li, Xiaofei Wang, Peng-Jun Wan, Zhu Han, and Victor CM Leung. Hierarchical edge caching in device-to-device aided mobile networks: Modeling, optimization, and design. *IEEE Journal on Selected Areas in Communications*, 2018a.

Yi Li, Chen Zhong, M. Cenk Gursoy, and Senem Velipasalar. Learning-based delay-aware caching in wireless D2D caching networks. *IEEE Access*, 6: 77250–77264, 2018b.

Dong Liu, Binqiang Chen, Chenyang Yang, and Andreas F. Molisch. Caching at the wireless edge: design aspects, challenges, and future directions. *IEEE Communications Magazine*, 54(9): 22–28, 2016.

Ryan Lowe, Yi Wu, Aviv Tamar, Jean Harb, OpenAI Pieter Abbeel, and Igor Mordatch. Multi-agent actor-critic for mixed cooperative-competitive environments. In *Advances in Neural Information Processing Systems*, pages 6382–6393, 2017.

Alireza Sadeghi, Fatemeh Sheikholeslami, and Georgios B. Giannakis. Optimal and scalable caching for 5G using reinforcement learning of space-time popularities. *IEEE Journal of Selected Topics in Signal Processing*, 12(1): 180–190, 2018.

Avik Sengupta, SaiDhiraj Amuru, Ravi Tandon, R. Michael Buehrer, and T. Charles Clancy. Learning distributed caching strategies in small cell networks. In *Wireless Communications Systems (ISWCS), 2014 11th International Symposium on*, pages 917–921. IEEE, 2014.

Jiongjiong Song, Min Sheng, Tony QS Quek, Chao Xu, and Xijun Wang. Learning-based content caching and sharing for wireless networks. *IEEE Transactions on Communications*, 65(10): 4309–4324, 2017.

Jihoon Sung, Kyounghye Kim, Junhyuk Kim, and June-Koo Kevin Rhee. Efficient content replacement in wireless content delivery network with cooperative caching. In *Machine Learning and Applications (ICMLA), 2016 15th IEEE International Conference on*, pages 547–552. IEEE, 2016.

Ravi Tandon and Osvaldo Simeone. Cloud-aided wireless networks with edge caching: Fundamental latency trade-offs in fog radio access networks. In *Information Theory (ISIT), 2016 IEEE International Symposium on*, pages 2029–2033. IEEE, 2016.

Tuyen X. Tran, Abolfazl Hajisami, and Dario Pompili. Cooperative hierarchical caching in 5G cloud radio access networks. *IEEE Network*, 31(4): 35–41, 2017.

Shuo Wang, Xing Zhang, Yan Zhang, Lin Wang, Juwo Yang, and Wenbo Wang. A survey on mobile edge networks: Convergence of computing, caching and communications. *IEEE Access*, 5: 6757–6779, 2017.

Wenbo Wang, Dusit Niyato, Ping Wang, and Amir Leshem. Decentralized caching for content delivery based on blockchain: A game theoretic perspective. *arXiv preprint arXiv:1801.07604*, 2018.

Yifei Wei, F. Richard Yu, Mei Song, and Zhu Han. Joint optimization of caching, computing, and radio resources for fog-enabled iot using natural actor-critic deep reinforcement learning. *IEEE Internet of Things Journal*, 2018.

Kai Xu, Xiang Li, Sanjay Kumar Bose, and Gangxiang Shen. Joint replica server placement, content caching, and request load assignment in content delivery networks. *IEEE Access*, 6: 17968–17981, 2018.

Qifa Yan, Minquan Cheng, Xiaohu Tang, and Qingchun Chen. On the placement delivery array design for centralized coded caching scheme. *IEEE Transactions on Information Theory*, 63(9): 5821–5833, 2017.

Qianqian Yang, Parisa Hassanzadeh, Deniz Gündüz, and Elza Erkip. Centralized caching and delivery of correlated contents over a Gaussian broadcast channel. In *Modeling and Optimization in Mobile, Ad Hoc, and Wireless Networks (WiOpt), 2018 16th International Symposium on*, pages 1–6. IEEE, 2018.

Shan Zhang, Peter He, Katsuya Suto, Peng Yang, Lian Zhao, and Xuemin Shen. Cooperative edge caching in user-centric clustered mobile networks. *IEEE Transactions on Mobile Computing*, 17 (8): 1791–1805, 2018.

Xiangyang Zhang, Ying Wang, Ruijin Sun, and Dong Wang. Clustered device-to-device caching based on file preferences. In *Personal, Indoor, and Mobile Radio Communications (PIMRC), 2016 IEEE 27th Annual International Symposium on*, pages 1–6. IEEE, 2016.

Chen Zhong, M. Cenk Gursoy, and Senem Velipasalar. A deep reinforcement learning-based framework for content caching. In *Information Sciences and Systems (CISS), 2018 52nd Annual Conference on*, pages 1–6. IEEE, 2018.

Bo Zhou,Ying Cui, and Meixia Tao. Optimal dynamic multicast scheduling for cache-enabled content-centric wireless networks. *IEEE Transactions on Communications*, 65(7): 2956–2970, 2017.

Hao Zhu, Yang Cao, Wei Wang, Tao Jiang, and Shi Jin. Deep reinforcement learning for mobile edge caching: Review, new features, and open issues. *IEEE Network*, 32(6): 50–57, 2018.

Index

Machine Learning for Future Wireless Communications, First Edition. Edited by Fa-Long Luo.
© 2020 John Wiley & Sons Ltd. Published 2020 by John Wiley & Sons Ltd.

d

e

f

g